D0760134

Laplace Transforms

$\bar{f}(s) = \int_0^\infty f(t)e^{-st}\,dt$	$f(t) = \frac{1}{2\pi i}\int_{\gamma-i\infty}^{\gamma+i\infty} \bar{f}(s)e^{st}\,ds$
$\frac{1}{s-a}$	e^{at}
$\frac{1}{s^2+a^2}$	$\frac{\sin at}{a}$
$\frac{s}{s^2+a^2}$	$\cos at$
$\frac{1}{s^2-a^2}$	$\frac{\sinh at}{a}$
$\frac{s}{s^2-a^2}$	$\cosh at$
$\frac{1}{s^a} \quad (a > 0)$	$\frac{t^{a-1}}{\Gamma(a)}$
$\frac{e^{-as}}{s} \quad (a \geq 0)$	$H(t-a)$
$\frac{1}{(s-b)^a} \quad (a > 0)$	$\frac{t^{a-1}e^{bt}}{\Gamma(a)}$
$-\frac{\gamma + \ln s}{s}$*	$\ln t$
$e^{-a\sqrt{s}} \quad (a > 0)$	$\frac{ae^{-a^2/4t}}{2\sqrt{\pi}\,t^{3/2}}$
$\frac{e^{-a\sqrt{s}}}{\sqrt{s}} \quad (a \geq 0)$	$\frac{e^{-a^2/4t}}{\sqrt{\pi t}}$
$\left[1 + \frac{1}{2} + \frac{1}{3} + \cdots + \frac{1}{n} - \gamma - \ln s\right]\frac{n!}{s^{n+1}}$ *	$t^n \ln t$
$s\bar{f}(s) - f(0)$	$f'(t)$
$s^2\bar{f}(s) - sf(0) - f'(0)$	$f''(t)$

* γ = Euler's constant $\doteq$ 0.577215665.

FOUNDATIONS OF APPLIED MATHEMATICS

Michael D. Greenberg
Professor Emeritus
Department of Mechanical and Aerospace Engineering
University of Delaware

Dover Publications, Inc.
Mineola, New York

Copyright

Bibliographical Note

This Dover edition, first published in 2013, is an unabridged republication of the work originally published by Prentice-Hall, Inc., Englewood Cliffs, N.J., in 1978. This book was previously published by Pearson Education, Inc.

Library of Congress Cataloging-in-Publication Data

Greenberg, Michael D., 1935–author.
Foundations of applied mathematics / Michael D. Greenberg, professor emeritus, Department of Mechanical and Aerospace Engineering, University of Delaware. — Dover edition.
p. cm.
Summary: "A longtime classic text in applied mathematics, this volume also serves as a reference for undergraduate and graduate students of engineering. Topics include real variable theory, complex variables, linear analysis, partial and ordinary differential equations, and other subjects. Answers to selected exercises are provided, along with Fourier and Laplace transformation tables and useful formulas. 1978 edition"—Provided by publisher.
Includes bibliographical references and index.
ISBN-13: 978-0-486-49279-7 (pbk.)
ISBN-10: 0-486-49279-6 (pbk.)
1. Engineering mathematics. I. Title.

TA330.G73 2013
620.001'51—dc23

2013015816

Manufactured in the United States by LSC Communications
49279606 2019
www.doverpublications.com

This book is dedicated,
with love, to Mim.

Contents by Part

Part

I	Real Variable Theory *1*
II	Complex Variables *229*
III	Linear Analysis *307*
IV	Ordinary Differential Equations *391*
V	Partial Differential Equations *521*

Contents

Preface

PART I REAL VARIABLE THEORY

Chapter 1 *The Important Limit Processes 3*

1.1. Functions and Functionals *3*
1.2. Limits, Continuity, and Uniform Continuity *4*
1.3. Differentiation *8*
1.4. Integration *9*
1.5. Asymptotic Notation and the "Big Oh" *11*
1.6. Numerical Integration *12*
1.7. Differentiation of Integrals Containing a Parameter; Leibnitz's Rule *17*
Exercises *19*

Chapter 2 *Infinite Series 23*

2.1. Sequences and Series; Fundamentals and Tests for Convergence *23*
2.2. Series with Terms of Mixed Sign *29*
2.3. Series with Terms That Are Functions; Power Series *31*
2.4. Taylor Series *34*
2.5. That's All Very Nice, But How Do We Sum the Series? Acceleration Techniques *38*
***2.6.** Asymptotic Expansions *43*
Exercises *46*

Chapter 3 *Singular Integrals* *51*

3.1. Choice of Summability Criteria; Convergence and Cauchy Principal Value *51*
3.2. Tests for Convergence *54*
3.3. The Gamma Function *56*
3.4. Evaluation of Singular Integrals *59*
Exercises *61*

Chapter 4 *Interchange of Limit Processes and the Delta Function* *66*

4.1. Theorems on Limit Interchange *67*
4.2. The Delta Function and Generalized Functions *68*
Exercises *73*

Chapter 5 *Fourier Series and the Fourier Integral* *77*

5.1. The Fourier Series of $f(x)$ *78*
5.2. Pointwise Convergence of the Series *80*
5.3. Termwise Integration and Differentiation of Fourier Series *85*
5.4. Variations: Periods Other than 2π, and Finite Interval *86*
5.5. Infinite Period; the Fourier Integral *88*
Exercises *92*

Chapter 6 *Fourier and Laplace Transforms* *98*

6.1. The Fourier Transform *99*
6.2. The Laplace Transform *103*
6.3. Other Transforms *112*
Exercises *112*

Chapter 7 *Functions of Several Variables* *120*

7.1. Chain Differentiation *120*
7.2. Taylor Series in Two or More Variables *125*
Exercises *127*

Chapter 8 *Vectors, Surfaces, and Volumes* *132*

8.1. Vectors and Elementary Operations *132*
8.2. Coordinate Systems, Base Vectors, and Components *136*
8.3. Angular Velocity of a Rigid Body *139*
8.4. Curvilinear Coordinate Representation of Surfaces *141*
8.5. Curvilinear Coordinate Representation of Volumes *144*
Exercises *145*

Chapter 9 Vector Field Theory 150

9.1. Line Integrals *150*
9.2. The Curves, Surfaces, and Regions Under Consideration *152*
9.3. Divergence, Gradient, and Curl *153*
9.4. The Divergence Theorem *160*
9.5. Stokes' Theorem *166*
9.6. Irrotational and Solenoidal Fields *170*
9.7. Noncartesian Systems *175*
9.8. Fluid Mechanics; Irrotational Flow *177*
9.9. The Gravitational Potential *182*
Exercises *187*

Chapter 10 The Calculus of Variations 196

10.1. Functions of One or More Variables *197*
10.2. Constraints; Lagrange Multipliers *199*
10.3. Functionals and the Calculus of Variations *202*
10.4. Two or More Dependent Variables; Hamilton's Principle *209*
10.5. Two or More Independent Variables; Vibrating Strings and Membranes *211*
10.6. The Ritz Method *216*
10.7. Optimal Control *218*
Exercises *220*

Additional References for Part I 227

PART II COMPLEX VARIABLES

Chapter 11 Complex Numbers 231

11.1. The Algebra of Complex Numbers *231*
11.2. The Complex Plane *232*
Exercises *234*

Chapter 12 Functions of a Complex Variable 235

12.1. Basic Notions *235*
12.2. Differentiation and the Cauchy–Riemann Conditions for Analyticity *237*
12.3. Harmonic Functions *239*
12.4. Some Elementary Functions and Their Singularities *240*
12.5. Multivaluedness and the Need for Branch Cuts *243*
Exercises *248*

Chapter 13 *Integration, Cauchy's Theorem, and the Cauchy Integral Formula* *252*

13.1. Integration in the Complex Plane *252*
13.2. Bounds on Contour Integrals *255*
13.3. Cauchy's Theorem *255*
13.4. An Important Little Integral *256*
13.5. The Cauchy Integral Formula *257*
Exercises *259*

Chapter 14 *Taylor and Laurent Series* *261*

14.1. Complex Series *261*
14.2. Taylor Series *263*
14.3. Laurent Series *265*
14.4. Classification of Isolated Singularities *269*
Exercises *272*

Chapter 15 *The Residue Theorem and Contour Integration* *276*

15.1. The Residue Theorem *276*
15.2. The Calculation of Residues *277*
15.3. Applications *279*
15.4. Cauchy Principal Value *285*
Exercises *286*

Chapter 16 *Conformal Mapping* *291*

16.1. The Fundamental Problem *291*
16.2. Conformality *292*
16.3. The Bilinear Transformation and Applications *294*
Exercises *298*

Additional References for Part II *304*

PART III *LINEAR ANALYSIS*

Chapter 17 *Linear Spaces* *309*

17.1. Introduction; Extension to *n* tuples *309*
17.2. The Definition of an Abstract Vector Space *311*
Introducing a Norm and Inner Product *312*
17.3. Linear Dependence, Dimension, and Bases *315*
17.4. Function Space *318*
***17.5.** Continuity of the Inner Product *324*
Exercises *324*

Chapter 18 *Linear Operators* *328*

18.1. Some Definitions *328*
18.2. Operator Algebra *331*
18.3. Further Discussion of Matrices *334*
18.4. The Adjoint Operator *335*
Exercises *337*

Chapter 19 *The Linear Equation* $Lx = c$ *340*

19.1. Introduction *340*
19.2. Existence and Uniqueness *342*
19.3. The Inverse Operator L^{-1} *344*
The Inverse of a Matrix *345*
*Inverse of $I + M$ where M is Small; the Neumann Series *349*
Exercises *354*

Chapter 20 *The Eigenvalue Problem* $Lx = \lambda x$ *359*

20.1. Statement of the Eigenvalue Problem *359*
20.2. Some Eigenhunts *360*
20.3. The Sturm–Liouville Theory *363*
20.4. Additional Discussion for the Matrix Case *367*
The Inertia Tensor *372*
20.5. The Inhomogeneous Problem $Lx = c$ *376*
20.6. Eigen Bounds and Estimates *377*
Exercises *380*

Additional References for Part III *389*

PART IV ORDINARY DIFFERENTIAL EQUATIONS

Chapter 21 *First-Order Equations* *393*

21.1. Standard Methods of Solution *393*
Variables Separable *394*
Homogeneous of Degree Zero *394*
Exact Differentials and Integrating Factors *395*
21.2. The Questions of Existence and Uniqueness *398*
Exercises *403*

Chapter 22 *Higher-Order Systems* *408*

***22.1.** Nonlinear Case; Existence and Uniqueness *408*
22.2. The Linear Equation; Some Theory *411*
22.3. The Linear Equation; Some Methods of Solution *415*
22.4. Series Solution; Bessel and Legendre Functions *423*
22.5. The Method of Green's Functions *434*

22.6. Some Nonlinear Problems and Techniques *438*
Nonlinear Algebraic Equations *440*
Application to Nonlinear Differential Equations *443*
Exercises *444*

Chapter 23 *Qualitative Methods; The Phase Plane* *454*

23.1. The Phase Plane and Some Simple Examples *455*
23.2. Singular Points *457*
23.3. Additional Examples *461*
Exercises *467*

Chapter 24 *Quantitative Methods* *472*

24.1. The Methods of Taylor and Euler *472*
24.2. Improvements: Midpoint Rule and Runge–Kutta *476*
24.3. Stability *479*
24.4. Application to Higher-Order Systems *483*
24.5. Boundary Value Problems *485*
*Invariant Imbedding *487*
24.6. The Method of Weighted Residuals *488*
24.7. The Finite-Element Method *491*
Exercises *494*

Chapter 25 *Perturbation Techniques* *500*

25.1. Introduction; the Regular Case *501*
25.2. Singular Perturbations: Straining Methods *502*
25.3. Singular Perturbations: Boundary Layer Methods *508*
Exercises *514*

Additional References for Part IV *519*

PART V *PARTIAL DIFFERENTIAL EQUATIONS*

Chapter 26 *Separation of Variables and Transform Methods* *523*

26.1. Introduction *523*
26.2. Some Background; the Sturm–Liouville Theory *525*
26.3. The Diffusion Equation *528*
26.4. The Wave Equation *540*
26.5. The Laplace Equation *547*
Exercises *553*

Chapter 27 *Classification and the Method of Characteristics* *564*

27.1. Characteristics and Classification *564*
27.2. The Hyperbolic Case *566*
27.3. Reduction of $A\phi_{xx} + 2B\phi_{xy} + C\phi_{yy} = F$ to Normal Form *573*
27.4. Comparison of Hyperbolic, Elliptic, and Parabolic Systems; Summary *574*
Exercises *576*

Chapter 28 *Green's Functions and Perturbation Techniques* *579*

28.1. The Green's Function Approach *579*
28.2. Perturbation Methods *585*
Exercises *589*

Chapter 29 *Finite-Difference Methods* *596*

29.1. The Heat Equation; An Explicit Method and Its Convergence and Stability *596*
29.2. The Heat Equation; Implicit Methods *604*
29.3. Hyperbolic and Elliptic Problems *607*
Exercises *608*

Additional References for Part V *612*

Survey-Type References *614*

Answers to Selected Exercises *616*

Index *625*

Fourier and Laplace Transform Tables, Inside Front Cover

Some Frequently Needed Formulas, Inside Back Cover

Preface

Purpose. This book is intended primarily as a text for a single- or multi-semester course in applied mathematics for first-year graduate students or seniors in engineering. It grew out of a set of lecture notes for a two-semester course that I taught both at Cornell University and at the University of Delaware, a course taken primarily by first-year graduate students in applied mechanics, materials science, mechanical, chemical, civil, and aerospace engineering, as well as some undergraduates and a number of people from electrical engineering, oceanography, chemistry, and astronomy.

Prerequisites. Prerequisites consist of the usual undergraduate four-semester sequence in calculus and ordinary differential equations, together with the general maturity and background of a senior or beginning graduate student. Knowledge of computer programming is not required.

Content and Organization. In books of this type it is tempting to try to include "something for everyone," that is, diverse physical applications from all branches of engineering, applied physics, and so on. Instead we have elected to concentrate on only a few areas of application, including fluid mechanics, heat conduction, and Newtonian mechanics. Discussion of such topics is self-contained; for instance, the governing equations of fluid mechanics and heat conduction are developed in the section on vector field theory before solutions are sought in the later sections on partial differential equations. As a result, these physical concepts weave through the mathematics and act as a unifying element. I try to emphasize not only how the mathematics permits us to clarify and understand the physics but also how our physical insight can support the mathematics and provide the key to finding the appropriate mathematical line of approach. Practical and numerical aspects are emphasized as well.

We begin in Part I with basic notions of functions and limit processes and proceed through various aspects of advanced calculus of real variables, with complex variables treated in Part II. In Part III we introduce linear vector space, operators, and the eigenvalue problem, with an emphasis on matrix operators, and an introduction to integral operators which is arranged in "parallel" so that it can be bypassed if desired. Ordinary and partial differential operators are so important that they are split off and treated separately in Parts IV and V.

Although the most natural sequence of topics is Parts I through V in that order, I have tried to maximize the book's adaptability to other formats of presentation. For example, in Part V on Partial Differential Equations a brief summary of the Sturm–Liouville theory is included, in Section 26.2, so that the reader can cope with the necessary eigenfunction expansions without necessarily having read Part III, which deals with that subject in more detail. I've also tried to refer the reader back to any specific formulas or results that are used and that appeared in an earlier part of the book.

It should be possible to *begin* with any of Parts I, II, or III. For Parts IV and V (not including the starred material), however, I suggest the following preliminary reading:

For Part IV: Section 4.2.

For Part V: Section 4.2, Chapters 5, 6, 7, 8, 9, Section 22.4.

In addition, some complex variable theory is used in Part V, and so it is recommended, although not essential, that Part II be read in advance as well. Perhaps it goes without saying that if the last two chapters of Part V (on Green's functions, perturbation methods, and finite-difference methods) are included, then it would be helpful to the reader first to have read the corresponding sections on *ordinary* differential equations in Part IV.

Suggested Courses.

TWO OR THREE SEMESTERS: In the two-semester course that I teach we cover about 90 percent of the material. This means that the pace is rather fast and does not permit class time to be used for both lecture and the discussion of assigned homework. Instead, I use class time for lecture and simply make the Solutions Manual available to the students to assist them in working out the suggested Exercises. The grade is based solely on two or three examinations with at least one class before each exam reserved for questions and review. At a more leisurely pace, the entire book could be covered in three semesters.

ONE SEMESTER: A natural one-semester course might cover Part I only, and would amount to a course in *Advanced Calculus for Scientists and Engineers.*

Starred Material. A certain amount of starred material is included. It can be bypassed with no loss of continuity, and is either of a more advanced or supplementary nature.

Exercises. The Exercises at the end of each chapter cover the material in that chapter and introduce additional material as well. They are arranged in the same order as the topics within the chapter and are *not* ordered in terms of their degree of difficulty. Starred problems, however, are generally more difficult and/or are based on material from starred sections within the chapter.

References. Besides the references cited in footnotes, a list of Additional References appears at the end of each part. Undoubtedly numerous excellent sources have not been cited, and I apologize in advance for these oversights.

Acknowledgment. I would like to extend special thanks to my department chairperson, Professor Jack R. Vinson, who has encouraged me in this time-consuming endeavor, to my students who are a continuing source of inspiration and friendship, and to Miss Leslie A. Month in particular. Finally, I am also grateful to Professors A. Eshett, Peter J. Gingo, Jerry Kazdan, Ronald J. Lomax, Ivar Stakgold, Dale W. Thoe, and Erich Zauderer for offering helpful comments on the manuscript.

Closure. In spite of all precautions, errors will no doubt be uncovered in using this book, and I would like to encourage you to let me know about them, if you have the time, so that they can be corrected. Beyond that, any comments, suggestions, or ideas that you may wish to share would be most appreciated. Applied mathematics is a beautiful subject, and I certainly hope that some of this comes through, in a quiet way, within this book.

MICHAEL D. GREENBERG

PART I

REAL VARIABLE THEORY

The first chapter, on LIMIT PROCESSES, sets the stage for much of what follows in Part I: Chapters 2 and 3 deal with the related limits INFINITE SERIES and SINGULAR INTEGRALS; Chapter 4 takes time out to consider questions concerning the interchange of the order of multiple limit processes, and this discussion leads naturally to the delta function and generalized functions; Chapter 5 returns to series and considers the so-called Fourier series of a periodic function (with mean-square convergence and vector space notions treated separately in Part III). It ends with the passage to the Fourier integral as the period becomes infinite, thereby leading to the Fourier and

Laplace transforms in Chapter 6. Besides discussing FUNCTIONS OF SEVERAL VARIABLES, VECTORS, SURFACES AND VOLUMES, VECTOR FIELD THEORY, and THE CALCULUS OF VARIATIONS, the final four chapters provide self-contained introductions to the physics and governing differential equations of heat transfer, fluid mechanics, strings and membranes, and the gravitational potential, so that these subject areas will be readily available for many of the applications that form an important part of the rest of the book.

Chapter 1

The Important Limit Processes

In this first chapter we review a number of fundamentals: functions, functionals, limits, continuity, and uniform continuity. Then we look at two important limit processes, differentiation and integration, the latter in some detail. Practical and computational aspects are emphasized as well.

1.1. FUNCTIONS AND FUNCTIONALS

Recall that a **function** f is a *mapping* from one set, called the *domain* of f, to another, called the *range* of f. Suppose, for instance, that f is defined by $f(x) = 8x(x - 1)^2 + 2x$ over the domain D, which is the segment $0.1 < x \leq 1$ of a real axis (the "x axis"). Clearly, this f is real valued so that the range R will also be some portion of a real axis (the "f axis"), as sketched in Fig. 1.1. Although the symbols f and $f(x)$ are generally used interchangeably without confusion, note that there is a difference and that occasions do arise in which it is important to respect this difference in the interest of clarity.

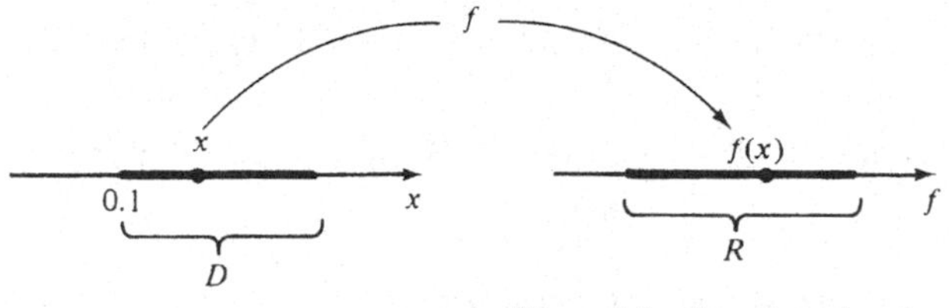

Figure 1.1. Domain and range.

Specifically, $f(x)$ actually denotes the range point associated with the domain point x, whereas the single letter f denotes the entire mapping, as can be displayed graphically if, following Descartes (1596–1650), we arrange the x and f axes at right angles (hence "Cartesian axes"), as shown in Fig. 1.2. In this example we find that R is the segment $0.848 < f \leq 2$.

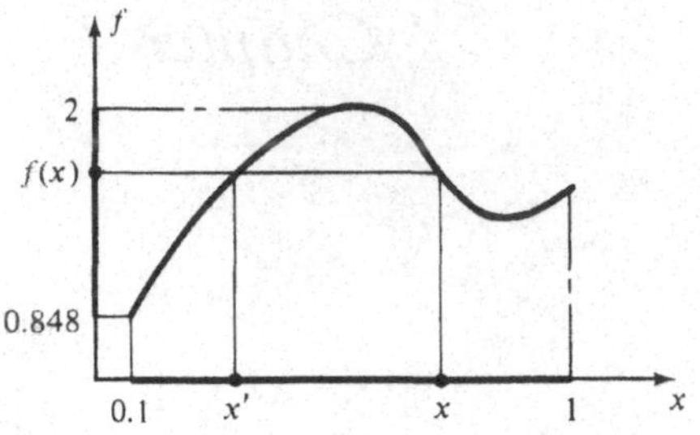

Figure 1.2. Graph of $f(x) = 8x(x-1)^2 + 2x$.

Similarly, a real-valued function $f(x, y)$ of two real variables x, y can be displayed (or at least visualized) as a *surface*, but when the number of dimensions of D plus the number of dimensions of R exceeds three, we're in difficulty in regard to graphical display.

We *always* require that f be *single valued*—that is, that each point in D be sent into a uniquely defined point in R. Yet a given point in R may correspond to more than one point in D. For instance, both x and x' in Fig. 1.2 get sent into the same range point. That's fine; but if each point in R were to correspond to a unique point in D, we would then say that f is *one-to-one*. Clearly, f must be one-to-one if an *inverse function* $x = x(f)$ is to exist as well.

In formulating a water wave problem, for example, and designating the free surface elevation by $y = \eta(x, t)$, where x is the horizontal coordinate and t is the time, we thereby make it impossible to consider waves that break, since η would then be multivalued. Of course, we can describe the profile of a breaking wave *parametrically* by $x = x(s, t), y = y(s, t)$, where s is a convenient parameter, perhaps arc length along the free surface, and where $x(s, t), y(s, t)$ *are* single-valued functions of s for each t.

A bit more exotic than a function, a **functional** has as its domain a set of *functions* and as its range a portion of the real axis; thus it is a "function of a function." To illustrate, consider the functional F defined by

$$F(u) = \int_0^1 u^2(x)\, dx, \tag{1.1}$$

where the domain D of F might be defined to be the set of all (real-valued) functions u defined over $0 \leq x \leq 1$, subject only to the restriction that the integral (1.1) does in fact exist. For instance, if $u(x) = 6x$, then $F(u) = 12$. In this example it is clear that the range R of F is the positive real axis, $0 \leq F(u) < \infty$.

Functionals are the objects of interest in the so-called *variational calculus*, which is the subject of Chapter 10.

1.2. *LIMITS, CONTINUITY, AND UNIFORM CONTINUITY*

We say that $f(x)$ has a **limit** L as x tends to x_0 and write

$$\lim_{x \to x_0} f(x) = L \quad \text{or} \quad f(x) \longrightarrow L \text{ as } x \longrightarrow x_0 \tag{1.2}$$

if *to each* $\epsilon > 0$ *(no matter how small) there corresponds a* $\delta(\epsilon, x_0) > 0$ *such that*

$|f(x) - L| < \epsilon$ *whenever* $0 < |x - x_0| < \delta$; in other words, $f(x)$ will be close to L if x is close to x_0.

Example 1.1. Consider $f(x) = 1/x$ over $0.1 \leq x \leq 1$. At the x_0 shown in Fig. 1.3, we claim that the limit exists and is equal to the L shown—namely, $1/x_0$. To prove it, we will put forward a suitable $\delta(\epsilon, x_0)$. First, we draw an arbitrarily small ϵ band

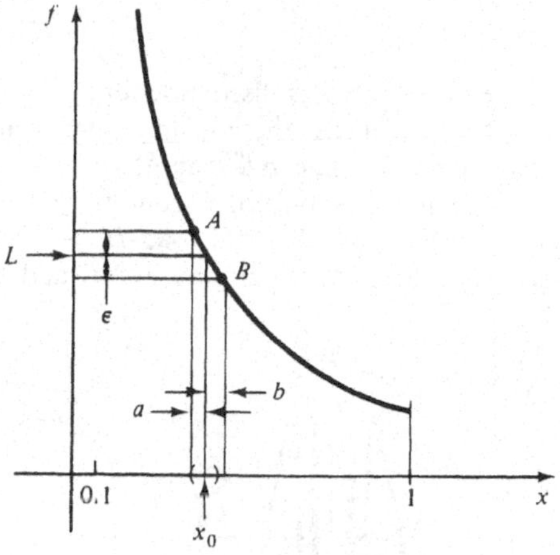

Figure 1.3. $\lim f(x)$ for $f(x) = 1/x$.

about $f(x) = L$ and, where it intersects the graph (at A and B), drop verticals to the x axis. Noting that $a < b$, we choose $\delta = a$; then $|x - x_0| < \delta$ is the centered interval denoted by the small parentheses. Surely, when x is closer to x_0 than δ—that is, $0 < |x - x_0| < \delta$, $f(x)$ will be closer to L than ϵ, as desired. To evaluate δ in terms of ϵ and x_0, we write $f(x_0 - \delta) - f(x_0) = \epsilon$—that is, $1/(x_0 - \delta) - 1/x_0 = \epsilon$. Solving,

$$\delta(\epsilon, x_0) = x_0 - \frac{x_0}{\epsilon x_0 + 1}. \tag{1.3}$$

Of course, any *smaller* (nonzero) value, say $\delta = [x_0 - x_0/(\epsilon x_0 + 1)]/5$, would do just as well. ▮

Note carefully that because of the $0 < |x - x_0| < \delta$ in our definition of $\lim_{x \to x_0} f(x)$, $f(x_0)$ need not *equal* L. For instance, if

$$f(x) = \begin{cases} x^2, & 0 \leq x \leq 4 \quad \text{but } x \neq 3 \\ 27, & x = 3 \end{cases} \tag{1.4}$$

then $\lim_{x \to 3} f(x) = 9$, *not* 27! If *in addition* to having $\lim_{x \to x_0} f(x) = L$ we also have $f(x_0) = L$, we say that $f(x)$ is **continuous** at x_0. In ϵ, δ language, *$f(x)$ is continuous at x_0 if to* *each* $\epsilon > 0$ *there corresponds a* $\delta(\epsilon, x_0) > 0$ *such that* $|f(x) - f(x_0)| < \epsilon$ *whenever* $|x - x_0| < \delta$. [Recalling our definition of lim, we see that the foregoing definition of continuity is equivalent to the statement $\lim_{\Delta x \to 0} f(x + \Delta x) = f(x)$.] Clearly, the function given in Example 1.1 is continuous over its domain, x^2 and $\sin x$ are continuous for all x (i.e., for $-\infty < x < \infty$), and so on. On the other hand, the so-called **Heaviside step function**, defined by

$$H(x) = \begin{cases} 1, & x > 0 \\ 0, & x < 0 \end{cases} \tag{1.5}$$

and $H(0) = \frac{1}{2}$, say, is discontinuous at $x = 0$; for any ϵ smaller than $\frac{1}{2}$ we see [from a graph of $H(x)$] that there is no δ such that $H(x)$ remains within an ϵ band of $H(0)$ whenever x is closer to 0 than δ.

A more heroic type of discontinuity is exhibited by the function $f(x) = \cos(1/x)$. To sketch its graph, note that $f(x) = \cos(1/x^2)x \equiv \cos \omega x$, where the frequency $\omega = 1/x^2$ tends to infinity as $x \to 0$ and to zero as $x \to \infty$, as shown in Fig. 1.4.

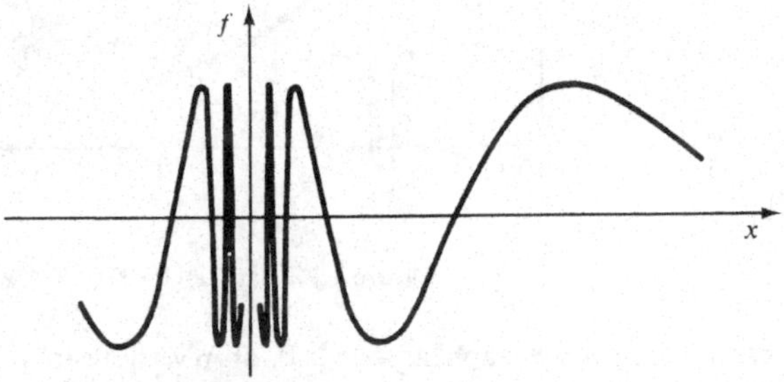

Figure 1.4. Graph of $f(x) = \cos(1/x)$.

Actually, $f(x)$ is not defined at $x = 0$, since $\cos \infty$ is undefined. So in setting up this example, let us define $f(x) = \cos(1/x)$ for all $x \neq 0$ and $f(0) \equiv A$, say. No matter what value is assigned to A, $f(x)$ is discontinuous at $x = 0$ because $\lim_{\Delta x \to 0} f(0 + \Delta x) = \lim_{\Delta x \to 0} \cos(1/\Delta x)$ cannot equal A.

Finally, we say that $f(x)$ is **uniformly** *continuous over D if to each* $\epsilon > 0$ *there corresponds a* $\delta(\epsilon)$ (i.e., *independent of* x_0) *such that* $|f(x) - f(x_0)| < \epsilon$ *whenever* $|x - x_0| < \delta$ *for all* x_0*'s in D*. Consider Example 1.1 again. For a given ϵ it is clear that $\delta(\epsilon, x_0)$ will be smallest, over $0.1 \leq x_0 \leq 1$, when $x_0 = 0.1$, since a is "skinniest" there. Thus $\delta(\epsilon, 0.1) = 0.1 - 1/(\epsilon + 10) \equiv \delta(\epsilon)$ will suffice for *all* x_0's in D, and hence $f(x) = 1/x$ is uniformly continuous over $0.1 \leq x \leq 1$. On the other hand, suppose that D were $0 < x \leq 1$ instead. Then $\delta(\epsilon, x_0)$ tends to zero as $x_0 \to 0$, since a merely gets skinnier and skinnier. Consequently, we are unable to single out one value to serve as our $\delta(\epsilon)$ over all D, and $f(x) = 1/x$ is *not* uniformly continuous over $0 < x \leq 1$.

Applying these ideas to *functionals*, what might be meant by a statement like "$F(u)$ is continuous at u_0"? Following the foregoing discussion on *functions*, we will mean that to each $\epsilon > 0$ there corresponds a $\delta(\epsilon, u_0)$ such that $|F(u) - F(u_0)| < \epsilon$ whenever u is closer to u_0 than δ, as indicated in Fig. 1.5. Note carefully that whereas the range R is part of a real axis, the domain D is a set of functions, indicated *schematically* in the figure, with each point in D representing a function. The only catch is, what do we mean by "u is *closer* to u_0 than δ"? Well, we need to introduce a notion of distance for the set D by defining a so-called **distance function**, or **metric**, $d(u, v)$. That is, for any points u, v in D we define a real-valued function $d(u, v)$ called "the distance from u to v." It seems entirely reasonable that we require it to be nonnegative and the same as

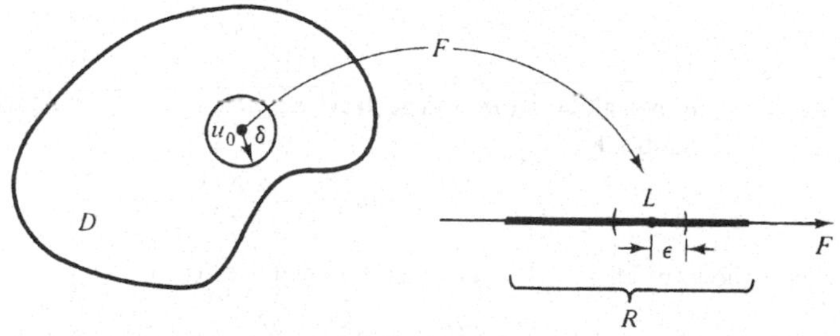

Figure 1.5. Schematic representation of a functional.

the distance from v to u. Also, the distance from u to itself should certainly be zero. So let us require of $d(u, v)$ that it satisfy the following conditions.

(i) *Symmetry:*

$$d(u, v) = d(v, u) \tag{1.6}$$

(ii) *Positiveness:*

$$\begin{aligned} d(u, v) &> 0 \quad \text{for} \quad u \neq v \\ &= 0 \quad \text{for} \quad u = v \end{aligned} \tag{1.7}$$

(iii) *Triangle Inequality:*

$$d(u, w) \leq d(u, v) + d(v, w), \tag{1.8}$$

where the triangle inequality, not mentioned above, is motivated by the simple schematic picture in Fig. 1.6 and is the abstract analog of the Euclidean proposition that the length of any one side of a triangle cannot exceed the sum of the lengths of the other two sides.

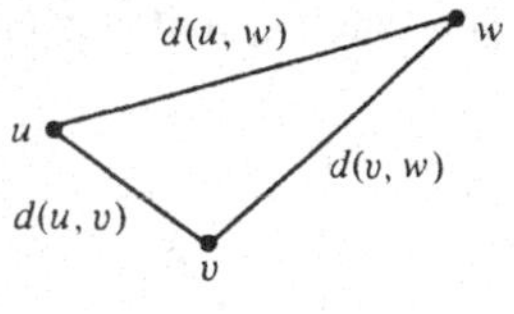

Figure 1.6. Schematic motivation for triangle inequality.

One distance function that can be assigned to D is simply

$$d(u, v) \equiv \begin{cases} 0, & u = v \\ 1, & u \neq v, \end{cases} \tag{1.9}$$

but this is not very discriminating. A more reasonable choice might be

$$d(u, v) \equiv \max_{0 \leq x \leq 1} |u(x) - v(x)|, \tag{1.10}$$

where the functions in D are defined over $0 \le x \le 1$, say; both choices satisfy the requirements (1.6) to (1.8). To illustrate, $d(5x, 4x^2)$ is 1 according to (1.9) and 25/16 according to (1.10).

With this done, we are now in a position to talk about $\lim F(u)$, continuity of F, and so on. By $\lim_{u \to u_0} F(u) = L$, for example, we mean that to each $\epsilon > 0$ (no matter how small) there corresponds a $\delta(\epsilon, u_0)$ such that $|F(u) - L| < \epsilon$ whenever $0 < d(u, u_0) < \delta$. That is (Fig. 1.5), $F(u)$ must be closer to L than ϵ if u is kept within a "neighborhood" of radius δ of u_0.

1.3. DIFFERENTIATION

Recall from the calculus that the **derivative** df/dx, or $f'(x)$, is defined as the limit of the difference quotient

$$\frac{df}{dx} = \lim_{\Delta x \to 0} \frac{f(x + \Delta x) - f(x)}{\Delta x} \tag{1.11}$$

when the limit exists. If $f(x) = x^2$ for instance, then

$$f'(x) = \lim_{\Delta x \to 0} \frac{(x + \Delta x)^2 - x^2}{\Delta x} = \lim_{\Delta x \to 0} (2x + \Delta x) = 2x \tag{1.12}$$

for all x.

It is easy to show that *if $f(x)$ is differentiable, then it must be continuous* because (Exercise 1.6)

$$\lim_{\Delta x \to 0} [f(x + \Delta x) - f(x)] = \lim_{\Delta x \to 0} \frac{f(x + \Delta x) - f(x)}{\Delta x} \Delta x = f'(x) \cdot 0 = 0. \tag{1.13}$$

For example, since the Heaviside function $H(x)$ is discontinuous at $x = 0$, it follows that $H'(0)$ does not exist; that is, $H(x)$ is not differentiable at $x = 0$. We say that it is *singular* there.

The converse of the preceding statement, however, is not true—continuity does *not* imply differentiability. To illustrate, the continuous function $f(x) = |x|$ is not differentiable at $x = 0$ because its difference quotient, $|\Delta x|/\Delta x \equiv \operatorname{sgn} \Delta x$ (read as "signum Δx" or "the sign of Δx"), is discontinuous at $\Delta x = 0$ (Fig. 1.7), so that $\lim_{\Delta x \to 0}$ of the difference quotient does not exist. In fact, Weierstrass (1815–1897) constructed an example of a function that is continuous everywhere and yet differentiable *nowhere*!

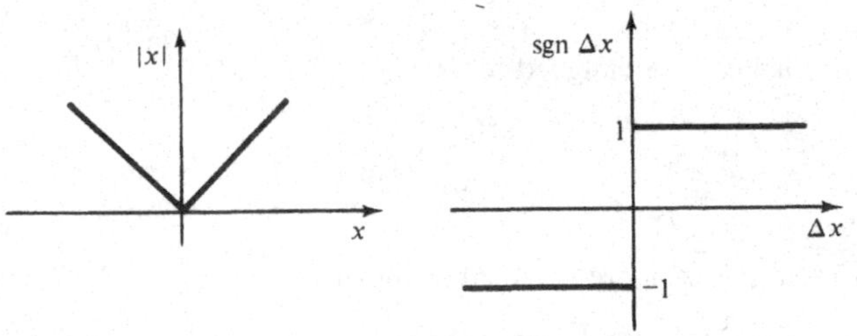

Figure 1.7. $|x|$ and its difference quotient at $x = 0$.

Differentiability can break down at various kinds of singularities: at a jump discontinuity [as in $H(x)$], at a kink (as in $|x|$), and at a point in the neighborhood of which the function wiggles too much (amplitudewise) and too fast (frequencywise) [as in $g(x)$ of Exercise 1.15]. As a result of the standard graphical interpretation of $f'(x)$ as the *slope* of f at x, it is clear that $f'(x)$ will also fail to exist at a point where the slope is infinite, as for $f(x) = \sqrt{x}$ (defined over $0 \leq x < \infty$) at $x = 0$ (Fig. 1.8).

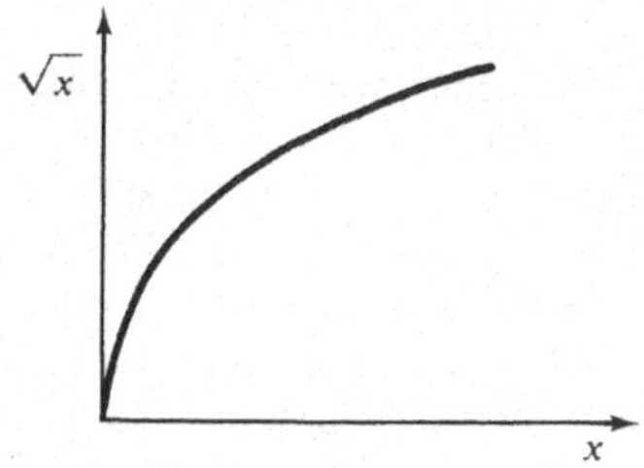

Figure 1.8. Singularity in $\sqrt{x}$ at $x = 0$.

1.4. *INTEGRATION*

Consider next the *integral* of $f(x)$, say from $x = a$ to $x = b$, where both a and b are finite; the case where one or both limits are infinite will be considered later. First, we partition the interval $a \leq x \leq b$ by choosing an integer n and designating a set of points x_0 to x_n, not necessarily evenly spaced, such that $x_0 = a < x_1 < x_2 < \cdots < x_{n-1} < x_n = b$, as shown in Fig. 1.9. Second, we designate a ξ_k point in each subinterval such that $x_{k-1} \leq \xi_k \leq x_k$ and compute the **Riemann sum**

$$\sum_{k=1}^{n} f(\xi_k)(x_k - x_{k-1})$$

corresponding to the partition P. Then we choose a *finer* partition—that is, one with a smaller *norm* (defined as the size of the largest subinterval and denoted as $|P|$)—and

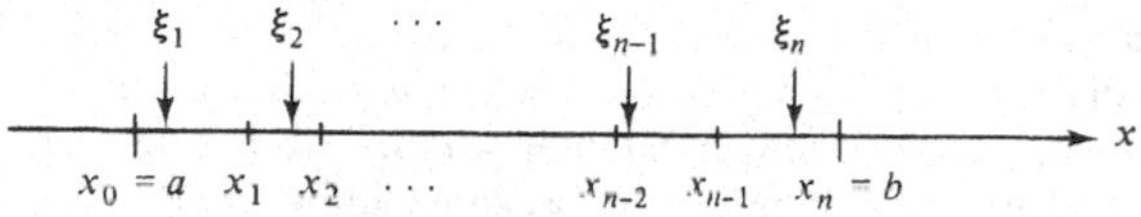

Figure 1.9. A typical partition P.

recompute the Riemann sum that corresponds to the new partition. Doing so for a sequence of partitions whose norms tend to zero, we define the **Riemann integral** as the limit of the sequence of Riemann sums,

$$I = \int_a^b f(x)\,dx \equiv \lim_{|P|\to 0} \sum_{k=1}^{n} f(\xi_k)(x_k - x_{k-1}), \tag{1.14}$$

provided, of course, that the limit does exist[1]; if so, we say that the integral *exists* or *converges*.

So far we've assumed that $a < b$. To complete our definition, we define the integral to be zero if $a = b$ and the negative of the integral from b to a if $b < a$.

---

[1]In ϵ, δ language, we say that to each $\epsilon > 0$ there corresponds a $\delta(\epsilon) > 0$ such that

$$\left| \int_a^b f(x)\,dx - \sum_{k=1}^{n} f(\xi_k)(x_k - x_{k-1}) \right| < \epsilon$$

whenever $|P| < \delta$.

There are theorems which guarantee existence of the integral if $f(x)$ is sufficiently well behaved—for instance, if it is *continuous* or of *bounded variation*[2] over $a \leq x \leq b$. Of more concern to us in applied mathematics, however, is the problem of actually *evaluating* integrals. First, however, let us consider a moment how we evaluate *derivatives*. We can compute the derivatives of a number of "elementary functions" (x^2, $\sin x$, e^x, etc.) directly from the limit definition (1.11) as we did in (1.12). Then we show that the derivative d/dx is a **linear operator**—that is,

$$\frac{d}{dx}(\alpha u + \beta v) = \alpha\frac{du}{dx} + \beta\frac{dv}{dx} \tag{1.15}$$

for any scalars α, β and any (differentiable) functions u, v; we show that

$$\frac{d}{dx}(uv) = u\frac{dv}{dx} + v\frac{du}{dx}, \qquad \frac{d}{dx}f[u(x)] = \frac{df}{du}\frac{du}{dx}, \tag{1.16}$$

and so on. With these operational formulas, as well as a fairly short list of derivatives like $d(x^2)/dx = 2x$, $d(\sin x)/dx = \cos x, \ldots$, we can then handle more complicated combinations, such as $3x^2 \sin 4x/(x^3 + 1)$.

The general procedure for evaluating integrals is about the same. We generate a short list of integrals of some simple functions and then add a few operational formulas, such as linearity,

$$\int_a^b (\alpha u + \beta v)\,dx = \alpha \int_a^b u\,dx + \beta \int_a^b v\,dx, \tag{1.17}$$

integration by parts,

$$\int_a^b u\,dv = (uv)\Big|_a^b - \int_a^b v\,du, \tag{1.18}$$

plus various tricks like "partial fractions" and "clever substitutions" that are undoubtedly familiar from calculus. Here and in later chapters other techniques will be developed, including numerical integration, Leibnitz differentiation, asymptotic expansions, contour integration, as well as some additional "tricks."

Note, however, that we *don't* evaluate the integral of even our simple functions (x^2, $\sin x, \ldots$) directly from the limit definition (1.14) because it is too unwieldy (exceptions like Exercise 1.16 notwithstanding). Instead we return to the (simpler) realm of differentiation by means of the **Fundamental Theorem of the Integral Calculus**:

If $f(x)$ is continuous over $a \leq x \leq b$ and

$$\int_a^x f(\xi)d\xi \equiv F(x) \tag{1.19}$$

for $a \leq x \leq b$, then $F'(x) = f(x)$ over $a \leq x \leq b$.[3] *Conversely, if $f(x)$ is continuous for $a \leq x \leq b$ and $F'(x) = f(x)$, then $\int_a^x f(\xi)\,d\xi = F(x)$.*

---

[2]By the **total variation** of $f(x)$ between two points a and b, denoted $\mathrm{TV}_a^b f$, we mean our total "up-down" travel as we move from $x = a$ to $x = b$ along the graph of $f(x)$. For example, $\mathrm{TV}_0^\pi \sin x = 2$, $\mathrm{TV}_{-2}^4 H(x) = 1$, and $\mathrm{TV}_0^1 \sin 1/x = \infty$. If $\mathrm{TV}_a^b f$ is finite, we say that f is of **bounded variation** over $a \leq x \leq b$.

[3]*Indefinite Integrals.* Sometimes we write $\int f(x)\,dx$ (i.e., without limits) or sometimes $\int^x f(x)\,dx$. Called an *indefinite integral* of f, it denotes a function whose derivative is f. For example, $\int 2x\,dx = x^2$ or $x^2 - 37$ or, as generally as possible, $x^2 + C$, where C is an arbitrary constant.

From this theorem and our list of derivatives it follows that the integral of $\sin x$ is $-\cos x$, . . . and so on.

Consider the integral

$$\frac{2}{\sqrt{\pi}}\int_0^x e^{-\xi^2}\,d\xi \equiv \operatorname{erf} x \qquad (0 \le x < \infty). \tag{1.20}$$

It turns out that this integral is not expressible in terms of elementary functions. Yet it appears quite often in applications and so has been given its own name, the **error function** $\operatorname{erf} x$, and has been tabulated. The $2/\sqrt{\pi}$ is introduced by way of normalization so that $\operatorname{erf} \infty = 1$, and

$$1 - \operatorname{erf} x = \frac{2}{\sqrt{\pi}}\int_x^\infty e^{-\xi^2}\,d\xi \equiv \operatorname{erfc} x \tag{1.21}$$

is called the **complementary error function** $\operatorname{erfc} x$. But tables, because of interpolation problems and storage, are not too convenient in computer applications, particularly if the domain is infinite as in (1.20). The trend in recent years has been away from tabulations and toward approximate expressions, not just for $\operatorname{erf} x$ but for virtually all of the important special functions. For instance, the expression

$$\operatorname{erf} x \approx 1 - (a_1 t + a_2 t^2 + a_3 t^3 + a_4 t^4 + a_5 t^5)e^{-x^2} \tag{1.22}$$

where

$$t = \frac{1}{1 + 0.3275911x}, \quad a_1 = 0.254829592, \quad a_2 = -0.284496736$$

$$a_3 = 1.421413741, \quad a_4 = -1.453152027, \quad a_5 = 1.061405429$$

due to Hastings[4] is uniformly accurate over $0 \le x < \infty$ to within $\pm 1.5 \times 10^{-7}$! A virtually indispensable reference for special functions, containing a wealth of computational formulas like (1.22) (as well as a variety of other information), is the *Handbook of Mathematical Functions.*[5]

Needless to say, we often encounter integrals that we cannot evaluate in terms of known functions—that is, elementary functions and/or special functions for which tables or approximate formulas are available. We may then wish to carry out the integration "numerically." Roughly speaking, this process means going back to the Riemann sum definition (1.14) or some variation of it. First, however, let us introduce some important notation.

1.5. ASYMPTOTIC NOTATION AND THE "BIG OH"

By $f(x) \sim g(x)$ *as* $x \to x_0$*; we mean that* $[f(x)/g(x)] \to 1$ *as* $x \to x_0$; we say that f is *asymptotic to* g as $x \to x_0$. For instance, $(3x^4 + x + 8)/(x^2 + 2)$ is asymptotic to $3x^2$

[4]C. Hastings, Jr., *Approximations for Digital Computers*, Princeton University Press, Princeton, N.J., 1955.

Note: Besides the footnoted references, a list of Additional References is included at the end of each of Parts I to V.

[5]M. Abramowitz and I. Stegun (Eds.), National Bureau of Standards Applied Math Series, 1964. Updated material concerning the special functions can be found in Y. L. Luke, *Mathematical Functions and their Approximation*, Academic, New York, 1975.

(i.e., $3x^4/x^2$) as $x \longrightarrow \infty$, to 4 (i.e., 8/2) as $x \longrightarrow 0$, and to 4 as $x \longrightarrow 1$, say. As another illustration, recalling the Taylor expansion

$$\sin x = x - \frac{x^3}{3!} + \frac{x^5}{5!} - \cdots,$$

it is apparent that $\sin x \sim x$ as $x \longrightarrow 0$ or, revealing a bit more[6], $\sin x \sim x - x^3/6$. But for the limit $x \longrightarrow \infty$ note that the asymptotic notation is of no help, since there is no "simpler" function to which $\sin x$ is asymptotic; in other words, the best that we can manage is the trivial statement that $\sin x \sim \sin x$.

It is often more convenient to use the *big oh* Bachmann–Landau *order of magnitude* symbol. *By $f(x) = O[g(x)]$ as $x \longrightarrow x_0$, we simply mean that $f(x)/g(x)$ is bounded as $x \longrightarrow x_0$.* For example, $(3x^4 + x + 8)/(x^2 + 2)$ is $O(x^2)$ as $x \longrightarrow \infty$[7] and $O(1)$ as $x \longrightarrow 0$. There is no need to worry about multiplicative constants like 3 or 100, since we are not stating asymptotic behavior but only the order of magnitude. For instance, $O(3x^2)$ and $O(x^2)$ are entirely equivalent, and so we simply omit the 3. As final illustrations, observe that

$$\sin x = \begin{cases} O(x) & \text{as } x \longrightarrow 0 \\ x + O(x^3) & \text{as } x \longrightarrow 0 \\ O(1) & \text{as } x \longrightarrow \infty. \end{cases}$$

Be sure to see Exercises 1.19 and 1.20 concerning e^x, e^{-x} and $\ln x$.

1.6. NUMERICAL INTEGRATION

Returning to (1.14), suppose that we break the interval into n equal parts of length $(b - a)/n \equiv h$ (i.e., $x_k = a + kh$) and choose $\xi_k = x_{k-1}$—that is, at the left endpoint of each interval. As $n \longrightarrow \infty$, $|P| \longrightarrow 0$ and the Riemann sum tends to I. But we're not going to let $n \longrightarrow \infty$ (in fact, doing so would be a neat trick!); we're going to choose a large n and accept whatever error results. Denoting $f(\xi_k) = f(x_{k-1}) \equiv f_{k-1}$, we have

$$I \approx h(f_0 + f_1 + \cdots + f_{n-1}) \equiv R_n. \tag{1.23}$$

This is known as the **rectangular rule,** and R_n corresponds to the cross-hatched area in Fig. 1.10, where we've set $n = 4$ for simplicity. If $f'(x)$ is continuous over $a \leq x \leq b$, it can be shown[8] that the discrepancy between I and R_n can be expressed as

$$I - R_n = \frac{(b - a)hf'(\xi)}{2} \equiv E_n^R \tag{1.24}$$

for some ξ between a and b. Although there's no way to determine ξ, (1.24) is still useful as an error estimate. Suppose, for instance, that $b - a = 1$ and $f'(\xi) \approx 1$. If

---

[6]Actually, we could also argue that $\sin x \sim x + x^3$, since $\sin x/(x + x^3)$ does $\longrightarrow 1$ as $x \longrightarrow 0$; but the understanding is that if we bother to include a third-order term, it should at least be correct! That is, transposing the x term, note that $\sin x - x \sim -x^3/6$ *is* true, but $\sin x - x \sim x^3$ is *not*.

[7]Actually, it is $O(x^p)$ for any $p \geq 2$, but the limiting value $p = 2$ is the most informative and surely the most natural choice.

[8]See, for example, the very nice book by S. D. Conte, *Elementary Numerical Analysis*, McGraw-Hill, New York, 1965. Incidentally, you may wonder how $f'(x)$ can exist and *not* be continuous. If so, see Exercise 1.15.

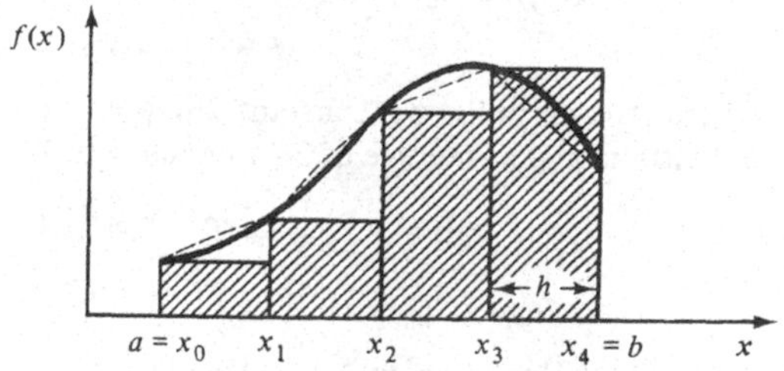

Figure 1.10. Rectangular and trapezoidal rules, for $n = 4$.

we desire 10^{-5} accuracy (i.e., to five decimal places), it is clear from (1.24) that we need $h \approx 2 \times 10^{-5}$ or $n \approx 50{,}000$. Not very encouraging!

Instead suppose that we had chosen $\xi_k = x_k$—that is, at the *right* endpoint of each interval instead of at the left. Then we'd have $I \approx h(f_1 + f_2 + \cdots + f_n)$. Adding this result to (1.23) and dividing by 2, we obtain the **trapezoidal rule**

$$I = h\left(\frac{f_0}{2} + f_1 + f_2 + \cdots + f_{n-1} + \frac{f_n}{2}\right) \equiv T_n, \tag{1.25}$$

where T_n can be interpreted as the area under the dashed lines in Fig. 1.10. This time [assuming that $f''(x)$ is continuous over $a \le x \le b$] it is found that

$$I - T_n = -\frac{(b-a)h^2 f''(\xi)}{12} \equiv E_n^T, \tag{1.26}$$

where ξ is some point between a and b, not the same as ξ of (1.24) except by coincidence. The important point to note is that whereas E_n^R was $O(h)$, E_n^T is $O(h^2)$, so that for a given (small) h (1.25) should be appreciably more accurate than (1.23). Or if the rectangular rule needs $n \approx 50{,}000$ to achieve a certain accuracy, the trapezoidal rule will need only $n \approx \sqrt{50{,}000} \approx 224$ for comparable accuracy!

Example 1.2. Suppose that we plan to use the trapezoidal rule to evaluate erf 1 and would like 10^{-6} accuracy. To determine how small h needs to be, note that

$$E_n^T = -(1-0)h^2 \frac{2}{\sqrt{\pi}} e^{-\xi^2} \frac{4\xi^2 - 2}{12}.$$

Not knowing ξ, let us be conservative and use $\xi = 0$, which gives the maximum value of $e^{-\xi^2}(4\xi^2 - 2)$ over $0 \le \xi \le 1$. Then $|E_n^T| \le h^2/3\sqrt{\pi}$. Equating this result to 10^{-6} gives $h = 0.002306$, or $n = (1-0)/h = 433.7$—say 434 to be safe. Using "extended-precision" arithmetic, $n = 434$ would probably suffice. Using "single-precision" arithmetic, however, machine roundoff might become significant and we may be tempted to increase n somewhat to allow for that. Unfortunately, increasing n is likely to increase the effects of roundoff. In fact, from the calculations presented in Conte[9] for this example, it appears that 10^{-6} accuracy simply cannot be achieved by using the trapezoidal rule and single-precision arithmetic, no matter how large we make n. ▮

But we can do much better than the trapezoidal rule. For instance, recall that

[9] *Ibid.*, p. 124.

$I = T_n + O(h^2)$. More precisely, it can be shown (e.g., Conte, p. 127) that

$$I = T_n + Ah^2 + O(h^4), \tag{1.27}$$

where A is an unknown constant. Suppose that we run *two* calculations,[10] one with n divisions and a step size of $2h$ and one with $2n$ divisions and a step size of h. Then

$$I = T_n + A(2h)^2 + O(h^4)$$
$$I = T_{2n} + Ah^2 + O(h^4).$$

Multiplying the second equation by 4 and subtracting give

$$I = \frac{4T_{2n} - T_n}{3} + O(h^4). \tag{1.28}$$

Since T_n and T_{2n} are already computed, (1.28) gives I to within an error that is now $O(h^4)$, compared with $O(h^2)$ for either T_n or T_{2n} individually. Furthermore, we can repeat the process. For instance, if we suspect (or can prove) that (1.27) proceeds in even powers of h, then (1.28) will actually be

$$I = \frac{4T_{2n} - T_n}{3} + Bh^4 + O(h^6),$$

and by repeating the preceding process, we can eliminate the Bh^4 term and be left with an error that is $O(h^6)$ (Exercise 1.23). This procedure is known as **Romberg integration.**

Romberg integration actually amounts to an *extrapolation* process. That is, knowing how the error decays with h in (1.27), basically as Ah^2, we are able to determine A empirically (or, what is equivalent, to eliminate it) by fitting two data points to (1.27)—by writing down (1.27) for two different n's. This basic idea is important and will be employed again with great success in the discussion on summation of series.

Before continuing, it is interesting to write out (1.28). Recalling (1.25), we have

$$\begin{aligned} I &= \frac{1}{3}\Big[4(h)\Big(\frac{f_0}{2} + f_1 + f_2 + \cdots + f_{2n-1} + \frac{f_{2n}}{2}\Big) \\ &\quad -(2h)\Big(\frac{f_0}{2} + f_2 + f_4 + \cdots + f_{2n-2} + \frac{f_{2n}}{2}\Big)\Big] + O(h^4) \\ &= \frac{h}{3}(f_0 + 4f_1 + 2f_2 + 4f_3 + \cdots + 4f_{2n-1} + f_{2n}) + O(h^4), \end{aligned} \tag{1.29}$$

which is the well-known **Simpson's rule.** Derivation of Simpson's rule usually follows other lines; here it occurs as the first Romberg extrapolation of the trapezoidal rule.

The numerical integration formulas considered thus far have all involved partitions consisting of equal subdivisions and are known as **Newton–Cotes**-type formulas. They are of the form

$$I \approx \sum_{k=0}^{n} A_k f(x_k), \tag{1.30}$$

where $x_k = a + kh$ and the A_k "weights" are selected (as in the discussion above) so

[10]Note that half the f_k function values needed in T_{2n} (i.e., every other one) have already been computed for T_n and need not be calculated again.

that the resulting error is of reasonably high order in h. Gauss (1777–1855), on the other hand, found that, for a given n, greater accuracy can generally be achieved if the x_k "abscissas" are not fixed a priori but rather are selected together with the A_k weights in an optimal way for a wide class of integrands. But "optimal" is a rather subjective term. What *Gauss* meant by optimal was that (1.30) should be *exact for all polynomial* $f(x)$*'s of as high a degree as possible*. Since we have $2n + 2$ parameters at our disposal $(A_0, \ldots, A_n, x_0, \ldots, x_n)$ and an Nth-degree polynomial contains $N + 1$ parameters (i.e., the coefficients), we expect to be able to obtain a formula that is exact for all polynomials up to and including degree $2n + 1$.

It is customary to first change the integration limits from a, b to $-1, 1$. This step is easily accomplished by means of the linear change of variables

$$x = \left(\frac{b-a}{2}\right)t + \left(\frac{b+a}{2}\right). \tag{1.31}$$

Then

$$\int_a^b f(x)\,dx = \frac{b-a}{2}\int_{-1}^{1} F(t)\,dt, \tag{1.32}$$

where $F(t) = f[x(t)]$,[11] and so it suffices to consider

$$\int_{-1}^{1} F(t)\,dt \approx A_0F(t_0) + \cdots + A_nF(t_n). \tag{1.33}$$

For example, let $n = 1$. Then we want to choose A_0, A_1, t_0, t_1 so that (1.33) is exact for all polynomials of degree 3 or less or, equivalently, for $F(t) = 1, t, t^2$, and t^3. Setting F equal to $1, t, t^2, t^3$ in (1.33) yields the algebraic equations

$$\left.\begin{aligned} 2 &= A_0 + A_1 \\ 0 &= A_0t_0 + A_1t_1 \\ \tfrac{2}{3} &= A_0t_0^2 + A_1t_1^2 \\ 0 &= A_0t_0^3 + A_1t_1^3 \end{aligned}\right\} \tag{1.34}$$

in A_0, A_1, t_0, t_1. Although these coupled equations are nonlinear, it's not hard to show that $A_0 = A_1 = 1$ and $t_0 = -t_1 = 1/\sqrt{3}$, so that (1.33) becomes

$$\int_{-1}^{1} F(t)\,dt \approx F\left(\frac{1}{\sqrt{3}}\right) + F\left(\frac{-1}{\sqrt{3}}\right). \tag{1.35}$$

For n *not* small, the nonlinear equations for the A_k's and t_k's are apparently quite unwieldy. However, it can be shown that the t_k's are the zeros of the so-called Legendre polynomials (Exercise 1.24). With the t_k's thus determined, the A_k's are easily found, since the governing equations are *linear* in the A_k's [see, for example, (1.34)]. We have listed the A_k's and t_k's in Table 1.1 for $n = 3$ and 9; 15-decimal-place tabulations, for n's as large as 95, can be found in Abramowitz and Stegun.[12]

---

[11] This is a point of nomenclature that sometimes causes confusion. Suppose, for example, that $f(x) = x^2$ and $x = 2t + 1$. Then $f[x(t)] = (2t + 1)^2 = 4t^2 + 4t + 1 \equiv F(t)$. Thus $f(\ ) = (\ )^2$, whereas $F(\ ) = 4(\ )^2 + 4(\ ) + 1$; that is, f and F are *different functions*, which is why we use different names, f and F.

[12] *Ibid.*, p. 916.

Example 1.3. As in Example 1.2, let us compute erf 1.

$$I = \text{erf } 1 = \frac{2}{\sqrt{\pi}} \int_0^1 e^{-x^2}\, dx = \frac{1}{\sqrt{\pi}} \int_{-1}^1 e^{-(t+1)^2/4}\, dt.$$

Using "extended-precision," four-point Gauss ($n = 3$), and ten-point Gauss ($n = 9$) integration yield $I \approx 0.84270117$ and 0.84270079, respectively, compared with the

TABLE 1.1
Abscissas and Weight Factors for Gaussian Integration

$n = 3$			$n = 9$		
k	A_k	t_k	k	A_k	t_k
0	0.6521451549	0.3399810436	0	0.2955242247	0.1488743390
1	0.3478548451	0.8611363116	1	0.2692667193	0.4333953941
2, 3	See note below		2	0.2190863625	0.6794095683
			3	0.1494513492	0.8650633667
			4	0.0666713443	0.9739065285
			5–9	See note below	

Note: The missing values (i.e., $k = 2, 3$ for $n = 3$ and $k = 5, \ldots, 9$ for $n = 9$) are given by $A_{n-k} = A_k$ and $t_{n-k} = -t_k$.

known (tabulated) value $I \doteq 0.84270079$. By way of comparison, we find that the trapezoidal rule (1.25) gives

$$T_{50} = 0.84267312$$
$$T_{100} = 0.84269387$$
$$T_{200} = 0.84269906$$

and a double Romberg extrapolation of these values (Exercise 1.23) yields the value 0.84270079. ∎

Example 1.4. Consider

$$I = \int_0^1 \frac{\sqrt{x}\,\cos x^2}{x+1}\, dx. \tag{1.36}$$

We know that N-point Gauss integration (i.e., $N = n + 1$) will be *exact* (except, of course, for roundoff error) if $f(x)$ is a polynomial of degree $2N - 1$ or less and is expected to be inexact but *very accurate* if $f(x)$ can be *closely approximated* by such a polynomial. Since $f(x)$ in (1.36) is continuous over $0 \le x \le 1$ we *know* that it can be uniformly approximated to any desired accuracy by a sufficiently high degree polynomial.[13] But as $x \longrightarrow 0$, observe that the integrand behaves like $\sqrt{x}$, which has an infinite slope at $x = 0$ (Fig. 1.8) and is not at all agreeable to polynomial approximation; thus a discouragingly large N will be needed in order to achieve good accuracy. One solution is to make a change of variables, say $x = \xi^2$, which

[13]This statement is guaranteed by the famous **Weierstrass approximation theorem**, which states that if $f(x)$ is real valued and continuous over $a \le x \le b$, then given any $\epsilon > 0$ (no matter how small) there exists a polynomial $p(x)$ such that $|f(x) - p(x)| < \epsilon$ for all x's in the interval.

stretches out the immediate neighborhood of the origin (e.g., $x = 0.01$ becomes $\xi = 0.1$) so that the slope at the origin is brought down to a finite value. In fact, (1.36) then becomes

$$I = 2\int_0^1 \frac{\xi^2 \cos \xi^4}{\xi^2 + 1}\, d\xi, \tag{1.37}$$

and we see that the new integrand behaves like $2\xi^2$ as $\xi \longrightarrow 0$ (since $\cos \xi^4 \sim 1$ and $\xi^2 + 1 \sim 1$), which is fine. Application of Gauss integration to (1.37) will undoubtedly be more successful than to (1.36).

Another interesting feature of (1.36) lies in the $\cos x^2$ factor. Its frequency increases linearly with x (i.e., $\cos x^2 = \cos \omega x$, where the frequency $\omega = x$), so that if our upper limit were 20, say, instead of 1, the integrand would then suffer *many* oscillations. The integration methods discussed, all of which are fundamentally based on the notion of polynomial approximation, would not be able to cope and special techniques would be necessary.[14] ▮

1.7. DIFFERENTIATION OF INTEGRALS CONTAINING A PARAMETER; LEIBNITZ'S RULE

The integral

$$I(\alpha) = \int_{a(\alpha)}^{b(\alpha)} f(x, \alpha)\, dx \tag{1.38}$$

is a function of the "parameter" α. (It is *not* a function of x, which is only the dummy variable of integration.) As we shall see, it is important to know how to take the derivative

$$I'(\alpha) = \frac{d}{d\alpha}\int_{a(\alpha)}^{b(\alpha)} f(x, \alpha)\, dx. \tag{1.39}$$

In principle, we can do so by evaluating the integral and then taking $d/d\alpha$ of it; in practice, however, it may be advantageous somehow to invert the order of integration and differentiation. Forming the difference quotient,

$$\begin{aligned}
\frac{I(\alpha + \Delta\alpha) - I(\alpha)}{\Delta\alpha} &= \frac{1}{\Delta\alpha}\left\{\int_{a+\Delta a}^{b+\Delta b} f(x, \alpha + \Delta\alpha)\, dx - \int_a^b f(x, \alpha)\, dx\right\} \\
&= \int_a^b \frac{f(x, \alpha + \Delta\alpha) - f(x, \alpha)}{\Delta\alpha}\, dx \\
&\quad + \frac{1}{\Delta\alpha}\int_b^{b+\Delta b} f(x, \alpha + \Delta\alpha)\, dx - \frac{1}{\Delta\alpha}\int_a^{a+\Delta a} f(x, \alpha + \Delta\alpha)\, dx \\
&\sim \int_a^b \frac{f(x, \alpha + \Delta\alpha) - f(x, \alpha)}{\Delta\alpha}\, dx + f(b, \alpha + \Delta\alpha)\frac{\Delta b}{\Delta\alpha} \\
&\quad - f(a, \alpha + \Delta\alpha)\frac{\Delta a}{\Delta\alpha},
\end{aligned} \tag{1.40}$$

since f is essentially constant (if it is continuous) over the infinitesimal intervals

[14] See, for example, I. M. Longman, A Method for the Numerical Evaluation of Finite Integrals of Oscillatory Functions, *Mathematics of Computation*, Vol. 14, 1960, pp. 53–59.

$b \leq x \leq b + \Delta b$ and $a \leq x \leq a + \Delta a$. Formally letting $\Delta\alpha \to 0$, we obtain the **Leibnitz rule,**[15]

$$I'(\alpha) = \int_{a(\alpha)}^{b(\alpha)} \frac{\partial f(x, \alpha)}{\partial \alpha}\, dx + f[b(\alpha), \alpha]\frac{db}{d\alpha} - f[a(\alpha), \alpha]\frac{da}{d\alpha}. \tag{1.41}$$

Suppose that $\alpha_1 \leq \alpha \leq \alpha_2$, and that $c_1 \leq a(\alpha) \leq c_2$ and $c_1 \leq b(\alpha) \leq c_2$ for each α in $\alpha_1 \leq \alpha \leq \alpha_2$. It can be shown that *a sufficient condition for the validity of (1.41) is that $\partial f/\partial\alpha$ be continuous in the rectangular x, α domain $c_1 \leq x \leq c_2$, $\alpha_1 \leq \alpha \leq \alpha_2$ (and, of course, that the derivatives $\partial f/\partial\alpha$, $db/d\alpha$ and $da/d\alpha$ all exist)*. For applications that we are likely to encounter, however, it is generally true that (1.41) is correct simply if the integral in (1.41) does converge.

For the special case where a and f are independent of α and $b(\alpha) = \alpha$, the Leibnitz rule reduces to the Fundamental Theorem of the Integral Calculus, (1.19).

Example 1.5. To illustrate the "mechanics" of (1.41), note that if

$$I(\alpha) = \int_{-\alpha}^{\alpha^2} \sin(\alpha x^2)\, dx,$$

then

$$I'(\alpha) = \int_{-\alpha}^{\alpha^2} x^2 \cos \alpha x^2\, dx + 2\alpha \sin \alpha^5 + \sin \alpha^3. \quad \blacksquare$$

Example 1.6. To illustrate how (1.41) can be of help in evaluating integrals, consider

$$I = \int_0^1 \frac{x - 1}{\ln x}\, dx. \tag{1.42}$$

To evaluate it, consider instead

$$J(\alpha) = \int_0^1 \frac{x^\alpha - 1}{\ln x}\, dx. \tag{1.43}$$

Then (if $\alpha \neq -1$)

$$J'(\alpha) = \int_0^1 \frac{x^\alpha \ln x}{\ln x}\, dx = \left.\frac{x^{\alpha+1}}{\alpha + 1}\right|_0^1 = \frac{1}{\alpha + 1}.$$

Integrating this simple differential equation for J,

$$J(\alpha) = \ln(\alpha + 1) + C.$$

To evaluate the constant of integration C, observe from (1.43) that $J(0) = 0$. Thus $0 = \ln 1 + C$, so $C = 0$. Finally,

$$I = J(1) = \ln(1 + 1) = \ln 2. \quad \blacksquare$$

A word of caution. To compute $I'(x)$, where

$$I(x) = \int_0^x x^2\, dx,$$

it is a good idea first to reexpress

$$I(x) = \int_0^x \xi^2\, d\xi$$

[15] When a mathematician uses "formally," what is really meant is "*in*formally"—that is, without worrying about questions of rigor at each step.

say, to emphasize that the ξ's and the x are entirely distinct; ξ is the dummy variable of integration and x is the right endpoint of the interval of integration. Thus $I'(x) = x^2$.

EXERCISES

Exercises 1.1 through 1.7 are on the algebra of limits. It would be best to read through them even if you plan to work only one or two.

1.1. Prove the simple inequality $|A + B| \leq |A| + |B|$, which will be useful in the following exercises. Show that it implies that

$$|A_1 + A_2 + \cdots + A_n| \leq |A_1| + \cdots + |A_n|.$$

1.2. Prove that if $\lim_{x\to a} f(x) = A$ and $\lim_{x\to a} f(x) = B$, then B *must* equal A; that is, the limit must be unique.

1.3. Is $\lim_{x\to a} f(x) = A$ equivalent to the statement $\lim_{x\to a} (f(x) - A) = 0$? Prove or disprove.

1.4. Prove (a) $\lim_{x\to a} Cf(x) = C \lim_{x\to a} f(x)$. (Allow for the possibility that $C = 0$.)

(b) $\lim_{x\to a} [f(x) + g(x)] = \lim_{x\to a} f(x) + \lim_{x\to a} g(x)$.

I'll do (b); you do (a). Let $\lim_{x\to a} f(x) \equiv A$ and $\lim_{x\to a} g(x) \equiv B$. Then for any $\epsilon > 0$ there is a δ_1 such that $|f(x) - A| < \epsilon$ whenever $0 < |x - a| < \delta_1$ and there is a δ_2 such that $|g(x) - B| < \epsilon$ whenever $0 < |x - a| < \delta_2$. Then

$$\begin{aligned} |[f(x) + g(x)] - (A + B)| &= |[f(x) - A] + [g(x) - B]| \\ &\leq |f(x) - A| + |g(x) - B| \\ &< \epsilon + \epsilon = 2\epsilon \equiv \epsilon' \end{aligned}$$

for any $\epsilon' > 0$ whenever $0 < |x - a| <$ the smaller of δ_1, δ_2.

1.5. Using 1.4(a) and (b), show that $\lim_{x\to a}$ is a *linear* operator—that is, that

$$\lim_{x\to a} [\alpha f(x) + \beta g(x)] = \alpha \lim_{x\to a} f(x) + \beta \lim_{x\to a} g(x)$$

for arbitrary scalars α, β, provided, of course, that the two limits on the right exist.

1.6. Does $\lim_{x\to a} [f(x)g(x)] = [\lim_{x\to a} f(x)][\lim_{x\to a} g(x)]$? Not always; for instance, if $a = 0$, $f(x) = 1/x$, and $g(x) = x$, then the left-hand side is clearly 1, whereas the right-hand side is undefined because $\lim_{x\to 0} 1/x$ does not exist! But it *is* true if both limits on the right-hand side exist. For suppose that $f(x) \longrightarrow A$ and $g(x) \longrightarrow B$. Then for any $\epsilon > 0$ there is a δ_1 such that $|f(x) - A| < \epsilon$ whenever $0 < |x - a| < \delta_1$ and there is a δ_2 such that $|g(x) - B| < \epsilon$ whenever $0 < |x - a| < \delta_2$. Let ϵ be < 1. Then $|f(x)| = |A + [f(x) - A]| < |A| + \epsilon < |A| + 1$; so

$$\begin{aligned} |f(x)g(x) - AB| &= |f(x)g(x) - Bf(x) + Bf(x) - AB| \\ &\leq |f(x)||g(x) - B| + |B||f(x) - A| \\ &< (|A| + 1)\epsilon + |B|\epsilon. \end{aligned}$$

Finish the story.

1.7. If $\lim_{x\to a} f(x) = A \neq 0$, prove that $\lim_{x\to a} 1/f(x) = 1/A$. This time no hints.

1.8. Does $x^2 + y^2 = 4$ define a function $y(x)$ over $|x| \leq 2$? Discuss.

1.9. Is $f(x) = x^2$ continuous over $0 \leq x < \infty$? Uniformly continuous? Why (not)? Interpret graphically.

1.10. Show that $f(x) = \sin x$ is uniformly continuous over $0 \leq x \leq \pi/2$. Give a suitable $\delta(\epsilon)$ and interpret graphically.

1.11. What is a linear functional? Is $F(u) = u(0)$ linear? $F(u) = u^2(0)$?

***1.12.** Show that the distance functions (1.9) and (1.10) satisfy the requirements (1.6) to (1.8).

***1.13.** Consider the functional $F(u) = u(0)$, whose domain is the set of all continuous functions $u(x)$ defined over $0 \leq x \leq 1$. Show whether $F(u)$ is continuous with respect to the distance function (1.10).

1.14. Show that the integral of the Heaviside function is $\int H(x)\,dx = xH(x) + C$. Sketch $H(x)$, $xH(x)$, and $H(x - a)$.

***1.15.** We showed in (1.13) that if $f'(x)$ exists at x, then f must be continuous there. But does the *derivative* $f'(x)$ need to be continuous there? Well, if $f'(x)$ had a *jump* discontinuity at x_0, then its integral $f(x)$ would have a kink there (e.g., recall the preceding exercise) and would not have been differentiable in the first place. But there are other kinds of discontinuities besides jump discontinuities. In fact, show that for

$$g(x) = \begin{cases} x^2 \sin \dfrac{1}{x}, & x \neq 0 \\ 0, & x = 0 \end{cases}$$

$g'(x)$ exists at $x = 0$ but is not continuous there.
(You may be interested in the "dictionary" of exceptional functions and examples compiled by B. Gelbaum and J. Olmsted in *Counterexamples in Analysis*, Holden-Day, San Francisco, 1964.)

1.16. Show, directly from the definition (1.14), that

$$\int_a^b x\,dx = \frac{b^2 - a^2}{2}.$$

Hint: Choose equal subintervals and choose each ξ_k at the center of the kth subinterval.

1.17. Show that if $m \leq f(x) \leq M$ over $a \leq x \leq b$, then

$$m(b - a) \leq \int_a^b f(x)\,dx \leq M(b - a).$$

1.18. (a) $xe^{-x}/(\cos x + x^3)$ is asymptotic to what, as $x \longrightarrow 0$?
(b) Give asymptotic expressions for $(3x - \sqrt{x^2 + 2})/(\cos x + x^3)$ as $x \longrightarrow \infty$ and $x \longrightarrow 0$.

1.19. (*Important*) Show that $e^{-x} = O(1)$ as $x \longrightarrow \infty$. As a more informative statement, show that $e^{-x} = O(x^{-n})$ for arbitrarily large n. Note that *no matter how large we make n, e^{-x} dies out faster than x^{-n}!* Since we can't seem to "get a handle" on e^{-x} in terms of powers of x (i.e., since n is arbitrarily large), we often don't attempt to find a simpler form and simply leave the e^{-x}'s intact. For instance, we might write $x^2e^{-x}/(\cos x + 3x)$

---

*Asterisked material, both in text and exercises, is either a little more difficult or of a supplementary nature and can be bypassed with no loss of continuity.

$= O(xe^{-x})$ as $x \longrightarrow \infty$ rather than "$O(x^{-n})$ for n arbitrarily large and positive." How about e^{+x} as $x \longrightarrow \infty$?

1.20. (*Important*) Sketch the graph of $\ln x$ for $0 < x < \infty$. Show that $\ln x = O(x^{\alpha})$ as $x \longrightarrow \infty$, where α is an *arbitrarily small* positive number. Thus $\ln x$ has an *extremely* weak singularity at infinity. Show that it has an equally weak singularity at the origin—that is, that $\ln x = O(x^{-\alpha})$ as $x \longrightarrow 0$, where α is again an arbitrarily small positive number. Since we can't seem to "get a handle" on $\ln x$ as $x \longrightarrow 0$ or ∞ (i.e., since α is arbitrarily small), frequently we simply leave $\ln x$ as $\ln x$. For example, we might write $(3x^2 + \sin x) \ln x = O(x^2 \ln x)$ as $x \longrightarrow \infty$ rather than "$O(x^{2+\alpha})$, where α is an arbitrarily small positive number."

1.21. Estimate n so that integration of $\int_1^2 x^{-1/2}e^{-x}\,dx$ by the trapezoidal rule is accurate to within $\pm 10^{-5}$.

1.22. Whereas the trapezoidal rule

$$\int_{x_0}^{x_1} f(x)\,dx \approx \frac{h(f_0 + f_1)}{2}$$

is exact if $f(x)$ is linear, show that Simpson's rule

$$\int_{x_0}^{x_2} f(x)\,dx \approx \frac{h(f_0 + 4f_1 + f_2)}{3}$$

is exact if $f(x)$ is a quadratic (i.e., a parabolic arc).

1.23. (*Romberg integration*) Starting with $I = (4T_{2n} - T_n)/3 + Bh^4 + O(h^6)$, obtain

$$I = \frac{64T_{4n} - 20T_{2n} + T_n}{45} + O(h^6).$$

1.24. The **Legendre polynomials** $P_m(x)$ are defined by the recursion formula

$$P_{m+1}(x) = \frac{(2m+1)xP_m(x) - mP_{m-1}(x)}{m+1}$$

for $m = 1, 2, \ldots$, where $P_0(x) \equiv 1$ and $P_1(x) \equiv x$. Recall that the t_k abscissas in (1.33) (i.e., for the $n + 1$-point Gauss) are the zeros of $P_{n+1}(x)$. Compute the t_k's for the four-point Gauss scheme, say, to slide rule accuracy and compare with Table 1.1 (left half). Then with the t_k's known solve the pertinent simultaneous equations for the weights A_0, A_1, A_2, A_3, to slide rule accuracy, and compare with Table 1.1.

1.25. Apply the Leibnitz rule to

(a) $I(\alpha) = \int_{-\alpha}^{\alpha} \sin \alpha x^2\,dx.$ (b) $I(x) = \int_0^{x^2} x^3 \ln x\,dx.$

1.26. Verify that

$$x(t) = \frac{1}{k}\int_0^t \sin k(t - \tau) f(\tau)\,d\tau \qquad (k = \text{constant})$$

satisfies the differential equation $d^2x/dt^2 + k^2x = f(t)$, plus the initial conditions $x(0) = (dx/dt)(0) = 0$.

1.27. Knowing that

$$\int_0^{\pi} \frac{dx}{1 + a\cos x} = \frac{\pi}{(1 - a^2)^{1/2}}$$

for $a^2 < 1$, show that

$$\int_0^\pi \frac{\ln(1 + 0.7\cos x)}{\cos x}\,dx = \pi \sin^{-1} 0.7.$$

1.28. Use the known integral $\int_0^\pi \cos ax\,dx = (\sin a\pi)/a$ to evaluate $\int_0^\pi x^2 \cos x\,dx$.

1.29. Evaluate $\int_0^1 x^{0.7} \ln x\,dx$.

Hint: Start with the known integral $\displaystyle\int_0^1 x^\alpha\,dx = \frac{1}{\alpha + 1}$ $(\alpha > -1)$.

1.30. The Bessel function of integral order n may be defined as

$$J_n(x) = \frac{1}{\pi}\int_0^\pi \cos(n\xi - x\sin\xi)\,d\xi.$$

Using this expression, show that

$$J_n'(x) = \tfrac{1}{2}[J_{n-1}(x) - J_{n+1}(x)].$$

1.31. Make up two examination-type questions on the material in Chapter 1. (Potentially, this can be a valuable and nontrivial exercise.)

Chapter 2

Infinite Series

Limit Processes: Differentiation, Integration, and Infinite Series might have been covered in Chapter 1, but the discussion of infinite series is long and best considered separately. As usual, we try to tie together the underlying principles and the practical and computational aspects. For instance, beyond determining if a given series is convergent, there is the crucial practical problem of actually summing it. Thus a convergent series may be so *slowly* convergent as to be essentially worthless insofar as computation is concerned, whereas merely the first few terms of certain *divergent* series may provide excellent results. Furthermore, we will see that it is often possible to take a given series and *accelerate* its convergence substantially by means of certain nonlinear transformations.

2.1. SEQUENCES AND SERIES; FUNDAMENTALS AND TESTS FOR CONVERGENCE

Whereas a finite sum

$$\sum_{k=1}^{N} a_k = a_1 + a_2 + \cdots + a_N, \tag{2.1}$$

is well defined (thanks to the associative law of addition), an *infinite* sum or **series**

$$\sum_{k=1}^{\infty} a_k = a_1 + a_2 + a_3 + \cdots \tag{2.2}$$

is not because it is clearly not possible actually to add up an infinite number of terms.

To give meaning to (2.2), we *define*

$$\sum_{k=1}^{\infty} a_k \equiv \lim_{n\to\infty} \sum_{k=1}^{n} a_k. \tag{2.3}$$

That is, we form the sequence of *partial sums* s_n, where

$$s_1 = a_1, \qquad s_2 = a_1 + a_2, \qquad s_3 = a_1 + a_2 + a_3, \cdots$$

and then define the sum of the infinite series as the limit of the sequence of partial sums.[1] If $\lim s_n = s$ exists,[2] we say that the series is *convergent* or *converges to* s; if not, we say that the series is *divergent* or *diverges*.

In "ϵ language"

$$\sum_{k=1}^{\infty} a_k = \lim s_n = s \tag{2.4}$$

means that to each ϵ (no matter how small) there corresponds an $N(\epsilon)$ such that $|s_n - s| < \epsilon$ for all $n > N$.

It is important to be aware that even though (2.3) is the standard definition, it is not the *only* one possible. Various "summability methods" are sometimes employed, of which (2.3) (often called *ordinary convergence*) is only one.[3] For instance, according to *Cesàro summability*, which is especially useful in the theory of Fourier series,

$$\sum_{k=1}^{\infty} a_k \equiv \lim \left(\frac{s_1 + s_2 + \cdots + s_n}{n}\right), \tag{2.5}$$

that is, the limit of the arithmetic means of the partial sums. It can be shown that if a series converges to s (i.e., according to "ordinary convergence"), then it will also be Cesàro-summable, in fact, to the same value. Yet there are series that diverge in the "ordinary" sense, that are, nevertheless, Cesàro-summable.

Example 2.1. Consider the well-known **geometric series** $1 + x + x^2 + \cdots$. Here the terms are functions of x, but we think of x as being fixed so that we actually have a series of constants. Now

$$\begin{aligned} s_n &= 1 + x + x^2 + \cdots + x^{n-1} \\ xs_n &= \phantom{1 +{}} x + x^2 + \cdots + x^{n-1} + x^n \\ \hline s_n - xs_n &= 1 - x^n \end{aligned}$$

and so

$$s_n = \frac{1 - x^n}{1 - x} \tag{2.6}$$

provided that $x \neq 1$; if $x = 1$, then clearly $s_n = n$. Letting $n \longrightarrow \infty$, $x^n \longrightarrow 0$ or diverges, depending on whether $|x| < 1$ or $|x| > 1$, respectively. Treat $|x| = 1$ separately. If $x = +1$, then $s_n = n$, which diverges (to infinity); and if $x = -1$, then the s_n sequence oscillates between 1 and 0 and hence again diverges (not to

---

[1]Time and time again, in mathematics, new concepts are introduced and defined as limits of old ones. Here, for example, an infinite sum is defined as the limit of a sequence of finite sums.

[2]In this section lim will be understood to mean $\lim\limits_{n\to\infty}$.

[3]For an interesting account of the various summability theories, see G. H. Hardy, *Divergent Series*, Oxford University Press, Oxford, 1949.

infinity this time but simply because it does not converge). So we can say that the geometric series $1 + x + x^2 + \cdots$ converges to

$$s = \frac{1}{1-x} \quad \text{if } |x| < 1 \tag{2.7}$$

and diverges if $|x| \geq 1$.

Next, reconsider the geometric series from the Cesàro point of view. We find (Exercise 2.1) that

$$\frac{s_1 + s_2 + \cdots + s_n}{n} = \frac{1}{1-x} - \frac{x}{n}\frac{1-x^n}{(1-x)^2} \tag{2.8}$$

if $x \neq 1$. As before, this result tends to $1/(1-x)$ as $n \longrightarrow \infty$ for $|x| < 1$ and diverges for $|x| > 1$ (Exercise 2.2). And as before, we have divergence if $x = +1$ (Exercise 2.3). The only difference is that, for $x = -1$, the limit of (2.8) *does* exist and equals $\frac{1}{2}$,[4] whereas *ordinary convergence* tells us the series diverges. Which is correct? Is $1 - 1 + 1 - 1 + \cdots = \frac{1}{2}$, or does it diverge? The answer depends on whether we define the series according to (2.5) or (2.3), and the definition adopted is up to the person who wrote the series down! From here on we adopt the notion of ordinary convergence (2.3). ∎

It should be clear that any finite number of (finite) terms of a series—for example, the first six million—in no way affects the convergence or divergence of the series. In other words, it is the "tail" of the series that counts. This idea is basic to the Cauchy convergence theorem. First,

DEFINITION. We say that s_n is a ***Cauchy sequence***[5] *if to each $\epsilon > 0$ (no matter how small), there corresponds an $N(\epsilon)$ such that $|s_m - s_n| < \epsilon$ for all m and $n > N$.*

Then we have

THEOREM 2.1. (Cauchy Convergence Theorem) A sequence s_n is convergent if and only if it is a Cauchy sequence.

This is unquestionably one of the central concepts of analysis, as is more apparent in Part III. Proof of the theorem consists of two parts: first, we show that if s_n is convergent, then it must be a Cauchy sequence; next, we show the converse—that if s_n is a Cauchy sequence, it must be convergent.

---

[4]Although we see this result directly from (2.8), let's actually look at some numbers. The series is $1 - 1 + 1 - 1 + \cdots$; $s_1 = 1$, $s_2 = 0$, $s_3 = 1$, $s_4 = 0, \ldots$; $s_1/1 = 1$, $(s_1 + s_2)/2 = 1/2$, $(s_1 + s_2 + s_3)/3 = 2/3$, $(s_1 + s_2 + s_3 + s_4)/4 = 1/2, \ldots$, and it *is* apparent that the limit is 1/2. We might note that the same result was obtained by Guido Grandi (1671–1742), a monk and professor at Pisa, using a somewhat different notion of summability. Considering the case of a father who bequeathes a gem to his two sons, who are to retain it one year each, in alternation, he argued that each son would possess the gem 1/2 of the time.

[5]The s_n's may be the *partial sums of a certain series* or simply some sequence without reference to any series; by a *sequence*, we simply mean an ordered set of numbers.

Proof. Suppose that s_n is convergent, say to s. Then for any $\epsilon/2 > 0$ there must correspond an N such that

$$|s_m - s| < \frac{\epsilon}{2} \quad \text{for all } m > N,$$

$$|s_n - s| < \frac{\epsilon}{2} \quad \text{for all } n > N.$$

Adding,

$$|s_m - s| + |s_n - s| < \epsilon \quad \text{for all } m, n > N.$$

But using the simple inequality of Exercise 1.1 (with $A \equiv s_m - s$ and $B = s - s_n$), we have

$$|s_m - s_n| \le |s_m - s| + |s_n - s| < \epsilon$$

for all $m, n > N$, and so s_n must, in fact, be a Cauchy sequence.

Proof of the converse is contained in the * section below.

*

First some background. Consider a general set S—for example, the set of points $0 < x < 1$ of an x axis, the points $0 < x < 1$ and $y = 0$ of an x, y plane, the set of all differentiable functions defined over $a \le x \le b$, and so on. And suppose that we introduce a notion of "distance between two points" by defining a distance function $d(u, v)$ according to (1.6) to (1.8).

Next, we define a **neighborhood** $N(x_0, R)$ *of a point* x_0 *as the set of all points* x *such that* $d(x_0, x) < R$; R is called the "radius" of the neighborhood. We say that x_0 is a **limit point** *of* S *if each* $N(x_0, R)$ *(no matter how small) contains at least one point of* S *other than* x_0 *itself.* [In fact, it follows that each $N(x_0, R)$ must then contain an *infinite* number of points of S. Suppose that, for some R, there were only a finite number of such points. We could find a point x' in $N(x_0, R)$ that is closest to x_0, and then with $d(x_0, x') \equiv R'$ the neighborhood $N(x_0, R')$ would contain only x_0, which contradicts the assumption that x_0 is a limit point.] As we'll see shortly, x_0 can be a limit point of S without actually belonging to S. If S does contain all its limit points, we say that it is **closed**.

Example 2.2. Let S be the set of points $x = 1/n$ ($n = 1, 2, 3, \ldots$) of an x axis with the usual Euclidean distance function $d(u, v) = |u - v|$. In this case, a neighborhood $N(x_0, R)$ is the one-dimensional interval $|x - x_0| < R$, centered at x_0. Clearly, $x = 0$ is a limit point of S (the only one, in fact), since every interval centered at $x = 0$, no matter how small, contains an infinite number of points of S. But note that $x = 0$ is not in S (n *approaches* infinity but never quite gets there, so that $1/n$ *approaches* zero but never quite gets there), and so S is not closed. Of course, if $x = 0$ were added to S, then the resulting set would be closed. ▮

Example 2.3. Let us reconsider the foregoing example with the Euclidean metric replaced by the simple metric (1.9)—namely,

$$d(u, v) = \begin{cases} 0, & u = v \\ 1, & u \neq v. \end{cases} \tag{2.9}$$

Then S has *no* limit points! In particular, $x = 0$ is no longer a limit point, since $N(0, R)$ contains only $x = 0$ for all Rs < 1. (Think of yourself as a tiny insect at the origin with an arrangement of lenses for eyes so that when you look to the right you perceive *all* the points $0 < x < \infty$ to be lumped at a single location that is a unit distance away, and similarly when you look to the left.) ▮

A set S is said to be **bounded** if there is a constant $M < \infty$ such that $d(u, v) < M$ for all u, v in S. [Note that the set $x = n$ ($n = 1, 2, \ldots$) is bounded with respect to the metric (2.9) but unbounded with respct to the Euclidean metric.]

THEOREM 2.2. (*Bolzano–Weierstrass*) *If a bounded set S (with Euclidean metric) contains an infinite number of points, then it must have at least one limit point.*

Finally, we are in a position to complete our proof of the Cauchy convergence theorem. We've already shown that if a sequence s_n is convergent, then it must be a Cauchy sequence, and now we want to prove the converse. Suppose, then, that s_n is a Cauchy sequence. With $\epsilon = 1$, say, it follows that there exists an N such that $|s_m - s_n| < 1$ for all $m, n > N$. For definiteness, let $n = N + 1$. Then s_m lies closer to s_{N+1} than unity for all $m > N$; therefore the sequence is bounded. According to the Bolzano–Weierstrass theorem, it follows that the s_n sequence has a limit point, and it is not hard to see that that limit point must be unique (Exercise 2.7).

An important and basic result that follows easily from the Cauchy convergence theorem is that *in order for the series* $\sum^{\infty} a_n$ *to converge, it is necessary that* $a_n \to 0$ *as* $n \to \infty$. For if we set $m = n - 1$, then to every $\epsilon > 0$ (no matter how small) there corresponds an N such that $|s_{n-1} - s_n| = |a_n| < \epsilon$ for all $n > N$, which is exactly the same (in ϵ language) as saying that $a_n \to 0$ as $n \to \infty$. Unfortunately, $a_n \to 0$ is not also *sufficient* to ensure convergence. (Our story would certainly be much simpler if it were.) This result is obvious from the well-known **harmonic series** $1 + \frac{1}{2} + \frac{1}{3} + \frac{1}{4} + \ldots$, since $a_n = 1/n \to 0$, and yet

$$|s_{2n} - s_n| = \frac{1}{n+1} + \frac{1}{n+2} + \cdots + \frac{1}{n+n}$$

$$> \frac{1}{n+n} + \frac{1}{n+n} + \cdots + \frac{1}{n+n} = \frac{n}{2n} = \frac{1}{2}$$

for all n. Consequently, s_n is not a Cauchy sequence and hence diverges.

Here we have just used the Cauchy convergence theorem to prove that the harmonic series diverges, but the procedure was designed specifically for the harmonic series and its generalization is not apparent; for instance, try applying it to the series with $a_n = 1/n^2$. The same is true for the geometric series of Example 2.1, which we tested by getting s_n into "closed form" and examining $\lim s_n$ directly; this step was especially fortuitous because not only did we answer the question of convergence/divergence, we also actually summed the series! Yet it is a rare series that is willing to let us get its s_n into closed form. What we need, then, is a handful of tests that can be easily applied to a wide variety of series. Before considering a few of them,[6] one other fundamental result should be mentioned.

THEOREM 2.3. (*Linearity*)

$$\sum_{k=1}^{\infty} (\alpha a_k + \beta b_k) = \alpha \sum_{k=1}^{\infty} a_k + \beta \sum_{k=1}^{\infty} b_k,$$

provided that both series on the right-hand side converge.

Proof. Left for Exercise 2.8.

Now a few of the standard tests for convergence.

THEOREM 2.4. (*Comparison Test*) *If for two series $\sum a_n$ and $\sum b_n$ of positive terms there exists a constant K such that $a_n < Kb_n$ for all sufficiently large n, then the convergence of $\sum b_n$ implies the convergence of $\sum a_n$; or if there is some $k > 0$ such that $kb_n < a_n$ for all sufficiently large n, then the divergence of $\sum b_n$ implies the divergence of $\sum a_n$.*

It follows that if we can find *both* constants k and K—that is, such that $k < a_n/b_n < K$—then $\sum a_n$ and $\sum b_n$ either converge together or diverge together. Surely this will be the case if $a_n/b_n \sim C$ as $n \to \infty$ (where $C > 0$, of course), since we can then choose $k = C/2$ and $K = 2C$, say.

THEOREM 2.5. (*Ratio Test*) *Consider a series $\sum a_n$ of positive terms, where $\lim a_{n+1}/a_n = r$—that is, $a_{n+1}/a_n \sim r$. If $r < 1$, the series converges; if $r > 1$, it diverges; and if $r = 1$, we obtain no information. If, for the case $r = 1$, we can further show that $a_{n+1}/a_n \sim 1 - r'/n$, then we can say that the series converges if $r' > 1$ and diverges if $r' < 1$. If $r' = 1$, we again obtain no information.*

Proof. We will prove only the first part ($r < 1$). Choosing any point q between r and 1—that is, $r < q < 1$—it follows from the statement $\lim a_{n+1}/a_n = r$ that we will have $a_{n+1}/a_n < q$ for all sufficiently large n's. In this case,

$$a_{n+1} < qa_n$$

$$a_{n+2} < qa_{n+1} < q^2 a_n$$

$$a_{n+3} < qa_{n+2} < q^3 a_n,$$

and so on, so that

$$a_n + a_{n+1} + \cdots < (1 + q + q^2 + \cdots)a_n,$$

which is convergent, according to Example 2.1, because $0 < q < 1$.

Example 2.4. Consider the so-called ***p* series**

$$\sum_1^{\infty} \frac{1}{n^p} \equiv \zeta(p), \tag{2.10}$$

[6]A more complete treatment can be found in the little book by J. M. Hyslop, *Infinite Series*, 5th ed., Oliver and Boyd, Edinburgh and London, 1954. A more advanced reference is the book by T. M. Apostol, *Mathematical Analysis*, Addison-Wesley, Reading, Mass., 1957.

whose sum (as a function of p) is known as the **Riemann zeta function**. Applying the ratio test, $a_{n+1}/a_n = n^p/(n+1)^p \sim 1$, which yields no information. But expanding *beyond* the leading term,

$$\frac{a_{n+1}}{a_n} = \left(1 + \frac{1}{n}\right)^{-p} = 1 - \frac{p}{n} + \cdots \sim 1 - \frac{p}{n}.$$

Thus $r' = p$, and so (2.10) converges for $p > 1$ and diverges for $p < 1$. For $p = 1$ we obtain no information; but in this case, (2.10) is the so-called harmonic series, which, as mentioned earlier, is divergent. ▮

Example 2.5. Consider

$$\sum_1^\infty \left(\frac{1}{n} - \frac{1}{n+10}\right). \tag{2.11}$$

It is tempting to split (2.11) into $\sum 1/n - \sum 1/(n+10)$ and claim that it diverges because these two series do. But Theorem 2.3 endorses this procedure only if "both series on the right-hand side converge," and here they do not. Instead note that

$$a_n = \frac{1}{n} - \frac{1}{n+10} = \frac{10}{n(n+10)} \sim \frac{10}{n^2} \equiv b_n,$$

and since $\sum b_n$ is a convergent p series ($p = 2$), it follows from the comparison test that $\sum a_n$ converges, too. ▮

Example 2.6.

$$\sum_1^\infty \left(\frac{\cos n}{2n-1}\right)^2. \tag{2.12}$$

Since $\cos^2 n < 1$ for all n, we have $a_n < 1/(2n-1)^2 \sim 1/4n^2$, which is again a convergent p series, and so (2.12) converges. More simply, we may note directly from (2.12) that $a_n = O(1/n^2)$ and conclude that the series converges. ▮

2.2. *SERIES WITH TERMS OF MIXED SIGN*

The last two theorems deal with series whose terms are all positive (actually, just of the same sign for all sufficiently large n's). More generally, suppose that the signs of the a_n's are mixed. As a first step, we establish a link between such series and the series of positive terms already considered.

THEOREM 2.6. (Absolute Convergence) If $\sum |a_n|$ converges, then so does $\sum a_n$, and we say that $\sum a_n$ converges "absolutely."

Proof. Defining

$$u_n = a_n, \quad v_n = 0 \quad \text{if } a_n \geq 0,$$

and

$$u_n = 0, \quad v_n = -a_n \quad \text{if } a_n \leq 0,$$

it follows that $u_n \geq 0$, $v_n \geq 0$ and

$$|a_n| = u_n + v_n, \qquad a_n = u_n - v_n.$$

The first implies that $u_n \leq |a_n|$ and $v_n \leq |a_n|$; and since $\sum |a_n|$ is convergent, then so are $\sum u_n$ and $\sum v_n$ according to the comparison test. Finally, $\sum u_n - \sum v_n = \sum (u_n - v_n) = \sum a_n$ converges according to the Linearity Theorem 2.3.

Example 2.7. Consider the series

$$\sum_1^\infty a_n = 1 - \tfrac{1}{2} + \tfrac{1}{3} - \tfrac{1}{4} + \cdots. \tag{2.13}$$

If $\sum |a_n|$ converges, then Theorem 2.6 assures us that $\sum a_n$ does. But $\sum |a_n| = 1 + \frac{1}{2} + \frac{1}{3} + \cdots$ is the harmonic series, which is divergent; so the theorem does not apply and we are in need of help. Which brings us to

THEOREM 2.7. (The Dirichlet Test) If $b_n \downarrow 0$[7] as $n \longrightarrow \infty$ and the sequence a_n has bounded partial sums, then the series $\sum a_n b_n$ converges.

Let us return to Example 2.7. Writing

$$1 - \frac{1}{2} + \frac{1}{3} - \cdots = \sum_1^\infty \frac{(-1)^{n+1}}{n},$$

let us take $1/n$ to be b_n and $(-1)^{n+1}$ to be a_n. Then $1/n \downarrow 0$ as required, and the partial sums

$$\sum_{k=1}^n a_k = 1, 0, 1, 0, \ldots \quad \text{for } n = 1, 2, 3, 4, \ldots$$

are surely bounded. For instance,

$$\left| \sum_{k=1}^n a_k \right| < 6, \text{ say, for all } n, \tag{2.14}$$

so that (2.13) is convergent. Since it is convergent but not absolutely convergent, we say that it **converges conditionally**.

Because of the $(-1)^{n+1}$ sign alternation, (2.13) is called an *alternating series*. The Dirichlet test implies that *any* alternating series $\sum (-1)^{n+1} b_n$ converges if $b_n \downarrow 0$. Of course, not all series with terms of mixed sign are alternating series. ▮

Consider the next example.

Example 2.8.

$$\sum_1^\infty \frac{\sin^3 (n\pi/4)}{n+1}. \tag{2.15}$$

Set $b_n = 1/(n+1)$, which does decrease monotonely to zero. The a_n's are

$$a_n = \sin^3 \frac{n\pi}{4} = \frac{1}{2\sqrt{2}},\quad 1,\quad \frac{1}{2\sqrt{2}},\quad 0,\quad -\frac{1}{2\sqrt{2}},\quad -1,\quad -\frac{1}{2\sqrt{2}},\quad 0, \ldots$$

for $n = 1, 2, \ldots$. Their partial sums build up to $1 + 1/\sqrt{2}$ when n is 3 and then decrease back to zero when n is 7 and so on. Thus (2.14) again applies, and so (2.15) converges. ▮

[7]By $b_n \downarrow 0$ as $n \longrightarrow \infty$, we mean that b_n is *monotone decreasing to zero*; that is, $m > n$ implies that $b_m \leq b_n$ and $b_n \longrightarrow 0$ as $n \longrightarrow \infty$.

The Dirichlet test can also be applied to series of positive terms. To illustrate, with $\sum 1/n^4$ we can set $a_n = 1/n^2$ and $b_n = 1/n^2$. Then $b_n \downarrow 0$, and the a_n's must have bounded partial sums, since $\sum 1/n^2$ is, in fact, convergent. Thus $\sum 1/n^4$ converges by the Dirichlet test. But this is a very "weak" application of the test because if we really knew that $\sum 1/n^2$ is convergent, then of course $\sum 1/n^4$ is, too, simply by the comparison test. In other words, the *power* of the Dirichlet test actually lies in its ability to cope with *conditionally* convergent series.

2.3. *SERIES WITH TERMS THAT ARE FUNCTIONS; POWER SERIES*

Thinking of x as fixed, so that the $a_n(x)$ terms are, in fact, constants, consideration of the series $\sum a_n(x)$ apparently involves nothing especially new except that convergence/divergence certainly depends on x now and we expect the x axis to be divisible into intervals of convergence and divergence.

Example 2.9. Consider, again, the geometric series[8]

$$\sum_{n=0}^{\infty} x^n. \tag{2.16}$$

Applying the ratio test (or, just as conveniently, the root test),

$$\lim \left| \frac{x^{n+1}}{x^n} \right| = \lim |x| = |x|,$$

and thus (2.16) is absolutely convergent for $|x| < 1$. For $|x| > 1$ it is at best conditionally convergent; but if $|x| > 1$, then the nth term does not tend to zero as $n \longrightarrow \infty$, which means that the series must, in fact, diverge. The cases $|x| = 1$ (i.e., $x = +1$ and -1) must be examined separately, and it's easily seen that (2.16) diverges in both cases. We conclude that the series converges absolutely over the interval $|x| < 1$ and diverges for $|x| \geq 1$, which, of course, coincides with our earlier conclusion in Example 2.1 (where we put s_n into closed form and took the limit as $n \longrightarrow \infty$). ∎

The geometric series (2.16) is a special case of the more general **power series**

$$\sum_{n=0}^{\infty} a_n(x - a)^n, \tag{2.17}$$

where a and the a_n's are constants. As a result of the preceding example, it is not surprising that we can make the following general statement.

THEOREM 2.8. If $\lim |a_{n+1}/a_n| = L$, *then the power series (2.17) converges absolutely for* $|x - a| < 1/L$ *and diverges for* $|x - a| > 1/L$.

Proof. Proof of absolute convergence follows easily from the ratio test. The interesting part is the proof of divergence for $|x - a| > 1/L$, since the ratio test merely

---

[8]For $x = 0$ we will understand x^0 to mean 1.

denies *absolute* convergence; it says nothing about the possibility of conditional convergence. Suppose that $|x - a| > 1/L$. Denoting $a_n(x - a)^n \equiv A_n$, we have

$$\left|\frac{A_{n+1}}{A_n}\right| \sim \rho > 1 \quad \text{or} \quad |A_{n+1}| \sim \rho|A_n|.$$

Incrementing n, it follows that

$$|A_{n+2}| \sim \rho|A_{n+1}| \sim \rho^2|A_n|, \qquad |A_{n+3}| \sim \rho|A_{n+2}| \sim \rho^3|A_n|,$$

and so on. Since $\rho > 1$, we see that $|A_m| \to \infty$ as $m \to \infty$, whereas a *necessary* condition for convergence is that $|A_m| \to 0$. This completes the proof.

The interval $|x - a| < 1/L$ is called the *interval of convergence*, and $1/L \equiv R$ is called the *radius of convergence*.

Note that (2.17) *always* converges for $x = a$, no matter what the a_n's are doing, for then each term (i.e., for $n \geq 1$) is simply zero. If, for example, $a_n = n!$, then L is infinite and $R = 0$, so that (2.17) converges *only* for $x = a$. On the other hand, if $a_n = 1/n!$, then $L = 0$ and R is infinite; that is, (2.17) converges for *all* x.

Example 2.10.

$$\sum_1^\infty \frac{\cos nx}{n^2}. \tag{2.18}$$

Of course, this happens not to be of power series form. But it is easy to see that $|\cos nx/n^2| \leq 1/n^2$, and so the comparison test tells us that (2.18) converges absolutely for all x. ∎

As we will see in Chapter 4, it is important to introduce a notion of "uniformity of convergence."

DEFINITION. We say that $\sum a_n(x)$ converges uniformly to $s(x)$ over $\alpha \leq x \leq \beta$ if to each $\epsilon > 0$ (no matter how small) there corresponds an $N(\epsilon)$, independent of x, such that $|s_n(x) - s(x)| < \epsilon$ for all $n > N$ and $\alpha \leq x \leq \beta$.[9]

This situation is illustrated graphically in Fig. 2.1. Supposing that $s(x)$ is as shown, choose an arbitrarily small $\epsilon > 0$ and draw an ϵ band about $s(x)$—that is, $s(x) - \epsilon < y < s(x) + \epsilon$. If $s_n(x)$ converges uniformly, then there must be some N such that $s_n(x)$ lies entirely within the ϵ band for all n's $> N$.

Consequently, it is often important to be able to test a given series $\sum a_n(x)$ not only for convergence but also for uniform convergence. With slight modifications, most tests for convergence can be converted to tests for *uniform* convergence. A few are given below.

THEOREM 2.9. If there is a number r (independent of x) such that $|a_{n+1}(x)/a_n(x)| \leq r < 1$ over some interval I ($\alpha \leq x \leq \beta$) for all sufficiently large n, then the series $\sum a_n(x)$ converges (absolutely and) uniformly over I.

---

[9]The concept of uniform convergence is due to the mathematical physicist G. G. Stokes (1819–1903).

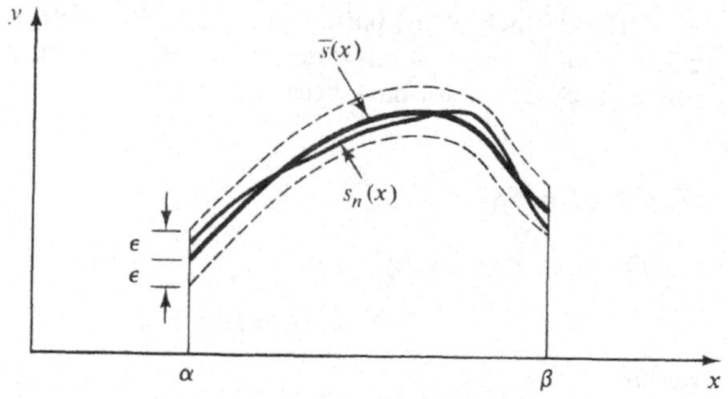

Figure 2.1. Uniform convergence.

THEOREM 2.10. If $b_n(x) \downarrow 0$ uniformly over some interval I as $n \longrightarrow \infty$ (i.e., there are constants $B_n \downarrow 0$ where $b_n(x) < B_n$ over I) and there is a number K (independent of x and n) such that

$$\left|\sum_{k=1}^{n} a_k(x)\right| < K$$

for all n and for all x in I, then the series $\sum a_n(x)b_n(x)$ converges uniformly over I.

We also have the comparison test. If $\sum b_n(x)$ converges uniformly over an interval I such that $|a_n(x)| \leq b_n(x)$ over I, then $\sum a_n(x)$ converges (absolutely and) uniformly over I. Surely any convergent series of *constants* is uniformly convergent (for all x); therefore as a special but important case, we have

THEOREM 2.11. (The Weierstrass M-Test) If $\sum M_n$ is a convergent series of constants such that $|a_n(x)| \leq M_n$ over I, then $\sum a_n(x)$ is (absolutely and) uniformly convergent over I.

For instance, with $M_n = 1/n^2$ in Example 2.10, we see that (2.18) is not only absolutely convergent but also uniformly convergent for all x.

Something That is Not True. It should be emphasized that convergence of $\sum a_n(x)$ for each x in some interval $\alpha \leq x \leq \beta$ does *not* in itself imply that the convergence is uniform! When we study Fourier series in Chapter 5, we will find that the series

$$\sum_{1}^{\infty} \frac{\sin (2n-1)x}{2n-1} \tag{2.19}$$

converges for *all* x to the "square wave" $s(x)$ shown in Fig. 2.2. where $s(x) = 0$ at $x = 0, \pm\pi, \pm 2\pi, \ldots$. But the convergence, because of the following theorem, cannot possibly be uniform over any interval that contains one or more of the jump discontinuities.

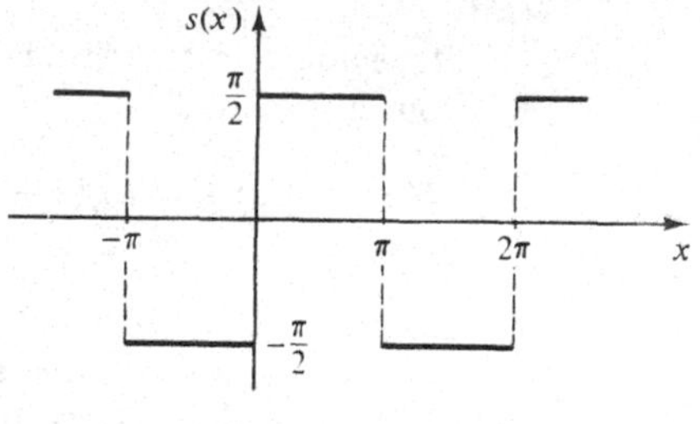

Figure 2.2. The square wave (2.19).

THEOREM 2.12. If a sequence of continuous functions $s_n(x)$ (perhaps the partial sums of a series) converges uniformly over some x interval I, then the limit function $s(x)$ will also be continuous over I.

Certainly each of the partial sums $s_n(x)$ of (2.19) is continuous over $-\pi/2 \leq x \leq \pi/2$, say; and since the limit function $s(x)$ is *not* continuous over that interval, then the convergence could not have been uniform.

2.4. *TAYLOR SERIES*

Starting with the identity

$$\int_a^x f'(x)\,dx = f(x) - f(a),$$

we have

$$f(x) = f(a) + \int_a^x f'(x)\,dx. \tag{2.20}$$

Since $f(x)$ is arbitrary, (2.20) should also hold with $f(x)$ replaced by the function $f'(x)$,[10]

$$f'(x) = f'(a) + \int_a^x f''(x)\,dx. \tag{2.21}$$

Using this expression for $f'(x)$ in the integral in (2.20), we obtain

$$\begin{aligned} f(x) &= f(a) + \int_a^x \left\{ f'(a) + \int_a^x f''(x)\,dx \right\} dx \\ &= f(a) + f'(a)(x-a) + \int_a^x \int_a^x f''(x)\,dx\,dx. \end{aligned} \tag{2.22}$$

But just as we obtained (2.21) from (2.20) by replacing $f(x)$ with $f'(x)$, we can replace $f'(x)$ in (2.21) by $f''(x)$, so that

$$f''(x) = f''(a) + \int_a^x f'''(x)\,dx. \tag{2.23}$$

Using this result in (2.22),

$$\begin{aligned} f(x) &= f(a) + f'(a)(x-a) + \int_a^x \int_a^x \left\{ f''(a) + \int_a^x f'''(x)\,dx \right\} dx\,dx \\ &= f(a) + f'(a)(x-a) + \frac{f''(a)}{2!}(x-a)^2 + \int_a^x \int_a^x \int_a^x f'''(x)\,dx\,dx\,dx. \end{aligned} \tag{2.24}$$

Repeating this process, under the assumption that $f(x)$ is sufficiently differentiable of course,[11] we find

$$f(x) = f(a) + f'(a)(x-a) + \frac{f''(a)}{2!}(x-a)^2 + \cdots + \frac{f^{(n-1)}(a)}{(n-1)!}(x-a)^{n-1} + R_n \tag{2.25a}$$

---

[10] This step is *not* the same as differentiating (2.20).

[11] Whereas the function $xH(x)$ (where H is the Heaviside function) is not even once differentiable over any interval containing the origin (because of its kink at $x = 0$), the function $x \sin(x^2)$, for example, is differentiable over $-\infty < x < \infty$ as many times as we like; we say that it is *infinitely differentiable*. As one more example, the function $x^3 H(x)$ is twice differentiable over any interval containing the origin (Exercise 2.23).

where
$$R_n = \int_a^x \cdots \int_a^x f^{(n)}(x)(dx)^n \tag{2.25b}$$

and $(dx)^n$ means $dx \ldots dx$, n times. Equation (2.25) is known as **Taylor's formula with remainder,** where the remainder is expressed in *integral* form by (2.25b).

Suppose next that $m \le f^{(n)}(x) \le M$ over $[a, x]$,[12] where m and M are constants. Then

$$\int_a^x \cdots \int_a^x m(dx)^n \le R_n \le \int_a^x \cdots \int_a^x M(dx)^n.$$

Integrating, we have

$$m\frac{(x-a)^n}{n!} \le R_n \le M\frac{(x-a)^n}{n!} \tag{2.26}$$

If we assume that $f^{(n)}(x)$ is continuous over $[a, x]$, then (it can be shown that) it must take on all values between its minimum m and its maximum M over the interval. It follows from (2.26) that we must be able to express

$$R_n = \frac{f^{(n)}(\xi)}{n!}(x-a)^n, \tag{2.27}$$

where ξ is *some* suitable point in $[a, x]$; this is the *Lagrange* form of R_n. For the special case $n = 1$, Taylor's formula with Lagrange remainder is simply

$$f(x) = f(a) + f'(\xi)(x-a)$$

or

$$\frac{f(x)-f(a)}{x-a} = f'(\xi) \qquad (a \le \xi \le x) \tag{2.28}$$

which is known as the **mean value theorem of the differential calculus.**

Assuming that f is infinitely differentiable at the point a, let us formally write down the infinite series

$$f(a) + f'(a)(x-a) + \frac{f''(a)}{2!}(x-a)^2 + \cdots + \frac{f^{(n)}(a)}{n!}(x-a)^n + \cdots \tag{2.29}$$

and call it **the Taylor series of** $f(x)$[13] about the point a. Hoping that (2.29) is *equal* to $f(x)$, we are tempted to make the tentative claim that if (2.29) converges, it converges to $f(x)$, and so we can write

$$f(x) = f(a) + f'(a)(x-a) + \frac{f''(a)}{2!}(x-a)^2 + \cdots \tag{2.30}$$

But

Example 2.11. The Taylor expansion of $f(x) = e^{-1/x^2}$ about $x = 0$,[14] also called

[12]It is standard practice to denote a *closed* interval $\alpha \le x \le \beta$ by the symbol $[\alpha, \beta]$ and an *open* interval $\alpha < x < \beta$ by (α, β). Similarly, $[\alpha, \beta)$ means $\alpha \le x < \beta$ and so on.

[13]Introduced by Brook Taylor (1685–1731) in 1715, although apparently known to James Gregory at least 45 years earlier, its importance was not fully recognized until 1755 when Euler applied it in his development of the differential calculus.

[14]If we were fussy, we would note that $f(0)$ is not actually defined by $f(x) = e^{-1/x^2}$, since 1/0 is not defined. To complete the definition of f let us define $f(x) = e^{-1/x^2}$ for $x \neq 0$, and $f(0) = 0$. Incidentally, you should sketch this function. Note how *extremely* flat it is in the neighborhood of $x = 0$, since $f'(0) = f''(0) = f'''(0) = \cdots = 0$, and yet is does eventually rise toward its asymptotic value of unity as $|x| \longrightarrow \infty$.

its *Maclaurin expansion* (i.e., the Taylor expansion about $x = 0$), is (Exercise 2.22)

$$0 + 0 + 0 + \cdots$$

which certainly converges (to zero) but *not* to the given $f(x)$ (except at $x = 0$, where $f(0) = 0$). The difficulty is that R_n, which gets "lost" when we go from (2.25) to (2.30), does not tend to zero as $n \longrightarrow \infty$ but is, in fact, equal to e^{-1/x^2} for all n! ∎

What we need to show is that $R_n \longrightarrow 0$ if we want to be certain of the equality in (2.30). Let us illustrate the procedure.

Example 2.12. Let $f(x) = e^x$ and $a = 0$. The resulting Taylor series

$$1 + x + \frac{x^2}{2!} + \cdots + \frac{x^n}{n!} + \cdots \tag{2.31}$$

is seen (e.g., by the ratio test) to converge for all x. Now examine R_n. Consider $x > 0$. From (2.27)

$$R_n = \frac{e^\xi}{n!}x^n,$$

Where $0 \leq \xi \leq x$. Actually, we don't know the value of ξ, but since $0 \leq \xi \leq x$, it follows that $R_n \leq e^x x^n/n!$, which tends to zero (for x fixed) as $n \longrightarrow \infty$. [If you're not sure why $x^n/n! \longrightarrow 0$ as $n \longrightarrow \infty$, recall that (2.31) converges, by the ratio test, and hence the nth term, $x^n/n!$, *must* $\longrightarrow 0$ as $n \longrightarrow \infty$.] A similar argument applies for $x < 0$. So not only does the Taylor series of e^x converge for all x, it also converges to e^x. We therefore have the equality

$$e^x = 1 + x + \frac{x^2}{2!} + \cdots \qquad (-\infty < x < \infty). \ ∎$$

Example 2.13. The expansion of $f(x) = 1/x$ about $x = a$ $(a \neq 0)$,

$$\sum_0^\infty \frac{(-1)^n}{a^{n+1}}(x-a)^n, \tag{2.32}$$

is seen (e.g., by the ratio test) to converge in $|x - a| < a$—that is, in $0 < x < 2a$. We state (this time without proof) that $R_n \longrightarrow 0$ as $n \longrightarrow \infty$, and so the sum function coincides, in fact, with $1/x$. ∎

If the equality (2.30) *holds over* $|x - a| < \delta$ *for some* $\delta > 0$, *we say that* $f(x)$ *is* **analytic** *at* $x = a$. For instance, recalling that the Taylor series (2.32) of $1/x$ about $x = a$ converges to $1/x$ within $|x - a| < a$ for all $a \neq 0$, it follows that $1/x$ is analytic for all $x \neq 0$. The Taylor series about $x = 0$, however, does *not* converge in any interval $|x| < \delta$ [in fact, the terms $f(0), f'(0), f''(0), \ldots$ in the series do not exist]; therefore $1/x$ is not analytic at $x = 0$. We say that it is **singular** there or that $x = 0$ is a **singular point** of $1/x$. The singular point $x = 0$ is significant not only in regard to the failure of the Taylor expansion about this point but also in that it limits the interval of convergence of the Taylor expansion about any other point $a \neq 0$. For example, the interval of convergence of the expansion about $a = 2$ is $|x - 2| < 2$ or $0 < x <$

4; that is, the interval of convergence "spreads out equally to the left and right until one end bumps up against a singularity," in this case at $x = 0$.

The function e^{-1/x^2} of Example 2.11 also happens to have a singular point at $x = 0$, since the Taylor series $0 + 0 + 0 + \ldots$ does not converge to e^{-1/x^2} in $|x| < \delta$ for any $\delta > 0$. Why? What's wrong? The function seems to be nicely behaved at $x = 0$; in fact, $f(0)$, $f'(0)$, $f''(0)$, ... all exist (and are zero). Although complex variable theory is not discussed until Part II, let us anticipate a little and consider $f(z) = e^{-1/z^2} = e^{-1/(x+iy)^2}$; in other words, we extend the domain off the real axis into the complex z plane. As $x \longrightarrow 0$ along the real axis ($y = 0$), $f = e^{-1/x^2} \longrightarrow 0$, which looks fine. But if we approach the origin along the *imaginary* axis ($x = 0$), then $f = e^{-1/(iy)^2} = e^{1/y^2} \longrightarrow \infty$ as $y \longrightarrow 0$, and the singular nature of $f(z)$ at $z = 0$ now comes into view!

The fact is that we are not going to understand Taylor series and analyticity fully until we study complex variable theory. As a final example, we note that the Taylor expansion $1/(1 + x^2) = 1 - x^2 + x^4 - x^6 + \ldots$ holds in the interval of convergence $|x| < 1$. Why the restriction $|x| < 1$? The function $1/(1 + x^2)$ is beautifully smooth and well behaved for *all* $-\infty < x < \infty$. Quite so, but $f(z) = 1/(1 + z^2)$ is singular at $z = \pm i$, and these singularities limit the region of convergence to the unit disk $|z| < 1$ or, putting our "blinders" back on and just looking at f on the x axis, to the interval $|x| < 1$. Again, a detailed discussion must wait until Part II.

Having discussed both power series and Taylor series, we point out (without proving it) the following fundamental connection: namely, that *every power series with nonzero radius of convergence is the Taylor series of its sum function*. Suppose that we have

$$\sum_0^\infty a_n(x - a)^n = \sum_0^\infty b_n(x - a)^n \tag{2.33}$$

over some common interval of convergence, say $|x - a| < R$; that is, the two series converge to the same sum function, say $f(x)$, over that interval. Then the previous statement (italics) implies that $a_n = f^{(n)}(a)/n!$ and that $b_n = f^{(n)}(a)/n!$, so that $a_n = b_n$. We rely on this important result quite often—in the power series solution of differential equations, for instance. It is stated below as a theorem for reference.

THEOREM 2.13. Equation (2.33) holds over some common interval of convergence $|x - a| < R$ if and only if $a_n = b_n$ for all n.

COROLLARY.

$$\sum_0^\infty a_n(x - a)^n = 0 \tag{2.34}$$

over $|x - a| < R$ if and only if all the a_n's $= 0$.

Let us close the present discussion with an elementary look at (2.30). Basically, the problem is one of extrapolation: given data at some point $x = a$—namely, the values of f and all its derivatives—can we predict what f will be at some other point x? The first partial sum provides the simplest extrapolation, simply $f(x) = f(a)$. The second and third partial sums provide tangent-line and parobolic fits and so on (Exercise 2.25).

2.5. THAT'S ALL VERY NICE, BUT HOW DO WE SUM THE SERIES? ACCELERATION TECHNIQUES

Suppose that we would like to sum the series

$$1 - \frac{1}{2} + \frac{1}{3} - \cdots = \sum_{1}^{\infty} \frac{(-1)^{k+1}}{k} \equiv s. \tag{2.35}$$

Recalling the Taylor series

$$\ln(1 + x) = x - \frac{x^2}{2} + \frac{x^3}{3} - \cdots$$

we can note, by comparison with (2.35), that $s = \ln(1 + 1) = \ln 2 \doteq 0.6931472.$[15] Generally, of course, we are not lucky or clever enough to identify a given series as the expansion of some known (and tabulated) function (indeed, it may not *be* the expansion of any such function), and we must face up to summing it numerically.

Since the signs alternate and the terms decrease in magnitude with n for our series (2.35), it is not hard to see that the error incurred by breaking off the series at any point must be less than the next term—that is, $|s - s_n| < 1/(n + 1)$. So in order to achieve $\pm 10^{-5}$ accuracy, we apparently need $n \approx 10^5$—that is, we must sum approximately 100,000 terms. Not a very happy thought, and we therefore turn our attention to the possibility of improving such a sad state of affairs.

Aitken–Shanks Transformation. Suppose that our series is such that

$$s \approx s_n + \kappa\rho^n \tag{2.36}$$

for some constants κ and ρ; that is, suppose that the remainder $s - s_n$ can be "fitted" approximately by the simple expression $\kappa\rho^n$ for some constants κ and ρ. If so, we see that by taking three different n's, (2.36) yields three equations in the three unknowns κ, ρ, s. Eliminating κ and ρ, we can solve for s. In other words, we fit three data points with the form (2.36) and "extrapolate to the limit." Thus

$$s - s_n \approx \kappa\rho^n \tag{2.37}$$

$$s - s_{n+1} \approx \kappa\rho^{n+1} \approx \rho(s - s_n) \tag{2.38}$$

$$s - s_{n+2} \approx \kappa\rho^{n+2} \approx \rho(s - s_{n+1}). \tag{2.39}$$

Dividing (2.38) by (2.39) and solving for s give

$$s \approx \frac{s_{n+2}s_n - s_{n+1}^2}{s_{n+2} + s_n - 2s_{n+1}}. \tag{2.40}$$

If (2.36) were *exactly* true, then the right side of (2.40) would be the same for all n—namely, s. If it is only approximately true, however, then the right side of (2.40) will vary with n, and this situation suggests using (2.40) to define a new sequence

$$s_n^{(1)} = \frac{s_{n+2}s_n - s_{n+1}^2}{s_{n+2} + s_n - 2s_{n+1}}, \tag{2.41}$$

---

[15]By $\doteq$ we mean "approximately equal to"—that is, to the number of decimal places indicated; for example, $\frac{2}{3} \doteq 0.67$.

which, hopefully, will converge to s more rapidly than the original s_n sequence (Exercise 2.29).

Returning to the example (2.35), we compute the first nine partial sums s_n and list them in Table 2.1. Observe that s_9 is correct to only one decimal place. Computing as many $s_n^{(1)}$'s as possible from these s_n's—namely, the first seven—we see that $s_7^{(1)}$ is correct to three decimal places. Encouraged, we repeat the process, computing

$$s_n^{(2)} = \frac{s_{n+2}^{(1)} s_n^{(1)} - s_{n+1}^{(1)2}}{s_{n+2}^{(1)} + s_n^{(1)} - 2s_{n+1}^{(1)}}$$

and so on. The terminal value, $s_1^{(4)}$, is seen to be correct to *seven* places. (Even if we did not have the exact answer for comparison, as in practical applications, the fact that seven-place accuracy is achieved is suggested by the fact that the results have apparently settled down to seven places by the time we reach $s_1^{(4)}$.) For comparable accuracy, straight calculation of the s_n's would have required about ten million terms!

TABLE 2.1
Aitken–Shanks Acceleration of the Series (2.35)

n	s_n	$s_n^{(1)}$	$s_n^{(2)}$	$s_n^{(3)}$	$s_n^{(4)}$
1	1.0	0.7	0.6932773	0.6931489	0.6931472
2	0.5	0.6904762	0.6931058	0.6931467	
3	0.8333333	0.6944444	0.6931633	0.6931474	
4	0.5833333	0.6924242	0.6931399		
5	0.7833333	0.6935897	0.6931508		
6	0.6166667	0.6928571			
7	0.7595238	0.6933473			
8	0.6345238				
9	0.7456349				

Next, consider the *geometric series* $1 + x + x^2 + \cdots$, which we know in advance sums to $1/(1 - x)$ for $|x| < 1$ and diverges for $|x| \geq 1$. For fun, let us take $x = 2$; then the series is $1 + 2 + 4 + 8 + \cdots$. Proceeding as before,

n	s_n	$s_n^{(1)}$
1	1	-1
2	3	-1
3	7	-1
4	15	
5	31	

so that whereas s_n diverges, $s_n^{(1)}$ is found to converge (in one step) to -1, which agrees with the value of $1/(1 - x)$ at $x = 2$. In fact, for *any* x we have $s_1 = 1$, $s_2 = 1 + x$, and $s_3 = 1 + x + x^2$, and thus (2.41) yields

$$s_1^{(1)} = \frac{(1 + x + x^2)(1) - (1 + x)^2}{(1 + x + x^2) + (1) - 2(1 + x)} = \frac{1}{1 - x}$$

for all $x \neq 1$. So although $1 + x + x^2 + \cdots$ is only a "partial representation" of $1/(1 - x) \equiv f(x)$—that is, only over its interval of convergence $|x| < 1$—we see that the transformed sequence $s_n^{(1)}$ offers a broader (in fact, a complete) representation of f![16]

Suppose, for example, that we seek the solution of the differential equation

$$(1 + x)y' + y = 0 \qquad (0 \leq x < \infty) \tag{2.42}$$

$$y(0) = 3 \tag{2.43}$$

in the form of a power series $\sum a_n x^n$. We find that

$$y(x) = 3(1 - x + x^2 - x^3 + \cdots). \tag{2.44}$$

Unfortunately, this series converges for $|x| < 1$ and hence only over the $0 \leq x < 1$ part of the domain; over the rest of the domain, $1 \leq x < \infty$, (2.44) is apparently useless. Nevertheless, observe that one application of the transformation (2.41) to the series (2.44) yields $y(x) = 3/(1 + x)$, which, in fact, is the "complete" solution! The difficulty with (2.44) is now clear: $3/(1 + x)$ is singular at $x = -1$. This situation is fine, since $x = -1$ is outside the domain of interest, but the singularity there limits the interval of convergence of (2.44) to $|x| < 1$. The Aitken–Shanks transformation provides the desired analytic continuation.

To see why the transformation works so beautifully for the geometric series, recall from (2.6) that

$$s_n = \frac{1 - x^n}{1 - x} = \frac{1}{1 - x} - \frac{x^n}{1 - x} = s - \left(\frac{1}{1 - x}\right)x^n, \tag{2.45}$$

which is exactly of the form (2.36), with $\kappa = 1/(1 - x)$ and $\rho = x$. (They *are* constants, since x is regarded as fixed in the summation process.) Thus the "closer" a given series is to a geometric series, the more effective will be the transformation (2.41). [How close is (2.35) to being a geometric series? Comparing it with $1 + x + x^2 + \cdots$, we see from the second terms that $x = -\frac{1}{2}$. But then x^2 is $\frac{1}{4}$, whereas the third term in (2.35) is $\frac{1}{3}$; x^3 is $-\frac{1}{8}$, whereas the fourth term in (2.35) is $-\frac{1}{5}$, and so on. Consequently, (2.35) is not exactly a geometric series, which is why Aitken–Shanks (Table 2.1) did not converge in one step.]

Historically, successive application of the transformation (2.41) is generally associated with Aitken, in connection with a paper published in 1926,[17] although (2.41) was noted by others both before and after Aitken's contribution; perhaps you've already observed that the basic extrapolation concept that led to (2.41) is the same as that employed in Romberg integration (Section 1.6) except that the form (1.27) was *known* to be true, whereas (2.36) was put forward in the absence of any advance information about the nature of the remainder $s - s_n$. I'm inclined to call (2.41) an **Aitken–Shanks transformation** in order to acknowledge the important 1955 paper by Shanks[18] in which a unified and systematic theory emerges for a whole class of nonlinear trans-

---

[16] In the language of complex variable theory (Part II), we say that $s_n^{(1)}$ provides the *analytic continuation* of the original geometric series.

[17] A. C. Aitken, On Bernoulli's Numerical Solution of Algebraic Equations, *Proc. Roy. Soc. Edinburgh*, Vol. 46, 1926, pp. 289–305.

[18] D. Shanks, Non-Linear Transformations of Divergent and Slowly Convergent Sequences, *Journal of Mathematics and Physics*, Vol. XXXIV, 1955, pp. 1–42.

formations, of which (2.41) is but one.[19] More generally than (2.36), Shanks allows for the presence of *several* "harmonics," and hence substantial deviation from "geometric behavior," by assuming the form

$$s \approx s_n + \kappa_1 \rho_1^n + \cdots + \kappa_N \rho_N^n.$$

Included in his paper are theorems that show the effectiveness of the transformations in accelerating the convergence of some slowly convergent sequences, as well as in inducing the convergence of some divergent ones.

In closing, note that Shanks' transformations are actually *summability methods*,[20] just like Cesàro summability (2.5). To see the similarity, it might be helpful to rewrite (2.5) in the form $s_n^{(1)} = (s_1 + s_2 + \cdots + s_n)/n$. However, Cesàro's transformation is linear, whereas Shanks' are nonlinear and considerably more powerful (Shanks, p. 40).

Of the various important transformation techniques, let us discuss one more, the so-called Euler transformation.

Euler's Transformation. First, we introduce some operator notation, the *displacement operator E* and the *forward difference operator* Δ:

$$Ef(x) \equiv f(x + h), \tag{2.46}$$

$$\Delta f(x) \equiv f(x + h) - f(x). \tag{2.47}$$

Observe that these operators are related, since

$$\Delta f = f(x + h) - f(x) = Ef - f = (E - 1)f, \tag{2.48}$$

so that $\Delta = E - 1$. Furthermore, by ABf, where A and B are any two operators, we always mean $A(Bf)$; that is, first B acts on f, yielding Bf, and then A acts on Bf. For instance, $E^2 f(x) = Ef(x + h) = f(x + 2h)$ and $E^k f(x) = f(x + kh)$.

Consider the alternating series $\sum (-1)^n u_n$. If we regard u_n as a function of the discrete variable n and let $h = 1$, then $Eu_n \equiv u_{n+1}$ and $\Delta u_n \equiv u_{n+1} - u_n$. Proceeding formally, we obtain

$$\begin{aligned}\sum_0^\infty (-1)^n u_n &= \sum_0^\infty (-E)^n u_0 = (1 + E)^{-1} u_0 \\ &= \frac{1}{2}\left(1 + \frac{\Delta}{2}\right)^{-1} u_0 = \frac{1}{2}\sum_0^\infty \left(-\frac{\Delta}{2}\right)^n u_0.\end{aligned}$$

The resulting linear transformation (Exercise 2.31)

$$\sum_0^\infty (-1)^n u_n = \sum_0^\infty \frac{(-1)^n}{2^{n+1}} \Delta^n u_0, \tag{2.49}$$

where $\Delta^0 u_0 \equiv u_0$, $\Delta u_0 = u_1 - u_0$, $\Delta^2 u_0 = \Delta(\Delta u_0) = \Delta u_1 - \Delta u_0$, and so on, is known as **Euler's transformation** and also known to be regular in the sense of footnote 20; application of (2.49) is often called "Eulerizing" the series on the left-hand side.

Be sure you see that our derivation of (2.49) was only formal. For instance, $(1 + E)^{-1} = 1 - E + E^2 - \cdots$ is correct if $|E| < 1$. But E is not a number, it is an

---

[19](2.41) is said to be *nonlinear* because $s_n^{(1)}$ is not a linear combination of the s_k's.

[20]Note that we generally require summability methods to be "regular" in the sense that they lead to the same results as ordinary convergence for all series that do converge in the ordinary sense.

operator, and so the statement "$|E| < 1$" is simply not meaningful (at least not *yet*; it will be made meaningful in Part III). Nevertheless, what we've done looks reasonable, and we are hopeful that the result can be made rigorous. What we'd like to prove is that whenever the left side of (2.49) converges, so does the right side, to the same value but more rapidly. If, in addition, the right side sometimes converges even when the left side does not, then we'd be happier still. Such a discussion, however, is beyond our present scope, and we close instead with a numerical illustration.

Example 2.14. Consider again the series (2.35). Summing the first six terms explicitly, $s_6 \doteq 0.616667$, so that

$$1 - \frac{1}{2} + \frac{1}{3} + \cdots \doteq 0.616667 + \frac{1}{7} - \frac{1}{8} + \frac{1}{9} - \cdots = 0.616667 + \sum_{n=0}^{\infty} \frac{(-1)^n}{n+7}. \tag{2.50}$$

Next, let us Eulerize the first five terms, say, of the series on the right side of (2.50). The calculations are shown in Table 2.2.

TABLE 2.2
The Difference Table for Example 2.14

n	u_n	Δu_n	$\Delta^2 u_n$	$\Delta^3 u_n$	$\Delta^4 u_n$	$\frac{(-1)^n \Delta^n u_0}{2^{n+1}}$
0	0.142857	−0.017857	0.003968	−0.001190	0.000433	0.071429
1	0.125000	−0.013889	0.002778	−0.000758		0.004464
2	0.111111	−0.011111	0.002020			0.000496
3	0.100000	−0.009091				0.000074
4	0.090909					0.000014
					Total =	0.076477

Adding, $s \approx 0.616667 + 0.076477 = 0.693144$, which is correct to five places. Even without knowing the exact answer, it is reasonably clear that we have five-place accuracy from the way the terms in the last column are dropping off; for instance, the term following 0.000014 would be around 0.000003 and would probably leave us with six-place accuracy. Note that the five underlined terms are the ones that enter directly into the calculation of the last column; they are $\Delta^0 u_0, \Delta u_0, \ldots, \Delta^4 u_0$, respectively.

The reason that we summed the first six terms explicitly (nothing special about six; could just as well have been ten) was to postpone the Eulerizing until farther along in the series, where successive u_n's are closer together and hence the differences ($\Delta u_n, \Delta^2 u_n, \ldots$) are smaller. By comparison, if we had Eulerized the first five terms of (2.35), only two-place accuracy would have been attained (Exercise 2.32). ▮

As with the Aitken–Shanks transforms, Eulerization may work even if the original series is divergent. A drawback, however, is that we must start with an alternating series $\sum (-1)^n u_n$, whereas our given series may not be of that form. An intermediate transform, due to van Wijngaarden,[21] can be used to produce an alternating series, but it is not always easy to apply.

*

To illustrate, consider the so-called **exponential integral,**[22]

$$\text{Ei}(x) = \int_x^\infty \frac{e^{-t}}{t}\,dt \qquad (0 < x < \infty). \tag{2.51}$$

This integral arises frequently, as does the related **sine integral** and **cosine integral,**

$$\text{Si}(x) = \int_0^x \frac{\sin t}{t}\,dt \quad \text{and} \quad \text{Ci}(x) = \int_x^\infty \frac{\cos t}{t}\,dt. \tag{2.52}$$

These integrals cannot be evaluated in terms of elementary functions, and so we give them names of their own. The right-hand side of (2.51) is an "integral representation" of Ei (x) in the same way that $\int_1^x dt/t$ is an integral representation of $\ln x$. But (2.51) is not a computational formula; that is, it is not a recipe for actually computing Ei (x) by some combination of numerical operations. Therefore it is important to develop alternative expressions, say in the form of series expansions. For instance, it can be shown that

$$\text{Ei}(x) = -\gamma - \ln x - \sum_1^\infty \frac{(-1)^n x^n}{n n!}, \tag{2.53}$$

where $\gamma \doteq 0.5772157$ is *Euler's constant*. The series converges over the entire domain $0 < x < \infty$—very rapidly, in fact, for small x. For large x, however, it converges more slowly due to the x^n's. Rather than rely on the use of a transform to speed the convergence, let us try to develop an expansion *specifically for large* x, to complement (2.53). If it is to converge more rapidly with increasing x, we might expect it to proceed in *inverse* powers of x, unlike (2.53), and this fact will guide the following development.

Integrating by parts repeatedly, with $e^{-t}\,dt =$ "dv" in each case, we obtain

$$\begin{aligned}
\text{Ei}(x) &= \int_x^\infty \frac{e^{-t}}{t}\,dt = -\frac{e^{-t}}{t}\bigg|_x^\infty - \int_x^\infty \frac{e^{-t}}{t^2}\,dt \\
&= \frac{e^{-x}}{x} - \frac{e^{-x}}{x^2} + 2!\int_x^\infty \frac{e^{-t}}{t^3}\,dt = \text{etc.} \\
&= \frac{e^{-x}}{x}\left[1 - \frac{1}{x} + \frac{2!}{x^2} - \cdots + \frac{(-1)^{n-1}(n-1)!}{x^{n-1}}\right] + (-1)^n n!\int_x^\infty \frac{e^{-t}}{t^{n+1}}\,dt \\
&\equiv s_n(x) + R_n(x).
\end{aligned} \tag{2.54}$$

[21] See, for instance, *Modern Computing Methods*, 2nd ed., Philosophical Library, London, 1961.

[22] Again a warning about notation. In Abramowitz and Stegun, for example, the integral (2.51) is called $E_1(x)$ and Ei (x) is defined differently.

Observe that the sequence $s_n(x)$ diverges as $n \longrightarrow \infty$, since $|a_{n+1}/a_n| = n/x > 1$ for all $n > x$; so we cannot simply let $n \longrightarrow \infty$ and claim that

$$\text{Ei}(x) = \frac{e^{-x}}{x} \sum_0^\infty \frac{(-1)^n n!}{x^n}.$$

Nevertheless, since $s_n(x) \sim e^{-x}/x$ as $x \longrightarrow \infty$ (for n fixed), and

$$|R_n(x)| < n! \int_x^\infty \frac{e^{-t}}{x^{n+1}}\, dt = \frac{n!\, e^{-x}}{x^{n+1}}, \tag{2.55}$$

it follows that for *each fixed n* the ratio $R_n(x)/s_n(x) \longrightarrow 0$ as $x \longrightarrow \infty$, and therefore $\text{Ei}(x) = s_n(x)[1 + R_n(x)/s_n(x)] \sim s_n(x)$.

Not only does $R_n(x)/s_n(x) \longrightarrow 0$ as $x \longrightarrow \infty$, in fact, $R_n(x) \longrightarrow 0$ faster than *every term* in $s_n(x)$, since $R_n(x) = O(e^{-x}/x^{n+1})$ from (2.55), whereas the terms in $s_n(x)$ are $O(e^{-x}/x), \ldots, O(e^{-x}/x^n)$. Thus it makes sense[23] to write

$$\text{Ei}(x) \sim \frac{e^{-x}}{x} \sum_0^n \frac{(-1)^k k!}{x^k} \quad \text{for each fixed } n. \tag{2.56}$$

This is the asymptotic expression that we sought. Actually, it is standard practice to write (2.56) in *series* form

$$\text{Ei}(x) \sim \frac{e^{-x}}{x} \sum_0^\infty \frac{(-1)^k k!}{x^k} \tag{2.57}$$

but with the understanding that the series is to be cut off at some finite n.

By definition, then, we say that $\sum_0^\infty a_k x^{-k}$ is an **asymptotic series** for $f(x)$, as $x \longrightarrow \infty$, and write

$$f(x) \sim \sum_0^\infty a_k x^{-k} \tag{2.58}$$

if for each positive integer n

$$\lim_{x\to\infty} x^n \left[f(x) - \sum_0^n a_k x^{-k} \right] = 0 \tag{2.59}$$

or

$$\lim_{x\to\infty} x^n [f(x) - s_{n+1}(x)] = \lim_{x\to\infty} x^n R_{n+1}(x) = 0. \tag{2.60}$$

That is, $R_{n+1}(x) \longrightarrow 0$ faster than $1/x^n$ and hence faster than all the terms in $s_{n+1}(x)$.

For instance, we see from the preceding discussion that

$$xe^x\, \text{Ei}(x) \sim \sum_0^\infty \frac{(-1)^k k!}{x^k}$$

satisfies this definition, since $|x^n R_{n+1}(x)| < |x^n (n+1)!/x^{n+1}| \longrightarrow 0$ as $x \longrightarrow \infty$, for each fixed n.

---

[23]To illustrate what we mean, note that $e^x \sim 1 - 2x + x^2$ is correct, as $x \longrightarrow 0$, since the ratio of the left- and right-hand sides tends to unity, but it violates the implicit understanding that if we bother to write down terms through $O(x^2)$, they should at least be correct. That is, we know that $e^x = 1 + x + x^2/2! + \cdots$, so that the statement $e^x \sim 1 - 2x + x^2$ is actually only correct through zeroth order, *not* second order.

As a numerical example, consider the calculation of Ei (10), say, or, equivalently, e^{10} Ei (10). Recall that the asymptotic series

$$e^{10}\,\mathrm{Ei}\,(10) \sim \sum_{0}^{\infty} \frac{(-1)^k k!}{10^{k+1}} \tag{2.61}$$

is to be cut off at some finite n. *Any* n? We expect that there is some *optimal* n that gives the best accuracy for any given x. Let us look at the successive terms and partial sums listed in Table 2.3. The terms decrease in magnitude until $k = 10$ and then begin to grow without bound.[24] Since $R_n(x)$ alternates sign with n, it follows that the exact value of e^{10} Ei (10) must lie between any two consecutive partial sums.[25] The "tightest" pair of partial sums occurs where the terms have reached their minimum magnitude and are about to start increasing again. Thus we can say that $0.0915456320 < e^{10}\,\mathrm{Ei}\,(10) < 0.09158192000$ or, probably, $e^{10}\,\mathrm{Ei}\,(10) \doteq 0.0916$.

It is important to see that it is the best we can do; retaining additional terms makes things worse, not better. This is a fundamental feature of divergent asymptotic series.[26] It does not mean, however, that we cannot squeeze out more information. For exam-

TABLE 2.3
Terms and Partial Sums of the Series in (2.61)

k	Terms	Partial Sums
0	0.1	0.1
1	−0.01	0.09
2	0.002	0.092
3	−0.0006	0.0914
4	0.00024	0.09164
5	−0.000120	0.091520
6	0.0000720	0.0915920
7	−0.00005040	0.09154160
8	0.000040320	0.091581920
9	−0.0000362880	0.0915456320
10	0.00003628800	0.09158192000
11	−0.000039916800	0.091542003200
12	0.0000479001600	etc.
13	−0.00006227020800	
14	0.000087178291200	
15	−0.0001307674368000	

ple, the first ten terms (i.e., $k = 0$ through 9) sum explicitly to 0.0915456320. Eulerizing the next five terms (i.e., $k = 10$ through 14) yields an additional contribution of 0.0000177203 (Exercise 2.33); so the total is 0.0915633523, compared with the tabulated value 0.0915633339!

[24]We already mentioned that $|a_{k+1}/a_k| = k/x$ is < 1 for $k < x$ and > 1 for $k > x$; so this situation should be no surprise.

[25]If $s = s_n + R_n$ and $s = s_{n+1} + R_{n+1}$, where $R_n > 0$ and $R_{n+1} < 0$, then $s > s_n$ and $s < s_{n+1}$; so $s_{n+1} < s < s_n$. This result holds whether the sequence s_n converges or not.

[26]As we'll see in Part II, not all asymptotic series are divergent.

Another path to (2.57) involves expanding the integrand and integrating termwise. Letting $\tau = t - x$ and recalling the identity

$$\frac{1}{1-z} = 1 + z + z^2 + \cdots + z^{n-1} + \frac{z^n}{1-z} \qquad (z \neq 1) \tag{2.62}$$

we have

$$\begin{aligned}
\mathrm{Ei}\,(x) &= \int_x^\infty \frac{e^{-t}}{t}\,dt = \int_0^\infty \frac{e^{-(\tau+x)}}{\tau + x}\,d\tau = \frac{e^{-x}}{x}\int_0^\infty \frac{e^{-\tau}}{1+\tau/x}\,d\tau \\
&= \frac{e^{-x}}{x}\int_0^\infty e^{-\tau}\left[1 - \frac{\tau}{x} + \frac{\tau^2}{x^2} - \cdots + \frac{(-1)^{n-1}\tau^{n-1}}{x^{n-1}} + \frac{(-1)^n\tau^n}{x^n(1+\tau/x)}\right]d\tau \\
&= \frac{e^{-x}}{x}\left\{1 - \frac{1}{x} + \frac{2!}{x^2} - \cdots + \frac{(-1)^{n-1}(n-1)!}{x^{n-1}}\right\} + \frac{(-1)^n e^{-x}}{x^n}\int_0^\infty \frac{\tau^n e^{-\tau}}{\tau + x}\,d\tau \\
&= s_n(x) + R_n(x)
\end{aligned} \tag{2.63}$$

as in (2.54), except that the remainder term is in slightly different form. If $|z| = \tau/x$ were < 1, we might expect the resulting series [$s_n(x)$ as $n \to \infty$] to be convergent, but τ goes from zero to infinity, and so $|z| = \tau/x > 1$ for all $\tau > x$; in a sense, this is the source of the divergence.

So far we've emphasized the *computational* importance of asymptotic formulas like (2.57). Equally important, however, is that they tell us the asymptotic nature of the function, and this information is often crucial. In the present example we have found, from the leading term of (2.57), that

$$\mathrm{Ei}\,(x) \sim \frac{e^{-x}}{x} \quad \text{as } x \longrightarrow \infty. \tag{2.64}$$

Furthermore, we see from (2.53) that

$$\mathrm{Ei}\,(x) \sim \ln x \text{ as } x \longrightarrow 0; \tag{2.65}$$

thus Ei (x) is singular at $x = 0$, since it tends to ∞ as $x \to 0$.

EXERCISES

2.1. Derive (2.8) for the geometric series $1 + x + x^2 + \cdots$.

2.2. Show that the right-hand side of (2.8) tends to $1/(1 - x)$ for $|x| < 1$ and diverges for $|x| > 1$.

2.3. Applying Cesàro summability, show that $1 + x + x^2 + \cdots$ diverges for $x = 1$.

***2.4.** Consider the point set $0 \leq x < 1$ of the x axis, subject to the Euclidean distance function $d(x_1, x_2) = |x_1 - x_2|$. Identify all the set's limit points. Is the set closed? Bounded? Repeat, using the distance function (2.9).

***2.5.** Consider the set of all differentiable functions defined over $0 \leq x \leq 1$, subject to the distance function (1.10). Is the set bounded? Show that it is not closed. *Hint:* Use the Weierstrass approximation theorem (footnote 13 of Chapter 1).

*2.6. Just before Theorem 2.2 we defined a set S to be bounded "if there is a constant $M < \infty$ such that $d(u, v) < M$ for all u, v in S." Is that equivalent to defining it to be bounded if $d(u, v) < \infty$ for all u, v in S? Explain.

*2.7. Show that a Cauchy sequence has exactly *one* limit point, whereas $s_n = 1, 0, 1, 0, 1, \ldots$, say, has two.

2.8. Prove Theorem 2.3. *Hint:* Recall Exercise 1.5.

2.9. Determine whether each of the following series converges or diverges. Use whatever test you like, preferably more than one test wherever possible.

(a) $\sum_1^\infty (n+3)^{-3/2}$ (b) $\sum_1^\infty \frac{n}{n^2 + 3\ln n}$

(c) $\sum_1^\infty e^{1/n}$ (d) $\sum_1^\infty \left(\frac{n^2+2n-1}{n^4+3}\right)^{3/2}$

(e) $\sum_1^\infty \frac{n!}{n^n}$ (f) $\sum_1^\infty n^{-100}$ (Use ratio test.)

(g) $\sum_1^\infty \left(1 + \frac{1}{n^2}\right)$ (h) $\sum_1^\infty \left[1 + n\ln\left(\frac{2n-1}{2n+1}\right)\right]$

2.10. Determine the absolute convergence, conditional convergence, or divergence of the following series.

(a) $\sum_1^\infty (-1)^n \ln\left(1 + \frac{1}{\sqrt{n}}\right)$ (b) $\sum_1^\infty \frac{\sin n\pi/4}{\ln(n+1)}$

(c) $\sum_1^\infty \frac{(-1)^n}{2^n}$ (d) $\sum_1^\infty (-1)^n\left(1 + \frac{1}{n^4}\right)$

2.11. Convergence/divergence of the series $\sum_2^\infty 1/(n\ln n)$ is generally established by means of the Integral Test (not included here). Instead use the following test to show that the series diverges:

CAUCHY CONDENSATION TEST. *If $\sum a_n$ is a series of positive terms such that $a_{n+1} \le a_n$ for all n, then the series converges if and only if the "condensed" series $\sum^\infty 2^j a_{2^j}$ converges.*

Note: Recalling Exercise 1.20, one might argue that $1/(n\ln n) = O(1/n^{1+\alpha})$, which, by comparison with the p series, implies convergence, since $1 + \alpha > 1$. The argument fails, however, since α is arbitrarily small. In fact, the divergence of $\sum 1/(n\ln n)$ emphasizes just how weak the singularity in $\ln n$ is at infinity, not even strong enough to give the n in the denominator the tiny boost that it needs for convergence.

2.12. Determine the x intervals of convergence of the series shown.

(a) $\sum_1^\infty \frac{n + \cos^2 nx}{n^2+4}$ (b) $\sum_1^\infty \frac{\cos^3 nx}{(2n+3)^2}$

(c) $\sum_1^\infty e^{-nx}$ (d) $\sum_1^\infty \sin\frac{x}{n^2}$

(e) $\sum_1^\infty \frac{1}{x^2+n^2}$ (f) $\sum_1^\infty \frac{n}{x^n}$

(g) $\sum_1^\infty \ln\left[1 + \left(\frac{x}{n}\right)^2\right]$ (h) $\sum_1^\infty \ln\left(2 + \frac{x}{n}\right)$

(i) $\sum_1^\infty \frac{(-1)^n}{x+n}$ (j) $\sum_1^\infty n^{-x}$

(k) $\sum_1^\infty (\sin x)^n$ (l) $\sum_1^\infty \frac{1}{n}\left(\frac{x}{x+4}\right)^n$

2.13. Determine the x intervals of convergence of the power series below.

(a) $\sum_1^\infty n!x^n$ (b) $\sum_2^\infty \frac{(x-3)^n}{\ln n}$

(c) $\sum_1^\infty [\ln(n^2+1)]x^n$ (d) $\sum_1^\infty e^n x^n$

2.14. Comment on the following excerpts from examination papers.

(a) "$1/n$ is monotone decreasing and $\sin nx$ is bounded, < 10 say; so

$$\sum_1^\infty \frac{\sin nx}{n}$$

is convergent by the Dirichlet test."

(b) "If $a_n > 0$ and $a_{n+1}/a_n < 1$ for each n, then $\sum a_n$ is convergent."

(c) "If $s_n(x)$ converges uniformly to $s(x)$, then $s(x)$ must be continuous."

2.15. Show whether the following series from Exercise 2.12 converge *uniformly* over the indicated intervals.

(b) For all x

(c) For $\alpha \le x < \infty$, where $\alpha > 0$

(d) For all x

(d) For $|x| < 100$

(i) For $-1 < x < \infty$

(i) For $-0.99 \le x < \infty$

2.16. If $s_n(x) = x^n$, evaluate $\lim_{n\to\infty} s_n(x)$ over $[0, 1]$. Is convergence *uniform* over $[0, 1]$? Prove that it is uniform over $[0, \frac{1}{2}]$ by actually demonstrating a suitable "$N(\epsilon)$" relation.

2.17. Is the sequence of partial sums s_n, for the series (a) of Exercise 2.9, a Cauchy sequence? Why (not)?

2.18. Prove the second part of Theorem 2.5. *Hint:* Suppose that $r' > 1$ and define $p = (1 + r')/2$. Consider the comparison series $b_n = An^{-p}$. Then

$$\frac{b_{n+1}}{b_n} = \left(1 + \frac{1}{n}\right)^{-p} \sim 1 - \frac{p}{n}.$$

Since $a_{n+1}/a_n \sim 1 - r'/n$, where $p < r'$, we see that, by choosing A large enough, we will have $a_n < b_n$ for all n. Moreover, since the b_n series converges, then so does the a_n series. Now suppose that $r' < 1$ and prove, in similar fashion, that the a_n series diverges.

2.19. Power series may be multiplied according to the rule

$$(\sum a_n x^n)(\sum b_n x^n) = \sum c_n x^n$$

for all x within the intervals of convergence of the two series on the left, where $c_n = a_0 b_n + a_1 b_{n-1} + \cdots + a_n b_0$. Thus work out $(\sum_0^\infty x^n)^2$, let's say through fourth-degree terms (e.g., $3x^5$ is "fifth degree"). Does $(1 + x + x^2)^2$ give the same result?

2.20. (a) Obtain the Taylor series of $\sin(3x^2)$ about $x = 0$ through tenth-degree terms. What is the radius of convergence of the series?

(b) Obtain the Taylor series of $x^2 - 3$ about $x = 0$ and $x = 1$.

(c) Expand $\ln x$ about $x = 3$. What is the radius of convergence?

(d) Expand the functions

$$\int_x^1 \ln(1 + x^3)\, dx \quad \text{and} \quad \int_0^1 \ln(1 + x^3)\, dx$$

about $x = 1$ through terms of second degree.

2.21. Show that

$$1-\frac{x^2}{6}\le\frac{\sin x}{x}\le 1 \quad \text{for all } x.$$

Hint: Use Taylor's formula with Lagrange remainder for each of the inequalities. (Can you do it by using Taylor's *series* instead?) Incidentally (and this is important), $(\sin x)/x \le 1$ for all x is *not* the same as saying $\sin x \le x$ for all x. That is, if we multiply an inequality $A \le B$ by C, we obtain $AC \le BC$ only if C is positive. If $C < 0$, then $AC \ge BC$.

***2.22.** Obtain the Maclaurin expansion of $f(x)$, defined as e^{-1/x^2} for $x \neq 0$ and as zero for $x = 0$. (Be careful.)

2.23. Sketch $x^3H(x)$ and its first and second derivatives and show that the third and higher derivatives do not exist.

2.24. Sketch the function $f(x) = x^4$ for $x \ge 0$ and $= x^2$ for $x < 0$. Is f differentiable at $x = 0$? Is it analytic there? Expand f in a Taylor series about $x = 3$. For which x's does it converge? For which does it converge to f? Comment on the following excerpt from an examination paper regarding this function. "$f(x)$ is differentiable at $x = 0$ because $\lim_{x\to 0-} f'(x) = \lim_{x\to 0+} f'(x) = 0$. $f(x)$ is not analytic at $x = 0$ because $\lim_{x\to 0-} f''(x) = \lim_{x\to 0+} f''(x)$; so $f''(0)$ does not exist."

2.25. Sketch the function e^x, as well as the first, second, and third partial sums of its Maclaurin expansion. State any moral that seems appropriate.

2.26. Evaluate the integral

$$I = \int_0^1 x^{1/3}e^x\,dx$$

to four decimal places by expanding the integrand according to Taylor's formula with Lagrange remainder and using a conservative estimate of the remainder term to guarantee the desired accuracy.

2.27. Obtain a polynomial representation of erf x [recall (1.20)] that is uniformly valid to ± 0.001 in $0 \le x \le 1$.

2.28. Derive the error estimate (1.24) for the rectangular rule of integration. *Hint:* Integrate

$$\int_{x_0}^{x_1} f(x)\,dx = \int_{x_0}^{x_1} [f_0 + f'(\xi_0)(x - x_0)]\,dx = f_0h + \frac{h^2}{2}f'(\xi_0).$$

Doing the same for each subinterval, show that

$$E_n^R = \frac{h^2}{2}[f'(\xi_0) + f'(\xi_1) + \cdots + f'(\xi_{n-1})]$$

and then use the fact that $f'(x)$ is assumed to be continuous to deduce (1.24).

2.29. Suppose that $s = s_n + \kappa\rho^n + \alpha_n$, where $|\rho| < 1$ and $\alpha_n/\kappa\rho^n \longrightarrow 0$ as $n \longrightarrow \infty$. Then show that $s = s_n^{(1)} + \beta_n$, where $\beta_n/\kappa\rho^n \longrightarrow 0$ as $n \longrightarrow \infty$, so that the $s_n^{(1)}$ sequence defined by (2.41) does converge more rapidly than the original sequence s_n.

2.30. Estimate e by applying the Aitken–Shanks transformation to the first nine terms of the Maclaurin expansion of e^x, for $x = 1$, as in Table 2.1.

2.31. Verify that the Euler transformation (2.49) is *linear* as claimed.

2.32. Reread Example 2.14. Instead of summing the first six terms explicitly and Eulerizing the next five, simply Eulerize the first five terms of the series (2.35) and compare the resulting accuracy with that achieved in Example 2.14.

2.33. Eulerize terms 11 through 15 of Table 2.3 (i.e., $k = 10$ through 14) and obtain the result 0.0000177203, as stated in the text.

2.34. Sometimes the Euler transformation is stated in power series form[27]

$$\sum_0^\infty (-1)^n u_n x^n = \frac{1}{1+x} \sum_0^\infty (-1)^n \Delta^n u_0 \left(\frac{x}{1+x}\right)^n \quad \text{for } x > 0. \qquad (2.66)$$

(This can also be applied to trigonometric series, which we will meet shortly, by means of the substitution $x = e^{i\theta}$, where $i = \sqrt{-1}$.) Provide a formal derivation of (2.66).

***2.35.** Test the series

$$\sum_1^\infty e^{5n} \operatorname{Ei}(n^2)$$

for convergence.

***2.36.** Deduce the asymptotic expansion

$$\int_0^\infty \frac{e^{-xt}}{1+t^2}\, dt \sim \frac{1}{x} - \frac{2!}{x^3} + \frac{4!}{x^5} - \cdots, \quad \text{as } x \longrightarrow \infty,$$

and verify that it *is* an asymptotic expansion—that is, that it satisfies condition (2.58).

***2.37.** Derive the asymptotic behavior $\operatorname{erfc} x \sim e^{-x^2}/\sqrt{\pi}\, x$ as $x \longrightarrow \infty$, where $\operatorname{erfc} x$ was defined by (1.21).

2.38. Make up two examination-type problems on this chapter.

---

[27]For several applications from the fluid mechanics literature, see M. Van Dyke, *Perturbation Methods in Fluid Mechanics*, Academic, New York, 1964, Sections 10.6–10.8.

Chapter 3

Singular Integrals

An integral is said to be **singular**, or **improper**, if one or both limits are infinite and/or if the integrand becomes unbounded ("blows up") at one or more points in the interval. Such integrals appear time and again in applied mathematics and deserve a section to themselves.

In many ways they are analogous to infinite series, and our story closely parallels the previous chapter on series. First, we decide on a summability (or "integrability") definition, then we consider tests for convergence, and, finally, we consider their calculation.

3.1. CHOICE OF SUMMABILITY CRITERIA; CONVERGENCE AND CAUCHY PRINCIPAL VALUE

Suppose first that the integrand is finite but that the upper limit is infinite. With the Riemann integral defined only for finite intervals, we *define* the improper integral

$$\int_a^\infty f(x)\,dx \equiv \lim_{b\to\infty} \int_a^b f(x)\,dx, \tag{3.1}$$

that is, as the limit of a sequence of Riemann integrals. It is analogous to the ordinary convergence definition (2.3) for infinite series.

Example 3.1.

$$\int_1^\infty \frac{1}{x^p}\,dx = \lim_{b\to\infty}\int_1^b x^{-p}\,dx = \lim_{b\to\infty}\begin{cases}\left.\dfrac{x^{1-p}}{1-p}\right|_1^b & \text{if } p \neq 1\\[2ex] \left.\ln x\right|_1^b & \text{if } p = 1.\end{cases} \tag{3.2}$$

As $b \longrightarrow \infty$, $b^{1-p} \longrightarrow \infty$ for $p < 1$, $b^{1-p} \longrightarrow 0$ for $p > 1$, and $\ln b \longrightarrow \infty$. Thus the limit exists only for $p > 1$, and so the integral *converges* for $p > 1$ and *diverges* for $p \leq 1$. ▮

Example 3.2.

$$\int_0^\infty \cos x\,dx = \lim_{b\to\infty} \sin x\Big|_0^b = \lim_{b\to\infty} \sin b \tag{3.3}$$

diverges because the limit does not exist. ▮

Similarly,

$$\int_{-\infty}^a f(x)\,dx \equiv \lim_{b\to\infty}\int_{-b}^a f(x)\,dx \tag{3.4}$$

and

$$\int_{-\infty}^\infty f(x)\,dx \equiv \lim_{\substack{a\to\infty\\ b\to\infty}}\int_{-b}^a f(x)\,dx. \tag{3.5}$$

The important point to note about (3.5) is that the right-hand side is *not*

$$\lim_{a\to\infty}\int_{-a}^a f(x)\,dx; \tag{3.6}$$

in other words, a and b in (3.5) are to tend to infinity *independently*. We might emphasize this fact by rewriting (3.5) in the equivalent form

$$\int_{-\infty}^\infty f(x)\,dx \equiv \lim_{a\to\infty}\int_0^a f(x)\,dx + \lim_{b\to\infty}\int_{-b}^0 f(x)\,dx. \tag{3.7}$$

Example 3.3.

$$\begin{aligned}\int_{-\infty}^\infty x\,dx &= \lim_{a\to\infty}\int_0^a x\,dx + \lim_{b\to\infty}\int_{-b}^0 x\,dx\\ &= \left\{\lim_{a\to\infty}\frac{a^2}{2}\right\} - \left\{\lim_{b\to\infty}\frac{b^2}{2}\right\}\end{aligned} \tag{3.8}$$

diverges because neither of these two limits exists, even though

$$\lim_{a\to\infty}\int_{-a}^a x\,dx = \lim_{a\to\infty}\left(\frac{a^2}{2} - \frac{a^2}{2}\right) = \lim_{a\to\infty} 0 = 0 \tag{3.9}$$

does exist. ▮

Finally, suppose that the integral is singular due to $f(x)$ being unbounded, either at some point c inside the interval or at one (or both) of the (finite) endpoints. By analogy with (3.1), we define

$$\int_a^b f(x)\,dx \equiv \lim_{\epsilon\to 0}\int_a^{b-\epsilon} f(x)\,dx \tag{3.10}$$

if f is unbounded at the right endpoint (similarly, if it is unbounded at the left endpoint); and by analogy with (3.5) and (3.7), we define

$$\int_a^b f(x)\,dx \equiv \lim_{\epsilon_1 \to 0} \int_a^{c-\epsilon_1} f(x)\,dx + \lim_{\epsilon_2 \to 0} \int_{c+\epsilon_2}^b f(x)\,dx \tag{3.11}$$

if f is unbounded at an interior point c—that is, if f "blows up" at c.

Example 3.4. For instance, noting that $1/x^p$ blows up at the origin for $p > 0$, we have

$$\int_0^1 \frac{1}{x^p}\,dx = \lim_{\epsilon \to 0} \int_\epsilon^1 x^{-p}\,dx = \lim_{\epsilon \to 0} \begin{cases} \left.\dfrac{x^{1-p}}{1-p}\right|_\epsilon^1 & \text{if } p \neq 1 \\ \left.\ln x\right|_\epsilon^1 & \text{if } p = 1, \end{cases} \tag{3.12}$$

which limit exists only if $p < 1$; so the integral converges for $p < 1$ and diverges for $p \geq 1$. It is important to understand that divergence occurs (for $p \geq 1$) *not* because x^{-p} is infinite at $x = 0$ but because of how fast it *approaches* infinity as $x \to 0$. In fact, the value of the integrand *at* $x = 0$ is completely immaterial, as should be clear from the definition (3.12). We could define the integrand to be zero *at* $x = 0$, and doing so would in no way affect the convergence/divergence or the value of the integral!

Alternatively, if we set $x = 1/t$, then

$$\int_0^1 \frac{1}{x^p}\,dx = \int_1^\infty \frac{1}{t^{2-p}}\,dt,$$

which, according to Example 3.1, converges if and only if $2 - p > 1$—that is, if and only if $p < 1$, as found before.

As a summary look at Examples 3.1 and 3.4, consider the graph of $1/x^p$ in Fig. 3.1. In the case (3.2) we need p to be sufficiently *large* (> 1) for the area A under the

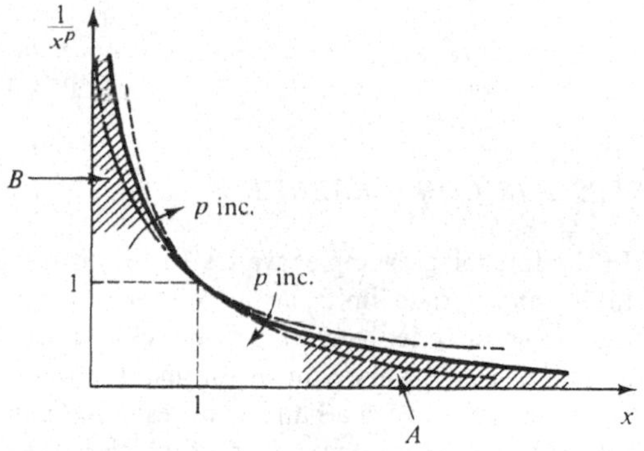

Figure 3.1. $1/x^p$.

horizontal "tail" to be finite, whereas in the case (3.12) we need p to be sufficiently *small* (< 1) for the area in the vertical "column" B to be finite. ▮

Example 3.5.

$$\int_0^3 \frac{dx}{x-1} = \lim_{\epsilon_1 \to 0} \int_0^{1-\epsilon_1} \frac{dx}{x-1} + \lim_{\epsilon_2 \to 0} \int_{1+\epsilon_2}^3 \frac{dx}{x-1}$$

$$= \lim_{\epsilon_1 \to 0} \ln|\epsilon_1| + \lim_{\epsilon_2 \to 0} \{\ln 2 - \ln|\epsilon_2|\}, \tag{3.13}$$

which diverges because the limits do not exist.[1] ▮

In contrast with (3.11), there is also the so-called **Cauchy Principal Value** definition,

$$⨍_a^b f(x)\,dx \equiv \lim_{\epsilon \to 0} \left\{ \int_a^{c-\epsilon} f(x)\,dx + \int_{c+\epsilon}^b f(x)\,dx \right\}, \tag{3.14}$$

which is denoted by various authors as $⨍$, $P\int$, $PV\int$, or $⨍$. It should be clear that if the integral exists in the "ordinary sense" (3.11), then it will also exist in the "Cauchy principal value sense" (3.14). On the other hand,

Example 3.6.

$$⨍_0^3 \frac{dx}{x-1} = \lim_{\epsilon \to 0} \{\ln|\epsilon| + \ln 2 - \ln|\epsilon|\} = \ln 2, \tag{3.15}$$

so that the integral *does* exist in the Cauchy sense, whereas it didn't (Example 3.5) in the ordinary sense. Roughly speaking, the Cauchy definition allows positive and negative infinite areas (shaded, in Fig. 3.2) to cancel. Similarly, if the integrand were $(x-1)^{-3}$, say. But if it were $(x-1)^{-2}$, then the infinite areas would both be positive, would not cancel, and the integral would diverge even in the Cauchy sense. ▮

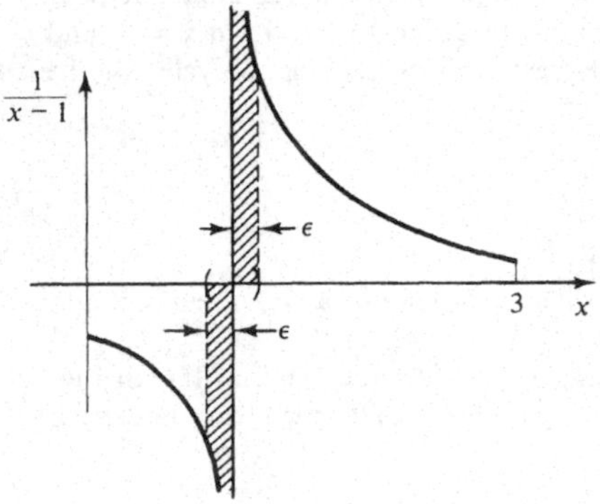

Figure 3.2. Cauchy principal value.

3.2. *TESTS FOR CONVERGENCE*

In the foregoing examples we tested for convergence by evaluating the integrals and taking appropriate limits, as in Example 2.1, where we determined the convergence of the geometric series by getting s_n into closed form and examining $\lim s_n$. Yet we are not always able to evaluate integrals, just as we are not always able to get s_n into closed form, and so we need a handful of tests that can be easily applied to a wide variety of integrals. Here are a few,[2] each of which is the analog of some test for series convergence already discussed in Chapter 2 and should come as no surprise.

---

[1] If the absolute values in (3.13) bother you, recall that $\int dx/x = \ln|x|$.

[2] For a more complete discussion, see, for example, Chapter 14 of Apostol.

THEOREM 3.1. (Comparison Test) If for two integrals

$$\int_a^{\infty} f(x)\,dx \quad \text{and} \quad \int_a^{\infty} g(x)\,dx$$

with positive integrands there exists a constant K such that $f(x) < Kg(x)$ for all sufficiently large x, then the convergence of the g integral implies the convergence of the f integral; or if there is some $k > 0$ such that $kg(x) < f(x)$ for all sufficiently large x, then the divergence of the g integral implies the divergence of the f integral.

It follows that if we can find *both* constants k and K—that is, such that $k < f(x)/g(x) < K$—then the two integrals either converge or diverge together. Surely such will be the case if $f/g \sim C$ as $x \longrightarrow \infty$ (where $C > 0$, of course), since we can choose $k = C/2$ and $K = 2C$, say.

THEOREM 3.2. (Absolute Convergence) If $\int_a^{\infty} |f(x)|\,dx$ converges, then so does $\int_a^{\infty} f(x)\,dx$, and we say that the latter converges "absolutely."

THEOREM 3.3. (The Dirichlet Test[3]) If $g(x) \downarrow 0$ as $x \longrightarrow \infty$ and there exists a constant $M < \infty$ such that

$$\left| \int_a^{x} f(x)\,dx \right| < M$$

for all $x > a$, then $\int_a^{\infty} f(x)g(x)\,dx$ converges.

Implicit in these three theorems is the fact that the integral from a to x does indeed exist for all *finite* $x > a$. The presence of interior singularities must be examined separately, as illustrated in the examples to follow.

Example 3.7.

$$\int_0^{\infty} \frac{(x^3 + 1)e^{-x}}{\sqrt{x}\,\ln x}\,dx. \tag{3.16}$$

Denoting the integrand as $f(x)$, we see that $f \sim x^{5/2}e^{-x}/\ln x < x^{5/2}e^{-x}$ as $x \longrightarrow \infty$. But

$$\frac{x^{5/2}}{e^x} = \frac{x^{5/2}}{1 + x + (x^2/2!) + \cdots} < \frac{x^{5/2}}{x^7/7!},$$

say, so that f is dominated by $7!x^{-9/2}$ as $x \longrightarrow \infty$. Comparing with the "p integral" of Example 3.1, we see that $p = \frac{9}{2} > 1$, and so (3.16) converges, provided that the integral from 0 to x exists for all finite $x > 0$. Well, there are two singularities, one at the left endpoint $x = 0$ and one at $x = 1$. As $x \longrightarrow 0, f \sim 1/\sqrt{x}\,\ln x = O(x^{-1/2+\alpha})$ for α arbitrarily small and positive. (Recall Exercise 1.20.) Since $\frac{1}{2} - \alpha < 1$, we see from Example 3.4 that this singularity is "integrable"; that is, it is not strong enough to cause divergence. Finally, consider $x = 1$, where the $\ln x$ in the denominator

[3] In connection with *integrals*, it is also referred to as *Chartier's test*.

becomes zero. To expose the strength and exact nature of the singularity, we *expand about that point*:

$$(x^3 + 1)e^{-x} = 2e^{-1} + \cdots$$
$$\sqrt{x} = 1 + \cdots$$
$$\ln x = (x - 1) + \cdots,$$

so that $f(x) \sim 2e^{-1}/(x - 1)$, which (Example 3.5) is *not* integrable, although it *would* be in the Cauchy principal value sense. We conclude that (3.16) is divergent, unless we're willing to interpret it in the Cauchy sense. How do we know which way to interpret it? That point is generally decided by the circumstances leading to the integral (e.g., see Exercise 4.8c); in the present case, I simply made up (3.16) as an illustration and am free to specify whichever criterion I wish. ∎

Example 3.8.

$$\int_0^\infty \sin x^2 \, dx. \tag{3.17}$$

Setting $x^2 = t$, (3.17) becomes $\int_0^\infty t^{-1/2} \sin t \, dt/2$, which converges by the Dirichlet test, since $t^{-1/2} \downarrow 0$ as $t \longrightarrow \infty$ and $\left|\int_0^t \sin t \, dt\right| = |1 - \cos t| < 3$, say, for all t. Observe that it would *not* be correct to state that $\left|\int_0^\infty \sin t \, dt\right| < 3$ because $\int_0^\infty \sin t \, dt$ does not exist!

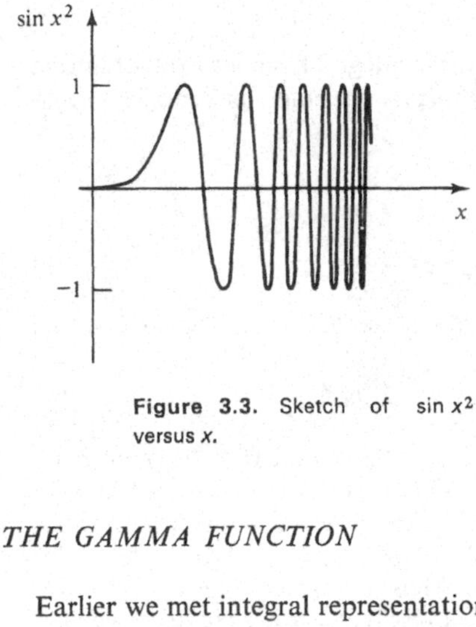

Figure 3.3. Sketch of $\sin x^2$ versus x.

Notice further that the convergence occurs despite the fact that the integrand $\sin x^2$ does not tend to zero as $x \longrightarrow \infty$, in contrast with series, where $a_n \longrightarrow 0$ is necessary for convergence! To see how this convergence is achieved, let us sketch $\sin x^2$. Writing $\sin x^2 = \sin \omega x$, observe that the frequency $\omega = x$ varies from zero to infinity as x varies from zero to infinity, as sketched in Fig. 3.3. Roughly speaking, then, the convergence of (3.17) is due to the considerable cancellation of succesive positive and negative areas, a tendency that becomes increasingly pronounced as $x \longrightarrow \infty$. (Another striking example is given in Exercise 3.1 where convergence is obtained without the benefit of the integrand tending to zero *or* a cancellation of areas.) ∎

3.3. *THE GAMMA FUNCTION*

Earlier we met integral representations of a few important special functions—namely, erf (x), erfc (x), Ei (x), Si (x), and Ci (x). We will find it helpful to introduce another, the **gamma function** $\Gamma(x)$, defined by

$$\Gamma(x) \equiv \int_0^\infty t^{x-1} e^{-t} \, dt \tag{3.18}$$

for $x > 0$; for $x \leq 0$ the integral is divergent and (3.18) does not apply. An important recursion formula can be derived for $\Gamma(x)$ by integration by parts. With "u" $= t^{x-1}$ and "dv" $= e^{-t}\,dt$,

$$\Gamma(x) = -t^{x-1}e^{-t}\Big|_0^\infty + (x-1)\int_0^\infty t^{x-2}e^{-t}\,dt. \tag{3.19}$$

The integral in (3.19) exists [and is $\Gamma(x-1)$] only if $x > 1$, in which case the boundary term is zero. Thus (3.19) becomes

$$\Gamma(x) = (x-1)\Gamma(x-1) \quad \text{for } x > 1. \tag{3.20}$$

If we tabulate $\Gamma(x)$ for $0 < x \leq 1$ only, the recursion formula (3.20) enables us to compute $\Gamma(x)$ for all $x > 1$. For instance,

$$\Gamma(3.7) = 2.7\Gamma(2.7) = (2.7)(1.7)\Gamma(1.7) = (2.7)(1.7)(0.7)\Gamma(0.7).$$

Note, in particular, that if x is a positive integer n, then

$$\Gamma(n) = (n-1)\Gamma(n-1) = \text{etc.} = (n-1)(n-2)\cdots(1)\Gamma(1) = (n-1)!, \tag{3.21}$$

since

$$\Gamma(1) = \int_0^\infty e^{-t}\,dt = \text{etc.} = 1.$$

Another x that permits direct integration is $x = \frac{1}{2}$. With the substitution $t = u^2$, (3.18) becomes

$$\Gamma(\tfrac{1}{2}) = 2\int_0^\infty e^{-u^2}\,du. \tag{3.22}$$

Since u is only a dummy variable, it is surely true that

$$[\Gamma(\tfrac{1}{2})]^2 = 4\int_0^\infty e^{-u^2}\,du \int_0^\infty e^{-v^2}\,dv = 4\int_0^\infty\int_0^\infty e^{-(u^2+v^2)}\,du\,dv. \tag{3.23}$$

Regarding u, v as rectangular Cartesian coordinates, (3.23) is an integral over the first quadrant of the u, v plane. Switching to polar coordinates r, θ, we have $u^2 + v^2 = r^2$ and $du\,dv = d(\text{area}) = r\,dr\,d\theta$, so that

$$\begin{aligned}[\Gamma(\tfrac{1}{2})]^2 &= 4\int_0^{\pi/2}\int_0^\infty e^{-r^2}r\,dr\,d\theta \\ &= 2\pi\int_0^\infty e^{-r^2}r\,dr = \pi\int_0^\infty e^{-r^2}\,dr^2 = \pi,\end{aligned} \tag{3.24}$$

and hence

$$\Gamma(\tfrac{1}{2}) = \sqrt{\pi} \tag{3.25}$$

Multiple integrals are covered later on, but some comments are in order here. We say that

$$I = \iint_S f(x, y)\,dA$$

is the **double integral** of f over the region S of the x, y plane. Roughly, this means: partition S into many small areas ΔA, multiply these ΔAs by the local values of f, add it all up, and then repeat the process for a finer partition. The limit (if it exists), as the norm of the partitions $\longrightarrow 0$, is I. *Numerical* integration of I involves just such a partition-summation process, but to evaluate I *analytically*, we convert to an **iterated**

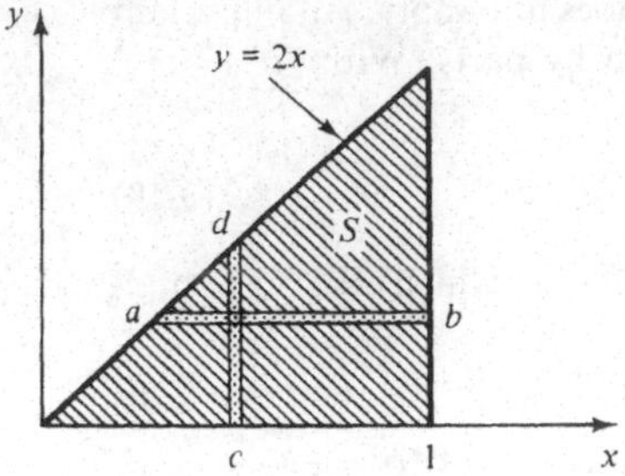

Figure 3.4. Iterated integral.

integral wherein we integrate first on x, with y fixed, and then on y, or vice versa. To illustrate, consider the simple case where $f(x, y) = xy^2$ and S is the triangular region shown in Fig. 3.4. Integrating first on x and then on y (i.e., integrating across the sliver ab, and then "sweeping" the sliver, which is bounded on the left by $y = 2x$ and on the right by $x = 1$, from $y = 0$ to $y = 2$), we write

$$I = \int_0^2 \left\{ \int_{y/2}^1 xy^2\, dx \right\} dy = \int_0^2 \left\{ \frac{x^2y^2}{2} \Big|_{y/2}^1 \right\} dy$$

$$= \int_0^2 \left(\frac{y^2}{2} - \frac{y^4}{8} \right) dy = \frac{8}{6} - \frac{32}{40} = \frac{8}{15}.$$

Or integrating first on y, over the sliver cd, and then sweeping from $x = 0$ to $x = 1$,

$$I = \int_0^1 \left\{ \int_0^{2x} xy^2\, dy \right\} dx = \int_0^1 \frac{8x^4}{3}\, dx = \frac{8}{15}.$$

Various notations are used. For instance,

$$\int_0^2 \left\{ \int_0^{3x} xe^{xy}\, dy \right\} dx, \qquad \int_0^2 \int_0^{3x} xe^{xy}\, dy\, dx, \qquad \int_0^2 x\, dx \int_0^{3x} e^{xy}\, dy$$

all denote the same thing. The last form does *not* mean $\int_0^2 x\, dx$ *times* $\int_0^{3x} e^{xy}\, dy$; rather it means that we first evaluate $\int_0^{3x} e^{xy}\, dy \equiv F(x)$, say, and then $\int_0^2 xF(x)\, dx$.

As another example, consider the evaluation of

$$I = \int_0^1 dy \int_y^{\sqrt{y}} xe^{y/x}\, dx.$$

Integration of $xe^{y/x}$ with respect to x is not a happy prospect, whereas with respect to y it would be quite simple, and so we wish to "invert" the order of integration:

$$I = \int_?^? x\, dx \int_?^? e^{y/x}\, dy.$$

To find the new limits, it is helpful to go back and determine S. Noting from the original limits that S is bounded on the left by $x = y$, on the right by $x = \sqrt{y}$, below by $y = 0$, and above by $y = 1$, we easily deduce the region S shown in Fig. 3.5, and we see that

$$I = \int_0^1 x\, dx \int_{x^2}^x e^{y/x}\, dy = \int_0^1 x^2(e - e^x)\, dx = \text{etc.}$$

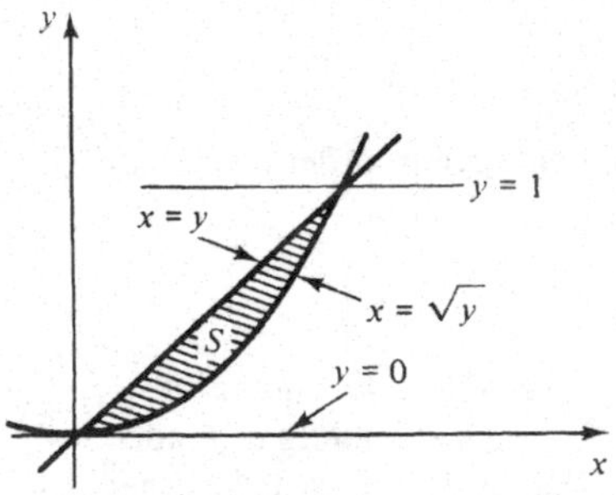

Figure 3.5. Determining S.

Concluding our discussion of the gamma function, recall that (3.18) defines $\Gamma(x)$ only for $x > 0$; for $x \leq 0$ the integral diverges and (3.18) is meaningless. What is a reasonable way to define $\Gamma(x)$ for $x < 0$? Recall that if we know $\Gamma(x)$, we can compute $\Gamma(x + 1)$

from the recursion formula $\Gamma(x+1) = x\Gamma(x)$. Instead of using this to march to the right, we can turn it around and use it to march to the left. Thus let us *define*

$$\Gamma(x) \equiv \frac{\Gamma(x+1)}{x} \quad \text{for } x < 0. \tag{3.26}$$

For instance,

$$\Gamma(-0.5) = \frac{\Gamma(0.5)}{(-0.5)} = -2\sqrt{\pi}$$

$$\Gamma(-1.3) = \frac{\Gamma(-0.3)}{(-1.3)} = \frac{\Gamma(0.7)}{(-1.3)(-0.3)}$$

$$\Gamma(-2.4) = \frac{\Gamma(-1.4)}{(-2.4)} = \frac{\Gamma(-0.4)}{(-2.4)(-1.4)}$$

$$= \frac{\Gamma(0.6)}{(-2.4)(-1.4)(-0.4)},$$

and, just as for $x > 1$, we see that $\Gamma(x)$ can be computed for all $x < 0$ (except for negative integers) in terms of $\Gamma(x)$ values for $0 < x < 1$. The singularity at $x = 0$ [i.e., $\Gamma(x)$, as given by the right-hand side of (3.18), tends to $+\infty$ as $x \to 0+$] propagates to the left via (3.26), so that $\Gamma(x)$ is singular at all negative integer x's. The resulting graph of $\Gamma(x)$ is shown in Fig. 3.6.

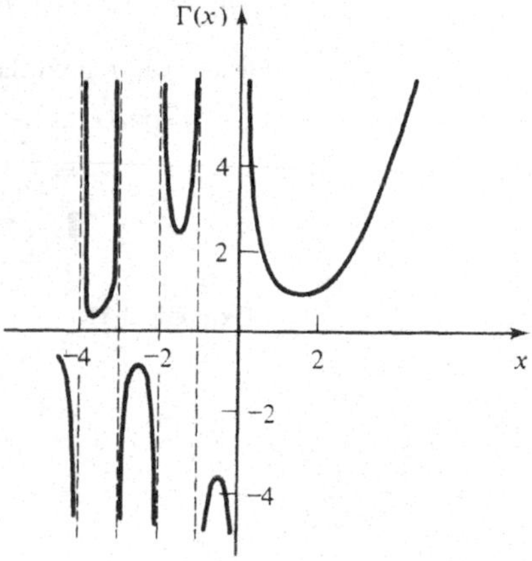

Figure 3.6. The gamma function.

3.4. EVALUATION OF SINGULAR INTEGRALS

There are many techniques for the analytical and numerical evaluation of singular integrals. We will illustrate a few by means of examples. The powerful method of contour integration in the complex plane will have to wait until Part II. Additional techniques and references are given by Abramowitz.[4]

Example 3.9.

$$I = \int_0^\infty e^{-x^4}\,dx = ? \tag{3.27}$$

With $x^4 = t$,

$$I = \int_0^\infty t^{-3/4} e^{-t} \frac{dt}{4} = \frac{\Gamma(1/4)}{4}. \quad \blacksquare \tag{3.28}$$

Example 3.10.

$$I = \int_0^\infty e^{-\alpha x} \sin x\,dx = ? \qquad (\alpha > 0) \tag{3.29}$$

---

[4] M. Abramowitz, On The Practical Evaluation of Integrals, *Journal of the Society of Industrial and Applied Mathematics*, Vol. 2, No. 1, March, 1954, pp. 20–35.

Using Euler's formula $e^{ix} = \cos x + i \sin x$, we can express[5]

$$\begin{aligned} I &= \int_0^\infty \operatorname{Im}(e^{-\alpha x}e^{ix})\,dx = \operatorname{Im}\int_0^\infty e^{-\alpha x}e^{ix}\,dx \\ &= \operatorname{Im}\int_0^\infty e^{(-\alpha+i)x}\,dx = \operatorname{Im}\frac{e^{(-\alpha+i)x}}{-\alpha+i}\bigg|_0^\infty = \operatorname{Im}\frac{1}{\alpha - i} \\ &= \operatorname{Im}\frac{1}{\alpha - i}\left(\frac{\alpha+i}{\alpha+i}\right) = \operatorname{Im}\frac{\alpha+i}{\alpha^2+1} = \frac{1}{\alpha^2+1}, \end{aligned} \tag{3.30}$$

where we've used the fact that

$$\begin{aligned} |e^{-\alpha x}e^{ix}| &= |e^{-\alpha x}||e^{ix}| = e^{-\alpha x}|\cos x + i\sin x| = e^{-\alpha x}\sqrt{\cos^2 x + \sin^2 x} \\ &= e^{-\alpha x} \longrightarrow 0 \end{aligned}$$

as $x \longrightarrow \infty$. ▮

Example 3.11.

$$I = \int_0^\infty \frac{\sin x}{x}\,dx = \,? \tag{3.31}$$

Let

$$J(\alpha) = \int_0^\infty \frac{e^{-\alpha x}\sin x}{x}\,dx, \tag{3.32}$$

so that $I = J(0)$. Proceeding formally (e.g., without worrying about the legitimacy of Leibnitz differentiation of a *singular* integral),

$$J'(\alpha) = -\int_0^\infty e^{-\alpha x}\sin x\,dx = -\frac{1}{\alpha^2+1} \tag{3.33}$$

per Example 3.10. Integrating (3.33), $J(\alpha) = -\tan^{-1}\alpha + C$. But $J(\infty) = 0$ implies that $C = \pi/2$, and so $I = J(0) = \pi/2$.

As a matter of fact, Abramowitz[6] mentions that more than half a dozen different methods of evaluation of this well-known integral appear in the literature. ▮

Finally, a few remarks concerning the *numerical* evaluation of singular integrals.

Example 3.12. To evaluate I of (3.31) *numerically*, suppose that we evaluate (say by Simpson's rule or Gauss integration) the integral of $\sin x/x$ over $[0, \pi]$, $[\pi, 2\pi]$, . . . , $[8\pi, 9\pi]$. To five decimals, they turn out to be[7] 1.85194, −0.43379, 0.25661, −0.18260, 0.14180, −0.11593, 0.09850, −0.08495, and 0.07495. This doesn't look very promising. But if we sum the first four terms, say, and Eulerize the next five, we arrive (Exercise 3.12) at $I \approx 1.57078$, compared with the exact value $I = \pi/2 \doteq 1.57080$. ▮

---

[5] For any complex number $z = x + iy$, we denote $\operatorname{Re} z = x$ and $\operatorname{Im} z = y$ (i.e., the real and imaginary parts of z).

[6] *Ibid.*

[7] *Handbook of Mathematical Functions*, ed. by M. Abramowitz and I. Stegun, National Bureau of Standards Applied Math Series, 1964, p. 22.

Example 3.13.

$$I = \int_0^1 \frac{\cos(x^2+2)}{\sqrt{x}}\,dx = \int_0^1 \frac{\cos(x^2+2) - \cos 2}{\sqrt{x}}\,dx + \cos 2 \int_0^1 \frac{dx}{\sqrt{x}}$$

$$= \int_0^1 \frac{\cos(x^2+2) - \cos 2}{\sqrt{x}}\,dx + 2\cos 2. \qquad (3.34)$$

The point is that whereas the original integral exists, the integrand's square root singularity (integrand $\sim \cos 2/\sqrt{x}$ as $x \longrightarrow 0$) at $x = 0$ makes accurate numerical evaluation difficult. By subtracting and adding $\cos 2/\sqrt{x}$, we split off the singular part (which is trivially integrable). The final integral in (3.34) is "desingularized" and therefore more amenable to numerical evaluation. Specifically, $\cos(x^2 + 2) = (\cos 2) - (\sin 2)x^2 + \cdots$, so that $[\cos(x^2 + 2) - \cos 2]/\sqrt{x} \sim -(\sin 2)x^{3/2} \longrightarrow 0$ as $x \longrightarrow 0$. (See Exercise 3.5 for an alternative approach.) ▮

Example 3.14. Cauchy integrals can be desingularized by "folding." For instance,

$$⨍_0^2 \frac{dx}{(1-x)\sqrt{x+1}} = \lim_{\epsilon\to 0}\left\{\int_0^{1-\epsilon} \frac{dx}{(1-x)\sqrt{x+1}} + \int_{1+\epsilon}^2 \frac{dx}{(1-x)\sqrt{x+1}}\right\}.$$

Setting $1 - x = t$ in the first integral on the right and $x - 1 = t$ in the second, we have

$$⨍_0^2 \frac{dx}{(1-x)\sqrt{x+1}} = \lim_{\epsilon\to 0}\left\{\int_\epsilon^1 \frac{dt}{t\sqrt{2-t}} - \int_\epsilon^1 \frac{dt}{t\sqrt{2+t}}\right\}$$

$$= \lim_{\epsilon\to 0}\int_\epsilon^1 \left\{\frac{1}{\sqrt{2-t}} - \frac{1}{\sqrt{2+t}}\right\}\frac{dt}{t}$$

$$= \int_0^1 \left\{\frac{1}{\sqrt{2-t}} - \frac{1}{\sqrt{2+t}}\right\}\frac{dt}{t},$$

where $\lim_{\epsilon\to 0}$ has been dropped in the last step since the final integrand is nonsingular; in particular, $(2 - t)^{-1/2} = 1/\sqrt{2} + t/4\sqrt{2} + \cdots$, and $(2 + t)^{-1/2} = 1/\sqrt{2} - t/4\sqrt{2} + ;\cdots$ therefore

$$\frac{(2-t)^{-1/2} - (2+t)^{-1/2}}{t} = \frac{t/2\sqrt{2} + \cdots}{t} \sim \frac{1}{2\sqrt{2}} \quad \text{as } t \longrightarrow 0.$$

Note that if we evaluate the final integral by Simpson's rule, say, then we need to *tell* the computer that the integrand is $1/2\sqrt{2}$ at $t = 0$, since, left to itself, it would have to deal with 0/0. For Gauss integration, however, doing so would be unnecessary because Gauss abscissas never coincide with the interval endpoints. ▮

EXERCISES

3.1. With $f(x) = n$ over $(n, n + n^{-3})$ for each $n = 1, 2, 3, \ldots$, and zero otherwise, show that the integral of $f(x)$ from zero to infinity converges even though $f(x)$ grows unboundedly and without oscillation. [Sketch $f(x)$.]

3.2. For what (real) values of α do the following integrals converge?

(a) $\displaystyle\int_0^\infty \frac{x^\alpha(1-\cos x)}{x^4+1}\,dx$ (b) $\displaystyle\int_1^\infty \frac{x^\alpha}{\sqrt{x^2-1}}\,dx$

(c) $\int_0^\infty \frac{(x+1)^\alpha e^{-x}}{\sqrt{x}}\,dx$ (d) $\int_0^\infty x^\alpha e^{-x^2}\,dx$

(e) $\int_0^\infty x^\alpha \sin x^3\,dx$ (f) $\int_{-\infty}^{\alpha} e^{(x+\cos x)}\,dx$

(g) $\int_0^\infty x^{-\alpha}\,dx$ (h) $⨍_{-\infty}^{\infty} \frac{|x|^\alpha}{x-1}\,dx$

3.3. Test for convergence: [Recall from Exercise 1.20 that $\ln x = O(x^\alpha)$ as $x \longrightarrow \infty$ and $O(x^{-\alpha})$ as $x \longrightarrow 0$, where α is arbitrarily small.)

(a) $\int_0^\infty \frac{\sin(x^2 - x)}{x}\,dx$ *(b) $\int_0^\infty \frac{(x-1)\,\mathrm{Ei}(x)}{(\ln x)^{3/2}}\,dx$

(c) $\int_0^\infty e^{-x}\sin\frac{1}{x}\,dx$ (d) $⨍_0^\infty \frac{\cos x}{\ln x}\,dx$

(e) $\int_0^2 \frac{1}{x^{3/2}}\sin\frac{1}{x}\,dx$ (f) $\int_1^\infty \sin^2\frac{1}{x}\,dx$

(g) $\int_1^\infty \cos^2\frac{1}{x}\,dx$ (h) $⨍_{\infty}^{\infty} \frac{\ln|1-x|}{x^2}\,dx$

3.4. Is $\int_0^a x\,dx/\sin x$ convergent for $a = \pi/2$? For $a = \pi$? For $a = 3\pi/2$? Is it convergent in the *Cauchy* sense for $a = 3\pi/2$? Explain.

3.5. Show that the convergent singular integrals

$$\int_0^1 \frac{\cos(x^2+2)}{\sqrt{x}}\,dx \quad \text{and} \quad \int_{-1}^1 \frac{x + 3\cos x}{\sqrt{1-x^2}}\,dx$$

are desingularized by the substitutions $x = t^2$ and $x = \sin t$, respectively.

3.6. (a) Show that

$$\int_0^\infty e^{-4(x^2+x)}\,dx = \frac{e\sqrt{\pi}}{4}\,\mathrm{erfc}\,(1).$$

Hint: "Complete the square," $4x^2 + 4x = (2x+1)^2 - 1$, and then set $2x + 1 = t$.

(b) Evaluate

$$\int_0^x e^{x-x^2}\,dx.$$

3.7. Desingularize

$$⨍_0^3 \frac{dx}{(1-x^2)\sqrt{x+1}}$$

by "folding," as in Example 3.14.

3.8. The following was proposed in class. "Setting $x = \sqrt{t}$,

$$I = \int_0^\infty \sin x\,dx = \frac{1}{2}\int_0^\infty \frac{\sin\sqrt{t}}{\sqrt{t}}\,dt.$$

Now $t^{-1/2} \downarrow 0$ and $\left|\int_0^T \sin\sqrt{t}\,dt\right| < 10$ for all T, and so I converges." On the other hand,

$$I = \lim_{b\to\infty}\int_0^b \sin x\,dx = \lim_{b\to\infty}(1 - \cos b)$$

does not exist. Discuss. (Eventually in your discussion you should be interested in sketching $\sin\sqrt{t}$. Express $\sin\sqrt{t} = \sin\omega t$ and note that the frequency $\omega = t^{-1/2}$ tends to zero as $t \longrightarrow \infty$.)

3.9. Evaluate

$$\int_0^\infty x^2 e^{-x^2}\, dx$$

from the known integral $\int_0^\infty e^{-x^2}\, dx = \sqrt{\pi}/2$ by differentiating with respect to a suitably inserted parameter.

3.10. Evaluate

(a) $\int_0^\infty e^{-\alpha x} \cos x\, dx \quad (\alpha > 0)$; (b) $\int_{-\infty}^\infty x^2 \sin x^5\, dx$.

3.11. (*The beta function*) We state without proof that

$$B(a, b) = \int_0^1 x^{a-1}(1-x)^{b-1}\, dx$$

$$= 2\int_0^{\pi/2} \cos^{2a-1} t \sin^{2b-1} t\, dt = \frac{\Gamma(a)\Gamma(b)}{\Gamma(a+b)},$$

where $a > 0$, $b > 0$, and $B(a, b)$ is known as the **beta function.**
Using this formula,
(a) show that

$$\int_0^1 \frac{dx}{\sqrt{1-x^n}} = \frac{\sqrt{\pi}\,\Gamma(1/n)}{n\Gamma[(n+2)/2n]}.$$

(b) evaluate

$$\int_0^{\pi/2} \sqrt{\tan x}\, dx.$$

3.12. Do the Eulerizing mentioned in Example 3.12, thus obtaining the quoted result $I \approx 1.57078$.

3.13. Evaluate the so-called **Dirichlet discontinuous integral**

$$\frac{1}{\pi}\int_{-\infty}^\infty \frac{\sin ax}{x}\, dx \qquad (-\infty < a < \infty)$$

with the help of Example 3.11.

3.14. Evaluate

$$\int_0^1 (\ln x)^3\, dx.$$

Hint: Differentiate the known result $\int_0^1 x^\alpha\, dx = (\alpha+1)^{-1}$ repeatedly with respect to α.

3.15. Show that

$$\int_0^\infty e^{-x^2-(a^2/x^2)}\, dx = \sqrt{\pi}\, e^{-2a}/2 \qquad (a \geq 0).$$

Hint: Denoting the integral as $I(a)$, obtain dI/da. Setting $x = a/t$ in the resulting integral, show that $dI/da = -2I$.

3.16. "Grade" the following solution (from an old examination paper) to Exercise 3.2(d) above: "$e^{-x^2} \downarrow 0$, and $\int_0^x x^\alpha \, dx = x^{\alpha+1}/(\alpha + 1)$ is bounded for $\alpha \le -1$; and so the integral converges by the Dirichlet test for all $\alpha \le -1$."

3.17. Evaluate

$$\int_0^\infty \int_y^\infty e^{-(x^2+y^2)} \, dx \, dy.$$

3.18. Invert the order of integration for the iterated integral

$$\int_0^2 \int_{x^2}^1 f(x, y) \, dy \, dx.$$

Hint: Be careful.

3.19. Inverting the order of integration is *not* always valid. For instance, show that

$$\int_1^\infty \int_1^\infty \frac{x - y}{(x + y)^3} \, dx \, dy = \frac{1}{2},$$

whereas the inverted integral equals $-\frac{1}{2}$. To clarify what's happening, it may help to evaluate the integral over the finite rectangle $1 \le x \le X$, $1 \le y \le Y$ and take $\lim_{Y\to\infty} \lim_{X\to\infty}$ and $\lim_{X\to\infty} \lim_{Y\to\infty}$ of this result.

3.20. Fill in the missing limits:

$$\int_0^1 \int_0^s \int_0^r \int_0^q f(p, q, r, s) \, dp \, dq \, dr \, ds$$
$$= \int_?^? \int_?^? \int_?^? \int_?^? f(p, q, r, s) \, ds \, dr \, dq \, dp.$$

Hint: It is not possible to deduce and sketch the region of integration in p, q, r, s space because it is *four dimensional.* Instead switch the order of integration in only two variables at a time; that is, starting with the integration order *pqrs*, switch to *qprs*, *qrps*, *qrsp*, *qsrp*, *sqrp*, and *srqp* in turn.

3.21. (a) Grade the following two proposals for evaluating

$$I = \int_0^\infty \frac{\sin^3 x}{x^2} \, dx = \int_0^\infty \frac{3 \sin x - \sin 3x}{4x^2} \, dx.$$

Proposal 1

$$I = \frac{3}{4} \int_0^\infty \frac{\sin x}{x^2} \, dx - \frac{1}{4} \int_0^\infty \frac{\sin 3x}{x^2} \, dx$$
$$= \frac{3}{4} \int_0^\infty \frac{\sin x}{x^2} \, dx - \frac{3}{4} \int_0^\infty \frac{\sin t}{t^2} \, dt = 0.$$

Proposal 2

$$I = \lim_{\epsilon\to 0} \left[\int_\epsilon^\infty \frac{3 \sin x}{4x^2} \, dx - \int_\epsilon^\infty \frac{\sin 3x}{4x^2} \, dx \right]$$
$$= \frac{3}{4} \lim_{\epsilon\to 0} \left[\int_\epsilon^\infty \frac{\sin x}{x^2} \, dx - \int_{3\epsilon}^\infty \frac{\sin t}{t^2} \, dt \right]$$
$$= \frac{3}{4} \lim_{\epsilon\to 0} \int_\epsilon^{3\epsilon} \frac{\sin x}{x^2} \, dx = \frac{3 \ln 3}{4}.$$

(b) Evaluate

$$I = \int_0^\infty \frac{2 \sin (x^3) - \sin (2x^3)}{x^4} \, dx.$$

3.22. Show that

(a) $\displaystyle\int_x^\infty \frac{\sin 3\xi}{\xi^2} \, d\xi \sim -3 \ln x \quad \text{as} \quad x \longrightarrow 0.$

(b) $\displaystyle\int_0^x \cos \frac{1}{\xi} \, d\xi \sim x \quad \text{as} \quad x \longrightarrow \infty.$

*(c) $\displaystyle\int_0^x \cos \frac{1}{\xi} \, d\xi = x + \frac{\pi}{2} - \frac{1}{2x} + \frac{2!}{4!\,3!} \frac{1}{x^3} - \frac{4!}{6!\,5!} \frac{1}{x^5} + \cdots$

for $x > 0$.

3.23. Make up two examination-type problems on this chapter.

Chapter 4

Interchange of Limit Processes and the Delta Function

Consider the calculation

$$\int_0^{\pi/2} \cos(\sin^{1/3} x)\, dx = \int_0^{\pi/2} \sum_0^{\infty} (-1)^n \frac{\sin^{2n/3} x}{(2n)!}\, dx$$
$$= \sum_0^{\infty} \frac{(-1)^n}{(2n)!} \int_0^{\pi/2} \sin^{2n/3} x\, dx$$
$$= \frac{1}{2} \sum_0^{\infty} \frac{(-1)^n}{(2n)!} \frac{\Gamma(1/2)\Gamma(n/3 + 1/2)}{\Gamma(n/3 + 1)}, \tag{4.1}$$

where we've used first the Maclaurin expansion of $\cos z$ and then the formula of Exercise 3.11. As will be seen, interchanging the order of limit processes is not always legitimate, and such an exchange is exactly what we've done with the infinite sum and the Riemann integral (each of which is defined in terms of a limit process) in the second equality.

Short of a rigorous test based on an appropriate theorem given later in this section, the least we can do is check the convergence of the final series in (4.1). To do so, we need to know the asymptotic behavior of the gamma function, and it is provided by **Stirling's formula,**[1]

$$\Gamma(x) \sim \sqrt{2\pi}\, x^{x-1/2} e^{-x} \quad \text{as } x \longrightarrow \infty. \tag{4.2}$$

Using (4.2), we find (Exercise 4.1) that (4.1) does indeed converge, which tends to

[1]See, for example, G. Carrier, M. Krook, and C. Pearson, *Functions of a Complex Variable*, McGraw-Hill, New York, 1966, Section 6-2.

inspire some confidence in its validity, together with the fact that the original integral in (4.1) is regular (i.e., not singular). In other words, we could rely on the following principle and let the matter rest. *When things* seem *to be going well, probably they* are *going well.* Alternatively, various theorems are available concerning the interchange of specific limit processes.

4.1. THEOREMS ON LIMIT INTERCHANGE

First, a theorem that covers the interchange of summation and integration, as in (4.1), and a companion theorem for summation and differentiation.

THEOREM 4.1.

$$\int_a^b \sum^{\infty} a_n(x)\, dx = \sum^{\infty} \int_a^b a_n(x)\, dx \tag{4.3}$$

if $\sum^{\infty} a_n(x)$ *converges uniformly over* $[a, b]$.

THEOREM 4.2.

$$\frac{d}{dx} \sum^{\infty} a_n(x) = \sum^{\infty} \frac{d}{dx} a_n(x) \tag{4.4}$$

for all x *in* (a, b) *if the series on the right converges uniformly over* (a, b).

It goes without saying that all integrals, derivatives, and series in these two equations are assumed to exist.

Let us prove Theorem 4.1. With $s_n(x)$ denoting the nth partial sum $a_1(x) + \cdots + a_n(x)$, (4.3) is equivalent (Why?) to

$$\lim_{n\to\infty} \int_a^b s_n(x)\, dx = \int_a^b \lim_{n\to\infty} s_n(x)\, dx = \int_a^b s(x)\, dx. \tag{4.5}$$

To prove (4.5), write

$$\lim_{n\to\infty} \int_a^b s_n(x)\, dx = \lim_{n\to\infty} \int_a^b [s_n(x) - s(x)]\, dx + \int_a^b s(x)\, dx. \tag{4.6}$$

But since $s_n(x) \longrightarrow s(x)$ uniformly, then to each $\epsilon > 0$ (no matter how small) there is an N such that $|s_n(x) - s(x)| < \epsilon$ over $[a, b]$ for all $n > N$. Thus the first integral on the right of (4.6) is smaller than $\epsilon(b - a)$ in magnitude for all $n > N$. With ϵ arbitrarily small, however, the first term on the right must, in fact, be zero, in which case (4.6) reduces to the desired result. Actually, (4.5) is not too surprising if we recall Fig. 2.1.

In applications, we sometimes have $b = \infty$; then the preceding proof breaks down because the bound $\epsilon(b - a)$ is no longer arbitrarily small.

There are many such theorems governing the interchange of various limit processes; interested readers might try the books by Apostol,[2] Buck,[3] or Rudin.[4]

---

[2]T. M. Apostol, *Mathematical Analysis*, Addison-Wesley, Reading, Mass., 1957.
[3]R. C. Buck, *Advanced Calculus*, 2nd ed., McGraw-Hill, New York, 1965.
[4]W. Rudin, *Principles of Mathematical Analysis*, 2nd ed., McGraw-Hill, New York, 1964.

4.2. *THE DELTA FUNCTION AND GENERALIZED FUNCTIONS*

In a very rough sense the so-called Dirac delta function, first introduced by the physicist P. Dirac,[5] can be regarded as a concentrated "spike" of unit area—for example, the limit of the rectangular pulse (Fig. 4.1)

$$w_k(x) = \begin{cases} \dfrac{k}{2}, & |x| \le \dfrac{1}{k} \\ 0, & |x| > \dfrac{1}{k} \end{cases} \tag{4.7}$$

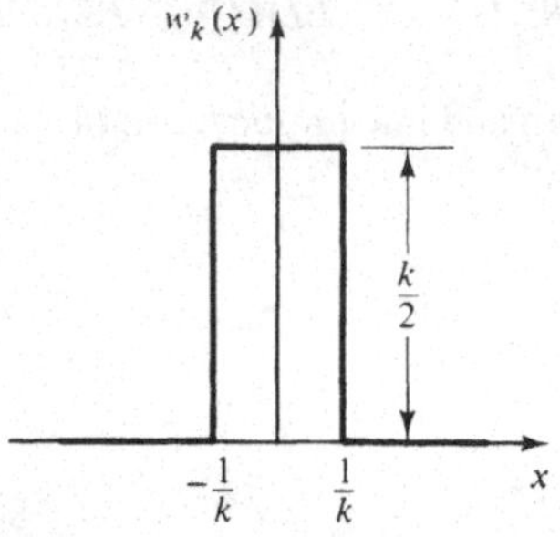

Figure 4.1. The $w_k(x)$ sequence (4.11).

as $k \longrightarrow \infty$. Note carefully how $w_k(x)$ gets taller and skinnier as k increases, so that its area remains equal to unity.

As a result of the spikelike nature of $w_k(x)$, as $k \longrightarrow \infty$, observe that for any function $g(x)$ that is continuous at $x = 0$,

$$\lim_{k\to\infty} \int_{-\infty}^{\infty} w_k(x)g(x)\,dx = \lim_{k\to\infty} \frac{k}{2} \int_{-1/k}^{1/k} g(x)\,dx$$
$$= \lim_{k\to\infty} \frac{k}{2} \cdot g(\xi) \cdot \frac{2}{k} = g(0), \tag{4.8}$$

where the next to last step follows from the mean value theorem of the integral calculus[6] with ξ some number between $-1/k$ and $+1/k$; clearly, ξ is forced to zero as $k \longrightarrow \infty$.

We now use (4.8) to provide our definition of the delta function; we *define* the **delta function** $\delta(x)$ by the relation

$$\int_{-\infty}^{\infty} \delta(x)g(x)\,dx = g(0). \tag{4.9}$$

That is, the "action" of $\delta(x)$ on $g(x)$ is to pick out its value at $x = 0$. We've already

---

[5]P. A. M. Dirac, The Physical Interpretation of the Quantum Mechanics, *Proc. Roy. Soc., London*, Sec. A113, 1926–1927, pp. 621–641. For an historical account, see G. Temple, *J. London Math. Soc.*, Vol. 28, 1953, pp. 134–148.

[6]Recall from Exercise 1.17 that if $m \le f(x) \le M$ over $[a, b]$, then

$$m(b-a) \le \int_a^b f(x)\,dx \le M(b-a).$$

If m and M are minimum and maximum values of f over $[a, b]$ and f is continuous, then there must be some point in $[a, b]$, say ξ, such that

$$\int_a^b f(x)\,dx = f(\xi)(b-a),$$

and this is known as the *mean value theorem of the integral calculus*. [For comparison, recall from (2.28) the mean value theorem of the *differential* calculus

$$f(b) - f(a) = f'(\xi)(b-a),$$

where $f'(x)$ is continuous over $[a, b]$ and ξ is some point in the interval.]

mentioned that $g(x)$ is to be continuous at $x = 0$. In addition, let us require that it be bounded.

If we could slip the limit past the integral sign in the left-hand side of (4.8), we could identify $\delta(x)$ simply as $\lim_{k\to\infty} w_k(x)$. But the limit interchange is *not* valid; in fact, $\lim_{k\to\infty} w_k(x)$ doesn't even exist, since $\lim_{k\to\infty} w_k(0) = \infty$. Instead we state the close connection between the $w_k(x)$ sequence and $\delta(x)$ as

$$\delta(x) \overset{\text{so}}{=} \lim_{k\to\infty} w_k(x), \tag{4.10}$$

where "so" means "sort of"; what we *actually* mean is that

$$\int_{-\infty}^{\infty} \delta(x)g(x)\,dx = \lim_{k\to\infty} \int_{-\infty}^{\infty} w_k(x)g(x)\,dx,$$

where the lim acts on the integral [as in (4.8)], *not* on $w_k(x)$ directly!

We call (4.7) a δ **sequence**. It is not the only one; there are many.

THEOREM 4.3. If $w(x)$ is nonnegative function satisfying $\int_{-\infty}^{\infty} w(x)\,dx = 1$, then $kw(kx) \equiv w_k(x)$ is a δ sequence.

*

Proof. Let us express

$$\begin{aligned}\lim_{k\to\infty} \int_{-\infty}^{\infty} w_k(x)g(x)\,dx &= \lim_{k\to\infty} \int_{-\infty}^{\infty} w_k(x)[g(x) - g(0)]\,dx \\ &\quad + \lim_{k\to\infty} \int_{-\infty}^{\infty} w_k(x)g(0)\,dx \qquad (4.11) \\ &= I + J,\end{aligned}$$

say. First consider J.

$$\begin{aligned} J &= g(0) \lim_{k\to\infty} \int_{-\infty}^{\infty} kw(kx)\,dx \\ &= g(0) \lim_{k\to\infty} \int_{-\infty}^{\infty} w(\xi)\,d\xi = g(0) \lim_{k\to\infty} 1 = g(0), \end{aligned}$$

and we now hope to show that $I = 0$ so that the right-hand side of (4.11) will be $g(0)$ as desired. Choose any number $\epsilon > 0$. Since g is assumed to be continuous at $x = 0$, there must exist a number $\gamma > 0$ such that $|g(x) - g(0)| < \epsilon$ whenever $|x - 0| = |x| < \gamma$.

Breaking up the I integral,

$$I = \lim_{k\to\infty} \int_{-\infty}^{-\gamma} + \lim_{k\to\infty} \int_{-\gamma}^{\gamma} + \lim_{k\to\infty} \int_{\gamma}^{\infty} \equiv I_1 + I_2 + I_3,$$

where $I_1 = I_3 = 0$. Because $g(x)$ is assumed to be bounded, $|g(x) - g(0)| < M$, say, and

$$|I_3| < M \lim_{k\to\infty} \int_{\gamma}^{\infty} w_k(x)\,dx = M \lim_{k\to\infty} \int_{\gamma}^{\infty} kw(kx)\,dx = M \lim_{k\to\infty} \int_{k\gamma}^{\infty} w(\xi)\,d\xi = 0,$$

and similarly for I_1. Turning to I_2,

$$|I_2| \leq \epsilon \lim_{k\to\infty} \int_{-\gamma}^{\gamma} w_k(x)\,dx = \epsilon \lim_{k\to\infty} \int_{-\gamma}^{\gamma} kw(kx)\,dx$$
$$= \epsilon \lim_{k\to\infty} \int_{-k\gamma}^{k\gamma} w(\xi)\,d\xi = \epsilon.$$

Finally, since ϵ can be chosen as small as we like, and $|I_2| < \epsilon$, it follows that I_2 must equal zero, which completes the proof.

Example 4.1. Since

$$w(x) = \frac{1}{\pi(1 + x^2)} \tag{4.12}$$

is nonnegative and has unit area, it follows from Theorem 4.3 that

$$w_k(x) \equiv kw(kx) = \frac{k}{\pi(1 + k^2x^2)} \tag{4.13}$$

is a δ sequence; see Fig. 4.2. ▮

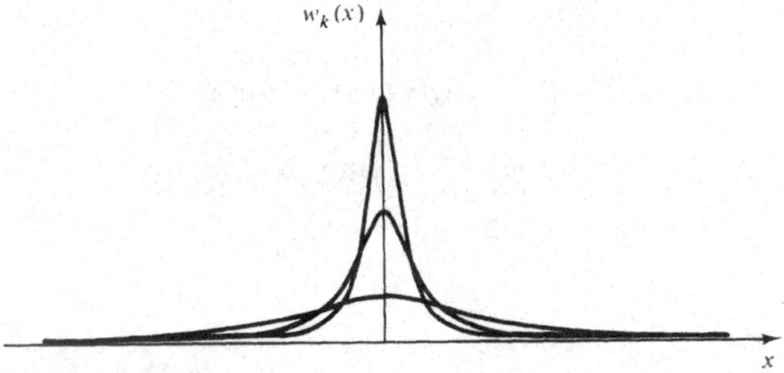

Figure 4.2. The δ-sequence (4.13).

We emphasize that $\delta(x)$ is thus defined by its action on a given function g, as spelled out by (4.9); *we will not draw a "graph" of a generalized function or talk about its "values"*! This point is important. If we insist on thinking in terms of an ordinary function representation of $\delta(x)$, the best we can do is recall the connection (4.10) with the sequence $w_k(x)$ of ordinary functions. Or, in a very formal way, we can visualize $\delta(x)$ as a concentrated spike of unit area at $x = 0$.

It is permissible, within the framework of generalized function theory, to "differentiate" generalized functions. To illustrate, consider the derivative $\delta'(x)$. Recalling the integral definition (4.9) of $\delta(x)$, note that defining "$\delta'(x)$" in similar fashion actually amounts to deciding on a functional $\mathfrak{F}(g)$ in the relation

$$\int_{-\infty}^{\infty} \delta'(x)g(x)\,dx = \mathfrak{F}(g). \tag{4.14}$$

Integrating by parts,

$$\int_{-\infty}^{\infty} \delta'(x)g(x)\,dx = \delta(x)g(x)\Big|_{-\infty}^{\infty} - \int_{-\infty}^{\infty} \delta(x)g'(x)\,dx$$

$$= \delta(x)g(x)\Big|_{-\infty}^{\infty} - g'(0). \tag{4.15}$$

It is tempting to say that the boundary term is zero, since $\delta(x)$ is zero for all $x \neq 0$, but remember that we do not speak of the *values* of a generalized function. Indeed, the integration by parts itself is only a formal carryover from classical function theory. So let us simply and formally drop the boundary term and *define* $\delta'(x)$ according to the relation

$$\int_{-\infty}^{\infty} \delta'(x)g(x)\,dx = -g'(0). \tag{4.16}$$

In similar fashion, the jth derivative $\delta^{(j)}(x)$ is defined by

$$\int_{-\infty}^{\infty} \delta^{(j)}(x)g(x)\,dx = (-1)^j g^{(j)}(0). \tag{4.17}$$

It should be clear, by the way, that just as $\delta(x)$ acts at $x = 0$, $\delta(x - \xi)$ acts at $x - \xi = 0$, that is, at $x = \xi$. Thus

$$\int_{-\infty}^{\infty} \delta(x - \xi)g(x)\,dx = g(\xi), \tag{4.18}$$

and similarly

$$\int_{-\infty}^{\infty} \delta^{(j)}(x - \xi)g(x)\,dx = (-1)^j g^{(j)}(\xi). \tag{4.19}$$

There is a simple relationship between the delta function $\delta(x)$ and the Heaviside step function $H(x)$. Integrating,

$$\int_{-\infty}^{\infty} H'(x)g(x)\,dx = H(x)g(x)\Big|_{-\infty}^{\infty} - \int_{-\infty}^{\infty} H(x)g'(x)\,dx$$

$$= g(\infty) - \int_0^{\infty} g'(x)\,dx$$

$$= g(\infty) - g(\infty) + g(0) = g(0), \tag{4.20}$$

and comparing this result to (4.9), we see that

$$H'(x) = \delta(x). \tag{4.21}$$

Observe that equations like (4.21) are not to be understood in a numerical sense. In other words, we do not say that for each x the value of the left-hand side is equal to the value of the right-hand side, since we do not talk about the values of generalized functions. Instead what is meant by (4.21) is that the action of $H'(x)$ on any $g(x)$ (which is bounded, and continuous at $x = 0$) is the same as the action of $\delta(x)$ on $g(x)$:

$$\int_{-\infty}^{\infty} H'(x)g(x)\,dx = \int_{-\infty}^{\infty} \delta(x)g(x)\,dx. \tag{4.22}$$

As one more example of this convention, consider the statement

$$x\,\delta(x) = 0. \tag{4.23}$$

To verify it, we need to show that

$$\int_{-\infty}^{\infty} x\,\delta(x)g(x)\,dx = \int_{-\infty}^{\infty} 0\cdot g(x)\,dx.$$

The left-hand side is $[xg(x)]|_{x=0} = 0$,[7] and the right-hand side is 0, too, so that (4.23) is correct. More generally [Exercise 4.16(a)], we have $f(x)\,\delta(x) = f(0)\,\delta(x)$.

Later on, in physical applications, we will find it natural to use delta functions to represent point forces, impulsive forces, and so on. For instance, suppose that we press a circular coin against the edge of a metal plate of the same thickness, which extends over $y > 0$ and $-\infty < x < \infty$, as shown in Fig. 4.3(a). We press with a force

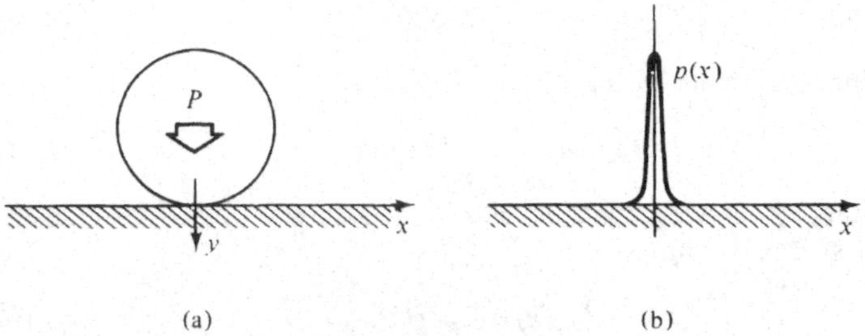

Figure 4.3. Force applied to edge of plate.

P and are interested in the resulting stress field induced in the plate. To set up the boundary value problem, we would need to know the force distribution, say $p(x)$ pounds per unit x-length, applied to the edge of the plate. However, the function p, which is probably of the general form shown in Fig. 4.3(b), is not known a priori. We do know that it will be quite concentrated and that

$$\int_{-\infty}^{\infty} p(x)\,dx = P \tag{4.24}$$

so that its net force is P, but its exact shape must be computed, together with the stress field, as part of the overall problem. (In the theory of elasticity, this is called a "contact problem.")

On the other hand, we expect the induced stress field to be quite insensitive to the

---

[7]You may be worried about our claim that $\int_{-\infty}^{\infty} [x\,\delta(x)]g(x)\,dx = \int_{-\infty}^{\infty} \delta(x)[xg(x)]\,dx$, since it involves breaking up the generalized function $x\,\delta(x)$. But what do we mean by "$x\,\delta(x)$"? Well, surely an $x\delta$ sequence should be $xw_k(x)$, where $w_k(x)$ is a δ sequence. Then

$$\lim_{k\to\infty} \int_{-\infty}^{\infty} xw_k(x)g(x)\,dx = \lim_{k\to\infty} \int_{-\infty}^{\infty} w_k(x)[xg(x)]\,dx$$

$$\equiv \int_{-\infty}^{\infty} \delta(x)[xg(x)]\,dx = [xg(x)]\Big|_{x=0} = 0,$$

where the first equality is trivial because x, $w_k(x)$, and $g(x)$ are all *ordinary* functions.

detailed shape of p,[8] and so we might just as well simplify the problem by assuming, a priori, that $p(x) = P\,\delta(x)$—that is, an idealized "point force" of strength P, acting at $x = 0$. Point *couples* are considered in Exercise 4.19.

EXERCISES

4.1. Using (4.2), show that the series in (4.1) converges, as we have claimed.

4.2. Is it true that

$$\lim_{k\to\infty} \int_0^1 f_k(x)\,dx = \int_0^1 \lim_{k\to\infty} f_k(x)\,dx$$

for the $f_k(x)$ sequence shown in Fig. 4.4? Consider separately the cases where $c_k = 2k$, 1, and $1/k$. Use appropriate theorem(s) where possible but also actually evaluate the left- and right-hand sides.

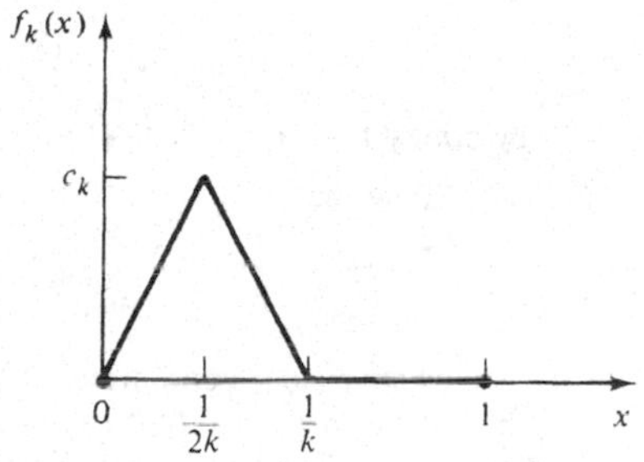

Figure 4.4

4.3. Sketch the sequence $s_n(x) = 1/(1 + n^2x^2)$ over $0 \leq x \leq 1$, showing the trend as n becomes large. What is (if it exists) $\lim_{n\to\infty} s_n(x)$? Is the convergence uniform? Why (not)? Discuss the validity of the claim

$$\lim_{n\to\infty} \int_0^1 s_n(x)\,dx = \int_0^1 \lim_{n\to\infty} s_n(x)\,dx$$

(*Hint:* You will see that Theorem 4.1 does not apply directly. Proceed by breaking the integral into two parts, 0 to γ and γ to 1. Apply Theorem 4.1 to the second part and, by a simple bound, show that the first part is arbitrarily small and hence zero, in the same way that we handled I in the proof of Theorem 4.3.) Finally, evaluate the left- and right-hand sides directly and show that the result is in keeping with your previous conclusion.

---

[8] More precisely, two highly concentrated force distributions will induce essentially identical stress fields except in the immediate neighborhood of their point of application, provided that they are *statically equivalent*—that is, provided that their resultant forces and couples are identical. (In the present example there will be no couple because of the symmetry about $x = 0$.) This is known as *Saint-Venant's principle*.

4.4. Evaluate

$$\lim_{\alpha \to 0+} \int_0^1 \frac{x^\alpha \, dx}{x^2 + 1}$$

by moving the limit past the integral sign. Justify the interchange by following the hint in the preceding exercise.

4.5. Proceeding formally, show that

$$\int_0^\infty \frac{x^{p-1}}{e^x - 1} \, dx = \zeta(p)\Gamma(p),$$

where $\zeta(p) \equiv \sum_1^\infty (1/n^p)$ is the **Riemann zeta function.**

4.6. Proceeding formally, evaluate $\int_0^\infty (\sin \sqrt{x}) e^{-x^2} \, dx$ as a series involving the gamma function.

4.7. Evaluate

$$\lim_{R \to \infty} \int_0^{\pi/2} \sqrt{R} \, e^{-R \sin \theta} \, d\theta$$

by noting that $|e^{-R \sin \theta}| \leq e^{-2R\theta/\pi}$ over $0 \leq \theta \leq \pi/2$.

4.8. (a) Show that

$$\lim_{\epsilon \to 0+} \frac{1}{\pi} \int_{-\infty}^\infty \frac{\epsilon f(x)}{x^2 + \epsilon^2} \, dx = f(0),$$

where $f(x)$ is continuous and $|f(x)| < M$ for all x. *Hint:* Recall Example 4.1.

*(b) Show that if the lower limit is changed to zero, then the result is $f(0)/2$.

*(c) In classical wing theory the normal velocity or "upwash" induced on the wing by the wing itself is found to be expressible in the form

$$v(x) = \lim_{y \to 0} \frac{1}{2\pi} \int_{-a}^a \frac{(x - \xi)\gamma(\xi)}{(\xi - x)^2 + y^2} \, d\xi,$$

where the wing lies on the x axis from $-a$ to $+a$ and $\gamma(\xi)$ is the "circulation per unit length" along the wing. Assuming that $\gamma(\xi)$ is "decent," say that it satisfies a so-called **Lipschitz condition** $|\gamma(x_2) - \gamma(x_1)| \leq C|x_2 - x_1|$ for all x_1's and x_2's in $-a \leq x \leq a$ for some finite constant C (that is, γ has a bounded difference quotient $[\gamma(x_2) - \gamma(x_1)]/(x_2 - x_1)$ over the interval), show that the preceding limit yields

$$v(x) = -\frac{1}{2\pi} ⨍_{-a}^a \frac{\gamma(\xi)}{\xi - x} \, d\xi.$$

Observe that the Cauchy principal value interpretation is not selected here on a whim but follows logically from the mathematical context. We mention this point because you may have wondered—when the Cauchy principal value was introduced in Chapter 3—when you should interpret in the Riemann sense and when in the Cauchy sense.

4.9. It can be shown that

$$\lim_{x \to x_0} \sum^\infty a_n(x) = \sum^\infty \lim_{x \to x_0} a_n(x)$$

for all x_0 in some interval I if the $a_n(x)$'s are all continuous over I and the series on the left converges uniformly over I. Is this theorem, not mentioned in the present section, equivalent to our earlier Theorem 2.12?

4.10. Prove the following important result. *A power series may be differentiated (or integrated) term by term (as many times as we like for any x inside its interval of convergence.* Furthermore, *the resulting series has the same interval of convergence as the original series and converges to the derivative (or integral) of the function to which the original series converges.*

4.11. Verify carefully that $y = \sum_1^\infty (-1)^{n+1}x^{2n}/(2n-1)!$ satisfies the differential equation $y'' + y = 2\cos x$ and the conditions $y(0) = y'(0) = 0$. (Use the result of Exercise 4.10.) Include in your answer verification of the term by term addition of the two power series on the left hand side of the differential equation.

4.12. Verify that

$$\lim_{n\to\infty} \int_\pi^{2\pi} \frac{\sin nx}{nx}\,dx = 0.$$

4.13. Verify that

$$\int_0^\pi \sum_1^\infty \frac{\sin nx}{n^2}\,dx = \sum_1^\infty \frac{2}{(2n-1)^3}.$$

4.14. Show that

$$\frac{d}{dx}\sum_1^\infty a_n \sin nx = \sum_1^\infty na_n \cos nx$$

for all x if $\sum_1^\infty |na_n|$ converges.

4.15. Sketch the following sequences and verify that they are δ sequences.

(a) $$w_k(x) = \begin{cases} 0, & |x| > \dfrac{1}{2k} \\ 4k^2x + 2k, & -\dfrac{1}{2k} \le x \le 0 \\ -4k^2x + 2k, & 0 \le x \le \dfrac{1}{2k}. \end{cases}$$

(b) $$w_k(x) = \frac{ke^{-k^2x^2}}{\sqrt{\pi}}.$$

(You may recognize that this is a **gaussian distribution**, which becomes more and more peaked as k increases.)

4.16. Prove that

(a) $f(x)\,\delta(x) = f(0)\,\delta(x)$ (b) $x\,\delta'(x) = -\delta(x)$

(c) $x^2\,\delta''(x) = 2\,\delta(x)$ (d) $\dfrac{d^4}{dx^4}|x|^3 = 12\,\delta(x)$

Hint for part (d): $|x| = x[2H(x) - 1]$ and $|x|^3 = x^3[2H(x) - 1]^3 = x^3[2H(x) - 1]$.

4.17. (a) Show that $\delta(\xi - x) = \delta(x - \xi)$.

*(b) More generally, show that if $f(\xi)$ is a strictly monotone increasing or decreasing function of ξ that vanishes for $\xi = x$, then $\delta[f(\xi)] = \delta(\xi - x)/|f'(x)|$. *Hint:* For a start, set $f(\xi)$ equal to a new variable, say u.

4.18. (a) Find the second derivative of $f(x)$, in terms of generalized functions, where $f(x)$ is defined to equal x over $|x| \leq 1$, $+1$ for $x > 1$, and -1 for $x < -1$. *Hint:* First draw a picture of $f(x)$ and express it in terms of a combination of Heaviside functions. Also, recall Exercises 4.16 and 4.17.

(b) The following alternative approach to part (a) was put forward in an examination paper:

$$\begin{aligned}\int_{-\infty}^{\infty} f''g\,dx &= -\int_{-\infty}^{\infty} f'g'\,dx = \int_{-\infty}^{\infty} fg''\,dx\\ &= -\int_{-\infty}^{-1} g''\,dx + \int_{-1}^{1} xg''\,dx + \int_{1}^{\infty} g''\,dx\\ &= -(g')\Big|_{-\infty}^{-1} + (xg')\Big|_{-1}^{1} - \int_{-1}^{1} g'\,dx + (g')\Big|_{1}^{\infty}\\ &= g(-1) - g(1),\end{aligned}$$

so that $f''(x) = \delta(x+1) - \delta(x-1)$. Is this correct? Discuss.

4.19. Following the lines of (4.18), Show that

$$w_k(x) = \begin{cases} 0, & |x| > \dfrac{1}{2k} \\ 4k^2, & -\dfrac{1}{2k} \leq x \leq 0 \\ -4k^2, & 0 \leq x \leq \dfrac{1}{2k} \end{cases}$$

constitutes a δ' sequence. [What does this sequence have to do with the one in Exercise 4.15(a)?] Show that $\delta'(x)$ may be interpreted physically, in terms of force distributions for example, as a counterclockwise point *couple* of unit strength acting at $x = 0$.

4.20. (*Finite limits*) Recalling (4.18), show that

$$\int_a^b \delta(x-\xi)g(x)\,dx = g(\xi),$$

provided that $a < \xi < b$. *Hint:* Rewrite the integral as

$$\int_{-\infty}^{\infty} [H(x-a) - H(x-b)]\,\delta(x-\xi)g(x)\,dx.$$

4.21. Make up two examination-type problems on this chapter.

Chapter 5

Fourier Series and the Fourier Integral

Although Fourier series won't be used frequently until we come to partial differential equations, its introduction seems to fit best at this point.

Since we have already discussed limits, convergence, and similar topics, the material that follows should come as no surprise. Yet, historically, when Fourier (1768–1830) announced in 1807, in connection with his work on the conduction of heat, that an "arbitrary" function defined over $(-\pi, \pi)$ could be represented in the form $\sum_0^\infty (a_n \cos nx + b_n \sin nx)$, he met with considerable opposition, even from Joseph Louis Lagrange (1736–1813), perhaps the greatest mathematician of the time. Remember, however, that this book is not a chronological development; even such fundamental notions as uniform convergence and Riemann integration were not introduced until the middle of the nineteenth century, and here Fourier was talking about representing arbitrary (e.g., even discontinuous) functions by means of beautifully smooth sines and cosines!

In fact, his claim turned out to be not quite true; mild restrictions are needed on the class of f's, and Dirichlet, Riemann, and others subsequently put forward various sets of sufficient conditions.

An enormous amount of research has been devoted to the subject, and important results continue to be obtained.[1]

[1]The little book by R. L. Jeffery (*Trigonometric Series*, University of Toronto Press, Toronto, 1956) gives an interesting historical (and mathematical) account, although important results have been obtained since that time.

5.1. THE FOURIER SERIES OF $f(x)$

It will be helpful to say something first about "odd functions" and "even functions." We say that $f(x)$ is **even** if $f(-x) = f(x)$ for all x; that is, the graph of $f(x)$ is *symmetric* about $x = 0$. And $f(x)$ is **odd** if $f(-x) = -f(x)$; in other words, its graph is *antisymmetric* about $x = 0$. For instance, x^2, $\cos x$, $\sin x^2$ are even, and x, $\sin x$, $x \cos x$ are odd. It is easy to show (Exercise 5.1) that "odd times odd is even" (e.g., $x \cdot x = x^2$), "even times even is even," and "odd times even is odd."

A given function is not necessarily one or the other; for example, e^x is neither. But any $f(x)$ can be decomposed into the sum of an even function plus an odd function according to[2]

$$f(x) = \frac{f(x) + f(-x)}{2} + \frac{f(x) - f(-x)}{2} \equiv f_{\text{even}} + f_{\text{odd}}. \tag{5.1}$$

To illustrate,

$$e^x = \frac{e^x + e^{-x}}{2} + \frac{e^x - e^{-x}}{2} \equiv \cosh x + \sinh x,$$

where $\cosh x$ is even and $\sinh x$ is odd.

Furthermore, we say that a function $f(x)$ is **periodic** with period τ if

$$f(x + \tau) = f(x) \tag{5.2}$$

for all x. For instance, $\sin x$ is periodic with period 2π, since $\sin(x + 2\pi) = \sin x \cos 2\pi + \sin 2\pi \cos x = \sin x$. Actually, $\sin x$ is also periodic with period $\tau = 4\pi, 6\pi, \ldots$, but in speaking of the period, we generally mean the smallest possible one.

Consider functions $f(x)$ that are periodic with period 2π; later on we will allow τ to be any value that we like. For example, consider the *square wave* shown in Fig. 5.1,

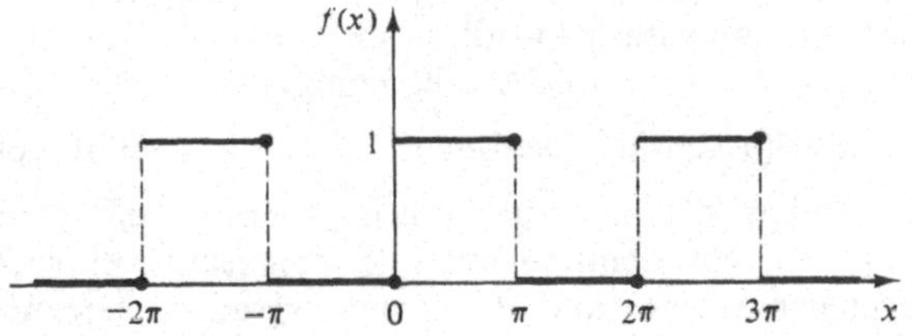

Figure 5.1. The square wave (5.3).

where the heavy dots indicate the values of f at the jump discontinuities. That is, over one full period, say $-\pi$ to π, we have

$$f(x) = \begin{cases} 0, & -\pi < x \leq 0 \\ 1, & 0 < x \leq \pi. \end{cases} \tag{5.3}$$

---

[2]Such decompositions appear time and again: we break a given function into the sum of two functions, one monotone increasing and the other monotone decreasing. A "matrix" can be split into a symmetric part plus an antisymmetric part. A vector field can be split into an "irrotational" part plus a "solenoidal" part, and so on (as we will see).

Although the representation (5.3), together with the periodic extension $f(x + 2\pi) = f(x)$, suffices to define f, we often need a single analytical expression—for example, a series $\sum a_n(x)$—that is valid for all x. Taylor series are of little help in the present example because f is not even continuous, let alone analytic! Let us try to generate an expansion like $\sum a_n(x)$ by *inspection*, if only approximately. It seems reasonable to choose for the first term $a_1(x) = \text{constant} = \frac{1}{2}$, that is, the average height of the square wave; thus the first partial sum is $s_1(x) = a_1(x) = \frac{1}{2}$, as shown in Fig. 5.2. To

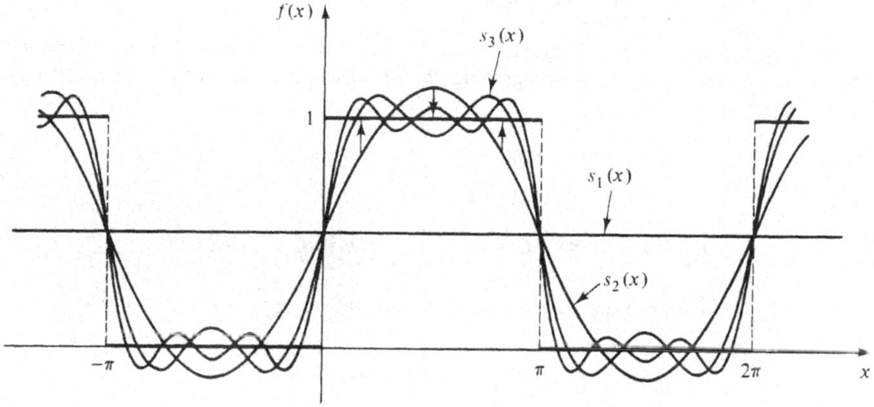

Figure 5.2. Series approximation to square wave.

this sum it seems reasonable to add $a_2(x) \approx 0.6 \sin x$, say, so that $s_2(x) \approx \frac{1}{2} + 0.6 \sin x$, as shown in the figure. $s_2(x)$ is beginning to take on the correct shape but needs to be "squared-off," as the arrows indicate. Consequently, let us choose $a_3(x) \approx 0.2 \sin 3x$, so that $s_3(x) \approx \frac{1}{2} + 0.6 \sin x + 0.2 \sin 3x$ as sketched. By suitable addition of higher harmonics ($\sin 5x$, $\sin 7x$, . . .), it looks as though we may be able to approach the desired square wave.

So let us consider **trigonometric series** $\sum_0^\infty (a_n \cos nx + b_n \sin nx)$ for the representation of a given 2π-periodic function $f(x)$ or, equivalently,

$$f(x) = \frac{a_0}{2} + \sum_1^\infty (a_n \cos nx + b_n \sin nx), \tag{5.4}$$

where the $\frac{1}{2}$ is introduced for later convenience. Note that the cosines (which are *even* functions) are to take care of the even part of $f(x)$, whereas the sines (which are *odd*) are to handle the odd part; that is,

$$f_{\text{even}} = \frac{a_0}{2} + \sum_1^\infty a_n \cos nx \quad \text{and} \quad f_{\text{odd}} = \sum_1^\infty b_n \sin nx.$$

Supposing, tentatively, that the representation (5.4) is possible, how are the a_n's and b_n's determined? We make use of the integrals

$$\left.\begin{aligned}\int_{-\pi}^{\pi} \cos mx \cos nx \, dx &= \begin{cases} 0, & m \neq n \\ \pi, & m = n \geq 1 \\ 2\pi, & m = n = 0 \end{cases} \\ \int_{-\pi}^{\pi} \sin mx \sin nx \, dx &= \begin{cases} 0, & m \neq n \\ \pi, & m = n \geq 1 \end{cases} \\ \int_{-\pi}^{\pi} \sin mx \cos nx \, dx &= 0 \quad \text{for all } m, n, \end{aligned}\right\} \tag{5.5}$$

where m and n are integers (Exercise 5.3). Multiply both sides of (5.4) by 1, $\cos mx$, and $\sin mx$ in turn. Formally integrating the resulting three equations term by term and using (5.5), we obtain

$$\int_{-\pi}^{\pi} f(x) \, dx = \frac{a_0}{2} 2\pi + 0 + 0 + \cdots$$

$$\int_{-\pi}^{\pi} f(x) \cos mx \, dx = 0 + \cdots + 0 + a_m \pi + 0 + \cdots$$

$$\int_{-\pi}^{\pi} f(x) \sin mx \, dx = 0 + \cdots + 0 + b_m \pi + 0 + \cdots$$

or

$$\begin{aligned} a_n &= \frac{1}{\pi} \int_{-\pi}^{\pi} f(x) \cos nx \, dx \qquad (n = 0, 1, 2, \ldots) \\ b_n &= \frac{1}{\pi} \int_{-\pi}^{\pi} f(x) \sin nx \, dx \qquad (n = 1, 2, 3, \ldots) \end{aligned} \tag{5.6}$$

where we've changed the dummy index from m back to n. Equations (5.6) are known as **Euler's formulas**.

The a_n's and b_n's are called the **Fourier coefficients** of $f(x)$, and $a_0/2 + \sum (a_n \cos nx + b_n \sin nx)$ is called **the Fourier series of $f(x)$**, whether or not it actually converges to $f(x)$ [just as we define *the Taylor series of* $f(x)$ as $\sum f^{(n)}(a)(x - a)^n/n!$ whether or not the series actually converges to $f(x)$]. Finally, we pull things together by proving that the Fourier series of $f(x)$ *does*, in fact, converge—to $f(x)$—almost everywhere, provided that f is sufficiently well behaved.

5.2. *POINTWISE CONVERGENCE OF THE SERIES*

In this section we consider the convergence of the series with x regarded as fixed, as we have in the past for series of the form $\sum a_n(x)$. Such convergence is local or *pointwise convergence*, in contrast to *mean square convergence*, which is a global type and which arises naturally within the context of the abstract, linear, vector space formulation developed in Part III.

First, two definitions. A function $f(x)$ is said to be **piecewise continuous** over a given interval if the interval can be divided into a finite number of subintervals, inside each of which f is continuous, with finite limits at the left and right endpoints. If, in addition, f is differentiable over each closed subinterval, then we say that it is **piecewise smooth.**

[By differentiable over a *closed* interval, say $\alpha \leq x \leq \beta$, it is understood that the derivatives at the left and right endpoints are the "right- and left-hand derivatives"

$$f'(\alpha) \equiv \lim_{\Delta x \to 0+} \frac{f(\alpha + \Delta x) - f(\alpha + 0)}{\Delta x}$$

and

$$f'(\beta) \equiv \lim_{\Delta x \to 0+} \frac{f(\beta - \Delta x) - f(\beta - 0)}{\Delta x},$$

respectively, where we recall that $f(\alpha + 0)$ means $\lim f(\alpha + \epsilon)$ as $\epsilon \longrightarrow 0$ through positive values; $f(\alpha + 0) = f(\alpha - 0) = f(\alpha)$ only if f happens to be continuous at α.]

THEOREM 5.1. (*Pointwise Convergence*) *If $f(x)$ is 2π-periodic and piecewise smooth over each full period, then its Fourier series converges to $f(x)$ at each point where f is continuous and to the average value $[f(x + 0) + f(x - 0)]/2$ at each jump discontinuity.*

Consider, for example, the square wave of Fig. 5.2. At $x = 0, \pm\pi, \pm 2\pi, \ldots$ it is clear that each $s_n(x)$, and hence the limit $s(x)$, is the average value $\frac{1}{2}$. It so happens that this value does *not* coincide with the function values, which are alternately zero and one. Here, then, the Fourier series converges to f everywhere except at the isolated points $x = 0, \pm\pi, \ldots$. But such a minor discrepancy is of no importance at all in practical applications. Some authors remind us of such possible isolated discrepancies by using $\overset{\text{ae}}{=}$ in (5.4), where the ae means "almost everywhere," instead of the strict equality sign.

*

Proof of Theorem 5.1. We will need the so-called Riemann–Lebesgue lemma. This lemma is discussed in Part III, and it will suffice here to state it without proof.

RIEMANN–LEBESGUE LEMMA.[3] *If a, b are finite, β is real, and f is Riemann integrable over $[a, b]$, then*

$$\lim_{\alpha \to \infty} \int_a^b f(x) \sin(\alpha x + \beta)\, dx = 0. \tag{5.7}$$

To prove Theorem 5.1, we first insert (5.6) into $s_n(x)$,

$$\begin{aligned} s_n(x) &= \frac{a_0}{2} + \sum_1^n (a_j \cos jx + b_j \sin jx) \\ &= \frac{1}{2\pi} \int_{-\pi}^{\pi} f(\xi)\, d\xi + \sum_1^n \left\{ \frac{1}{\pi} \int_{-\pi}^{\pi} f(\xi) \cos j\xi\, d\xi \cdot \cos jx \right. \\ &\qquad \left. + \frac{1}{\pi} \int_{-\pi}^{\pi} f(\xi) \sin j\xi\, d\xi \cdot \sin jx \right\} \\ &= \frac{1}{\pi} \int_{-\pi}^{\pi} f(\xi) \left\{ \frac{1}{2} + \sum_1^n \cos j(x - \xi) \right\} d\xi, \end{aligned} \tag{5.8}$$

[3] A *lemma* is a helping theorem, or subsidiary proposition, as well as a bract in a grass spikelet just below the pistil and stamens.

where the dummy integration variable in (5.6) has been changed to ξ to avoid confusion with x. But the kernel can be expressed in "closed form" (Exercise 5.4) as

$$\frac{1}{2} + \sum_1^n \cos j(x - \xi) = \frac{\sin (n + 1/2)(x - \xi)}{2 \sin (x - \xi)/2}, \tag{5.9}$$

and so

$$s_n(x) = \frac{1}{\pi} \int_{-\pi+x}^{\pi+x} f(\xi) \frac{\sin (n + 1/2)(x - \xi)}{2 \sin (x - \xi)/2} \, d\xi,$$

where we've also shifted the limits by x (which is all right, since the integrand is 2π-periodic). Letting $x - \xi \equiv \xi'$ and discarding the primes, for simplicity,

$$\begin{aligned} s_n(x) &= \frac{1}{\pi} \int_{-\pi}^{\pi} f(x - \xi) \frac{\sin (n + 1/2)\xi}{2 \sin (\xi/2)} \, d\xi \\ &= \frac{1}{\pi} \int_0^{\pi} [f(x + \xi) + f(x - \xi)] \frac{\sin (n + 1/2)\xi}{2 \sin (\xi/2)} \, d\xi \\ &= \frac{1}{\pi} \int_0^{\pi} [f(x + \xi) + f(x - \xi)] \left\{ \frac{\sin (n + 1/2)\xi}{2 \sin (\xi/2)} - \frac{\sin (n + 1/2)\xi}{\xi} \right\} d\xi \\ &\quad + \frac{1}{\pi} \int_0^{\pi} [f(x + \xi) + f(x - \xi)] \frac{\sin (n + 1/2)\xi}{\xi} \, d\xi \\ &\sim \frac{1}{\pi} \int_0^{\pi} [f(x + \xi) + f(x - \xi)] \frac{\sin (n + 1/2)\xi}{\xi} \, d\xi \end{aligned} \tag{5.10}$$

as $n \longrightarrow \infty$, applying the Riemann–Lebesgue lemma to the first integral, since

$$[f(x + \xi) + f(x - \xi)] \left\{ \frac{1}{2 \sin (\xi/2)} - \frac{1}{\xi} \right\} \tag{5.11}$$

is well enough behaved to be Riemann integrable over the interval (Exercise 5.5). Again adding and subtracting terms,

$$\begin{aligned} s_n(x) &\sim \frac{1}{\pi} \int_0^{\pi} \{[f(x + \xi) + f(x - \xi)] - [f(x + 0) + f(x - 0)]\} \frac{\sin (n + 1/2)\xi}{\xi} \, d\xi \\ &\quad + \frac{f(x + 0) + f(x - 0)}{\pi} \int_0^{\pi} \frac{\sin (n + 1/2)\xi}{\xi} \, d\xi \\ &\sim \frac{f(x + 0) + f(x - 0)}{2} \end{aligned} \tag{5.12}$$

as $n \longrightarrow \infty$, since

$$\int_0^{\pi} \frac{\sin (n + 1/2)\xi}{\xi} \, d\xi = \int_0^{(n+1/2)\pi} \frac{\sin \zeta}{\zeta} \, d\zeta \longrightarrow \int_0^{\infty} \frac{\sin \zeta}{\zeta} \, d\zeta,$$

which (Example 3.10) is $\pi/2$, and the first integral approaches zero by the Riemann–Lebesgue lemma. The only possible source of trouble in the first integral is the apparent singularity at $\xi = 0$ due to the ξ in the denominator; but since f has been assumed to be piecewise smooth, we are assured that the difference quotients

$$\frac{f(x+\xi)-f(x+0)}{\xi} \quad \text{and} \quad \frac{f(x-\xi)-f(x-0)}{\xi}$$

are bounded as $\xi \longrightarrow 0+$, which completes the proof.

Example 5.1. The square wave of Fig. 5.1 is 2π-periodic and is surely piecewise smooth. We compute

$$a_n = \frac{1}{\pi}\int_{-\pi}^{\pi} f(x)\cos nx\,dx = \frac{1}{\pi}\int_0^{\pi}\cos nx\,dx = \begin{cases} 1, & n=0 \\ 0, & n>0 \end{cases}$$

$$b_n = \frac{1}{\pi}\int_{-\pi}^{\pi} f(x)\sin nx\,dx = \frac{1}{\pi}\int_0^{\pi}\sin nx\,dx$$

$$= -\frac{1}{n\pi}\cos nx\Big|_0^{\pi} = \frac{1-(-1)^n}{n\pi} = \begin{cases} \dfrac{2}{n\pi}, & n \text{ odd} \\ 0, & n \text{ even} \end{cases}$$

so that

$$f(x) = \frac{1}{2} + \frac{2}{\pi}\sum_{n=1,3,\ldots}^{\infty}\frac{\sin nx}{n}, \tag{5.13}$$

or

$$f(x) = \frac{1}{2} + \frac{2}{\pi}\sum_{n=1}^{\infty}\frac{\sin(2n-1)x}{2n-1}, \tag{5.13a}$$

the equality being guaranteed by Theorem 5.1 for all x except $0, \pm\pi, \pm 2\pi, \ldots,$ where the Fourier series should be converging to the average value $\frac{1}{2}$; clearly, it does, since all the $\sin nx$'s in (5.13) are zero at those points, just leaving the $\frac{1}{2}$.

Observe that in this example we need not have actually computed the a_n's, since

$$\frac{a_0}{2} = \frac{1}{2\pi}\int_{-\pi}^{\pi} f(x)\,dx$$

is, by definition, the *average height* of the function, which, by inspection of the graph, is obviously $\frac{1}{2}$; also, by inspection, we see that $f(x) = \frac{1}{2} +$ an odd function, so that we need no cosine terms, and hence $a_1 = a_2 = \cdots = 0$.

We know that the right-hand side of (5.13) converges, as well as what it converges to, because we know that it is the Fourier series of the square wave $f(x)$, which satisfies the conditions of Theorem 5.1. The reverse is not so simple. That is, given the series (5.13) ("out of the blue"), does it converge and, if so, to what function $f(x)$? We can at least handle the first question as follows. The $1/n$ decay is not sufficient for convergence without some help from the sign alternations of the $\sin nx$ factor, and this situation suggests the Dirichlet test. Recalling that

$$1 + z + z^2 + \cdots + z^{n-1} = \frac{1-z^n}{1-z}$$

for all $z \neq 1$ and that

$$\operatorname{Im} e^{ikx} = \text{the imaginary part of } (\cos kx + i\sin kx)$$

namely, $\sin kx$,

$$\begin{aligned}\sum_{j=1}^{n} \sin (2j-1)x &= \operatorname{Im}\,[e^{ix} + e^{i3x} + \cdots + e^{i(2n-1)x}] \\ &= \operatorname{Im} e^{ix}[1 + (e^{i2x}) + \cdots + (e^{i2x})^{n-1}] \\ &= \operatorname{Im} e^{ix}\left(\frac{1 - e^{i2nx}}{1 - e^{i2x}}\right) = \operatorname{Im}\left(\frac{1 - e^{i2nx}}{e^{-ix} - e^{ix}}\right) \\ &= \operatorname{Im}\left(\frac{e^{i2nx} - 1}{2i \sin x}\right) = \frac{1 - \cos 2nx}{2 \sin x} \end{aligned} \tag{5.14}$$

except for $x = 0, \pm\pi, \pm 2\pi, \ldots$, where the right-hand side is indeterminate. For these x's, however, the original sum [i.e., the left-hand side of (5.14)] is clearly zero; and so if we define $(1 - \cos 2nx)/2 \sin x \equiv 0$ for them, then (5.14) is valid for *all* x. Moreover, since $|1 - \cos nx| \leq 2$, it follows that, for any fixed value of x,

$$\begin{aligned}\left|\sum_{j=1}^{n} \sin (2j-1)x\right| &= 0 \quad \text{for } x = 0, \pm\pi, \pm 2\pi, \ldots \\ &\leq \frac{1}{|\sin x|} \quad \text{for all other } x, \end{aligned} \tag{5.15}$$

independent of n. In addition, $1/(2n-1) \downarrow 0$ as $n \to \infty$; thus the Fourier series (5.13) or (5.13a) converges for all x by the Dirichlet test.

Although (5.13) converges for *all x*, it follows from Theorem 2.12 that the convergence will not be uniform over any interval containing one or more of the jump discontinuities. Yet it is not hard to show (Exercise 5.6) that it will be uniform over any interval not containing any of these jumps—for instance, over $0.0001 \leq x \leq 2$. (Be sure to see Exercise 5.18 also.) ▮

Example 5.2. For the $f(x)$ of Fig. 5.3, $a_0/2$ = average height $= \pi/2$ and the b_n's $= 0$, since $f(x)$ is even. Also, for $n \geq 1$,

$$\begin{aligned} a_n &= \frac{1}{\pi}\int_{-\pi}^{\pi} f(x) \cos nx\, dx = \frac{2}{\pi}\int_{0}^{\pi} f(x) \cos nx\, dx \\ &= \frac{2}{\pi}\int_{0}^{\pi} x \cos nx\, dx = \text{etc.} = \frac{2[(-1)^n - 1]}{n^2\pi} \\ &= \begin{cases} -\dfrac{4}{n^2\pi}, & n \text{ odd} \\ 0, & n \text{ even,} \end{cases} \end{aligned} \tag{5.16}$$

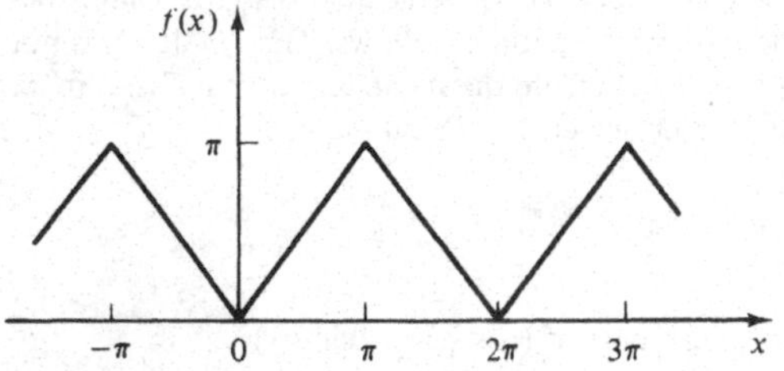

Figure 5.3. $f(x)$.

where the second equality follows from the fact that $f(x) \cos nx = |x| \cos nx =$ (even)(even) = even, and the fact that the interval of integration is symmetric about the origin. Thus

$$f(x) = \frac{\pi}{2} - \frac{4}{\pi} \sum_{n=1,3,\ldots}^{\infty} \frac{\cos nx}{n^2}. \tag{5.17}$$

COMMENT. Note that the terms of (5.13) drop off only as $1/n$, so that the series converges with painful slowness. This situation reflects the fact that the beautifully smooth sines are having difficulty representing the *discontinuous* function f. (As we'll see in Exercise 5.7, the convergence is fastest near $x = \pm\pi/2, \pm 3\pi/2, \ldots$ and slowest near the jumps.) In contrast, f in the present example is *continuous* (although not differentiable) and the terms in (5.17) drop off as $1/n^2$, which is somewhat better. *More generally, if the Nth derivative $f^{(N)}(x)$ is continuous, then the Fourier coefficients of f are $O(1/n^{N+2})$ as $n \longrightarrow \infty$.* ▮

Theorem 5.1 is not the only such theorem. Various other conditions are known to be sufficient for the pointwise convergence of $f(x)$. One well-known set, the **Dirichlet conditions**, is that *f be bounded, with only a finite number of maxima and minima and only a finite number of discontinuities over any one period.*

What about the question of *uniqueness*? In our discussion of Taylor series we noted that if $\sum a_n(x - a)^n = \sum b_n(x - a)^n$ over some common interval of convergence, then $a_n = b_n$; that is, the coefficients are uniquely determined by the sum function. A corresponding statement can be made for trigonometric series (although the proof is not as simple as for Taylor series and will be omitted). *If*

$$\sum_0^{\infty} (a_n \cos nx + b_n \sin nx) = \sum_0^{\infty} (c_n \cos nx + d_n \sin nx) \tag{5.18}$$

for all x, then $a_n = c_n$ and $b_n = d_n$ for all n.

5.3. TERMWISE INTEGRATION AND DIFFERENTIATION OF FOURIER SERIES

Although the integration and differentiation of series have already been discussed (Section 4.1), it will be helpful to say a few words specifically in regard to Fourier series.

THEOREM 5.2. Any Fourier series (convergent or not) can be integrated term by term (i.e., $\int \sum = \sum \int$) between any two limits and the resulting series converges (to the integral of the periodic function corresponding to the original series.)

Example 5.3. Consider the (convergent) series

$$\sum_1^{\infty} \frac{\sin nx}{\ln (n + 1)}. \tag{5.19}$$

Integrating from $-\pi$ to x, say,

$$\int_{-\pi}^{x} \sum_1^{\infty} \frac{\sin nx}{\ln (n + 1)}\, dx = \sum_1^{\infty} \frac{\cos n\pi - \cos nx}{n \ln (n + 1)}. \tag{5.20}$$

Observe that for $x = 0$, for instance, this becomes

$$-2 \sum_{n=1,3,\ldots}^{\infty} \frac{1}{n \ln (n+1)}, \tag{5.21}$$

which is *divergent* (recall Exercise 2.11). Since the integrated series was, according to Theorem 5.2, supposed to converge for all x, we conclude that (5.19) *could not have been a Fourier series*! That is, there exists no function $f(x)$, periodic with period 2π, such that

$$\frac{1}{\pi} \int_{-\pi}^{\pi} f(x) \cos nx \, dx = 0$$

for $n = 0, 1, 2, \ldots$ and

$$\frac{1}{\pi} \int_{-\pi}^{\pi} f(x) \sin nx \, dx = \frac{1}{\ln (n+1)}$$

for $n = 1, 2, \ldots$. ∎

Thus *trigonometric series* [i.e., of the form $\sum (a_n \cos nx + b_n \sin nx)$] *are not necessarily Fourier series even if they are convergent*. This situation is in contrast to Taylor series, where (Section 2.4) every power series with nonzero radius of convergence is the Taylor series of its sum function.

What about termwise *differentiation* of Fourier series? Trying this procedure on (5.13), for example, gives (Exercise 5.8)

$$f'(x) = \frac{2}{\pi} \sum_{n=1,3,\ldots}^{\infty} \cos nx = \lim_{N\to\infty} \frac{2}{\pi} \sum_{j=1}^{N} \cos (2j-1)x$$

$$= \frac{2}{\pi} \lim_{N\to\infty} \begin{cases} N & \text{for } x = 0, \pm 2\pi, \pm 4\pi, \ldots \\ -N & \text{for } x = \pm\pi, \pm 3\pi, \ldots \\ \dfrac{\sin 2Nx}{2 \sin x} & \text{for all other } x\text{'s.} \end{cases} \tag{5.22}$$

Since the limit does not exist for any x, it follows that the series diverges for all x, and hence the tentative first equality in (5.22) must be incorrect.

In this example $f(x)$ was not continuous. On the other hand,

THEOREM 5.3. If $f(x)$ is periodic with period 2π and ***continuous for all x*** *and $f'(x)$ is piecewise continuous, then termwise differentiation of the Fourier series of $f(x)$ produces the Fourier series of $f'(x)$.*

5.4. VARIATIONS: PERIODS OTHER THAN 2π, AND FINITE INTERVAL

Thus far we've considered f to be periodic over $-\infty < x < \infty$, with period 2π. Suppose, more generally, that the period is $2l$. Then the simple change of variables $x = lx'/\pi$ changes the period back to 2π, as illustrated in Fig. 5.4, and we can write

$$f = \frac{a_0}{2} + \sum_{1}^{\infty} (a_n \cos nx' + b_n \sin nx')$$

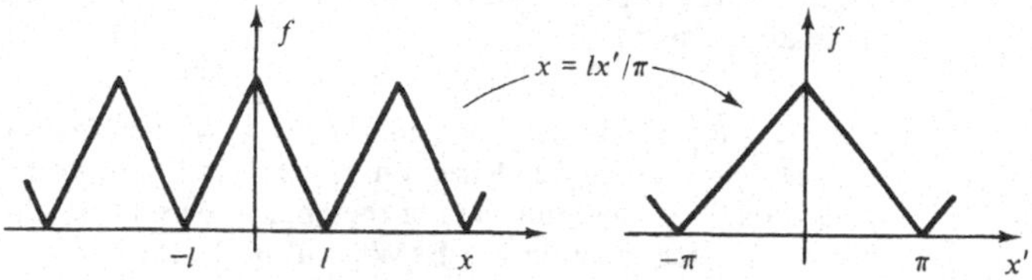

Figure 5.4. Period not 2π.

as before, where

$$a_n = \frac{1}{\pi}\int_{-\pi}^{\pi} f \cos nx'\, dx' \qquad (n = 0, 1, 2, \ldots)$$

$$b_n = \frac{1}{\pi}\int_{-\pi}^{\pi} f \sin nx'\, dx' \qquad (n = 1, 2, \ldots).$$

Changing back to the original x variable,

$$f(x) = \frac{a_0}{2} + \sum_1^\infty \left(a_n \cos\frac{n\pi x}{l} + b_n \sin\frac{n\pi x}{l}\right) \tag{5.23a}$$

where

$$\left.\begin{aligned} a_n &= \frac{1}{l}\int_{-l}^{l} f(x) \cos\frac{n\pi x}{l}\, dx \qquad (n = 0, 1, 2, \ldots) \\ b_n &= \frac{1}{l}\int_{-l}^{l} f(x) \sin\frac{n\pi x}{l}\, dx \qquad (n = 1, 2, \ldots). \end{aligned}\right. \tag{5.23b}$$

Observe that $a_0/2$ is still the average height.

Finally, suppose that the domain of f is only a *finite interval.* We can still represent f by a Fourier series if we first extend the definition of f over the rest of the x axis so that the resulting "extension" of f is periodic.

Example 5.4. *Infinite Beam on Elastic Foundation.* Consider an infinitely long beam on an elastic foundation, as sketched in Fig. 5.5, where the constant k is the "foundation modulus" (i.e., the spring stiffness per unit x length) and $w(x)$ is a prescribed periodic loading (force per unit x length). Physically, the beam might be a train track, with the elastic foundation used to model the track bed.

According to classical *Euler beam theory,*[4] the vertical deflection $u(x)$ resulting

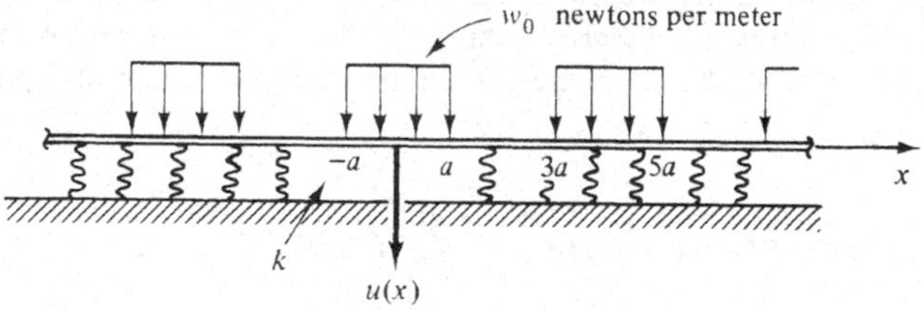

Figure 5.5. Infinite beam on elastic foundation.

[4]See, for example, S. Timoshenko, *Strength of Materials*, Part I, D. Van Nostrand, Princeton, N.J., 1955.

from a load distribution $p(x)$ newtons per unit length satisfies the fourth-order differential equation

$$EIu'''' = p(x), \tag{5.24}$$

where E and I are physical constants; EI is called the "flexural rigidity" of the beam, since u'''', and hence the deflection u, is inversely proportional to EI. Now $p(x)$ is the *net* loading, consisting here of the applied periodic loading $w(x)$ downward and the spring force $ku(x)$ upward. (We will neglect the weight of the beam for simplicity.) Thus $p(x) = w(x) - ku(x)$, and (5.24) becomes

$$EIu'''' + ku = w(x). \tag{5.25}$$

To solve for $u(x)$, let us expand

$$w(x) = \frac{w_0}{2} + \frac{2w_0}{\pi}\sum_1^\infty \frac{\sin(n\pi/2)}{n}\cos\frac{n\pi x}{2a} \tag{5.26}$$

and seek $u(x)$ in the form

$$u(x) = \sum_0^\infty a_n \cos\frac{n\pi x}{2a}, \tag{5.27}$$

where sines will be unnecessary because $u(x)$ is certainly symmetric. Putting (5.26) and (5.27) into (5.25) and equating coefficients of each cosine harmonic, we obtain

$$ka_0 = \frac{w_0}{2}$$

and

$$\left(EI\frac{n^4\pi^4}{16a^4} + k\right)a_n = \frac{2w_0}{\pi}\frac{\sin(n\pi/2)}{n} \quad \text{for } n \geq 1,$$

so that (5.27) becomes

$$u(x) = \frac{w_0}{2k} + \frac{2w_0}{\pi}\sum_1^\infty \frac{\sin(n\pi/2)}{n[EI(n^4\pi^4/16a^4) + k]}\cos\frac{n\pi x}{2a}. \tag{5.28}$$

Actually, to this "particular solution" we should also add the "complementary solution"[5]

$$e^{\beta x}(A\sin\beta x + B\cos\beta x) + e^{-\beta x}(C\sin\beta x + D\cos\beta x),$$

where $\beta \equiv \sqrt[4]{EI}/\sqrt{2}$ but $e^{\beta x} \longrightarrow \infty$ as $x \longrightarrow +\infty$, and $e^{-\beta x} \longrightarrow \infty$ as $x \longrightarrow -\infty$. Requiring that our solution be bounded, we therefore set $A = B = C = D = 0$ and are left only with the particular solution (5.28). ▮

On seeing the words "Fourier series," we generally think immediately of sines and cosines. It turns out, however, that there are also analogous expansions in terms of Bessel functions, Legendre functions, and so on, and they are called *Fourier–Bessel series, Fourier–Legendre series,* . . . . Such series are important and we will return to them in Part III when we reexamine Fourier series from quite a different and more general point of view.

5.5. *INFINITE PERIOD; THE FOURIER INTEGRAL*

Can the function $f(x) = e^{-x^2}$ $(-\infty < x < \infty)$ be expanded in a Fourier series? Apparently not; it is not periodic. On the other hand, we might argue that it *is* periodic

[5] These terms are reviewed in Sections 22.2 and 22.3 of Part IV; in particular, see Example 22.5.

but with infinite period. Let us therefore examine the representation (5.23) as $l \longrightarrow \infty$. We do not simply set $l = \infty$, which would yield nonsense; instead we let l *tend* to infinity [in the same way that (sin 0)/0 is meaningless, whereas $(\sin x)/x \longrightarrow 1$ as x *tends* to zero].

As motivation, note first that the $n\pi/l$'s in (5.23) are the *frequencies*. For instance, if x denotes time, then the units of $n\pi/l$ are radians per unit time; if x is a space variable, then $n\pi/l$ is a spatial frequency. We call the set of all possible frequencies

$$\frac{\pi}{l}, \frac{2\pi}{l}, \frac{3\pi}{l}, \ldots$$

the *frequency spectrum*. To illustrate, with $l = \pi, 2\pi$, and 10π, the frequency spectra are as follows.

l	Frequencies				
π	1,	2,	3,	4,	...
2π	0.5,	1,	1.5,	2,	...
10π	0.1,	0.2,	0.3,	0.4,	...

It is important to see that as l increases, the spectrum becomes finer and finer and tends to a *continuous* spectrum (from 0 to ∞) as $l \longrightarrow \infty$. It is therefore reasonable to anticipate that the (discrete) summation in (5.23a) will give way to an integral from 0 to ∞. Let us proceed.

If we change the dummy integration variable in (5.23b) to ζ, say, insert (5.23b) into (5.23a), and use the identity $\cos A \cos B + \sin A \sin B = \cos(A - B)$ to simplify, we have

$$f(x) = \frac{1}{2l}\int_{-l}^{l} f(\zeta)\,d\zeta + \sum_{k=1}^{\infty}\frac{1}{l}\int_{-l}^{l} f(\zeta)\cos\frac{k\pi}{l}(\zeta - x)\,d\zeta \equiv I + J. \qquad (5.29)$$

Certainly

$$|I| \leq \frac{1}{2l}\int_{-\infty}^{\infty} |f(\zeta)|\,d\zeta < \frac{M}{2l} \longrightarrow 0 \qquad (5.30)$$

as $l \longrightarrow \infty$, if we are willing to require of f that

$$\int_{-\infty}^{\infty} |f(\zeta)|\,d\zeta < M, \qquad (5.31)$$

where M is some finite constant. Proceeding rather formally with J,

$$J = \frac{1}{\pi}\int_{-l}^{l} f(\zeta)\,d\zeta \sum_{1}^{\infty} \Delta\omega \cos k\,\Delta\omega(\zeta - x)$$

where $\Delta\omega \equiv \pi/l$. Recalling Section 1.4, we see that the sum is a *Riemann sum*. The interval is $0 \leq \omega < \infty$, partitioned according to $\omega_0 = 0, \omega_1 = \pi/l, \omega_2 = 2\pi/l, \ldots$, and the "$\xi_k$" points (recall Fig. 1.9) are at the right endpoint of each subinterval—that is, $\xi_1 = \pi/l = \Delta\omega$, $\xi_2 = 2\pi/l = 2\Delta\omega$, and so on. Thus

$$\sum_{1}^{\infty} \Delta\omega \cos k\,\Delta\omega(\zeta - x) = \sum_{1}^{\infty} [\cos \xi_k(\zeta - x)](\omega_k - \omega_{k-1}) \longrightarrow \int_{0}^{\infty} \cos \omega(\zeta - x)\,d\omega$$

as $l \longrightarrow \infty$ (i.e., as the norm of the partition $\pi/l \longrightarrow 0$), so that

$$J \longrightarrow \frac{1}{\pi}\int_{-\infty}^{\infty} f(\zeta)\,d\zeta \int_0^{\infty} \cos\omega(\zeta - x)\,d\omega. \tag{5.32}$$

Switching the order of integration in (5.32), we finally have

$$f(x) = \frac{1}{\pi}\int_0^{\infty} d\omega \int_{-\infty}^{\infty} f(\zeta)\cos\omega(\zeta - x)\,d\zeta \tag{5.33}$$

for the case where f's period is infinite.

In fact, we have the following important theorem.

THEOREM 5.4. (The Fourier Integral Theorem) If $f(x)$ is piecewise smooth[6] *over $(-l, l)$ for every l and*[7]

$$\int_{-\infty}^{\infty} |f(\zeta)|\,d\zeta < M \tag{5.34}$$

for some finite constant M, then

$$\frac{1}{\pi}\int_0^{\infty} d\omega \int_{-\infty}^{\infty} f(\zeta)\cos\omega(\zeta - x)\,d\zeta = f(x) \tag{5.35}$$

almost everywhere; specifically, if f has a jump discontinuity at x, then the Fourier integral of f [i.e., the left-hand side of (5.35)] converges to the average value $[f(x+0) + f(x-0)]/2$.

Rigorous proof of this important result (in contrast with our formal derivation) is similar to our proof of the analogous Theorem 5.1 for Fourier *series* and will be omitted.[8]

To emphasize the analogy between the Fourier series (5.23) and the Fourier integral theorem, it is helpful to rewrite (5.35) in the less compact form

$$f(x) = \frac{1}{\pi}\int_0^{\infty} d\omega \int_{-\infty}^{\infty} f(\zeta)(\cos\omega\zeta\cos\omega x + \sin\omega\zeta\sin\omega x)\,d\zeta$$

$$= \int_0^{\infty} [a(\omega)\cos\omega x + b(\omega)\sin\omega x]\,d\omega, \tag{5.36a}$$

where

$$a(\omega) = \frac{1}{\pi}\int_{-\infty}^{\infty} f(x)\cos\omega x\,dx$$
$$b(\omega) = \frac{1}{\pi}\int_{-\infty}^{\infty} f(x)\sin\omega x\,dx \tag{5.36b}$$

and where we have switched the dummy variable of integration back from ζ to x in the integrals (5.36b). Compare the form of (5.36) with that of (5.23).

---

[6]Or satisfies the Dirichlet conditions, etc.

[7]These conditions are sufficient, not necessary. For example, (5.35) holds for $f(x) = \sin x/x$ even though the latter is not absolutely integrable—that is, does not satisfy (5.34).

[8]See, for example, E. C. Titchmarsh, *The Theory of Functions*, 2nd ed., Oxford University Press, Oxford, 1937.

Example 5.5. The rectangular pulse shown in Fig. 5.6 can certainly be expressed as a combination of Heaviside functions,

$$f(x) = H(x + 1) - H(x - 1), \tag{5.37}$$

but in some applications it is important to have a *Fourier* representation of f, in this case a Fourier *integral*, since f's period is infinite. Computing the Fourier coefficients,

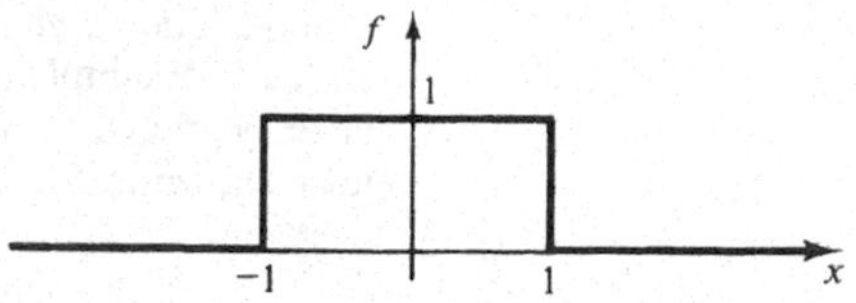

Figure 5.6. Rectangular pulse.

$$a(\omega) = \frac{1}{\pi}\int_{-\infty}^{\infty} f(x)\cos\omega x\,dx = \frac{1}{\pi}\int_{-1}^{1}\cos\omega x\,dx = \frac{2}{\pi}\frac{\sin\omega}{\omega}, \tag{5.38}$$

and $b(\omega) = 0$, since $f(x)$ is symmetric. Thus the Fourier integral representation (5.36a) is

$$f(x) = \frac{2}{\pi}\int_{0}^{\infty}\frac{\sin\omega}{\omega}\cos\omega x\,d\omega. \tag{5.39}$$

What use is it? After all, it seems much more unwieldy than (5.37). We will illustrate its utility in Example 5.6.

Finally, just as it was illuminating to plot the partial sums $s_n(x)$ of Fourier *series*, it should be interesting to see how the "partial integral"

$$f_\Omega(x) = \frac{2}{\pi}\int_{0}^{\Omega}\frac{\sin\omega}{\omega}\cos\omega x\,d\omega \tag{5.40}$$

tends to $f(x)$ as $\Omega \longrightarrow \infty$. Recalling the identity $2\sin A\cos B = \sin(A + B) + \sin(A - B)$ and the sine integral (2.52),

$$\begin{aligned} f_\Omega(x) &= \frac{1}{\pi}\int_{0}^{\Omega}\frac{\sin\omega(1 + x)}{\omega}\,d\omega + \frac{1}{\pi}\int_{0}^{\Omega}\frac{\sin\omega(1 - x)}{\omega}\,d\omega \\ &= \frac{1}{\pi}\int_{0}^{\Omega(1+x)}\frac{\sin\alpha}{\alpha}\,d\alpha + \frac{1}{\pi}\int_{0}^{\Omega(1-x)}\frac{\sin\alpha}{\alpha}\,d\alpha \\ &= \frac{\text{Si}\,[\Omega(1 + x)] + \text{Si}\,[\Omega(1 - x)]}{\pi}, \end{aligned} \tag{5.41}$$

which we've plotted in Fig. 5.7 for $\Omega = 4$, 16, and 128. ∎

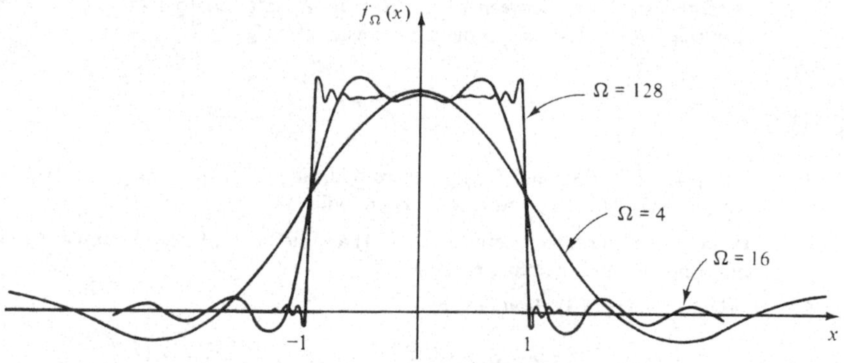

Figure 5.7. The "partial integral" (5.40).

Example 5.6. *Infinite Beam on Elastic Foundation.* Let us consider the same problem as in Example 5.4 but one in which the applied load $w(x)$ is not periodic or, more precisely, in which it has an *infinite* period—for instance, the rectangular pulse shown in Fig. 5.8.

Again,

$$EIu'''' + ku = w(x), \tag{5.42}$$

but this time we use Fourier *integral* representations of $w(x)$ and $u(x)$:

$$w(x) = \frac{2w_0}{\pi} \int_0^\infty \frac{\sin \omega}{\omega} \cos \omega x \, d\omega \quad \text{per (5.39)}$$

and

$$u(x) = \int_0^\infty a(\omega) \cos \omega x \, d\omega. \tag{5.43}$$

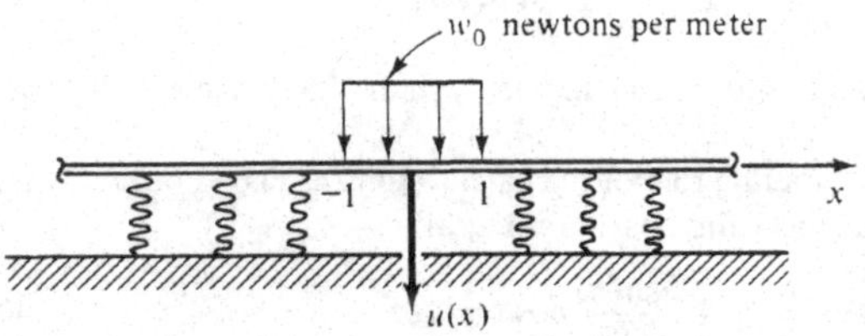

Figure 5.8. Finite rectangular pulse.

Putting them into (5.42) and equating cosine harmonics, we obtain

$$[EI\omega^4 + k]a(\omega) = \frac{2w_0}{\pi} \frac{\sin \omega}{\omega},$$

so that (5.43) becomes

$$u(x) = \frac{2w_0}{\pi} \int_0^\infty \frac{\sin \omega}{\omega(EI\omega^4 + k)} \cos \omega x \, d\omega. \tag{5.44}$$

The integral can be evaluated by using the residue theorem of Part II, but for our purposes it will suffice to leave the answer in integral form. Be sure that you see the essential similarity between the Fourier *series* solution to Example 5.4 and the Fourier *integral* solution to the present example. ∎

EXERCISES

5.1. If $f_1(x)$, $f_2(x)$ are even and $f_3(x)$, $f_4(x)$ are odd, show that $f_1(x)f_2(x)$ and $f_3(x)f_4(x)$ are even and that $f_1(x)f_3(x)$ and $f_1(x)/f_3(x)$ are odd.

5.2. Decompose each of the functions $1/(x + 1)$ and $\sin (x + a)$ into the sum of two functions, one odd and the other even.

5.3. Derive formulas (5.5). *Hint:* Express

$$\cos mx \cos nx = \frac{\cos (m + n)x + \cos (m - n)x}{2}$$

and similarly for $\sin mx \sin nx$ and $\sin mx \cos nx$.

5.4. Derive (5.9). *Hint:* Proceed as in (5.14).

***5.5.** Show that the expression in curly braces in (5.11) is continuous and bounded over $(0, \pi)$.

5.6. Show that the series (5.13) is *uniformly* convergent over $0.0001 \leq x \leq 2$, say, as claimed in Example 5.1.

***5.7.** **(Gibbs' phenomenon)** From (5.10), plus the Riemann–Lebesgue lemma, show that

$$s_n(x) \sim \frac{1}{\pi}\int_0^\gamma [f(x+\xi) + f(x-\xi)]\frac{\sin(n+1/2)\xi}{\xi}\,d\xi$$

for $0 < \gamma < \pi$. Choosing $\gamma = x$, which will be small and positive, and taking f to be the square wave of Fig. 5.9, show that

$$s_n(x) \sim \frac{2}{\pi}\int_0^x \frac{\sin(n+1/2)\xi}{\xi}\,d\xi = \frac{2}{\pi}\,\mathrm{Si}\left[\left(n+\frac{1}{2}\right)x\right]$$

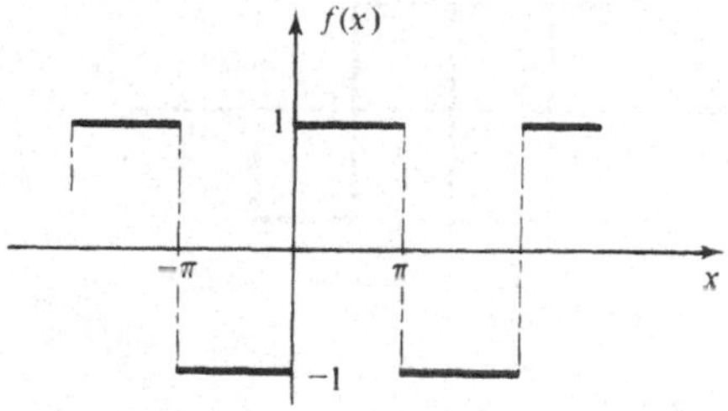

Figure 5.9. Square wave.

(which is sketched in Fig. 5.10). Show that the maxima and minima occur at $\pi/(n+\frac{1}{2})$, $2\pi/(n+\frac{1}{2}), \ldots$, as indicated in the figure. Observe that as we increase n, the curve squeezes to the left, as indicated by the dotted curve and the arrow, so that at a fixed positive x we have $s_n(x) \sim 1$ as $n \longrightarrow \infty$ (as it should). But clearly the closer x is to zero, the slower the convergence is, since n must be increased until enough of the bumps get moved to the left of x. In particular, note that the 0.179 . . . overshoot does

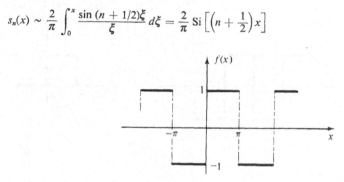

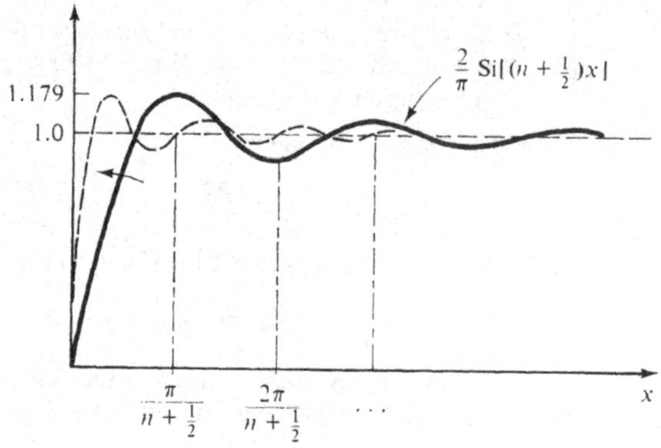

Figure 5.10. Gibbs phenomenon.

not diminish with n; it simply gets skinnier and skinnier, tending to a "spikelet" at $x = 0+$. Known as **Gibbs' phenomenon**, this overshoot occurs at any jump discontinuity, not just for the $f(x)$ of Fig. 5.9. It has been shown by Lanzcos that the Gibbs phenomenon can be suppressed through the introduction of certain smoothing factors. For a summary of this story see, for example, the book by Arfken.[9]

5.8. Derive the closed-form expression

$$\sum_{j=1}^{N} \cos(2j-1)x = \frac{\sin 2Nx}{2 \sin x}$$

for all $x \neq 0, \pm\pi, \pm 2\pi, \pm 3\pi, \ldots$, used in (5.22). *Hint:* See Example 5.1.

5.9. Obtain the Fourier series of the functions shown in Fig. 5.11 directly from the expansion (5.13) of the square wave shown in Fig. 5.1.

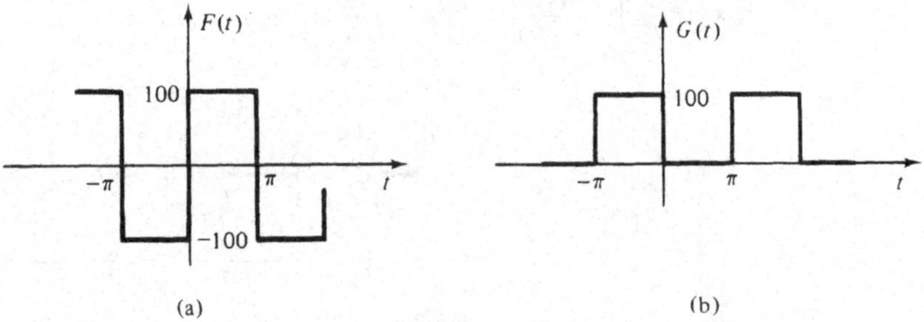

Figure 5.11.

5.10. Obtain the Fourier series of the periodic functions below, defined over one period as follows.

(a) $f(x) = x^2\ (-\pi \le x \le \pi)$
(b) $f(x) = \sin x\ (0 \le x \le \pi)$
(c) $f(x) = \sin x\ (-\pi \le x \le \pi)$
(d) $f(x) = 0$ over $-\pi \le x \le 0$ and x over $0 \le x < \pi$
(e) $f(x) = 0$ over $-\pi \le x \le 0$ and x^2 over $0 \le x < \pi$
(f) $f(x) = 100$ over $-\pi/2 \le x < 0$ and -100 over $0 \le x < \pi/2$
(g) $f(x) = \cos x\ (-\pi/2 \le x \le \pi/2)$
(h) $f(x) = 0$ over $-\pi \le x \le 0$ and $\sin x$ over $0 \le x \le \pi$
(i) $f(x) = 0$ over $-2 \le x < 0$, 25 at $x = 0$, and 50 over $0 < x < 1$
(j) $f(x) = 50$ over $-\pi \le x < \pi$.

5.11. Show that for α not an integer

$$\cos \alpha x \frac{\sin \pi\alpha}{\pi\alpha} + \sum_{1}^{\infty} (-1)^n \frac{2\alpha \sin \pi\alpha}{\pi(\alpha^2 - n^2)} \cos nx$$

over $-\pi \le x \le \pi$ and deduce from this result the formula

$$\cot \pi\alpha = \frac{1}{\pi}\left(\frac{1}{\alpha} - \sum_{1}^{\infty} \frac{2\alpha}{n^2 - \alpha^2}\right).$$

5.12. At what points (if any) does the Fourier series for the $f(x)$ defined in Exercise 5.10(i) fail to converge to $f(x)$? To what values *does* it converge at those points?

---

[9]G. Arfken, *Mathematical Methods for Physicists*, 2nd ed., Academic, New York, 1970, p. 665.

5.13. Sketch $f(x)$ of Example 5.2, together with the first four partial sums of its Fourier series (5.17)—that is, $\pi/2$, $\pi/2 - 4\cos x$, and so on.

5.14. Find a_n's and b_n's such that

$$\cos^2 x = \sum_0^\infty (a_n \cos nx + b_n \sin nx)$$

for all x. Are they *uniquely* determined?

5.15. Given $f(x) = x + 1$ over $0 \leq x \leq 1$, define (sketch) both odd and even periodic extensions with period $\tau = 2$. Sketch at least one other possible periodic extension with $\tau = 2$ and at least one with $\tau = 3$.

5.16. Obtain the Fourier series of the even periodic extension (with period $\tau = 2$) of the function $f(x) = 100$ over $0 \leq x \leq 1$.

5.17. The edge temperature $u(a, \theta)$ of the metal disk shown in Fig. 5.12 is 100 over the top half (i.e., $0 < \theta < \pi$) and 0 over the bottom half. Expand $u(a, \theta)$ in a Fourier series in θ.

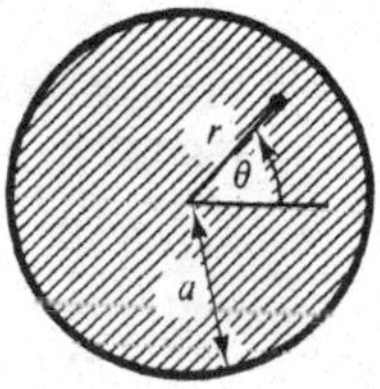

Figure 5.12

***5.18.** Consider the following *claim*. "Surely $(1 - \cos 2nx)/2 \sin x$ in (5.14) can 'blow up' only at the zeros of $\sin x$—that is, at $x = 0, \pm 2\pi, \ldots$. But recall that we *defined* it to be zero at these points. Furthermore, it tends to zero at these points (by L'Hospital's rule), so that it must be continuous and bounded for *all* x. Clearly, then, there is some finite constant M such that

$$\left| \sum_{j=1}^{n} \sin (2j - 1)x \right| < M$$

for all n and *for all* x. Furthermore, $1/(2n - 1) \downarrow 0$ as $n \longrightarrow \infty$, and so the series (5.13a) converges *uniformly* for all x." On the other hand, we know that the convergence cannot be uniform over any interval containing one or more of the jump discontinuities of the sum function; therefore the claim appears to be incorrect. *Explain.*

***5.19.** Consider the mechanical system of Fig. 5.13 and suppose that the mass $m = 1$, the spring stiffness $k = 4$, $x = 0$ is the equilibrium position, and the driving force $F(t)$ is the same as the $F(t)$ in Exercicise 5.9.

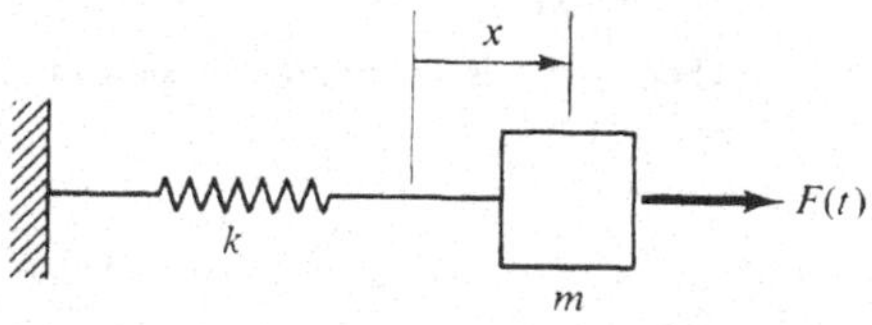

Figure 5.13. Mechanical oscillator.

(a) Show that the equation of motion (Newton's second law) is the second-order differential equation $\ddot{x} + 4x = F(t)$, where dots denote time derivatives. Express $F(t)$ as a Fourier series, solve for the "response" $x(t)$, and verify your solution (insofar as possible).

(b) Re-solve for $x(t)$ for the case where $k = 9$ and make any remarks that seem cogent.

5.20. The **complex Fourier series**[10]

$$f(x) = \sum_{-\infty}^{\infty} c_n e^{in\pi x/l} \tag{5.45}$$

is sometimes a more convenient form than the trigonometric form (5.23a). Recalling Euler's formula $e^{ix} = \cos x + i \sin x$, find the relationship between the c_n's of (5.45) and the a_n's and b_n's of (5.23a) and thus establish the equivalence of the two series. Give an integral expression for c_n analogous to (5.23b).

5.21. Show that

$$\int_{-l}^{l} e^{-im\pi x/l} e^{in\pi x/l}\, dx = \begin{cases} 2l & \text{if } m = n \\ 0 & \text{if } m \neq n. \end{cases} \tag{5.46}$$

Multiply both sides of (5.45) by $e^{-im\pi x/l}$, integrate, and thus (formally) obtain the expression

$$c_n = \frac{1}{2l} \int_{-l}^{l} f(x) e^{-in\pi x/l}\, dx \tag{5.47}$$

for the c_n's. (If should coincide with your result in Exercise 5.20.)

5.22. Obtain the complex form of the Fourier integral formula

$$f(x) = \frac{1}{2\pi} \int_{-\infty}^{\infty} e^{i\omega x}\, d\omega \int_{-\infty}^{\infty} f(\xi) e^{-i\omega\xi}\, d\xi \tag{5.48}$$

two ways.

(a) First, obtain it directly from (5.33) by inserting

$$\cos \omega(\xi - x) = \frac{e^{i\omega(\xi - x)} + e^{-i\omega(\xi - x)}}{2}$$

and "fooling around."

(b) Secondly, obtain it by starting from scratch with the Fourier series, (5.45) and (5.47), and formally taking the limit as $l \longrightarrow \infty$.

5.23. The pulse of Fig. 5.7 obviously has zero average height over the interval $-\infty < x < \infty$ and yet $a(0) \neq 0$. Explain why this is not a contradiction.

5.24. Recalling that $\int_0^\infty (\sin x)/x\, dx = \pi/2$, let $\Omega \longrightarrow \infty$ in (5.40) and show that

$$\frac{2}{\pi} \int_0^\infty \frac{\sin \omega \cos \omega x}{\omega}\, d\omega = \frac{\operatorname{sgn}(1 + x) + \operatorname{sgn}(1 - x)}{2} = f(x) \text{ of Fig. 5.6,}$$

where sgn x is shorthand for **signum** x—that is, the sign of x; sgn $x = +1$ if $x > 0$ and -1 if $x < 0$.

---

[10] By $\sum_{-\infty}^{\infty}$ we mean the limit of $\sum_{-M}^{N}$ as M and $N \longrightarrow \infty$ independently, as you might have guessed. Note further that the terms in (5.45) are *complex*. We say that $\sum (a_n + ib_n) = A + iB$ if and only if $\sum a_n = A$ and $\sum b_n = B$.

***5.25.** Without necessarily attempting a rigorous proof, put forward some simple comments and/or calculations that help to explain why the Riemann–Lebesgue lemma (5.7) should be true.

5.26. Make up two examination-type problems on Chapter 5.

Chapter 6

Fourier and Laplace Transforms

The so-called Fourier transform is actually a restatement of the Fourier integral formula, which, in complex exponential form, is as follows (recall Exercise 5.22).

$$f(x) = \frac{1}{2\pi} \int_{-\infty}^{\infty} e^{i\omega x} \, d\omega \int_{-\infty}^{\infty} f(\xi) e^{-i\omega\xi} \, d\xi, \tag{6.1}$$

where the equal sign holds "almost everywhere" and f is expected to satisfy the conditions of Theorem 5.4. Specifically, if we define

$$\hat{f}(\omega) \equiv \int_{-\infty}^{\infty} f(x) e^{-i\omega x} \, dx, \tag{6.2}$$

then (6.1) becomes[1]

$$f(x) = \frac{1}{2\pi} \int_{-\infty}^{\infty} \hat{f}(\omega) e^{i\omega x} \, d\omega, \tag{6.3}$$

and the pair (6.2), (6.3) is equivalent to the original statement (6.1).

Calling $\hat{f}(\omega)$ the **Fourier transform of $f(x)$**, we refer to (6.3) as the corresponding *inversion formula* because it gives us back $f(x)$. Sometimes it is convenient to use the operator notation $\mathbf{F}$ and $\mathbf{F}^{-1}$ for the transform and its inverse—that is, $\mathbf{F}f = \hat{f}$ and $\mathbf{F}^{-1}\hat{f} = f$.

[1]Although the dummy variable ξ is helpful in (6.1)—to avoid confusion with x—it is not really needed once we break (6.1) apart into (6.2) and (6.3). Thus we change ξ back to x in (6.2) in order to minimize the nomenclature and to emphasize the symmetry between (6.2) and (6.3). Some authors further enhance the symmetry by using $1/\sqrt{2\pi}$ factors in both (6.2) and (6.3) rather than lumping the entire $1/2\pi$ into (6.3).

Aside from its definition (6.2), is there any special meaning or significance that we can attach to the transform $\hat{f}(\omega)$? Comparing (6.3) with (5.45), we see that (give or take a factor of $1/2\pi$) the transform $\hat{f}(\omega)$ plays the role of the Fourier coefficients in the case of the continuous spectrum in the same way that the c_n's do in the case of the discrete spectrum.

In this section we first look at the principal properties of the Fourier transform and show how it can be applied in the solution of ordinary differential equations. Then after introducing the *Laplace* transform and considering its properties and application, we list several other transforms and their inversion formulas.[2]

It should be mentioned that this chapter is concerned more with introducing the transforms and the technique of their application than with illustrating their considerable range and power, a fact that will become more apparent when we use them later on to solve a variety of ordinary and partial differential equations, as well as a few "integral equations."

6.1. *THE FOURIER TRANSFORM*

Most of the main ideas are illustrated by the following example.

Example 6.1. *Infinite Beam on Elastic Foundation.* Let us return to the problem posed in Example 5.4—namely,

$$EIu'''' + ku = w(x), \tag{6.4}$$

or

$$u'''' + \alpha^4 u = W(x), \tag{6.5}$$

where $\alpha^4 \equiv k/EI$, $W(x) \equiv w(x)/EI$, and $w(x)$ is some prescribed loading with infinite period (see Fig. 6.1). This time, however, let us solve by the *Fourier transform*, which is actually only a formalization of the Fourier integral method.

To solve (6.5) by the Fourier transform, we "transform the whole equation";

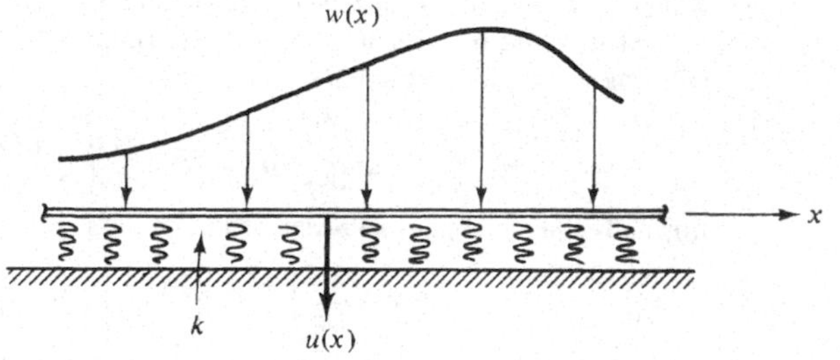

Figure 6.1. Infinite beam on elastic foundation.

---

[2]Noting that the Fourier transform derives, in a sense, from the trigonometric Fourier series and recalling our statement (at the end of Section 5.4) that there are various *non*trigonometric Fourier series as well (Fourier–Bessel series), it is not surprising that there are also various other integral transforms.

that is, we multiply by the Fourier kernel $e^{-i\omega x}$ and integrate on x from $-\infty$ to ∞:

$$\int_{-\infty}^{\infty} (u'''' + \alpha^4 u)e^{-i\omega x}\,dx = \int_{-\infty}^{\infty} We^{-i\omega x}\,dx \tag{6.6}$$

or

$$\int_{-\infty}^{\infty} u''''e^{-i\omega x}\,dx + \alpha^4\hat{u} = \hat{W}. \tag{6.7}$$

Integrating the first term by parts,

$$\int_{-\infty}^{\infty} u''''e^{-i\omega x}\,dx = u'''e^{-i\omega x}\Big|_{-\infty}^{\infty} + i\omega\int_{-\infty}^{\infty} u'''e^{-i\omega x}\,dx. \tag{6.8}$$

So far we have not mentioned any boundary conditions. Let us suppose tentatively that $W(x)$ is localized enough so that u, u', u'', and u''' all $\longrightarrow 0$ as $x \longrightarrow \pm\infty$. Then $u'''e^{-i\omega x} \longrightarrow 0$ as $x \longrightarrow \pm\infty$, since $e^{-i\omega x} = \cos\omega x - i\sin\omega x$ is bounded [in fact, $|e^{-i\omega x}| = \sqrt{\cos^2\omega x + \sin^2\omega x} = 1$] and $u''' \longrightarrow 0$, and so the boundary term in (6.8) drops out. Integrating by parts three more times, the boundary terms drop out, and we have

$$\int_{-\infty}^{\infty} u''''e^{-i\omega x}\,dx = \text{etc.} = (i\omega)^4\int_{-\infty}^{\infty} ue^{-i\omega x}\,dx = (i\omega)^4\hat{u} = \omega^4\hat{u}, \tag{6.9}$$

so that (6.7) is simply

$$(\omega^4 + \alpha^4)\hat{u} = \hat{W}. \tag{6.10}$$

Thus what was originally a (linear) *differential* equation has been reduced to a (linear) *algebraic* equation in the "transform domain." Solving,

$$\hat{u} = \frac{\hat{W}}{\omega^4 + \alpha^4}. \tag{6.11}$$

Having found $\hat{u}$, the final step is the inversion, by means of (6.3), from $\hat{u}$ back to u,

$$u(x) = \frac{1}{2\pi}\int_{-\infty}^{\infty} \frac{\hat{W}}{\omega^4 + \alpha^4}e^{i\omega x}\,d\omega. \tag{6.12}$$

Even though an integration must still be carried out, we regard (6.12) as "the solution." For any given (Fourier transformable) $W(x)$, we can compute $\hat{W}(\omega)$ and then work out the integral in (6.12), at least in principle; in practice, we might have to resort to numerical integration and/or we might be able to extract asymptotic behavior from the integral. Or we might be lucky enough to find the inverse of $\hat{W}/(\omega^4 + \alpha^4)$ in a *table of Fourier transforms*.[3]

Yet it would be nicer if our solution contained W, or w, rather than $\hat{W}$. Substituting for $\hat{W}$, (6.12) becomes

$$u(x) = \frac{1}{2\pi}\int_{-\infty}^{\infty} \frac{e^{i\omega x}}{\omega^4 + \alpha^4}\,d\omega \int_{-\infty}^{\infty} W(\xi)e^{-i\omega\xi}\,d\xi. \tag{6.13}$$

Formally interchanging the order of integration, we have

$$u(x) = \frac{1}{2\pi}\int_{-\infty}^{\infty} W(\xi)\,d\xi \int_{-\infty}^{\infty} \frac{e^{i(x-\xi)\omega}}{\omega^4 + \alpha^4}\,d\omega \tag{6.14}$$

or

$$u(x) = \int_{-\infty}^{\infty} w(\xi)G(\xi, x)\,d\xi, \tag{6.15}$$

[3]There are a number of extensive tables of Fourier transforms—for example, see *Tables of Integral Transforms*, Vols. 1 and 2, ed. by A. Erdèlyi, McGraw-Hill, New York, 1954. Volume 1 contains Fourier, Laplace, and Mellin transforms, and Volume 2 contains Hankel and various other transforms. A very brief table of Fourier transforms is included in the inside front cover of this book.

where the kernel is[4]

$$G(\xi, x) = \frac{1}{2\pi EI}\int_{-\infty}^{\infty} \frac{e^{i(x-\xi)\omega}}{\omega^4 + \alpha^4}\,d\omega = \frac{\alpha^4}{\pi k}\int_0^{\infty} \frac{\cos(\xi - x)\omega}{\omega^4 + \alpha^4}\,d\omega. \tag{6.16}$$

In Part II on complex variable theory we discuss the evaluation of such integrals, using *contour integration*. For the present, we merely state the result,

$$G(\xi, x) = \frac{\alpha}{2k} e^{-\alpha|\xi - x|/\sqrt{2}} \sin\left(\frac{\alpha|\xi - x|}{\sqrt{2}} + \frac{\pi}{4}\right). \tag{6.17}$$

COMMENT 1. Suppose, for example, that $w(x)$ is a point load of unit strength acting at some point x_0—that is, $w(x) = \delta(x - x_0)$. Then the solution (6.15) becomes

$$u(x) = \int_{-\infty}^{\infty} \delta(\xi - x_0) G(\xi, x)\,d\xi = G(x_0, x), \tag{6.18}$$

and it follows that the kernel $G(\xi, x)$ in (6.15) has the physical significance of being the deflection, as a function of x, due to a point unit load at ξ (see Fig. 6.2). Thus the deflection (as a function of x) due to a point load of strength $w(\xi)\,d\xi$ at ξ is $w(\xi)\,d\xi\;G(\xi, x)$, and (6.15) is exposed as a *superposition* of all these incremental deflections.

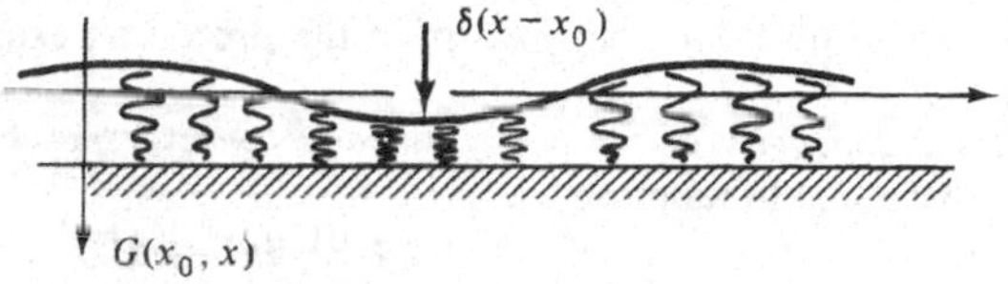

Figure 6.2. Physical significance of $G(\xi, x)$.

Certain questions of rigor exist in connection with the delta function, but we sidestep them for the moment. The same questions arise in discussing Green's functions, and at that time we will be in a position to face up to them.

Finally, observe that for $w(x) = \delta(x - x_0)$, and undoubtedly for other w's as well, $u(x)$ is not everywhere positive. Actually, if we *are* talking about a track on a bed, as mentioned in Example 5.4, then we should have only a "one-way" spring, as shown in Fig. 6.3(a), as opposed to the "two-way" spring of Fig. 6.3(b) and assumed

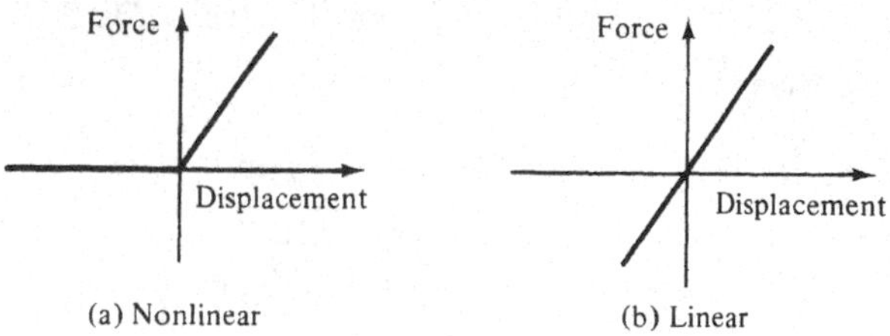

Figure 6.3. Nonlinear versus linear spring models.

---

[4] G is called the **Green's function** for this particular problem. Green's functions are discussed in Parts IV and V.

in our model. Thus the ku term in our original differential equation (6.4) should have been the *nonlinear* function $F(u) = kuH(u)$, where $H(u)$ is the Heaviside step function—that is, $H(u) = 0$ for $u < 0$ and 1 for $u > 0$. This term would resist the Fourier transform, and we'd have a more difficult problem on our hands.

COMMENT 2. Suppose next that $w(x) = \text{constant} \equiv w$. Integrating (6.15), we find (Exercise 6.5) that $u(x) = w/k$, which is obviously correct. A pleasant surprise, since the transform $\hat{W}$ of $w(x) = w$ doesn't even exist. In other words, (6.13) is meaningless because the ξ integral does not exist, and yet (6.14) is all right. The reason lies in the interchange in the order of integration, since the ω integral in (6.14) results in a function of ξ that dies out very strongly (exponentially) at ∞, thereby helping out in the ξ integration. Furthermore, $u(x)$ does *not* $\longrightarrow 0$ as $x \longrightarrow \pm\infty$, as tentatively assumed. If we had worried about the validity of each step, however, we would not have come as far as (6.14). The important point is that it is often best to proceed formally until the answer is in hand. *If it can then be shown that the answer does indeed satisfy the original requirements* (generally a differential equation and boundary conditions), there is no need to worry about the rigor of intermediate steps. ∎

Without calling attention to it, we have used the following three important properties of the Fourier transform in the preceding example.

(F1) *Linearity of the transform and its inverse.* For α, β any scalars and f, g any transformable functions,

$$\mathbf{F}(\alpha f + \beta g) = \alpha \mathbf{F} f + \beta \mathbf{F} g \tag{6.19}$$

and

$$\mathbf{F}^{-1}(\alpha \hat{f} + \beta \hat{g}) = \alpha \mathbf{F}^{-1} \hat{f} + \beta \mathbf{F}^{-1} \hat{g}. \tag{6.20}$$

These properties follow easily from the definition of the transform and its inverse and the linearity of the Riemann integral (Exercise 6.6).

(F2) *Transform of nth derivative.* If $f^{(n-1)}(x), f^{(n-2)}(x), \ldots, f(x)$ all $\longrightarrow 0$ as $x \longrightarrow \pm\infty$, then

$$F[f^{(n)}(x)] = (i\omega)^n \hat{f}(\omega). \tag{6.21}$$

(F3) *Fourier convolution.* Defining the **Fourier convolution** $f * g$ of $f(x)$ and $g(x)$ as

$$f * g \equiv \int_{-\infty}^{\infty} f(x - \xi) g(\xi) \, d\xi, \tag{6.22}$$

we have

$$\begin{aligned} \mathbf{F}^{-1}(\hat{f}\hat{g}) &= \frac{1}{2\pi} \int_{-\infty}^{\infty} \hat{f}(\omega) e^{i\omega x} \, d\omega \int_{-\infty}^{\infty} g(\xi) e^{-i\omega\xi} \, d\xi \\ &= \frac{1}{2\pi} \int_{-\infty}^{\infty} g(\xi) \, d\xi \int_{-\infty}^{\infty} \hat{f}(\omega) e^{i\omega(x-\xi)} \, d\omega \\ &= \int_{-\infty}^{\infty} g(\xi) f(x - \xi) \, d\xi = f * g \end{aligned} \tag{6.23}$$

or

$$\mathbf{F}(f * g) = \hat{f}\hat{g}. \tag{6.24}$$

The second equality in (6.23) involves interchanging the order of integration. Sufficient conditions for the validity of the interchange are that both $|\hat{f}|$ and $|g|$ be integrable over $(-\infty, \infty)$. (But recall the moral of Comment 2 above.)

Applying (6.23) to Example 6.1, for instance, we see from (6.11) that $u(x)$ is the convolution of $W(x)$ and the inverse of $(\omega^4 + \alpha^4)^{-1}$. The latter can be found either by direct integration according to the inversion formula (6.3) or by looking it up in a table of Fourier transforms. Observe that our derivation of (6.23) is simply a generalization of the steps between (6.11) and (6.15). Incidentally, if (6.15) does not appear to be of convolution form, notice from (6.17) that $G(\xi, x)$ is actually a function of the combination $x - \xi$, and so (6.15) is of the form $\int_{-\infty}^{\infty} w(\xi)G(x - \xi)\, d\xi = w * G$.

Our introduction to the Fourier transform has been brief; additional properties are discussed in Exercise 6.8, and applications to partial differential equations are contained in Part V.

6.2. *THE LAPLACE TRANSFORM*

In contrast to the preceding infinite beam problem, the independent variable is often only *semi*-infinite. To illustrate, a system may be in a state of rest until a certain instant, say $t = 0$, when a disturbance is applied, and we wish to know the response for $0 < t < \infty$. Furthermore, the functions involved often fail to decay fast enough, as $t \rightarrow \infty$, to be Fourier transformable—for example, the Heaviside step function $H(t - t_0)$ and $H(t) \sin t$.

Although the Fourier transform can still be applied sometimes, it is best to develop a transform that is tailored specifically to such problems.

Suppose that $f(t)$ satisfies the following conditions.

(i) f is piecewise smooth over every finite interval.
(ii) f is of **exponential order**; that is, there exist (real) constants K, c, and T such that

$$|f(t)| < Ke^{ct}$$

for all $t > T$. (6.25)

This class of functions is quite extensive and even allows for exponential *growth*. For example, $f(t) = 1/(t + 2)$, t^6, and e^{27t} are all of exponential order, although $f(t) = e^{t^2}$ is not (Exercise 6.13).

Although an f satisfying (6.25) is not necessarily Fourier transformable, the constructed function

$$F(t) \equiv \begin{cases} 0, & t < 0 \\ e^{-\gamma t}f(t), & t > 0 \end{cases} \tag{6.26}$$

is if we choose $\gamma > c$. Inserting this $F(t)$ into (6.1), we have, for $t > 0$ (which is the domain of interest),

$$e^{-\gamma t}f(t) = \frac{1}{2\pi}\int_{-\infty}^{\infty} e^{i\omega t}\, d\omega \int_0^{\infty} e^{-\gamma\tau}f(\tau)e^{-i\omega\tau}\, d\tau$$

or

$$f(t) = \frac{1}{2\pi}\int_{-\infty}^{\infty} e^{(\gamma+i\omega)t}\, d\omega \int_0^{\infty} f(\tau)e^{-(\gamma+i\omega)\tau}\, d\tau.$$

With $\gamma + i\omega \equiv s$,

$$f(t) = \frac{1}{2\pi i} \int_{\gamma - i\infty}^{\gamma + i\infty} e^{st}\, ds \int_0^\infty f(\tau) e^{-s\tau}\, d\tau, \tag{6.27}$$

which yields the **Laplace transform** and its inverse,[5]

$$\bar{f}(s) \equiv \int_0^\infty f(t) e^{-st}\, dt, \tag{6.28}$$

$$f(t) = \frac{1}{2\pi i} \int_{\gamma - i\infty}^{\gamma + i\infty} \bar{f}(s) e^{st}\, ds. \tag{6.29}$$

Sometimes it is convenient to use the operator notation $\mathbf{L}$ and $\mathbf{L}^{-1}$ for the transform and its inverse—that is, $\mathbf{L}f = \bar{f}$ and $\mathbf{L}^{-1}\bar{f} = f$.

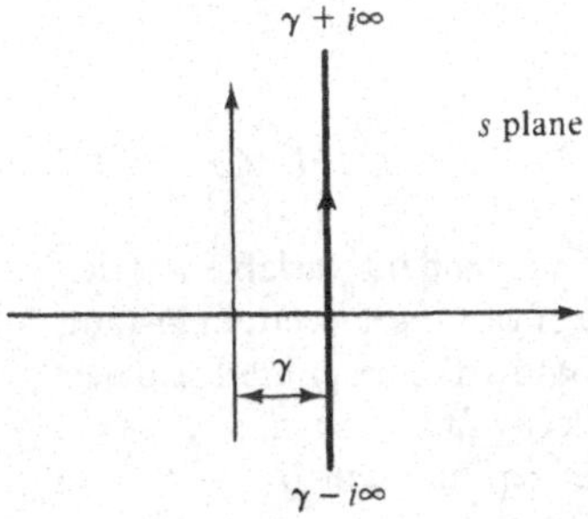

Figure 6.4. Path of integration for inversion integral.

The fact that γ and ω are real means that the transform variable $s = \gamma + i\omega$ is *complex*. The path of integration in the inversion integral (6.29) is the straight line from $\gamma - i\infty$ to $\gamma + i\infty$ in the complex s plane, as shown in Fig. 6.4. So in order to carry out the actual Laplace inversion, we need some knowledge of complex variable methods, specifically "contour integration" in the complex plane; in fact, the Fourier inversion integral (6.3) also is generally carried out by contour integration, as will be seen later on. This topic will be developed in Part II; in the present section we are content to rely on the use of tables for the Laplace inversion.[6]

Meanwhile, you may be wondering how we decide what value to assign to γ in (6.29). In our derivation γ was simply assumed to be arbitrarily greater than c, but, of course, we don't know c in advance because we don't (in practical applications) know (the solution) $f(t)$ in advance. The fact is that it does not matter, since γ will not appear in the final result.

Note that the conditions (6.25) are sufficient, although not necessary, for the existence of $\bar{f}(s)$. For example, condition (i) of (6.25) is not satisfied by $f(t) = t^{-1/2}$, and yet $\bar{f}$ does exist (Exercise 6.15):

$$\mathbf{L}(t^{-1/2}) = \sqrt{\frac{\pi}{s}}. \tag{6.30}$$

Let us look at the transform of a few of the most common functions.

Example 6.2.

$$f(t) = H(t-a) = \begin{cases} 0, & t < a \\ 1, & t > a. \end{cases}$$

[5]As with (6.1), the equality in (6.27) is understood to hold "almost everywhere."

[6]Besides contour integration and tables, specialized methods for the *numerical* integration of (6.29) are also available. See, for example, V. Krylov and N. Skoblya, *Handbook of Numerical Inversion of Laplace Transforms*, Israel Program for Scientific Translations, Jerusalem, 1969.

Observe that conditions (6.25) are satisfied. In particular, note that $c = 0$, so that we need to choose $\gamma > 0$. Thus

$$\mathbf{L}[H(t-a)] = \int_0^\infty H(t-a)e^{-st}\,dt$$
$$= \int_a^\infty e^{-st}\,dt = \frac{e^{-st}}{-s}\Big|_a^\infty = \frac{e^{-as}}{s}, \tag{6.31}$$

where $e^{-st} \longrightarrow 0$ as $t \longrightarrow \infty$, since $\operatorname{Re} s = \gamma > 0$. ∎

Similarly, we find easily (Exercise 6.15) that

$$\mathbf{L}(t^a) = \frac{\Gamma(a+1)}{s^{a+1}} \quad \text{for } a > -1, \tag{6.32}$$

$$\mathbf{L}(e^{at}) = \frac{1}{s-a}, \tag{6.33}$$

$$\mathbf{L}(\sin at) = \frac{a}{s^2 + a^2}, \tag{6.34}$$

and

$$\mathbf{L}(\cos at) = \frac{s}{s^2 + a^2}. \tag{6.35}$$

In particular, when a is an integer, (6.32) becomes $\mathbf{L}(1) = 1/s$, $\mathbf{L}(t) = 1/s^2$, $\mathbf{L}(t^2) = 2!/s^3$, and so on.

As with the Fourier transform, the following three operational properties are crucial.

(L1) *Linearity of the transform and its inverse.* For α, β any scalars and f, g any transformable functions,

$$\mathbf{L}(\alpha f + \beta g) = \alpha \mathbf{L} f + \beta \mathbf{L} g \tag{6.36}$$

and

$$\mathbf{L}^{-1}(\alpha f + \beta g) = \alpha \mathbf{L}^{-1} f + \beta \mathbf{L}^{-1} g. \tag{6.37}$$

(L2) *Transform of nth derivative.* Integrating by parts,

$$\int_0^\infty f^{(n)}(t)e^{-st}\,dt = e^{-st}f^{(n-1)}(t)\Big|_0^\infty + s\int_0^\infty f^{(n-1)}(t)e^{-st}\,dt.$$

Assuming that $f^{(n-1)}(t)$ is of exponential order, it follows that we can choose γ positive enough so that the boundary term tends to zero at the upper limit—that is,

$$|e^{-st}f^{(n-1)}(t)| = |e^{-\gamma t}|\,|e^{-i\omega t}|\,|f^{(n-1)}(t)| = e^{-\gamma t}|f^{(n-1)}(t)| \longrightarrow 0$$

as $t \longrightarrow \infty$. Thus we have

$$\mathbf{L} f^{(n)} = s\mathbf{L} f^{(n-1)} - f^{(n-1)}(0). \tag{6.38}$$

For $n = 1$ and 2, for instance,

$$\mathbf{L}[f'(t)] = s\bar{f}(s) - f(0) \tag{6.39}$$

and

$$\mathbf{L}[f''(t)] = s[s\bar{f}(s) - f(0)] - f'(0)$$
$$= s^2\bar{f}(s) - sf(0) - f'(0). \tag{6.40}$$

In some applications (e.g., Exercise 6.23), $f^{(n-1)}$ has a jump discontinuity at $t = 0$, so that "$f^{(n-1)}(0)$" is not meaningful. In this case, we have

$$\mathbf{L}f^{(n)} = \int_0^\infty f^{(n)}(t)e^{-st}\,dt = \lim_{\epsilon\to 0}\int_\epsilon^\infty f^{(n)}(t)e^{-st}\,dt \qquad (\epsilon > 0)$$

$$= \lim_{\epsilon\to 0}\left\{e^{-st}f^{(n-1)}(t)\Big|_\epsilon^\infty + s\int_\epsilon^\infty f^{(n-1)}(t)e^{-st}\,dt\right\}$$

$$= s\mathbf{L}f^{(n-1)} - f^{(n-1)}(0+);$$

that is, (6.38) is correct if we interpret $f^{(n-1)}(0)$ as $\lim_{\epsilon\to 0} f^{(n-1)}(\epsilon)$ or $f^{(n-1)}(0+)$.

(L3) *Laplace convolution.* Defining the **Laplace convolution** $f * g$ of $f(t)$ and $g(t)$ as

$$f * g \equiv \int_0^t f(\tau)g(t-\tau)\,d\tau, \tag{6.41}$$

we have

$$\mathbf{L}(f * g) = \int_0^\infty e^{-st}\,dt \int_0^t f(\tau)g(t-\tau)\,d\tau$$

$$= \int_0^\infty f(\tau)\,d\tau \int_\tau^\infty e^{-st}g(t-\tau)\,dt$$

$$= \int_0^\infty f(\tau)\,d\tau \int_0^\infty e^{-s(\mu+\tau)}g(\mu)\,d\mu$$

$$= \int_0^\infty f(\tau)e^{-s\tau}\,d\tau \int_0^\infty g(\mu)e^{-s\mu}\,d\mu$$

$$= \bar{f}(s)\bar{g}(s) \tag{6.42}$$

or

$$\mathbf{L}^{-1}(\bar{f}\bar{g}) = f * g. \tag{6.43}$$

If the mechanics of the interchange in the order of integration is not clear, recall footnote 4 of Chapter 3. (The legitimacy of the interchange is left for Exercise 6.30.)

Additional properties are contained in Exercise 6.16.

Example 6.3. Consider the harmonic oscillator of Fig. 6.5, where the wall is given some prescribed displacement $f(t)$ and we would like to know the response $x(t)$, assuming that the system is initially at rest. According to Newton's second law,

$$m\ddot{x} = k[f(t) - x]$$

and so

$$m\ddot{x} + kx = kf(t) \equiv F(t), \qquad x(0) = \dot{x}(0) = 0, \tag{6.44}$$

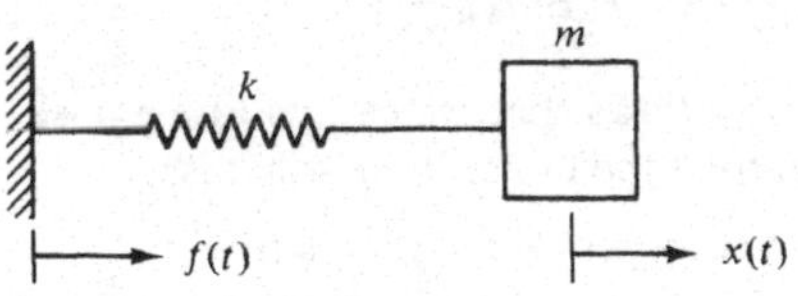

Figure 6.5. Driven harmonic oscillator.

where, as usual, the dots denote time derivatives. [Interestingly, the governing equation is the same as if the wall were fixed and a force $F(t)$ were applied directly to the mass m. Thus from a design point of view, if we wish to drive m with some prescribed force $F(t)$, a good way to do so would be through the wall-spring arrangement of Fig. 6.5, since displacements are more easily arranged than forces.] The domain being semi-infinite (i.e., $0 < t < \infty$) suggests the Laplace transform. To transform the differential equation, we multiply through by the Laplace kernel e^{-st}

and integrate from zero to infinity. Using (6.36) and (6.40), we obtain the (linear) algebraic equation

$$m[s^2\bar{x}(s) - sx(0) - \dot{x}(0)] + k\bar{x}(s) = \bar{F}(s) \tag{6.45}$$

for $\bar{x}(s)$. Solving,

$$\bar{x}(s) = \bar{G}(s)\bar{I}(s), \tag{6.46}$$

where
$$\bar{G}(s) = \frac{1}{m(s^2 + \omega^2)}, \qquad \omega^2 \equiv \frac{k}{m} \tag{6.47}$$

and
$$\bar{I}(s) = \bar{F}(s) + m[sx(0) + \dot{x}(0)]. \tag{6.48}$$

In the language of *control theory*, $\bar{G}(s)$ is called the **transfer function** of the system and $\bar{I}(s)$ is the **input**. $\bar{G}(s)$ reflects the nature of the system—that is, the left-hand side $m\ddot{x} + kx$ of the differential equation. $\bar{I}(s)$ is the total input; in this case, it consists of two parts, the driving force $\bar{F}(s)$ plus an input $m[s\dot{x}(0) + x(0)]$ from the initial conditions. In fact, *note carefully* how the initial conditions are thus built right into (6.46).

Inverting, with the help of (6.37), (6.43), and Table A.2 of Appendix A, we have

$$\begin{aligned} x(t) &= \mathbf{L}^{-1}(\bar{G}\bar{F}) + x(0)\mathbf{L}^{-1}\Big(\frac{s}{s^2+\omega^2}\Big) + \dot{x}(0)\mathbf{L}^{-1}\Big(\frac{1}{s^2+\omega^2}\Big) \\ &= G(t) * F(t) + x(0)\cos\omega t + \frac{\dot{x}(0)}{\omega}\sin\omega t \\ &= \omega\int_0^t \sin\omega(t-\tau)f(\tau)\,d\tau + x(0)\cos\omega t + \frac{\dot{x}(0)}{\omega}\sin\omega t, \end{aligned} \tag{6.49}$$

where we've replaced $F(\tau)$ by $kf(\tau)$ in the last step. Actually, the last two terms drop out, since the initial conditions were $x(0) = \dot{x}(0) = 0$.

For instance, if f is the Heaviside step function

$$f(t) = H(t - T), \tag{6.50}$$

then (6.49) becomes

$$\begin{aligned} x(t) &= \begin{cases} 0, & t < T \\ 1 - \cos\omega(t - T), & t > T \end{cases} \\ &= H(t - T)[1 - \cos\omega(t - T)]. \end{aligned} \tag{6.51}$$

COMMENT 1. Clearly, (6.51) satisfies the initial conditions $x(0) = \dot{x}(0) = 0$. To verify its satisfaction of the differential equation, note from (6.51) [and Exercise 4.16(a)] that

$$\begin{aligned} \dot{x}(t) &= \delta(t - T)[1 - \cos\omega(t - T)] + \omega H(t - T)\sin\omega(t - T) \\ &= \omega H(t - T)\sin\omega(t - T) \end{aligned}$$

and
$$\begin{aligned} \ddot{x}(t) &= \omega\delta(t - T)\sin\omega(t - T) + \omega^2 H(t - T)\cos\omega(t - T) \\ &= \frac{k}{m}H(t - T)\cos\omega(t - T), \end{aligned}$$

so that

$$\begin{aligned} m\ddot{x} + kx &= m\frac{k}{m}H(t - T)\cos\omega(t - T) + kH(t - T)[1 - \cos\omega(t - T)] \\ &= kH(t - T) = kf(t). \end{aligned}$$

COMMENT 2. To understand the convolution integral in (6.49), let us set $f(t) = a\,\delta(t - t_0)$—that is, an impulsive force of strength a acting at time t_0. Then

$$\omega \int_0^t \sin \omega(t - \tau)\, a\, \delta(\tau - t_0)\, d\tau = \begin{cases} 0, & t < t_0 \\ a\omega \sin \omega(t - t_0), & t > t_0 \end{cases}$$

$$= H(t - t_0)\, a\omega \sin \omega(t - t_0).$$

That is, nothing happens until t_0 when the impulse is applied; subsequently, the motion is $a\omega \sin \omega(t - t_0)$. Thus the deflection x at time t, due to an incremental force of strength $f(\tau)\, d\tau$ acting at an earlier time τ, will be $\omega \sin \omega(t - \tau) f(\tau)\, d\tau$. Adding them up from $\tau = 0$ to $\tau = t$ gives the convolution integral in (6.49); the cutoff at t (the upper limit) makes sense because only forces that act at τ's $< t$ will be felt at time t. As with the Fourier convolution (recall Comment 1 of Example 6.1), we therefore see through the *superposition* nature of the Laplace convolution.

COMMENT 3. We've already mentioned the pertinence of the semi-infinite domain to the Laplace transform method. We should also point out that the method applies most naturally to *initial value* problems. The present example is such a type, since both boundary conditions are at $t = 0$—that is, $x(0) = 0$ and $\dot{x}(0) = 0$. This situation is most convenient because it is *exactly* that data that is needed in (6.45). Summarizing, the Laplace transform is most natural for initial value problems on semi-infinite domains, which is generally the context in which it is used. The fact that it can sometimes be used under other circumstances, however, is illustrated by the following simple example. ▮

Example 6.4. Find the deflection $u(x)$ of a cantilever beam of length l subjected to a distributed load $w(x) = x$, as shown in Fig. 6.6. According to the Euler beam theory,

$$EIu'''' = x \tag{6.52a}$$

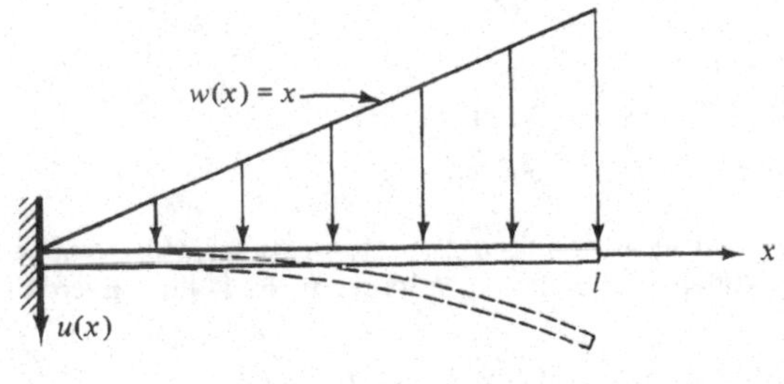

Figure 6.6. Boundary value problem on finite domain.

subject to the boundary conditions[7]

$$u(0) = u'(0) = u''(l) = u'''(l) = 0. \tag{6.52b}$$

Clearly, the domain is not semi-infinite; furthermore, it is a *boundary value* rather than an initial value problem because the boundary conditions are applied at *both*

[7]At $x = l$, $u''(l) = 0$ because there's no "bending moment" there, and $u'''(l) = 0$ because there's no "shear" there. If these terms are not meaningful to you, don't worry about it; simply try to accept the system (6.52) on mathematical terms.

ends of the interval. But there is no harm in extending the beam (fictitiously) from l to ∞, provided that we define $w(x) = 0$ for $x > l$. Then (6.52a) becomes $EIu'''' = [1 - H(x - l)]x$, and Laplace transforming this, with $u(0) = u'(0) = 0$, gives

$$EI[s^4\bar{u}(s) - su''(0) - u'''(0)] = \frac{1}{s^2} - \frac{e^{-sl}}{s^2} - \frac{le^{-sl}}{s}$$

or

$$\bar{u}(s) = \frac{1}{EIs^6} + \frac{u''(0)}{s^3} + \frac{u'''(0)}{s^4} - \frac{e^{-sl}}{EIs^6} - \frac{le^{-sl}}{EIs^5}.$$

Inverting, with the help of Exercise 6.16(b),

$$u(x) = \frac{x^5}{120EI} + \frac{u''(0)x^2}{2} + \frac{u'''(0)x^3}{6} - \frac{H(x-l)(x-l)^5}{120EI} - \frac{lH(x-l)(x-l)^4}{24EI}. \tag{6.53}$$

Unfortunately, $u''(0)$ and $u'''(0)$ are not prescribed. We have not yet used the data $u''(l) = u'''(l) = 0$, however. Applying these conditions,

$$u''(l) = \frac{l^3}{6EI} + u''(0) + u'''(0)l = 0$$

$$u'''(l) = \frac{l^2}{2EI} + u'''(0) = 0,$$

so that

$$u'''(0) = -\frac{l^2}{2EI}, \qquad u''(0) = \frac{l^3}{3EI},$$

and, since the last two terms in (6.53) are zero over $0 \le x \le l$,

$$u(x) = (x^5 + 20l^3x^2 - 10l^2x^3)/120EI. \quad ▮ \tag{6.54}$$

Example 6.5. *Bessel Equation.* The initial value problem

$$t\ddot{x} + \dot{x} + tx = 0, \tag{6.55a}$$

$$x(0) = 1, \qquad \dot{x}(0) = 0 \tag{6.55b}$$

is more complicated than the preceding examples because of the nonconstant coefficients in the differential equation. Laplace transforming (6.55a),

$$\int_0^\infty t\ddot{x}e^{-st}\,dt + s\bar{x} - x(0) + \int_0^\infty txe^{-st}\,dt = 0$$

or (note carefully, recalling the Leibnitz rule from Section 1.7)

$$-\frac{d}{ds}\int_0^\infty \ddot{x}e^{-st}\,dt + s\bar{x} - 1 - \frac{d}{ds}\int_0^\infty xe^{-st}\,dt = 0,$$

$$-\frac{d}{ds}[s^2\bar{x} - sx(0) - \dot{x}(0)] + s\bar{x} - 1 - \frac{d\bar{x}}{ds} = 0,$$

$$(s^2+1)\frac{d\bar{x}}{ds} + s\bar{x} = 0. \tag{6.56}$$

There is still a differential equation to deal with, but it is a much simpler one. Separating variables, we have

$$\frac{d\bar{x}}{\bar{x}} = -\frac{s\,ds}{s^2+1}.$$

Integrating and simplifying give

$$\bar{x}(s) = \frac{C}{\sqrt{s^2+1}}. \tag{6.57}$$

Not finding $(s^2 + 1)^{-1/2}$ in our brief tables, let us expand it in powers of s:

$$(1 + s^2)^{-1/2} = 1 - \tfrac{1}{2}s^2 + \cdots. \tag{6.58}$$

The only problem is that we cannot find the inverse of $1, s^2, s^4, \ldots$ in our tables or any other tables, as a matter of fact, since the inversion integral diverges in these cases![8] Yet *inverse* powers of s *are* all right, and so we discard (6.58) and try again:

$$\frac{1}{\sqrt{s^2+1}} = \frac{1}{s}\left(1 + \frac{1}{s^2}\right)^{-1/2} = \frac{1}{s}\left(1 - \frac{1}{2s^2} + \cdots\right).$$

Thus

$$\bar{x}(s) = C \sum_0^\infty \frac{(-1)^n(2n)!}{(2^n n!)^2} \frac{1}{s^{2n+1}}, \tag{6.59}$$

and since $\mathbf{L}^{-1}(s^{-m}) = t^{m-1}/(m-1)!$, term-by-term inversion of (6.59) yields

$$x(t) = C \sum_0^\infty \frac{(-1)^n}{(2^n n!)^2} t^{2n}.$$

Finally, applying $x(0) = 1$ again, we see that the integration constant C is unity; therefore

$$x(t) = \sum_0^\infty \frac{(-1)^n}{(2^n n!)^2} t^{2n}. \tag{6.60}$$

Although our derivation has been rather formal, it can be verified that the series converges for all t and does satisfy the *system* (6.55)—that is, the differential equation plus the initial conditions (Exercise 6.31).

In fact, (6.55a) is an extremely important equation, **Bessel's equation of order zero**; and the solution (6.60) is denoted as $J_0(t)$, the **Bessel function of the first kind and order zero**. More about Bessel functions later. ∎

Quite often we are confronted with the inversion of a rational fraction—that is, $\bar{f}(s) = P(s)/Q(s)$, where P and Q are polynomials. For $\mathbf{L}^{-1}\bar{f}$ to exist (recall footnote 8), we will need Q to be of higher degree than P. Suppose that we are able to determine the roots of $Q(s) = 0$, which may be real or complex. If a_1 is a root of multiplicity m_1, a_2 is a root of multiplicity m_2, and so on, then

$$\bar{f}(s) = \frac{P(s)}{\alpha(s-a_1)^{m_1}(s-a_2)^{m_2}\cdots(s-a_n)^{m_n}}, \tag{6.61}$$

which (we state without proof) admits the **partial fractions** representation

$$\frac{A_1^1 + A_2^1 s + \cdots + A_{m_1}^1 s^{m_1-1}}{(s-a_1)^{m_1}} + \frac{A_1^2 + A_2^2 s + \cdots + A_{m_2}^2 s^{m_2-1}}{(s-a_2)^{m_2}} + \cdots + \frac{A_1^n + A_2^n s + \cdots + A_{m_n}^n s^{m_n-1}}{(s-a_n)^{m_n}}. \tag{6.62}$$

[8] A *necessary* condition for the inverse of a given $\bar{f}(s)$ to exist is that $\bar{f}(s) \to 0$ as $s \to \infty$ through real values [at least for the inverse to exist as a classical function; for instance, $\bar{f}(s) = 1$ does not $\to 0$, but its inverse exists as the generalized function $\delta(t)$].

Example 6.6.

$$\bar{f}(s) = \frac{1}{2s^2 + s - 1} = \frac{1}{2(s - 1/2)(s + 1)}$$

$$= \frac{A}{s - 1/2} + \frac{B}{s + 1}. \tag{6.63}$$

To evaluate A and B, we multiply through by $(s - 1/2)(s + 1)$:

$$\frac{1}{2} = (s + 1)A + \left(s - \frac{1}{2}\right)B. \tag{6.64}$$

Here (6.63) [and hence (6.64)] is to be *identically* true—that is, for *all* s. Setting $s = \frac{1}{2}$, we see that $A = \frac{1}{3}$; and setting $s = -1$, we see that $B = -\frac{1}{3}$. So

$$f(t) = \mathbf{L}^{-1}\left(\frac{1/3}{s - 1/2} - \frac{1/3}{s + 1}\right) = \frac{1}{3}e^{t/2} - \frac{1}{3}e^{-t}.$$

The point is that the terms in the partial fraction representation are simpler and more easily inverted than the original rational fraction. ▮

Example 6.7.

$$\bar{f}(s) = \frac{s + 1}{s^3(s + 2)} = \frac{A + Bs + Cs^2}{s^3} + \frac{D}{s + 2}. \tag{6.65}$$

Multiplying through by $s^3(s + 2)$,

$$s + 1 = (s + 2)(A + Bs + Cs^2) + Ds^3$$

or

$$1 + s = 2A + (A + 2B)s + (B + 2C)s^2 + (C + D)s^3. \tag{6.66}$$

Equating coefficients of like powers of s,[9] we find

$$2A = 1, \qquad A + 2B = 1, \qquad B + 2C = 0, \qquad C + D = 0$$

so that $A = \frac{1}{2}$, $B = \frac{1}{4}$, $C = -\frac{1}{8}$, $D = \frac{1}{8}$. Thus inverting (6.65),

$$f(t) = \frac{t^2}{4} + \frac{t}{4} - \frac{1}{8} + \frac{e^{-2t}}{8}. \quad ▮$$

These two examples will suffice for our purposes. More systematic recipes for the inversion of (6.61) are given by the so-called **Heaviside expansion theorems**; see, for example, Wylie.[10] Systematic development of the operational calculus (i.e., the "calculus" of integral transforms) and its adoption as a standard tool of modern applied mathematics are due largely to the electrical engineer, Oliver Heaviside (1850–1925), although the Laplace transform was actually introduced about a hundred years earlier, by Laplace, in connection with probability theory.

So ends our introduction to Fourier and Laplace transforms. The topic also appears in Part II, where we discuss the evaluation of the inversion integrals by means

[9]Equation (6.65) is to hold for all s. Setting $s = 0$ gives $2A = 1$; differentiating (6.66) once with respect to s and then setting $s = 0$ give $A + 2B = 1$; differentiating (6.66) twice and setting $s = 0$ give $B + 2C = 0$ and so on. Or recall Theorem 2.13.

[10]C. R. Wylie, Jr., *Advanced Engineering Mathematics*, 3rd ed., McGraw-Hill, New York, 1966, p. 255.

of contour integration, and in Part V, where we apply transform techniques to *partial* differential equations.

As usual, we encourage the reader at least to read through the exercises, even those you don't intend to work, especially Exercises 6.8 and 6.16, which contain various additional properties of the Fourier and Laplace transforms. For further reference, we recommend the accounts by Churchill,[11] Miles,[12] and Tranter.[13] Other sources are included in the Additional References at the end of Part I.

6.3. *OTHER TRANSFORMS*

In addition to the Laplace and Fourier transforms defined here, several other transforms are more natural in certain applications—for instance, the **Hankel transform**

$$\left.\begin{aligned} \bar{f}(p) &\equiv \int_0^\infty f(x)xJ_n(px)\,dx \qquad &\text{(transform)} \\ f(x) &= \int_0^\infty \bar{f}(p)pJ_n(px)\,dp \qquad &\text{(inversion)}, \end{aligned}\right\} \tag{6.67}$$

where J_n is a *Bessel function of order n*, and the **Mellin transform**

$$\left.\begin{aligned} \bar{f}(p) &\equiv \int_0^\infty f(x)x^{p-1}\,dx \qquad &\text{(transform)} \\ f(x) &= \frac{1}{2\pi i}\int_{-i\infty}^{i\infty} \bar{f}(p)x^{-p}\,dp \qquad &\text{(inversion)}. \end{aligned}\right\} \tag{6.68}$$

Moreover, there are variations of the Laplace and Fourier transforms: in contrast to the (one-sided) Laplace transform, there is also a two-sided or **bilateral Laplace transform**; and in contrast to the (two-sided) Fourier transform, there are also one-sided versions called the **Fourier sine and cosine transforms** (Exercise 6.12). In addition, there are *finite* Fourier transforms, but they are actually only a formalization of the classical Fourier series approach.

EXERCISES

The Fourier Transform

6.1. Find the Fourier transform of the functions below.

(a) $f(x) = e^{-\alpha|x|}$ (b) $f(x) = \begin{cases} 0, & x < 0 \\ e^{-\alpha x}, & x \geq 0 \end{cases}$

(c) $f(x) = \delta(x - x_0)$

***6.2.** Given $f(x) = 1$ for $|x| < 1$ and 0 for $|x| \geq 1$, find $\hat{f}(\omega)$ and then verify the inversion formula; that is, compute the inverse of $\hat{f}$ and show that it is the same as the original $f(x)$. (Everywhere or almost everywhere? Explain.)

---

[11] R. V. Churchill, *Operational Mathematics*, 2nd ed., McGraw-Hill, New York, 1958.

[12] *Modern Mathematics for the Engineer*, Second Series, ed. by E. F. Beckenbach, McGraw-Hill, 1961. See Part 1, Chapter 3 on Integral Transforms, by J. W. Miles.

[13] C. J. Tranter, *Integral Transforms in Mathematical Physics*, Methuen, London, 1956.

6.3. We could have identified the first term in (6.7) simply as $\widehat{u''''}$; so why did we bother to integrate by parts four times?

***6.4.** Show that the deflection of the beam in Example 6.1, due to a counterclockwise point couple of unit strength at x_0, is $u(x) = \partial G(x_0, x)/\partial x$. Sketch $u(x)$. *Hint:* Recall Exercise 4.19 and the fact that $G(\xi, x)$ is actually a function of $|\xi - x|$.

6.5. Integrating (6.15) for the case $w(x) = \text{constant} \equiv w$, where G is given by (6.17), show that $u(x) = \text{constant} = w/k$ as claimed in Comment 2.

6.6. Verify the linearity of the Fourier transform and its inverse, (6.19) and (6.20). *Hint:* Recall (6.2), (6.3), and (1.17). [Actually, the limits a, b were *finite* in (1.17). Show that $\int_{-\infty}^{\infty} (\alpha u + \beta v)\, dx = \alpha \int_{-\infty}^{\infty} u\, dx + \beta \int_{-\infty}^{\infty} v\, dx$, too.]

6.7. Recalling the definition (6.22), show that $f * g = g * f$.

6.8. **(Additional properties of the Fourier transform)** Show that

(a)

$$\mathbf{F}[(-ix)^n f(x)] = \hat{f}^{(n)}(\omega).$$

[See (6.21).] You need not justify your steps.

(b) x shift:

$$\mathbf{F}^{-1}[e^{-ia\omega}\hat{f}(\omega)] = f(x-a).$$

(c) ω shift:

$$\mathbf{F}^{-1}[\hat{f}(\omega - a)] = e^{iax} f(x).$$

(d) If

$$h(x) = \int_{-\infty}^{x} f(\xi)\, d\xi \quad \text{and} \quad h \longrightarrow 0$$

as $x \longrightarrow \infty$, then $\mathbf{F}[h(x)] = \hat{f}(\omega)/i\omega$. *Hint:* Integrate by parts.

6.9. The circuit of Fig. 6.7 is governed by the "integrodifferential" equation

$$L\frac{dI}{dt} + RI + \frac{1}{C}\int_{-\infty}^{t} I\, dt = E_i(t). \tag{6.69}$$

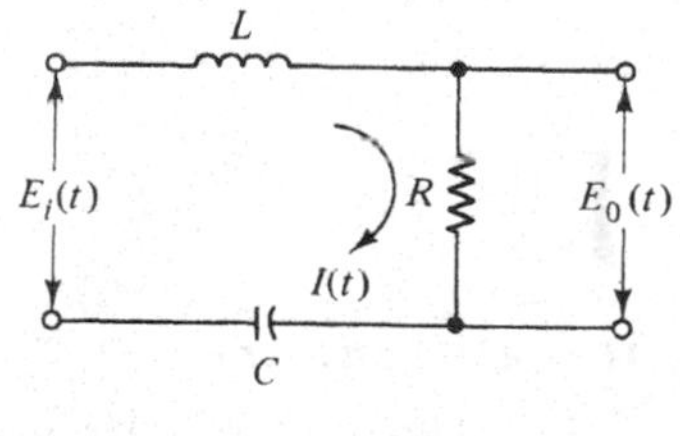

Figure 6.7. Circuit.

Fourier transforming (6.69), show that

$$E_0(t) = I(t)R = \int_{-\infty}^{\infty} E_i(t-\tau)\mathbf{F}^{-1}\left(\frac{i\omega RC}{1 + i\omega RC - \omega^2 LC}\right) d\tau.$$

Show that we obtain the same result if we first take d/dt of (6.69) in order to eliminate the integral sign.

6.10. Expanding in "partial fractions," find $\alpha, \beta, \gamma, \delta$ so that

$$\frac{i\omega RC}{1 + i\omega RC - \omega^2 LC} = \frac{\alpha}{\omega - \beta} + \frac{\gamma}{\omega - \delta}.$$

Noting from Appendix A that

$$\mathbf{F}^{-1}\left(\frac{1}{\omega - a}\right) = \begin{cases} ie^{iat}, & t > 0 \\ 0, & t < 0 \end{cases}$$

if Im $a > 0$, show that

$$\mathbf{F}^{-1}\left(\frac{i\omega RC}{1 + i\omega RC - \omega^2 LC}\right) = 0 \quad \text{for } t < 0$$

and

$$\frac{2e^{-Rt/2L}}{A\sqrt{LC}} \sin\left(\frac{RA}{2L} t - \tan^{-1} A\right) \quad \text{for } t > 0,$$

where $A = \sqrt{(4L/R^2C) - 1}$, and so the solution to Exercise 6.9 becomes

$$E_0(t) = \int_0^\infty E_i(t - \tau)\frac{2e^{-R\tau/2L}}{A\sqrt{LC}} \sin\left(\frac{RA}{2L}\tau - \tan^{-1} A\right) d\tau.$$

Hint: See the next exercise.

6.11. Show that

$$a \cos \omega t + b \sin \omega t = c \sin (\omega t + \phi),$$

where
$$c = \sqrt{a^2 + b^2} \quad \text{and} \quad \phi = \tan^{-1} \frac{a}{b}.$$

6.12. If $f(x)$ is even (or if it is only defined over $0 < x < \infty$, in which case we can extend it so as to be even), show that (6.2) and (6.3) become

$$\hat{f}(\omega) = 2 \int_0^\infty f(x) \cos \omega x \, dx$$

$$f(x) = \frac{1}{\pi} \int_0^\infty \hat{f}(\omega) \cos \omega x \, d\omega$$

This is called the **Fourier cosine transform** of f and its inverse. Similarly, obtain the **Fourier sine transform** and its inverse

$$\hat{f}(\omega) = 2 \int_0^\infty f(x) \sin \omega x \, dx$$

$$f(x) = \frac{1}{\pi} \int_0^\infty \hat{f}(\omega) \sin \omega x \, d\omega.$$

The Laplace Transform

6.13. Which of the following functions satisfy (6.25)? For those that don't, explain why not.

(a) t^{100}
(b) $\sin (e^{t^2})$; How about its derivative?
(c) te^t
(d) $e^t \ln t$
(e) $te^t \ln t$
(f) $\sin (\alpha t + \beta)$
(g) $[H(t - 1) - H(t - 2)]e^{t^2}$

6.14. Show that if $f_1(t)$ and $f_2(t)$ are of exponential order, then $f_1(t) + f_2(t)$ and $f_1(t)f_2(t)$ are, too.

6.15. Compute the Laplace transforms of the following functions directly; where possible, compare with Table A.2.

(a) t^a
(b) e^{at}
(c) $\sin at$
(d) $\cos at$

(e) $\sinh at$ (f) $\cosh at$

(g) $tH(t-1)$ (h) $\delta(t-a)$

*(i) Si (t) *Hint:* Recall the definition (2.63) and see Exercise 6.16(g).

6.16. **(Additional properties of the Laplace transform)** Provide formal derivations of the following properties.

(a)

$$\mathbf{L}[(-t)^n f(t)] = \bar{f}^{(n)}(s);$$

for example,

$$\mathbf{L}[tf(t)] = -\frac{d\bar{f}}{ds}.$$

(Note that this was used in Example 6.5.)

(b) t shift:

$$\mathbf{L}^{-1}[e^{-as}\bar{f}(s)] = H(t-a)f(t-a).$$

(c) s shift:

$$\mathbf{L}^{-1}[\bar{f}(s+a)] = e^{-at}f(t).$$

(d)

$$\mathbf{L}^{-1}\frac{\bar{f}(s)}{s} = \int_0^t f(\tau)\,d\tau.$$

(e)

$$\mathbf{L}\frac{f(t)}{t} = \int_s^\infty \bar{f}(s)\,ds.$$

(f) If $f(t)$ is periodic with period a over $0 \le t < \infty$, then

$$\bar{f}(s) = \frac{1}{1-e^{-as}}\int_0^a f(t)e^{-st}\,dt.$$

(g)

$$\lim_{s\to\infty} s\bar{f}(s) = \lim_{t\to 0+} f(t) = f(0+), \tag{6.70}$$

where $s \to \infty$ through real values. *Hint:* Recalling that $\mathbf{L}[f'(t)] = s\bar{f}(s) - f(0+)$, take the limit as $s \to \infty$ and note that $\mathbf{L}[f'(t)] \to 0$, since

$$\begin{aligned}\left|\int_0^\infty f'(t)e^{-st}\,dt\right| &\le \int_0^\infty \left|f'(t)e^{-st}\right| dt\\ &= \int_0^\infty \left|f'(t)\right| e^{-st}\,dt \qquad (s \text{ is real here})\\ &< K\int_0^\infty e^{-(s-c)t}\,dt = \frac{K}{s-c}\end{aligned}$$

[if conditions (6.25) are required of f']. As a companion result to (6.70), we state (but you need not prove) that if $\bar{f}(s)$ is bounded for all Re $s \ge 0$, then

$$\lim_{s\to 0} s\bar{f}(s) = \lim_{t\to\infty} f(t). \tag{6.71}$$

6.17. Evaluate C in (6.57) directly, using (6.70).

6.18. Invert the following transforms, using the Laplace Transform Table A.2, together with properties L1 to L3 and (a) to (f) of Exercise 6.16. (Do not use partial fractions.)

*(a) $\dfrac{e^{-s}}{s(s+3)^{3/2}}$ (b) $\dfrac{1}{s^2+2s+5}$

(c) $\dfrac{1}{s(s^2+4)}$ (d) $\dfrac{e^{-\pi s}+e^{-2\pi s}}{s^2+1}$ [Sketch $f(t)$.]

6.19. (*Abel's integral equation*)

(a) Suppose that a bead of mass m, starting from rest at time $t = 0$, slides down a frictionless slope under the action of gravity. If the initial height is x, the height at any $t > 0$ is $\xi(t)$, and $s(t)$ is the arc length traveled, then from conservation of energy the speed is $\dot{s}(t) = \sqrt{2g(x-\xi)}$, independent of the shape of the curve. But the *time* of descent T does depend on the shape; in particular, show that

$$T = -\frac{1}{\sqrt{2g}}\int_0^x \frac{s'(\xi)}{\sqrt{x-\xi}}\,d\xi. \tag{6.72}$$

Regarding T as prescribed, this is an *integral equation* on s', a *singular* integral equation because the integral is singular (due to the singularity in the *kernel* $1/\sqrt{x-\xi}$ at $\xi = x$). Studied by Abel, it is known as **Abel's integral equation.** In fact, Abel also considered the more general case

$$f(x) = \int_0^x \frac{y(\xi)}{(x-\xi)^\alpha}\,d\xi \qquad (0 < \alpha < 1). \tag{6.73}$$

(b) Observing that the integral in (6.73) is of Laplace convolution form, show that

$$\bar{y}(s) = \frac{s^{1-\alpha}\bar{f}(s)}{\Gamma(1-\alpha)}.$$

Inverting it by convolution is impossible, since $\mathbf{L}^{-1}(s^{1-\alpha})$ does not exist. To remedy the situation, divide both sides by s and obtain the result

$$y(x) = \frac{1}{\Gamma(\alpha)\Gamma(1-\alpha)}\frac{d}{dx}\int_0^x \frac{f(\xi)}{(x-\xi)^{1-\alpha}}\,d\xi. \tag{6.74}$$

6.20. Recalling the definition (6.41), show that $f * g = g * f$.

6.21. The deflection $u(x)$ of the "simply supported" beam shown in Fig. 6.8, due to the point load P at midspan, is governed by the system

$$EIu'''' = P\,\delta\left(x - \frac{l}{2}\right)$$

$$u(0) = u''(0) = u(l) = u''(l) = 0,$$

where EI is constant. Solve for $u(x)$ by the Laplace transform.

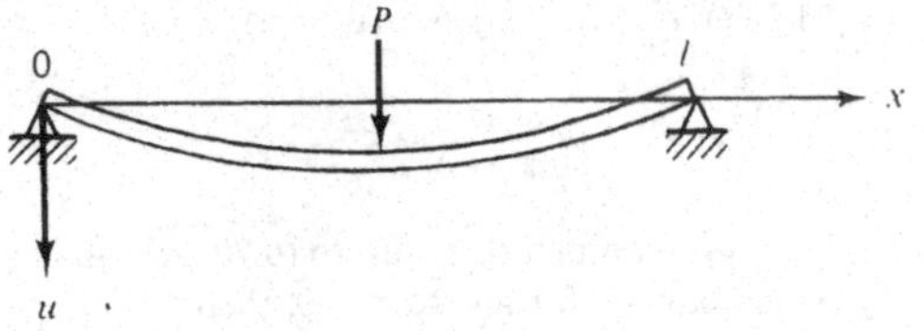

Figure 6.8. Simply supported beam.

6.22. Find the Laplace transform of the following periodic functions, using property (f) of Exercise 6.16.

(a) $f(t) = |\sin t|$

(b) $f(t) = t, \quad 0 \le t < 2$
$f(t+2) = f(t), \quad t > 0$

(c) $f(t) = \begin{cases} 1, & 0 \le t < 1 \\ 0, & 1 \le t < 3 \end{cases}$

$f(t+3) = f(t), \qquad t > 0$

6.23. (a) The mass m in Fig. 6.9 is initially at rest. At $t = 0$ a unit impulsive force is applied. Thus $m\ddot{x} + kx = \delta(t)$, subject to the initial conditions $x(0) = \dot{x}(0) = 0$. Solve for $x(t)$ by Laplace transform. Is $\dot{x}(t)$ continuous at $t = 0$? Comment.

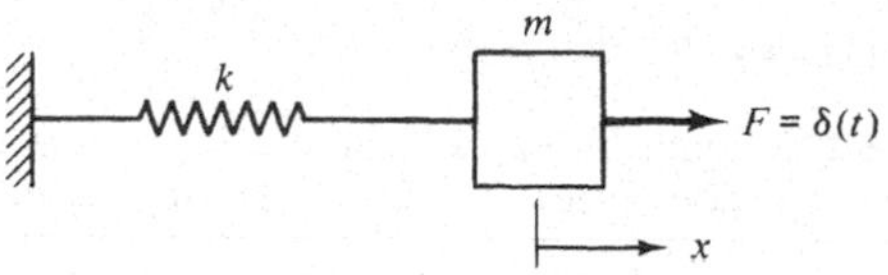

Figure 6.9. Impulsive excitation.

[*Note:* To avoid any confusion about $\delta(t)$ acting *at* the endpoint $t = 0$, consider it to act at $t = 0+$.]

(b) Find $x(t)$ if $F = H(t)$ instead.

(c) Show that the response $x(t)$ due to the unit impulse is the derivative of the response due to the unit step function.

6.24. If $\bar{f}(s) = P(s)/[(s - a_1)(s - a_2) \cdots (s - a_n)]$, where $P(s)$ is a polynomial of degree $m < n$, so that we can express

$$\bar{f}(s) = \frac{A_1}{s - a_1} + \cdots + \frac{A_n}{s - a_n}$$

[per (6.62), where $\alpha = 1$ and all the m_j's are unity; that is, where the roots are all nonrepeated], show that

$$A_j = \lim_{s \to a_j} [(s - a_j)\bar{f}(s)].$$

For example, if

$$\bar{f}(s) = \frac{s^2 + 2}{s(s - 1)(s + 3)} = \frac{A_1}{s} + \frac{A_2}{s - 1} + \frac{A_3}{s + 3},$$

then

$$A_1 = \lim_{s \to 0} \frac{s(s^2 + 2)}{s(s - 1)(s + 3)} = \frac{(2)}{(-1)(3)} = -\frac{2}{3}.$$

Finish the evaluation of A_2 and A_3 in this way.

6.25. Invert the following, using partial fractions.

(a) $\dfrac{1}{s(s^2 + 4)}$ (b) $\dfrac{1}{s^2 + 2s + 5}$

(c) $\dfrac{s}{(s^2 + 9)(s^2 + 1)}$ (d) $\dfrac{s + 2}{s^3(s - 1)}$

(e) $\dfrac{s}{(s + 1)(s + 2)^3}$ (f) $\dfrac{s^2}{s^4 - 2s^2 + 1}$

6.26. If the switch K is closed at time $t = 0$, the current $I(t)$ for the circuit of Fig. 6.10 is governed by the system $L\ddot{I} + R\dot{I} + I/C = \dot{E}_i(t)$ plus the initial conditions $I(0) = \dot{I}(0) = 0$. Regarding the applied voltage $E_i(t)$ as known, solve by Laplace transform.

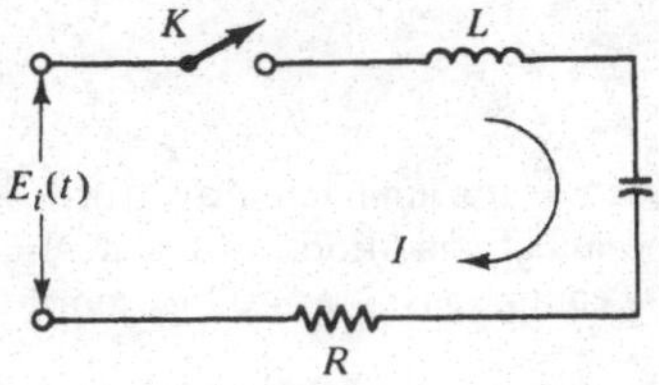

Figure 6.10. RLC series circuit.

6.27. Find, by Laplace transform, the displacements $x_1(t), x_2(t)$ due to the initial conditions $x_1(0) = x_2(0) = 1$, $\dot{x}_1(0) = \dot{x}_2(0) = 0$ (see Fig. 6.11). Let $m = k$.

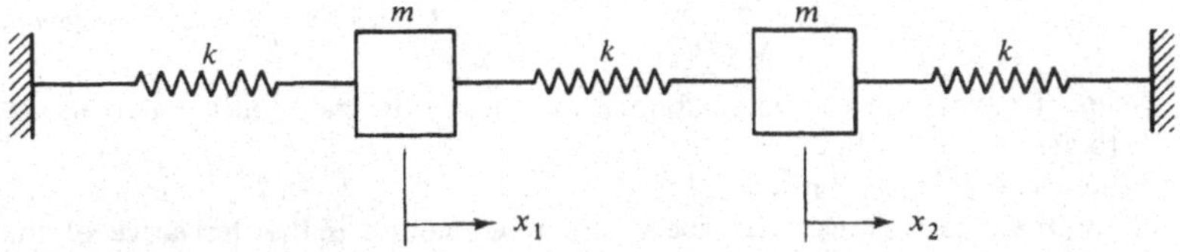

Figure 6.11. Mass-spring system.

6.28. Solve the following problems by Laplace transforms.

(a) $\ddot{x} + 2\dot{x} = H(t - 1);\ x(0) = 0,\ \dot{x}(0) = 1$

(b) $\ddot{x} + x = |\sin t|;\ x(0) = \dot{x}(0) = 0$

(c) $\dddot{x} - \dot{x} = f(t);\ x(0) = 1,\ \dot{x}(0) = \ddot{x}(0) = 0$

(d) $t\ddot{x} + (1 - t)\dot{x} = 0;\ x(0) = 1,\ \dot{x}(0) = 0$

(e) $\left.\begin{aligned} \dot{x} &= x + y \\ \dot{y} &= 4x + y \end{aligned}\right\}\ x(0) = 1,\ y(0) = 0$

(f) $\left.\begin{aligned} \dot{x} - 2y &= e^t - 4t \\ \dot{y} + x - y &= e^t - 2t + 2 \end{aligned}\right\}\ x(0) = 1,\ y(0) = 0$

(g) $t\ddot{x} + \dot{x} = 0;\ x(0) = 1,\ \dot{x}(0) = 0.$

6.29. From (6.38) show that

$$\mathbf{L}f^{(n)} = s^n\bar{f} - s^{n-1}f(0) - s^{n-2}f'(0) - s^{n-3}f''(0) - \cdots - f^{(n-1)}(0).$$

***6.30.** The interchange

$$\int_0^\infty \int_0^\infty F(t, \tau)\, d\tau\, dt = \int_0^\infty \int_0^\infty F(t, \tau)\, dt\, d\tau$$

is valid if at least one side converges absolutely—that is, if

$$\int_0^\infty \int_0^\infty |F(t, \tau)|\, d\tau\, dt < \infty \quad \text{or} \quad \int_0^\infty \int_0^\infty |F(t, \tau)|\, dt\, d\tau < \infty.$$

Using this result, verify the legitimacy of the interchange in the derivation of the convolution theorem (6.43); assume that f and g satisfy conditions (6.25). *Hint:* The limits on both τ and t will be 0 to ∞ if we insert the factor $H(t - \tau)$ into the integrand.

6.31. Verify rigorously that (6.60) does satisfy conditions (6.55).

6.32. Discuss the evaluation of $x = 2^{1/3}$, say, by the use of natural logarithms, from an "operational" point of view—that is, in terms of transform concepts.

***6.33.** The integral equation

$$C(T) = \int_0^\infty e^{-0.0744\nu^2/T^2} \rho(\nu)\, d\nu$$

is an approximate relation between the frequency spectrum $\rho(\nu)$ and the specific heat $C(T)$ of a crystal. Solve for $\rho(\nu)$ for the cases

(a) $C(T) = T$; (b) $C(T) = Te^{-1/T}$

6.34. (*Use of Laplace transform in evaluating integrals*) If

$$I = \int_0^\infty f(s)g(s)\, ds \tag{6.75}$$

and $f(s) = L\{\phi(t)\}$, then

$$I = \int_0^\infty g(s)\, ds \int_0^\infty \phi(t)e^{-st}\, dt = \int_0^\infty \phi(t)\, dt \int_0^\infty g(s)e^{-st}\, ds, \tag{6.76}$$

where (6.76) may be easier to evaluate than (6.75). Use this idea (without justifying the interchange in the order of integration) to evaluate

$$I = \int_0^\infty \frac{\sin x}{x}\, dx.$$

6.35. Make up two examination-type problems on this chapter.

Chapter 7

Functions of Several Variables

Here we extend concepts already established for functions of a single variable to functions of two or more variables. As before, both the variables and the functions are assumed to be real valued.

7.1. CHAIN DIFFERENTIATION

We will consider only two independent variables, since the extension to the case of three or more should be transparent.

If $x = x(t)$ and $y = y(t)$, then $u(x, y) = u[x(t), y(t)]$ can be regarded as a function of the single variable t; we say that u is a **composite function** of t. Recalling the "chain rule" from calculus, we expect that

$$\frac{du}{dt} = \frac{\partial u}{\partial x}\frac{dx}{dt} + \frac{\partial u}{\partial y}\frac{dy}{dt}, \tag{7.1}$$

assuming, of course, that the four derivatives on the right-hand side all exist. Let us have a closer look, however.

Example 7.1.

$$u(x, y) = \begin{cases} x & \text{on } y = x^2 \\ 0 & \text{for } y \neq x^2. \end{cases} \tag{7.2}$$

Introduce a curve C in the x, y plane according to parametric equations $x = x(t)$, $y = y(t)$, as sketched in Fig. 7.1. Then du/dt is the rate of change of u with respect to t along C. For instance, let us choose $x = 3t$ and $y = 9t^2$, so that C happens to coincide with the curve $y = x^2$, and let us compute du/dt at the origin. The simplest way to do so is to note that on C we have $u(x, y) = x = 3t$; therefore du/dt is clearly 3. Alternatively, (7.1) gives

$$\frac{du}{dt} = (0)(3) + (0)(0) = 0,$$

which is incorrect! ∎

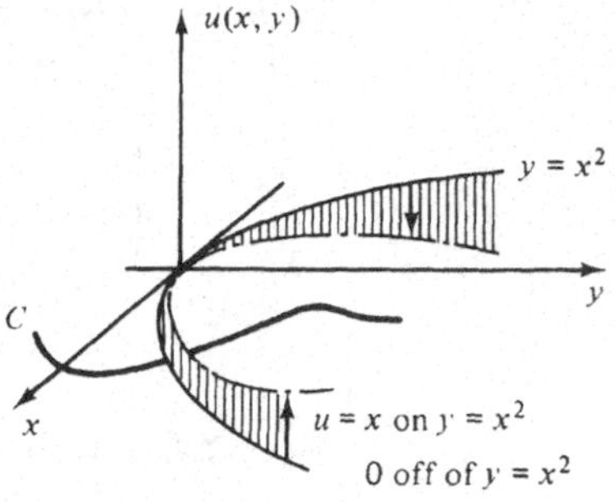

Figure 7.1. Chain differentiation, Example 7.1.

THEOREM 7.1. (The Chain Rule) If $u(x, y)$ has continuous partials u_x and u_y and x and y are differentiable functions of t, then[1]

$$\frac{du}{dt} = \frac{\partial u}{\partial x}\frac{dx}{dt} + \frac{\partial u}{\partial y}\frac{dy}{dt}. \tag{7.3}$$

The difficulty within Example 7.1 is that u_x and u_y are *not* continuous at the point in question, $x = y = 0$, since they don't even *exist* throughout an arbitrarily small disk centered at the origin; in particular, u_x and u_y do not exist at any points on the curve $y = x^2$ (except for the origin) due to the jump discontinuity in u along that curve.

Why is the requirement of continuity on u_x and u_y in Theorem 7.1 entirely reasonable? It's reasonable because, after all, (7.3) is basically an *interpolation* formula, where we are interpolating between the x and y directions, and interpolation is fundamentally dependent on continuity.

Suppose that we parameterize, in Example 7.1, according to $x(t) = t^2$, $y(t) = t^4$ instead; C again coincides with $y = x^2$. On C we have $u(x, y) = x = t^2$, so that $du/dt = 2t = 0$ at the origin. This time, however, (7.3) gives $du/dt = (0)(0) + (0)(0) = 0$, which *is* correct, even though u_x and u_y are still discontinuous at the point in question. We conclude that the conditions of the theorem are *sufficient* but not necessary.

Do you agree that $u_x = 0$ at $x = y = 0$, as we've been claiming, or, looking at (7.2), do you think it should be 1? Well,

$$u_x(0, 0) = \lim_{\Delta x \to 0} \frac{u(\Delta x, 0) - u(0, 0)}{\Delta x} = \lim_{\Delta x \to 0} \frac{0 - 0}{\Delta x} = 0!$$

Example 7.2. Express the quantity

$$u_{xx} + u_{yy} \tag{7.4}$$

[1] By "continuous" and "differentiable" we mean at the point, or points, of application of the formula, not necessarily everywhere. Incidentally, exactly what do we mean by these terms for functions of *several* variables, say $f(x_1, \ldots, x_n)$? Extending the basic idea from Chapter 1, we say that $f(x_1, \ldots, x_n)$ is *continuous* at X_0 if $\lim_{X \to X_0} f(X) = f(X_0)$; that is, if to each $\epsilon > 0$ there corresponds a $\delta(\epsilon, X_0)$ such that $|f(X) - f(X_0)| < \epsilon$ for all X such that $d(X, X_0) < \delta$, where X is shorthand for the "point" $(x_1, \ldots, x_n)$ and d is some distance function, for example the *n-dimensional Euclidean distance* $d(X, Y) = \sqrt{(x_1 - y_1)^2 + \cdots + (x_n - y_n)^2}$. Similarly for differentiability.

in terms of **polar coordinates** r, θ, which are related to the original cartesian x, y variables according to

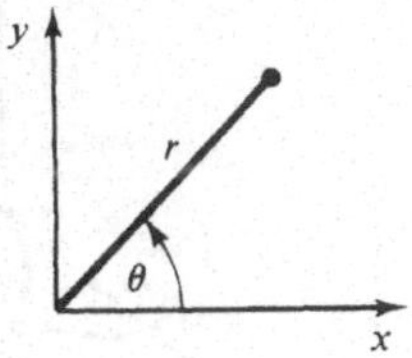

Figure 7.2. Polar variables.

$$x = r\cos\theta \tag{7.5a}$$

$$y = r\sin\theta, \tag{7.5b}$$

as shown in Fig. 7.2. First, let us transform $\partial/\partial x$ and $\partial/\partial y$. Applying the chain rule,

$$\frac{\partial}{\partial x} = \frac{\partial}{\partial r}\frac{\partial r}{\partial x} + \frac{\partial}{\partial \theta}\frac{\partial \theta}{\partial x}, \tag{7.6}$$

$$\frac{\partial}{\partial y} = \frac{\partial}{\partial r}\frac{\partial r}{\partial y} + \frac{\partial}{\partial \theta}\frac{\partial \theta}{\partial y}. \tag{7.7}$$

[It is important to keep the bookkeeping straight. There are two alternative sets of independent variables, x, y and r, θ. By $\partial(\quad)/\partial x$, for example, we mean the derivative of ( ), considered as a function of the x, y variables, with respect to x, with all the other variables in the x group held fixed—namely, y. Similarly, $\partial/\partial\theta$ means r is held fixed and so on.]

We need to know $r_x, \theta_x, r_y, \theta_y$ in (7.6) and (7.7).

Although $x_r = \cos\theta$, $x_\theta = -r\sin\theta$, $y_r = \sin\theta$, $y_\theta = r\cos\theta$ are easily obtained directly from (7.5), the "reverse" derivatives $r_x, \ldots, \theta_y$ are a little more elusive. To obtain r_x and θ_x, let (7.6) act on the functions x and y, say, in turn:

$$\frac{\partial x}{\partial x} = 1 = \frac{\partial x}{\partial r}\frac{\partial r}{\partial x} + \frac{\partial x}{\partial \theta}\frac{\partial \theta}{\partial x} = cr_x - rs\theta_x \tag{7.8a}$$

$$\frac{\partial y}{\partial x} = 0 = \frac{\partial y}{\partial r}\frac{\partial r}{\partial x} + \frac{\partial y}{\partial \theta}\frac{\partial \theta}{\partial x} = sr_x + rc\theta_x, \tag{7.8b}$$

where c and s are shorthand for $\cos\theta$ and $\sin\theta$, respectively. Equations (7.8) are two simultaneous equations in the desired quantities r_x and θ_x. Solving,

$$r_x = \cos\theta \tag{7.9}$$

$$\theta_x = -\frac{\sin\theta}{r}. \tag{7.10}$$

Thus[2]

$$\begin{aligned} u_{xx} &= \left(\frac{\partial}{\partial x}\right)\left(\frac{\partial}{\partial x}\right)u = \left(c\frac{\partial}{\partial r} - \frac{s}{r}\frac{\partial}{\partial\theta}\right)\left(c\frac{\partial}{\partial r} - \frac{s}{r}\frac{\partial}{\partial\theta}\right)u \\ &= \left(c\frac{\partial}{\partial r} - \frac{s}{r}\frac{\partial}{\partial\theta}\right)\left(cu_r - \frac{s}{r}u_\theta\right) \end{aligned} \tag{7.11}$$

$$u_{xx} = c^2u_{rr} + \frac{cs}{r^2}u_\theta - \frac{cs}{r}u_{\theta r} + \frac{s^2}{r}u_r - \frac{sc}{r}u_{r\theta} + \frac{sc}{r^2}u_\theta + \frac{s^2}{r^2}u_{\theta\theta}. \tag{7.12}$$

To obtain r_y and θ_y, we let (7.7) act on the functions x and y,

$$\frac{\partial x}{\partial y} = 0 = \frac{\partial x}{\partial r}\frac{\partial r}{\partial y} + \frac{\partial x}{\partial \theta}\frac{\partial \theta}{\partial y} = cr_y - rs\theta_y \tag{7.13a}$$

$$\frac{\partial y}{\partial y} = 1 = \frac{\partial y}{\partial r}\frac{\partial r}{\partial y} + \frac{\partial y}{\partial \theta}\frac{\partial \theta}{\partial y} = sr_y + rc\theta_y. \tag{7.13b}$$

---

[2]Note that (7.6) really means $\partial(\quad)/\partial x = [\partial(\quad)/\partial r]r_x + [\partial(\quad)/\partial\theta]\theta_x$; that is, the $\partial/\partial r$ and $\partial/\partial\theta$ act only on ( ) and *not* on their respective neighbors r_x and θ_x. We sometimes emphasize this point by writing (7.6) as $\partial/\partial x = r_x\partial/\partial r + \theta_x\partial/\partial\theta$ instead, as in (7.11). Similarly for (7.7).

Solving, we have

$$r_y = \sin\theta \tag{7.14}$$

$$\theta_y = \frac{\cos\theta}{r}. \tag{7.15}$$

Then

$$u_{yy} = \left(\frac{\partial}{\partial y}\right)\left(\frac{\partial}{\partial y}\right)u = \text{etc.}$$

$$= s^2 u_{rr} - \frac{sc}{r^2}u_\theta + \frac{sc}{r}u_{\theta r} + \frac{c^2}{r}u_r + \frac{cs}{r}u_{r\theta} - \frac{cs}{r^2}u_\theta + \frac{c^2}{r^2}u_{\theta\theta}. \tag{7.16}$$

Adding (7.12) and (7.16) gives

$$\frac{\partial^2 u}{\partial x^2} + \frac{\partial^2 u}{\partial y^2} = \frac{\partial^2 u}{\partial r^2} + \frac{1}{r}\frac{\partial u}{\partial r} + \frac{1}{r^2}\frac{\partial^2 u}{\partial \theta^2} \tag{7.17}$$

or in terms of the differential operator

$$\frac{\partial^2}{\partial x^2} + \frac{\partial^2}{\partial y^2} = \frac{\partial^2}{\partial r^2} + \frac{1}{r}\frac{\partial}{\partial r} + \frac{1}{r^2}\frac{\partial^2}{\partial \theta^2}, \tag{7.18}$$

which is the well-known **Laplace operator**, or **Laplacian**, expressed in both cartesian and polar coordinates; we'll be seeing much more of it later on.

COMMENT 1. In fact, r_x, θ_x, r_y, and θ_y could have been determined more easily by solving (7.5) for $r(x, y)$ and $\theta(x, y)$: squaring and adding give $r = \sqrt{x^2 + y^2}$, and dividing gives $\theta = \tan^{-1}(y/x)$. Thus $r_x = x/\sqrt{x^2 + y^2} = (r\cos\theta)/r = \cos\theta$, and so on. Had the $x = x(r, \theta)$, $y = y(r, \theta)$ relations been more complicated, however, we might not have been able to solve for $r(x, y)$ and $\theta(x, y)$, and we might then have been glad to fall back on the more systematic procedure demonstrated above.

COMMENT 2. Observe that $\partial x/\partial r$ and $\partial r/\partial x$, for instance, are *not* algebraic reciprocals of each other the way that ordinary derivatives are; that is, $\partial x/\partial r = \cos\theta$ and $\partial r/\partial x = \cos\theta$ (identical by coincidence), *not* $1/\cos\theta$. This situation occurs because $\partial x/\partial r$ is computed with θ fixed (i.e., the other member of the r, θ group), whereas $\partial r/\partial x$ is computed with y fixed (i.e., the other member of the x, y group), as displayed in Fig. 7.3. Since different triangles are involved, there's no reason why $\Delta x/\Delta r$ and $\Delta r/\Delta x$ should be reciprocals of each other (see Exercise 7.18, however).

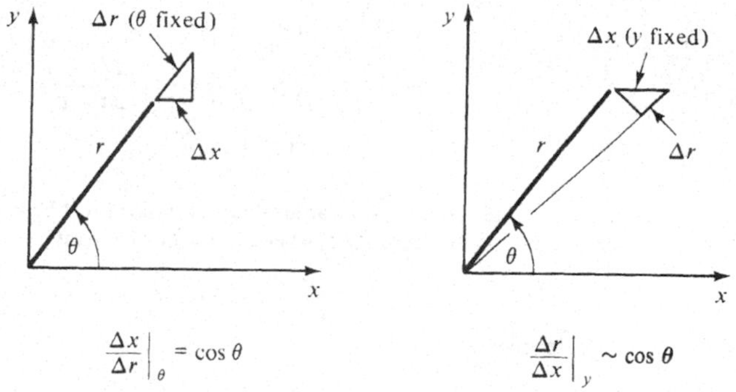

$\left.\dfrac{\Delta x}{\Delta r}\right|_\theta = \cos\theta$ $\qquad$ $\left.\dfrac{\Delta r}{\Delta x}\right|_y \sim \cos\theta$

Figure 7.3. Graphical interpretation of $\frac{\partial x}{\partial r}$ and $\frac{\partial r}{\partial x}$.

On the other hand, consider the equation of state of an ideal gas, $pv = RT$, where p, v, T are the pressure, volume, and absolute temperature and R is a physical constant. In this case,

$$\frac{\partial p}{\partial T} = \frac{R}{v} \quad \text{and} \quad \frac{\partial T}{\partial p} = \frac{v}{R},$$

for instance, *are* reciprocals. The reason that they are is that both partials are computed with the *same* variable (v) held constant:

$$\frac{\partial p}{\partial T} = \frac{1}{\partial T/\partial p}.$$

(As soon as we write down $\partial p/\partial T$, say, it's implied that we are regarding p as the dependent variable and T, v as independent. Similarly, $\partial T/\partial p$ implies that T is to be regarded as a function of p and v.) ▮

Generalizing a bit, equations (7.5) are of the form

$$F(x, y, u, v) = 0 \tag{7.19a}$$

$$G(x, y, u, v) = 0 \tag{7.19b}$$

For example, (7.5a) is equivalent to $F = x - u \cos v = 0$. In (7.19) x, y, u, v are all independent variables. However, we may wish to change our point of view and regard u and v as functions of x and y, say, as defined *implicitly* by (7.19). How might we compute the partials u_x, u_y, v_x, v_y? Well,

$$\frac{\partial}{\partial x} F[x, y, u(x, y), v(x, y)] = F_x + F_u u_x + F_v v_x = 0 \tag{7.20a}$$

$$\frac{\partial}{\partial x} G[x, y, u(x, y), v(x, y)] = G_x + G_u u_x + G_v v_x = 0 \tag{7.20b}$$

and solving in terms of *determinants*,[3]

$$u_x = -\frac{\begin{vmatrix} F_x & F_v \\ G_x & G_v \end{vmatrix}}{\begin{vmatrix} F_u & F_v \\ G_u & G_v \end{vmatrix}} = -\frac{J(x, v)}{J(u, v)}, \tag{7.21a}$$

$$v_x = -\frac{\begin{vmatrix} F_u & F_x \\ G_u & G_x \end{vmatrix}}{\begin{vmatrix} F_u & F_v \\ G_u & G_v \end{vmatrix}} = -\frac{J(u, x)}{J(u, v)}, \tag{7.21b}$$

---

[3]Determinants and simultaneous linear algebraic equations will be discussed in depth in Part III. For the present, let us merely define the so-called 2×2 determinant

$$\begin{vmatrix} a & b \\ c & d \end{vmatrix} \equiv ad - bc$$

and the 3×3 determinant

$$\begin{vmatrix} a & b & c \\ d & e & f \\ g & h & i \end{vmatrix} \equiv a(ei - fh) - b(di - fg) + c(dh - eg)$$

where the J determinants are known as **Jacobians.** For instance, $J(x, v)$ is *the Jacobian of F and G with respect to x and v.* The alternative notations

$$\frac{\partial(F, G)}{\partial(x, v)} \quad \text{and} \quad J\left(\frac{F, G}{x, v}\right)$$

contain reference to F and G and are sometimes preferred.

Similarly,

$$u_y = -\frac{J(y, v)}{J(u, v)} \quad \text{and} \quad v_y = -\frac{J(u, y)}{J(u, v)}. \tag{7.22}$$

7.2. TAYLOR SERIES IN TWO OR MORE VARIABLES

Suppose that we know "all about" some function $f(x, y)$ at a point (a, b) of the x, y plane; that is, we know all the values $f(a, b)$, $f_x(a, b)$, $f_y(a, b)$, $f_{xx}(a, b)$, $f_{xy}(a, b)$, $f_{yy}(a, b)$, and so on. Based on these data, can we predict the value of f at some other point, say x_0, y_0? That is, can we extrapolate away from (a, b) throughout some or all the x, y plane?

Having already explored this question for functions $f(x)$ of a *single* variable, it is natural to try to reduce the two-dimensional case to one dimension; if we can do so, we're "home free." We can accomplish it by running a curve C from (a, b) to (x_0, y_0). Defining C parametrically by $x = x(t)$, $y = y(t)$, we then have $f(x, y) = f[x(t), y(t)] \equiv F(t)$, say, a function of the single variable t. This line of approach is left for Exercise 7.15.

---

and note that the solution of

$$\begin{aligned} ax + by &= p \\ cx + dy &= q \end{aligned}$$

is given by

$$x = \frac{\begin{vmatrix} p & b \\ q & d \end{vmatrix}}{D}, \qquad y = \frac{\begin{vmatrix} a & p \\ c & q \end{vmatrix}}{D},$$

where

$$D = \begin{vmatrix} a & b \\ c & d \end{vmatrix},$$

provided that $D \neq 0$, and that the solution of

$$\begin{aligned} ax + by + cz &= p \\ dx + ey + fz &= q \\ gx + hy + iz &= r \end{aligned}$$

is

$$x = \frac{\begin{vmatrix} p & b & c \\ q & e & f \\ r & h & i \end{vmatrix}}{D}, \qquad y = \frac{\begin{vmatrix} a & p & c \\ d & q & f \\ g & r & i \end{vmatrix}}{D}, \qquad z = \frac{\begin{vmatrix} a & b & p \\ d & e & q \\ g & h & r \end{vmatrix}}{D},$$

where

$$D = \begin{vmatrix} a & b & c \\ d & e & f \\ g & h & i \end{vmatrix},$$

provided again that $D \neq 0$.

Instead note that we can reduce the problem to one dimension, alternatively, simply by expanding in one variable at a time. Expanding in x first (holding y fixed),

$$f(x, y) = f(a, y) + f_x(a, y)(x - a) + \frac{f_{xx}(a, y)}{2!}(x - a)^2 + \cdots. \tag{7.23}$$

Next, we expand the coefficients $f(a, y), f_x(a, y), \ldots$ in terms of y,

$$\begin{aligned} f(x, y) = &\left[f(a, b) + f_y(a, b)(y - b) + \frac{f_{yy}(a, b)}{2!}(y - b)^2 + \cdots\right] \\ &+ [f_x(a, b) + f_{xy}(a, b)(y - b) + \cdots](x - a) \\ &+ [f_{xx}(a, b) + \cdots]\frac{(x - a)^2}{2!} + \cdots. \end{aligned} \tag{7.24}$$

Finally, we arrange the terms according to ascending degree,[4]

$$\begin{aligned} f(x, y) = &f(a, b) + f_x(a, b)(x - a) + f_y(a, b)(y - b) \\ &+ \frac{f_{xx}(a, b)}{2!}(x - a)^2 + \frac{2f_{xy}(a, b)}{2!}(x - a)(y - b) \\ &+ \frac{f_{yy}(a, b)}{2!}(y - b)^2 + \cdots, \end{aligned} \tag{7.25}$$

which, it turns out, is expressible as

$$f(x, y) = \sum_{m=0}^{\infty} \sum_{n=0}^{\infty} \frac{1}{m!\, n!} \frac{\partial^{m+n} f}{\partial x^m\, \partial y^n}(a, b)(x - a)^m (y - b)^n. \tag{7.26}$$

Example 7.3. Expand $f(x, y) = y^x$ about (1, 1), up to and including second-degree terms. Instead of simply plugging into (7.26), it may be more informative to proceed by expanding in one variable at a time, as mentioned above. Since y^x is a simpler function of y than of x, let us expand in y first, about $y = 1$, with x considered as *fixed.*

$$\begin{aligned} f(x, y) &= f(x, 1) + f_y(x, 1)(y - 1) + \frac{f_{yy}(x, 1)}{2!}(y - 1)^2 + \cdots \\ y^x &= [1]^x + [x \cdot 1^{x-1}](y - 1) + [x(x - 1) \cdot 1^{x-2}]\frac{(y - 1)^2}{2} + \cdots \\ &= [1] + [x](y - 1) + [x(x - 1)]\frac{(y - 1)^2}{2} + \cdots. \end{aligned} \tag{7.27}$$

The coefficients of this expansion (in the square brackets) are functions of x; we now expand each of them about $x = 1$. Since we plan to go only as far as second degree and $(y - 1)^2$ is already second degree, we need $[x(x - 1)]$ only through *zeroth* degree; similarly, we need $[x]$ through first degree and [1] through second degree. Thus

$$x(x - 1) = 0 + \cdots, \qquad x = 1 + (x - 1), \qquad 1 = 1,$$

so that (7.27) becomes

$$y^x = 1 + [1 + (x - 1)](y - 1) + \cdots.$$

Finally, we rearrange so that the terms are in ascending degree:

$$y^x = 1 + (y - 1) + (x - 1)(y - 1) + \cdots.$$

[4] $(x - a)^m (y - b)^n$ is said to be of degree (or "order") $m + n$.

We do *not* "simplify" per

$$y^x = 1 + y - 1 + xy - x - y + 1 + \cdots = 1 - x + xy + \cdots;$$

in expanding about (a, b), we leave the quantities $(x - a)$ and $(y - b)$ intact! ▮

You may have noticed that in going from (7.24) to (7.25), we tacitly assumed that $f_{yx}(a, b) = f_{xy}(a, b)$, and similarly for higher-order partials, such as f_{xxy} and f_{xyx}. That this assumption is not necessarily true should come as no surprise in view of our discussion of limit interchange in Chapter 4. It can be shown, for instance, that *if f_x, f_y, and f_{xy} are continuous in some neighborhood of a, b, then we do have $f_{yx} = f_{xy}$ there*, and similar statements can be made about mixed partials of higher order as well.[5] In engineering applications, however, our functions turn out to be sufficiently well behaved with such regularity that we generally don't worry about the order of differentiation—although you may worry if you wish.

EXERCISES

7.1. Defining $u(x, y) = xy/(x^2 + y^2)$ except at $x = y = 0$, where it is defined to be zero, show that $u(x, y)$ is discontinuous at $x = y = 0$. Nevertheless, $u(x, y)$ is a continuous function of x for all x and all y, and a continuous function of y for all x and all y. Explain how this fact is compatible with the fact that u is discontinuous at the origin.

7.2. Defining $u(x, y) = (x^3 - y^3)/(x^2 + y^2)$ except at $x = y = 0$, where it is defined to be zero, show that $u_x(x, y)$ exists at $x = y = 0$ (Evaluate it.) but is not continuous there.

7.3. (a) Verify that $u_{xy} = u_{yx}$ for all values of x and y, for the cases $u = e^{x-2\sin y}$ and $u = 4x \sin x^3 y$. *Note:* By u_{xy} we mean $(u_x)_y$.

(b) Given that $u(x, y) = xy^3/(x^2 + y^2)$ for $x^2 + y^2 \neq 0$, and $u(0, 0) = 0$, compute $u_{xy}(0, 0)$ and $u_{yx}(0, 0)$.

7.4. The *second law of thermodynamics* may be expressed in differential form as $T\,ds = dh - v\,dp$. Since $s = s(p, T)$ and $h = h(p, T)$, we have $ds = s_T\,dT + s_p\,dp$ and $dh = h_T\,dT + h_p\,dp$. Thus show that $s_T = h_T/T$ and $s_p = (h_p - v)/T$. Eliminating s between these two equations, show that $h_{pT} = T[h_p - v)/T]_T$. So for the case of a *perfect gas* (i.e., where $pv = RT$ with R a constant), show that h is, in fact, a function of T alone; that is, $h = h(T)$.

7.5. Show that $u = f(x + at) + g(x - at)$ satisfies the partial differential ("wave") equation $a^2 u_{xx} = u_{tt}$, where a is a constant and f and g are arbitrary twice-differentiable functions. With $a = \sqrt{-1} = i$, it follows that $u = f(x + iy) + g(x - iy)$ satisfies the partial differential ("Laplace") equation $u_{xx} + u_{yy} = 0$.

7.6. The differential equation $xy'' + y' + xy = 0$, known as **Bessel's equation of order zero,** has the general solution $y = AJ_0(x) + BY_0(x)$, where J_0 is the **Bessel function of the first kind and order zero** and Y_0 is the **Bessel function of the second kind and order zero.**

(a) Solve $xy'' + y' + k^2xy = 0$ in terms of Bessel functions. *Hint:* Set $x = \alpha t$ and choose α so that the new differential equation is the Bessel equation.

(b) Solve $xy'' + y' + k^2y = 0$ in terms of Bessel functions. *Hint:* Set $x = \alpha t^\beta$.

7.7. $f(x_1, \ldots, x_n)$ is said to be **homogeneous of degree k** if

$$f(\lambda x_1, \ldots, \lambda x_n) = \lambda^k f(x_1, \ldots, x_n). \tag{7.28}$$

---

[5]T. M. Apostol, *Mathematical Analysis*, Addison-Wesley, Reading, Mass., 1957, pp. 120–123.

(a) Are the functions $f = x^2 + 3xy, f = \ln(x^2 + y^2), f = (x^2 - xy)/(2x + y), f = x^2 e^{x/2y}$ homogeneous? Of what degree?

(b) Show that if $f(x, y, z)$ is homogeneous of degree k, then $\partial f/\partial x$, $\partial f/\partial y$, $\partial f/\partial z$ are homogeneous of degree $k - 1$.

(c) (*Euler's theorem on homogeneous functions*) Show that if $f(x_1, \ldots, x_n)$ is homogeneous of degree k, then

$$x_1 \frac{\partial f}{\partial x_1} + \cdots + x_n \frac{\partial f}{\partial x_n} = kf.$$

Hint: Differentiate (7.28) with respect to λ and then set $\lambda = 1$. *Note:* There's an important point of notation to straighten out. By $f_x(\lambda x, \lambda y)$, say, we mean $f_x(x, y)$ with x and y *then* replaced by λx and λy, respectively. For instance, if $f(x, y) = x^2 y + 3y$, then $f_x(\lambda x, \lambda y) = 2(\lambda x)(\lambda y) = 2\lambda^2 xy$.

(d) Verify Euler's theorem for the case

$$f = \sqrt{x^4 + y^4} \sin^{-1} \frac{y}{2x}.$$

7.8. **(Similarity transformation)** The one-dimensional diffusion in an isotropic medium, where the diffusion coefficient is a function of the concentration C, is governed by the (nonlinear) partial differential equation

$$\frac{\partial C}{\partial t} = \frac{\partial}{\partial x}\left[D(C)\frac{\partial C}{\partial x}\right]. \tag{7.29}$$

Suppose that the x, t dependence of a solution $C(x, t)$ is such that C is actually a function of some combination of x and t, say $x/\sqrt{t}$; then with $\eta \equiv x/\sqrt{t}$, we have $C = C(\eta)$, which is fortunate because of the reduction from two independent variables to one. Are there such combinations? Let us try, for example, the form $\eta = xt^\alpha$, where α is at our disposal. Then

$$\frac{\partial C}{\partial x} = \frac{\partial C[\eta(x, t)]}{\partial x} = C'(\eta)\frac{\partial \eta}{\partial x} = t^\alpha C',$$

and similarly for C_t and $[D(C)C_x]_x$ in (7.29). Thus show that (7.29) becomes $\alpha\eta C' = t^{2\alpha+1}(D'C'^2 + DC'')$, so that x and t can be suppressed, in favor of η, if we choose $2\alpha + 1 = 0$, $\alpha = -\frac{1}{2}$.[6] Then (7.29) reduces to the *ordinary* differential equation $(D'C'^2 + DC'') = -\eta C'/2$, where $\eta = x/\sqrt{t}$. (Of course, there may also be boundary and initial conditions attached to (7.29), which we've ignored. More about this in Part V.)

7.9. **(Von Mises transformation)** With ν the constant kinematic viscosity and the dp/dx term omitted, the equations governing the boundary layer on a semi-infinite flat plate (Fig. 7.4) are

$$uu_x + vu_y = \nu u_{yy} \qquad (x \text{ momentum})$$

$$u_x + v_y = 0 \qquad (\text{continuity})$$

With the "stream function" ψ defined by $u = \psi_y$, $v = -\psi_x$, so that the continuity equation is automatically satisfied, the Von Mises transformation consists of replacing the independent variable y by ψ. Thus $u(x, y) = U(x, \psi)$, say, so that $u_x = U_x + U_\psi \psi_x = U_x - vU_\psi$, and similarly for u_y and u_{yy}. Thus show that the x-mo-

[6] D' means dD/dC. By a prime, we will *always* mean differentiation with respect to the argument, whatever the argument happens to be. In the present case, D is a function of C, and so D' must be dD/dC.

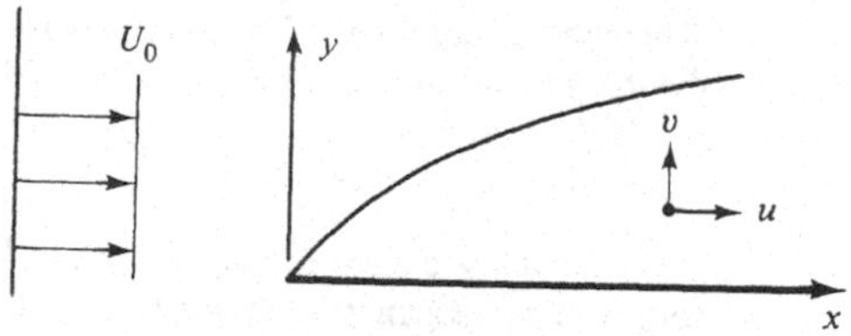

Figure 7.4. Boundary layer on flat plate.

mentum equation becomes

$$\frac{\partial U}{\partial x} = \frac{\partial}{\partial \psi}\left[\nu U \frac{\partial U}{\partial \psi}\right],$$

which is of the type (7.29) with $D(U) = \nu U$. The transformations referred to Exercises 7.8, 7.9, and 7.23, as well as many other techniques of this type, are discussed in the book by W. F. Ames, *Nonlinear Partial Differential Equations in Engineering*, Academic, New York, 1965.

7.10. If $u = x^2 + y^2$ and $v = x + y$, evaluate u_x, x_u, x_v and x_{uu}.

7.11. If $u = u(x, y, z)$, and x, y, z are given in terms of **spherical polar coordinates** ρ, θ, ϕ (Fig. 7.5), according to

$$x = \rho \sin\theta \cos\phi, \qquad y = \rho \sin\theta \sin\phi, \qquad z = \rho\cos\theta, \tag{7.30}$$

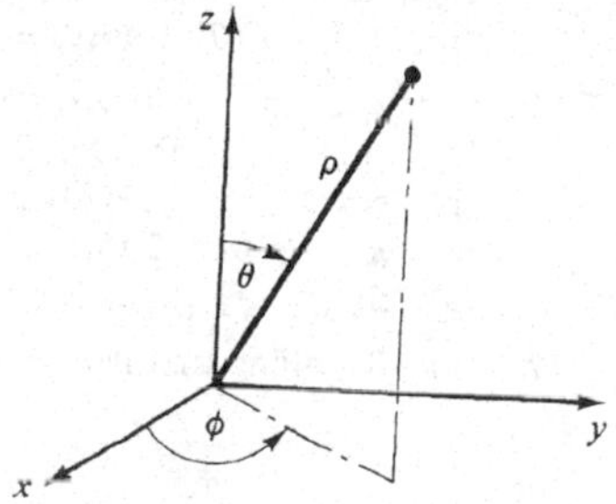

Figure 7.5. Spherical polar coordinates.

(a) show that

$$\frac{\partial^2 u}{\partial x^2} + \frac{\partial^2 u}{\partial y^2} + \frac{\partial^2 u}{\partial z^2} = \frac{1}{\rho^2 \sin\theta}\left[\frac{\partial}{\partial \rho}\left(\rho^2 \sin\theta \frac{\partial u}{\partial \rho}\right) + \frac{\partial}{\partial \theta}\left(\sin\theta \frac{\partial u}{\partial \theta}\right) + \frac{\partial}{\partial \phi}\left(\frac{1}{\sin\theta}\frac{\partial u}{\partial \phi}\right)\right] \tag{7.31}$$

(b) show that the Jacobian

$$\frac{\partial(x, y, z)}{\partial(\rho, \theta, \phi)} = \begin{vmatrix} x_\rho & x_\theta & x_\phi \\ y_\rho & y_\theta & y_\phi \\ z_\rho & z_\theta & z_\phi \end{vmatrix} \text{ is } = \rho^2 \sin\theta.$$

7.12. Consider the change of variables $u = u(x, y)$, $v = v(x, y)$, subject to the restrictions $u_x = v_y$ and $u_y = -v_x$ (the so-called **Cauchy-Riemann conditions**). Show that under this change of variables

$$f_{xx} + f_{yy} = (u_x^2 + u_y^2)(f_{uu} + f_{vv}).$$

(This fact is actually the key to the *conformal mapping* technique of complex variable theory, as we will see in Part II.) *Hint:* Note that

$$f_{xx} = \left(u_x \frac{\partial}{\partial u} + v_x \frac{\partial}{\partial v}\right)\left(u_x \frac{\partial}{\partial u} + v_x \frac{\partial}{\partial v}\right) f,$$

which expands to eight terms, and similarly for f_{yy}. Quantities like $(\partial/\partial u)(u_x)$ that arise may well be puzzling. It is *not* true that $(u_x)_u = (u_u)_x = (1)_x = 0$ [e.g., if $u = x^{1/2}$, then $u_x = 1/(2x^{1/2}) = 1/(2u)$, so that $(u_x)_u = -1/(2u^2) \neq 0$.] Instead note that

$$\frac{\partial}{\partial u}(u_x)u_x + \frac{\partial}{\partial v}(u_x)v_x = u_{xx},$$

and similarly for various other combinations of such terms.

7.13. If $u = e^x \cos y$ and $v = e^x \sin y$, show that

$$f_{xx} + f_{yy} = e^{2x}(f_{uu} + f_{vv}).$$

Hint: Recall Exercise 7.12.

7.14. If $f(x, y, z) = 0$, show that

$$\frac{\partial y}{\partial x} = -\frac{f_x}{f_y}, \qquad \frac{\partial x}{\partial y}\frac{\partial y}{\partial x} = 1, \qquad \frac{\partial x}{\partial y}\frac{\partial y}{\partial z}\frac{\partial z}{\partial x} = -1.$$

7.15. Derive the Taylor series (7.25), following the line of approach suggested in the second paragraph of Section 7.2. *Hint:* Setting $x = a + (x_0 - a)t$, $y = b + (y_0 - b)t$, expand per $f(x, y) = f[x(t), y(t)] = F(t) = F(0) + F'(0)t + \cdots$ or, with $t = 1$, $f(x_0, y_0) = F(0) + F'(0) + \cdots$. Finally, change "$x_0, y_0$" to x, y.

7.16. Carry out the Taylor expansions below.
(a) $\sin(x + y)$ about 0, 0 through third degree.
(b) $\sin(x + y + z^2)$ about 0, 0, 1 through second degree.
(c) $x^2 y$ about 1, 2 through 27th degree.
(d) $\ln(x + y)$ about 0, 1 through second degree.

7.17. (a) Recalling that if $x = x[u(s)]$, then

$$\frac{dx}{du}\frac{du}{ds} = \frac{dx}{ds},$$

generalize this result to

$$\frac{\partial(x, y)}{\partial(u, v)}\frac{\partial(u, v)}{\partial(r, s)} = \frac{\partial(x, y)}{\partial(r, s)} \quad \text{and} \quad \frac{\partial(x, y, z)}{\partial(u, v, w)}\frac{\partial(u, v, w)}{\partial(r, s, t)} = \frac{\partial(x, y, z)}{\partial(r, s, t)}.$$

Note: The three-dimensional Jacobian notation is defined in Exercise 7.11.
(b) Verify for the case where $x = u \cos v$, $y = u \sin v$ and $u = r + s$, $v = r^2 + s^2$.

7.18. (a) Show that

$$\frac{\partial(x, y)}{\partial(u, v)} = \left[\frac{\partial(u, v)}{\partial(x, y)}\right]^{-1} \quad \text{and} \quad \frac{\partial(x, y, z)}{\partial(u, v, w)} = \left[\frac{\partial(u, v, w)}{\partial(x, y, z)}\right]^{-1} \tag{7.32}$$

that is

$$J(u, v) = [J(x, y)]^{-1} \quad \text{and} \quad J(u, v, w) = [J(x, y, z)]^{-1}.$$

Recall that in Comment 2 of Example 7.2 we noted that if $x = x(u)$, then dx/du and du/dx *are* inverses, i.e.,

$$\frac{dx}{du} = \left(\frac{du}{dx}\right)^{-1}, \tag{7.33}$$

whereas, if $x = x(u, v)$, and $y = y(u, v)$, say, then (in general) pairs such as $\partial x/\partial u$ and $\partial u/\partial x$ are *not* inverses. Nevertheless, we now see that there *is* an analog of (7.33), namely, the first of equations (7.32). Similarly, for the case where $x = x(u, v, w)$, $y = y(u, v, w)$, $z = z(u, v, w)$ the analog of (7.33) is provided by the second of equations (7.32).

Note: The three-dimensional Jacobian notation is defined in Exercise 7.11.

(b) Verify for the case $x = u \cos v$, $y = u \sin v$.

7.19. (a) Find z_x and z_y if $xy + \sin(x + z) - z^2 = 5$.
(b) Find y_x and y_z if $xe^y - y^2 - z^2 \sin z = 0$.

7.20. Expand $y(x)$ about a point x of your choosing, through second order, if $(y - 1)e^y = \sqrt{x} - 1$.

7.21. Expand $z(x, y)$ about $x = y = 0$, through second order, if $x^2 + y^2 + z^2 - 4$ and $z \geq 0$. Calling the second-order approximation $z_{approx}(x, y)$, sketch the original and approximating surfaces, $z(x, y)$ and $z_{approx}(x, y)$, respectively.

7.22. Expand $t(q, r, s)$ about the point 1, 1, 1 through second order if $q^2 + r^2 + s^2 + t^2 - 2t = 3$; choose the root $t(1, 1, 1) = 0$.

7.23. **(Hodograph transformation)** A variety of problems in fluid mechanics, plasticity, and vibration theory can be modeled by the partial differential equations

$$\left.\begin{aligned} Au_x + Bu_y + Cv_x + Dv_y &= P \\ Eu_x + Fu_y + Gv_x + Hv_y &= Q. \end{aligned}\right\} \tag{7.34}$$

If $P = Q = 0$ and A through H are functions only of u, v, then we say that equations (7.33) are *reducible*; suppose that such is the case. Let us switch the dependent and independent variables; that is, we now regard x and y as functions of u and v:

$$x = x(u, v) \quad \text{and} \quad y = y(u, v). \tag{7.35}$$

This is called the hodograph transformation. Show that for any region in which the Jacobian $J = \partial(u, v)/\partial(x, y) \neq 0$, (7.34) converts the original *nonlinear* system to the *linear* system

$$\begin{aligned} Ay_v - Bx_v - Cy_u + Dx_u &= 0 \\ Ey_v - Fx_v - Gy_u + Hx_u &= 0. \end{aligned}$$

Again, we recommend the book by Ames mentioned in Exercise 7.9.

7.24. Make up two cogent exercises on Chapter 7.

Chapter 8

Vectors, Surfaces, and Volumes

This chapter was intended to be about surfaces and volumes with vectors postponed until Chapter 9 on vector field theory, but vectors and vector products will be helpful in our discussion of surfaces and volumes. Thus we start the present chapter with an introduction to vectors and their manipulation.

8.1. VECTORS AND ELEMENTARY OPERATIONS

Although our notion of what constitutes a vector will be considerably broadened in Part III, when we introduce the concept of an abstract linear vector space, for the present we will regard a **vector** simply as a directed magnitude.

Graphically, it may be represented by an arrow, the length of which is adjusted to some scale to indicate the desired magnitude. Physically, it may denote a force, the position (with respect to some reference point) of some particle in space, the particle's velocity, and so on. But note that the position of the vector in space is immaterial; for instance, the vectors $\mathbf{A}$, $\mathbf{B}$, $\mathbf{C}$ in Fig. 8.1(a) are identical. That is not to say that the *effect* of a vector is independent of its position; it should be apparent that the motion of the body $\mathcal{B}$ induced by a force $\mathbf{F}$ [Fig. 8-1(b)] will certainly depend on its point of application (as will the stress field induced in $\mathcal{B}$, and so on).

In this book we denote vectors by boldface type, such as $\mathbf{A}$; in writing, it is customary to underline instead. The **length** or **magnitude** of $\mathbf{A}$ is denoted $\|\mathbf{A}\|$ or simply A.

We define the following operations on and between vectors.

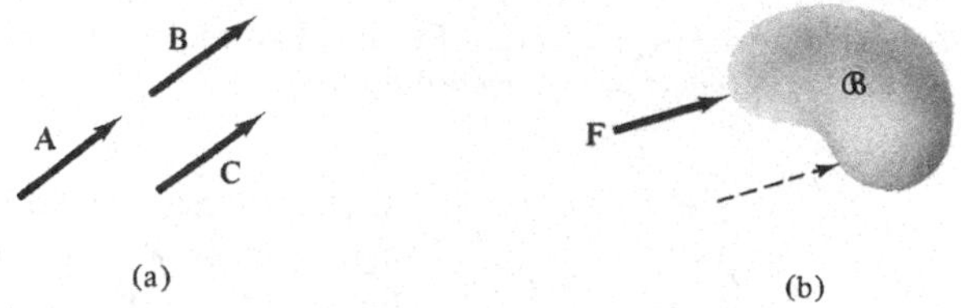

Figure 8.1. Position of a vector.

Scalar Multiplication. By $c\mathbf{A}$ (or $\mathbf{A}c$), where c is any real number, we mean the vector that is $|c|$ times as long as $\mathbf{A}$, in the same direction as $\mathbf{A}$ if c is positive and in the opposite direction if c is negative. If $c = 0$, then $c\mathbf{A} = 0\mathbf{A} \equiv \mathbf{0}$, the so-called *null vector* or *zero vector*, having zero length. In particular, if $c = -1$, then we denote $c\mathbf{A} = (-1)\mathbf{A} \equiv -\mathbf{A}$, read as "minus **A**."

In contrast with the vector quantity **A**, we call c a **scalar**.

Addition and Subtraction. Given any vectors **A** and **B** [Fig. 8.2(a)], we define $\mathbf{A} + \mathbf{B}$, called the *sum* of **A** and **B**, according to any of the (equivalent) constructions shown in Fig. 8.2(b), (c), or (d). In view of the construction of Fig. 8.2(d), we refer to this definition of addition as the **parallelogram law**. With vector addition defined, we also define the *difference* of **A** and **B** as $\mathbf{A} - \mathbf{B} \equiv \mathbf{A} + (-\mathbf{B})$, as shown, for example, in Fig. 8.2(b). Note that in Fig. 8.2 **A** and **B** happen to be coplanar; both lie in the plane of the paper. Even if they were not, however, one or both could be shifted (as discussed above) so that they become coplanar and the parallelogram law could then be applied.

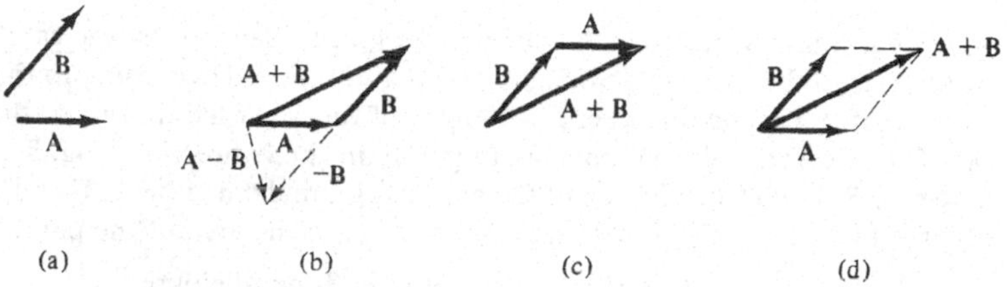

Figure 8.2. Vector addition and subtraction.

From our definitions of the scalar multiplication and addition of vectors it follows (Exercise 8.1) that

$$a(b\mathbf{A}) = (ab)\mathbf{A}, \qquad (a + b)\mathbf{A} = a\mathbf{A} + b\mathbf{A}$$
$$c(\mathbf{A} + \mathbf{B}) = c\mathbf{A} + c\mathbf{B} \tag{8.1a}$$

and

$$\mathbf{A} + \mathbf{B} = \mathbf{B} + \mathbf{A}$$
$$\mathbf{A} + (\mathbf{B} + \mathbf{C}) = (\mathbf{A} + \mathbf{B}) + \mathbf{C}. \tag{8.1b}$$

Scalar (or Dot) Product. For any two vectors **A** and **B** we define their **scalar product** or **dot product** $\mathbf{A} \cdot \mathbf{B}$ according to

$$\mathbf{A} \cdot \mathbf{B} \equiv AB \cos \theta, \tag{8.2}$$

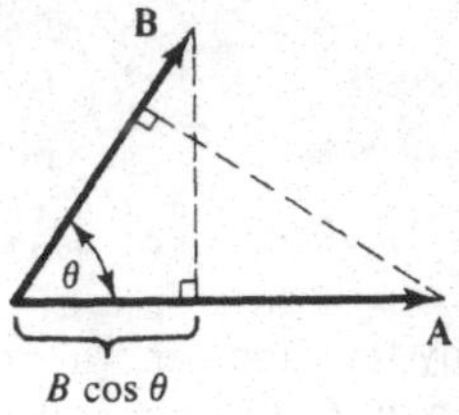

Figure 8.3. The scalar or dot product.

where A, B are the magnitudes of **A**, **B** and θ is the angle ($\leq \pi$) between **A** and **B** when they are arranged "tail to tail," as in Fig. 8.3. Observe that $B \cos \theta$ is the projection of **B** on **A**, so that $\mathbf{A} \cdot \mathbf{B} = AB \cos \theta$ is A times the projection of **B** on **A**. Alternatively, it is B times the projection of **A** on **B**. In particular, if $\theta = \pi/2$, the projection is zero and $\mathbf{A} \cdot \mathbf{B} = 0$; we then say that **A** and **B** are **orthogonal.** ($\mathbf{A} \cdot \mathbf{B} = 0$ implies that $\mathbf{A} = \mathbf{0}$ and/or $\mathbf{B} = \mathbf{0}$ and/or $\theta = \pi/2$.) On the other hand, $\mathbf{A} \cdot \mathbf{A} = A^2$, since $\theta = 0$ in this case.

From its definition it follows that the scalar product is *commutative*

$$\mathbf{A} \cdot \mathbf{B} = \mathbf{B} \cdot \mathbf{A} \tag{8.3}$$

and *linear*

$$(\alpha \mathbf{A} + \beta \mathbf{B}) \cdot \mathbf{C} = \alpha(\mathbf{A} \cdot \mathbf{C}) + \beta(\mathbf{B} \cdot \mathbf{C}), \tag{8.4}$$

where **A, B, C** are any vectors and α, β any scalars. For instance,

$$(\mathbf{A} + 2\mathbf{B}) \cdot (\mathbf{C} - \mathbf{D}) = \mathbf{A} \cdot \mathbf{C} - \mathbf{A} \cdot \mathbf{D} + 2\mathbf{B} \cdot \mathbf{C} - 2\mathbf{B} \cdot \mathbf{D} \tag{8.5}$$

as in "ordinary algebra."

Vector Product. Analogous to (8.2), we also define a **vector product** or **cross product** of any two vectors **A**, **B** according to

$$\mathbf{A} \times \mathbf{B} \equiv AB \sin \theta \hat{\mathbf{e}}, \tag{8.6}$$

where $\hat{\mathbf{e}}$ is a unit vector (i.e., it is of unit length, as denoted by the caret) that is perpendicular to the plane containing **A** and **B** and directed according to the "right-hand rule"; that is, curling the other four fingers of our right hand from **A** (the first vector) into **B** (the second vector), our thumb points in the $\hat{\mathbf{e}}$ direction (Fig. 8.4).

Since $B \sin \theta$ is the altitude of the parallelogram defined by **A**, **B** and A is the base, observe that $\|\mathbf{A} \times \mathbf{B}\| = AB \sin \theta$ corresponds to the area of the parallelogram.

$$\|\mathbf{A} \times \mathbf{B}\| = \text{area of the } \mathbf{A}, \mathbf{B} \text{ parallelogram.} \tag{8.7}$$

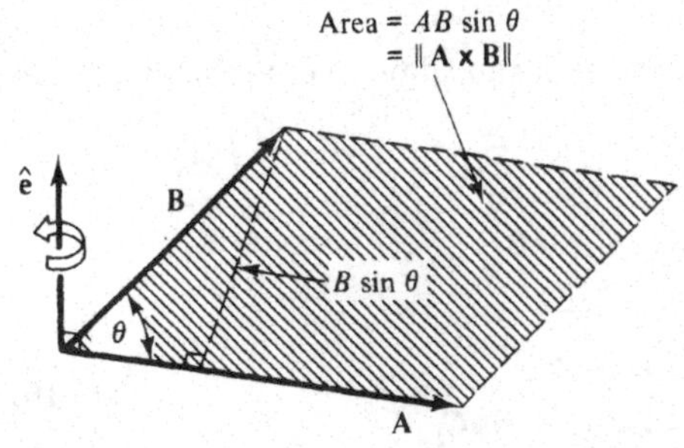

Figure 8.4. The vector or cross product.

[Alternatively, we can regard $A \sin\theta$ as the altitude and B as the base; the conclusion (8.7) is the same.] $\mathbf{A} \times \mathbf{B} = \mathbf{0}$ implies that $\mathbf{A} = \mathbf{0}$ and/or $\mathbf{B} = \mathbf{0}$ and/or $\theta = 0$ or π; for example, $\mathbf{A} \times \mathbf{A} = \mathbf{0}$.

Note carefully that the vector product is *not* commutative; $\mathbf{A} \times \mathbf{B}$ is not the same as $\mathbf{B} \times \mathbf{A}$. Rather,

$$\mathbf{A} \times \mathbf{B} = -\mathbf{B} \times \mathbf{A}, \tag{8.8}$$

since $\|\mathbf{A} \times \mathbf{B}\| = \|\mathbf{B} \times \mathbf{A}\| = AB \sin\theta$, but the $\hat{\mathbf{e}}$'s for $\mathbf{A} \times \mathbf{B}$ and $\mathbf{B} \times \mathbf{A}$ are oppositely oriented. It *is* true (Exercise 8.12), however, that the vector product is *linear*—that is,

$$(\alpha\mathbf{A} + \beta\mathbf{B}) \times \mathbf{C} = \alpha(\mathbf{A} \times \mathbf{C}) + \beta(\mathbf{B} \times \mathbf{C}) \tag{8.9}$$

for any vectors **A**, **B**, **C** and any scalars α, β.

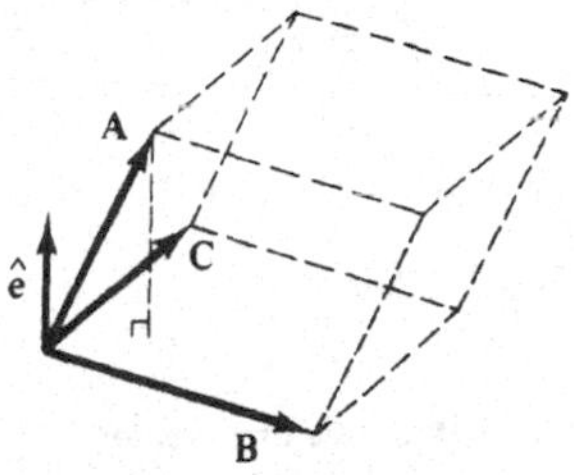

Figure 8.5. Volume interpretation of $\mathbf{A} \cdot (\mathbf{B} \times \mathbf{C})$.

Consider next the so-called **scalar triple product** $\mathbf{A} \cdot (\mathbf{B} \times \mathbf{C})$. Actually, the parentheses can be omitted without confusion, since $\mathbf{A} \cdot \mathbf{B} \times \mathbf{C}$ can't possibly be interpreted as $(\mathbf{A} \cdot \mathbf{B}) \times \mathbf{C}$ because the latter is a scalar crossed into a vector, which is meaningless. Since $\mathbf{B} \times \mathbf{C}$ is the area of the **B**, **C** parallelogram times the unit vector $\hat{\mathbf{e}}$ (Fig. 8.5) and $\mathbf{A} \cdot \hat{\mathbf{e}}$ is the altitude of the **A**, **B**, **C** parallelepiped, it follows that

$$\mathbf{A} \cdot \mathbf{B} \times \mathbf{C} = \text{volume of the } \mathbf{A}, \mathbf{B}, \mathbf{C} \text{ parallelepiped.} \tag{8.10}$$

The volume is also given by $\mathbf{C} \times \mathbf{A} \cdot \mathbf{B}$ and $\mathbf{A} \times \mathbf{B} \cdot \mathbf{C}$, so that we have

$$\mathbf{A} \cdot \mathbf{B} \times \mathbf{C} = \mathbf{C} \cdot \mathbf{A} \times \mathbf{B} = \mathbf{B} \cdot \mathbf{C} \times \mathbf{A}; \tag{8.11}$$

that is, any cyclic permutation of **A**, **B**, **C** in (8.10) leaves the scalar triple product unchanged.

Besides $\mathbf{A} \cdot \mathbf{B} \times \mathbf{C}$ there is also the **vector triple product** $\mathbf{A} \times (\mathbf{B} \times \mathbf{C})$. We should be able to express it in the form $\alpha\mathbf{B} + \beta C$, since it must clearly be perpendicular to both **A** and $\mathbf{B} \times \mathbf{C}$; in other words, it must lie in the **B**, **C** plane such that it is perpendicular to **A**. To obtain this form, let $\hat{\mathbf{e}}_1$ be a unit vector aligned with **B** and let $\hat{\mathbf{e}}_2$ be a unit vector lying in the **B**, **C** plane perpendicular to $\hat{\mathbf{e}}_1$ such that $\hat{\mathbf{e}}_1 \times \hat{\mathbf{e}}_2 = \hat{\mathbf{e}}_3$ coincides with the $\hat{\mathbf{e}}$ shown in Fig. 8.5. Then we can express

$$\mathbf{B} = B\hat{\mathbf{e}}_1, \qquad \mathbf{C} = C_1\hat{\mathbf{e}}_1 + C_2\hat{\mathbf{e}}_2, \qquad \mathbf{A} = A_1\hat{\mathbf{e}}_1 + A_2\hat{\mathbf{e}}_2 + A_3\hat{\mathbf{e}}_3$$

so that $\mathbf{B} \times \mathbf{C} = BC_2\hat{\mathbf{e}}_3$, and

$$\begin{aligned}\mathbf{A} \times (\mathbf{B} \times \mathbf{C}) &= (A_1\hat{\mathbf{e}}_1 + A_2\hat{\mathbf{e}}_2 + A_3\hat{\mathbf{e}}_3) \times BC_2\hat{\mathbf{e}}_3 \\ &= -A_1BC_2\hat{\mathbf{e}}_2 + A_2BC_2\hat{\mathbf{e}}_1 \\ &= (A_1C_1 + A_2C_2)B\hat{\mathbf{e}}_1 - A_1B(C_1\hat{\mathbf{e}}_1 + C_2\hat{\mathbf{e}}_2) \\ &= (\mathbf{A} \cdot \mathbf{C})\mathbf{B} - (\mathbf{A} \cdot \mathbf{B})\mathbf{C}\end{aligned} \tag{8.12}$$

as claimed. That is, α turns out to be $\mathbf{A} \cdot \mathbf{C}$ and β to be $-\mathbf{A} \cdot \mathbf{B}$.

Do we need the parentheses around $\mathbf{B} \times \mathbf{C}$? Is $\mathbf{A} \times (\mathbf{B} \times \mathbf{C}) = (\mathbf{A} \times \mathbf{B}) \times \mathbf{C}$, in which case we could drop the parentheses? Well,

$$(\mathbf{A} \times \mathbf{B}) \times \mathbf{C} = -\mathbf{C} \times (\mathbf{A} \times \mathbf{B}) = \mathbf{C} \times (\mathbf{B} \times \mathbf{A}) = (\mathbf{A} \cdot \mathbf{C})\mathbf{B} - (\mathbf{C} \cdot \mathbf{B})\mathbf{A},$$

which, in general, is *not* the same as the right-hand side of (8.12), and so the parentheses are necessary.

Derivative. A vector need not be constant; it may vary over space and/or the time t. For instance, suppose that $\mathbf{A} = \mathbf{A}(t)$. We define the *derivative* of $\mathbf{A}$ with respect to t by the limit

$$\frac{d\mathbf{A}}{dt} \equiv \lim_{\Delta t \to 0} \frac{\mathbf{A}(t + \Delta t) - \mathbf{A}(t)}{\Delta t} \tag{8.13}$$

if, of course, the limit exists. We see from its definition that $d\mathbf{A}/dt$ is a vector quantity.

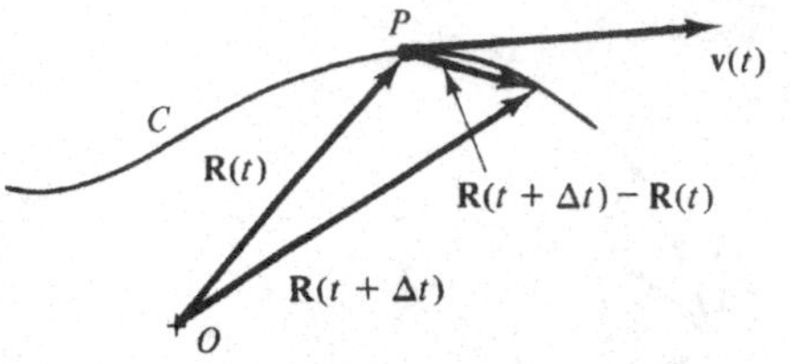

Figure 8.6. Velocity vector.

Consider, for example, the "position vector" $\mathbf{R}(t)$ from some fixed point O to a particle that is moving along some curve C. Then $d\mathbf{R}/dt$ is the limit of $[\mathbf{R}(t + \Delta t) - \mathbf{R}(t)]/\Delta t$ as $\Delta t \to 0$, and is called the *velocity* of the particle, $\mathbf{v}(t)$. Clearly, $\mathbf{v}(t)$ is tangent to C, since $\mathbf{R}(t + \Delta t) - \mathbf{R}(t)$ *defines* the tangent to C at P as $\Delta t \to 0$ (Fig. 8.6). The next derivative, $d\mathbf{v}/dt = d^2\mathbf{R}/dt^2$, is called the *acceleration*, $\mathbf{a}(t)$. In general, $\mathbf{a}$ will *not* be tangent to C; see Exercise 8.21.

What can we say about $d(\mathbf{A} \cdot \mathbf{B})/dt$ and $d(\mathbf{A} \times \mathbf{B})/dt$? Expressing $\mathbf{A}(t + \Delta t) = \mathbf{A}(t) + [\mathbf{A}(t + \Delta t) - \mathbf{A}(t)] \equiv \mathbf{A} + \Delta\mathbf{A}$, and similarly for $\mathbf{B}(t + \Delta t)$, we have [with the help of (8.3) and (8.4)]

$$\begin{aligned} \frac{d}{dt}(\mathbf{A} \cdot \mathbf{B}) &= \lim_{\Delta t \to 0} \frac{(\mathbf{A} + \Delta\mathbf{A}) \cdot (\mathbf{B} + \Delta\mathbf{B}) - \mathbf{A} \cdot \mathbf{B}}{\Delta t} \\ &= \lim_{\Delta t \to 0} \left(\frac{\Delta\mathbf{A}}{\Delta t} \cdot \mathbf{B} + \mathbf{A} \cdot \frac{\Delta\mathbf{B}}{\Delta t} + \Delta\mathbf{A} \cdot \frac{\Delta\mathbf{B}}{\Delta t} \right) = \frac{d\mathbf{A}}{dt} \cdot \mathbf{B} + \mathbf{A} \cdot \frac{d\mathbf{B}}{dt} \end{aligned} \tag{8.14}$$

as we might have guessed. Similarly,

$$\frac{d}{dt}(\mathbf{A} \times \mathbf{B}) = \frac{d\mathbf{A}}{dt} \times \mathbf{B} + \mathbf{A} \times \frac{d\mathbf{B}}{dt}. \tag{8.15}$$

8.2. *COORDINATE SYSTEMS, BASE VECTORS, AND COMPONENTS*

In calculations it is often convenient to turn to a specific reference frame—a coordinate system and a corresponding set of unit base vectors. Three of the most widely used systems will now be discussed.

Cartesian. Here we refer to the rectangular x, y, z coordinates and the corresponding unit base vectors $\hat{\mathbf{i}}, \hat{\mathbf{j}}, \hat{\mathbf{k}}$ shown in Fig. 8.7. We say that the base vectors $\hat{\mathbf{i}}, \hat{\mathbf{j}}, \hat{\mathbf{k}}$ are **orthonormal**, since they are both mutually orthogonal and also "normalized"; that is, they're of unit length. We have

$$\left.\begin{aligned} &\hat{\mathbf{i}} \cdot \hat{\mathbf{j}} = \hat{\mathbf{i}} \cdot \hat{\mathbf{k}} = \hat{\mathbf{j}} \cdot \hat{\mathbf{k}} = 0 \\ &\hat{\mathbf{i}} \cdot \hat{\mathbf{i}} = \hat{\mathbf{j}} \cdot \hat{\mathbf{j}} = \hat{\mathbf{k}} \cdot \hat{\mathbf{k}} = 1 \\ &\hat{\mathbf{i}} \times \hat{\mathbf{j}} = \hat{\mathbf{k}}, \quad \hat{\mathbf{i}} \times \hat{\mathbf{k}} = -\hat{\mathbf{j}}, \quad \hat{\mathbf{j}} \times \hat{\mathbf{k}} = \hat{\mathbf{i}}, \end{aligned}\right\} \tag{8.16}$$

and so on.

To illustrate, the vector **A** running from the point $(-1, 4, 1)$ to $(1, 3, 4)$ in Fig. 8.7 is (by the definition of vector addition) $\mathbf{A} = \mathbf{a} + \mathbf{b} + \mathbf{c} = 2\hat{\mathbf{i}} - \hat{\mathbf{j}} + 3\hat{\mathbf{k}}$. More generally, we can express *any* vector **A** in the three-dimensional x, y, z space in the form

$$\mathbf{A} = A_x\hat{\mathbf{i}} + A_y\hat{\mathbf{j}} + A_z\hat{\mathbf{k}}, \tag{8.17}$$

where the A_x, A_y, A_z are called the x, y, z **components** of **A**, respectively. "Dotting" $\hat{\mathbf{i}}, \hat{\mathbf{j}}$ and $\hat{\mathbf{k}}$ into (8.17) and recalling (8.16) yield

$$A_x = \mathbf{A} \cdot \hat{\mathbf{i}}, \qquad A_y = \mathbf{A} \cdot \hat{\mathbf{j}}, \qquad A_z = \mathbf{A} \cdot \hat{\mathbf{k}}; \tag{8.18}$$

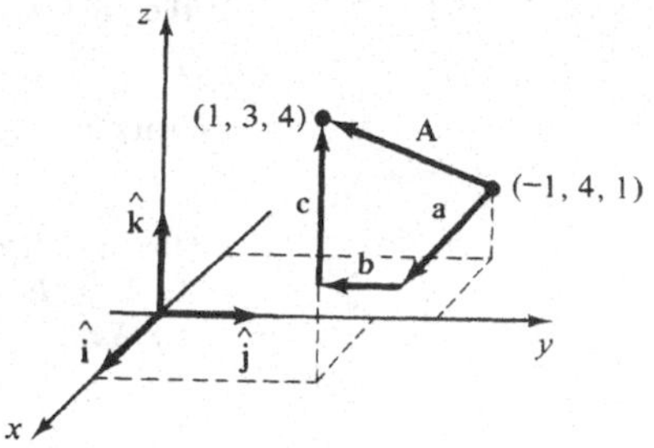

Figure 8.7. Cartesian system.

that is, A_x, A_y, A_z are the projections of **A** on the x, y, z axes, respectively.

Observe that $\hat{\mathbf{i}}, \hat{\mathbf{j}}, \hat{\mathbf{k}}$ are *constant* vectors, and therefore

$$\frac{d}{dt}\mathbf{A}(t) = \frac{dA_x}{dt}\hat{\mathbf{i}} + \frac{dA_y}{dt}\hat{\mathbf{j}} + \frac{dA_z}{dt}\hat{\mathbf{k}}. \tag{8.19}$$

Cylindrical. The cylindrical coordinates r, θ, z and their respective unit base vectors $\hat{\mathbf{e}}_r, \hat{\mathbf{e}}_\theta, \hat{\mathbf{e}}_z$ are shown in Fig. 8.8; r, θ, z are related to x, y, z according to

$$x = r\cos\theta, \qquad y = r\sin\theta, \qquad z = z. \tag{8.20}$$

Any vector in the space can be expressed in the form

$$\mathbf{A} = A_r\hat{\mathbf{e}}_r + A_\theta\hat{\mathbf{e}}_\theta + A_z\hat{\mathbf{e}}_z. \tag{8.21}$$

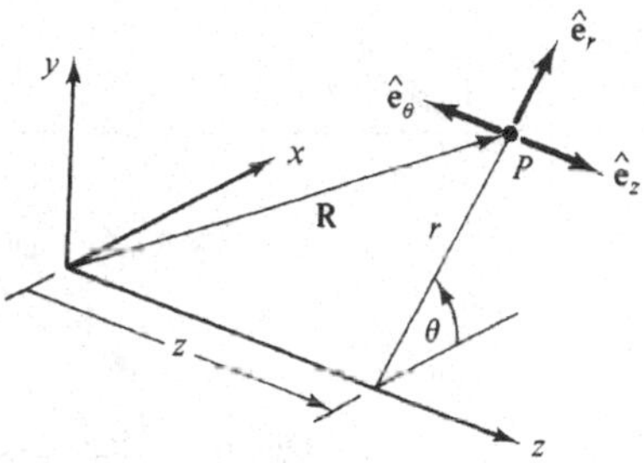

Figure 8.8. Cylindrical system.

Since the set $\hat{\mathbf{e}}_r, \hat{\mathbf{e}}_\theta, \hat{\mathbf{e}}_z$ is orthonormal (i.e., $\hat{\mathbf{e}}_r \cdot \hat{\mathbf{e}}_\theta = \hat{\mathbf{e}}_r \cdot \hat{\mathbf{e}}_z = \hat{\mathbf{e}}_\theta \cdot \hat{\mathbf{e}}_z = 0$ and $\hat{\mathbf{e}}_r \cdot \hat{\mathbf{e}}_r = \hat{\mathbf{e}}_\theta \cdot \hat{\mathbf{e}}_\theta = \hat{\mathbf{e}}_z \cdot \hat{\mathbf{e}}_z = 1$), it follows that the r, θ, z components are given by

$$A_r = \mathbf{A} \cdot \hat{\mathbf{e}}_r, \qquad A_\theta = \mathbf{A} \cdot \hat{\mathbf{e}}_\theta, \qquad A_z = \mathbf{A} \cdot \hat{\mathbf{e}}_z. \tag{8.22}$$

Example 8.1. Obtain general expressions for the velocity and acceleration of a particle with respect to a cylindrical reference system. We start with the "position vector"

$$\mathbf{R} = z\hat{\mathbf{e}}_z + r\hat{\mathbf{e}}_r \tag{8.23}$$

from the origin out to the r, θ, z point, say P (Fig. 8.8). Note that **R** is *not* $r\hat{\mathbf{e}}_r + \theta\hat{\mathbf{e}}_\theta + z\hat{\mathbf{e}}_z$; the *units* of this expression aren't even consistent, since θ is dimensionless (it's measured in radians), whereas r and z have the correct units of length.

Even though $\hat{\mathbf{i}}, \hat{\mathbf{j}}, \hat{\mathbf{k}}$ are *constant* vectors, note that $\hat{\mathbf{e}}_r, \hat{\mathbf{e}}_\theta, \hat{\mathbf{e}}_z$ are not. More specifically, $\hat{\mathbf{e}}_r$ and $\hat{\mathbf{e}}_\theta$ vary with θ, although not with r, whereas $\hat{\mathbf{e}}_z$ is, in fact, constant. For instance, if we imagine sliding the $\hat{\mathbf{e}}_r, \hat{\mathbf{e}}_\theta, \hat{\mathbf{e}}_z$ triad to a new r location, say $r + \Delta r$, with both θ and z held *fixed*, we see that the three vectors remain unchanged because both their magnitude (unity) and direction remain unchanged, even though their *location* is different. But if we vary θ, with r and z held fixed, then $\hat{\mathbf{e}}_r$ and $\hat{\mathbf{e}}_\theta$ *rotate*. The upshot is that $\hat{\mathbf{e}}_r = \hat{\mathbf{e}}_r(\theta)$, $\hat{\mathbf{e}}_\theta = \hat{\mathbf{e}}_\theta(\theta)$, $\hat{\mathbf{e}}_z =$ constant.

If the point P moves about according to $r = r(t)$, $\theta = \theta(t)$, $z = z(t)$, then

$$\mathbf{R}(t) = z(t)\hat{\mathbf{e}}_z + r(t)\hat{\mathbf{e}}_r[\theta(t)]. \tag{8.24}$$

Its velocity is

$$\mathbf{v}(t) = \dot{\mathbf{R}}(t) = \dot{z}\hat{\mathbf{e}}_z + \dot{r}\hat{\mathbf{e}}_r + r\dot{\hat{\mathbf{e}}}_r, \tag{8.25}$$

where the dots denote time derivatives as usual. Yet it remains to express $\dot{\hat{\mathbf{e}}}_r$ in terms of $\hat{\mathbf{e}}_r$, $\hat{\mathbf{e}}_\theta$, $\hat{\mathbf{e}}_z$. (*All* vectors are to be expressed as linear combinations of the base vectors.) Well,

$$\frac{d\hat{\mathbf{e}}_r}{dt} = \frac{d\hat{\mathbf{e}}_r}{d\theta}\frac{d\theta}{dt}, \tag{8.26}$$

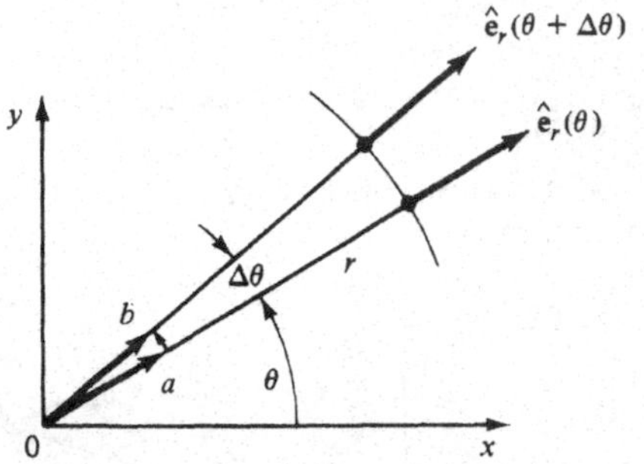

Figure 8.9. Calculation of $d\hat{\mathbf{e}}_r/d\theta$.

so that what we actually need is $d\hat{\mathbf{e}}_r/d\theta$. By definition,

$$\frac{d\hat{\mathbf{e}}_r}{d\theta} = \lim_{\Delta\theta\to 0}\frac{\hat{\mathbf{e}}_r(\theta + \Delta\theta) - \hat{\mathbf{e}}_r(\theta)}{\Delta\theta}. \tag{8.27}$$

The vector difference in the numerator is easily evaluated if we slide the two vectors back to O (Fig. 8.9) so that they are tail to tail. Then $\hat{\mathbf{e}}_r(\theta + \Delta\theta) - \hat{\mathbf{e}}_r(\theta)$ is the little vector from a to b; its length is $1\,\Delta\theta$ and its direction is $\hat{\mathbf{e}}_\theta$. Thus

$$\frac{d\hat{\mathbf{e}}_r}{d\theta} = \lim_{\Delta\theta\to 0}\frac{\Delta\theta\hat{\mathbf{e}}_\theta}{\Delta\theta} = \hat{\mathbf{e}}_\theta, \tag{8.28}$$

and so (8.25) becomes

$$\mathbf{v}(t) = \dot{r}\hat{\mathbf{e}}_r + r\dot{\theta}\hat{\mathbf{e}}_\theta + \dot{z}\hat{\mathbf{e}}_z. \tag{8.29}$$

Differentiating once more, the acceleration $\mathbf{a}(t)$ is

$$\begin{aligned}\mathbf{a}(t) &= \ddot{r}\hat{\mathbf{e}}_r + \dot{r}\dot{\hat{\mathbf{e}}}_r + (\dot{r}\dot{\theta} + r\ddot{\theta})\hat{\mathbf{e}}_\theta + r\dot{\theta}\dot{\hat{\mathbf{e}}}_\theta + \ddot{z}\hat{\mathbf{e}}_z \\ &= (\ddot{r} - r\dot{\theta}^2)\hat{\mathbf{e}}_r + (r\ddot{\theta} + 2\dot{r}\dot{\theta})\hat{\mathbf{e}}_\theta + \ddot{z}\hat{\mathbf{e}}_z,\end{aligned} \tag{8.30}$$

where we've used the fact that (Exercise 8.13)

$$\frac{d\hat{\mathbf{e}}_\theta}{d\theta} = -\hat{\mathbf{e}}_r \tag{8.31}$$

to express $\dot{\hat{\mathbf{e}}}_\theta = (d\hat{\mathbf{e}}_\theta/d\theta)(d\theta/dt) = -\dot{\theta}\hat{\mathbf{e}}_r$.

The derivatives of the base vectors, (8.28) and (8.31), are important because they inevitably arise whenever we differentiate vectors that are expressed in terms of cylindrical coordinates.

Of the five terms constituting $\mathbf{a}(t)$, in (8.30), $-r\dot{\theta}^2\hat{\mathbf{e}}_r$ and $2\dot{r}\dot{\theta}\hat{\mathbf{e}}_\theta$ are called the **centripetal** and **Coriolis** accelerations, respectively. ∎

Spherical. The spherical polar ρ, θ, ϕ system and its corresponding base vectors are shown in Fig. 8.10. The spherical polar and cartesian coordinates are related according to

$$x = \rho\sin\theta\cos\phi, \qquad y = \rho\sin\theta\sin\phi, \qquad z = \rho\cos\theta. \tag{8.32}$$

Observe that $\hat{\mathbf{e}}_\rho = \hat{\mathbf{e}}_\rho(\theta, \phi)$, $\hat{\mathbf{e}}_\theta = \hat{\mathbf{e}}_\theta(\theta, \phi)$, and $\hat{\mathbf{e}}_\phi = \hat{\mathbf{e}}_\phi(\phi)$. We leave it for you

(Exercise 8.14) to show that

$$\left.\begin{aligned}
&\frac{\partial \hat{\mathbf{e}}_\rho}{\partial \theta} = \hat{\mathbf{e}}_\theta, \qquad \frac{\partial \hat{\mathbf{e}}_\rho}{\partial \phi} = \sin\theta\, \hat{\mathbf{e}}_\phi \\
&\frac{\partial \hat{\mathbf{e}}_\theta}{\partial \theta} = -\hat{\mathbf{e}}_\rho, \qquad \frac{\partial \hat{\mathbf{e}}_\theta}{\partial \phi} = \cos\theta\, \hat{\mathbf{e}}_\phi \\
&\frac{\partial \hat{\mathbf{e}}_\phi}{\partial \phi} = -\sin\theta\, \hat{\mathbf{e}}_\rho - \cos\theta\, \hat{\mathbf{e}}_\theta.
\end{aligned}\right\} \tag{8.33}$$

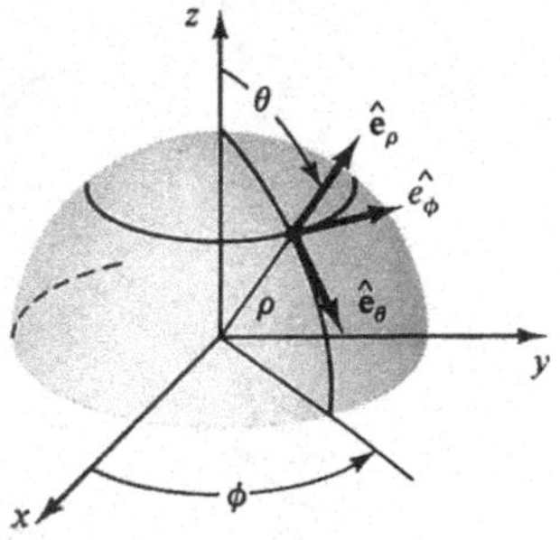

Figure 8.10. Spherical polar system.

Vector Products and Scalar Triple Products as Determinants. Let $\hat{\mathbf{e}}_1, \hat{\mathbf{e}}_2, \hat{\mathbf{e}}_3$ be any right-handed orthonormal set of base vectors—in other words, mutually orthogonal unit vectors such that $\hat{\mathbf{e}}_1 \times \hat{\mathbf{e}}_2 = \hat{\mathbf{e}}_3$, $\hat{\mathbf{e}}_2 \times \hat{\mathbf{e}}_3 = \hat{\mathbf{e}}_1$, and $\hat{\mathbf{e}}_3 \times \hat{\mathbf{e}}_1 = \hat{\mathbf{e}}_2$. For example, they might be $\hat{\mathbf{i}}, \hat{\mathbf{j}}, \hat{\mathbf{k}}$ (cartesian), $\hat{\mathbf{e}}_r, \hat{\mathbf{e}}_\theta, \hat{\mathbf{e}}_z$ (cylindrical), $\hat{\mathbf{e}}_\rho, \hat{\mathbf{e}}_\theta, \hat{\mathbf{e}}_\phi$ (spherical), or any other such set.

For any vectors **A**, **B** observe that we can express

$$\mathbf{A} \times \mathbf{B} = \begin{vmatrix} \hat{\mathbf{e}}_1 & \hat{\mathbf{e}}_2 & \hat{\mathbf{e}}_3 \\ A_1 & A_2 & A_3 \\ B_1 & B_2 & B_3 \end{vmatrix}. \tag{8.34}$$

This expression is verified easily by expanding both $(A_1\hat{\mathbf{e}}_1 + A_2\hat{\mathbf{e}}_2 + A_3\hat{\mathbf{e}}_3) \times (B_1\hat{\mathbf{e}}_1 + B_2\hat{\mathbf{e}}_2 + B_3\hat{\mathbf{e}}_3)$ and the determinant and noting that the results are the same. Furthermore,

$$\mathbf{A} \cdot \mathbf{B} \times \mathbf{C} = (A_1\hat{\mathbf{e}}_1 + A_2\hat{\mathbf{e}}_2 + A_3\hat{\mathbf{e}}_3) \cdot \begin{vmatrix} \hat{\mathbf{e}}_1 & \hat{\mathbf{e}}_2 & \hat{\mathbf{e}}_3 \\ B_1 & B_2 & B_3 \\ C_1 & C_2 & C_3 \end{vmatrix} = \begin{vmatrix} A_1 & A_2 & A_3 \\ B_1 & B_2 & B_3 \\ C_1 & C_2 & C_3 \end{vmatrix}. \tag{8.35}$$

8.3. ANGULAR VELOCITY OF A RIGID BODY

Consider a rigid body $\mathcal{B}$ rotating about a fixed z axis with *angular velocity* Ω (radians per unit time), as sketched in Fig. 8.11. That is, each point in $\mathcal{B}$ moves in a circular path with $\dot{\theta} = \Omega$; the center of the circle is on the axis of rotation, and the plane of the circle is perpendicular to the axis. According to the right-hand rule, we define the angular velocity vector $\mathbf{\Omega}$ as shown in the figure.

We claim that the velocity of each point in $\mathcal{B}$ can be expressed as

$$\mathbf{v} = \mathbf{\Omega} \times \mathbf{R}, \tag{8.36}$$

where **R** is the position vector from some point on the axis, say O (the origin of our cylindrical coordinate system), to the point in question. This statement is easily verified:

$$\mathbf{R} = z\hat{\mathbf{e}}_z + r\hat{\mathbf{e}}_r$$

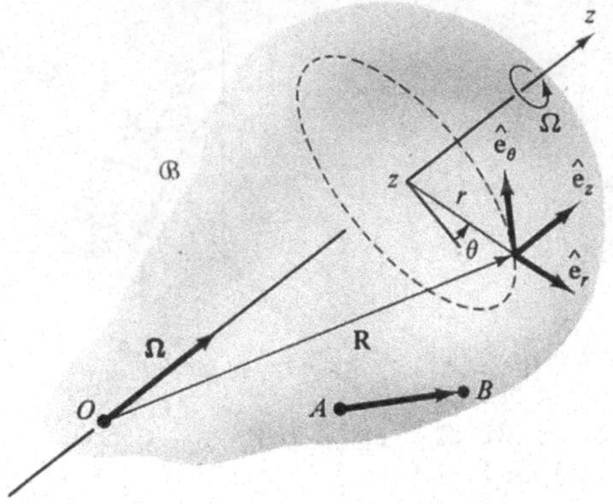

Figure 8.11. Rotation about a fixed axis.

and so

$$\mathbf{v} = \dot{\mathbf{R}} = r\dot{\hat{\mathbf{e}}}_r = r\dot{\theta}\hat{\mathbf{e}}_\theta = \Omega r\hat{\mathbf{e}}_\theta.$$

On the other hand,

$$\boldsymbol{\Omega} \times \mathbf{R} = \Omega\hat{\mathbf{e}}_z \times (z\hat{\mathbf{e}}_z + r\hat{\mathbf{e}}_r) = \Omega r\hat{\mathbf{e}}_\theta.$$

In fact, if A and B are any two points in ℬ and **AB** denotes the vector from A to B (Fig. 8.11), it follows that

$$\begin{aligned}\dot{\mathbf{AB}} &= \frac{d}{dt}(\mathbf{OB} - \mathbf{OA}) = \dot{\mathbf{OB}} - \dot{\mathbf{OA}} \\ &= \boldsymbol{\Omega} \times \mathbf{OB} - \boldsymbol{\Omega} \times \mathbf{OA} = \boldsymbol{\Omega} \times (\mathbf{OB} - \mathbf{OA}) = \boldsymbol{\Omega} \times \mathbf{AB}. \qquad (8.37)\end{aligned}$$

Generalizing further, suppose that the body is undergoing an *arbitrary* motion in space, not necessarily a rotation about a fixed axis. According to **Chasle's theorem** from elementary mechanics, the motion can be decomposed into a *translation* plus a *rotation*. That is, consider any point P in ℬ with velocity $\mathbf{v}_P$. Then at any given instant the motion of ℬ may be regarded as a translation with velocity $\mathbf{v}_P$ (i.e., where every point in ℬ has the velocity $\mathbf{v}_P$) plus a rotation with angular velocity $\boldsymbol{\Omega}$ about some axis through P. (Furthermore, $\boldsymbol{\Omega}$ is independent of the choice of P.)[1]

Thus if A and B are points in ℬ, then

$$\mathbf{v}_A = \mathbf{v}_P + \boldsymbol{\Omega} \times \mathbf{PA}, \qquad \mathbf{v}_B = \mathbf{v}_P + \boldsymbol{\Omega} \times \mathbf{PB},$$

so that

$$\mathbf{v}_B - \mathbf{v}_A = \boldsymbol{\Omega} \times (\mathbf{PB} - \mathbf{PA}) = \boldsymbol{\Omega} \times \mathbf{AB}$$

or

$$\dot{\mathbf{AB}} = \boldsymbol{\Omega} \times \mathbf{AB} \qquad (8.38)$$

as before [equation (8.37)]. Note carefully that **AB** is a vector of *fixed length*, since ℬ is a rigid body. This situation is reflected in the fact that $\dot{\mathbf{AB}}$ is perpendicular to **AB**, per (8.38), since a component along **AB** would correspond to a stretching or contraction of the **AB** vector.

[1] For a proof of Chasle's theorem, see, for example, the appendix in I. Shames, *Engineering Mechanics; Dynamics*, Prentice-Hall, Englewood Cliffs, N.J., 1960.

Example 8.2. Equation (8.38) is quite helpful in differentiating base vectors. Consider, for instance, the spherical polar system shown in Fig. 8.10. Let us regard the $\hat{\mathbf{e}}_\rho$, $\hat{\mathbf{e}}_\theta$, $\hat{\mathbf{e}}_\phi$ triad as a rigid body; in other words, think of the three vectors as metal rods that are welded together. Their angular velocity, for any given $\rho(t)$, $\theta(t)$, and $\phi(t)$, is

$$\begin{aligned}\mathbf{\Omega} &= \dot{\theta}\hat{\mathbf{e}}_\phi + \dot{\phi}\hat{\mathbf{e}}_z = \dot{\theta}\hat{\mathbf{e}}_\phi + \dot{\phi}[(\hat{\mathbf{e}}_z \cdot \hat{\mathbf{e}}_\rho)\hat{\mathbf{e}}_\rho + (\hat{\mathbf{e}}_z \cdot \hat{\mathbf{e}}_\theta)\hat{\mathbf{e}}_\theta + (\hat{\mathbf{e}}_z \cdot \hat{\mathbf{e}}_\phi)\hat{\mathbf{e}}_\phi] \\ &= \dot{\theta}\hat{\mathbf{e}}_\phi + \dot{\phi}\cos\theta\,\hat{\mathbf{e}}_\rho + \dot{\phi}\cos\left(\theta + \frac{\pi}{2}\right)\hat{\mathbf{e}}_\theta \\ &= \dot{\phi}\cos\theta\,\hat{\mathbf{e}}_\rho - \dot{\phi}\sin\theta\,\hat{\mathbf{e}}_\theta + \dot{\theta}\hat{\mathbf{e}}_\phi.\end{aligned}$$

(Seeking to express all vectors in terms of $\hat{\mathbf{e}}_\rho$, $\hat{\mathbf{e}}_\theta$, $\hat{\mathbf{e}}_\phi$, we have expanded $\hat{\mathbf{e}}_z$ in the form $\hat{\mathbf{e}}_z = a\hat{\mathbf{e}}_\rho + b\hat{\mathbf{e}}_\theta + c\hat{\mathbf{e}}_\phi$. Dotting both sides with $\hat{\mathbf{e}}_\rho$ yields $\hat{\mathbf{e}}_z \cdot \hat{\mathbf{e}}_\rho = a$, and similarly for b and c, as expressed above.)

Then since $\hat{\mathbf{e}}_\rho$ is of fixed length, (8.38) applies:

$$\begin{aligned}\dot{\hat{\mathbf{e}}}_\rho &= \mathbf{\Omega} \times \hat{\mathbf{e}}_\rho = (\dot{\phi}\cos\theta\,\hat{\mathbf{e}}_\rho - \dot{\phi}\sin\theta\,\hat{\mathbf{e}}_\theta + \dot{\theta}\hat{\mathbf{e}}_\phi) \times \hat{\mathbf{e}}_\rho \\ &= \dot{\phi}\sin\theta\,\hat{\mathbf{e}}_\phi + \dot{\theta}\hat{\mathbf{e}}_\theta,\end{aligned} \tag{8.39}$$

and similarly for $\hat{\mathbf{e}}_\theta$ and $\hat{\mathbf{e}}_\phi$.

The space derivatives are also easily available. For example,

$$\dot{\hat{\mathbf{e}}}_\rho[\theta(t), \phi(t)] = \frac{\partial \hat{\mathbf{e}}_\rho}{\partial \theta}\dot{\theta} + \frac{\partial \hat{\mathbf{e}}_\rho}{\partial \phi}\dot{\phi}$$

and comparing this equation with (8.39), we see that

$$\frac{\partial \hat{\mathbf{e}}_\rho}{\partial \theta} = \hat{\mathbf{e}}_\theta \quad \text{and} \quad \frac{\partial \hat{\mathbf{e}}_\rho}{\partial \phi} = \sin\theta\,\hat{\mathbf{e}}_\phi,$$

as claimed earlier in (8.33). ∎

8.4. *CURVILINEAR COORDINATE REPRESENTATION OF SURFACES*[2]

Just as a space curve can be represented by parametric equations $x = x(u)$, $y = y(u)$, $z = z(u)$, we expect that a *surface* can be represented by a *two*-parameter family

$$x = x(u, v), \qquad y = y(u, v), \qquad z = z(u, v). \tag{8.40}$$

In general, (8.40) yields a space curve for each fixed v; and as v varies continuously, we anticipate that this family of curves will generate a surface.

Example 8.3. Suppose that

$$x = a \sin v \cos u, \qquad y = a \sin v \sin u, \qquad z = a \cos v. \tag{8.41}$$

It follows that

$$x^2 + y^2 + z^2 = a^2, \tag{8.42}$$

[2]One especially important application of the differential geometry of curved surfaces is in the theory of shells and curved plates; see, for example, Chapter 12 in C. Wang, *Applied Elasticity*, McGraw-Hill, New York, 1953. For a more complete treatment of the mathematics of curves and surfaces, we recommend D. J. Struik, *Differential Geometry*, 2nd ed., Addison-Wesley, Reading, Mass., 1961.

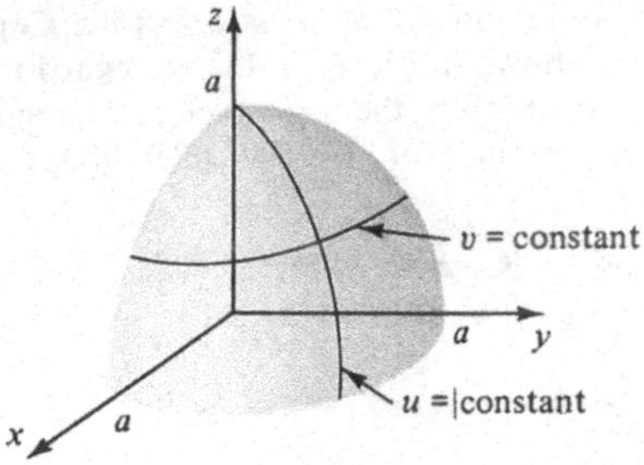

Figure 8.12. The spherical surface defined by (8.41).

and so the parametric equations (8.41) define the surface of a sphere of radius a, part of which is shown in Fig. 8.12. If $v =$ constant, then $z = a \cos v$ is constant; therefore the $v =$ constant curves are lines of "latitude" (if we regard 0, 0, a and 0, 0, $-a$ as the "north and south poles"). On the other hand, if $u =$ constant, then $y = (\tan u)x$, and thus the $u =$ constant curves are lines of "longitude" or "meridians."[3] [If this discussion looks familiar, recall the spherical polar system (8.32) and note that u and v actually correspond to the angles ϕ and θ, respectively.]

To develop the full spherical surface, we confine u, v to the domain $0 \leq u < 2\pi$, say, and $0 \leq v \leq \pi$.

It is not hard to imagine that the u, v coordinates are more natural for calculations involving the surface of the sphere than the original cartesian variables. The "calculations" might, for instance, consist of the dynamics of a bead on the surface, a calculation of the surface area, and so on. Finally, note that the u, v coordinates in this example happen to be *orthogonal*; that is, the $u =$ constant and $v =$ constant curves intersect at right angles. The special significance of orthogonality will be discussed later in Chapter 9. ▮

Area of a Curved Surface. Part of the problem of computing the area of a curved surface is first deciding what is meant by the area of a curved surface. We know what the area of a *flat* surface means. For example, the area of a rectangle is defined to be the product of its dimensions. Extending this idea, we define a differential area dA of a *curved* surface as the area of its tangent plane approximation, as sketched in Fig. 8.13. The point of tangency is immaterial; we've chosen the corner P. The tangent vectors $\mathbf{ds}_1$ and $\mathbf{ds}_2$, which define the tangent plane parallelogram, may be expressed as follows.

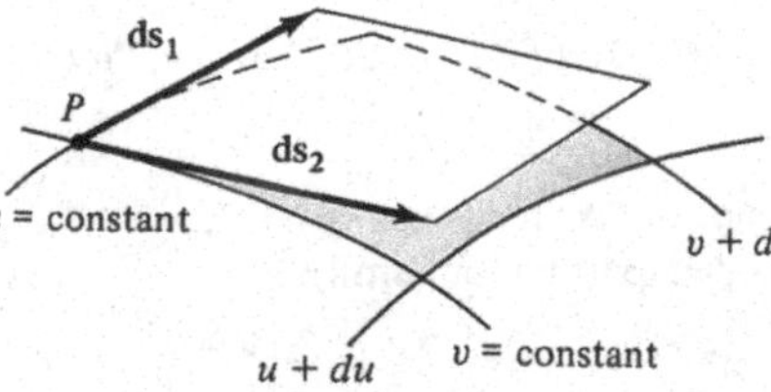

Figure 8.13. Tangent-plane definition of dA.

$$\mathbf{ds}_1 = dx\hat{\mathbf{i}} + dy\hat{\mathbf{j}} + dz\hat{\mathbf{k}}$$

where

$$dx = x_u\,du + x_v\,dv = x_v\,dv$$
$$dy = y_u\,du + y_v\,dv = y_v\,dv$$
$$dz = z_u\,du + z_v\,dv = z_v\,dv,$$

since $\mathbf{ds}_1$ is along the $u =$ constant curve, and thus $du = 0$. So

$$\mathbf{ds}_1 = (x_v\hat{\mathbf{i}} + y_v\hat{\mathbf{j}} + z_v\hat{\mathbf{k}})\,dv \equiv \mathbf{V}\,dv, \tag{8.43}$$

and similarly

$$\mathbf{ds}_2 = (x_u\hat{\mathbf{i}} + y_u\hat{\mathbf{j}} + z_u\hat{\mathbf{k}})\,du \equiv \mathbf{U}\,du. \tag{8.44}$$

[3] A *meridian* is a curve on a surface of revolution formed by the intersection of the surface with a plane containing the axis of revolution.

Next, recalling (8.7),

$$dA = \|\mathbf{ds}_2 \times \mathbf{ds}_1\| = \|(\mathbf{U}\,du) \times (\mathbf{V}\,dv)\| = \|\mathbf{U} \times \mathbf{V}\|\,du\,dv$$
$$= \sqrt{(\mathbf{U} \times \mathbf{V}) \cdot (\mathbf{U} \times \mathbf{V})}\,du\,dv.$$

With **U** and **V** as defined above, this yields

$$dA = \sqrt{EG - F^2}\,du\,dv, \tag{8.45a}$$

where

$$E = x_u^2 + y_u^2 + z_u^2$$
$$F = x_u x_v + y_u y_v + z_u z_v \tag{8.45b}$$
$$G = x_v^2 + y_v^2 + z_v^2.$$

Example 8.4. Returning to Example 8.3, let us use (8.45) to compute the surface area of the sphere. With $x(u, v)$, $y(u, v)$, and $z(u, v)$ given by (8.41), we find that

$$E = a^2 \sin^2 v, \qquad F = 0, \qquad G = a^2$$

so that

$$A = \int_0^{\pi} \int_0^{2\pi} a^2\,|\sin v|\,du\,dv. \tag{8.46}$$

Dropping the absolute value signs, since $\sin v \geq 0$ over the domain of integration, and integrating, we easily obtain $A = 4\pi a^2$, which, of course, we recognize as being correct.

The point to notice is the happy simplicity of (8.46); the integrand is quite simple and the limits are independent (i.e., the u limits are not functions of v). This situation merely reflects the fact that the curvilinear coordinates (8.41) "fit" the spherical geometry so naturally.

Finally, note the significance of $F = 0$. In general, $F = x_u x_v + y_u y_v + z_u z_v = \mathbf{U} \cdot \mathbf{V}$. Thus $F = 0$ means that $\mathbf{U} \cdot \mathbf{V} = 0$—that is, the constant u and constant v curves are *orthogonal.* ▮

Special cases of (8.45).

(i) Suppose that $z = 0$—that is,

$$x = x(u, v), \qquad y = y(u, v), \qquad z = 0, \tag{8.47}$$

so that our surface is *flat* and lies in the x, y plane. Then

$$E = x_u^2 + y_u^2, \qquad F = x_u x_v + y_u y_v, \qquad G = x_v^2 + y_v^2;$$

therefore

$$EG - F^2 = x_u^2 y_v^2 - 2x_u y_v x_v y_u + x_v^2 y_u^2 = (x_u y_v - x_v y_u)^2,$$

which is none other than the *Jacobian* $J(u, v)$ squared! So we have the neat result

$$dA = |J(u, v)|\,du\,dv, \tag{8.48}$$

the absolute value signs being needed to ensure that we have the *positive* square root of $EG - F^2$, since dA must surely be positive.

For instance, if

$$x = r\cos\theta, \qquad y = r\sin\theta,$$

then $|J| =$ etc. $= r$ (independent of whether we identify r as u and θ as v or vice versa) and $dA = r\,dr\,d\theta$. Of course, this result might have been deduced instead from the

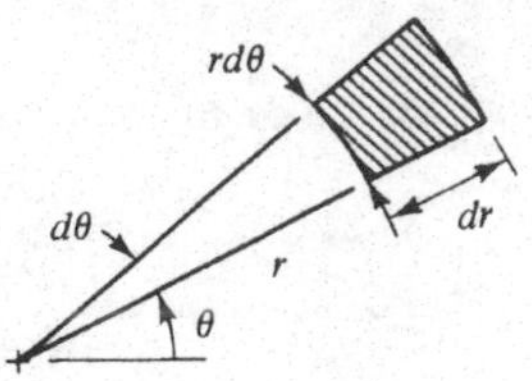

Figure 8.14. Area element for polar coordinates.

simple sketch in Fig. 8.14, but the "method of sketches" may not be practicable in more complicated cases, whereas (8.48) is always straightforward.

(ii) Suppose that we are given the equation of the surface in the form $z = f(x, y)$. That's equivalent to the system

$$x = x(u, v) = u, \qquad y = y(u, v) = v,$$
$$z = z(u, v) = f(u, v). \tag{8.49}$$

Thus

$$E = 1 + f_u^2, \qquad F = f_u f_v, \qquad G = 1 + f_v^2$$

which means that

$$dA = \sqrt{f_u^2 + f_v^2 + 1}\, du\, dv$$

or since $u = x$ and $v = y$,

$$dA = \sqrt{f_x^2 + f_y^2 + 1}\, dx\, dy. \tag{8.50}$$

Example 8.5. Again returning to our sphere $x^2 + y^2 + z^2 = a^2$, let us compute its surface area, using (8.50). With $z = f(x, y) = +\sqrt{a^2 - x^2 - y^2}$, on the top half, say, we have (Exercise 8.26)

$$A = 2\int_{-a}^{a}\int_{-\sqrt{a^2-y^2}}^{\sqrt{a^2-y^2}} \frac{a}{\sqrt{a^2 - x^2 - y^2}}\, dx\, dy = 4\pi a^2. \quad \blacksquare$$

8.5. *CURVILINEAR COORDINATE REPRESENTATION OF VOLUMES*

You will probably not be surprised that we now consider the representation of a *volume* by a *three*-parameter family,

$$x = x(u, v, w), \qquad y = y(u, v, w), \qquad z = z(u, v, w). \tag{8.51}$$

In general, (8.51) yields a surface for each fixed w; and as w varies continuously, we expect this family of surfaces to generate a volume.

Proceeding along essentially the same lines as earlier, we have from (8.10) and Fig. 8.15 the expression

$$dV = |\mathbf{U}\, du \cdot \mathbf{V}\, dv \times \mathbf{W}\, dw| = |\mathbf{U} \cdot \mathbf{V} \times \mathbf{W}|\, du\, dv\, dw$$

for the volume element, where $\mathbf{U}$ and $\mathbf{V}$ are the same as above [(8.43) and (8.44)] and $\mathbf{W}$ is $x_w\hat{\mathbf{i}} + y_w\hat{\mathbf{j}} + z_w\hat{\mathbf{k}}$. Recalling (8.35), we can express

$$dV = \left\| \begin{matrix} x_u & x_v & x_w \\ y_u & y_v & y_w \\ z_u & z_v & z_w \end{matrix} \right\| du\, dv\, dw,$$

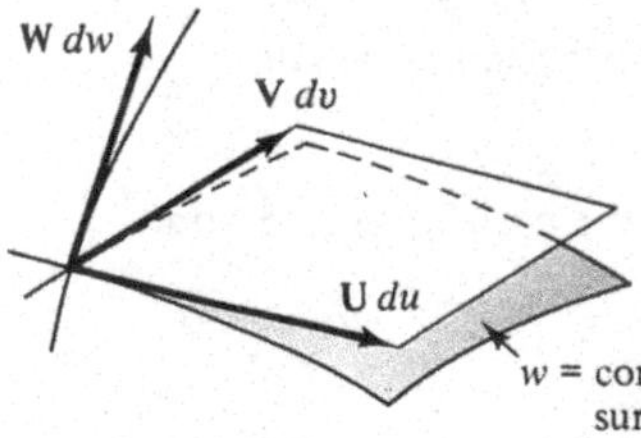

Figure 8.15. Forming the parallelepiped for dV.

where the inner set of vertical lines denotes determinant and the outer set indicates absolute value.

But the determinant is none other than the Jacobian $J(u, v, w)$, and so

$$dV = |J(u, v, w)|\, du\, dv\, dw. \tag{8.52}$$

[Note the similarity to (8.48).]

Example 8.6. To compute the *volume* of our sphere of radius a, we introduce the more natural curvilinear coordinates

$$x = w \sin v \cos u, \qquad y = w \sin v \sin u, \qquad z = w \cos v.$$

Then $|J(u, v, w)| = \text{etc.} = |w^2 \sin v|$, and therefore

$$V = \int_0^a \int_0^\pi \int_0^{2\pi} w^2 \sin v \, du\, dv\, dw = \tfrac{4}{3}\pi a^3. \quad \blacksquare$$

EXERCISES

8.1. Show that equations (8.1) follow from our definitions of addition and scalar multiplication of vectors.

8.2. Using (8.1) and the definition of **0**, show carefully that $\mathbf{A} + \mathbf{0} = \mathbf{A}$ (for all **A**s) and that $\mathbf{A} + \mathbf{B} = \mathbf{C}$ implies $\mathbf{A} = \mathbf{C} - \mathbf{B}$.

8.3. Exactly how does (8.5) follow from (8.3) and (8.4)?

8.4. If $\|\mathbf{A}\| = 1$, $\|\mathbf{B}\| = 2$, and $\|\mathbf{C}\| = 5$, can $\mathbf{A} + \mathbf{B} + \mathbf{C} = \mathbf{0}$? How about if $\|\mathbf{A}\| = 4$ instead? Explain.

8.5. If $\mathbf{A} = \hat{\imath} - \hat{k}$, $\mathbf{B} = 2\hat{\imath} + \hat{\jmath} + \hat{k}$, and $\mathbf{C} = -\hat{\imath} + \hat{\jmath}$, evaluate $2\mathbf{A} - 3\mathbf{B}$, $\mathbf{A} \cdot \mathbf{B}$, $\mathbf{B} \cdot \mathbf{A}$, $\mathbf{A} \times \mathbf{B}$, $\mathbf{B} \times (3\mathbf{A})$, $\mathbf{A} \times (\mathbf{B} \times \mathbf{C})$, the (acute) angle between **A** and **B**, the projection of **B** on **A**, and the projection of **A** on **B**. Find a scalar α such that $\mathbf{A} + \alpha\mathbf{B}$ is orthogonal to **A**. Find scalars α, β such that $\mathbf{A} + \alpha\mathbf{B} + \beta\mathbf{C}$ is orthogonal to both **A** and **B**.

8.6. What (acute) angle does $\mathbf{A} = \hat{\imath} - 2\hat{k}$ make with the normal to the plane containing the vectors $\mathbf{B} = \hat{\jmath} - \hat{k}$ and $\mathbf{C} = \hat{\imath} + \hat{\jmath} + \hat{k}$?

8.7. Given distinct vectors **A**, **B**, **C**, **D**, which are tail to tail, give a necessary and sufficient condition(s) for their heads all to lie in a common plane.

8.8. To show that the diagonals of a parallelogram bisect each other, proceed as follows. Note from Fig. 8.16 that $\mathbf{A} + \mathbf{B} = \mathbf{C}$, $\mathbf{A} - \alpha\mathbf{D} = \beta\mathbf{C}$, $\mathbf{A} = \mathbf{B} + \mathbf{D}$. Eliminating **A** and **B**, say, we obtain $(2\beta - 1)\mathbf{C} = (1 - 2\alpha)\mathbf{D}$. Since **C** and **D** are not aligned, it must be true that $2\beta - 1 = 1 - 2\alpha = 0$ — that is, $\alpha = \beta = \frac{1}{2}$. Use this same general procedure to show that a line from one vertex of a parallelogram to the midpoint of a nonadjacent side trisects a diagonal.

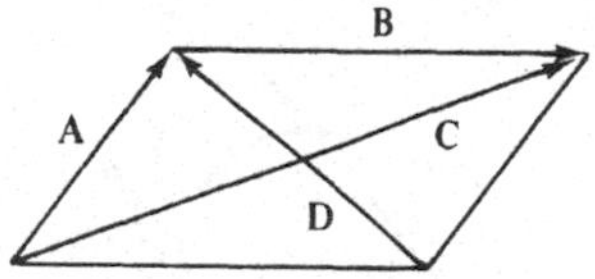

Figure 8.16. Parallelogram.

8.9. In mechanics we call $\mathbf{OA} \times \mathbf{F} = \mathbf{M}$ the *moment* of the force **F** (which acts through the point A) about the point O, where **OA** is the vector from O to A.

(a) If $\mathbf{F}_1 = \hat{\mathbf{i}} + \hat{\mathbf{j}}$ acts through (0, 3, 2) and $\mathbf{F}_2 = \hat{\mathbf{i}} - 2\hat{\mathbf{j}} - \hat{\mathbf{k}}$ acts through (1, −1, 4), find (if possible) the force $\mathbf{F}_3$ needed to act through (2, 2, 1) such that the system has no net moment about the point (1, 1, 0). Show that the net moment induced by $\mathbf{F}_1$ and $\mathbf{F}_2$ about (1, 1, 0) is not orthogonal to the position vector from (1, 1, 0) to (2, 2, 1). How is this fact pertinent?

(b) If $\mathbf{F}_1$ and $\mathbf{F}_2$ act through distinct points A_1 and A_2, derive a condition on $\mathbf{F}_1$ and $\mathbf{F}_2$ that will ensure that their resultant moment about O is independent of the location of O. Such a system of forces is said to constitute a pure *couple*. Can you give a specific example of *three* (nonzero) forces through distinct points that constitute a couple?

8.10. (a) If $\mathbf{A}, \mathbf{B}, \mathbf{C}$ form a triangle such that $\mathbf{A} = \mathbf{B} + \mathbf{C}$, and α, β, γ are the angles opposite the sides $\mathbf{A}, \mathbf{B}, \mathbf{C}$, respectively, derive the *law of cosines*, $C^2 = A^2 + B^2 - 2AB\cos\alpha$, by starting with the identity $\mathbf{C} \cdot \mathbf{C} = (\mathbf{A} - \mathbf{B}) \cdot (\mathbf{A} - \mathbf{B})$.

(b) Derive the *law of sines*, $(\sin\alpha)/A = (\sin\beta)/B$, by starting with $\mathbf{C} \times \mathbf{C} = \mathbf{C} \times (\mathbf{A} - \mathbf{B})$.

8.11. Derive **Lagrange's identity**,

$$(\mathbf{A} \times \mathbf{B}) \cdot (\mathbf{A} \times \mathbf{B}) = \begin{vmatrix} \mathbf{A} \cdot \mathbf{A} & \mathbf{A} \cdot \mathbf{B} \\ \mathbf{A} \cdot \mathbf{B} & \mathbf{B} \cdot \mathbf{B} \end{vmatrix}$$

by noting that $(\mathbf{A} \times \mathbf{B}) \cdot (\mathbf{A} \times \mathbf{B}) = \mathbf{A} \cdot \mathbf{B} \times (\mathbf{A} \times \mathbf{B})$ (why?) and using (8.11) and (8.12).

8.12. (a) Show that the statement of linearity, (8.9), is equivalent to the two separate statements $(\alpha\mathbf{A}) \times \mathbf{B} = \alpha(\mathbf{A} \times \mathbf{B})$ and $(\mathbf{A} + \mathbf{B}) \times \mathbf{C} = \mathbf{A} \times \mathbf{C} + \mathbf{B} \times \mathbf{C}$.

(b) Prove that each of these two statements is correct.

8.13. Derive (8.31) by taking the limit of the difference quotient as we did in deriving (8.28).

8.14. Derive (8.33) by taking the limit of the difference quotient as we did in deriving (8.28).

8.15. Derive (8.28) and (8.31) by the method employed in Example 8.2.

8.16. Finish Example 8.2; that is, compute $\partial\hat{\mathbf{e}}_\theta/\partial\theta$, $\partial\hat{\mathbf{e}}_\theta/\partial\phi$, and $\partial\hat{\mathbf{e}}_\phi/\partial\phi$ and compare your results with (8.33).

8.17. Compute $d\mathbf{A}/dt$ if $\mathbf{A} = \sin\theta\,\hat{\mathbf{e}}_r - r\theta^2\hat{\mathbf{e}}_\theta + r\hat{\mathbf{e}}_z$, with $r = t^2$ and $\theta = 3t$.

8.18. A particle moves in the x, y plane according to $x = t^2$, $y = 2$. Compute its velocity and acceleration in terms of $\hat{\mathbf{e}}_r$ and $\hat{\mathbf{e}}_\theta$.

8.19. Consider a particle P moving under the influence of a *central force field*—that is, where the force $\mathbf{F}$ on P is always aligned with the position vector $\mathbf{R}$ from some fixed point, say O, as shown in Fig. 8.17. Starting with Newton's law $\mathbf{F} = m\ddot{\mathbf{R}}$ and noting that the moment $\mathbf{R} \times \mathbf{F}$ about O is zero, show that the so-called *moment of momentum*

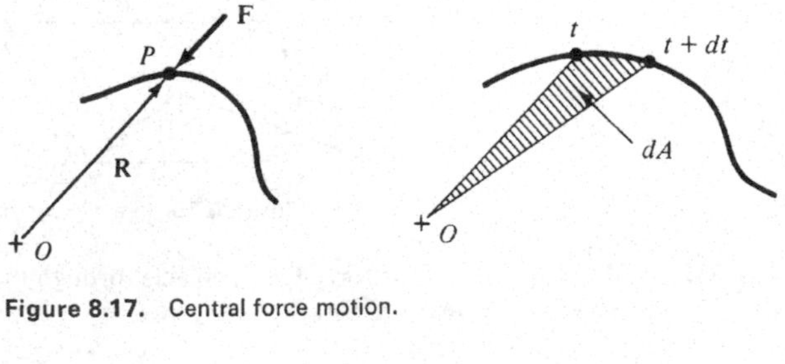

Figure 8.17. Central force motion.

$\mathbf{R} \times m\dot{\mathbf{R}}$ is a constant vector. Using this fact, show that the motion of P must lie in a *plane* and that the *areal velocity* dA/dt (i.e., the rate at which area is swept out by the $\mathbf{R}$ vector) must be constant.

8.20. (**Frenet's formulas**) Consider a space curve C defined by $\mathbf{R} = \mathbf{R}(s)$, where s is the arc length along C, measured from some reference point. The length of the tangent vector $d\mathbf{R} = dx\hat{\mathbf{i}} + dy\hat{\mathbf{j}} + dz\hat{\mathbf{k}}$ is $\sqrt{(dx)^2 + (dy)^2 + (dz)^2}$, or ds, so that $d\mathbf{R}/ds \equiv \hat{\mathbf{t}}$ is a *unit tangent* vector. Since $\hat{\mathbf{t}}$ is of unit length, show that $d\hat{\mathbf{t}}/ds$ must be normal to it. Thus we write

$$\frac{d\hat{\mathbf{t}}}{ds} \equiv \kappa\hat{\mathbf{n}} \qquad (\kappa \geq 0), \tag{8.53}$$

where the unit vector $\hat{\mathbf{n}}$ is called the **principal normal**; the plane of $\hat{\mathbf{t}}$ and $\hat{\mathbf{n}}$ is called the **osculating plane.** With the help of suitable sketches show that the multiplier κ is the **curvature** of C—that is, $1/\rho$, where ρ is the local **radius of curvature** of C, and that $\hat{\mathbf{n}}$ points toward the center of curvature. (If C is straight, then ρ is infinite and $\kappa = 0$.) We introduce a third unit vector, the **binormal** $\mathbf{b} \equiv \hat{\mathbf{t}} \times \hat{\mathbf{n}}$. (Sketch all this.) Taking d/ds of $\hat{\mathbf{b}} = \hat{\mathbf{t}} \times \hat{\mathbf{n}}$, show that $d\hat{\mathbf{b}}/ds$ is parallel to $\hat{\mathbf{n}}$, so that

$$\frac{d\hat{\mathbf{b}}}{ds} = \tau\hat{\mathbf{n}}. \tag{8.54}$$

The factor τ is called the **torsion**, since it is a measure of the rate at which C twists out of its osculating plane. Next, show that

$$\frac{d\hat{\mathbf{n}}}{ds} = -\kappa\hat{\mathbf{t}} - \tau\hat{\mathbf{b}}. \tag{8.55}$$

Equations (8.53) to (8.55) are called the *Frenet–Serret formulas*. Finally, show that

$$\kappa = \|\hat{\mathbf{t}}'\| = \|\mathbf{R}''\| \tag{8.56}$$

and that $\hat{\mathbf{t}} \cdot \hat{\mathbf{t}}' \times \hat{\mathbf{t}}'' = -\kappa^2\tau$, and thus

$$\tau = -\frac{\mathbf{R}' \cdot \mathbf{R}'' \times \mathbf{R}'''}{\mathbf{R}'' \cdot \mathbf{R}''}, \tag{8.57}$$

where the primes denote d/ds.

8.21. (First read the previous exercise.) If we move along the curve C defined by $\mathbf{R} = \mathbf{R}(s)$ according to $s = s(t)$, where t is the time, show that our velocity and acceleration are

$$\mathbf{v} = \dot{s}\hat{\mathbf{t}} \quad \text{and} \quad \mathbf{a} = \ddot{s}\hat{\mathbf{t}} + \frac{\dot{s}^2}{\rho}\hat{\mathbf{n}}.$$

8.22. A tiny elephant moves along the path shown in Fig. 8.18 with speed $\dot{s} = 6t$. Using the results of the previous exercise, give an expression for its acceleration $\mathbf{a}$ over each

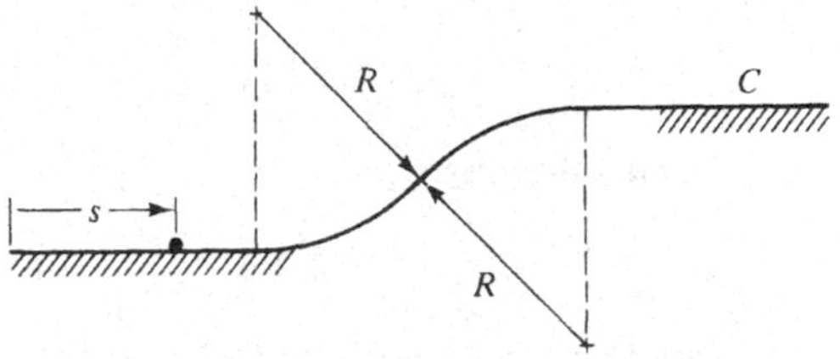

Figure 8.18.

segment of C—that is, the two linear parts and the two circular arcs. Where does the maximum acceleration occur?

8.23. For the helix defined parametrically by

$$x = a\cos\zeta, \qquad y = a\sin\zeta, \qquad z = \omega\zeta \quad (a, \omega \text{ constants})$$

(a) Compute the arc length of the helix from $\zeta = 0$ out to an arbitrary ζ.
(b) Give expressions for $\hat{\mathbf{t}}$, $\hat{\mathbf{n}}$ and $\hat{\mathbf{b}}$ (see Exercise 8.20) in terms of $\hat{\mathbf{i}}, \hat{\mathbf{j}}, \hat{\mathbf{k}}$. *Hint:* $\hat{\mathbf{t}} = (d\mathbf{R}/d\zeta)(d\zeta/ds)$, where $d\zeta/ds$ is found simply as $1/\|d\mathbf{R}/d\zeta\|$, since $\hat{\mathbf{t}}$ is to be of unit length, or from your solution to part (a).
(c) Evaluate κ, ρ and τ. Check your solution for the special limiting cases where $\omega \longrightarrow 0$ and $\omega \longrightarrow \infty$. Do your answers look right? For example, would you expect κ, ρ, τ to vary along the helix or to be constant? Would you expect τ to be positive or negative? Why?

8.24. For the curve of intersection of the circular cylinder $x^2 + y^2 = 1$ with the circular cylinder $y^2 + z^2 = 1$, find the radius of curvature at the points (0, 1, 0) and (1, 0, 1). (See Exercise 8.20.)

***8.25.** In Section 8.3 we defined the angular velocity vector $\boldsymbol{\Omega}$. Can we also define an angular *rotation* vector, say $\boldsymbol{\theta}$? That is, $\boldsymbol{\theta}$ would be the amount of angular rotation θ about some axis, times a unit vector along that axis in the direction assigned by the "right-hand rule." *Hint:* Show that vector addition would not be commutative.

8.26. Fill in the omitted steps in Example 8.5 without using an integral table.

8.27. For polar coordinates, we can bypass the expression $dA = |J|\,dr\,d\theta$ and obtain $dA = r\,dr\,d\theta$ simply from Fig. 8.14. Try this for the curvilinear u, v coordinates defined by $x = u + v$, $y = u - v$ and compare with the result given by $|J|\,du\,dv$.

8.28. Derive the volume element $dV = \rho^2 \sin\theta\, d\rho\, d\theta\, d\phi$ for spherical polars first from a simple three-dimensional sketch and then from the expression $dV = |J|\,d\rho\,d\theta\,d\phi$.

8.29. Evaluate the integrals shown.

(a) $I = \int_0^{1/2}\int_0^{1-2y} e^{x/(x+2y)}\,dx\,dy.$

Hint: Set $u = x$, $v = x + 2y$. In computing the needed Jacobian $J(u, v) = x_u y_v - x_v y_u$, note that $J(u, v) = [J(x, y)]^{-1} = (u_x v_y - u_y v_x)^{-1}$, per Exercise 7.18, is more convenient, since we know $u(x, y)$ and $v(x, y)$, not $x(u, v)$ and $y(u, v)$. [In this example there's little difference because we can easily solve for $x(u, v)$ and $y(u, v)$, but this inversion is not always simple.] The same comment applies to the *volume* element, where we need the Jacobian $J(u, v, w)$. Again, it's true (Exercise 7.18) that $J(u, v, w) = [J(x, y, z)]^{-1}$.

(b) $I = \int_0^1\int_0^{1-z}\int_0^{1-y-z} e^{x/(x+y)}\,dx\,dy\,dz.$

(c) $I = \int_0^\infty\int_y^\infty e^{-(x^2+y^2)}\,dx\,dy.$

(d) $I = \int_0^1\int_0^{1-y} \sin\left(\dfrac{x-y}{x+y}\right)dx\,dy.$

***8.30.** We meet the integral

$$I = -\frac{1}{8\pi^3}\int_{-\infty}^{\infty}\int_{-\infty}^{\infty}\int_{-\infty}^{\infty}\frac{e^{i(\xi x+\eta y+\zeta z)}}{\xi^2+\eta^2+\zeta^2}\,d\xi\,d\eta\,d\zeta$$

when we solve the so-called Poisson equation by the Fourier transform. Show that $I = -1/4\pi r$, where $r = \sqrt{x^2 + y^2 + z^2}$. *Hint:* Note that $\exp i(\xi x + \eta y + \zeta z) =$

exp $i\mathbf{R} \cdot \mathbf{r} = \exp irR\cos\theta$, where $\mathbf{R}$ and $\mathbf{r}$ are shown in Fig. 8.19, and change over to spherical polars R, θ, ϕ with $\mathbf{r}$ as the polar axis.

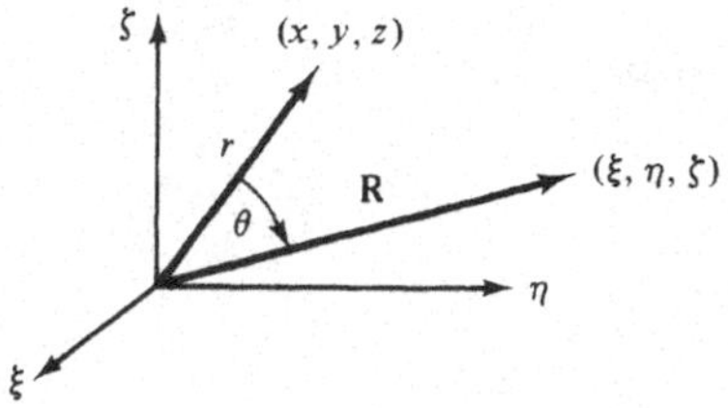

Figure 8.19. The coordinates.

8.31. Compute the moment of inertia about the y axis, $I_y = \iint x^2\,dA$, of the area enclosed by the circle $(x - a)^2 + y^2 = a^2$ by changing to polar coordinates.

8.32. The gravitational force of attraction induced per unit mass at the origin by the right circular cone of height h and (uniform) mass density σ shown in Fig. 8.20 is given by

$$F = \iiint_V \frac{\sigma z\,dx\,dy\,dz}{(x^2 + y^2 + z^2)^{3/2}},$$

where V is the volume of the cone.

(a) Evaluate F by changing to cylindrical coordinates.
(b) Repeat, using spherical polars.
(c) Evaluate F if V is instead the *hemisphere* $x^2 + y^2 + z^2 < a^2$ for $0 \leq z \leq a$.

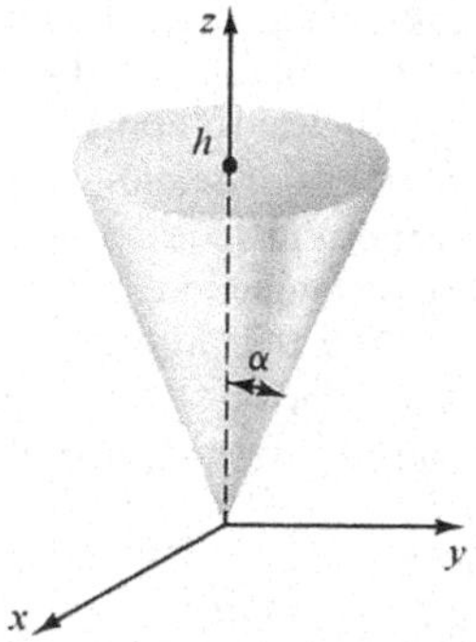

Figure 8.20. Right circular cone.

8.33. Show that the paraboloid $x^2 + y^2 = a^2z$ can be represented by the two-parameter family $x = av\cos u$, $y = av\sin u$, $z = v^2$. Sketch the paraboloid and the constant u and constant v curves on it.

8.34. Make up two examination-type problems on this chapter.

Chapter 9

Vector Field Theory

In this chapter we introduce the so-called gradient, divergence, and curl vector differential operators and then develop the several transformation theorems of Gauss, Green, and Stokes that relate volume, surface, and line integrals. These results, which are central to applied analysis, play an especially important role in the development of electrodynamics, thermodynamics, and the continuum mechanics. In fact, our illustrations include derivations of some of the governing equations of fluid mechanics and heat conduction, as well as a discussion of the gravitational potential. That is, here we see how the physical laws (e.g., conservation of mass, Fourier's law of heat conduction, and Newton's law of gravitational attraction) plus the vector integral theorems lead to the governing partial differential equations; later on we discuss their solution.

9.1. LINE INTEGRALS

To illustrate, suppose that we have defined throughout some region R of x, y, z space a vector force field $\mathbf{F}(x, y, z)$ per unit mass at x, y, z. If we denote by $\mathbf{r}$ the position vector from some fixed point O to a unit mass that moves along some space curve C (Fig. 9.1), then the work done by the force field on the particle in traversing C is defined by the **line integral**

$$W = \int_C \mathbf{F} \cdot d\mathbf{r}. \tag{9.1}$$

By the right-hand side we mean (as might be guessed from our discussion of the

Riemann integral in Section 1.4) the limit

$$\lim_{|P|\to 0} \sum_{k=1}^{n} \mathbf{F}(P_{k-1}) \cdot [\mathbf{r}(P_k) - \mathbf{r}(P_{k-1})], \tag{9.2}$$

where P_k is shorthand for the point x_k, y_k, z_k. The points $P_0(= A), P_1, \ldots, P_{n-1}, P_n(= B)$ define a "partition" P of C, and $|P| \equiv \max_{1\le k\le n} \|\mathbf{r}(P_k) - \mathbf{r}(P_{k-1})\|$ is called the "norm" of this partition. This definition is exactly analogous to the Riemann definition (1.14), where in this case we have decided to choose "ξ_k" = "x_{k-1}"—that is, at the left endpoint of each segment.

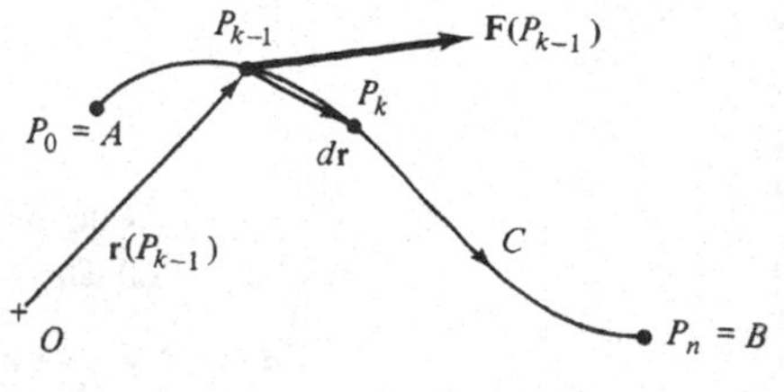

Figure 9.1. The line integral (9.1).

Note that there is an *orientation* to C, as indicated in the figure by the arrowhead. If we denote by $-C$ the same curve but with opposite orientation, then

$$\int_{-C} \mathbf{F} \cdot d\mathbf{r} = -\int_C \mathbf{F} \cdot d\mathbf{r}, \tag{9.3}$$

since the $d\mathbf{r}$'s are oppositely directed.

Line integrals also come in forms other than (9.1); for instance,

$$\int_C f\,ds, \quad \int_C f\,dx, \quad \int_C f\,dy, \quad \text{and} \quad \int_C f\,dz,$$

where $f = f(x, y, z)$, the Cs are given space curves, and s is arc length along C, are all line integrals and are defined as limits of Riemann-type sums. Practically speaking, however, they are simply special cases of (9.1). For instance, if $\mathbf{r} = \mathbf{r}(s)$, then

$$\int_C \mathbf{F} \cdot d\mathbf{r} = \int_C \left(\mathbf{F} \cdot \frac{d\mathbf{r}}{ds}\right) ds = \int_C f\,ds,$$

where

$$f \equiv \mathbf{F} \cdot \frac{d\mathbf{r}}{ds}.$$

Example 9.1. Consider the two-dimensional case where $\mathbf{F} = 2xy\hat{\mathbf{i}} + (x^2 - 1)\hat{\mathbf{j}}$ and C is the spiral $r = 2\theta$, from $\theta = 0$ to $\theta = 5\pi/2$, as shown in Fig. 9.2. In this case, $\mathbf{F}$ is given in terms of cartesian coordinates and C in terms of polar coordinates. Anticipating that it may be more convenient to work this problem in polar coordinates, let us take a step in that direction by expressing $d\mathbf{r} = dr\,\hat{\mathbf{e}}_r + r\,d\theta\,\hat{\mathbf{e}}_\theta$. Then

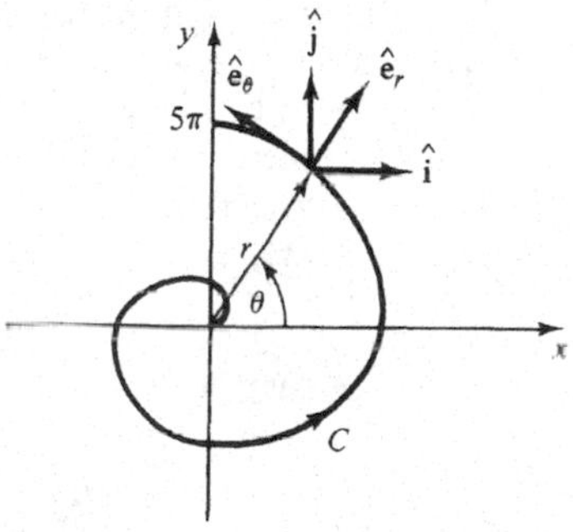

Figure 9.2. The spiral $r = 2\theta$.

$$\begin{aligned}
\int_C \mathbf{F}\cdot d\mathbf{r} &= \int_C [2xy\hat{\mathbf{i}} + (x^2 - 1)\hat{\mathbf{j}}] \cdot (dr\,\hat{\mathbf{e}}_r + r\,d\theta\,\hat{\mathbf{e}}_\theta) \\
&= \int_C [2xy(\hat{\mathbf{i}} \cdot \hat{\mathbf{e}}_r)\,dr + 2xy(\hat{\mathbf{i}} \cdot \hat{\mathbf{e}}_\theta)r\,d\theta \\
&\qquad + (x^2 - 1)(\hat{\mathbf{j}} \cdot \hat{\mathbf{e}}_r)\,dr + (x^2 - 1)(\hat{\mathbf{j}} \cdot \hat{\mathbf{e}}_\theta)r\,d\theta] \\
&= \int_C [2r^2 cs(c)\,dr + 2r^2 cs(-s)r\,d\theta \\
&\qquad + (r^2c^2 - 1)(s)\,dr + (r^2c^2 - 1)(c)r\,d\theta],
\end{aligned}$$

where $c \equiv \cos\theta$ and $s \equiv \sin\theta$ for short. Finally, on C $r = 2\theta$, so that

$$\int_C \mathbf{F} \cdot d\mathbf{r} = \int_0^{5\pi/2} (16\theta^2 c^2 s - 16\theta^3 c s^2 + 8\theta^2 c^2 s - 2s + 8\theta^3 c^3 - 2\theta c)\, d\theta$$

$$= \text{etc.} = -5\pi. \tag{9.4}$$

As in Chapter 8, we emphasize that $d\mathbf{r}$ in polar coordinates is $dr\,\hat{\mathbf{e}}_r + r\,d\theta\,\hat{\mathbf{e}}_\theta$, *not* $dr\,\hat{\mathbf{e}}_r + d\theta\,\hat{\mathbf{e}}_\theta$! This situation results from expressing $\mathbf{r} = r\hat{\mathbf{e}}_r$, so that

$$d\mathbf{r} = dr\,\hat{\mathbf{e}}_r + r\,d\hat{\mathbf{e}}_r = dr\,\hat{\mathbf{e}}_r + r\left[\left(\frac{\partial\hat{\mathbf{e}}_r}{\partial r}\right) dr + \left(\frac{\partial\hat{\mathbf{e}}_r}{\partial\theta}\right) d\theta\right] = dr\,\hat{\mathbf{e}}_r + r\,d\theta\,\hat{\mathbf{e}}_\theta,$$

as can also be seen graphically by sketching $\mathbf{r}$ and $\mathbf{r} + d\mathbf{r}$, identifying $d\mathbf{r}$, and breaking it into its $\hat{\mathbf{e}}_r$ and $\hat{\mathbf{e}}_\theta$ components. ▮

9.2. *THE CURVES, SURFACES, AND REGIONS UNDER CONSIDERATION*

Throughout this chapter we assume that our curves and surfaces are "suitably decent." In particular, we assume that they are **piecewise smooth**.

A curve C, given by $r = r(s)$, is said to be **smooth** if the unit tangent vector $\hat{\mathbf{t}} = d\mathbf{r}/ds$ is a continuous function of s over C, or, equivalently, if dx/ds, dy/ds, and dz/ds are continuous over C. If it is smooth except at a finite number of points, such that "left-" and "right-hand derivatives" $d\mathbf{r}/ds$ exist at these points, then we say that it is **sectionally** or **piecewise smooth**. Certainly the curves in Examples 9.1 and 9.2 are piecewise smooth. It can be shown that every piecewise smooth curve between endpoints is *rectifiable*—that is, has finite length. Finally, if the initial and final points of C (A and B in Fig. 9.1) coincide, we say that C is a *closed* curve.

Turning to surfaces, we say that a surface S is **smooth** if its unit normal $\hat{\mathbf{n}}$ is a continuous function of position over S. If S can be divided by smooth curves into a finite number of segments, each of which is smooth, then we say that S is **piecewise smooth**. For instance, a cube is not smooth, but it is piecewise smooth.

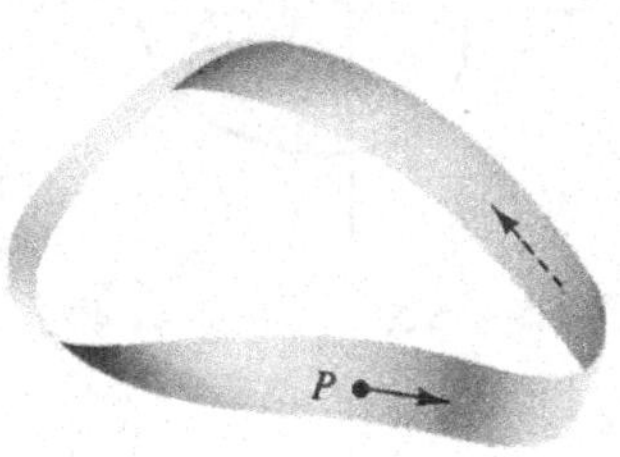

Figure 9.3. Möbius band.

Actually, we will also require our surfaces to be **orientable**—that is, having two sides—so that a skink could live on one side and a small klipspringer on the other without ever meeting. The classic example of a *non*orientable surface is the Möbius band, which can be constructed by taking a rectangular strip of paper and gluing the upper surface of one end to the lower surface of the other end (Fig. 9.3). If we start at any point P and follow the arrows, we eventually visit the entire surface, so that the Möbius band has only one side.

Our three-dimensional regions are to be bounded (by one or more piecewise smooth surfaces) and connected. By **connected** we mean that every pair of points in the region R can be joined by a piecewise smooth curve lying entirely in R. Furthermore, a connected region R is said to be **simply connected** if every *simple closed curve* (i.e., a piecewise smooth, closed curve) in R can be shrunk down to a point without leaving R; otherwise R is **multiply connected**.

To illustrate, consider the regions shown in Fig. 9.4; R_1 is the interior of a sphere, R_2 is the region *between* concentric spherical surfaces, and R_3 is the interior of a bagel. R_1 and R_2 are both simply connected, since every simple closed curve C (which we might think of as made of string) can be shrunk to a point; this fact is less obvious for R_2, since it appears that the inner spherical boundary gets in the way, but observe that the string can be slipped over this sphere and shrunk to the point P, for example. R_3 on the other hand, is multiply connected, since curves such as the one shown can*not* be shrunk to a point.

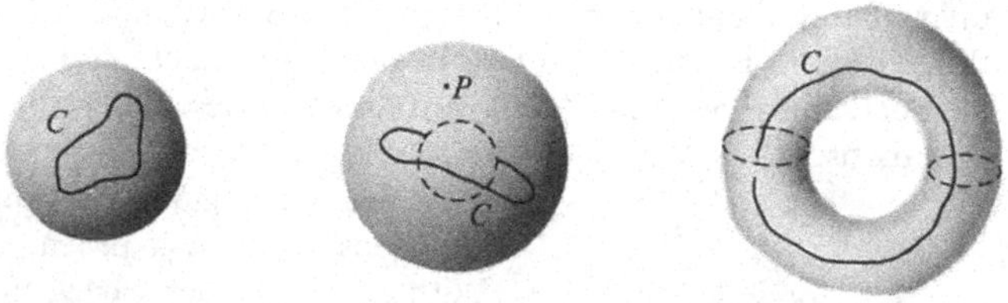

Figure 9.4. Simple and multiple connectedness.

9.3. *DIVERGENCE, GRADIENT, AND CURL*

In the following discussion we consider both scalar and vector fields, say $u(x, y, z)$ and $\mathbf{v}(x, y, z)$, or $u(P)$ and $\mathbf{v}(P)$ for short, defined throughout various three-dimensional regions R. Physically, u might be a temperature field and $\mathbf{v}$ a fluid velocity field, for example. They may vary with time as well, but the present discussion involves spatial considerations, and so we will simply ignore any t dependence except in applications where relevant.

We will assume of u and the three components of $\mathbf{v}$ that their first partial derivatives with respect to x, y and z all exist and are continuous throughout R.

Divergence. Associated with the vector field $\mathbf{v}$, we define a scalar field called div $\mathbf{v}$—that is, *the divergence of* $\mathbf{v}$—by

$$\text{div}\,\mathbf{v}(P) \equiv \lim_{\tau \to 0} \left\{ \frac{\int_\sigma \hat{\mathbf{n}} \cdot \mathbf{v}\, d\sigma}{\tau} \right\}, \tag{9.5}$$

where τ and σ are the volume and surface of a little "control volume," having an outward unit normal vector $\hat{\mathbf{n}}$ defined over σ[1] and containing the point P (see Fig. 9.5).

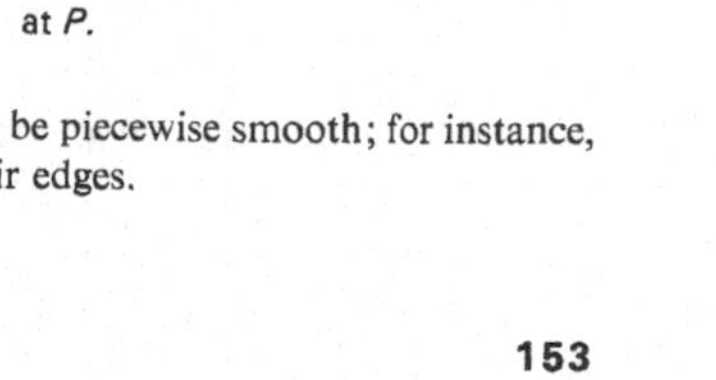

Figure 9.5. The control volume at P.

[1]That is, $\hat{\mathbf{n}}$ is defined over *most* of σ, since σ is only required to be piecewise smooth; for instance, for a cube, $\hat{\mathbf{n}}$ is defined over the six faces, although not along their edges.

For definiteness, let us think of $\mathbf{v}$ as a *fluid velocity field*, as indicated by the streamlines in Fig. 9.5. Note that the control volume τ is only a "mathematical" surface; it is completely permeable, and its presence in no way affects the flow.

Let us now interpret the right-hand side of (9.5). The outflow through the surface element $d\sigma$ is the normal velocity component $\hat{\mathbf{n}} \cdot \mathbf{v}$ (meters per second, say) times $d\sigma$ (meters squared)—that is, $\hat{\mathbf{n}} \cdot \mathbf{v}\, d\sigma$ m³/sec. (Note that the component of $\mathbf{v}$ that is *tangential* to $d\sigma$ produces no outflow.) Integrating over σ, $\int_\sigma \hat{\mathbf{n}} \cdot \mathbf{v}\, d\sigma$ is the *net* outflow, in cubic meters per second, say, across σ. And dividing by the volume τ, we have outflow per unit volume. Letting τ tend to 0, we see that div $\mathbf{v}(P)$ is the *outflow per unit volume at P*.

An important asset of (9.5) is that it provides a so-called *intrinsic* definition of div $\mathbf{v}$; that is, it contains no reference to any particular coordinate system. The price of this generality is that it is not particularly convenient computationally.

For definiteness, then, let us introduce a cartesian coordinate system, for instance, with the usual $\hat{\mathbf{i}}, \hat{\mathbf{j}}, \hat{\mathbf{k}}$ base vectors, and let us attempt to carry out the limit indicated in (9.5). The rectangular coordinate system suggests that we choose τ to be a prism, as sketched in Fig. 9.6. With x, y, z as our cartesian variables, let P be the point x_0, y_0, z_0 at the center of the prism.

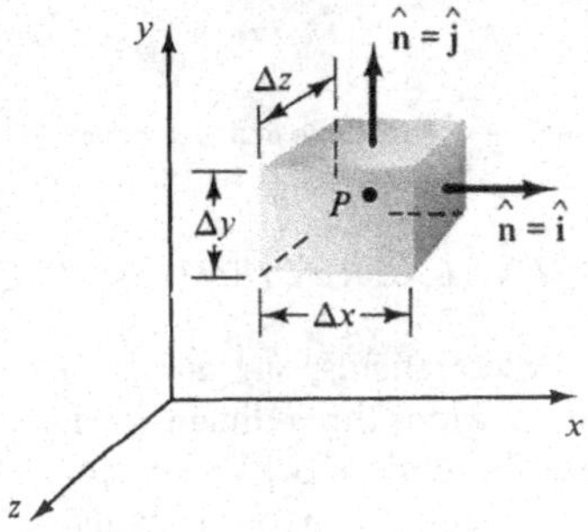

Figure 9.6. Cartesian coordinates.

Consider first the contribution from the right- and left-hand faces, on which $\hat{\mathbf{n}} = +\hat{\mathbf{i}}$ and $-\hat{\mathbf{i}}$, respectively. Since $\mathbf{v} = v_x\hat{\mathbf{i}} + v_y\hat{\mathbf{j}} + v_z\hat{\mathbf{k}}$, we have, with some help from the mean value theorem of the integral calculus,[2,3]

$$\int_{\text{right face}} \hat{\mathbf{n}} \cdot \mathbf{v}\, d\sigma = \int v_x\left(x_0 + \frac{\Delta x}{2}, y, z\right) d\sigma$$
$$= v_x\left(x_0 + \frac{\Delta x}{2}, y_1, z_1\right) \Delta y\, \Delta z$$

---

[2]If $f(x)$ has minimum and maximum values of m and M over $a \leq x \leq b$, then clearly

$$m(b - a) \leq \int_a^b f\, dx \leq M(b - a),$$

so that the integral equals $c(b - a)$, where $m \leq c \leq M$. But if f is continuous over $[a, b]$, then there is some point, say ξ, such that $f(\xi) = c$. Then

$$\int_a^b f\, dx = f(\xi)(b - a) \tag{9.6}$$

for some ξ in $[a, b]$, and this is the **mean value theorem of the integral calculus**. Similarly, if $f(x, y)$ is continuous over a bounded area A, then there is some point ξ, η in A such that

$$\iint_A f\, dx\, dy = f(\xi, \eta)A. \tag{9.7}$$

[3]Often we use only a single integral sign for multiple integrals.

and
$$\int_{\text{left face}} \hat{\mathbf{n}} \cdot \mathbf{v}\, d\sigma = \int -v_x\left(x_0 - \frac{\Delta x}{2}, y, z\right) d\sigma$$
$$= -v_x\left(x_0 - \frac{\Delta x}{2}, y_2, z_2\right) \Delta y\, \Delta z,$$

where $x_0 + (\Delta x/2), y_1, z_1$ is some point on the right-hand face, and $x_0 - (\Delta x/2), y_2, z_2$ is some point on the left hand face. Since $\tau = \Delta x\, \Delta y\, \Delta z$,

$$\lim_{\tau \to 0} \left\{ \frac{\int_{\text{left+right faces}} \hat{\mathbf{n}} \cdot \mathbf{v}\, d\sigma}{\tau} \right\}$$
$$= \lim_{\Delta x, \Delta y, \Delta z \to 0} \left\{ \frac{v_x[x_0 + (\Delta x/2), y_1, z_1] - v_x[x_0 - (\Delta x/2), y_2, z_2]}{\Delta x} \right\}$$
$$= \lim_{\Delta x \to 0} \frac{v_x[x_0 + (\Delta x/2), y, z] - v_x[x_0 - (\Delta x/2), y, z]}{\Delta x} = \frac{\partial v_x}{\partial x}. \tag{9.8}$$

Similarly, the top and bottom faces contribute $\partial v_y/\partial y$, and the front and back faces contribute $\partial v_z/\partial z$, and so
$$\operatorname{div} \mathbf{v} = \frac{\partial v_x}{\partial x} + \frac{\partial v_y}{\partial y} + \frac{\partial v_z}{\partial z}. \tag{9.9}$$

Thus if $\mathbf{v}$ is given in cartesian form—that is, as $v_x\hat{\mathbf{i}} + v_y\hat{\mathbf{j}} + v_z\hat{\mathbf{k}}$, then its divergence is obtained easily according to (9.9). Noncartesian systems will be treated in Section 9.7.

Rewriting (9.9) as
$$\operatorname{div} \mathbf{v} = \left(\hat{\mathbf{i}}\frac{\partial}{\partial x} + \hat{\mathbf{j}}\frac{\partial}{\partial y} + \hat{\mathbf{k}}\frac{\partial}{\partial z}\right) \cdot (v_x\hat{\mathbf{i}} + v_y\hat{\mathbf{j}} + v_z\hat{\mathbf{k}}),$$
let us define the vector differential operator
$$\hat{\mathbf{i}}\frac{\partial}{\partial x} + \hat{\mathbf{j}}\frac{\partial}{\partial y} + \hat{\mathbf{k}}\frac{\partial}{\partial z} \equiv \nabla \tag{9.10}$$
called *del* or *nabla*, so that
$$\operatorname{div} \mathbf{v} = \nabla \cdot \mathbf{v}. \tag{9.11}$$

Gradient. Starting instead with a *scalar* field $u(P)$, it will be fruitful to define an associated vector field called grad u, that is, *the gradient of u*—by
$$\operatorname{grad} u(P) \equiv \lim_{\tau \to 0} \left\{ \frac{\int_\sigma \hat{\mathbf{n}} u\, d\sigma}{\tau} \right\}. \tag{9.12}$$

If we introduce cartesian coordinates, take τ to be a prism, and carry out the limit as before, we find (Exercise 9.5) that
$$\operatorname{grad} u = \frac{\partial u}{\partial x}\hat{\mathbf{i}} + \frac{\partial u}{\partial y}\hat{\mathbf{j}} + \frac{\partial u}{\partial z}\hat{\mathbf{k}}. \tag{9.13}$$

Recalling (9.10), we see that this is none other than ∇ acting on u:
$$\operatorname{grad} u = \nabla u. \tag{9.14}$$

Physical interpretation of grad u does not seem to fall out of the definition (9.12) easily. Instead let us consider a field point P and the $u =$ constant surface (e.g., an *isothermal surface* if u is temperature) that passes through P (Fig. 9.7). To facilitate our story, let us introduce a space curve C through P according to $x = x(s)$, $y = y(s)$, $z = z(s)$ and examine the rate of change du/ds along C, in particular, at the point P on C.

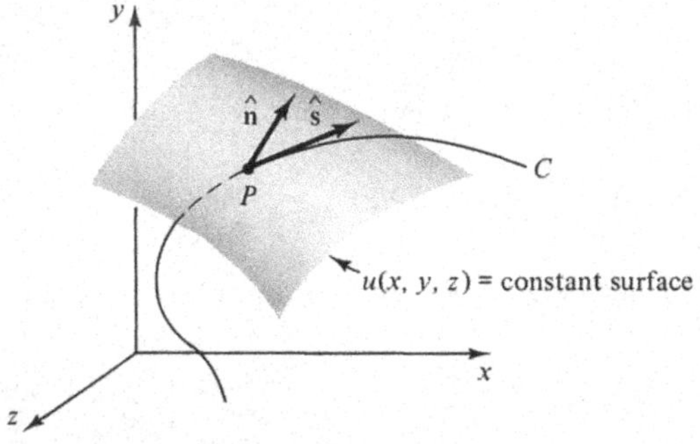

Figure 9.7. Interpretation of gradient.

By chain differentiation,

$$\frac{d}{ds}u[x(s), y(s), z(s)] = \frac{\partial u}{\partial x}\frac{dx}{ds} + \frac{\partial u}{\partial y}\frac{dy}{ds} + \frac{\partial u}{\partial z}\frac{dz}{ds}. \tag{9.15}$$

Recalling from Exercise 8.20 that the unit tangent vector to C is[4] $\hat{\mathbf{s}} = (dx/ds)\hat{\mathbf{i}} + (dy/ds)\hat{\mathbf{j}} + (dz/ds)\hat{\mathbf{k}}$, we observe that the right-hand side of (9.15) can be expressed more compactly as $\nabla u \cdot \hat{\mathbf{s}}$:

$$\frac{du}{ds} = \nabla u \cdot \hat{\mathbf{s}}. \tag{9.16}$$

This is an important little result. It says that the "**directional derivative**" du/ds at any point P is computed as the gradient of u at P dotted into a unit vector in the desired direction.

Equation (9.16) is very informative. Suppose, for instance, that we choose C to lie *in* the $u = c$ surface, so that $\hat{\mathbf{s}}$ is *tangent* to the surface at P. Since u is thereby constant on C, $du/ds = 0$. Thus $0 = \nabla u \cdot \hat{\mathbf{s}}$ for *all* $\hat{\mathbf{s}}$'s that are tangent to the surface at P, and it follows that *the* ∇u *vector at* P *must be perpendicular to the* $u =$ *constant surface at* P.

So much for the *direction* of ∇u. What about its *magnitude*? To answer, consider the problem of choosing the orientation of $\hat{\mathbf{s}}$ so as to achieve the largest possible value of du/ds. From (9.16) $du/ds = \|\nabla u\|\,\|\hat{\mathbf{s}}\| \cos\theta = \|\nabla u\| \cos\theta$, where θ is the angle between $\hat{\mathbf{s}}$ and ∇u. Thus du/ds is a maximum, namely, $\|\nabla u\|$, when $\cos\theta = 1$—that is, $\theta = 0$; so *the magnitude* $\|\nabla u\|$ *equals the maximum possible* du/ds, *and this value occurs when* $\hat{\mathbf{s}}$ *is normal to the* $u = c$ *surface.*

---

[4]We will use the designation $\hat{\mathbf{s}}$ here rather than $\hat{\mathbf{t}}$.

Example 9.2. Find the directional derivative of $u = x^2y + 3z$ at the point (1, 2, 0) in the direction of the vector $\hat{\mathbf{i}} - 4\hat{\mathbf{k}}$.

Solution.

$$\nabla u = 2xy\hat{\mathbf{i}} + x^2\hat{\mathbf{j}} + 3\hat{\mathbf{k}} = 4\hat{\mathbf{i}} + \hat{\mathbf{j}} + 3\hat{\mathbf{k}} \quad \text{at } (1, 2, 0)$$

$$\hat{\mathbf{s}} = \frac{\hat{\mathbf{i}} - 4\hat{\mathbf{k}}}{\sqrt{17}}$$

$$\frac{du}{ds} = \nabla u \cdot \hat{\mathbf{s}} = (4\hat{\mathbf{i}} + \hat{\mathbf{j}} + 3\hat{\mathbf{k}}) \cdot \frac{\hat{\mathbf{i}} - 4\hat{\mathbf{k}}}{\sqrt{17}} = -\frac{8}{\sqrt{17}}. \quad ▮$$

Curl. Finally, we also define for a vector field $\mathbf{v}$ an associated vector field called curl $\mathbf{v}$, *the curl of* $\mathbf{v}$, according to

$$\text{curl } \mathbf{v}(P) \equiv \lim_{\tau \to 0} \left\{ \frac{\int_\sigma \hat{\mathbf{n}} \times \mathbf{v}\, d\sigma}{\tau} \right\}. \tag{9.17}$$

Introducing cartesian coordinates and taking τ to be a prism (Fig. 9.6), we can carry out the limit, as we did for div $\mathbf{v}$. But it is shorter and neater to proceed as follows. Since $\mathbf{A} = (\mathbf{A} \cdot \hat{\mathbf{i}})\hat{\mathbf{i}} + (\mathbf{A} \cdot \hat{\mathbf{j}})\hat{\mathbf{j}} + (\mathbf{A} \cdot \hat{\mathbf{k}})\hat{\mathbf{k}}$ for any $\mathbf{A}$, we can express

$$\begin{aligned}\hat{\mathbf{n}} \times \mathbf{v} &= (\hat{\mathbf{n}} \times \mathbf{v} \cdot \hat{\mathbf{i}})\hat{\mathbf{i}} + (\hat{\mathbf{n}} \times \mathbf{v} \cdot \hat{\mathbf{j}})\hat{\mathbf{j}} + (\hat{\mathbf{n}} \times \mathbf{v} \cdot \hat{\mathbf{k}})\hat{\mathbf{k}} \\ &= (\hat{\mathbf{n}} \cdot \mathbf{v} \times \hat{\mathbf{i}})\hat{\mathbf{i}} + (\hat{\mathbf{n}} \cdot \mathbf{v} \times \hat{\mathbf{j}})\hat{\mathbf{j}} + (\hat{\mathbf{n}} \cdot \mathbf{v} \times \hat{\mathbf{k}})\hat{\mathbf{k}}.\end{aligned}$$

Thus

$$\begin{aligned}\text{curl } \mathbf{v}(P) &= \lim_{\tau \to 0} \left\{ \frac{\int_\sigma \hat{\mathbf{n}} \cdot (\mathbf{v} \times \hat{\mathbf{i}})\, d\sigma}{\tau} \right\}\hat{\mathbf{i}} + \lim_{\tau \to 0} \left\{ \frac{\int_\sigma \hat{\mathbf{n}} \cdot (\mathbf{v} \times \hat{\mathbf{j}})\, d\sigma}{\tau} \right\}\hat{\mathbf{j}} \\ &\quad + \lim_{\tau \to 0} \left\{ \frac{\int_\sigma \hat{\mathbf{n}} \cdot (\mathbf{v} \times \hat{\mathbf{k}})\, d\sigma}{\tau} \right\}\hat{\mathbf{k}} \\ &= [\nabla \cdot (\mathbf{v} \times \hat{\mathbf{i}})]\hat{\mathbf{i}} + [\nabla \cdot (\mathbf{v} \times \hat{\mathbf{j}})]\hat{\mathbf{j}} + [\nabla \cdot (\mathbf{v} \times \hat{\mathbf{k}})]\hat{\mathbf{k}} \\ &= \left(\frac{\partial v_z}{\partial y} - \frac{\partial v_y}{\partial z}\right)\hat{\mathbf{i}} + \left(\frac{\partial v_x}{\partial z} - \frac{\partial v_z}{\partial x}\right)\hat{\mathbf{j}} + \left(\frac{\partial v_y}{\partial x} - \frac{\partial v_x}{\partial y}\right)\hat{\mathbf{k}}.\end{aligned} \tag{9.18}$$

We can also express curl $\mathbf{v}$ in terms of ∇ as

$$\text{curl } \mathbf{v} = \nabla \times \mathbf{v}, \tag{9.19}$$

since, from (8.34),

$$\begin{aligned}\nabla \times \mathbf{v} &= \begin{vmatrix} \hat{\mathbf{i}} & \hat{\mathbf{j}} & \hat{\mathbf{k}} \\ \partial/\partial x & \partial/\partial y & \partial/\partial z \\ v_x & v_y & v_z \end{vmatrix} \\ &= \left(\frac{\partial v_z}{\partial y} - \frac{\partial v_y}{\partial z}\right)\hat{\mathbf{i}} + \left(\frac{\partial v_x}{\partial z} - \frac{\partial v_z}{\partial x}\right)\hat{\mathbf{j}} + \left(\frac{\partial v_y}{\partial x} - \frac{\partial v_x}{\partial y}\right)\hat{\mathbf{k}},\end{aligned}$$

in agreement with the last line of (9.18). Or backing up to the next to last line of (9.18), we could have noted that

$$\begin{aligned}&[\nabla \cdot (\mathbf{v} \times \hat{\mathbf{i}})]\hat{\mathbf{i}} + [\nabla \cdot (\mathbf{v} \times \hat{\mathbf{j}})]\hat{\mathbf{j}} + [\nabla \cdot (\mathbf{v} \times \hat{\mathbf{k}})]\hat{\mathbf{k}} \\ &\qquad = [(\nabla \times \mathbf{v}) \cdot \hat{\mathbf{i}}]\hat{\mathbf{i}} + [(\nabla \times \mathbf{v}) \cdot \hat{\mathbf{j}}]\hat{\mathbf{j}} + [(\nabla \times \mathbf{v}) \cdot \hat{\mathbf{k}}]\hat{\mathbf{k}} = \nabla \times \mathbf{v}\end{aligned}$$

again.

Example 9.3. For instance, if $\mathbf{v} = y^2\hat{\mathbf{i}} + 3xy\hat{\mathbf{j}} - x^2z\hat{\mathbf{k}}$, then

$$\text{curl}\,\mathbf{v} = (0-0)\hat{\mathbf{i}} + (0+2xz)\hat{\mathbf{j}} + (3y-2y)\hat{\mathbf{k}} = 2xz\hat{\mathbf{j}} + y\hat{\mathbf{k}}. \quad \blacksquare$$

To interpret curl $\mathbf{v}$ physically, consider the case of fluid flow, where $\mathbf{v}$ is chosen to be $\mathbf{q}$, the fluid velocity vector:

$$\mathbf{q} = u\hat{\mathbf{i}} + v\hat{\mathbf{j}} + w\hat{\mathbf{k}};$$

that is, u, v, w are the x, y, z velocity components. (With the letter v used for one of the velocity components, it would be confusing to use $\mathbf{v}$ for the velocity vector, and so we use $\mathbf{q}$ instead.) Let us examine curl $\mathbf{q}$, which is called the **vorticity** and denoted $\mathbf{\Omega}$. For simplicity, let $\mathbf{q}$ be a *plane flow*, that is, where $w = 0$ and u, v are functions of x and y but not z; thus $\mathbf{q} = u(x, y)\hat{\mathbf{i}} + v(x, y)\hat{\mathbf{j}}$. Then

$$\text{curl}\,\mathbf{q} = \mathbf{\Omega} = \left(\frac{\partial v}{\partial x} - \frac{\partial u}{\partial y}\right)\hat{\mathbf{k}}. \tag{9.20}$$

Consider the motion of the little rectangular element of fluid shown in Fig. 9.8. At time t it is in position 1. If the element were *rigid*, its motion would consist of a translation plus a rotation. At $t + \Delta t$ it might be in configuration 2, say, due to a translation OO' plus a rotation $\Delta\alpha$, and its angular velocity (taken as positive counterclockwise) would be $\omega \sim (\Delta\alpha/\Delta t)\hat{\mathbf{k}}$.

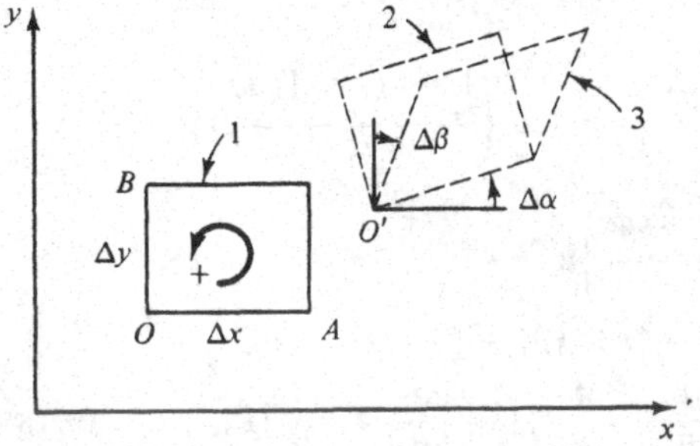

Figure 9.8. Plane motion of fluid element.

A fluid element, however, is *deformable*, not rigid, and it may also suffer a "shear" deformation, as indicated by 3. What, then, does the element's "angular velocity" mean? We *define* it as the *average* angular velocity of the initially perpendicular edges OA and OB. To illustrate, if $\Delta\beta = \Delta\alpha$, we would have $\boldsymbol{\omega} = \mathbf{0}$, as seems fair and reasonable. Then

$$\text{Angular velocity of } OA = \frac{v(A) - v(O)}{\Delta x}\hat{\mathbf{k}} \longrightarrow \frac{\partial v}{\partial x}\hat{\mathbf{k}} \quad \text{as } \Delta x \longrightarrow 0$$

$$\text{Angular velocity of } OB = \frac{u(O) - u(B)}{\Delta y}\hat{\mathbf{k}} \longrightarrow -\frac{\partial u}{\partial y}\hat{\mathbf{k}} \quad \text{as } \Delta y \longrightarrow 0$$

so that the

$$\text{Average of the two} \equiv \boldsymbol{\omega} = \frac{1}{2}\left(\frac{\partial v}{\partial x} - \frac{\partial u}{\partial y}\right)\hat{\mathbf{k}}. \tag{9.21}$$

Comparing (9.21) and (9.20), we see that $\mathbf{\Omega} = 2\boldsymbol{\omega}$; *the vorticity curl* $\mathbf{q}(P)$ *is twice the angular velocity of the fluid at P*. Although our discussion was for plane flows, it's not hard to show that the conclusion is the same for nonplanar flows as well.

The Laplace Operator. So far we've introduced the div, grad, and curl operators and found that all three can be expressed in terms of ∇. Specifically, div $= \nabla\cdot$, grad $= \nabla$, and curl $= \nabla\times$.

Finally, we also introduce the Laplacian operator "∇^2" as

$$\nabla^2 \equiv \nabla \cdot \nabla = \left(\hat{\mathbf{i}}\frac{\partial}{\partial x} + \hat{\mathbf{j}}\frac{\partial}{\partial y} + \hat{\mathbf{k}}\frac{\partial}{\partial z}\right) \cdot \left(\hat{\mathbf{i}}\frac{\partial}{\partial x} + \hat{\mathbf{j}}\frac{\partial}{\partial y} + \hat{\mathbf{k}}\frac{\partial}{\partial z}\right)$$
$$= \frac{\partial^2}{\partial x^2} + \frac{\partial^2}{\partial y^2} + \frac{\partial^2}{\partial z^2}. \tag{9.22}$$

Read as *del square*, we emphasize that it is not really ∇ times ∇ but rather ∇ *dot* ∇, as given above for cartesian coordinates. It can act on either scalar or vector fields:

$$\nabla^2 u = \frac{\partial^2 u}{\partial x^2} + \frac{\partial^2 u}{\partial y^2} + \frac{\partial^2 u}{\partial z^2} \tag{9.23a}$$

$$\nabla^2 \mathbf{v} = \hat{\mathbf{i}}\,\nabla^2 v_x + \hat{\mathbf{j}}\,\nabla^2 v_y + \hat{\mathbf{k}}\,\nabla^2 v_z \tag{9.23b}$$
$$= \hat{\mathbf{i}}\left(\frac{\partial^2 v_x}{\partial x^2} + \frac{\partial^2 v_x}{\partial y^2} + \frac{\partial^2 v_x}{\partial z^2}\right) + \hat{\mathbf{j}}(\text{etc.}) + \hat{\mathbf{k}}(\text{etc.}).$$

The importance of the div, grad, curl, and ∇^2 operators will become apparent shortly.

Some Important Vector Identities. The following properties and formulas will prove quite useful.

For α, β arbitrary scalars,

$$\text{grad}\,(\alpha u + \beta v) = \nabla(\alpha u + \beta v) = \alpha\,\nabla u + \beta\,\nabla v \tag{9.24}$$
$$\text{div}\,(\alpha \mathbf{u} + \beta \mathbf{v}) = \nabla \cdot (\alpha \mathbf{u} + \beta \mathbf{v}) = \alpha\,\nabla \cdot \mathbf{u} + \beta\,\nabla \cdot \mathbf{v} \tag{9.25}$$
$$\text{curl}\,(\alpha \mathbf{u} + \beta \mathbf{v}) = \nabla \times (\alpha \mathbf{u} + \beta \mathbf{v}) = \alpha\,\nabla \times \mathbf{u} + \beta\,\nabla \times \mathbf{v}; \tag{9.26}$$

that is, grad, div, and curl are *linear* operators (Exercise 9.9). Furthermore,

$$\text{div curl}\,\mathbf{v} = \nabla \cdot \nabla \times \mathbf{v} = 0 \tag{9.27}$$
$$\text{curl grad}\,u = \nabla \times \nabla u = 0 \tag{9.28}$$
$$\text{div}\,(u\mathbf{v}) = \nabla \cdot (u\mathbf{v}) = \nabla u \cdot \mathbf{v} + u\,\nabla \cdot \mathbf{v} \tag{9.29}$$
$$\text{curl}\,(u\mathbf{v}) = \nabla \times (u\mathbf{v}) = u\,\nabla \times \mathbf{v} + \nabla u \times \mathbf{v} \tag{9.30}$$
$$\text{div}\,(\mathbf{u} \times \mathbf{v}) = \nabla \cdot (\mathbf{u} \times \mathbf{v}) = \mathbf{v} \cdot \text{curl}\,\mathbf{u} - \mathbf{u} \cdot \text{curl}\,\mathbf{v} \tag{9.31}$$
$$\text{curl}\,(\mathbf{u} \times \mathbf{v}) = \nabla \times (\mathbf{u} \times \mathbf{v}) = \mathbf{u}\,\nabla \cdot \mathbf{v} - \mathbf{v}\,\nabla \cdot \mathbf{u} + (\mathbf{v} \cdot \nabla)\mathbf{u} - (\mathbf{u} \cdot \nabla)\mathbf{v} \tag{9.32}$$
$$\text{curl}^2\,\mathbf{v} = \nabla \times (\nabla \times \mathbf{v}) = \nabla\,\text{div}\,\mathbf{v} - \nabla^2\mathbf{v}, \tag{9.33}$$

where $(\mathbf{u} \cdot \nabla)\mathbf{v}$ in (9.32) means

$$\left(u_x\frac{\partial}{\partial x} + u_y\frac{\partial}{\partial y} + u_z\frac{\partial}{\partial z}\right)(v_x\hat{\mathbf{i}} + v_y\hat{\mathbf{j}} + v_z\hat{\mathbf{k}})$$
$$= \left(u_x\frac{\partial v_x}{\partial x} + u_y\frac{\partial v_x}{\partial y} + u_z\frac{\partial v_x}{\partial z}\right)\hat{\mathbf{i}} + (\text{etc.})\hat{\mathbf{j}} + (\text{etc.})\hat{\mathbf{k}},$$

and similarly for $(\mathbf{v} \cdot \nabla)\mathbf{u}$.

Let us prove just one of these formulas, say (9.29), and leave the others for Exercise 9.10.

Example 9.4. To verify (9.29), we simply expand[5]

$$\begin{aligned}
\nabla \cdot (u\mathbf{v}) &= \left(\hat{\mathbf{i}}\frac{\partial}{\partial x} + \hat{\mathbf{j}}\frac{\partial}{\partial y} + \hat{\mathbf{k}}\frac{\partial}{\partial z}\right) \cdot (uv_x\hat{\mathbf{i}} + uv_y\hat{\mathbf{j}} + uv_z\hat{\mathbf{k}}) \\
&= \frac{\partial}{\partial x}(uv_x) + \frac{\partial}{\partial y}(uv_y) + \frac{\partial}{\partial z}(uv_z) \\
&= v_x\frac{\partial u}{\partial x} + v_y\frac{\partial u}{\partial y} + v_z\frac{\partial u}{\partial z} + u\frac{\partial v_x}{\partial x} + u\frac{\partial v_y}{\partial y} + u\frac{\partial v_z}{\partial z} \\
&= \nabla u \cdot \mathbf{v} + u\nabla \cdot \mathbf{v}. \quad \blacksquare
\end{aligned}$$

Note carefully that all scalars, vectors, and operators in (9.25) through (9.32) are defined intrinsically—that is, without reference to any special coordinate system. Thus verification in any one coordinate system (e.g., cartesian) is equivalent to verification of *all* coordinate systems!

9.4. THE DIVERGENCE THEOREM

Consider a vector field $\mathbf{v}$ defined throughout some region V. Let us partition V into a large number of tiny subregions, say N of them, as suggested in Fig. 9.9(a), and define a point P_i in each of them. Then from definition (9.5) we have

$$\operatorname{div} \mathbf{v}(P_i)(d\tau)_i \approx \int_{\sigma_i} \hat{\mathbf{n}}_i \cdot \mathbf{v}\, d\sigma \tag{9.34}$$

for each little subregion.[6] Letting $i = 1, 2, \ldots, N$, we have N equations. Adding them,

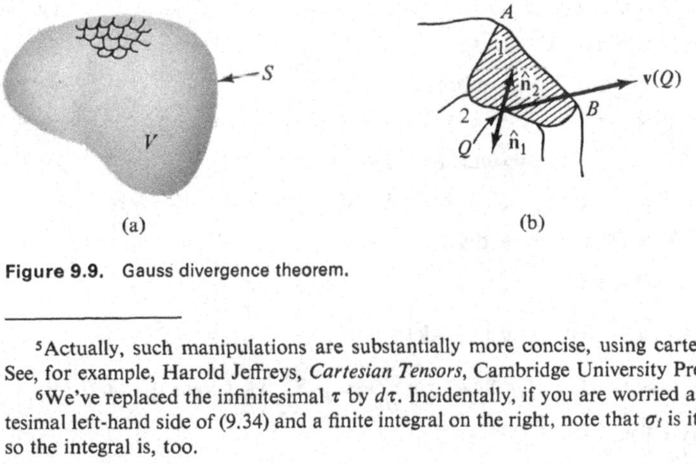

Figure 9.9. Gauss divergence theorem.

---

[5]Actually, such manipulations are substantially more concise, using cartesian tensor notation. See, for example, Harold Jeffreys, *Cartesian Tensors*, Cambridge University Press, Cambridge, 1952.

[6]We've replaced the infinitesimal τ by $d\tau$. Incidentally, if you are worried about having an infinitesimal left-hand side of (9.34) and a finite integral on the right, note that σ_i is itself infinitesimal, and so the integral is, too.

the left-hand side is

$$\sum_{i=1}^{N} \operatorname{div} \mathbf{v}(P_i)(d\tau)_i,$$

which $\longrightarrow \int_V \operatorname{div} \mathbf{v}\, d\tau$ as the partition becomes infinitely fine.

The sum of the right-hand sides, however, is less obvious. The idea is that contributions from all abutting surfaces cancel. Consider, for instance, subregions 1 and 2 in Fig. 9.9(b), which have part of their boundaries in common. At any point Q of their interface note that their outward unit normals are negatives of each other, $\hat{\mathbf{n}}_2 = -\hat{\mathbf{n}}_1$, whereas $\mathbf{v}(Q)$ is uniquely defined, so that $\hat{\mathbf{n}}_1 \cdot \mathbf{v}(Q)\, d\sigma = -\hat{\mathbf{n}}_2 \cdot \mathbf{v}(Q)\, d\sigma$. Thus we have cancellation from all abutting surfaces, and the only contributions that survive are those from the surface S of V (e.g., the portion AB of subregion 1), since they suffer no such cancellation. So

$$\sum_{i=1}^{n} \int_{\sigma_i} \hat{\mathbf{n}}_i \cdot \mathbf{v}\, d\sigma = \int_S \hat{\mathbf{n}} \cdot \mathbf{v}\, d\sigma.$$

Equating these two results, we have

$$\int_V \nabla \cdot \mathbf{v}\, d\tau = \int_S \hat{\mathbf{n}} \cdot \mathbf{v}\, d\sigma, \tag{9.35}$$

which is the **Gauss divergence theorem**. Of course, our derivation has been quite formal.[7]

Example 9.5. In order to illustrate the manipulations involved in (9.35), consider the case where V is the right circular cone shown in Fig. 9.10, and $\mathbf{v} = z^2\hat{\mathbf{k}}$. Then $\nabla \cdot \mathbf{v} = 2z$, and therefore the volume integral in (9.35) is

$$\int_V \nabla \cdot \mathbf{v}\, d\tau = \int_V 2z\, d\tau = \int_0^1 (2z)(\pi r^2\, dz)$$
$$= 2\pi \int_0^1 z(1-z)^2\, dz = \text{etc.} = \frac{\pi}{6}, \tag{9.36}$$

where r is the radius of a coin-shaped element of volume.

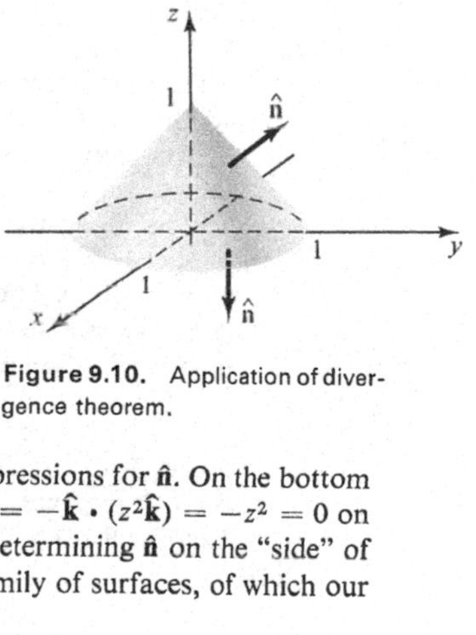

Figure 9.10. Application of divergence theorem.

In evaluating the surface integral, we first need expressions for $\hat{\mathbf{n}}$. On the bottom face ($x^2 + y^2 < 1$ and $z = 0$) $\hat{\mathbf{n}} = -\hat{\mathbf{k}}$, and so $\hat{\mathbf{n}} \cdot \mathbf{v} = -\hat{\mathbf{k}} \cdot (z^2\hat{\mathbf{k}}) = -z^2 = 0$ on $z = 0$ and there is no contribution from that face. Determining $\hat{\mathbf{n}}$ on the "side" of the cone is less obvious. Imagine a one-parameter family of surfaces, of which our cone is one member. If the equation of this family is

$$F(x, y, z; c) = 0,$$

where c_0 is the value of c corresponding to our particular cone, then

$$\hat{\mathbf{n}} = \pm \frac{\nabla F(x, y, z; c)}{\|\nabla F(x, y, z; c)\|}\bigg|_{c=c_0} = \pm \frac{\nabla F(x, y, z; c_0)}{\|\nabla F(x, y, z; c_0)\|}, \tag{9.37}$$

[7]For details, see, for example, O. D. Kellogg's *Foundations of Potential Theory*, Springer-Verlag, Berlin, 1929, p. 84.

where the sign is chosen so that $\hat{\mathbf{n}}$ is the *out*ward normal, as desired. In the present example,

$$F(x, y, z; c_0) = x^2 + y^2 - (1 - z)^2 = 0;$$

hence

$$\hat{\mathbf{n}} = \frac{x\hat{\mathbf{i}} + y\hat{\mathbf{j}} + (1 - z)\hat{\mathbf{k}}}{\sqrt{x^2 + y^2 + (1 - z)^2}} = \frac{x\hat{\mathbf{i}} + y\hat{\mathbf{j}} + (1 - z)\hat{\mathbf{k}}}{\sqrt{2}\,(1 - z)}, \tag{9.38}$$

where the last step follows from the fact that $x^2 + y^2 = (1 - z)^2$ on our particular cone. Then $\hat{\mathbf{n}} \cdot \mathbf{v} = z^2/\sqrt{2}$; therefore (see Fig. 9.11)

$$\int_S \hat{\mathbf{n}} \cdot \mathbf{v}\, d\sigma = \int_0^1 \frac{z^2}{\sqrt{2}} 2\pi(1 - z)(\sqrt{2}\, dz)$$
$$= \text{etc.} = \frac{\pi}{6},$$

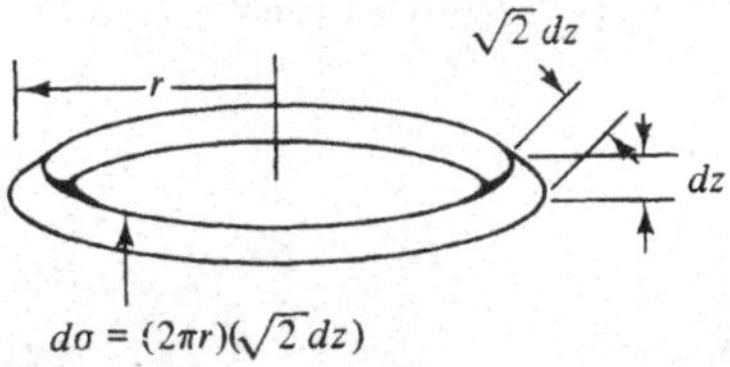

Figure 9.11. Area element $d\sigma$.

in agreement with (9.36).

Actually, you may have noticed that we've made the problem too hard; that is, $\hat{\mathbf{n}} \cdot \mathbf{v} = z^2(\hat{\mathbf{n}} \cdot \hat{\mathbf{k}}) = z^2/\sqrt{2}$, since $\hat{\mathbf{n}} \cdot \hat{\mathbf{k}} = \text{constant} = 1/\sqrt{2}$ over the whole side of the cone. But this simplification is fortuitous, and we chose to follow a more general approach. ∎

Example 9.6. *The Continuity Equation of Fluid Mechanics.* Let us *make up* a fluid flow. To illustrate, let the velocity field be $\mathbf{q} = -x\hat{\mathbf{i}}$ and the mass density $\rho = \text{constant} = \rho_0$. Is such a flow *possible*? Consider a control volume consisting of the slab $-1 < x < 1$. Fluid flows in through the left face ($x = -1$) at the rate 1 m/sec and in through the right face ($x = 1$) at the same rate. Since the amount of fluid in the control volume is constantly increasing, it follows from the principle of conservation of mass that the average density therein must be increasing with time. Yet we have prescribed that $\rho = \text{constant}$, which means that such a flow is not possible! So instead of $\mathbf{q}$ and ρ being independent, there is apparently some relationship between them that is implied by the fundamental requirement of conservation of mass. Let us find this relationship.

Consider a stationary control volume V of arbitrary shape and location within the flow. The amount of mass inside V at any time t is

$$M = \int_V \rho\, d\tau, \tag{9.39}$$

where $\rho = \rho(x, y, z, t)$ is the fluid mass density. Thus the rate of increase of M is (Exercise 9.13)

$$\frac{dM}{dt} = \frac{d}{dt}\int_V \rho(x, y, z, t)\, d\tau = \int_V \frac{\partial \rho}{\partial t}\, d\tau. \tag{9.40}$$

But if mass is to be conserved, it must be true that dM/dt also equals the rate at which mass enters V through its surface S; that is,

$$\frac{dM}{dt} = -\int_S \rho\mathbf{q} \cdot \hat{\mathbf{n}}\, d\sigma, \tag{9.41}$$

where the minus sign is needed, since $\hat{\mathbf{n}}$ is the *out*ward normal, and so $\rho\mathbf{q} \cdot \hat{\mathbf{n}}\, d\sigma$ is *out*flow, whereas we want the *in*flow. Consequently, we must have

$$\int_V \frac{\partial \rho}{\partial t}\, d\tau = -\int_S \rho \mathbf{q} \cdot \hat{\mathbf{n}}\, d\sigma. \tag{9.42}$$

In order to combine the two integrals, let us convert the surface integral to a volume integral by means of the divergence theorem:

$$\int_S \rho \mathbf{q} \cdot \hat{\mathbf{n}}\, d\sigma = \int_V \nabla \cdot (\rho \mathbf{q})\, d\tau;$$

then (9.42) becomes.

$$\int_V \left[\frac{\partial \rho}{\partial t} + \nabla \cdot (\rho \mathbf{q})\right] d\tau = 0. \tag{9.43}$$

The only way that this result can be true for *all* choices of V within the flow (recall that V is arbitrary) is if the integrand is identically zero throughout the flow—for all t, since (9.43) holds at every instant. Suppose the integrand were *positive*, say, at some point P and thus (by continuity) throughout some sufficiently small sphere N centered at P. Then choosing V to be N, the integral would be positive (i.e., nonzero), contradicting (9.43). Therefore

$$\frac{\partial \rho}{\partial t} + \nabla \cdot (\rho \mathbf{q}) = 0 \tag{9.44}$$

throughout the flow and for all time t; (9.44) is a partial differential equation and is called the **continuity equation** of fluid mechanics.

In particular, if $\rho =$ constant (i.e., the fluid is **incompressible**[8]), it reduces to

$$\nabla \cdot \mathbf{q} = 0, \tag{9.45}$$

or in terms of a cartesian reference frame

$$\frac{\partial u}{\partial x} + \frac{\partial v}{\partial y} + \frac{\partial w}{\partial z} = 0, \tag{9.46}$$

where u, v, w are the x, y, z velocity components.

COMMENT. Here we have elected to have V stationary, in which case dM/dt equals the rate of mass flux across the surface S of V. Alternatively, suppose that we define a control volume V at some instant t_0 and then require that it continue to enclose the same set of fluid particles for all $t > t_0$. In other words, V is to "drift" with the fluid. In this case, the statement of conservation of mass is simply

$$\frac{dM}{dt} = 0. \tag{9.47}$$

But this time the Leibnitz-type differentiation of

$$M = \int_{V(t)} \rho(x, y, z, t)\, d\tau$$

[8]Just as the *Reynolds number* turns out to be a measure of the effects of *viscosity*, the nondimensional parameter that provides a measure of the effects of *compressibility* is the so-called *mach number* M, defined as the ratio of the fluid velocity to the local speed of sound. If $M \ll 1$, then we can, to good accuracy, make the simplifying assumption that $\rho =$ constant. For example, the speed of sound in air, at 20°C and sea level, is 344 m/sec, so that if $|\mathbf{q}| < 30$ m/sec, say, the flow may be considered as incompressible. Do you think we could take $\rho =$ constant in the theory of water waves on the ocean?

is more difficult because the region V varies with t. Proceeding formally,[9] along the same lines as in our derivation of the Leibnitz rule (1.41), we have

$$\frac{M(t+\Delta t)-M(t)}{\Delta t}=\frac{1}{\Delta t}\left[\int_{V(t+\Delta t)}\rho(P,t+\Delta t)\,d\tau-\int_{V(t)}\rho(P,t)\,d\tau\right]$$
$$=\int_{V(t)}\frac{\rho(P,t+\Delta t)-\rho(P,t)}{\Delta t}\,d\tau$$
$$+\frac{1}{\Delta t}\left\{\int_{V(t+\Delta t)}\rho(P,t+\Delta t)\,d\tau-\int_{V(t)}\rho(P,t+\Delta t)\,d\tau\right\},$$

where P is short for x, y, z. From the sketch in Fig. 9.12 we see that the last term, in braces, is

$$\frac{1}{\Delta t}\{\quad\}\sim\frac{1}{\Delta t}\int_{S(t)}\rho(P,t+\Delta t)[\mathbf{q}(P,t)\,\Delta t]\cdot\hat{\mathbf{n}}\,d\sigma,$$

where $S(t)$ is the surface of $V(t)$. Letting $\Delta t\longrightarrow 0$,

$$\frac{dM}{dt}=\int_V\frac{\partial\rho}{\partial t}\,d\tau+\int_S\rho\mathbf{q}\cdot\hat{\mathbf{n}}\,d\sigma=0,$$

which coincides with our previous result (9.42), and we end up with the same continuity equation (9.44) (as we had better!), independent of whether our derivation considers V to be fixed or drifting with the fluid. ▮

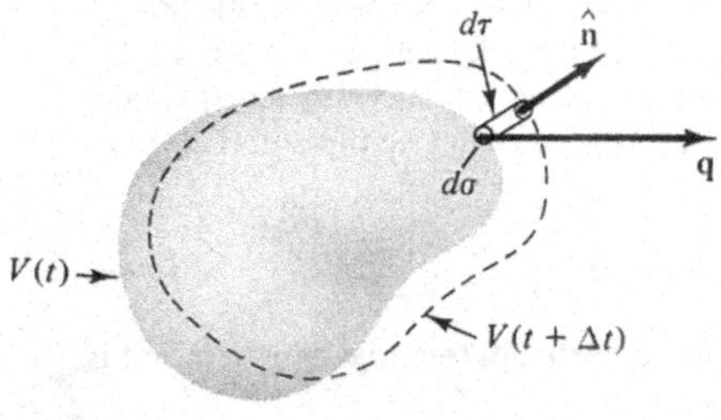

Figure 9.12. Leibnitz differentiation of $M(t)$.

Example 9.7. *Unsteady Heat Conduction.* Consider the unsteady conduction of heat within some region R. Our derivation of a differential equation governing the temperature field $u(x, y, z, t)$ will closely parallel our derivation of the continuity equation in the previous example.

First, we choose a fixed arbitrary control volume V within R. Instead of a "mass balance," this time we carry out a heat balance; that is, we equate the rate at which heat enters V through its surface S to the rate at which it is stored.

The **Fourier law of heat conduction** states that the heat flux q (calories per unit time, say) across a plane face is proportional to its area and the temperature gradient $\partial u/\partial n$ (i.e., $\nabla u\cdot\hat{\mathbf{n}}$) normal to the face, with a constant of proportionality k, called the *thermal conductivity* of the material.

Thus the flux through S is

$$q=\int_S k\frac{\partial u}{\partial n}\,d\sigma, \tag{9.48}$$

where $\hat{\mathbf{n}}$ is the outward normal on S. Note that the sign is correct because a positive $\partial u/\partial n$ causes a flux *into* V, as assumed.

Alternatively, since the amount of heat (in calories, say) in a mass m at tempera-

---

[9]For a rigorous derivation, see W. Kaplan, *Advanced Calculus*, Addison-Wesley, Reading, Mass., 1953, pp. 223, 299, and 300; also see the detailed discussion by H. Flanders, Differentiation under the Integral Sign, *American Mathematical Monthly*, Vol. 80, June–July, 1973, pp. 615–627.

ture u is mcu, where c is the *specific heat* of the material, we can express

$$q = \frac{d}{dt}\int_V cu\rho\, d\tau, \tag{9.49}$$

where ρ is the mass density.

Assuming, for simplicity, that the material is homogeneous so that k, c, ρ are all constant, we therefore have

$$\int_V \frac{\partial u}{\partial t}\, d\tau = \alpha^2 \int_S \frac{\partial u}{\partial n}\, d\sigma \qquad \left(\alpha^2 \equiv \frac{k}{c\rho}\right)$$

$$= \alpha^2 \int_S \nabla u \cdot \hat{\mathbf{n}}\, d\sigma = \alpha^2 \int_V \nabla \cdot \nabla u\, d\tau.$$

Thus

$$\int_V \left(\frac{\partial u}{\partial t} - \alpha^2 \nabla^2 u\right) d\tau = 0; \tag{9.50}$$

and since V is arbitrary, we must have

$$\alpha^2 \nabla^2 u = \frac{\partial u}{\partial t} \tag{9.51}$$

throughout the region R and for all t.

The differential equation (9.51) is called the equation of heat conduction or sometimes simply the **heat equation**, and α is the *coefficient of diffusivity*; (9.51) is also known as the **diffusion equation** because it applies not only to heat conduction but also to diffusion processes in general, such as material diffusion due to variation in concentration and the diffusion of vorticity in viscous flows.

Of special importance is the case of *steady-state* heat conduction—that is, where $\partial u/\partial t = 0$—since then (9.51) becomes the **Laplace equation**

$$\nabla^2 u = 0. \tag{9.52}$$

The importance of the diffusion equation, Laplace's equation, and the wave equation (not yet discussed) cannot be overstated. Most of our discussion of these equations is contained in Part V.

COMMENT. We have assumed that there is no generation or disappearance of heat in the material. Such is often the case, but not always. For instance, a chemical reaction might be taking place throughout the material so that heat is generated at the rate $Q(x, y, z, t)$ calories per unit volume per unit time ($Q > 0$ if the reaction is exothermic; $Q < 0$ if it is endothermic), or we might have a mechanical specimen in fatigue being heated by internal hysteresis losses. If this effect is included, it is not hard to see that the result is $c\rho u_t = k\,\nabla^2 u + Q$ in place of (9.51). ∎

Example 9.8. *Green's Identities.* If u and v are scalar fields, then $u\,\nabla v$ is a vector field. Applying the divergence theorem to the field $u\,\nabla v$ gives

$$\int_V \nabla \cdot (u\,\nabla v)\, d\tau = \int_S \hat{\mathbf{n}} \cdot (u\,\nabla v)\, d\sigma.$$

But [recall (9.29)] $\nabla \cdot (u\,\nabla v) = \nabla u \cdot \nabla v + u\,\nabla^2 v$ and $\hat{\mathbf{n}} \cdot (u\,\nabla v) = u\hat{\mathbf{n}} \cdot \nabla v = u\,\partial v/\partial n$, and therefore

$$\int_V (\nabla u \cdot \nabla v + u\,\nabla^2 v)\, d\tau = \int_S u \frac{\partial v}{\partial n}\, d\sigma. \tag{9.53}$$

This useful formula is known as **Green's first identity**. Interchanging u and v,

$$\int_V (\nabla v \cdot \nabla u + v \nabla^2 u) \, d\tau = \int_S v \frac{\partial u}{\partial n} \, d\sigma,$$

and expunging this equation from (9.53), we arrive at **Green's second identity**,

$$\int_V (u \nabla^2 v - v \nabla^2 u) \, d\tau = \int_S \left(u \frac{\partial v}{\partial n} - v \frac{\partial u}{\partial n} \right) d\sigma. \tag{9.54}$$

Recall that we required the three components of $\mathbf{v}$ in the divergence theorem to have continuous first-order partials with respect to x, y, and z. In deriving Green's identities, we set "$\mathbf{v}$" $= u \nabla v = uv_x\hat{\mathbf{i}} + uv_y\hat{\mathbf{j}} + uv_z\hat{\mathbf{k}}$ and then "$\mathbf{v}$" $= v \nabla u = vu_x\hat{\mathbf{i}} + vu_y\hat{\mathbf{j}} + vu_z\hat{\mathbf{k}}$. Thus we require uv_x, uv_y, uv_z, vu_x, vu_y, vu_z to have continuous first-order partials with respect to x, y, z. It follows that we require the continuity of u, v and all their first- and second-order partials throughout V in Green's identities (9.53) and (9.54). ▮

9.5. *STOKES' THEOREM*

Consider a caplike surface S bounded by a simple, closed, oriented curve C (Fig. 9.13) and with a unit normal $\hat{\mathbf{n}}$ defined at each point at which S is smooth. We will not say that $\hat{\mathbf{n}}$ is to be directed "outward" or "inward," since S is only a cap, not the surface of a closed volume, and such terms are not meaningful. By *convention*, we will take the direction of the unit normals $\hat{\mathbf{n}}$ to be dictated by the "right-hand rule"; that is, curling the fingers of our right hand around $\hat{\mathbf{n}}$ in the direction indicated by the orientation of C, our thumb points in the direction of $\hat{\mathbf{n}}$.

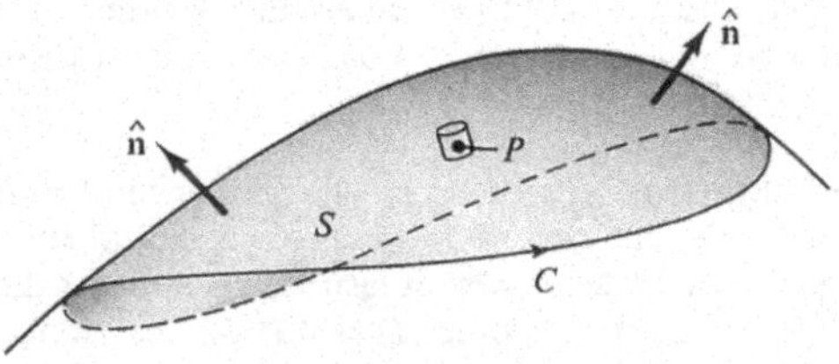

Figure 9.13. Caplike surface S.

At any point P on S let us construct a tiny cylindrical house (not necessarily of circular cross section), sitting on S as shown in Fig. 9.13. If S is immersed in some vector field $\mathbf{v}$, then from the definition (9.17) of the curl

$$\text{curl } \mathbf{v}(P) \approx \frac{1}{\tau} \int_\sigma \hat{\mathbf{N}} \times \mathbf{v} \, d\sigma, \tag{9.55}$$

with equality holding in the limit as the volume τ of the house tends to zero. $\hat{\mathbf{N}}$ is the outward unit normal over the surface σ of τ and should not be confused with $\hat{\mathbf{n}}$, which is the normal to S; in fact, $\hat{\mathbf{N}} = \hat{\mathbf{n}}$ on the top of σ, $\hat{\mathbf{N}} = -\hat{\mathbf{n}}$ on the bottom, and $\hat{\mathbf{N}} = \hat{\mathbf{e}}_n$, say, on the side, as shown in the blowup of the house in Fig. 9.14. Dotting $\hat{\mathbf{n}}$ into (9.55),

$$\hat{\mathbf{n}} \cdot \text{curl } \mathbf{v} \approx \frac{1}{\tau}\left\{\int_{\text{top}} \hat{\mathbf{n}} \cdot (\hat{\mathbf{n}} \times \mathbf{v})\, d\sigma + \int_{\text{bottom}} \hat{\mathbf{n}} \cdot (-\hat{\mathbf{n}} \times \mathbf{v})\, d\sigma + \int_{\text{side}} \hat{\mathbf{n}} \cdot (\hat{\mathbf{e}}_n \times \mathbf{v})\, d\sigma\right\}.$$

But $\hat{\mathbf{n}} \cdot (\hat{\mathbf{n}} \times \mathbf{v}) = \hat{\mathbf{n}} \cdot (-\hat{\mathbf{n}} \times \mathbf{v}) = 0$, since $\hat{\mathbf{n}} \times \mathbf{v}$ is orthogonal to both $\hat{\mathbf{n}}$ and $\mathbf{v}$, and [recalling (8.11)] $\hat{\mathbf{n}} \cdot (\hat{\mathbf{e}}_n \times \mathbf{v}) = (\hat{\mathbf{n}} \times \hat{\mathbf{e}}_n) \cdot \mathbf{v} = \hat{\mathbf{e}}_t \cdot \mathbf{v}$; therefore

$$\hat{\mathbf{n}} \cdot \text{curl } \mathbf{v} \approx \frac{1}{\tau} \int_{\text{side}} \hat{\mathbf{e}}_t \cdot \mathbf{v}\, d\sigma.$$

Expressing $d\sigma = h\, ds$ and $\tau = h\sigma_B$, where σ_B is the (cross-hatched) base area of the house,

$$\hat{\mathbf{n}} \cdot \text{curl } \mathbf{v} \approx \frac{1}{h\sigma_B} \int_c \hat{\mathbf{e}}_t \cdot \mathbf{v}\, h\, ds = \frac{1}{\sigma_B} \int_c \mathbf{v} \cdot d\mathbf{r}, \qquad (9.56)$$

where the tiny closed curve c is the edge of σ_B, oriented in accordance with $\hat{\mathbf{n}}(P)$ and the right-hand rule, and $d\mathbf{r}$ has entered through the combination $ds\hat{\mathbf{e}}_t$, as shown in Fig. 9.14.[10]

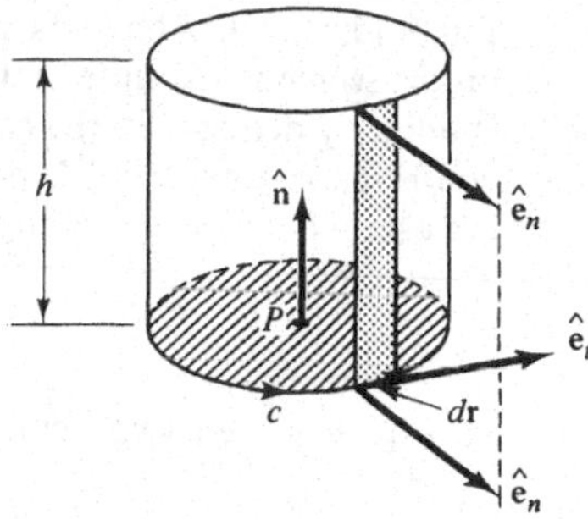

Figure 9.14. Blowup of the house at P.

Actually, σ_B constitutes an element of area "$d\sigma$" of S, and so (9.56) yields

$$\hat{\mathbf{n}} \cdot \text{curl } \mathbf{v}\, d\sigma \approx \int_c \mathbf{v} \cdot d\mathbf{r} \qquad (9.57)$$

with equality holding in the limit as $d\sigma \longrightarrow 0$. Thus if we partition S into a large number of tiny subregions, say N of them, as suggested in Fig. 9.15(a), and define a point P_i in each, then from (9.57)

$$\hat{\mathbf{n}} \cdot \text{curl } \mathbf{v}(P_i)(d\sigma)_i \approx \int_{c_i} \mathbf{v} \cdot d\mathbf{r} \quad \text{for } i = 1, 2, \ldots, N. \qquad (9.58)$$

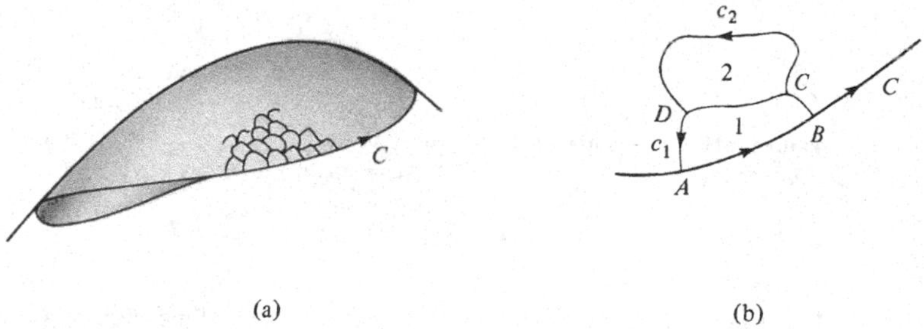

Figure 9.15. Tortoise-shell partition of S.

[10]Note that our derivation is rather formal. For instance, taking $d\sigma$ to be the shaded sliver $h\,ds$ really assumes that the integrand $\hat{\mathbf{e}}_t \cdot \mathbf{v}$ is constant over the sliver. Well, $\hat{\mathbf{e}}_t$ is, but $\mathbf{v}$ varies throughout space and thus over the height of the sliver. But we are letting h and σ_B tend to zero. Letting $h \longrightarrow 0$ first, the variation in $\mathbf{v}$ over the height of the sliver has no effect.

Adding these N equations, the left-hand side is

$$\sum_{i=1}^{N} \hat{\mathbf{n}} \cdot \operatorname{curl} \mathbf{v}(P_i)(d\sigma)_i,$$

which $\longrightarrow \int_S \hat{\mathbf{n}} \cdot \operatorname{curl} \mathbf{v}\, d\sigma$ as the partition becomes infinitely fine. The sum of the right-hand sides is less obvious, however. Consider, for instance, subregions 1 and 2 in Fig. 9.15(b), which have the portion DC of their boundaries in common. At any point on the segment DC note that the $d\mathbf{r}$'s for c_1 and c_2 are oppositely oriented, whereas $\mathbf{v}$ is uniquely defined. So the contribution from the DC portion of c_2 exactly cancels the contribution from the CD portion of c_1. Similarly, we have cancellation from *all* the c_i's except for segments along the boundary curve C, such as AB, which are not shared. Therefore

$$\sum_{i=1}^{N} \int_{c_i} \mathbf{v} \cdot d\mathbf{r} = \int_C \mathbf{v} \cdot d\mathbf{r}.$$

Equating these two results, we have the well-known and important **Stokes theorem,**

$$\int_S \hat{\mathbf{n}} \cdot \nabla \times \mathbf{v}\, d\sigma = \int_C \mathbf{v} \cdot d\mathbf{r}. \tag{9.59}$$

Notice how our derivation parallels that of the divergence theorem, starting with equation (9.57).

Example 9.9. Evaluate

$$I = \int_S \hat{\mathbf{n}} \cdot \nabla \times \mathbf{v}\, d\sigma,$$

where $\mathbf{v} = x^3\hat{\mathbf{j}} - (z+1)\hat{\mathbf{k}}$, S is the surface $z = 4 - 4x^2 - y^2$ between $z = 0$ and $z = 4$, and $\hat{\mathbf{n}}$ is directed as shown in Fig. 9.16.

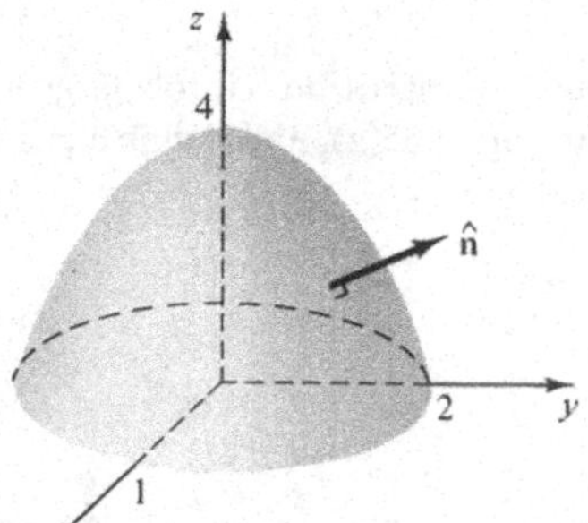

Figure 9.16. The surface $z = 4 - 4x^2 - y^2$.

Solution. Rather than evaluate I directly, it is easier to apply Stokes' theorem and deal with the line integral

$$I = \int_C \mathbf{v} \cdot d\mathbf{r} = \int_C [x^3\, dy - (z+1)\, dz],$$

where C is the edge curve $4x^2 + y^2 = 4$ in the x, y plane. (What should its orientation be?) Then $dz = 0$ on C, and parameterizing C according to $x = \cos t$, $y = 2 \sin t$, we have

$$I = 2 \int_0^{2\pi} \cos^4 t\, dt = \frac{3\pi}{2}. \quad \blacksquare$$

Example 9.10. *Green's Theorem.* Consider the two-dimensional case, where $\mathbf{v} = M(x, y)\hat{\mathbf{i}} + N(x, y)\hat{\mathbf{j}}$ and S is in the x, y plane. Let C be oriented *counterclockwise*, say, in which case $\hat{n} = \hat{\mathbf{k}}$. Then Stokes' theorem (9.59) becomes

$$\int_S \hat{\mathbf{k}} \cdot \left[0\hat{\mathbf{i}} + 0\hat{\mathbf{j}} + \left(\frac{\partial N}{\partial x} - \frac{\partial M}{\partial y}\right)\hat{\mathbf{k}}\right] d\sigma = \int_C (M\hat{\mathbf{i}} + N\hat{\mathbf{j}}) \cdot (dx\hat{\mathbf{i}} + dy\hat{\mathbf{j}})$$

or

$$\int_S \left(\frac{\partial N}{\partial x} - \frac{\partial M}{\partial y}\right) d\sigma = \int_C (M\, dx + N\, dy), \tag{9.60}$$

which is known as **Green's theorem** and should not be confused with "Green's identities," (9.53) and (9.54). Recall, from the first sentence of this section, that C is to be a *simple closed curve* or, equivalently, that S is to be a *simply connected region.*

COMMENT. Actually, it is also possible to bypass Stokes' theorem, in deriving Green's theorem, and to proceed instead along elementary lines. For instance, consider the N terms in (9.60). Referring to Fig. 9.17, we see that

$$\begin{aligned}\int_S \frac{\partial N}{\partial x}\,d\sigma &= \int_{y_1}^{y_2}\int_{x_L(y)}^{x_R(y)} \frac{\partial N}{\partial x}\,dx\,dy \\ &= \int_{y_1}^{y_2} N(x_R, y)\,dy - \int_{y_1}^{y_2} N(x_L, y)\,dy \\ &= \int_{C_R} N(x, y)\,dy + \int_{C_L} N(x, y)\,dy = \int_C N(x, y)\,dy,\end{aligned}$$

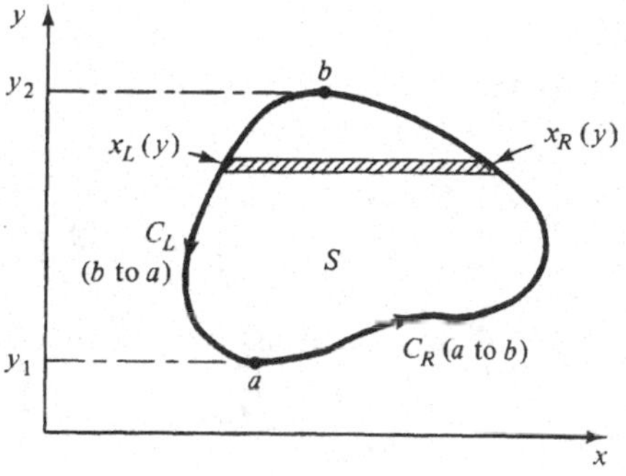

Figure 9.17. Green's theorem.

where C_R and C_L are the right and left halves of C, as shown in the figure.[11] We leave it for the reader to show, in similar fashion, that (Exercise 9.24)

$$\int_S \frac{\partial M}{\partial y}\,d\sigma = -\int_C M\,dx, \tag{9.61}$$

completing our alternative derivation of (9.60). ▮

Example 9.11. *A Multiply Connected Region.* Consider

$$I = \int_S \left(\frac{\partial N}{\partial x} - \frac{\partial M}{\partial y}\right) d\sigma, \tag{9.62}$$

where S is the annulus $1 < r < 2$ in the x, y plane, $N = x/(x^2 + y^2)$, and $M = -y/(x^2 + y^2)$. Differentiating, we find that $\partial N/\partial x - \partial M/\partial y = 0$ in S, so that $I = 0$ and we're done. But the reason that we pose this problem is to discuss the conversion of (9.62) to a line integral by means of Green's theorem (9.60). The catch is that the boundary C in Green's theorem is to be a simple closed curve, whereas that's not the case here, since S is multiply connected.

[11] Note that we've tacitly assumed that the region is *convex* in the x direction; that is, each cross-hatched "sliver" running from $x_L(y)$ to $x_R(y)$ lies entirely within S. Extension to the nonconvex case is not hard, and the final result (9.60) is the same.

To fix things, let us "cut" the annulus S, say along the x axis from $x = 1$ to $x = 2$. The cut annulus, S' say, is simply connected and *is* bounded by a simple closed curve C', as shown in Fig. 9.18(a). The two straight portions of C' are shown as separated for clarity; actually, they abut each other on $y = 0$.

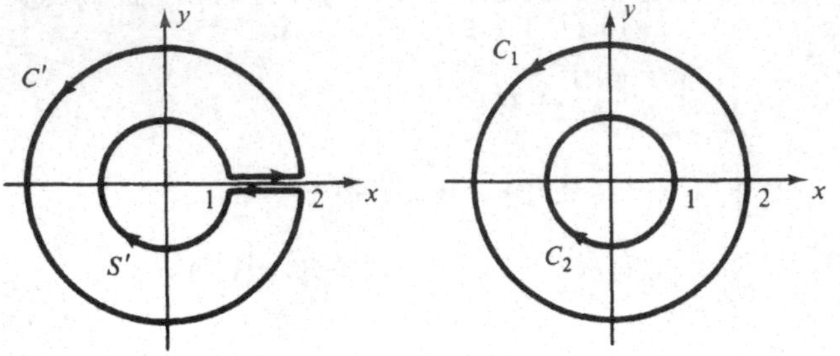

Figure 9.18. Extension of Green's theorem to a multiply connected region.

Surely

$$\int_S \left(\frac{\partial N}{\partial x} - \frac{\partial M}{\partial y}\right) d\sigma = \int_{S'} \left(\frac{\partial N}{\partial x} - \frac{\partial M}{\partial y}\right) d\sigma,$$

since S and S' differ only by an infinitely thin cut. *Now* applying Green's theorem,

$$I = \int_{S'} \left(\frac{\partial N}{\partial x} - \frac{\partial M}{\partial y}\right) d\sigma = \int_{C'} (M\,dx + N\,dy)$$

$$= \int_{C_1} (M\,dx + N\,dy) + \int_{C_2} (M\,dx + N\,dy),$$

since the two straight segments cancel. We leave it for the reader to show that the two integrals on the right are 2π and -2π, respectively, so that $I = 2\pi - 2\pi = 0$ as before. ▮

9.6. *IRROTATIONAL AND SOLENOIDAL FIELDS*

Irrotational Fields. If $\nabla \times \mathbf{v} = 0$ throughout R, then we say that $\mathbf{v}$ is ***irrotational*** in R.[12]

THEOREM 9.1. $\nabla \times \mathbf{v} = 0$ *throughout a simply connected region R if and only if there exists a scalar field u(P) such that* $\mathbf{v} = \nabla u$ *in R.* (Note that u is to have continuous *second* partials, $u_{xx}, u_{yy}, u_{zz}, u_{xy}, u_{xz}$, and u_{yz}, since the components of $\mathbf{v}$ are to have continuous first partials and $\mathbf{v} = \nabla u$.)

[12]Recall (Section 9.3) that for all scalar fields $u(P)$ and vector fields $\mathbf{v}(P)$ we assume continuity of the first partials throughout R. And from section 9.2 R is understood always to be bounded and connected; our surfaces are to be piecewise smooth and orientable, and our curves are to be piecewise smooth. These conditions will be assumed, without reiteration, throughout this chapter. Only when *more* is required, will these conditions be stated—for example, in Theorem 9.1, where R is required to be *simply* connected.

Proof. First, assume that there *is* a u such that $\mathbf{v} = \nabla u$. Then $\nabla \times \mathbf{v} = \nabla \times \nabla u = (u_{zy} - u_{yz})\hat{\mathbf{i}} + (u_{xz} - u_{zx})\hat{\mathbf{j}} + (u_{yx} - u_{xy})\hat{\mathbf{k}} = 0$. Next, we must show that $\nabla \times \mathbf{v} = 0$ implies the existence of a u such that $\mathbf{v} = \nabla u$.

Consider $\int \mathbf{v} \cdot d\mathbf{r}$ from any fixed point x_0, y_0, z_0 to a field point x, y, z along any two paths C_1 and C_2 that lie within R (Fig. 9.19). Let us denote by $C = C_1 + (-C_2)$ the closed path from P_0 to P along C_1 and then back to P_0 along $-C_2$. Next, let us stretch a caplike surface S across from C_1 to C_2. Then the edge of S is C. (Note that if R were not simply connected, S might need to have holes in it, so that we could not say that the edge of S is C.) Applying Stokes' theorem,

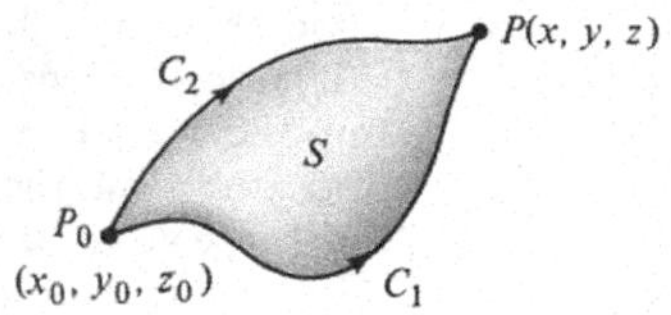

Figure 9.19. Getting ready to apply Stokes' theorem.

$$\int_C \mathbf{v} \cdot d\mathbf{r} = \int_S \hat{\mathbf{n}} \cdot \nabla \times \mathbf{v} \, d\sigma = 0,$$

since $\nabla \times \mathbf{v}$ is zero in R, and hence on S. But

$$\int_C \mathbf{v} \cdot d\mathbf{r} = \int_{C_1} \mathbf{v} \cdot d\mathbf{r} + \int_{-C_2} \mathbf{v} \cdot d\mathbf{r} = \int_{C_1} \mathbf{v} \cdot d\mathbf{r} - \int_{C_2} \mathbf{v} \cdot d\mathbf{r},$$

and therefore

$$\int_{C_1} \mathbf{v} \cdot d\mathbf{r} = \int_{C_2} \mathbf{v} \cdot d\mathbf{r}; \tag{9.63}$$

that is, the integral is *independent of the path* and depends only on the endpoints P_0 and P. Thus we can express[13]

$$\int_{x_0, y_0, z_0}^{x, y, z} \mathbf{v} \cdot d\mathbf{r} \equiv f(x, y, z), \tag{9.64}$$

where f is uniquely determined by the point x, y, z[14]; if the integral *did* depend on path, then it would not define a single-valued function of x, y, z. Then if we hold y and z fixed and express (see Fig. 9.20)

$$f(x, y, z) = \int_{x_0, y_0, z_0}^{x_1, y, z} (v_x\, dx + v_y\, dy + v_z\, dz) + \int_{x_1, y, z}^{x, y, z} (v_x\, dx + v_y\, dy + v_z\, dz), \tag{9.65}$$

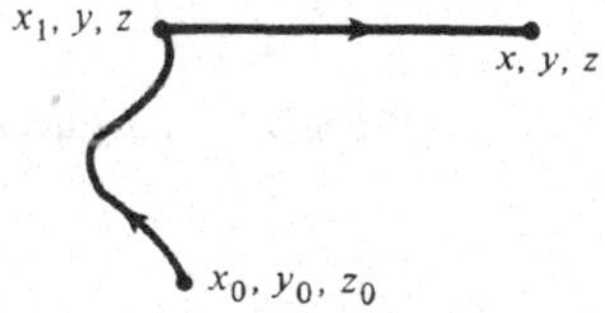

Figure 9.20. Getting ready for the Fundamental Theorem of the Integral Calculus.

we see that the first integral is constant and that the second reduces to

$$\int_{x_1}^{x} v_x(x, y, z)\, dx \quad \text{or} \quad \int_{x_1}^{x} v_x(\xi, y, z)\, d\xi$$

and so, by the Fundamental Theorem of the Integral Calculus (1.19), we have $\partial f/\partial x = v_x(x, y, z)$. Similarly, we have $\partial f/\partial y = v_y(x, y, z)$ and $\partial f/\partial z = v_z(x, y, z)$; consequently, $\mathbf{v} = \nabla f$, and f is the scalar field u that we've been looking for. This completes the proof.

[13]When a line integral is independent of path, we will denote this fact by dropping the subscripted C and indicating the endpoints as upper and lower limits.

[14]We will regard P_0 as fixed and P as variable, so that f is a function of the "active" variables x, y, z.

Example 9.12. *Conservative Force Fields.* A force field is said to be **conservative** if the work

$$W = \int_{P_0}^{P} \mathbf{F} \cdot d\mathbf{r},$$

is independent of the path for any points P_0 and P in the region. We've just seen that a necessary and sufficient condition for $\mathbf{F}$ to be conservative is that $\nabla \times \mathbf{F} = 0$. In this event, we can associate with $\mathbf{F}$ a *scalar* field, ϕ say, such that $\mathbf{F} = -\nabla\phi$, where the minus sign is inserted for convenience; ϕ is called the scalar potential or simply the **potential** of $\mathbf{F}$.

As a simple illustration, consider the uniform gravitational field

$$\mathbf{F} = -g\hat{\mathbf{k}} \tag{9.66}$$

per unit mass, where g is the acceleration of gravity. Clearly, $\nabla \times \mathbf{F} = 0$, since $\mathbf{F}$ is a constant. To find ϕ, note from $\mathbf{F} = -g\hat{\mathbf{k}} = -\nabla\phi$ that

$$\frac{\partial\phi}{\partial x} = 0, \qquad \frac{\partial\phi}{\partial y} = 0, \qquad \frac{\partial\phi}{\partial z} = g,$$

and so

$$\phi = gz + C, \tag{9.67}$$

where C is an arbitrary constant of integration; (9.67) is the familiar "gravitational potential energy" from elementary physics. There is *always* an arbitrary additive constant in ϕ that has no physical significance because it drops out when we compute the "physical" quantity $\mathbf{F} = -\nabla\phi$; we can take C to be zero or any other convenient value.

As a final example, let us reconsider the two-dimensional case discussed in Example 9.1. Noting that $\nabla \times \mathbf{F}$ happens to be zero, we have $\mathbf{F} = -\nabla\phi$; as a result,

$$\begin{aligned} W &= \int \mathbf{F} \cdot d\mathbf{r} = \int (F_x\, dx + F_y\, dy) \\ &= -\int \left(\frac{\partial\phi}{\partial x}\, dx + \frac{\partial\phi}{\partial y}\, dy\right) = -\int d\phi = -\phi(x, y)\Big|_{0,0}^{0,5\pi}. \end{aligned}$$

To find ϕ, let us integrate $\mathbf{F} = -\nabla\phi$—that is, $2xy\hat{\mathbf{i}} + (x^2 - 1)\hat{\mathbf{j}} = -\frac{\partial\phi}{\partial x}\hat{\mathbf{i}} - \frac{\partial\phi}{\partial y}\hat{\mathbf{j}}$—or

$$\frac{\partial\phi}{\partial x} = -2xy \tag{9.68a}$$

$$\frac{\partial\phi}{\partial y} = 1 - x^2. \tag{9.68b}$$

Integrating (9.68a),[15]

$$\phi = \int (-2xy)\, \partial x = -x^2y + A(y), \tag{9.69}$$

and inserting this expression in (9.68b) gives

$$-x^2 + A'(y) = 1 - x^2;$$

[15]By $\int f(x, y)\, \partial x$ we mean the integral of f with respect to x *with y held fixed.* So in order to be as general as possible, the constant of integration A must be permitted to depend on y, since y has been regarded as constant. If unhappy about $A(y)$ work backward: taking $\partial/\partial x$ of (9.69), we have $\partial\phi/\partial x = -2xy$, in agreement with (9.68a).

hence $A'(y) = 1$ and $A = y + C$. Thus $\phi = -x^2y + y + C$ and

$$W = -(-x^2y + y + C)\Big|_{0,0}^{0,5\pi} = -5\pi$$

in agreement with our result in Example 9.1.

Alternatively, recall from our proof of Theorem 9.1 that $\nabla \times \mathbf{F} = 0$ implies that the line integral $\int_C \mathbf{F} \cdot d\mathbf{r}$ is independent of the path C and depends only on the endpoints (which is consistent with the fact that $\int_C \mathbf{F} \cdot d\mathbf{r} = -\phi\Big|_{\text{initial point}}^{\text{final point}}$). Thus we can *simplify* C from the given spiral to a straight-line path $x = 0$, from $y = 0$ to $y = 5\pi$, so that

$$W = \int \mathbf{F} \cdot d\mathbf{r} = \int [2xy\,dx + (x^2 - 1)\,dy] = -\int_0^{5\pi} dy = -5\pi$$

again.

Of these two approaches, the first has the advantage that it yields ϕ, and once we have ϕ, we are able to compute W between *any* two points P_0 and P simply as $-\phi(x, y)\Big|_{P_0}^{P}$ with no further calculation. ▮

Solenoidal Fields. If $\nabla \cdot \mathbf{v} = 0$ throughout R, we say that $\mathbf{v}$ is ***solenoidal*** in R.

THEOREM 9.2. $\nabla \cdot \mathbf{v} = 0$ *throughout a simply connected region R if and only if there exists a vector field $\mathbf{w}$ such that $\mathbf{v} = \nabla \times \mathbf{w}$ in R.* (Note that $\mathbf{w}$ is to have continuous *second* partials in R, since $\mathbf{v} = \nabla \times \mathbf{w}$, and $\mathbf{v}$ is to have continuous first partials.)

Proof. First, suppose that there *is* a $\mathbf{w}$ such that $\mathbf{v} = \nabla \times \mathbf{w}$. Then

$$\begin{aligned} \nabla \cdot \mathbf{v} &= \frac{\partial}{\partial x}\left(\frac{\partial w_z}{\partial y} - \frac{\partial w_y}{\partial z}\right) + \frac{\partial}{\partial y}\left(\frac{\partial w_x}{\partial z} - \frac{\partial w_z}{\partial x}\right) + \frac{\partial}{\partial z}\left(\frac{\partial w_y}{\partial x} - \frac{\partial w_x}{\partial y}\right) \\ &= \text{etc.} = 0. \end{aligned}$$

It remains to prove the converse—namely, that if $\mathbf{v}$ is given such that $\nabla \cdot \mathbf{v} = 0$, then there is a solution $\mathbf{w}$ of $\nabla \times \mathbf{w} = \mathbf{v}$. But let us skip the proof, which is actually a generalization of the method outlined in the following example.

Example 9.13. If $\mathbf{v} = (xy - 1)\hat{\mathbf{i}} - xz\hat{\mathbf{j}} + (2 - yz)\hat{\mathbf{k}}$, then $\nabla \cdot \mathbf{v} = y - y = 0$, and there exists a $\mathbf{w}$ such that $\mathbf{v} = \nabla \times \mathbf{w}$. Equating components,

$$\frac{\partial w_z}{\partial y} - \frac{\partial w_y}{\partial z} = xy - 1 \tag{9.70a}$$

$$\frac{\partial w_x}{\partial z} - \frac{\partial w_z}{\partial x} = -xz \tag{9.70b}$$

$$\frac{\partial w_y}{\partial x} - \frac{\partial w_x}{\partial y} = 2 - yz. \tag{9.70c}$$

To find $\mathbf{w}$, we are faced with the three coupled, first-order, partial differential equations (9.70) in the unknowns w_x, w_y, and w_z. As a start, let us (tentatively) try setting

$w_x \equiv 0$. Then from (9.70b)

$$\frac{\partial w_z}{\partial x} = xz$$

$$w_z = \int xz\, \partial x = \frac{x^2 z}{2} + f(y, z) \tag{9.71}$$

and from (9.70c)

$$\frac{\partial w_y}{\partial x} = 2 - yz$$

$$w_y = \int (2 - yz)\, \partial x = 2x - xyz + g(y, z). \tag{9.72}$$

Inserting (9.71) and (9.72) into (9.70a), we need merely find an f and a g such that

$$\frac{\partial f}{\partial y} + xy - \frac{\partial g}{\partial z} = xy - 1.$$

Doing so is easy; for instance, we can set $g \equiv 0$ and $f = -y$; then

$$\mathbf{w} = 0\hat{\mathbf{i}} + (2x - xyz)\hat{\mathbf{j}} + \left(\frac{x^2 z}{2} - y\right)\hat{\mathbf{k}} \equiv \mathbf{w}_0.$$

In fact, $\mathbf{w}_0$ is not the *only* possible $\mathbf{w}$. We can add to it any gradient, ∇u, since the curl of a gradient is zero (per Theorem 9.1); that is,

$$\nabla \times \mathbf{w} = \nabla \times (\mathbf{w}_0 + \nabla u) = \nabla \times \mathbf{w}_0 + \nabla \times \nabla u = \mathbf{v} + \mathbf{0} = \mathbf{v}.$$

For example, with $u = yz^3$ we have $\mathbf{w} = \mathbf{w}_0 + \nabla u = \mathbf{w}_0 + z^3\hat{\mathbf{j}} + 3yz^2\hat{\mathbf{k}}$.

COMMENT 1. A nice way to remember the important formulas $\nabla \times \nabla u = 0$ for all u and $\nabla \cdot \nabla \times \mathbf{v} = 0$ for all $\mathbf{v}$ contained in Theorems 9.1 and 9.2 is to think of the letters d, g, and c (for divergence, gradient, and curl). The only combinations that come to mind (for me at least) are **cg** (for center of gravity) and **dc** (for direct current); that is, the **c**url of a **g**radient is zero and the **d**ivergence of a **c**url is zero.

COMMENT 2. Observe that the *scalar* potential ϕ of an irrotational field is unique to within an additive arbitrary *constant*, whereas the *vector* potential $\mathbf{w}$ of a solenoidal field is unique to within an additive arbitrary *gradient*.

COMMENT 3. It is interesting to look at some of the formulas we've developed as generalizations of the Fundamental Theorem of the Integral Calculus, which states that if $f(x)$ is continuous over $a \leq x \leq b$, we can express

$$\int_a^x f(\xi)\, d\xi = F(x) - F(a) \tag{9.73}$$

for $a \leq x \leq b$, where $F'(x) = f(x)$. [This statement is equivalent to our earlier statement (1.19); "$F(x)$" in (1.19) is our $F(x) - F(a)$ in (9.73).] To illustrate, in the course of proving Theorem 9.1, we found that

$$\int_{a,b,c}^{x,y,z} [v_x(\xi, \eta, \zeta)\, d\xi + v_y(\xi, \eta, \zeta)\, d\eta + v_z(\xi, \eta, \zeta)\, d\zeta]$$

$$= F(x, y, z) - F(a, b, c), \tag{9.74}$$

where $\nabla F = \mathbf{v}$; that is $\partial F/\partial x = v_x$, $\partial F/\partial y = v_y$, and $\partial F/\partial z = v_z$. This is the generalization of (9.73) to line integrals. But whereas we merely assumed f to be continuous in (9.73), here we require v_x, v_y, and v_z to be not only continuous but also related so that $\nabla \times \mathbf{v} = 0$; that is,

$$\frac{\partial v_z}{\partial y} - \frac{\partial v_y}{\partial z} = 0, \qquad \frac{\partial v_x}{\partial z} - \frac{\partial v_z}{\partial x} = 0, \qquad \frac{\partial v_y}{\partial x} - \frac{\partial v_x}{\partial y} = 0.$$

Just as (9.74) relates a *line* integral to certain *end* values, Stokes' theorem relates a *surface* integral to an integral around the *edge* of the surface, and the divergence theorem relates a *volume* integral to an integral over the *surface* of the volume.[16] ▌

9.7. NONCARTESIAN SYSTEMS

Having introduced the gradient, divergence, and curl operators through intrinsic definitions, we've worked them out thus far only with respect to cartesian coordinates. Many situations, however, are more naturally referred to curvilinear systems, and it is therefore important to obtain more general representations of these operators.

In order to derive such expressions, we can start with definitions (9.5), (9.12), (9.17) and introduce a curvilinear u, v, w system, as defined in Section 8.5. The derivation would follow essentially the same lines as our derivation of (9.9), for example, so let us simply state the final results.

First, note (as in Sections 8.4 and 8.5) that the "line element" $d\mathbf{s}$ is

$$\begin{aligned} d\mathbf{s} &= dx\hat{\mathbf{i}} + dy\hat{\mathbf{j}} + dz\hat{\mathbf{k}} \\ &= (x_u\,du + x_v\,dv + x_w\,dw)\hat{\mathbf{i}} + (y_u\,du + y_v\,dv + y_w\,dw)\hat{\mathbf{j}} \\ &\quad + (z_u\,du + z_v\,dv + z_w\,dw)\hat{\mathbf{k}} \\ &= (x_u\hat{\mathbf{i}} + y_u\hat{\mathbf{j}} + z_u\hat{\mathbf{k}})\,du + (x_v\hat{\mathbf{i}} + y_v\hat{\mathbf{j}} + z_v\hat{\mathbf{k}})\,dv \\ &\quad + (x_w\hat{\mathbf{i}} + y_w\hat{\mathbf{j}} + z_w\hat{\mathbf{k}})\,dw \\ &= \mathbf{U}\,du + \mathbf{V}\,dv + \mathbf{W}\,dw. \end{aligned} \tag{9.75}$$

Limiting ourselves to the case of *orthogonal* curvilinear coordinates—that is, where $\mathbf{U}$, $\mathbf{V}$, $\mathbf{W}$ are mutually orthogonal at each point in the space—it follows that

$$\begin{aligned} (ds)^2 = d\mathbf{s} \cdot d\mathbf{s} &= (\mathbf{U} \cdot \mathbf{U})(du)^2 + (\mathbf{V} \cdot \mathbf{V})(dv)^2 + (\mathbf{W} \cdot \mathbf{W})(dw)^2 \\ &= h_1^2(du)^2 + h_2^2(dv)^2 + h_3^2(dw)^2, \end{aligned} \tag{9.76}$$

where[17]

$$\begin{aligned} h_1 &= \sqrt{\mathbf{U} \cdot \mathbf{U}} = \sqrt{x_u^2 + y_u^2 + z_u^2} \\ h_2 &= \sqrt{\mathbf{V} \cdot \mathbf{V}} = \sqrt{x_v^2 + y_v^2 + z_v^2} \\ h_3 &= \sqrt{\mathbf{W} \cdot \mathbf{W}} = \sqrt{x_w^2 + y_w^2 + z_w^2}. \end{aligned} \tag{9.77}$$

---

[16]For a generalized development, see, for example, R. C. Buck, *Advanced Calculus*, 2nd ed., McGraw-Hill, New York, 1965, Section 7.4.

[17]Many authors use the symbols g_{11}, g_{22}, g_{33} in place of our h_1^2, h_2^2, h_3^2; they are called **metric coefficients.**

Then we state without derivation

$$\nabla f = \frac{1}{h_1}\frac{\partial f}{\partial u}\hat{\mathbf{i}}_1 + \frac{1}{h_2}\frac{\partial f}{\partial v}\hat{\mathbf{i}}_2 + \frac{1}{h_3}\frac{\partial f}{\partial w}\hat{\mathbf{i}}_3 \tag{9.78}$$

$$\nabla \cdot \mathbf{Q} = \frac{1}{h_1 h_2 h_3}\left[\frac{\partial}{\partial u}(h_2 h_3 Q_1) + \frac{\partial}{\partial v}(h_3 h_1 Q_2) + \frac{\partial}{\partial w}(h_1 h_2 Q_3)\right] \tag{9.79}$$

$$\begin{aligned}\nabla \times \mathbf{Q} &= \frac{1}{h_1 h_2 h_3}\begin{vmatrix} h_1\hat{\mathbf{i}}_1 & h_2\hat{\mathbf{i}}_2 & h_3\hat{\mathbf{i}}_3 \\ \partial/\partial u & \partial/\partial v & \partial/\partial w \\ h_1 Q_1 & h_2 Q_2 & h_3 Q_3 \end{vmatrix} \\ &= \frac{1}{h_2 h_3}\left[\frac{\partial(h_3 Q_3)}{\partial v} - \frac{\partial(h_2 Q_2)}{\partial w}\right]\hat{\mathbf{i}}_1 + \frac{1}{h_1 h_3}\left[\frac{\partial(h_1 Q_1)}{\partial w} - \frac{\partial(h_3 Q_3)}{\partial u}\right]\hat{\mathbf{i}}_2 \\ &\quad + \frac{1}{h_1 h_2}\left[\frac{\partial(h_2 Q_2)}{\partial u} - \frac{\partial(h_1 Q_1)}{\partial v}\right]\hat{\mathbf{i}}_3,\end{aligned} \tag{9.80}$$

where $\hat{\mathbf{i}}_1, \hat{\mathbf{i}}_2, \hat{\mathbf{i}}_3$ are the normalized $\mathbf{U}, \mathbf{V}, \mathbf{W}$ base vectors:

$$\begin{aligned}\hat{\mathbf{i}}_1 &= \frac{\mathbf{U}}{\|\mathbf{U}\|} = \frac{\mathbf{U}}{\sqrt{\mathbf{U}\cdot\mathbf{U}}} = \frac{\mathbf{U}}{h_1} \\ \hat{\mathbf{i}}_2 &= \text{etc.} = \frac{\mathbf{V}}{h_2} \\ \hat{\mathbf{i}}_3 &= \text{etc.} = \frac{\mathbf{W}}{h_3}.\end{aligned} \tag{9.81}$$

Combining (9.79) and (9.78), we have for the important *Laplacian operator* (Exercise 9.41)

$$\nabla^2 f = \nabla \cdot \nabla f = \frac{1}{h_1 h_2 h_3}\left[\frac{\partial}{\partial u}\left(\frac{h_2 h_3}{h_1}\frac{\partial f}{\partial u}\right) + \frac{\partial}{\partial v}\left(\frac{h_1 h_3}{h_2}\frac{\partial f}{\partial v}\right) + \frac{\partial}{\partial w}\left(\frac{h_1 h_2}{h_3}\frac{\partial f}{\partial w}\right)\right]. \tag{9.82}$$

Naturally, for the simple cartesian case—that is, where $u = x$, $v = y$, and $w = z$—these expressions should reduce to our previous results. In this case, $h_1 = \sqrt{x_x^2 + y_x^2 + z_x^2} = \sqrt{1 + 0 + 0} = 1$, $h_2 = \text{etc.} = 1$, and $h_3 = \text{etc.} = 1$, so that (9.82), for example, becomes

$$\nabla^2 f = \frac{\partial^2 f}{\partial u^2} + \frac{\partial^2 f}{\partial v^2} + \frac{\partial^2 f}{\partial w^2} = \frac{\partial^2 f}{\partial x^2} + \frac{\partial^2 f}{\partial y^2} + \frac{\partial^2 f}{\partial z^2}$$

as before.

For **cylindrical coordinates**, with $u = r, v = \theta, w = z$ say,[18] $\hat{\mathbf{i}}_1 = \hat{\mathbf{e}}_r$, $\hat{\mathbf{i}}_2 = \hat{\mathbf{e}}_\theta$, $\hat{\mathbf{i}}_3 = \hat{\mathbf{e}}_z$, and according to (9.77),

$$\begin{aligned}h_1 &= \sqrt{x_r^2 + y_r^2 + z_r^2} = \sqrt{\cos^2\theta + \sin^2\theta + 0} = 1 \\ h_2 &= \sqrt{x_\theta^2 + y_\theta^2 + z_\theta^2} = \sqrt{r^2\sin^2\theta + r^2\cos^2\theta + 0} = r \\ h_3 &= \sqrt{x_z^2 + y_z^2 + z_z^2} = \sqrt{0 + 0 + 1} = 1.\end{aligned}$$

---

[18] We could just as well have taken $u = z$, $v = r$, and $w = \theta$, for example; it doesn't matter.

It follows that

$$\nabla f = \frac{\partial f}{\partial r}\hat{\mathbf{e}}_r + \frac{1}{r}\frac{\partial f}{\partial \theta}\hat{\mathbf{e}}_\theta + \frac{\partial f}{\partial z}\hat{\mathbf{e}}_z, \tag{9.83}$$

$$\nabla \cdot \mathbf{Q} = \frac{1}{r}\frac{\partial}{\partial r}(rQ_r) + \frac{1}{r}\frac{\partial}{\partial \theta}Q_\theta + \frac{\partial}{\partial z}Q_z, \tag{9.84}$$

$$\nabla \times \mathbf{Q} = \left[\frac{1}{r}\frac{\partial Q_z}{\partial \theta} - \frac{\partial Q_\theta}{\partial z}\right]\hat{\mathbf{e}}_r + \left[\frac{\partial Q_r}{\partial z} - \frac{\partial Q_z}{\partial r}\right]\hat{\mathbf{e}}_\theta + \frac{1}{r}\left[\frac{\partial (rQ_\theta)}{\partial r} - \frac{\partial Q_r}{\partial \theta}\right]\hat{\mathbf{e}}_z, \tag{9.85}$$

and

$$\nabla^2 f = \frac{1}{r}\frac{\partial}{\partial r}\left(r\frac{\partial f}{\partial r}\right) + \frac{1}{r^2}\frac{\partial^2 f}{\partial \theta^2} + \frac{\partial^2 f}{\partial z^2} = \frac{\partial^2 f}{\partial r^2} + \frac{1}{r}\frac{\partial f}{\partial r} + \frac{1}{r^2}\frac{\partial^2 f}{\partial \theta^2} + \frac{\partial^2 f}{\partial z^2}, \tag{9.86}$$

where $\mathbf{Q} = Q_r\hat{\mathbf{e}}_r + Q_\theta\hat{\mathbf{e}}_\theta + Q_z\hat{\mathbf{e}}_z$.

For **spherical polars** ρ, θ, ϕ (defined in Exercise 7.11) we leave it for you to show (Exercise 9.38) that

$$\nabla f = \frac{\partial f}{\partial \rho}\hat{\mathbf{e}}_\rho + \frac{1}{\rho}\frac{\partial f}{\partial \theta}\hat{\mathbf{e}}_\theta + \frac{1}{\rho \sin\theta}\frac{\partial f}{\partial \phi}\hat{\mathbf{e}}_\phi, \tag{9.87}$$

$$\nabla \cdot \mathbf{Q} = \frac{1}{\rho^2}\frac{\partial}{\partial \rho}(\rho^2 Q_\rho) + \frac{1}{\rho \sin\theta}\frac{\partial}{\partial \theta}(\sin\theta Q_\theta) + \frac{1}{\rho\sin\theta}\frac{\partial Q_\phi}{\partial \phi}, \tag{9.88}$$

$$\begin{aligned}\nabla \times \mathbf{Q} &= \frac{1}{\rho \sin\theta}\left[\frac{\partial Q_\theta}{\partial \phi} - \frac{\partial}{\partial \theta}(\sin\theta Q_\phi)\right]\hat{\mathbf{e}}_\rho \\ &\quad + \frac{1}{\rho}\left[\frac{\partial(\rho Q_\phi)}{\partial \rho} - \frac{1}{\sin\theta}\frac{\partial Q_\rho}{\partial \phi}\right]\hat{\mathbf{e}}_\theta + \frac{1}{\rho}\left[\frac{\partial Q_\rho}{\partial \theta} - \frac{\partial(\rho Q_\theta)}{\partial \rho}\right]\hat{\mathbf{e}}_\phi,\end{aligned} \tag{9.89}$$

and

$$\nabla^2 f = \frac{1}{\rho^2}\left[\frac{\partial}{\partial \rho}\left(\rho^2\frac{\partial f}{\partial \rho}\right) + \frac{1}{\sin\theta}\frac{\partial}{\partial \theta}\left(\sin\theta\frac{\partial f}{\partial \theta}\right) + \frac{1}{\sin^2\theta}\frac{\partial^2 f}{\partial \phi^2}\right], \tag{9.90}$$

where $\mathbf{Q} = Q_\rho\hat{\mathbf{e}}_\rho + Q_\theta\hat{\mathbf{e}}_\theta + Q_\phi\hat{\mathbf{e}}_\phi$.

Of all the formulas (9.83) through (9.90), two that should definitely be memorized are (9.83) and (9.86).

9.8. FLUID MECHANICS; IRROTATIONAL FLOW

Some flows are irrotational—that is the vorticity $\mathbf{\Omega} = \nabla \times \mathbf{q}$ is identically zero—and some are not; here we consider only the class of flows that are. To understand which flows are irrotational and which are not, we state the following crucial result, which goes back to Helmholtz and Lord Kelvin (Sir William Thomson).

If the fluid is barotropic and inviscid, then the vorticity vector $\mathbf{\Omega}$ *of any fluid particle* (which we saw in Section 9.3 is actually twice the particle's angular velocity) *is constant in time*. By *barotropic* we mean that the fluid mass density ρ depends on the pressure p only, that is, $\rho = \rho(p)$. For instance, if the fluid is a perfect gas (i.e., $p = \rho RT$, where R is a constant and T is the temperature) and heat is transferred so that the flow is isothermal, then RT is a constant and ρ is proportional to p. Or if the fluid can be considered as incompressible (recall footnote 8), then, of course, $\rho =$ constant. And by *inviscid* we mean that the fluid has no viscosity.

It follows from the Kelvin–Helmholtz statement that if, for such a fluid, a particle once has zero vorticity, then it will *always* have zero vorticity. Physically, this situation makes sense, since the way to impart some spin (and hence vorticity) to the particle is by applying a torque through *shear stresses* at its surface; but an inviscid fluid, by definition, cannot support any shear stresses.

If, in fact, $\mathbf{\Omega} = 0$ throughout the flow, it follows from Theorem 9.1 (since $\nabla \times \mathbf{q} = 0$) that there exists a fluid **velocity potential** ϕ such that[19]

$$\mathbf{q} = \nabla\phi. \tag{9.91}$$

If, *in addition*, the fluid is incompressible, the continuity equation (9.45) holds:

$$\nabla \cdot \mathbf{q} = 0. \tag{9.92}$$

Inserting $\nabla\phi$ for $\mathbf{q}$, (9.92) becomes $\nabla \cdot \nabla\phi = 0$ or

$$\nabla^2\phi = 0, \tag{9.93}$$

which is the well-known Laplace equation. This partial differential equation and its solutions constitute the rich and well-developed subject of **potential theory**. Most of our discussion of potential theory is contained in Part V on partial differential equations.

Example 9.14. To illustrate, consider the steady two-dimensional flow of an inviscid, incompressible fluid over a rigid, stationary, circular cylinder of radius a, as sketched in Fig. 9.21. Except for the presence of the cylinder, the flow would simply be the uniform "free stream" $\mathbf{q} = U\hat{\mathbf{i}}$; the deviation from this free stream, due to the cylinder, is called the "disturbance" velocity field.

Now as $x \longrightarrow -\infty$, we have $\mathbf{q} \sim U\hat{\mathbf{i}}$. (In fact, $\mathbf{q} \sim U\hat{\mathbf{i}}$ as $r \longrightarrow \infty$ in *any* direction.) Moreover, since $\nabla \times U\hat{\mathbf{i}} = 0$, we conclude that $\nabla \times \mathbf{q} = 0$ initially (i.e., far upstream) and hence $\nabla \times \mathbf{q} = 0$ *throughout* the flow. Thus $\mathbf{q} = \nabla\phi$, and putting this expression into $\nabla \cdot \mathbf{q} = 0$ yields the governing partial differential equation

$$\nabla^2\phi = 0. \tag{9.94}$$

In addition, we have *boundary conditions* at infinity and on the cylinder.

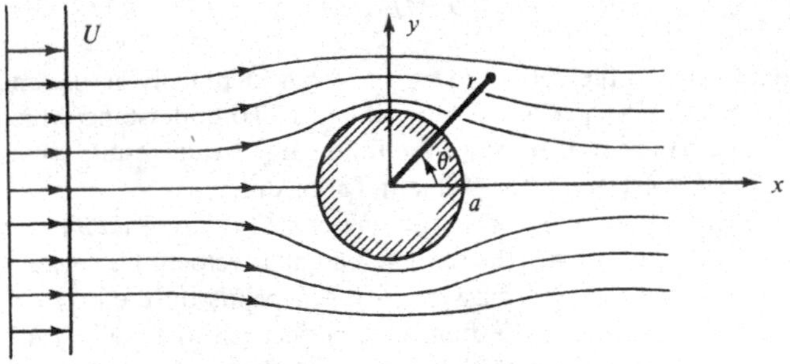

Figure 9.21. Potential flow over a cylinder.

[19]Some authors use $-\nabla\phi$ instead of $\nabla\phi$; it doesn't matter as long as we are consistent.

At infinity, $\mathbf{q} = \nabla\phi \sim U\hat{\imath}$; that is, $\phi_x \sim U$ and $\phi_y \longrightarrow 0$. Integrating, $\phi \sim Ux$ (plus an arbitrary constant, which we might as well set equal to zero).

On the cylinder $r = a$, we must have the normal velocity equal to zero, since the cylinder is rigid and stationary; that is, $\mathbf{q} \cdot \hat{\mathbf{e}}_r = \nabla\phi \cdot \hat{\mathbf{e}}_r = \partial\phi/\partial r = 0$ on $r = a$.

The complete boundary value problem is therefore as follows.

Partial differential equation:

$$\nabla^2\phi = 0 \qquad \text{in } r > a$$

Boundary conditions:

$$\phi \sim Ux \quad \text{as } r \longrightarrow \infty \qquad (9.95)$$

$$\frac{\partial\phi}{\partial r} = 0 \qquad \text{on } r = a.$$

We should decide whether we want to use cartesian or polar coordinates. The free stream $\mathbf{q} \sim U\hat{\imath}$ seems to suggest a cartesian system, whereas the circular shape of the cylinder strongly suggests polar coordinates. We will find later on that it is best to use the r, θ variables shown in the figure. Then (9.95) becomes

$$\frac{\partial^2\phi}{\partial r^2} + \frac{1}{r}\frac{\partial\phi}{\partial r} + \frac{1}{r^2}\frac{\partial^2\phi}{\partial\theta^2} = 0 \quad \text{in } r > a$$

$$\phi \sim Ur\cos\theta \quad \text{as } r \longrightarrow \infty \qquad (9.96)$$

$$\frac{\partial\phi}{\partial r} = 0 \quad \text{on } r = a.$$

The *solution* of (9.96) is the subject of Exercise 26.39 in Part V. Here we are concerned only with the problem formulation.

COMMENT 1. Sometimes people prefer to split $\mathbf{q}$ into two parts, the free stream plus the disturbance velocity $\mathbf{q}'$, say: $\mathbf{q} = U\hat{\imath} + \mathbf{q}'$. Then $\mathbf{q}' = \mathbf{q} - U\hat{\imath} = \nabla\phi - \nabla(Ux) = \nabla(\phi - Ux) \equiv \nabla\phi'$, where ϕ' is the "disturbance potential." To formulate the problem in terms of ϕ', take the divergence of the equation $\mathbf{q} = U\hat{\imath} + \mathbf{q}'$: $\nabla \cdot \mathbf{q} = \nabla \cdot (U\hat{\imath}) + \nabla \cdot \mathbf{q}'$ or $0 = 0 + \nabla \cdot \mathbf{q}'$. Thus $\nabla\cdot\nabla\phi' = 0$ or

$$\nabla^2\phi' = \frac{\partial^2\phi'}{\partial r^2} + \frac{1}{r}\frac{\partial\phi'}{\partial r} + \frac{1}{r^2}\frac{\partial^2\phi'}{\partial\theta^2} = 0, \qquad (9.97a)$$

that is, ϕ' satisfies the same (Laplace) equation as ϕ. We leave it for the reader to show (Exercise 9.38) that the associated boundary conditions are

$$\phi' \longrightarrow 0 \quad \text{as } r \longrightarrow \infty \qquad (9.97b)$$

$$\frac{\partial\phi'}{\partial r} = -Ua\cos\theta \quad \text{on } r = a. \qquad (9.97c)$$

The boundary value problem (9.97) is actually no easier to solve than (9.96); the choice is primarily a matter of personal preference.

COMMENT 2. Observe that in this problem we had both $\nabla \times \mathbf{q} = 0$ and $\nabla \cdot \mathbf{q} = 0$. We used the first to imply the existence of a scalar potential ϕ, so that $\mathbf{q} = \nabla\phi$, and the second to yield the differential equation (9.94) on ϕ. Alternatively, we could have reversed the procedure and used the continuity equation $\nabla \cdot \mathbf{q} = 0$ to imply the existence of a *vector potential* $\mathbf{A}$ (from Theorem 9.2), so that

$$\mathbf{q} = \nabla \times \mathbf{A}. \qquad (9.98)$$

Inserting (9.98) into $\nabla \times \mathbf{q} = 0$ then yields the governing partial differential equation

$$\nabla \times \nabla \times \mathbf{A} = 0 \tag{9.99}$$

on **A**. Nevertheless, it is standard to use the scalar potential approach instead. Why? Well, it's certainly preferable to deal with the single unknown ϕ than a vector **A** with two or three components. Furthermore, whereas the Laplace equation on ϕ is well known and susceptible to standard methods of solution, equation (9.99) is less tractable (Exercise 9.40). ▮

Example 9.15. Next, consider the plane axisymmetric, irrotational, incompressible flow in the annulus $a < r < b$ between stationary, rigid, concentric, circular cylinders, defined by the potential $\phi = \kappa\theta$, where κ is any given constant. First checking to see that such a flow is *possible*, note that $\kappa\theta$ does satisfy the Laplace equation $\nabla^2\phi = 0$. Also, the velocity

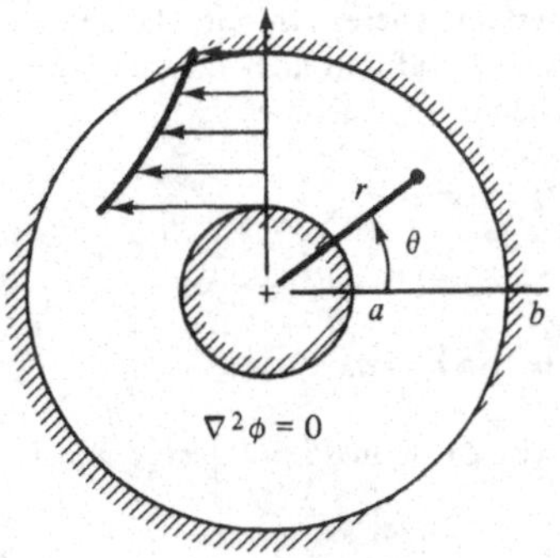

Figure 9.22. Axisymmetric potential flow between cylinders.

$$\mathbf{q} = \nabla\phi = \frac{\partial\phi}{\partial r}\hat{\mathbf{e}}_r + \frac{1}{r}\frac{\partial\phi}{\partial\theta}\hat{\mathbf{e}}_\theta = 0\hat{\mathbf{e}}_r + \frac{\kappa}{r}\hat{\mathbf{e}}_\theta \tag{9.100}$$

does satisfy the boundary conditions $\mathbf{q} \cdot \hat{\mathbf{e}}_r = 0$ at $r = a$ and $r = b$. In fact, the radial velocity component $\mathbf{q} \cdot \hat{\mathbf{e}}_r$ is zero for *all* $a < r < b$; the velocity is purely tangential and independent of θ (hence the adjective "axisymmetric"). The variation in the tangential component $\mathbf{q} \cdot \hat{\mathbf{e}}_\theta = \kappa/r$ is sketched in Fig. 9.22. (Anyone who has studied fluid mechanics probably recognizes this flow as the flow induced by a fictitious "vortex" of strength $\Gamma = 2\pi\kappa$ at $r = 0$.)

Let us calculate the kinetic energy of the fluid per unit depth (i.e., perpendicular to the paper). Surely,

$$\begin{aligned} \text{KE} &= \int \frac{1}{2}q^2\,dm = \frac{\rho}{2}\int_V (\mathbf{q} \cdot \mathbf{q})\,d\tau \\ &= \frac{\rho}{2}\int_0^1\int_0^{2\pi}\int_a^b \frac{\kappa^2}{r^2}\,r\,dr\,d\theta\,dz = \pi\kappa^2\rho\ln\frac{b}{a}. \end{aligned} \tag{9.101}$$

But suppose instead that we convert to a surface integral by Green's first identity (9.53) with "u" = "v" = ϕ:

$$\text{KE} = \frac{\rho}{2}\int_V \nabla\phi\cdot\nabla\phi\,d\tau = \frac{\rho}{2}\int_S \phi\frac{\partial\phi}{\partial n}\,d\sigma - \frac{\rho}{2}\int_V \phi\nabla^2\phi\,d\tau = \frac{\rho}{2}\int_S \phi\frac{\partial\phi}{\partial n}\,d\sigma, \tag{9.102}$$

since $\nabla^2\phi = 0$ throughout V. Now V is the "washer"-shaped region $a \le r \le b$, $0 \le z \le 1$, and so its surface S has four parts: an outer edge $r = b$, an inner edge $r = a$, a front face $z = 1$, and a back face $z = 0$. On each of these four parts $\partial\phi/\partial n = \nabla\phi \cdot \hat{\mathbf{n}} = \mathbf{q} \cdot \hat{\mathbf{n}}$ is zero. (On $z = 0$ and $z = 1$ it is zero because there is no velocity in the z direction.) Thus the right side of (9.102) is zero, which disagrees with (9.101) and which is obviously incorrect. What's wrong? The difficulty is that ϕ and its first-

and second-order partials are not continuous in V as required in Green's identity (9.53). That is, if we consider the θ domain to be $0 \leq \theta < 2\pi$, say, then ϕ is not even *continuous* in V, since $\phi \longrightarrow 0$ as we approach the line $\theta = 0$ from above and $\phi \longrightarrow 2\pi\kappa$ as we approach it from below!

We overcome this difficulty by redefining our region. Specifically, we take a hacksaw (with an infinitely thin blade) and cut the washer as in Fig. 9.23. Inside the *cut* washer ϕ *is* continuous, as are its partials of all orders, which means that (9.102) should now be correct. But note carefully that S now has two additional faces, the top of the cut ($\theta = 0$, say) and the bottom of the cut ($\theta = 2\pi$). The top part gives no contribution because $\phi = \kappa\theta = 0$ there. On the bottom, however,

$$\phi = \kappa\theta = 2\pi\kappa,$$

and $$\frac{\partial\phi}{\partial n} = \nabla\phi \cdot \hat{\mathbf{n}} = \mathbf{q} \cdot \hat{\mathbf{e}}_\theta = \frac{\kappa}{r}$$

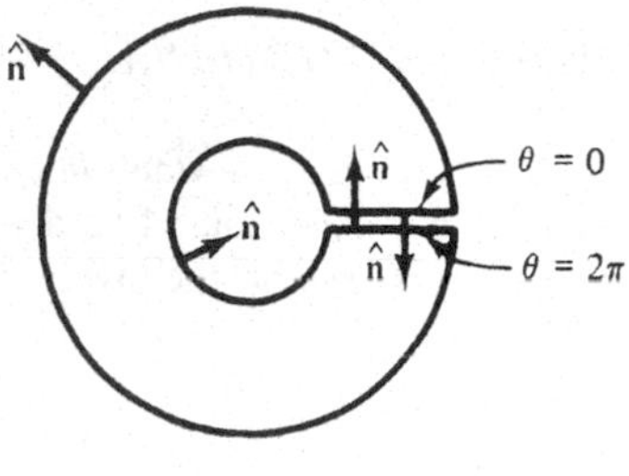

Figure 9.23. The cut washer.

from (9.100). So from the six parts of S we obtain

$$\text{KE} = 0 + 0 + 0 + 0 + 0 + \frac{\rho}{2}\int_0^1 \int_a^b (2\pi\kappa)\left(\frac{\kappa}{r}\right) dr\, dz = \pi\rho\kappa^2 \ln\frac{b}{a},$$

which is now in agreement with (9.101).

Note that the same result is obtained if we decide instead to let $\theta = 6\pi$, for example, on the top of the cut; then it is 8π on the bottom (Exercise 9.40).

COMMENT. Recalling that the vorticity $\mathbf{\Omega} = \nabla \times \mathbf{q}$ is twice the angular velocity of the fluid particles, the reader may well wonder how the flow around an annulus can possibly be irrotational, since it certainly *looks* as though the fluid particles must have some rotation. Of course, we're certain that it *is* irrotational because $\mathbf{q} = (\kappa/r)\hat{\mathbf{e}}_\theta$ is derived from the scalar potential $\phi = \kappa\theta$ and so, by Theorem 9.1, $\nabla \times \mathbf{q}$ must be zero. Or we can work out $\nabla \times [(\kappa/r)\hat{\mathbf{e}}_\theta]$ directly from (9.85) if you like, and again we obtain zero.

The trick is that the fluid particles simply move about like seats on a ferris wheel as shown in Fig. 9.24(a)—that is, without rotating! (Does this situation seem to fit together with the way that the tangential velocity varies with r?)

In contrast, consider the flow $\mathbf{q} = \omega r\hat{\mathbf{e}}_\theta$, where ω is any given constant. Using

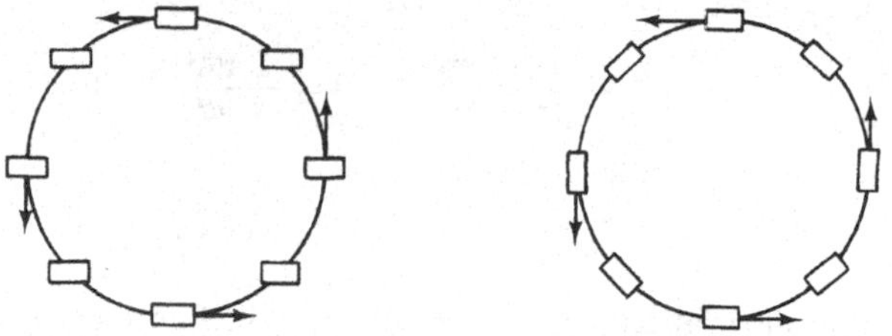

(a) (b)

Figure 9.24. Irrotational and rotational flow in an annulus.

(9.85), we find that $\mathbf{\Omega} = \nabla \times \mathbf{q} = 2\omega\hat{\mathbf{e}}_z \neq 0$; therefore the flow is *not* irrotational; it is rotational. In fact, since $\mathbf{\Omega}$ is a constant vector, it follows that *all* fluid particles have the same angular velocity—namely, $\omega\hat{\mathbf{e}}_z$ (i.e., half of $\mathbf{\Omega}$)—so that $\mathbf{q} = \omega r\hat{\mathbf{e}}_\theta$ happens to correspond to a state of "solid-body rotation" as shown in Fig. 9.24(b). ∎

9.9. THE GRAVITATIONAL POTENTIAL

According to *Newton's law of gravitation*, the attractive force F exerted by a point mass m on another mass at a distance ρ from m, per unit mass at that location, is proportional to m and the inverse square of ρ (Fig. 9.25):

$$\mathbf{F} = -\gamma\frac{m}{\rho^2}\hat{\mathbf{e}}_\rho, \tag{9.103}$$

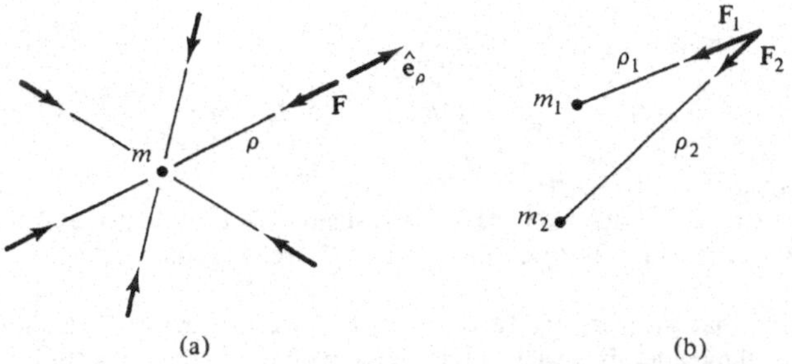

Figure 9.25. Gravitational force field.

where γ is a suitable constant of proportionality that can (and will) be set equal to unity by an appropriate choice of the units of mass. Note that we've adopted a spherical polar coordinate system, rather than cylindrical say, in order to take advantage of the spherical symmetry; that is, $F_\theta = F_\phi = 0$, and $F_\rho = -m/\rho^2$ is a function of ρ only.

From (9.89) we see that $\nabla \times \mathbf{F} = 0$; thus $\mathbf{F}$ is conservative and can be expressed as $\nabla\Phi$, where Φ is the so-called gravitational or Newtonian potential. To determine Φ, we equate the components of $\mathbf{F}$ to the corresponding components of $\nabla\Phi$. Recalling (9.87),

$$-\frac{m}{\rho^2} = \frac{\partial\Phi}{\partial\rho}, \qquad 0 = \frac{1}{\rho}\frac{\partial\Phi}{\partial\theta}, \qquad 0 = \frac{1}{\rho\sin\theta}\frac{\partial\Phi}{\partial\phi},$$

from which it follows that

$$\Phi = \frac{m}{\rho} \tag{9.104}$$

(plus an arbitrary constant, which we will choose to be zero).

If there are *two* point masses, m_1 and m_2 [Fig. 9.25(b)], the force at any point P is $\mathbf{F} = \mathbf{F}_1 + \mathbf{F}_2 = \nabla\Phi_1 + \nabla\Phi_2 = \nabla(\Phi_1 + \Phi_2) = \nabla\Phi$, where

$$\Phi = \Phi_1 + \Phi_2 = \frac{m_1}{\rho_1} + \frac{m_2}{\rho_2}$$

is the total potential at P. (Note how vector addition of the $\mathbf{F}_j$'s is equivalent to the scalar addition of their potentials.) Generalizing to a *distribution* of mass, of density $\sigma(P)$, we can say that

$$\Phi = \int_M \frac{dm}{\rho} = \int_V \frac{\sigma\, d\tau}{\rho}, \tag{9.105}$$

where M is the total mass and V is the region that it occupies. Or in terms of some cartesian reference frame (Fig. 9.26)

$$\Phi(x, y, z) = \int_V \frac{\sigma(x', y', z')\, dx'\, dy'\, dz'}{\sqrt{(x - x')^2 + (y - y')^2 + (z - z')^2}}. \tag{9.106}$$

Note carefully that although the "field point" x, y, z is arbitrary, it is regarded as *fixed* in the calculation (9.106).

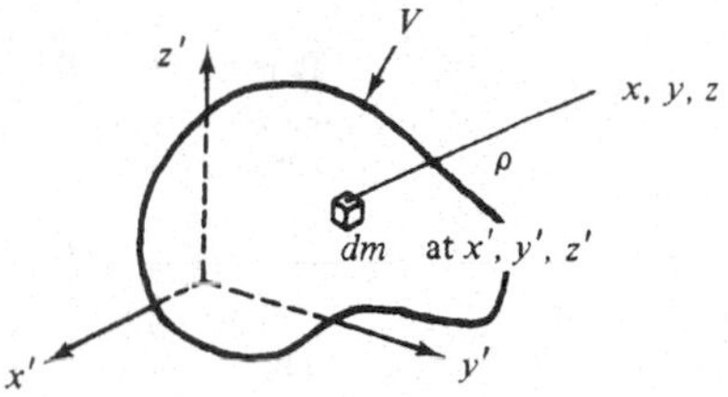

Figure 9.26. Distribution of mass.

Example 9.16. *A Hollow Sphere.* Suppose that V is a spherical shell with inner radius a, outer radius b, its center at $x' = y' = z' = 0$, and constant density σ. Because of the spherical symmetry, we can consider our field point to be on the z axis, say, with no loss of generality; in other words, set $x = y = 0$ in (9.106). Switching to the more convenient spherical polars,

$$x' = \rho \sin\theta \cos\phi, \qquad y' = \rho \sin\theta \sin\phi, \qquad z' = \rho\cos\theta,$$

and $$dx'\, dy'\, dz' \longrightarrow |J|\, d\rho\, d\theta\, d\phi = |\rho^2 \sin\theta|\, d\rho\, d\theta\, d\phi.$$

Therefore

$$\Phi(0, 0, z) = \sigma \int_0^{2\pi}\int_0^{\pi}\int_a^b \frac{\rho^2 \sin\theta\, d\rho\, d\theta\, d\phi}{\sqrt{\rho^2 + z^2 - 2\rho z \cos\theta}}$$

$$= 2\pi\sigma \int_a^b \rho^2\, d\rho \int_0^{\pi} \frac{\sin\theta\, d\theta}{\sqrt{\rho^2 + z^2 - 2\rho z\cos\theta}}.$$

Letting the terms inside the radical be u, say,

$$\Phi(0, 0, z) = \frac{\pi\sigma}{z}\int_a^b \rho\, d\rho \int_{(\rho-z)^2}^{(\rho+z)^2} u^{-1/2}\, du = \frac{2\pi\sigma}{z}\int_a^b \rho\, d\rho \left\{\sqrt{u}\,\Big|_{(\rho-z)^2}^{(\rho+z)^2}\right\}. \tag{9.107}$$

Next comes a key point. Tracing the square root back to (9.106), we see that it is actually a distance, and so the *positive* root is to be used. Thus

$$\sqrt{(\rho - z)^2} = |\rho - z| = \begin{cases} \rho - z & \text{if } z < \rho \\ z - \rho & \text{if } z > \rho. \end{cases} \tag{9.108}$$

Case 1: $z > b$.

That is, the field point lies outside the sphere. Thus $z > \rho$ for all $a < \rho < b$, and (9.107) becomes

$$\Phi(0, 0, z) = \frac{2\pi\sigma}{z}\int_a^b \rho[(\rho + z) - (z - \rho)]\, d\rho = \frac{4\pi\sigma}{3z}(b^3 - a^3). \tag{9.109}$$

But since $M = V\sigma = [4\pi(b^3 - a^3)/3]\sigma$, (9.109) is simply

$$\Phi(0, 0, z) = \frac{M}{z}, \tag{9.110}$$

and the force [per unit mass at (0, 0, z)] is

$$F = \frac{\partial \Phi}{\partial z} = -\frac{M}{z^2}. \tag{9.111}$$

It is remarkable that (9.111) *is precisely the same as if all the mass were concentrated at the origin*! The "averaging" process that leads to this result is not at all transparent.

Case 2: $z < a$.

That is, the field point lies within the cavity. Thus $z < \rho$ for all $a < \rho < b$, and (9.107) becomes

$$\Phi(0, 0, z) = \frac{2\pi\sigma}{z} \int_a^b \rho[(\rho + z) - (\rho - z)]\, d\rho = 2\pi\sigma(b^2 - a^2), \tag{9.112}$$

which is a *constant*, so that the force in the cavity is

$$F = \frac{\partial \Phi}{\partial z} = 0, \tag{9.113}$$

which is no less remarkable than (9.111).

Case 3: $a < z < b$.

That is, the field point is somewhere in the material. Then

$$\Phi(0, 0, z) = \frac{2\pi\sigma}{z}\left\{\int_a^z \rho[(\rho + z) - (z - \rho)]\, d\rho + \int_z^b \rho[(\rho + z) - (\rho - z)]\, d\rho\right\}$$
$$= \text{etc.}, \tag{9.114}$$

but it is easier simply to apply what we have already learned from Cases 1 and 2: there should be no force at z due to the $z < \rho < b$ portion of the shell, and the $a < \rho < z$ portion should be felt exactly the same as if it were lumped as a point mass at the origin. Therefore

$$F = -\frac{\text{mass}}{z^2} = -\frac{(4/3)\pi(z^3 - a^3)\sigma}{z^2}, \tag{9.115}$$

which, of course, agrees with (9.113) as $z \to a$ and with (9.111) as $z \to b$.

COMMENT 1. Let us return to (9.106) and pursue an entirely different line of approach.

Recall from the Leibnitz rule (1.41) that if

$$I(\alpha) = \int_a^b f(x, \alpha)\, dx,$$

where a and b are independent of α, then

$$\frac{dI}{d\alpha} = \int_a^b \frac{\partial f}{\partial \alpha}(x, \alpha)\, dx$$

if $\partial f/\partial \alpha$ is a continuous function of both x and α over the pertinent x and α intervals. Similarly, it is a fact that if V is independent of α, then

$$\frac{d}{d\alpha}\int_V f(P, \alpha)\, d\tau = \int_V \frac{\partial f}{\partial \alpha}(P, \alpha)\, d\tau \tag{9.116}$$

if $\partial f/\partial \alpha$ is continuous for all Ps (i.e., x, y, z) in V and α's in the given α interval.

Using this result to obtain $\partial/\partial x$ of (9.106), we have

$$\frac{\partial \Phi}{\partial x}(x, y, z) = \int_V \frac{(-1/2)(2)(x - x')\sigma(x', y', z')\, dx'\, dy'\, dz'}{[(x - x')^2 + (y - y')^2 + (z - z')^2]^{3/2}}.$$

But the integrand is *not* continuous, as desired, if x, y, z is inside V because it "blows up" at that point—that is, when $x' = x$, $y' = y$, and $z' = z$. So we can't be sure that the differentiation under the integral sign is legitimate, and from here on we will restrict our x, y, z field point to be *outside* V.

Differentiating again with respect to x yields an expression for $\partial^2\Phi/\partial x^2$; computing $\partial^2\Phi/\partial y^2$ and $\partial^2\Phi/\partial z^2$ in similar fashion and adding, we have much cancellation and find (Exercise 9.42) that

$$\frac{\partial^2\Phi}{\partial x^2} + \frac{\partial^2\Phi}{\partial y^2} + \frac{\partial^2\Phi}{\partial z^2} = 0. \tag{9.117}$$

That is,

$$\nabla^2\Phi = 0 \tag{9.118}$$

for all x, y, z outside V![20] Writing (9.118) in spherical polars per (9.90) and noting that Φ is a function of ρ only, for the spherical shell case,

$$\nabla^2\Phi = \frac{1}{\rho^2}\frac{d}{d\rho}\left(\rho\,\frac{2d\Phi}{d\rho}\right) = 0,$$

so that

$$\rho^2\frac{d\Phi}{d\rho} = A, \qquad \frac{d\Phi}{d\rho} = A\rho^{-2},$$

and

$$\Phi = -\frac{A}{\rho} + B, \qquad F = \frac{d\Phi}{d\rho} = \frac{A}{\rho^2}, \tag{9.119}$$

where A and B are the constants of integration.

Note carefully that we are solving (9.118) in two distinct regions, the exterior $\rho > b$ and the cavity $\rho < a$, and that there is no reason why the constants A and B for the exterior region need be the same as for the cavity. In $\rho < a$, (9.119) says that F is unbounded; that is, it tends to infinity as $\rho \longrightarrow 0$. Certainly this is ridiculous, and so we must choose $A = 0$. Thus we obtain $F = 0$ in the cavity as before.

---

[20] We cannot claim that (9.118) holds for points *within* V because the Leibnitz differentiations are then apparently incorrect. In fact, it turns out that within the material Φ satisfies the so-called **Poisson equation** $\nabla^2\Phi = -4\pi\sigma(x, y, z)$; of course, this reduces to (9.118) outside V, where $\sigma = 0$. Although we will not derive the Poisson equation (see, for example, Philip Franklin's *Methods of Advanced Calculus*, McGraw-Hill, New York, 1944, pp. 319–322), let us at least check it for the present example. Recalling the results of case 3,

$$F = -\frac{4\pi\sigma(\rho^3 - a^3)}{3\rho^2} = \frac{\partial\Phi}{\partial\rho},$$

so that

$$\nabla^2\Phi = \frac{1}{\rho^2}\frac{\partial(\rho^2\partial\Phi/\partial\rho)}{\partial\rho} = \frac{1}{\rho^2}\frac{-4\pi\sigma}{3}(3\rho^2) = -4\pi\sigma,$$

as claimed.

Having applied a "boundedness" condition in order to evaluate A for the region $\rho < a$, what boundary condition can we come up with so as to evaluate A in the exterior region $\rho > b$? Well, we must have

$$F \sim -\frac{M}{\rho^2} \quad \text{as } \rho \longrightarrow \infty, \tag{9.120}$$

because as $\rho \longrightarrow \infty$, the shell will look more and more like a *point* mass. Comparing (9.120) and (9.119), we see that we must choose $A = -M$ for the exterior region. Consequently, (9.119) becomes $F = -M/\rho^2$ for all $\rho > b$ (not just as $\rho \longrightarrow \infty$), in agreement with our previous result (9.111).

COMMENT 2. In Section 9.8 we noted that irrotationality ($\nabla \times \mathbf{q} = 0$) implied that $\mathbf{q} = \nabla\phi$. The *additional* assumption of incompressibility ($\nabla \cdot \mathbf{q} = 0$) then led to the Laplace equation $\nabla \cdot \nabla\phi = \nabla^2\phi = 0$. In the present section we verified directly that $\nabla \times \mathbf{F} = 0$ for the gravitational force field (9.103), so that $\mathbf{F} = \nabla\Phi$. Finally, in Comment 1, we showed by Leibnitz differentiation that $\nabla^2\Phi = 0$ (at points not in the material). More directly, however, this result follows on taking the divergence of (9.103). Using (9.88), we have

$$\nabla \cdot \mathbf{F} = \frac{1}{\rho^2}\frac{\partial}{\partial\rho}(\rho^2 F_\rho) = \frac{1}{\rho^2}\frac{\partial}{\partial\rho}(-m) = 0$$

for all points $\rho \neq 0$. So $\nabla \cdot \mathbf{F} = \nabla \cdot \nabla\Phi = 0$. It is also interesting to observe that whereas $\nabla \times \mathbf{F} = 0$ for *any* central force field $\mathbf{F} = F_\rho(\rho)\hat{\mathbf{e}}_\rho$, it is only for an *inverse-square law* $F_\rho(\rho) = \text{constant}/\rho^2$ that $\nabla \cdot \mathbf{F} = 0$ as well and hence that $\nabla^2\Phi = 0$.

*

COMMENT 3. Recall from Chapter 3 that if $f(x) \longrightarrow \infty$ as $x \longrightarrow x_0$, where x_0 is inside an interval I ($a \leq x \leq b$), then the improper integral $\int_a^b f\,dx$ is defined as the limit of the integral with an arbitrary neighborhood N_0 of x_0 deleted [Fig. 9.27(a)] as the length of N_0 shrinks to zero in any manner; symbolically,

$$\int_I f\,dx \equiv \lim_{N_0 \to 0} \int_{I-N_0} f\,dx. \tag{9.121}$$

In particular, we saw that a singularity $f(x) = O[1/(x - x_0)^p]$ causes divergence if and only if $p \geq 1$.

Improper *multiple* integrals are defined essentially the same way. For instance, if $f(x, y, z) \longrightarrow \infty$ as $x, y, z \longrightarrow x_0, y_0, z_0$, where x_0, y_0, z_0 is inside a volume V,

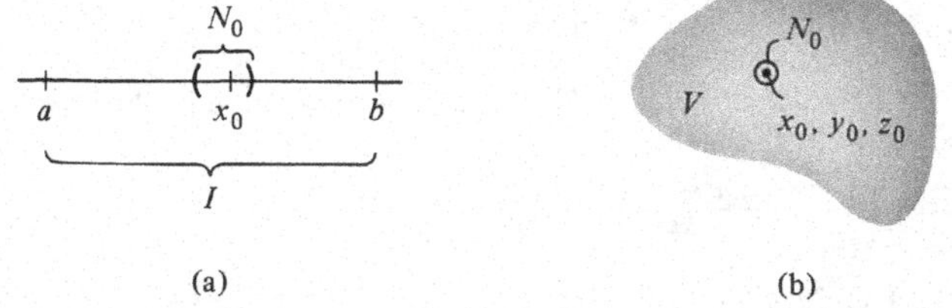

Figure 9.27. Deleting a neighborhood of singular point.

then [Fig. 9.27(b)]

$$\int_V f\,d\tau \equiv \lim_{N_0\to 0}\int_{V-N_0} f\,d\tau, \tag{9.122}$$

where the N_0s are a sequence of neighborhoods of x_0, y_0, z_0 of arbitrary shape that shrink to zero; that is, their largest dimension tends to zero.

We mention this definition because a number of integrals in the "shell example" were, in fact, singular; for instance, assuming that σ is nonzero and continuous at the field point x, y, z in (9.106), observe that x, y, z is a singular point of the integral.

More generally, consider the integral

$$I = \int_V \frac{g(x, y, z)\,dx\,dy\,dz}{[(x-x_0)^2+(y-y_0)^2+(z-z_0)^2]^p}, \tag{9.123}$$

where P_0 (i.e., x_0, y_0, z_0) is inside V and $\lim_{P\to P_0} g(P) = g(P_0) \neq 0$. Switching to a spherical polar system ρ, θ, ϕ with its origin at P_0, in order to expose the nature of the singularity, we have

$$dx\,dy\,dz \longrightarrow \rho^2 \sin\theta\,d\rho\,d\theta\,d\phi$$

and

$$[(x-x_0)^2+(y-y_0)^2+(z-z_0)^2]^p = \rho^{2p}.$$

Note how the ρ^2, which enters through the Jacobian, helps out. Instead of having an $O(1/\rho^{2p})$ behavior as $\rho \longrightarrow 0$, we have an $O(\rho^2/\rho^{2p})$ behavior, and so we have convergence of I if $2p - 2 < 1$—that is, if $p < \frac{3}{2}$.

The reader should, I think, be reasonably comfortable with this claim. Anyone who wishes to see a careful proof, plus some further discussion of the gravitational potential, should refer to the book by Tychonov and Samarski.[21] ∎

EXERCISES

9.1. Evaluate the line integral $\int_C \mathbf{F}\cdot d\mathbf{r}$, where C is a straight line from (0, 1, 0) to (1, 2, 3) and

(a) $\mathbf{F} = xz\hat{\mathbf{i}} - y^2z\hat{\mathbf{k}}$.
(b) $\mathbf{F} = x^2y\hat{\mathbf{i}} + 3\hat{\mathbf{j}} - x\hat{\mathbf{k}}$.
(c) $\mathbf{F} = xyz\hat{\mathbf{k}}$.

9.2. Evaluate $\int_C \mathbf{F}\cdot d\mathbf{r}$, where C is the cardioid $r = a(1+\cos\theta)$ from $\theta = 0$ to 2π and

(a) $\mathbf{F} = y\hat{\mathbf{i}}$, (b) $\mathbf{F} = 6\hat{\mathbf{e}}_r + r\hat{\mathbf{e}}_\theta$.

9.3. Evaluate $\int_C (xy^2\,dx - y\,dz)$, where C is given parametrically by $x = a\cos\zeta$, $y = a\sin\zeta$, $z = \zeta$, as ζ goes from 0 to 5π.

9.4. Evaluate the following line integrals, where C is the ellipse $(x/a)^2 + (y/b)^2 = 1$ clockwise. *Hint*: The parameterization $x = a\cos t$, $y = b\sin t$ may be of help in some of them.

[21] A. N. Tychonov and A. A. Samarski, *Partial Differential Equations of Mathematical Physics*, Vol. I, Holden-Day, San Francisco, 1964, pp. 292–303.

(a) $\int_C (y\,dx - x\,dy)$ (b) $\int_C (y\,dx + x\,dy)$

(c) $\int_C (x^6 \sin x\,dx + dy)$ (d) $\int_C x^3 y\,dy$

(e) $\int_C x^7\,dx$ (f) $\int_C (x + y)\,dy$

9.5. Taking τ to be a prism, as in Fig. 9.6, derive the expression (9.13) for grad u from definition (9.12).

9.6. Find the derivative of $f = xyz$ at the point (1, 3, 2) in the direction of the vector $2\hat{\mathbf{i}} - \hat{\mathbf{k}}$. What is the *maximum* possible directional derivative of f at that point, and what is its direction? What is the equation of the tangent plane to the surface $f =$ constant at the point (1, 3, 2)?

9.7. Find the derivative of $u = xy^2 - 3z^3$ at the point (1, − 2, 4) in the direction of the normal to the surface $xy + xz + yz = -6$.

9.8. Sketch the velocity profile for the one-dimensional flow $\mathbf{q} = (100 + 5y)\hat{\mathbf{i}}$. Compute $\nabla \times \mathbf{q}$ and explain qualitatively why the result "makes sense" in light of our physical interpretation of the vorticity vector.

9.9. Why are the linearity relations (9.24) to (9.26) true?

9.10. Verify the formulas (a) (9.27) and (9.28), (b) (9.30) and (9.31), (c) (9.32) and (9.33).

9.11. If $\mathbf{A}$ is a constant vector and $r = \sqrt{x^2 + y^2 + z^2}$, show that

$$\mathbf{A} \cdot \nabla\left(\frac{1}{r}\right) = -\frac{\mathbf{A} \cdot \mathbf{r}}{r^3} \quad \text{and} \quad \nabla(\mathbf{A} \cdot \mathbf{r}) = \mathbf{A}.$$

9.12. Just as we derived the divergence theorem (9.35) from definition (9.5), obtain the companion results

$$\int_V \nabla u\,d\tau = \int_S \hat{\mathbf{n}} u\,d\sigma \quad \text{and} \quad \int_V \nabla \times \mathbf{v}\,d\tau = \int_S \hat{\mathbf{n}} \times \mathbf{v}\,d\sigma. \tag{9.124}$$

9.13. Previously we applied Leibnitz differentiation only to single integrals. Can you explain why the Leibnitz differentiation of the *triple* integral in (9.40) is valid?

9.14. According to the fluid mechanics version of the **Biot–Savart law**, the velocity $\mathbf{q}$ induced in an incompressible fluid at a point P by a "vortex filament" of strength Γ is given by the line integral

$$\mathbf{q}(P) = -\frac{\Gamma}{4\pi} \int_C \frac{\mathbf{r} \times d\mathbf{s}}{r^3}, \tag{9.125}$$

where Γ is a constant, $\mathbf{r}$ is the vector from the $d\mathbf{s}$ element to the point P, as shown in Fig. 9.28, and the orientation of Γ and $d\mathbf{s}$ are related by the right-hand rule. For the case where C is an infinitely long helix, defined parametrically by $x = a \cos \theta$, $y =$

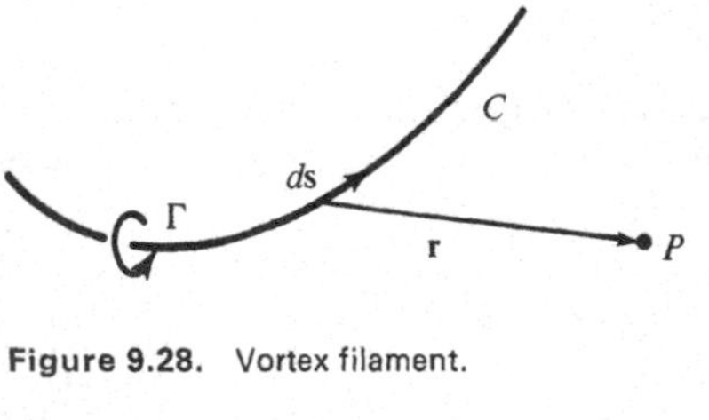

Figure 9.28. Vortex filament.

$a \sin\theta$, and $z = \kappa\theta$, for $-\infty < \theta < \infty$, use this integral to show that the z component of $\mathbf{q}$ at any point on the z axis is equal to $\Gamma/(2\pi\kappa)$. As a partial check, does this result make sense for the limiting cases $\kappa \longrightarrow \infty$ and $\kappa \longrightarrow 0$? How about the fact that it's independent of the helix radius a? (Note that our formula is called the Biot–Savart law by analogy with the expression for the magnetic flux induced by a current-carrying wire.)

9.15. (*Convective derivative*) Suppose that we have a fluid velocity field $\mathbf{q}(x, y, z, t)$. Consider some property of the flow, such as the temperature $T(x, y, z, t)$. If we swim along any desired path according to $x = x(t)$, $y = y(t)$, $z = z(t)$, then the rate of change observed is

$$\frac{d}{dt} T[x(t), y(t), z(t), t] = \frac{\partial T}{\partial t} + \frac{\partial T}{\partial x}\frac{dx}{dt} + \frac{\partial T}{\partial y}\frac{dy}{dt} + \frac{\partial T}{\partial z}\frac{dz}{dt}.$$

Note carefully that if we simply choose to *drift* with the fluid, then dx/dt, dy/dt, and dz/dt are the fluid velocity components u, v, and w. In this case, show that

$$\frac{dT}{dt} = \frac{\partial T}{\partial t} + \mathbf{q} \cdot \nabla T.$$

This is called the **convective derivative** because it corresponds to the case where we drift, or convect, with the fluid. The special notation D/Dt is often used:

$$\frac{D(\ )}{Dt} = \frac{\partial(\ )}{\partial t} + \mathbf{q} \cdot \nabla(\ ). \tag{9.126}$$

9.16. (*Equations of motion of fluid mechanics*) Considering a fluid velocity field $\mathbf{q}$, let us apply Newton's second law to an arbitrary control volume V with surface S that drifts with the fluid. If ρ and p are the fluid mass density and pressure, $\mathbf{F}$ is a body force per unit mass (e.g., a gravitational field), and $\mathfrak{M}$ is the total momentum of the fluid in V, then

$$\mathfrak{M} = \int_V \rho\mathbf{q}\, d\tau$$

and according to Newton's law,

$$\frac{d}{dt}\int_V \rho\mathbf{q}\, d\tau = -\int_S p\hat{\mathbf{n}}\, d\sigma + \int_V \rho\mathbf{F}\, d\tau \tag{9.127}$$

if we assume the fluid to be inviscid, so that there is no viscous shear force on S to include. Noting that both the integrand and V depend on t, show (with the help of Fig. 9.12 and the continuity equation) that the Leibnitz differentiation on the left of (9.127) yields

$$\frac{d}{dt}\int_V \rho\mathbf{q}\, d\tau = \int_V \frac{\partial}{\partial t}(\rho\mathbf{q})\, d\tau + \int_S (\hat{\mathbf{n}} \cdot \mathbf{q})(\rho\mathbf{q})\, d\sigma$$

$$= \text{etc.} = \int_V \left(\frac{\partial \mathbf{q}}{\partial t} + \mathbf{q} \cdot \nabla\mathbf{q}\right)\rho\, d\tau.$$

Finally, converting the pressure term in (9.127) to a volume integral with the help of (9.124), show that

$$\frac{\partial \mathbf{q}}{\partial t} + \mathbf{q} \cdot \nabla\mathbf{q} = -\frac{1}{\rho}\nabla p + \mathbf{F} \tag{9.128}$$

or using (9.126),

$$\frac{D\mathbf{q}}{Dt} = -\frac{1}{\rho}\nabla p + \mathbf{F}. \tag{9.129}$$

Here is the important vector equation of motion for an inviscid fluid. Whereas the continuity equation was a statement of conservation of mass (i.e., *kinematics*), the equation of motion (9.129) is a statement of Newton's second law (i.e., *dynamics*).

9.17. To obtain a work-energy equation for the flow of an inviscid, incompressible fluid, it is reasonable to dot an arbitrary displacement vector ds into (9.128), since $\mathbf{F} \cdot ds$, for example, is clearly a work term. Is the following derivation correct? If not, correct it.

Suppose that the flow $\mathbf{q}$ is irrotational and that the force field $\mathbf{F}$ is conservative. Then with $\mathbf{q} \equiv \nabla\phi$ and $\mathbf{F} \equiv -\nabla V$,

$$\frac{\partial}{\partial t}(\nabla\phi) \cdot ds + (\mathbf{q} \cdot \nabla\mathbf{q}) \cdot ds = -\frac{1}{\rho}\nabla p \cdot ds - \nabla V \cdot ds,$$

$$\nabla\left(\frac{\partial\phi}{\partial t}\right) \cdot ds + \mathbf{q} \cdot (\nabla\mathbf{q} \cdot ds) = -\frac{1}{\rho}\frac{\partial p}{\partial s}\,ds - \frac{\partial V}{\partial s}\,ds,$$

$$\frac{\partial}{\partial s}\left(\frac{\partial\phi}{\partial t}\right) + \mathbf{q} \cdot \frac{\partial\mathbf{q}}{\partial s} = -\frac{\partial}{\partial s}\left(\frac{p}{\rho}\right) - \frac{\partial}{\partial s}V,$$

$$\frac{\partial}{\partial s}\left(\frac{\partial\phi}{\partial t} + \frac{1}{2}\mathbf{q} \cdot \mathbf{q} + \frac{p}{\rho} + V\right) = 0,$$

and we obtain the *energy equation*

$$\frac{\partial\phi}{\partial t} + \frac{q^2}{2} + \frac{p}{\rho} + V = C(t) \tag{9.130}$$

throughout the flow, where the "constant" of integration $C(t)$ is often determined by considering the value of the terms on the left at "infinity" if the flow is unbounded. For instance, suppose that $V = 0$ and $\mathbf{q}$ and p approach the "free stream" values $U\hat{\imath}$ and p_0 at infinity. Then $C(t)$ must equal $U^2/2 + p_0/\rho$.

9.18. The following scalar equations occur in fluid and solid mechanics. Reexpress them more concisely in terms of vector and vector differential operator notation.

(a)
$$\rho\ddot{u}_x = \mu\left(\frac{\partial^2 u_x}{\partial x^2} + \frac{\partial^2 u_x}{\partial y^2} + \frac{\partial^2 u_x}{\partial z^2}\right) + (\lambda + \mu)\frac{\partial\Theta}{\partial x} + F_x$$
$$\rho\ddot{u}_y = \mu\left(\frac{\partial^2 u_y}{\partial x^2} + \frac{\partial^2 u_y}{\partial y^2} + \frac{\partial^2 u_y}{\partial z^2}\right) + (\lambda + \mu)\frac{\partial\Theta}{\partial y} + F_y$$
$$\rho\ddot{u}_z = \mu\left(\frac{\partial^2 u_z}{\partial x^2} + \frac{\partial^2 u_z}{\partial y^2} + \frac{\partial^2 u_z}{\partial z^2}\right) + (\lambda + \mu)\frac{\partial\Theta}{\partial z} + F_z,$$

where the x, y, z subscripts denote the respective components of a given vector, the dots denote partial derivatives with respect to t, and ρ, μ, λ are constants.

(b) $$\frac{\partial^4\psi}{\partial x^4} + 2\frac{\partial^4\psi}{\partial x^2\,\partial y^2} + \frac{\partial^4\psi}{\partial y^4} = 0.$$

9.19. Evaluate $\int_V \nabla \cdot \mathbf{v}\,d\tau$ both directly and as a surface integral, where

(a) $\mathbf{v} = -z^2\hat{\mathbf{k}}$ and V is bounded by four flat faces with corners at (0, 0, 0), (0, 1, 1), (0, 0, 1), and (1, 0, 1).

(b) $\mathbf{v} = 3xy^2\hat{\mathbf{j}}$ and V is the region $0 \le x \le 1, 0 \le y \le 2, 0 \le z \le 3$.

9.20. Evaluate $\int_S \mathbf{v} \cdot \hat{\mathbf{n}}\,d\sigma$, where S is the surface of the cylinder $x^2 + y^2 \le 4, 0 \le z \le 3$, and

(a) $\mathbf{v} = x\hat{\mathbf{\imath}} + y\hat{\mathbf{\jmath}} + z\hat{\mathbf{k}}$, (b) $\mathbf{v} = 3z^5\hat{\mathbf{k}} + 2\hat{\mathbf{\imath}}$.

9.21. Suppose that $\phi(x, y, z)$ has continuous first-order partials and satisfies the Poisson equation $\nabla^2\phi = f(x, y, z)$ throughout some simply connected region V, together with the boundary condition $\phi = g(x, y, z)$ on the surface S of V.

(a) Show that the solution (if it exists) is unique. *Hint:* Suppose that there are *two* solutions, say ϕ_1 and ϕ_2. With $\phi_1 - \phi_2 \equiv \Phi$, show that $\nabla^2\Phi = 0$ in V; $\Phi = 0$ on S. Setting $u = v = \Phi$ in (9.53), show that

$$\int_V (\Phi_x^2 + \Phi_y^2 + \Phi_z^2)\, d\tau = 0$$

and conclude that $\Phi_x = \Phi_y = \Phi_z = 0$ throughout V, so that Φ = constant $= 0$ and $\phi_1 = \phi_2$.

(b) Repeat part (a) with the boundary condition $\phi = g$ replaced by $\partial\phi/\partial n = g$.

(c) Repeat part (a) but with a "mixed" boundary condition, where $\phi = g$ over part of S and $\partial\phi/\partial n = h$ over the rest of S.

9.22. Evaluate the surface integral in Example 9.9 directly—that is, without using Stokes' theorem.

9.23. Verify Stokes' theorem if S is the hemispherical cap $z = +\sqrt{a^2 - x^2 - y^2}$ and
(a) $\mathbf{v} = y\hat{\mathbf{i}} - x\hat{\mathbf{j}} + (2z + 3)\hat{\mathbf{k}}$.
(b) $\mathbf{v} = x^3y\hat{\mathbf{i}} + (y^2 + 4)\hat{\mathbf{j}}$.
(c) $\mathbf{v} = x\hat{\mathbf{i}} + y\hat{\mathbf{j}} + z\hat{\mathbf{k}}$.

9.24. Derive (9.61) by means of integration by parts.

9.25. Evaluate the line integrals of Exercise 9.4(b), (c), (d), and (e) with the help of Green's theorem.

9.26. If we set $M = 0$ and $N = x$ in Green's theorem, we obtain the line integral representation

$$A = \int_C x\, dy$$

for the area enclosed by C. Or with $N = 0$ and $M = -y$, we obtain an alternative representation

$$A = -\int_C y\, dx.$$

Interpret these two formulas graphically and apply them to the following areas.
(a) The triangle with vertices at (0, 0), (0, 2), and (2, 2) counterclockwise.
(b) The triangle with vertices at (0, 0), (0, 2) and (4, 2) counterclockwise.

9.27. Express

$$\int_S \left(\frac{\partial N}{\partial x} - \frac{\partial M}{\partial y}\right) dx\, dy$$

as one or more line integrals, where S is the shaded region in Fig. 9.29.

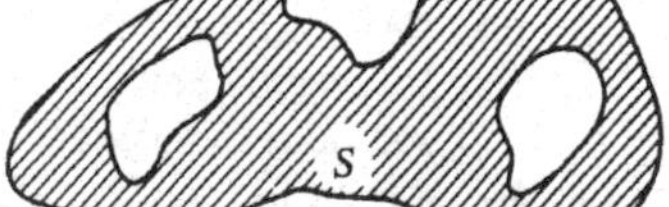

Figure 9.29. Green's theorem.

9.28. Show carefully that for the *two*-dimensional case [i.e., where $\mathbf{v} = \mathbf{v}(x, y)$, S is a region in the x, y plane, and C is its boundary], the *divergence theorem* (9.35) becomes

$$\int_S \nabla \cdot \mathbf{v}\, d\sigma = \int_C \hat{\mathbf{n}} \cdot \mathbf{v}\, |ds|,$$

where $|ds|$ emphasizes that all ds increments are to be positive. Similarly, write out the two-dimensional version of Green's first and second identities, (9.53) and (9.54).

9.29. Show that $\mathbf{F}$ in Example 9.1, is conservative. Use this information to evaluate the line integral much more simply.

9.30. Evaluate the line integrals shown.

(a) $\int_C (3x^2 \cos 2y\, dx - 2x^3 \sin 2y\, dy)$.

(b) $\int_C [5x^{3/2}e^{y^2}\, dx + (4yx^{5/2}e^{y^2} + 5y^2)\, dy]$,
where C is the curve $r = 2\theta^3$ as θ goes from 0 to 6π.

(c) $\int_C [3\, dx + 6y^2z^{5/2}\, dy + (5y^3z^{3/2} + 2)\, dz]$,
where C is the curve $x = \sin t$, $y = \cos 3t$, $z = 2t$ as t goes from 0 to 5π. *Hint:* They are *not* hard.

9.31. Show that the following vector fields are irrotational and in each case determine a scalar potential $u(x, y, z)$ such that $\mathbf{v} = \nabla u$.
(a) $\mathbf{v} = 2x^2\hat{\mathbf{i}} - 2yz\hat{\mathbf{j}} - (y^2 + 3)\hat{\mathbf{k}}$ (b) $\mathbf{v} = \hat{\mathbf{i}} - 4\hat{\mathbf{j}}$
(c) $\mathbf{v} = 3x^2 \cos 2y\hat{\mathbf{i}} - 2x^3 \sin 2y\hat{\mathbf{j}}$ (d) $\mathbf{v} = z^3\hat{\mathbf{k}}$

9.32. Show that the vector fields below are solenoidal and in each case find a vector potential $\mathbf{w}(x, y, z)$ such that $\mathbf{v} = \nabla \times \mathbf{w}$:
(a) $\mathbf{v} = \hat{\mathbf{i}} - 4\hat{\mathbf{j}}$ (b) $\mathbf{v} = z\hat{\mathbf{i}} + x\hat{\mathbf{j}} + y\hat{\mathbf{k}}$
(c) $\mathbf{v} = 2y\hat{\mathbf{i}} + x^2y^2\hat{\mathbf{j}} - 2x^2yz\hat{\mathbf{k}}$

9.33. (a) Consider the decomposition, or splitting, of a given vector field $\mathbf{q}$ into the sum $\mathbf{q} = \mathbf{v}_1 + \mathbf{v}_2$, where $\mathbf{v}_1$ is irrotational (so that there exists a ϕ such that $\mathbf{v}_1 = \nabla\phi$) and $\mathbf{v}_2$ is solenoidal (so that there exists a $\mathbf{w}$ such that $\mathbf{v}_2 = \nabla \times \mathbf{w}$). The problem is as follows. Given $\mathbf{q}$, find $\mathbf{v}_1$ and $\mathbf{v}_2$ or, equivalently, their potentials ϕ and $\mathbf{w}$. Show that ϕ is found as any solution of the Poisson equation

$$\nabla^2\phi = \operatorname{div} \mathbf{q} \tag{9.131}$$

and that $\mathbf{w}$ is found as any solution of

$$\nabla \operatorname{div} \mathbf{w} - \nabla^2\mathbf{w} = \operatorname{curl} \mathbf{q}$$

(if indeed such solutions exist). Note that since the gradient of an arbitrary scalar function, say f, can be included in $\mathbf{w}$ (Why?), it may be possible to adjust f so that $\operatorname{div} \mathbf{w} = 0$, in which case $\mathbf{w}$, like ϕ, satisfies a Poisson equation,

$$\nabla^2\mathbf{w} = -\operatorname{curl} \mathbf{q}. \tag{9.132}$$

(b) Carry out the decomposition described in part (a) for $\mathbf{q} = x\hat{\mathbf{i}} + x^2\hat{\mathbf{j}}$. (Although solving the Poisson equation has yet to be discussed, you should be able to find solutions by inspection.)

9.34. (*The stream function for plane steady flow*) Consider the plane flow [i.e., $\mathbf{q} = u(x, y)\hat{\mathbf{i}} + v(x, y)\hat{\mathbf{j}}$] of an incompressible fluid. The volume flow rate per unit depth (in cubic meters per second, say) crossing the curve OA from left to right (Fig. 9.30) is

$$Q = \int_C (u\,dy - v\,dx) \equiv \int_C \mathbf{F} \cdot d\mathbf{r}$$

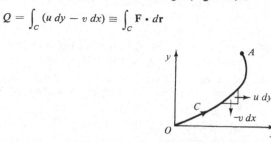

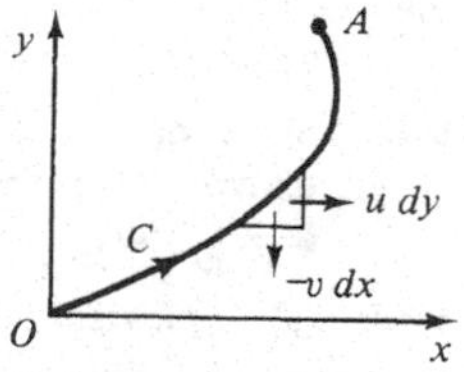

Figure 9.30. Stream function.

where $\mathbf{F} = -v\hat{\mathbf{i}} + u\hat{\mathbf{j}}$. Since $\nabla \times \mathbf{F} = (\partial u/\partial x + \partial v/\partial y)\hat{\mathbf{k}} = 0$ by the continuity equation, it follows that

$$Q = \int_{0,0}^{x,y} (u\,dy - v\,dx)$$

is a (single-valued) function of x, y, say $\psi(x, y)$, where

$$\frac{\partial \psi}{\partial x} = -v, \qquad \frac{\partial \psi}{\partial y} = u, \tag{9.133}$$

right? The $\psi(x, y) =$ constant lines are called **streamlines**.

(a) Show that

$$\nabla^2 \psi = -\Omega(x, y) \tag{9.134}$$

where $\boldsymbol{\Omega} = \Omega\hat{\mathbf{k}}$ is the vorticity. In particular, $\nabla^2\psi = 0$ if the flow is irrotational (i.e., if $\boldsymbol{\Omega} = 0$).

(b) Reformulate the problem of the flow over a circular cylinder (Example 9.14) in terms of ψ instead of ϕ. What is the resulting boundary value problem on ψ, analogous to the problem (9.96) on ϕ?

(c) Show that for plane, steady, *compressible* flow we can again introduce a stream function ψ according to

$$\frac{\partial \psi}{\partial x} = -\rho v, \qquad \frac{\partial \psi}{\partial y} = \rho u,$$

where this time ψ is the *mass* flow rate per unit depth across OA.

9.35. Show how (9.82) results from the combination of (9.79) and (9.78) as claimed in the text.

9.36. Write out the continuity equation (9.44) in cartesian, cylindrical, and spherical polar coordinates.

9.37. Write out the equations of motion (9.129) as three scalar equations in cartesian and cylindrical coordinates.

9.38. Derive equations (9.87) to (9.90) from (9.78) to (9.82).

9.39. Show how (9.97b) and (9.97c) were derived. Interpret (9.97c) in physical terms. (A picture would help.)

9.40. Write out the equation $\nabla \times \nabla \times \mathbf{A} = 0$ from Comment 2 of Example 9.14 as three scalar equations in terms of cartesian coordinates.

9.41. (a) In Example 9.15 show that if we set $\theta \equiv 6\pi$, say, on the top of the cut (Fig. 9.23), we still arrive at the same result, $\text{KE} = \pi\rho\kappa^2 \ln (b/a)$.

(b) Show also that the same answer results if we use a different cut, say along the y axis instead.

9.42. Verify that the Leibnitz differentiation of (9.106) does lead to (9.117). Is this result surprising in view of Exercise 9.44?

9.43. Show that *any* central force field $\mathbf{F} = f(\rho)\hat{\mathbf{e}}_\rho$ is conservative, where $f(\rho)$ is continuously differentiable [i.e., $f'(\rho)$ exists and is continuous]. What is the potential Φ such that $\mathbf{F} = -\nabla\Phi$?

9.44. (a) *Spherical Polars.* With $\rho = \sqrt{x^2 + y^2 + z^2}$, show that

$$\nabla^2(\rho^\alpha) = 0 \qquad (\text{for all } 0 < \rho < \infty)$$

only for $\alpha = -1$ (and, of course, $\alpha = 0$).

(b) *Cylindrical Coordinates.* With $r = \sqrt{x^2 + y^2}$, show that there are *no* α's (except, of course, $\alpha = 0$) such that

$$\nabla^2(r^\alpha) = 0$$

but that

$$\nabla^2(\ln r) = 0 \qquad (\text{for all } 0 < r < \infty).$$

(c) In order to attach a physical significance to the solutions $1/\rho$ and $\ln r$ of Laplace's equation, consider the context of the gravitational potential (Section 9.9). From (9.104) we see that $1/\rho$ may be regarded as the potential field induced by a unit mass at $\rho = 0$. Next, consider a straight, infinite wire with a uniform mass density of $\frac{1}{2}$ (for convenience) per unit length (Fig. 9.31). Integrating over $-\infty < z < \infty$,

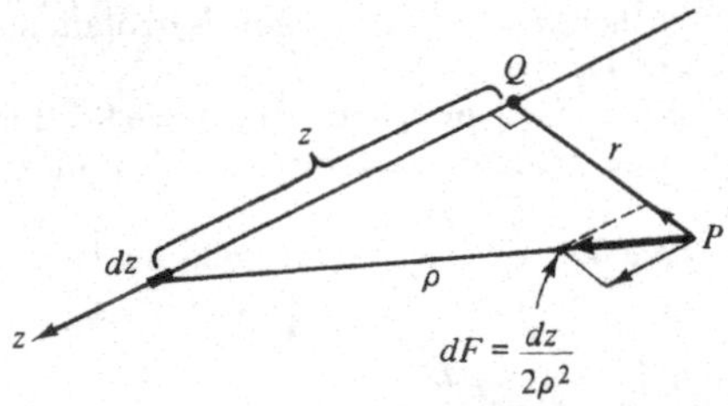

Figure 9.31. Potential due to an infinite wire.

show that the resulting force at P is $1/r$ and is normal to the wire—that is, toward Q. Thus deduce that the potential field induced by the wire is $\Phi = -\ln r$. This potential is called the two-dimensional or **logarithmic potential**. (These results will be important in our discussion of Green's functions in Part V.)

***9.45.** Consider the solution of $\nabla \times \mathbf{F} = \Phi(x, y, z)\mathbf{r}$ for $\mathbf{F}$, where

$$\Phi = \frac{1}{\pi}\frac{lx + my + nz}{(x^2 + y^2 + z^2)^2} \quad \text{and} \quad \mathbf{r} = x\hat{\mathbf{i}} + y\hat{\mathbf{j}} + z\hat{\mathbf{k}}.$$

It will help to define $\mathbf{A} \equiv l\hat{\mathbf{i}} + m\hat{\mathbf{j}} + n\hat{\mathbf{k}}$ and introduce spherical polars r, θ, ϕ with $\mathbf{A}$ as polar axis (Fig. 9.32), for then $\mathbf{F}$ will be independent of ϕ because of the symmetry about the polar axis (Right?). Thus

$$\nabla \times \mathbf{F} = \frac{1}{\pi}\frac{\mathbf{A} \cdot \mathbf{r}}{r^4}\mathbf{r} = \frac{1}{\pi}\frac{A\cos\theta}{r^2}\hat{\mathbf{e}}_r.$$

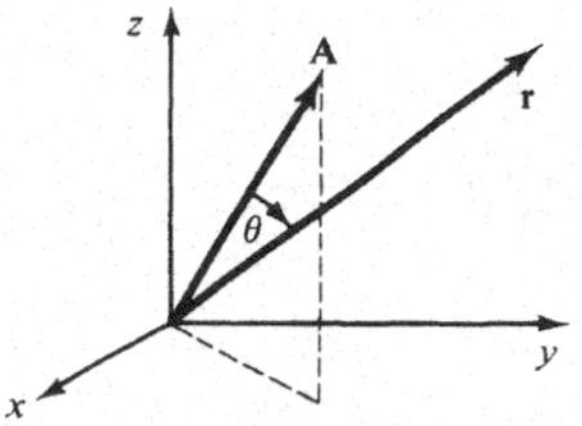

Figure 9.32. A as polar axis.

Integrating, obtain the solution

$$\mathbf{F} = \frac{(ny - mz)\hat{\mathbf{i}} + (lz - nx)\hat{\mathbf{j}} + (mx - ly)\hat{\mathbf{k}}}{2\pi r^2}.$$

(Note the similarity between this problem and Exercise 8.30.)

***9.46.** (*Improper multiple integrals*) In Comment 3 following Example 9.16 we briefly discussed improper triple integrals. For the *double* integral

$$I = \int_S \frac{g(x, y)\, dx\, dy}{[(x - x_0)^2 + (y - y_0)^2]^p}, \tag{9.135}$$

where x_0, y_0 is inside S, put forward a convergence definition, analogous to (9.122), and discuss the values of p for which (9.135) converges.

9.47. Consider the following claims: $\mathbf{F} = \theta\hat{\mathbf{e}}_r + \hat{\mathbf{e}}_\theta$ is conservative and $\mathbf{F} = -\nabla\Phi$, where $\Phi = \quad r\theta$ plus a constant, which can be taken to be zero. If C is the *ccw* unit circle, say, then

$$W = \int_C \mathbf{F} \cdot d\mathbf{r} = -\Phi\Big|_{\text{initial}}^{\text{final}} - 2\pi.$$

On the other hand, W should be zero by Stokes' theorem, since $\mathbf{F}$ is conservative and C is a closed curve. Explain the apparent contradiction.

9.48. Make up two examination-type problems on this chapter.

Chapter 10

The Calculus of Variations

One of the first problems discussed in a beginning course in calculus is the determination of the maxima and minima of a given function $F(x)$. Historically, too, such determination was an early problem in the development of the calculus. Newton, in a paper written in 1671 (published in 1736), argued that at a maximum or minimum the rate of change of $f(x)$ must be zero, and Leibnitz in 1684 made the equivalent statement that the tangent line must be horizontal there, statements that seem all too elementary to us now and that, to be appreciated, must be considered in the context of seventeenth-century mathematics.

A few years later Newton investigated a difficult problem concerning the drag on an axisymmetric body moving through a resisting medium. With x as the axial coordinate and the radius of the body denoted as $y(x)$, the problem was to find the function $y(x)$ connecting two given endpoints

$$y(x_1) = y_1 \quad \text{and} \quad y(x_2) = y_2, \tag{10.1}$$

such that the integral

$$I = \int_{x_1}^{x_2} \frac{yy'^3}{1 + y'^2}\, dx, \tag{10.2}$$

which (according to Newton's simplified physical model) is proportional to the drag, is a minimum.

Note carefully that Newton's problem involved the minimization not of a function but of a *functional* $I(y)$ over a certain domain, say the class of functions $y(x)$ that satisfy the end conditions (10.1) and are (at least) differentiable over $x_1 \leq x \leq x_2$. Although he found the solution, Newton expressed no interest in pursuing a generalization to a wider class of problems. (This situation reflects the fact that more interest attached to specific problems in those days than to the systematic development of a broad area.)

Not until 1696 was attention focused on these problems. In that year John Bernoulli put before the scholars of his time his *brachistochrone problem*—that is, finding the path from one given point to another, in the same vertical plane, along which a particle will descend (due to gravity) in the shortest time. The problem generated considerable interest and solutions were offered by Newton, Leibnitz, L'Hospital, Bernoulli himself, and his older brother James. Finally, a subsequent student of John Bernoulli, Leonhard Euler (1707–83), took James' solution and developed the *calculus of variations* as it is known today.

10.1 FUNCTIONS OF ONE OR MORE VARIABLES

Before looking at *functionals*, consider the simplest case—that is, a *function* of one variable, $F(x)$. As usual, x is a real variable and F is to be real valued.

We say that $F(x)$ has an *absolute maximum* over an interval I if there exists a point x_0 such that

$$F(x) \leq F(x_0) \tag{10.3}$$

for all x in I and a *relative* or *local maximum* at x_0 if there is some neighborhood $N(x_0)$, which is a subset of I, such that (10.3) holds in $N(x_0)$. The same is true for absolute and relative *minima* with (10.3) changed to $F(x) \geq F(x_0)$.

We are concerned here only with *relative* maxima and minima and so will dispense with the adjective "relative"; from here on "max" and "min" will mean relative maximum and relative minimum, respectively. Furthermore, since $N(x_0)$ is contained in I, it follows that we will consider only cases where x_0 is an interior point of I.

THEOREM 10.1. In order for $F(x)$ to attain a max or min at an interior point x_0 of I, it is necessary that $F'(x_0)$ either vanish or fail to exist.

That these conditions are not also sufficient is apparent from Fig. 10.1. F' fails to exist at P and Q and vanishes at O and R; yet the function has neither maximum nor minimum at the points O and Q.

As *sufficient* conditions, we state without proof[1]

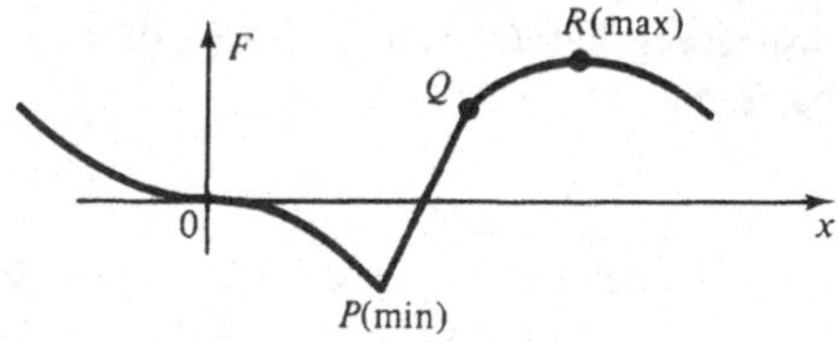

Figure 10.1. Max, min, and neither.

[1] For proof of Theorems 10.1, 10.2, and 10.4, as well as further details, see T. M. Apostol, *Mathematical Analysis*, Addison-Wesley, Reading, Mass., 1957, pp. 91, 92, 148–157.

THEOREM 10.2. Suppose that

$$F'(x_0) = F''(x_0) = \cdots = F^{(n-1)}(x_0) = 0 \quad \text{but} \quad F^{(n)}(x_0) \neq 0,$$

where $F^{(n)}(x)$ is continuous in the interval and x_0 is an interior point. Then if n is even, F has a min at x_0 if $F^{(n)}(x_0) > 0$ and a max at x_0 if $F^{(n)}(x_0) < 0$. If n is odd, F has neither a max nor a min at x_0.

Extending the discussion to the case of n independent variables, $F(x_1, \ldots, x_n)$, we state (without proof) two additional theorems. First, some *necessary* conditions.

THEOREM 10.3. In order for $F(x_1, \ldots, x_n) \equiv$ "$F(P)$" to attain a max or min at an interior point P_0 of the region, it is necessary that $\partial F/\partial x_1, \ldots, \partial F/\partial x_n$ all vanish at P_0 or that some of them fail to exist there.

That these conditions are not also sufficient follows from the examples $F(x, y) = xy$ and $G(x, y) = x^3 + |y|$, say; $F_x(0, 0) = F_y(0, 0) = G_x(0, 0) = 0$ and $G_y(0, 0)$ fails to exist; yet both F and G have neither max nor min at (0, 0).

In stating *sufficient* conditions, we limit ourselves to functions of *two* variables.

THEOREM 10.4. Suppose that

$$F_x(a, b) = F_y(a, b) = 0 \quad \text{and} \quad F_{xy}^2(a, b) - F_{xx}(a, b)F_{yy}(a, b) \equiv D,$$

where $F(x, y)$ has continuous second-order partials in the region and (a, b) is an interior point. If

(i) $F_{xx}(a, b) < 0$ and $D < 0$, then F has a max at (a, b).
(ii) $F_{xx}(a, b) > 0$ and $D < 0$, then F has a min at (a, b).
(iii) $D > 0$, then F has neither a max nor a min at (a, b).

The determinant

$$\begin{vmatrix} F_{xx}(a, b) & F_{xy}(a, b) \\ F_{yx}(a, b) & F_{yy}(a, b) \end{vmatrix} = -D$$

is sometimes called the **Hessian** of F.

Note that if $D = 0$, Theorem 10.4 provides no information and the nature of F at (a, b) will be dictated by the values of third- or higher-order partials. Sometimes the situation is transparent. For instance, $F(x, y) = x^4 + y^4$ has $F_x(0, 0) = F_y(0, 0) = 0$, but $D = 0$, so that Theorem 10.4 does not apply. Nevertheless, it is obvious from inspection that F has a min at (0, 0). Similarly, it is apparent that $F = -(x^4 + y^4)$ has a max at (0, 0).

Example 10.1. Consider $F(x, y) = x^2 - y^2$. Then $F_x = 2x = 0$ on the line $x = 0$ and $F_y = -2y = 0$ on the line $y = 0$. Thus $F_x = F_y = 0$ only at the origin (0, 0). Since the second-order partials are continuous everywhere and $D = 0 - (2)(-2) = 4 > 0$, it follows from Theorem 10.4 that the "flat spot" at the origin is neither a max nor a min. Note that $F(x, 0) = x^2$ is a "valley" and that $F(0, y) = -y^2$ is a "hill"; therefore the F surface is shaped somewhat like a "saddle" in the neighborhood of the origin, which, in fact, is called a **saddlepoint** (Fig. 10.2).

COMMENT. Note that our $F(x, y) = x^2 - y^2$ happens to be **harmonic**; that is, it has continuous second-order partials and satisfies Laplace's equation, $F_{xx} + F_{yy} = 0$. More generally, suppose that $F(x, y)$ is *any* nonconstant harmonic function in a given region R and consider any point a, b in R at which $F_x = F_y = 0$. Then since $F_{xx} + F_{yy} = 0$, it follows that F_{xx} and F_{yy} are of opposite sign, so that $D > 0$, and hence a, b is neither a max nor a min (unless $F_{xy} = F_{xx} = F_{yy} = 0$ at a, b, in which case no conclusion can be drawn). Consider this result in the context of *heat conduction*, for instance. From (9.51) we see that $F(x, y)$ can be regarded as a steadystate temperature field in R. The fact that F cannot have a max (or min) inside R makes sense physically. Suppose that F did have a max at some point in R. Then surely heat would flow away from this point and the temperature F there would drop, thereby contradicting the assumption that F is in steady state. ∎

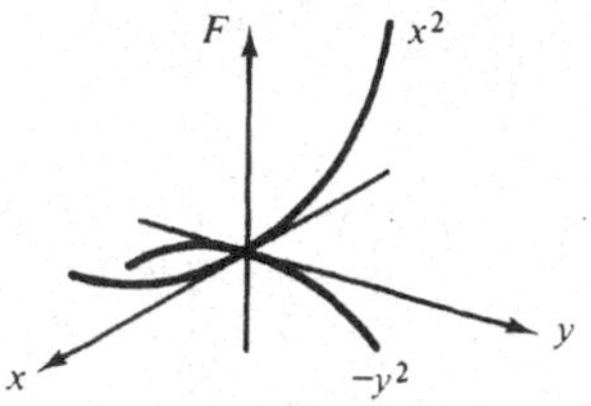

Figure 10.2. Saddlepoint.

10.2 CONSTRAINTS; LAGRANGE MULTIPLIERS

Example 10.2. Find the point(s) on the ellipse $5x^2 - 6xy + 5y^2 = 8$ that is closest to the origin; that is, find x, y such that $(x^2 + y^2)^{1/2}$ is a min, subject to the *constraint* $5x^2 - 6xy + 5y^2 = 8$. Equivalently, we want

$$F(x, y) = x^2 + y^2 = \min, \tag{10.4}$$

subject to the constraint

$$g(x, y) = 5x^2 - 6xy + 5y^2 = 8. \tag{10.5}$$

Notice that if we seek x, y by setting

$$F_x = 2x = 0, \qquad F_y = 2y = 0,$$

we arrive at $x = y = 0$, which certainly minimizes F but which does not satisfy the constraint (10.5)! That is, the point $x = y = 0$ does not lie on the ellipse $5x^2 - 6xy + 5y^2 = 8$.

The key fact is that whether or not the x, y variables are independent, the fundamental necessary condition for a differentiable function $F(x, y)$ to have a max or min at a particular point x, y is that the differential $dF = 0$ there. Now

$$dF = \nabla F \cdot ds. \tag{10.6}$$

If the variables x, y are *independent*, the orientation of ds is arbitrary, and so from (10.6) $dF = 0$ implies $\nabla F = 0$ or $F_x = F_y = 0$.[2] Then $dF = 0$ and $\nabla F = 0$ are equivalent and can be used interchangeably. But if x, y are *dependent*, the orientation of ds is *not* arbitrary and the condition $dF = 0$ for a max or min does *not* imply that $\nabla F = 0$.

[2] If the variables x, y are independent and $dF = F_x\,dx + F_y\,dy = 0$ at a particular point, it must be true that $F_x = F_y = 0$ there, since dx, dy are independent increments; if, for example, we choose $dx = 0$ and $dy \neq 0$, it follows from $dF = F_x\,dx + F_y\,dy = 0$ that $F_y = 0$, whereas the choice $dx \neq 0$, $dy = 0$ reveals that $F_x = 0$.

To illustrate, let us return to the present example. We have

$$dF = 2x\,dx + 2y\,dy = 0. \tag{10.7}$$

Instead of dx and dy being independent (i.e., ds being arbitrary), in which case it would follow that $F_x = 2x = 0$ and $F_y = 2y = 0$ (i.e., $\nabla F = 0$), dx and dy are related through the constraint (10.5), or

$$dg = (10x - 6y)\,dx + (10y - 6x)\,dy = 0. \tag{10.8}$$

Assuming, tentatively, that $10y - 6x$ is not zero (at the desired minimum), it follows from (10.8) that

$$dy = \frac{10x - 6y}{6x - 10y}\,dx,$$

and inserting this result into (10.7),

$$dF = \left[2x + 2y\left(\frac{5x - 3y}{3x - 5y}\right)\right] dx = 0.$$

Since dx all by itself *is* arbitrary, then

$$2x + 2y\left(\frac{5x - 3y}{3x - 5y}\right) = 0$$

or

$$3x^2 - 5xy + 5xy - 3y^2 = 0$$

or

$$y = \pm x.$$

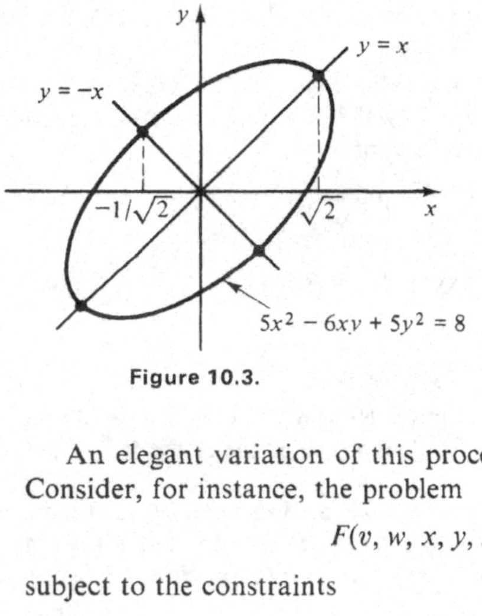

Figure 10.3.

Thus the points that we seek lie on the lines $y = \pm x$. Yet they must also lie on the ellipse (10.5), and so they are the intersections $(\sqrt{2}, \sqrt{2})$, $(-\sqrt{2}, -\sqrt{2})$, $(1/\sqrt{2}, -1/\sqrt{2})$, and $(-1/\sqrt{2}, 1/\sqrt{2})$, as indicated in Fig. 10.3.

It is clear from the figure that $(\sqrt{2}, \sqrt{2})$ and $(-\sqrt{2}, -\sqrt{2})$ correspond to maxima of F (namely, $F = 4$) and that $(1/\sqrt{2}, -1/\sqrt{2})$ and $(-1/\sqrt{2}, 1/\sqrt{2})$ correspond to minima (namely, $F = 1$) (Exercise 10.1). ∎

An elegant variation of this procedure, due to Lagrange, will be described next. Consider, for instance, the problem

$$F(v, w, x, y, z) = \max \qquad \text{(or min)}, \tag{10.9}$$

subject to the constraints

$$g(v, w, x, y, z) = a, \qquad h(v, w, x, y, z) = b. \tag{10.10}$$

First, we form the combination

$$F^* \equiv F - \lambda_1 g - \lambda_2 h, \tag{10.11}$$

where the λ's are constants. Now $dF = 0$ at a max and $dg = dh = 0$, since g and h are constant; consequently, $dF^* = dF - \lambda_1\,dg - \lambda_2\,dh = 0$ there, too; that is,

$$\begin{aligned} dF^* = (F_v - \lambda_1 g_v - \lambda_2 h_v)\,dv + (F_w - \lambda_1 g_w - \lambda_2 h_w)\,dw + \cdots \\ + (F_z - \lambda_1 g_z - \lambda_2 h_z)\,dz = 0. \end{aligned} \tag{10.12}$$

Suppose that we try to choose λ_1 and λ_2 so that $F_y - \lambda_1 g_y - \lambda_2 h_y = F_z - \lambda_1 g_z -$

$\lambda_2 h_z = 0$ at the desired max; that is,

$$g_y\lambda_1 + h_y\lambda_2 = F_y, \qquad g_z\lambda_1 + h_z\lambda_2 = F_z. \tag{10.13}$$

From Chapter 7 we note that this situation is possible [i.e., equations (10.13) have a solution] if

$$\begin{vmatrix} g_y & h_y \\ g_z & h_z \end{vmatrix} \neq 0 \tag{10.14}$$

at the desired max—in other words, if the Jacobian $\partial(g, h)/\partial(y, z) \neq 0$ there. Assuming that such is the case, (10.12) reduces to

$$dF^* = (F_v - \lambda_1 g_v - \lambda_2 h_v)\,dv + (F_w - \lambda_1 g_w - \lambda_2 h_w)\,dw + (F_x - \lambda_1 g_x - \lambda_2 h_x)dx = 0. \tag{10.15}$$

Having eliminated dy and dz, the remaining increments—dv, dw, and dx—are independent,[3] and therefore the three coefficients in parentheses in (10.15) must equal zero. This result, together with (10.13), says that

$$\begin{aligned} F_v{}^* &= F_v - \lambda_1 g_v - \lambda_2 h_v = 0 \\ F_w{}^* &= F_w - \lambda_1 g_w - \lambda_2 h_w = 0 \\ F_x{}^* &= F_x - \lambda_1 g_x - \lambda_2 h_x = 0 \\ F_y{}^* &= F_y - \lambda_1 g_y - \lambda_2 h_y = 0 \\ F_z{}^* &= F_z - \lambda_1 g_z - \lambda_2 h_z = 0 \end{aligned}$$

at the max. They are five equations in the seven unknowns $v, w, x, y, z, \lambda_1, \lambda_2$, and so we adjoin the two additional equations (10.10) to complete the system.

If (10.14) were *not* satisfied, it would still be all right if, for instance,

$$\begin{vmatrix} g_x & h_x \\ g_z & h_z \end{vmatrix} \neq 0$$

at the desired max; then we could eliminate the dx and dz terms instead of dy and dz.

Generalizing, suppose that

$$F(x_1, \ldots, x_m) = \max \qquad \text{(or min)}, \tag{10.16}$$

subject to the constraints

$$\begin{aligned} g_1(x_1, \ldots, x_m) &= c_1 \\ &\vdots \\ g_n(x_1, \ldots, x_m) &= c_n, \end{aligned} \tag{10.17}$$

where $F, g_1, \ldots, g_n$ have continuous first partials and $n < m$. To solve by the method of Lagrange multipliers, we form

$$F^* \equiv F - \lambda_1 g_1 - \cdots - \lambda_n g_n \tag{10.18}$$

(where the λ's are the so-called *Lagrange multipliers*) and seek $x_1, \ldots, x_m, \lambda_1, \ldots, \lambda_n$

[3]The condition (10.14) implies that equations (10.10) can be solved for y and z in terms of v, w, x. Thus $F = F[v, w, x, y(v, w, x), z(v, w, x)]$, and in place of the five dependent variables we have the three independent variables v, w, x.

from the $m + n$ equations

$$\frac{\partial F^*}{\partial x_j} = 0 \qquad (j = 1, 2, \ldots, m) \tag{10.19}$$

$$g_j = c_j \qquad (j = 1, \ldots, n.) \tag{10.20}$$

Note carefully that (10.19) amounts to *minimizing F^* subject to no constraints.*

Example 10.3. Let us re-solve Example 10.2 by the method of Lagrange multipliers. With $F^* = F - \lambda g = x^2 + y^2 - \lambda(5x^2 - 6xy + 5y^2)$, (10.19) and (10.20) are

$$2x - 10\lambda x + 6\lambda y = 0$$

$$2y + 6\lambda x - 10\lambda y = 0$$

$$5x^2 - 6xy + 5y^2 = 8.$$

Eliminating λ from the first two gives $y = \pm x$, which, together with the last equation, yields the same result as before. ▮

Finally, a few words about an important although slightly different kind of problem. Suppose that we seek $x_1, \ldots, x_m$ so that the *linear* function

$$F(x_1, \ldots, x_m) = b_1 x_1 + \cdots + b_m x_m = \max \qquad \text{(or min)}, \tag{10.21}$$

subject to the *linear* constraints

$$\begin{aligned} g_1(x_1, \ldots, x_m) &= a_{11}x_1 + \cdots + a_{1m}x_m = c_1 \\ &\vdots \\ g_n(x_1, \ldots, x_m) &= a_{n1}x_1 + \cdots + a_{nm}x_m = c_n, \end{aligned} \tag{10.22}$$

where the x_j's *must all be nonnegative*. Or the constraints may be *inequalities*, such as

$$a_{11}x_1 + \cdots + a_{1m}x_m \leq c_1. \tag{10.23}$$

But this situation is equivalent because we can always introduce an additional variable x_{m+1} and rewrite (10.23) as

$$a_{11}x_1 + \cdots + a_{1m}x_m + x_{m+1} = c_1$$

with the proviso that x_{m+1} is also nonnegative; x_{m+1} is called a *slack variable.*

Such problems are of interest in the broad area known as operations research. It turns out that their solutions cannot be found by setting various derivatives equal to zero; special techniques like the *simplex method* are needed. These various techniques come under the heading of **linear programming**, which is outside the scope of this book.[4]

10.3 FUNCTIONALS AND THE CALCULUS OF VARIATIONS

Finally, we come to the extremization (i.e., maximization or minimization of *functionals* or at least of certain classes of functionals. Let us start with the following problem.

---

[4]See, for example, R. M. Stark and R. L. Nichols, *Mathematical Foundations for Design*, McGraw-Hill, New York, 1972.

Find the function $y(x)$, defined over $x_1 \leq x \leq x_2$ and passing through the given endpoints

$$y(x_1) = y_1, \qquad y(x_2) = y_2, \tag{10.24a}$$

such that

$$I = \int_{x_1}^{x_2} f(x, y, y')\, dx = \text{extremum}. \tag{10.24b}$$

We ask that f have as many continuous partial derivatives with respect to its arguments x, y, y' as needed and seek $y(x)$ from within a certain class of admissible functions, say twice continuously differentiable. That is, y'' is to exist and be continuous over $x_1 \leq x \leq x_2$.[5]

To illustrate, suppose that we wish to find the arc of minimum length that passes through given endpoints x_1, y_1 and x_2, y_2. With $ds = [(dx)^2 + (dy)^2]^{1/2}$, the problem is as follows.

$$I = \int ds = \int_{x_1}^{x_2} \sqrt{1 + y'^2}\, dx = \min \tag{10.25a}$$

with

$$y(x_1) = y_1 \quad \text{and} \quad y(x_2) = y_2. \tag{10.25b}$$

As a general rule, note how new situations are often handled by recasting them in terms of old and familiar ones. For instance, in Chapter 7 we extended the idea of Taylor series to functions $f(x, y)$ of *two* variables by expressing x and y in terms of a parameter t so that $f(x, y) = f[x(t), y(t)] \equiv F(t)$ was reduced to a function of *one* variable; alternatively, we also considered expanding in *one* variable at a time.

In the same spirit we reduce problem (10.24) to the extremization of a *function* as follows. Denoting the *solution* (as yet unknown) as $y(x)$, let us consider a one-parameter family of "comparison functions" $Y(x)$, defined by

$$Y(x) = y(x) + \epsilon\eta(x), \tag{10.26}$$

where $\eta(x)$ is an arbitrary, twice continuously differentiable[6] function satisfying the end conditions

$$\eta(x_1) = \eta(x_2) = 0. \tag{10.27}$$

For example, if $y(x)$ and $\eta(x)$ are as shown in Fig. 10.4, then assigning various (positive and negative) values to the parameter ϵ produces the $Y(x)$ family indicated; of course, an infinite number of members actually exist, only a few of which have been sketched here.

Note that no matter how $\eta(x)$ is defined, the true solution $y(x)$ will certainly be a member of the Y family—namely, for $\epsilon = 0$. Note also that $Y(x_1) = y(x_1) + \epsilon\eta(x_1) = y_1 + \epsilon \cdot 0 = y_1$ and $Y(x_2) =$ etc. $= y_2$; thus all members of the family satisfy the end conditions (10.24a) and are qualified candidates.

Consider next the quantity

$$I(\epsilon) \equiv \int_{x_1}^{x_2} f(x, Y, Y')\, dx,$$

[5]Extension to a broader class of admissible y's, including y's with "corners," is discussed in J. C. Clegg, *Calculus of Variations*, Oliver & Boyd, Edinburgh, 1968.

[6]$f(x)$ is said to be **continuously differentiable** (smooth) wherever $f'(x)$ exists and is continuous.

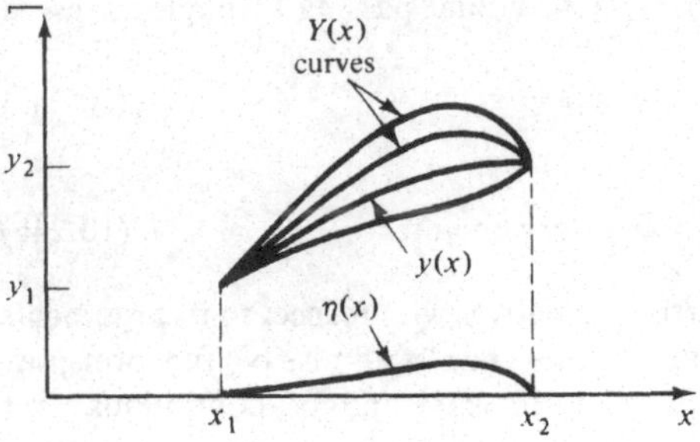

Figure 10.4. The comparison functions.

which is a *function* of the parameter ϵ (through Y and Y'). To extremize, we set $I'(\epsilon) = 0$. Unlike most extremization problems, however, we know beforehand where the minimum occurs—namely, at $\epsilon = 0$. Thus

$$I'(0) = 0 = \int_{x_1}^{x_2} \left(\frac{\partial f}{\partial Y} \frac{dY}{d\epsilon} + \frac{\partial f}{\partial Y'} \frac{dY'}{d\epsilon} \right)\bigg|_{\epsilon=0} dx.$$

But $Y = y + \epsilon\eta$ and $Y' = y' + \epsilon\eta'$, and so $dY/d\epsilon = \eta$ and $dY'/d\epsilon = \eta'$. Furthermore, all Y's and Y''s revert to y's and y''s when we set $\epsilon = 0$; consequently,

$$0 = \int_{x_1}^{x_2} \left(\frac{\partial f}{\partial y} \eta + \frac{\partial f}{\partial y'} \eta' \right) dx.$$

Integrating the second term by parts,

$$0 = \frac{\partial f}{\partial y'} \eta \bigg|_{x_1}^{x_2} + \int_{x_1}^{x_2} \left[\frac{\partial f}{\partial y} - \frac{d}{dx}\left(\frac{\partial f}{\partial y'}\right) \right] \eta(x)\, dx$$

$$= \int_{x_1}^{x_2} \left[\frac{\partial f}{\partial y} - \frac{d}{dx}\left(\frac{\partial f}{\partial y'}\right) \right] \eta(x)\, dx, \tag{10.28}$$

since $\eta(x_1) = \eta(x_2) = 0$.

Equation (10.28) must hold for *all* η's [which satisfy (10.27) and are twice continuously differentiable], and we suspect that this condition can be true only if the square bracket is zero over the interval. In fact, we have the following basic lemma.

LEMMA 10.1. If $F(x)$ is continuous and

$$\int_{x_1}^{x_2} F(x)\eta(x)\, dx = 0 \tag{10.29}$$

for **all** *continuously differentiable functions $\eta(x)$ for which $\eta(x_1) = \eta(x_2) = 0$, then it must be true that*

$$F(x) = 0$$

for all $x_1 \le x \le x_2$.

Outline of Proof. Suppose that $F(x)$ is *not* zero at some point ξ in the interval. Since F is continuous, there must exist an $\epsilon > 0$ such that F is of the same sign in the subinterval $(\xi - \epsilon, \xi + \epsilon)$ as at the point ξ. Using the arbitrariness of η, let η be zero outside this subinterval and positive inside (Fig. 10.5). Then the left-hand side of (10.29) will be nonzero, and we have proof by contradiction.

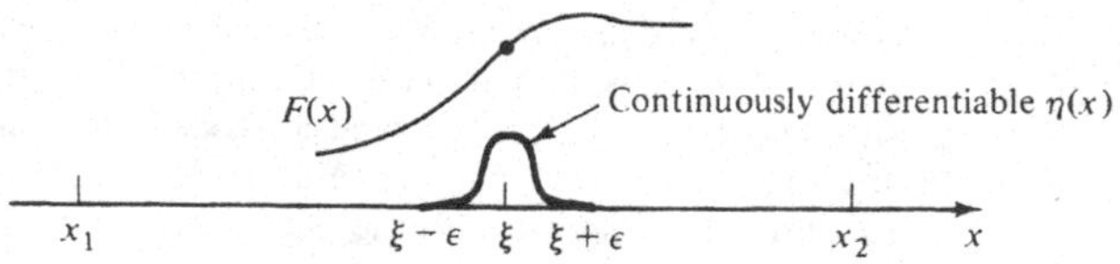

Figure 10.5. Proof of the basic lemma.

Returning to (10.28), we conclude that the desired $y(x)$ must satisfy the differential equation

$$\frac{\partial f}{\partial y} - \frac{d}{dx}\left(\frac{\partial f}{\partial y'}\right) = 0, \tag{10.30}$$

plus, of course, the prescribed end conditions (10.25b); (10.30) is the so-called **Euler equation** associated with the functional (10.24b). Note that the second term is *not* $\partial^2 f(x, y, y')/\partial x\,\partial y'$ but rather

$$\frac{d}{dx} f_{y'}[x, y(x), y'(x)] = f_{xy'} + f_{yy'}y' + f_{y'y'}y''.$$

In general, the Euler equation (10.30) will be a nonlinear, second-order, differential equation in the unknown $y(x)$.

Example 10.4. *Shortest Arc.* We are now able to solve the shortest arc problem, stated by (10.25). In this case, $f(x, y, y') = [1 + y'^2]^{1/2}$ happens to depend only on y'. The Euler equation (10.30) becomes

$$0 - \frac{d}{dx}\left(\frac{y'}{\sqrt{1 + y'^2}}\right) = 0.$$

Integrating once gives

$$\frac{y'}{\sqrt{1 + y'^2}} \equiv C.$$

Solving for y',

$$y' = \sqrt{\frac{C^2}{1 - C^2}} \equiv A,$$

say, and integrating once more, we obtain the straight line

$$y = Ax + B, \tag{10.31}$$

where A and B are determined by the end conditions (10.25b).

COMMENT 1. Our approach has been a little formal in several aspects. For instance, all we have actually shown is that the straight line provides a relative extremum. It's true that we can "sense" that it happens to be a *minimum*, and, in fact, an *absolute* minimum, but so far our analysis provides no grounds for such claims. Furthermore, note that in expressing $ds = [(dx)^2 + (dy)^2]^{1/2}$ as $(1 + y'^2)^{1/2}\,dx$ in (10.25a), we drop the view of x and y as independent variables and consider y as a function of x. Yet the desired curve is not necessarily expressible in the form $y = y(x)$; that is, perhaps it is not single valued. Several of these points are examined in the exercises, but generally such details lie outside our present scope.[7]

COMMENT 2. The present example is the simplest case of the more general problem of geodesics—that is, to find the arc of minimum length connecting two points on a given surface. Such an arc is called a **geodesic** for that particular surface. As an illustration, the geodesics for a sphere turn out to be great-circle arcs—that is, the intersection of the sphere with the plane containing the two given points and the center of the sphere.

If, for example, the surface is defined by $z = z(x, y)$ and we seek the geodesic between $x_1, y_1, z(x_1, y_1)$ and $x_2, y_2, z(x_2, y_2)$, we have the variational problem

$$I = \int \sqrt{(dx)^2 + (dy)^2 + (dz)^2}$$

$$= \int_{x_1}^{x_2} \sqrt{1 + y'^2 + (z_x + z_y y')^2}\, dx = \min, \tag{10.32a}$$

subject to the end conditions

$$y(x_1) = y_1, \qquad y(x_2) = y_2, \tag{10.32b}$$

which is of the form (10.24).

Of course, for a *sphere*, it would be more natural to use a parametric representation of the surface, $x = x(u, v)$, $y = y(u, v)$, $z = z(u, v)$. (See Exercises 10.18 and 10.19) ▮

Example 10.5. *Undetermined Endpoints.* Find the shape $y(x)$ of a hanging flexible cable that is looped around a frictionless, vertical wire at $x = a$, as sketched in Fig. 10.6.

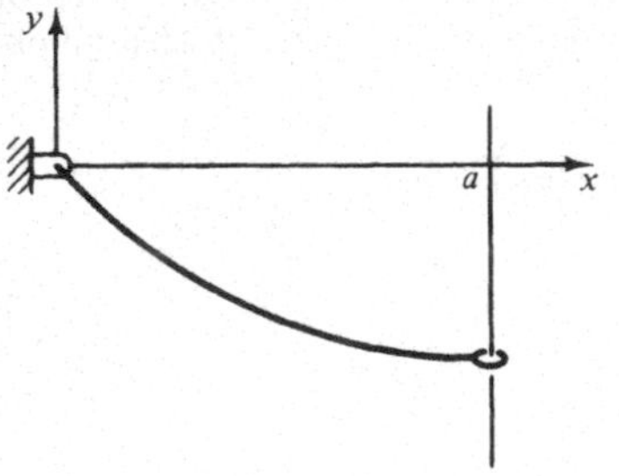

Figure 10.6. Hanging flexible cable.

The governing physical principle is that the cable will assume the shape that minimizes its potential energy V. In saying that the cable is "flexible," we mean that there is a negligible amount of elastic "strain energy" within the cable; for instance, it might be a chain or string. So there is only gravitational potential to consider. If ρ is the lineal mass density (mass per unit length), g is the acceleration of gravity, and $y = 0$ is the zero potential reference level, we have the variational problem

$$V = \int \rho g y\, ds = \int_0^a \rho g y \sqrt{1 + y'^2}\, dx = \min, \tag{10.33a}$$

where

$$y(0) = 0 \tag{10.33b}$$

but where $y(a)$ is not prescribed. Thus (10.33) is not quite of the form (10.24), and our derivation of the Euler equation needs to be reexamined.

[7]A rigorous and detailed treatment can be found in the monograph by G. A. Bliss, *Calculus of Variations*, Open Court Publishing Company, LaSalle, Ill., 1925.

Consider, then,

$$I = \int_{x_1}^{x_2} f(x, y, y')\, dx = \text{extremum}, \qquad (10.34)$$

$$y(x_1) = y_1, \qquad y(x_2) \text{ unspecified.}$$

As before, we form the family $Y = y + \epsilon\eta$, where $\eta(x_1) = 0$, but this time $\eta(x_2)$ need not equal zero; thus our comparison functions $Y(x)$ can "try out" a variety of right-end values, as sketched in Fig. 10.7. And, as earlier, we obtain

$$\begin{aligned} I'(0) = 0 &= \frac{\partial f}{\partial y'}\eta\bigg|_{x_1}^{x_2} + \int_{x_1}^{x_2} \left[\frac{\partial f}{\partial y} - \frac{d}{dx}\left(\frac{\partial f}{\partial y'}\right)\right]\eta(x)\, dx, \\ &= \frac{\partial f}{\partial y'}\bigg|_{x_2} \eta(x_2) + \int_{x_1}^{x_2} \left[\frac{\partial f}{\partial y} - \frac{d}{dx}\left(\frac{\partial f}{\partial y'}\right)\right]\eta(x)\, dx, \end{aligned} \qquad (10.35)$$

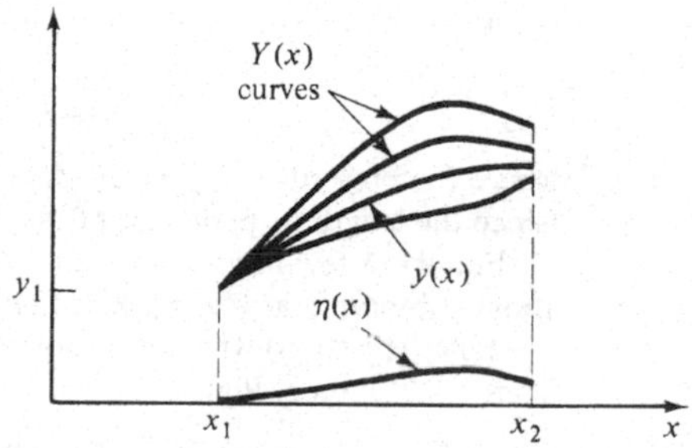

Figure 10.7. Undetermined right-end conditions.

since $\eta(x_1) = 0$. Because (10.35) is to hold for *all* (twice continuously differentiable) η's such that $\eta(x_1) = 0$, it must obviously hold for the subclass of η's that vanish at x_2 as well. Then the remaining boundary term in (10.35) drops out, and Lemma 10.1 leads to the same Euler equation as before,

$$\frac{\partial f}{\partial y} - \frac{d}{dx}\left(\frac{\partial f}{\partial y'}\right) = 0. \qquad (10.36)$$

So (10.35) reduces to

$$0 = \frac{\partial f}{\partial y'}\bigg|_{x_2} \eta(x_2);$$

and since $\eta(x_2)$ *need* not equal zero, it also follows that

$$\frac{\partial f}{\partial y'}\bigg|_{x_2} = 0. \qquad (10.37)$$

We call (10.37) a **natural boundary condition.**

For our hanging cable, the Euler equation and boundary conditions are therefore

$$yy'' - y'^2 - 1 = 0 \qquad (10.38)$$

(after algebraic simplification) and

$$y(0) = 0, \qquad \frac{\partial f}{\partial y'} = \frac{yy'}{\sqrt{1 + y'^2}} = 0 \quad \text{at } x = a. \qquad (10.39)$$

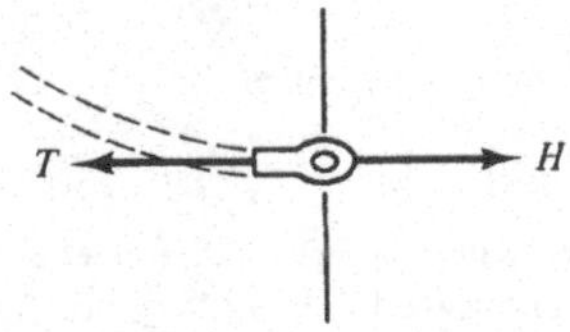

Figure 10.8. Equilibrium of the loop at $x = a$.

Apparently, from the right-hand side of (10.39), either $y(a) = 0$ or $y'(a) = 0$. Of these two, we accept the latter on the basis of a simple equilibrium argument. That is, consider an infinitesimal bit of the cable at the loop. Since the vertical wire is frictionless, the force H that it exerts on the loop must be horizontal (Fig. 10.8). This force is to be balanced by the tension T, which must therefore also be horizontal, so that $y'(a) = 0$ (Exercise 10.7).

Unfortunately, the (nonlinear) differential equation (10.38), plus the boundary conditions $y(0) = y'(a) = 0$, has no (real) solutions; for instance, at $x = 0$ (10.38) reduces to $y'^2 = -1$! Why? The difficulty is that the problem is not properly posed until we specify how *long* the cable is. That is, the problem should have been to minimize the potential subject to the end condition $y(0) = 0$ *and the constraint*

$$\int_0^a \sqrt{1 + y'^2}\, dx = l,$$

where l is prescribed ($\geq a$). Thus we have what is called an **isoperimetric problem** (since the length or perimeter of the curve is prescribed).

Finally, a few more words about natural boundary conditions. As discussed above, observe that the bracketed terms in (10.35) necessarily vanish (thus yielding the Euler equation) whether or not $y(x)$ is prescribed at x_1 and/or x_2. It follows from (10.35), then, that

$$0 = \frac{\partial f}{\partial y'}\eta \Big|_{x_1}^{x_2} = \frac{\partial f}{\partial y'}\Big|_{x_2} \eta(x_2) - \frac{\partial f}{\partial y'}\Big|_{x_1} \eta(x_1). \tag{10.40}$$

If $y(x_1)$ and/or $y(x_2)$ are unspecified, then $\eta(x_1)$ and/or $\eta(x_2)$ are arbitrary, and it follows from (10.40) that $\partial f/\partial y'|_{x_1} = 0$ and/or $\partial f/\partial y'|_{x_2} = 0$. These are referred to as *natural boundary conditions.* ∎

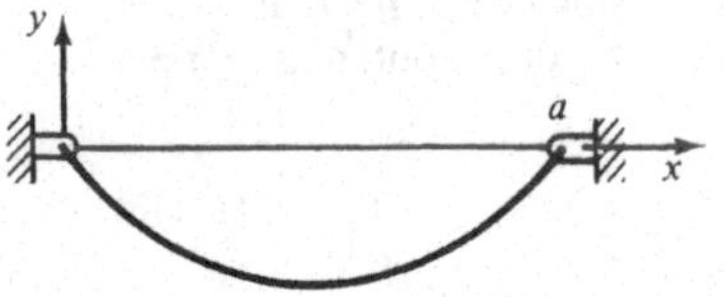

Figure 10.9.

Let us take the properly posed problem as our next example but with the end conditions $y(0) = y(a) = 0$ (Fig. 10.9), so that we need not worry about an undetermined end condition at the same time.

Example 10.6. *An Isoperimetric Problem.*

$$I = \int_0^a y\sqrt{1 + y'^2}\, dx = \min, \qquad y(0) = y(a) = 0, \tag{10.41a}$$

subject to the constraint

$$J = \int_0^a \sqrt{1 + y'^2}\, dx = l. \tag{10.41b}$$

Although the constraint may seem to complicate matters considerably, just recall the analogous situation for the extremization of a *function* F subject to a constraint $g =$ constant. We saw that the problem could be restated equivalently as $F^* = F - \lambda g =$ extremum subject to *no* constraint, where λ is the so-called Lagrange multiplier.

Considering the way that we are able to convert functionals to functions through the introduction of families of comparison functions, it is not surprising that the Lagrange method can be applied to problems like (10.41) as well.[8] So we have

$$I^* = I - \lambda J = \int_0^a (y - \lambda)\sqrt{1 + y'^2}\, dx = \min \tag{10.42a}$$

with

$$y(0) = y(a) = 0. \tag{10.42b}$$

Since $f^* = (y - \lambda)\sqrt{1 + y'^2}$ does not contain any explicit x dependence, the Euler equation has the first integral (Exercise 10.9)

$$\frac{y - \lambda}{\sqrt{1 + y'^2}} = A; \tag{10.43}$$

and integrating once more (Exercise 10.11),

$$y = \lambda - A \cosh\left(\frac{x - B}{A}\right). \tag{10.44}$$

Finally, $y(0) = y(a) = 0$ and (10.41b) yield three nonlinear algebraic equations in the unknowns A, B, and λ. Their unique solution is always possible, although perhaps numerically difficult. ▮

10.4. TWO OR MORE DEPENDENT VARIABLES; HAMILTON'S PRINCIPLE

Generalizing a little, consider the case of two or more dependent variables,

$$I = \int_{t_1}^{t_2} f(x, y, \ldots, z, \dot{x}, \dot{y}, \ldots, \dot{z}, t)\, dt = \text{extremum}, \tag{10.45}$$

where the dots denote d/dt and t is often, although not necessarily, the time.

In the discussion to follow, the variational problem will be supplied by what is known in classical mechanics as **Hamilton's principle**.

Suppose that our system is *conservative*; that is, (Chapter 9) each force that acts is derivable from a potential. Let V and T denote the *total* potential and kinetic energies, respectively. Introducing the so-called **Lagrangian,** $L \equiv T - V$, Hamilton's principle states that *the actual motion connecting two known states of the system, say at times t_1 and t_2, is the one that minimizes the integral*

$$I = \int_{t_1}^{t_2} (T - V)dt = \int_{t_1}^{t_2} L\, dt. \tag{10.46}$$

Example 10.7. Derive the equations of motion for the two-mass system shown in Fig. 10.10. Only lateral motion is considered—the allowed displacements of the two masses from their equilibrium positions are $x(t)$ and $y(t)$—and the three springs of stiffness k are assumed to be neither stretched nor compressed when $x = y = 0$

[8] See, for example, R. Weinstock, *Calculus of Variations*, McGraw-Hill, New York, 1952, pp. 48–50. A good general reference for the present chapter, it contains many important applications to engineering and physics.

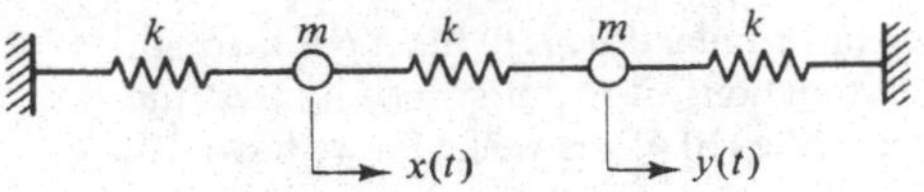

Figure 10.10. Vibration of a two-mass system.

(although the equations of motion would be the same even if there were some initial tension or compression).

Clearly, the kinetic and potential energies are

$$T = \tfrac{1}{2}m\dot{x}^2 + \tfrac{1}{2}m\dot{y}^2 \quad \text{and}$$

$$V = \tfrac{1}{2}kx^2 + \tfrac{1}{2}k(y - x)^2 + \tfrac{1}{2}ky^2,$$

and so, according to Hamilton's principle,

$$I = \int_{t_1}^{t_2} [\tfrac{1}{2}m(\dot{x}^2 + \dot{y}^2) - k(x^2 - xy + y^2)]\, dt = \text{min}. \tag{10.47}$$

In order to obtain the Euler equation(s) for the variational problem

$$I = \int_{t_1}^{t_2} f(x, y, \dot{x}, \dot{y}, t)\, dt = \text{extremum}, \tag{10.48}$$

of which our problem (10.47) is an example, we follow essentially the same lines as for the case of only one dependent variable discussed above. First, we form the comparison functions

$$X(t) = x(t) + \epsilon\xi(t), \qquad Y(t) = y(t) + \epsilon\eta(t),$$

where $\xi(t)$ and $\eta(t)$ are arbitrary, twice continuously differentiable functions satisfying the end conditions

$$\xi(t_1) = \xi(t_2) = \eta(t_1) = \eta(t_2) = 0. \tag{10.49}$$

Considering

$$I(\epsilon) = \int_{t_1}^{t_2} f(X, Y, \dot{X}, \dot{Y}, t)\, dt,$$

we set

$$\begin{aligned} I'(0) = 0 &= \int_{t_1}^{t_2} \left(\frac{\partial f}{\partial x}\xi + \frac{\partial f}{\partial y}\eta + \frac{\partial f}{\partial \dot{x}}\dot{\xi} + \frac{\partial f}{\partial \dot{y}}\dot{\eta}\right) dt \\ &= \left(\frac{\partial f}{\partial \dot{x}}\xi + \frac{\partial f}{\partial \dot{y}}\eta\right)\Bigg|_{t_1}^{t_2} + \int_{t_1}^{t_2} \left[\frac{\partial f}{\partial x} - \frac{d}{dt}\left(\frac{\partial f}{\partial \dot{x}}\right)\right]\xi(t)\, dt \\ &\quad + \int_{t_1}^{t_2} \left[\frac{\partial f}{\partial y} - \frac{d}{dt}\left(\frac{\partial f}{\partial \dot{y}}\right)\right]\eta(t)\, dt. \end{aligned}$$

The boundary terms drop out because of (10.49), and the arbitrariness of ξ and η implies (with the help of Lemma 10.1) that each of the square brackets must be zero over the interval:

$$\frac{\partial f}{\partial x} - \frac{d}{dt}\left(\frac{\partial f}{\partial \dot{x}}\right) = 0 \tag{10.50a}$$

$$\frac{\partial f}{\partial y} - \frac{d}{dt}\left(\frac{\partial f}{\partial \dot{y}}\right) = 0. \tag{10.50b}$$

Thus we have the same Euler equation as before—that is, (10.30)—but governing *each* of the dependent variables.

In particular, for our case (10.47), the *equations of motion* (10.50) become

$$\left.\begin{aligned} m\ddot{x} + 2kx - ky &= 0 \\ m\ddot{y} + 2ky - kx &= 0. \end{aligned}\right\} \tag{10.51}$$

Let us ignore the *solution* of equations (10.51) for the moment, since it is not central to the present discussion.

COMMENT. Naturally, (10.51) could have been derived much more directly from Newton's second law without resorting to Hamilton's principle and the calculus of variations. Nevertheless, we emphasize that variational methods are of the greatest importance to us for at least three reasons.

First, for complicated problems in dynamics and continuum mechanics, Hamilton's principle may be somewhat simpler to apply because we need only be able to write expressions for kinetic and potential energies; for instance, we never need to know what a "Coriolis force" is! This situation is illustrated by a problem from the literature in Exercise 10.22.

Secondly, for many problems like (10.25) there *is* no (apparent) alternative *non*-variational statement, analogous to Newton's law in the present example.

Finally, the Euler equation(s) for a given problem may prove too difficult to solve, whereas the variational principle itself provides an excellent starting point for various *approximate* techniques, such as the Ritz method, which is discussed briefly in Section 10.6. ∎

10.5. *TWO OR MORE INDEPENDENT VARIABLES; VIBRATING STRINGS AND MEMBRANES*

Starting with the case of *two* independent variables x, y, we seek $w(x, y)$ in some region D with boundary C such that

$$I = \iint_D f(x, y, w, w_x, w_y)\,dx\,dy = \text{extremum}, \tag{10.52a}$$

subject to the boundary values

$$w \text{ given on } C \tag{10.52b}$$

(Fig. 10.11).

We introduce the comparison functions

$$W(x, y) = w(x, y) + \epsilon\eta(x, y),$$

where η has continuous second-order partials and satisfies

$$\eta = 0 \text{ on } C \tag{10.53}$$

but is otherwise arbitrary in D. Then

$$I(\epsilon) = \iint_D f(x, y, W, W_x, W_y)\,dx\,dy$$

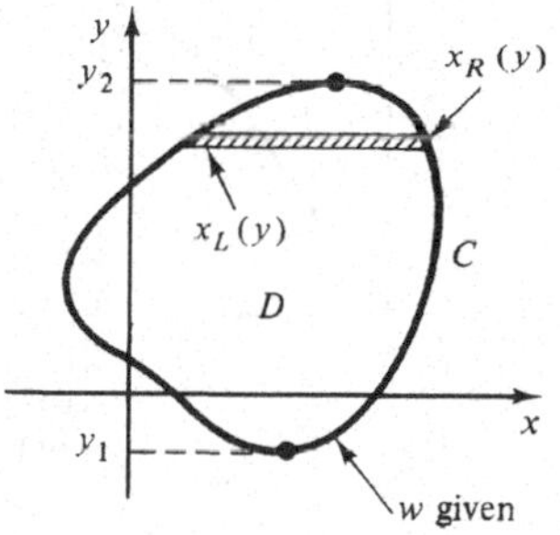

Figure 10.11. The region under consideration.

is an extremum for $\epsilon = 0$; that is,

$$I'(0) = 0 = \iint_D \left(\frac{\partial f}{\partial w}\eta + \frac{\partial f}{\partial w_x}\eta_x + \frac{\partial f}{\partial w_y}\eta_y\right) dx\,dy. \tag{10.54}$$

Integrating the second term by parts,

$$\iint_D \frac{\partial f}{\partial w_x}\eta_x \, dx \, dy = \int_{y_1}^{y_2}\left\{\int_{x_L(y)}^{x_R(y)} \frac{\partial f}{\partial w_x}\eta_x \, dx\right\} dy$$

$$= \int_{y_1}^{y_2}\left\{\frac{\partial f}{\partial w_x}\eta \Big|_{x_L(y)}^{x_R(y)} - \int_{x_L(y)}^{x_R(y)} \frac{\partial}{\partial x}\left(\frac{\partial f}{\partial w_x}\right)\eta \, dx\right\} dy$$

$$= \int_C \frac{\partial f}{\partial w_x}\eta \frac{dy}{ds}\, ds - \iint_D \frac{\partial}{\partial x}\left(\frac{\partial f}{\partial w_x}\right)\eta \, dx\, dy, \tag{10.55}$$

where s is the arc length along C measured counterclockwise.[9] Integrating the third term in the integrand of (10.54) in the same manner, (10.54) becomes

$$0 = \int_C \left(\frac{\partial f}{\partial w_x}\frac{dy}{ds} - \frac{\partial f}{\partial w_y}\frac{dx}{ds}\right)\eta \, ds + \iint_D \left[\frac{\partial f}{\partial w} - \frac{\partial}{\partial x}\left(\frac{\partial f}{\partial w_x}\right) - \frac{\partial}{\partial y}\left(\frac{\partial f}{\partial w_y}\right)\right]\eta \, dx \, dy$$

and as a result of (10.53) and (a two-dimensional version of) Lemma 10.1, we arrive at the Euler equation

$$\frac{\partial f}{\partial w} - \frac{\partial}{\partial x}\left(\frac{\partial f}{\partial w_x}\right) - \frac{\partial}{\partial y}\left(\frac{\partial f}{\partial w_y}\right) = 0. \tag{10.56}$$

Example 10.8. *Vibrating String.* Starting with a perfectly flexible string of length l_0, a plot of applied tension versus length is apt to be somewhat as sketched in Fig. 10.12(a). Suppose that we tie the string as shown in Fig. 10.12(b), under tension τ, and set it in motion in the x, y plane by means of some initial disturbance. We would like the equation of motion governing the displacement $y(x, t)$.

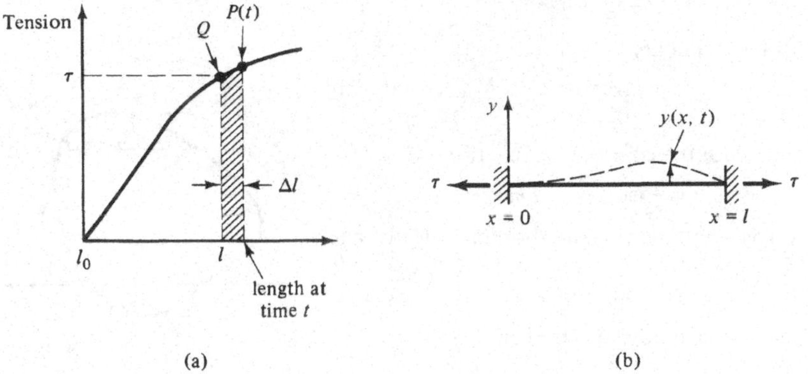

Figure 10.12. Vibrating string.

[9]We've tacitly assumed that D is *convex* in the x direction—that is, that each "sliver" running from $x_L(y)$ to $x_R(y)$ lies entirely within D. The minor modifications needed for the nonconvex case will not be pursued, since they are probably obvious and, in any case, do not affect the final result (10.55). Incidentally, rather than use integration by parts, we could have obtained (10.55) by using Green's theorem (9.60) with $M = 0$ and $N = \eta\, \partial f/\partial w_x$.

Let us suppose that the amplitude of vibration is so small that the slope $y_x(x, t)$ is small compared to unity for $0 \leq x \leq l$ and all t, that the tension is uniform in x at each instant, and that each particle of the string undergoes negligible x motion.[10]

Because of the displacement $y(x, t)$, the length of the string (and the tension in it) will vary slightly with t, as indicated by the movement of the representative point $P(t)$ in Fig. 10.12(a). The associated work is equal to the area of the shaded sliver and is stored as potential energy. Defining the equilibrium state $y = 0$ as our zero potential (reference state), we therefore have

$$\begin{aligned} V \approx \tau \Delta l &= \tau \left\{ \int_0^l \sqrt{1 + y_x^2}\, dx - l \right\} \\ &\approx \tau \left\{ \int_0^l \left(1 + \frac{1}{2} y_x^2\right) dx - l \right\} = \frac{\tau}{2} \int_0^l y_x^2\, dx. \end{aligned} \tag{10.57}$$

Furthermore, the kinetic energy is clearly

$$T = \frac{1}{2} \int_0^l \rho(x) y_t^2\, dx, \tag{10.58}$$

where $\rho(x)$ is the lineal mass density (mass per unit length).

According to Hamilton's principle (10.46), then,

$$\int_{t_1}^{t_2} (T - V)\, dt = \frac{1}{2} \int_{t_1}^{t_2} \int_0^l (\rho y_t^2 - \tau y_x^2)\, dx\, dt = \text{min}, \tag{10.59}$$

and so the Euler equation (10.56), with w corresponding to y and y to t, becomes the second-order partial differential equation of motion

$$\tau y_{xx} - \rho(x) y_{tt} = 0. \tag{10.60}$$

Methods of *solution* of partial differential equations are discussed in Part V.

COMMENT. Note that we have *not* assumed a linear tension-displacement curve [Fig. 10.12(a)]. In fact, its precise shape is immaterial because the process is assumed to take place in the neighborhood of the point Q, so that tension $\approx \tau$ is all that survives in the final equations. ∎

Finally, consider the following problem in *three* independent variables x, y, z. Find $w(x, y, z)$ in some region R bounded by a surface S such that

$$I = \iiint_R f(x, y, z, w, w_x, w_y, w_z)\, dx\, dy\, dz = \text{extremum}, \tag{10.61a}$$

subject to the boundary values

$$w \text{ given on } S. \tag{10.61b}$$

Referring to the one- and two-dimensional results, (10.30) and (10.56), it is not hard to imagine that the Euler equation relevant to (10.61) is (Exercise 10.23)

$$\frac{\partial f}{\partial w} - \frac{\partial}{\partial x}\left(\frac{\partial f}{\partial w_x}\right) - \frac{\partial}{\partial y}\left(\frac{\partial f}{\partial w_y}\right) - \frac{\partial}{\partial z}\left(\frac{\partial f}{\partial w_z}\right) = 0. \tag{10.62}$$

---

[10] For purposes of this example, the fact that these assumptions are "quite reasonable" will suffice. Rigorous development by means of perturbation methods (Parts IV and V) is possible but not really needed here.

Example 10.9. *Vibrating Membrane.* We consider a flexible membrane,[11] such as a drumhead, which is stretched over some region D of the x, y plane bounded by a curve C. Denote the elevation of the membrane above the x, y plane at the point x, y and the time t by $w(x, y, t)$ and suppose that w is prescribed along C for all t. The membrane is set in motion by some initial disturbance, and we seek the equation of motion governing the displacement $w(x, y, t)$.

Let us suppose that the boundary values of w on C and the initial disturbance are so small that the slope of the membrane with respect to the x, y plane in any direction is small compared to unity throughout D and for all t, that the tension τ per unit length is essentially constant throughout D and for all t, and that each particle undergoes negligible motion parallel to the x, y plane.

Let us elaborate on our assumption that the tension per unit length is essentially uniform throughout D. Consider a rectangular membrane held along its left and right edges, stretched as shown (Fig. 10.13) with a tension τ per unit length (thus with

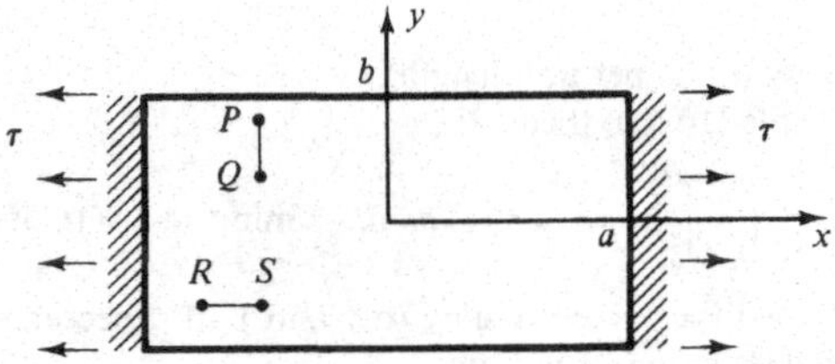

Figure 10.13. *Non*uniform membrane tension.

a total force $2b\tau$), and then also clamped along the top and bottom edges $y = \pm b$. In this case, the tension per unit length will *not* be uniform; for example, it will be τ along each vertical line (e.g., PQ) and 0 along each horizontal line (e.g., RS). Thus the present analysis would not apply to this case unless we undertook to render the tension uniform by also stretching suitably in the y direction.

To compute the potential V, first note that the work done by a constant and uniform tension per unit length τ in stretching the rectangular membrane element shown in Fig. 10.14 by the amounts da and db can be expressed formally as

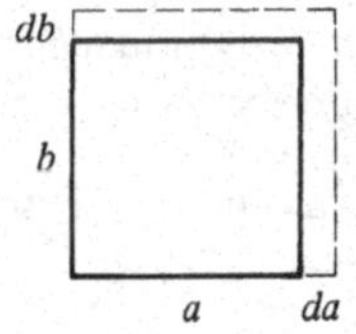

Figure 10.14. Work done in stretching.

$$\begin{aligned} d(\text{work}) &= (\tau b)\, da + (\tau a)\, db \\ &= \tau d(ab) = \tau d(\text{area}), \end{aligned} \tag{10.63}$$

and it is stored as an increase in potential, dV.

So if we recall from (8.50) that the surface area of the membrane is given at any instant by

$$A = \iint_D \sqrt{1 + w_x^2 + w_y^2}\, dx\, dy$$

[11]As with the "flexible string," the words "flexible membrane" imply that the membrane offers negligible resistance to bending.

and use the hypothetical state $w \equiv 0$ as our zero reference potential, then

$$V = \tau\left\{\iint_D \sqrt{1 + w_x^2 + w_y^2}\, dx\, dy - \iint_D dx\, dy\right\}$$

$$\approx \tau\left\{\iint_D \left(1 + \frac{w_x^2 + w_y^2}{2}\right) dx\, dy - \iint_D dx\, dy\right\} = \frac{\tau}{2}\iint_D (w_x^2 + w_y^2)\, dx\, dy. \tag{10.64}$$

Furthermore, the kinetic energy is

$$T = \frac{1}{2}\iint_D \rho(x, y) w_t^2\, dx\, dy, \tag{10.65}$$

where $\rho(x, y)$ is the membrane mass density (mass per unit area); consequently, Hamilton's principle becomes

$$\frac{1}{2}\int_{t_1}^{t_2}\iint_D [\rho w_t^2 - \tau(w_x^2 + w_y^2)]\, dx\, dy\, dt = \text{min}, \tag{10.66}$$

and the equation of motion follows from (10.62) as

$$\tau(w_{xx} + w_{yy}) - \rho(x, y) w_{tt} = 0,$$

or

$$\tau \nabla^2 w = \rho(x, y) w_{tt}. \tag{10.67}$$

COMMENT 1. Observe that, for the *static* case, (10.67) reduces to the *Laplace* equation, and therefore we have the so-called **Dirichlet problem**

$$\nabla^2 w = 0 \text{ in } D,\ w \text{ given on } C. \tag{10.68}$$

Now, for the static case, our variational principle is simply $V = \text{min}$ or

$$\iint_D (w_x^2 + w_y^2)\, dx\, dy = \text{min}; \tag{10.69}$$

that is, the Laplace equation $\nabla^2 w = w_{xx} + w_{yy} = 0$ is the Euler equation for the variational principle (10.69)!

COMMENT 2. Again, for the static case, suppose that the membrane were "pumped up" by a uniform net pressure p on its underside. In order to incorporate this feature into our formulation, we ask: Is there a *potential* associated with p? Well, consider a little element dA of the membrane (Fig. 10.15), subjected to a pressure force $p\, dA$. In place of the pressure we can induce exactly the same effect on the element by attaching a weightless string to it and hanging a weight of magnitude $p\, dA$ over some frictionless pulleys (Fig. 10.15); the w displacement of the membrane element and the height of the weight are related simply as shown. The gravitational potential is $-(p\, dA)w$. Integrating over the membrane and adding the result to (10.64), we have

$$V = \iint_D \left[\frac{\tau}{2}(w_x^2 + w_y^2) - pw\right] dx\, dy = \text{min} \tag{10.70}$$

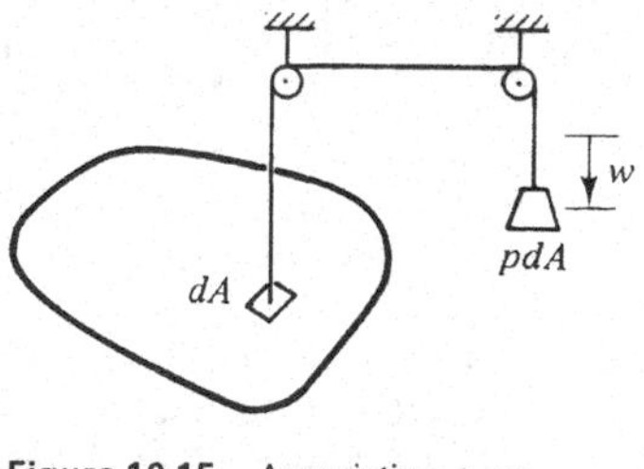

Figure 10.15. Associating a potential with p.

or from (10.62) the Poisson equation[12]

$$\nabla^2 w = -\frac{p}{\tau}. \quad \blacksquare \tag{10.71}$$

10.6. THE RITZ METHOD

Consider, for definiteness, the variational problem (10.52). The partial differential Euler equation (10.56), plus boundary conditions (10.52b), may, depending on f and the shape of D, be intractable to the point where we would gladly settle for a good *approximate* solution, or a good approximation may be all that we actually need in the first place. The *Ritz method* is a simple and powerful technique for obtaining such approximate solutions.

We seek

$$w \approx c_1\phi_1(x, y) + \cdots + c_n\phi_n(x, y), \tag{10.72}$$

where the $\phi_j(x, y)$'s are "suitably chosen functions" and the c_j's are constants to be determined. Generally the ϕ_j's are chosen so that (10.72) satisfies the given boundary conditions for all possible c_j's; for instance, if $w = 0$ on C, then $\phi_1(x, y) = \cdots = \phi_n(x, y) = 0$ on C. Then, we have $I = I(c_1, \ldots, c_n)$, and we extremize I with respect to the c_j's by setting $\partial I/\partial c_1 = \cdots = \partial I/\partial c_n = 0$. This process provides n algebraic equations in the n c_j's.

Example 10.10. *Pumped-Up Membrane.* Let D in (10.70) be the rectangle $-a \leq x \leq a$, $-b \leq y \leq b$ and suppose that $w = 0$ on its boundary. Let us try

$$w \approx (x^2 - a^2)(y^2 - b^2)(c_1 + c_2x^2 + c_3y^2 + c_4x^4 + c_5x^2y^2 + c_6y^4 + \cdots). \tag{10.73}$$

These ϕ's, which are of the form $(x^2 - a^2)(y^2 - b^2)x^my^n$, are a good choice for three reasons. They are zero on the boundary, as desired, and they are *simple*, so that the integration of V will be straightforward. Finally, note that $(x^2 - a^2)(y^2 - b^2)$ provides what we expect to be the right basic shape (dotted lines in Fig. 10.16), and the power series factor $c_1 + c_2x^2 + c_2y^2 + \cdots$ allows for the necessary "adjustments." Odd powers like $x, y, xy, x^2y, \ldots$ have been omitted because it is obvious that the solution should be symmetric in both x and y.

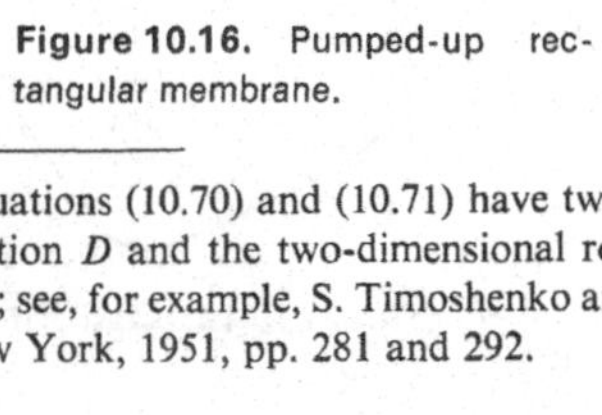

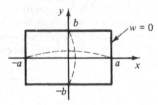

Figure 10.16. Pumped-up rectangular membrane.

Depending on the accuracy desired and how ambitious we feel, we cut (10.73) off at some point. Suppose that we simply keep the first term, $w \approx (x^2 - a^2)(y^2 - b^2)c_1$. Putting it into (10.70) and integrating, we obtain

$$V(c_1) = \frac{16a^3b^3[4\tau(a^2 + b^2)c_1^2 - 5pc_1]}{45}.$$

---

[12]Equations (10.70) and (10.71) have two well-known analogs in mechanics; torsion of a bar of cross section D and the two-dimensional rotational flow of an inviscid incompressible fluid in the region D; see, for example, S. Timoshenko and J. N. Goodier, *Theory of Elasticity*, 2nd ed., McGraw-Hill, New York, 1951, pp. 281 and 292.

Minimizing, we set

$$\frac{dV}{dc_1} = \frac{16a^3b^3[8\tau(a^2 + b^2)c_1 - 5p]}{45} = 0,$$

so that $c_1 = 5p/8\tau(a^2 + b^2)$ and

$$w \approx \frac{5p(x^2 - a^2)(y^2 - b^2)}{8\tau(a^2 + b^2).} \tag{10.74}$$

As a partial measure of the accuracy of this result, we note that if the volume under the membrane, $\iint w\,dx\,dy$ over D, is computed by using (10.74) and compared with the exact value [using equation (26.109a) of Part V], it is found that the error is 1.2% for $b/a = 1$ and 11.9% for $b/a = 10$. Why does the accuracy drop off as b/a becomes large (Exercise 10.26)?

Seeking better accuracy, let us try

$$w \approx (x^2 - 1)(y^2 - 1)(c_1 + c_2x^2 + c_3y^2), \tag{10.75}$$

where we've set $a = b = 1$ for definiteness. Setting $\partial V/\partial c_1 = \partial V/\partial c_2 = \partial V/\partial c_3 = 0$ leads to the simultaneous equations (Exercise 10.27)

$$\begin{aligned} \frac{16}{45}c_1 + \frac{32}{525}c_2 + \frac{32}{525}c_3 &= \frac{p}{9\tau} \\ \frac{32}{105}c_1 + \frac{304}{945}c_2 + \frac{16}{329}c_3 &= \frac{p}{9\tau} \\ \frac{32}{105}c_1 + \frac{16}{329}c_2 + \frac{304}{945}c_3 &= \frac{p}{9\tau}, \end{aligned} \tag{10.76}$$

and so $c_1 = 0.2920p/\tau$ (compared with $0.3125p/\tau$ before) and $c_2 = c_3 = 0.0597p/\tau$. This time it turns out that $\iint w\,dx\,dy$ is in error by only 0.2%. It seems plausible that *any* desired accuracy can be achieved, in this example at least, by including enough terms in (10.73). This fundamental question of convergence was examined by Ritz and is discussed in the book by Mikhlin.[13] ▮

Observe that the Ritz method proceeds from a variational principle. Often, however, we have an ordinary or partial differential equation to solve that was not obtained as the Euler equation of some variational problem. In this case, it *may* be possible to find a corresponding variational principle to which the Ritz method can then be applied; this topic is considered briefly in Exercise 10.28 and at length in the book by Finlayson.[14] Alternatively, a **method of weighted residuals**, such as the *Galerkin method*, can be applied directly to the governing differential equation(s). The methods of weighted residuals and the Ritz method are among the most important *direct methods*—that is, methods of approximate solution that reduce the problem to a finite system of *algebraic* equations. The subject is discussed further in Parts IV and V. For a detailed account, including many applications to problems in fluid mechanics and heat and mass transfer, we recommend Finlayson.[15]

---

[13] S. G. Mikhlin, *Variational Methods in Mathematical Physics*, Macmillan, New York, 1964.

[14] B. A. Finlayson, *The Method of Weighted Residuals and Variational Principles*, Academic, New York, 1972. See especially Chapters 8 and 9.

[15] *Ibid.*

Let us illustrate with an example.

Example 10.11. *Liquid Level Control System.* Suppose that a certain liquid is pumped into a tank of uniform cross-sectional area, say unity, at a steady rate of Q m^3/min and that it flows out the bottom at a rate that is proportional to the square root of the depth x (Fig. 10.17). (For instance, if the efflux at the bottom is at atmospheric pressure, then the exit velocity is known to be $\sqrt{2gx}$ where g is the acceleration of gravity, or perhaps some fraction of this value due to various losses.)

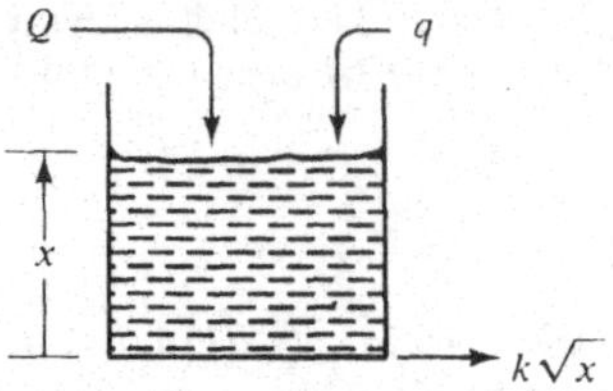

Figure 10.17. Liquid level control system.

Equating inflow to outflow, it is clear that the steady-state level x_s is given by $Q = k\sqrt{x_s}$. Although it is desired to operate at the equilibrium value x_s, x may, for various reasons, sometimes fall below this value, and we can attempt to bring it back in an optimal way by an additional input at a suitable rate $q(t)$.

The optimal control problem, then, consists of deciding how to manage the *control variable* $q(t)$. Naturally, doing so will depend on exactly what we mean by "optimal." Since we are told that it is important to operate close to equilibrium, it is reasonable to require that the square deviation $\int_0^T [x(t) - x_s]^2\,dt$ over the control time T be small. Additionally, we would not like the control expenditure, which we might measure by $\int_0^T q^2(t)\,dt$, to be uneconomically large. So given an initial state $x(0) \neq x_s$, we seek the control "policy" $q(t)$ such that

$$\int_0^T [(x - x_s)^2 + \alpha^2 q^2]\,dt = \min, \tag{10.77}$$

where the weighting factor α^2 (squared to emphasize that it is positive) is a measure of the relative importance that we attach to the square error and control expenditure.

Of course, we cannot expect to proceed much further without a statement of the dynamics of the system, and this statement is given by the mass balance

$$\dot{x} = Q + q - k\sqrt{x}\,; \tag{10.78}$$

that is d(volume)/dt equals the rate in minus the rate out.

Before continuing, let us simplify subsequent steps (Exercise 10.30) by "linearizing" (10.78)—that is, the nonlinear $\sqrt{x}$ term—about the equilibrium state x_s. In other words, we expand

$$\sqrt{x} = \sqrt{x_s} + \frac{1}{2\sqrt{x_s}}(x - x_s) + \cdots$$

and break the series off after the linear term on the grounds that x deviates little from x_s ($x - x_s$ is reasonably small), so that the linear or "tangent line" approximation

$$\sqrt{x} \approx \sqrt{x_s} + \frac{1}{2\sqrt{x_s}}(x - x_s)$$

should suffice (Fig. 10.18). Putting this approximation into (10.78), recalling that $Q = k\sqrt{x_s}$, and defining the error $x - x_s \equiv y$, we have the simpler (linear) equation

$$\dot{y} = -Ay + q(t), \qquad (10.79)$$

where $A \equiv k/2\sqrt{x_s}$.

So by inserting $q = \dot{y} + Ay$ into (10.77), we find

$$\int_0^T [y^2 + \alpha^2(\dot{y} + Ay)^2]\, dt = \min, \qquad (10.80a)$$

$$y(0) = x(0) - x_s \equiv y_0, \qquad y(T) \text{ unspecified}; \qquad (10.80b)$$

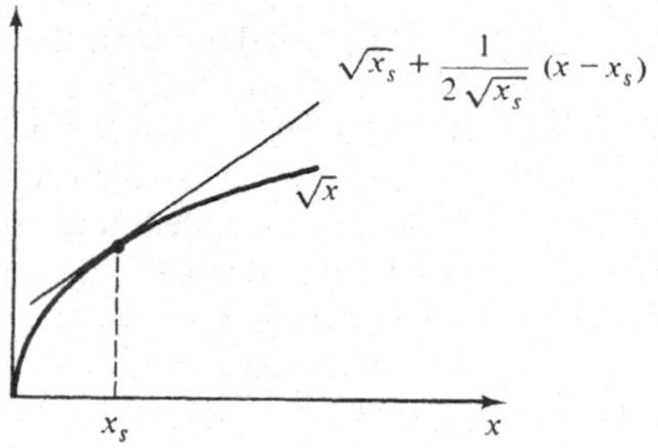

Figure 10.18. Tangent line linearization.

hence the Euler equation and boundary conditions are

$$\ddot{y} - \left(\frac{1 + \alpha^2 A^2}{\alpha^2}\right) y = 0; \qquad (10.81a)$$

$$y(0) = y_0, \qquad \dot{y}(T) + Ay(T) = 0 \quad \text{(natural boundary condition)}. \qquad (10.81b)$$

It is not hard to solve (10.81) for $y(t)$ and thus to find the desired policy $q(t)$ from $q(t) = \dot{y} + Ay$.

COMMENT 1. What is the "physical significance" of the natural boundary condition in (10.81b)? Recalling (10.79), we see that $\dot{y}(T) + Ay(T) = 0$ is the same as saying that $q(T) = 0$, which simply says that the control time is T!

COMMENT 2. Actually, a **feedback control system** would be operationally more convenient—that is, where the deviation $y = x - x_s$ is monitored and the control q is some optimally prescribed function of the state y; thus $q = q(y)$ rather than $q(t)$.

To illustrate, consider the simple case of *proportional control* $q(y) = My$, where the constant M is called the *gain*.[16] Then (10.79) becomes $\dot{y} + (A - M)y = 0$, and so

$$y = y_0 e^{-(A-M)t},$$

where we choose M so that

$$\int_0^T (y^2 + \alpha^2 q^2)\, dt = (1 + \alpha^2 M^2) \int_0^T y^2\, dt$$
$$= (1 + \alpha^2 M^2) y_0^2 \int_0^T e^{-2(A-M)t}\, dt = \text{etc.}$$

is a minimum. Setting d/dM of this equation equal to zero, we obtain (for the case where T is considered to be quite large) the optimal gain

$$M = \frac{A\alpha - \sqrt{1 + A^2\alpha^2}}{\alpha}. \quad ∎$$

[16]The present example involves a *first*-order system; its state is prescribed by the single quantity y. Suppose, however, that you are to drive a car on a straight one-lane road so as to follow a "lead car" by a constant distance, say 20 m. If y is the deviation of the distance between cars from the desired value, then (roughly speaking) we have a second-order system with feedback control. That is, your eyes monitor the two state variables y, $\dot{y}$ (the error in spacing and its rate of change), your brain evaluates some $q(y, \dot{y})$, and your foot activates the brake or gas pedal accordingly. Perhaps $q(y, \dot{y}) \approx M(y - 20) + N\dot{y}$, where M and N are determined from experience.

The foundations of the mathematical theory of optimal control date back to the late 1950s and include major contributions from L. S. Pontryagin, L. W. Neustadt, J. P. LaSalle, and a group of mathematicians led by R. Bellman.

The subject is sophisticated, and even the central results are outside our present scope. If interested, a good place to start might be the 36-page introduction in the book by Boltyanskii,[17] which includes the method of **dynamic programming, Pontryagin's maximum principle**, and **discontinuous control**. In addition, we recommend the very readable account by Bellman[18] and the book by Denn,[19] which includes numerous practical applications of the theory to various chemical engineering processes.

EXERCISES

10.1. Verify, by suitable examination of second derivative(s), that $(\sqrt{2}, \sqrt{2})$ and $(-\sqrt{2}, -\sqrt{2})$ correspond to maxima of F and that $(-1/\sqrt{2}, 1/\sqrt{2})$ and $(1/\sqrt{2}, -1/\sqrt{2})$ correspond to minima of F in Example 10.2 as claimed.

10.2. Where possible, verify that the extremum is indeed a minimum.
(a) Find the point on the curve of intersection of $z - xy = 10$ and $x + y + z = 1$ that is closest to the origin.
(b) Find the point(s) on the ellipsoid $x^2 - xy + y^2 + 2z^2 = 4$ that is closest to the origin.
(c) Find the point on the plane $ax + by + cz = d$ that is closest to the origin.

10.3. If $x_1 x_2 \cdots x_n = 1$, where each x_i is positive, show that $x_1 + x_2 + \cdots + x_n \geq n$.

10.4. (a) Show that of all parallelograms with a prescribed area A, the one with the least perimeter is a square.
(b) Furthermore, show that if their perimeter is prescribed instead, then the one with the largest area is a square.

10.5. (*Steepest descent*) (a) In discussing (necessary) conditions for the existence of extrema of functions of several variables, we have thus far said nothing about *computational* aspects; for instance, consider

$$f(x, y, \ldots) = \min \tag{10.82}$$

or

$$f_x(x, y, \ldots) = f_y(x, y, \ldots) = \ldots = 0. \tag{10.83}$$

Equation (10.83) involves the solution of simultaneous algebraic equations, which may be nonlinear and intractable. Generally we resort to one of several standard iterative schemes for their solution. The methods of *successive approximation* and *Newton–Raphson* are discussed in Part IV; here we will consider only the so-called **method of steepest descent** for the (10.82) version of the problem. Although refinements are possible, it goes basically like this. Consider, for definiteness, the two-dimensional case $f(x, y) = \min$. Choose an initial guess-estimate x^0, y^0 to the desired solution. To move toward the minimum, it is reasonable to strike out in the direction of steepest descent of $f(x, y)$—that

---

[17] V. G. Boltyanskii, *Mathematical Methods of Optimal Control*, Holt, Rinehart and Winston, New York, 1971.

[18] R. Bellman, *Introduction to the Mathematical Theory of Control Processes*, Vol. 1, *Linear Equations and Quadratic Criteria*, Academic, New York, 1967.

[19] M. Denn, *Optimization by Variational Methods*, McGraw-Hill, New York, 1969.

is, in the direction of ∇f at the point x^0, y^0. The question is how *far* should we move; in other words, what is the scale factor α so that

$$f(x^0 + \alpha f_x^0, y^0 + \alpha f_y^0) \approx f^0 + \alpha[(f_x^0)^2 + (f_y^0)^2]$$
$$+ \frac{\alpha^2}{2}[f_{xx}^0(f_x^0)^2 + 2f_{xy}^0 f_x^0 f_y^0 + f_{yy}^0(f_y^0)^2]$$

is a minimum, where $(\ )^0$ means $(\ )$ evaluated at x^0, y^0? Setting $d/d\alpha$ of the right-hand side equal to zero, solve for α and deduce the iterative scheme

$$x^{n+1} = x^n + \alpha f_x^n, \qquad y^{n+1} = y^n + \alpha f_y^n, \tag{10.84}$$

where
$$\alpha = -\frac{(f_x^n)^2 + (f_y^n)^2}{f_{xx}^n(f_x^n)^2 + 2f_{xy}^n f_x^n f_y^n + f_{yy}^n(f_y^n)^2}.$$

(b) Use (10.84) to locate the equilibrium point x, y that is near the origin, in the gravitational field induced by point unit masses at $(1, 0)$, $(0, 5)$, and $(-1, 0)$. Is it stable or unstable?

(c) Use (10.84) to solve the simultaneous equations $x = \sin(x + y)$, $y = \cos(x - y)$ by defining $f = [x - \sin(x + y)]^2 + [y - \cos(x - y)]^2$ and starting with $x^0 = y^0 = 0.5$, say.

10.6. (a) Verify that the straight-line solution (10.31) is, in fact, a *minimum* by showing that $I''(0) > 0$.

(b) Lifting the assumption that y be a single-valued function of x, re-solve the shortest arc Example 10.4 by describing the arc parametrically by $x = x(t)$, $y = y(t)$.

10.7. In our discussion of the equilibrium of the loop shown in Fig. 10.8 we omitted the weight of the element. Is this omission an approximation or is it strictly correct? Explain.

10.8. Derive the general solution $y = A\cosh[(x - B)/A]$ of the Euler equation (10.38). *Hint:* Set $y' = p$ and hence $y'' = p\,dp/dy$.

10.9. (*First integral of Euler equation*) Show that if f in (10.24b) does not depend explicitly on x—that is, $f = f(y, y')$—then the Euler equation (10.30) admits the first integral

$$f - y'\frac{\partial f}{\partial y'} = \text{constant}.$$

10.10. (a) Asked in an examination to find the Euler equation for the problem

$$\int_{x_1}^{x_2} f(x, y')\,dx = \min; \qquad y(x_1),\quad y(x_2) \text{ given},$$

one student showed that $0 = \int_{x_1}^{x_2} f_{y'}\eta'\,dx$ and concluded that the Euler equation is $f_{y'} = 0$. Evaluate these claims.

(b) Determine the Euler equation and natural boundary conditions for the problem

$$\int_{x_1}^{x_2} f(x, y, y', y'')\,dx = \min.$$

10.11. Integrate (10.43) to obtain (10.44).

10.12. (*John Bernoulli's brachistochrone problem*) We seek the curve $y(x)$ from $(0, 0)$ to (a, b) along which a bead of mass m will descend under the action of gravity (no friction) in the shortest time (Fig. 10.19). Noting, from conservation of energy, that

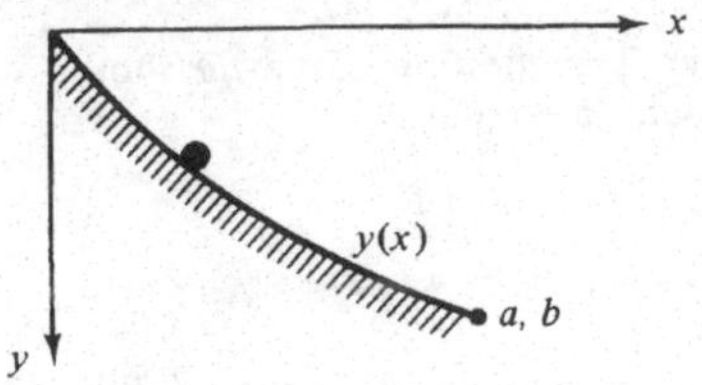

Figure 10.19. Brachistochrone.

the bead's velocity is $v = \sqrt{2gy}$, show that the variational problem may be stated as

$$\int_0^a \sqrt{\frac{1+y'^2}{2gy}}\,dx = \min; \qquad y(0) = 0, \quad y(a) = b.$$

Show (recall Exercise 10.9) that a first integral of the Euler equation is

$$y' = \sqrt{\frac{1-Cy}{Cy}}.$$

[The next integration is more tedious; it would reveal that $y(x)$ is a *cycloid.*]

10.13. (*Fermat's principle*) Consider two different, optically homogeneous media above and below the plane P (Fig. 10.20) and suppose that there is a point light source at

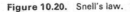

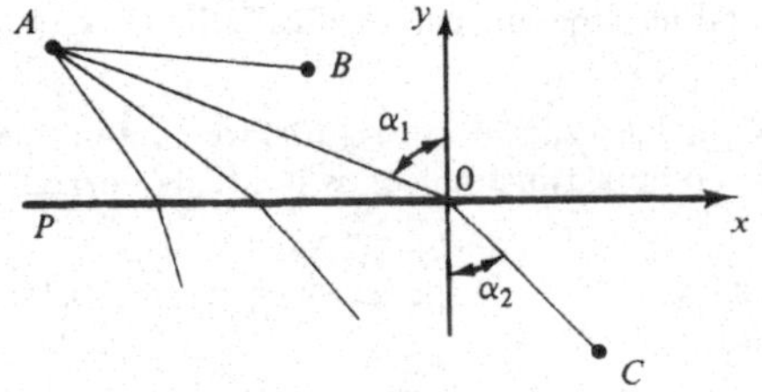

Figure 10.20. Snell's law.

A. Fermat's principle of optics states that *the path taken by a ray of light will be such that the travel time t is a minimum.* If the speed of light at each point of the upper medium, in any direction, is v_1 (a constant) and v_2 in the lower medium, show that

(a) the path taken from A to any point B in the upper medium is a straight line.

(b) the path from A to any point C in the lower medium is the broken straight line with the (Snell's) law of refraction

$$\frac{\sin\alpha_1}{\sin\alpha_2} = \frac{v_1}{v_2}.$$

10.14. (*More about Fermat's principle*) First read Exercise 10.13. Now instead of the simple two-layer medium, suppose that $v = v(y)$ varies continuously.

(a) Show that the paths $y(x)$ are determined by

$$v(y)y'' + v'(y)(1 + y'^2) = 0.$$

(b) Find the path from (0, 0) to (1, 0) if $v(y) = y$. (First use your intuition to sketch the solution and then see how good your intuition is.)

(c) Repeat part (b) for $v(y) = \sqrt{y}$.

***10.15.** Suppose that the length of the cable in Fig. 10.6 is l and that a weight W is hung from the loop. Derive the Euler equation and the natural boundary condition at $x = a$.

10.16. (*An isoperimetric exercise*) Find a curve $y(x)$ of length L over $0 \leq x \leq 1$, with $y(0) = y(1) = 0$, so that the area under it is maximized. Show that the result is the circle $(x - A)^2 + (y - B)^2 = \lambda^2$, where A and B are integration constants and λ is a Lagrange multiplier. Comment on the state of affairs when $L > \pi/2$.

10.17. (*Nonintegral constraints*) Consider the problem

$$I = \int_{t_1}^{t_2} f(t, x, y, \dot{x}, \dot{y})\, dt = \text{extremum},$$

with x and y prescribed at t_1 and t_2, subject to the nonintegral constraint $g(t, x, y) = c$; see (10.41b). Introducing $X = x + \epsilon\xi$ and $Y = y + \epsilon\eta$, as in Section 10.4, show that

$$0 = \int_{t_1}^{t_2} \left\{ \left[f_x - \frac{d}{dt}(f_{\dot{x}}) - \lambda g_x \right] \xi + \left[f_y - \frac{d}{dt}(f_{\dot{y}}) - \lambda g_y \right] \eta \right\} dt.$$

Thus show that

$$f_x - \frac{d}{dt}(f_{\dot{x}}) - \lambda(t) g_x = 0, \qquad f_y - \frac{d}{dt}(f_{\dot{y}}) - \lambda(t) g_y = 0.$$

Note carefully that here the Lagrange multiplier λ is a function of t unlike the case of an integral constraint, where it was a constant.

10.18. (*Geodesics for a sphere*) Show that the geodesics for a sphere of radius R are great circles. Do it two ways.

(a) With the help of Exercise 10.17, start with the problem statement

$$\int_{t_1}^{t_2} \sqrt{\dot{x}^2 + \dot{y}^2 + \dot{z}^2}\, dt = \min$$

with the constraint

$$g(t, x, y, z) = x^2 + y^2 + z^2 = R^2.$$

Hint: Show that $y\, d(dx/ds) = x\, d(dy/ds)$, so that $d(y\, dx/ds) = d(x\, dy/ds)$, and hence $y\, dx - x\, dy = A\, ds$. Similarly, obtain $z\, dx - x\, dz = B\, ds$. Eliminate ds and divide by x^2.

(b) In terms of spherical polars ρ, θ, ϕ, we have $(ds)^2 = (d\rho)^2 + \rho^2(d\theta)^2 + \rho^2 \sin^2\theta (d\phi)^2$, and so on $\rho = R$ we have

$$\int ds = \int_{\theta_1}^{\theta_2} R\sqrt{1 + (\sin^2\theta)(d\phi/d\theta)^2}\, d\theta = \min,$$

subject to *no* constraint. *Hint:* This solution can be quite short.

10.19. (*Geodesics for a cylinder*) Find the geodesics for the circular cylinder $x^2 + y^2 = R^2$. Do it two ways.

(a) With the help of Exercise 10.17, start with

$$\int_{t_1}^{t_2} \sqrt{\dot{x}^2 + \dot{y}^2 + \dot{z}^2}\, dt = \min$$

with the constraint

$$g(t, x, y, z) = x^2 + y^2 = R^2.$$

(b) With $x = R\cos\theta$, $y = R\sin\theta$, $z = z$, start with

$$\int ds = \int \sqrt{R^2(d\theta)^2 + (dz)^2} = \int_{\theta_1}^{\theta_2} \sqrt{R^2 + z'^2}\, d\theta = \text{min},$$

subject to *no* constraints, and recall Exercise 10.9.

10.20. Derive the equations of motion (not necessarily small) on $x(t)$, $y(t)$ for the system shown in Fig. 10.21 by means of Hamilton's principle. The springs are of length L and are neither stretched nor compressed when $x = y = 0$. Suppose that the x, y plane is horizontal, so that gravity is not relevant.

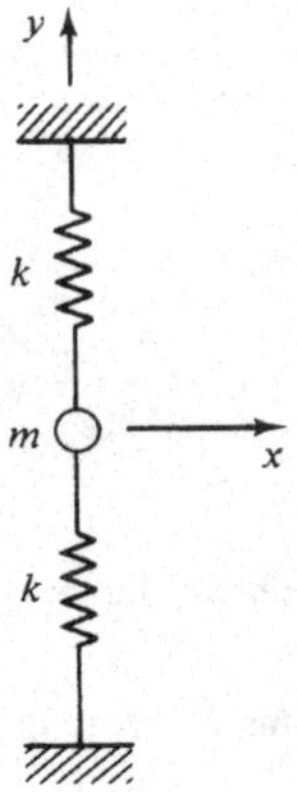

Figure 10.21. Two-degree-of-freedom oscillation.

10.21. Derive the equations of motion on $r(t)$, $\theta(t)$ for the system shown in Fig. 10.22 by means of Hamilton's principle. The unstretched length of the spring is L.

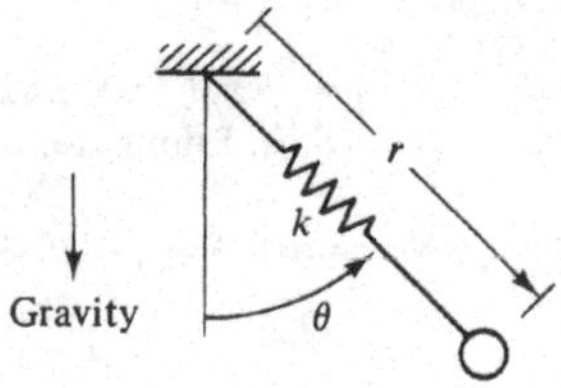

Figure 10.22. Spring pendulum.

10.22. (*Vibrating pipeline*) A nice problem from the literature[20] concerns the transverse vibrations of an above-ground oil pipeline that is supported at intervals of L feet. Let v be the (constant) velocity of the oil within the pipe, ρ the mass per unit length of the pipe, μ the mass per unit length of the fluid, E Young's modulus for the pipe, and I the moment of inertia of the pipe cross section about its neutral axis. Assume

[20] G. W. Housner, Bending Vibrations of a Pipe Line Containing Flowing Liquid, *Journal of Applied Mechanics*, Vol. 19, June 1952, p. 205.

small displacements of the pipe, that the pipe can be considered as a pinned-end beam (Fig. 10.23), and that the gravitational potential can be neglected compared with the strain energy of the pipe, so that

$$V = \int_0^L \tfrac{1}{2}EIy_{xx}^2\,dx,$$

where $y(x, t)$ is the deflection of the pipe's centerline. Write an expression for T and thus deduce the equation of motion

$$EIy_{xxxx} + (\rho + \mu)y_{tt} + \mu v^2 y_{xx} + 2\mu v y_{xt} = 0. \tag{10.85}$$

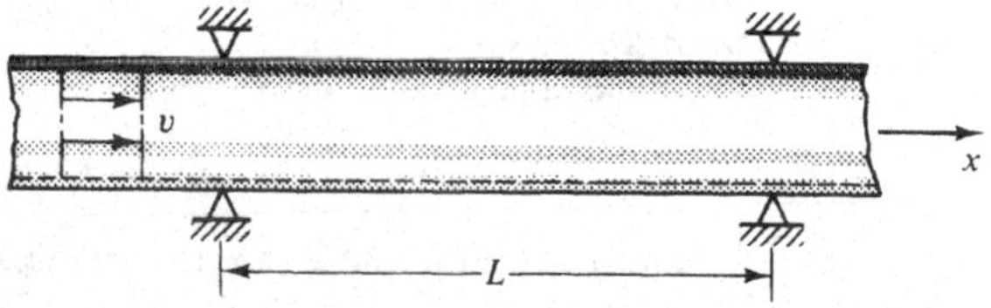

Figure 10.23. Vibrating pipeline.

Note that if $\mu \longrightarrow 0$, (10.85) reduces to the classical vibrating-beam equation (which you may have already encountered in mechanics); the third and fourth terms can be interpreted as centripetal and Coriolis terms, respectively. Note also that the only difficulty in this problem is in writing down the kinetic energy T, the expression for V being "standard" (in the *Euler beam theory*). By comparison, you will probably find that a nonvariational approach, based directly on Newton's law, is more difficult.

10.23. Provide a derivation of (10.62).

10.24. Reexpress (10.69) in terms of polar coordinates r, θ and write out the corresponding Euler equation.

10.25. (*The Ritz method*) Apply the Ritz method to find approximate solutions to the problems below.

(a) $\nabla^2 w = 3x^2 y$ in the triangle with corners at (0, 0), (1, 0), and (0, 1), where $w = 0$ on the boundary.

(b) The same as in part (a) but for the rectangular region $0 < x < 1, 0 < y < 1$.

10.26. Why does the accuracy of (10.74) drop off as b/a becomes large (or small)?

10.27. Go through the calculations that lead to equations (10.76). Incidentally, if b/a had been 10, say, instead of unity, might the three-term approximation

$$w \approx (x^2 - a^2)(y^2 - b^2)(c_1 + c_2 y^2 + c_3 y^4)$$

be more promising than

$$w \approx (x^2 - a^2)(y^2 - b^2)(c_1 + c_2 x^2 + c_3 y^2)?$$

Explain.

10.28. (*Working backward*) (a) In discussing the Ritz method, we pointed out that it would be nice to be able to take a given ordinary or partial differential equation and "work backward" to a corresponding variational principle. For instance, given the equation $y'' + y = F(x)$ over $x_1 \le x \le x_2$, a look at the derivation of (10.30) suggests that we proceed as follows.

$$0 = \int_{x_1}^{x_2} (y'' + y - F)\eta\, dx = y'\eta\Big|_{x_1}^{x_2} + \int_{x_1}^{x_2} (-y'\eta' + y\eta - F\eta)\, dx$$

$$= \int_{x_1}^{x_2} [(y - F)\eta + (-y')\eta']\, dx = \int_{x_1}^{x_2} \left[\frac{\partial f}{\partial y}\eta + \frac{\partial f}{\partial y'}\eta'\right] dx$$

so that $\partial f/\partial y = y - F$ and $\partial f/\partial y' = -y'$. Integrating, $f = (y^2/2) - Fy - (y'^2/2)$, and thus the variational principle is

$$\int_{x_1}^{x_2} \left(\frac{y^2}{2} - Fy - \frac{y'^2}{2}\right) dx = \text{extremum}$$

with $y'\eta = 0$ at x_1 and x_2—that is, $y' = 0$ (natural boundary condition) or y prescribed at each endpoint.

In the same manner, derive

$$\int_{x_1}^{x_2} (py'^2 - qy^2 + 2Fy)\, dx = \text{extremum}, \tag{10.86}$$

plus $py'\eta = 0$ at x_1 and x_2, for the more general differential equation

$$\frac{d}{dx}\left(p\frac{dy}{dx}\right) + qy = F(x), \tag{10.87}$$

or

$$py'' + p'y' + qy = F. \tag{10.88}$$

But if our given equation is

$$ay'' + by' + cy = d, \tag{10.89}$$

where $b \neq a'$, then it is not of the ("self-adjoint") form (10.88). Nevertheless, show that we can multiply (10.89) through by s so that $(sb) = (sa)'$—that is, $s = \exp \int [(b - a')/a]\, dx$—and hence throw it into the form (10.88).

(b) Thus show that the problem

$$xy'' + 2y' + y = 1; \qquad y(0) = 0, \quad y(1) = 1$$

corresponds to

$$\int_0^1 (x^2y'^2 - xy^2 + 2xy)\, dx = \text{extremum}; \qquad y(0) = 0, \quad y(1) = 1.$$

Using the Ritz method, seek an approximate solution of the form

$$y \approx x(x - 1)(c_1 + c_2x + c_3x^2 + \cdots) + x.$$

[This is slightly different from (10.72) due to the x term at the end, which is inserted because $y(1) = 1$, not 0.) Cut the $c_1 + c_2x + \cdots$ factor off after c_2x, say.

(c) Again "working backward," derive the variational principle corresponding to the two-dimensional Poisson equation $\nabla^2 w = \phi(x, y)$.

10.29. Instead of appealing to Lemma 10.1 to obtain the Euler equation from (10.28), suppose that we argue as follows. Set

$$\eta(x) = (x - x_1)^2(x - x_2)^2\left[\frac{\partial f}{\partial y} - \frac{d}{dx}\left(\frac{\partial f}{\partial y'}\right)\right].$$

Then (10.28) becomes

$$0 = \int_{x_1}^{x_2} (x - x_1)^2(x - x_2)^2\left[\frac{\partial f}{\partial y} - \frac{d}{dx}\left(\frac{\partial f}{\partial y'}\right)\right]^2 dx,$$

and since the integrand is nonnegative (and continuous), it follows that the square bracket must be identically zero over $x_1 < x < x_2$. Is this assumption correct? Discuss.

10.30. Is it really necessary that we linearize (10.78)? What eventual difficulties would arise if we did not?

10.31. Rework Example 10.11, using proportional control $q = My$, for the case where the tank is as shown in Fig. 10.24; in other words, the cross-sectional area varies as $a + bx$.

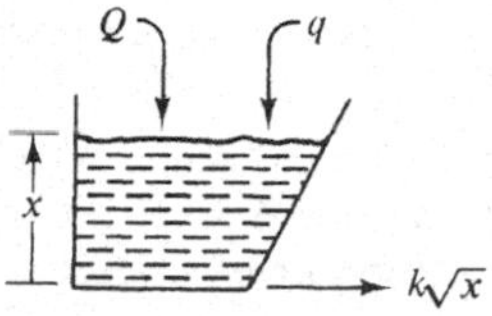

Figure 10.24. Nonconstant tank cross section.

10.32. (*Terminal process control*) Suppose that the optimal control criterion (10.80) is changed to

$$\int_0^T [y^2 + \alpha^2(\dot{y} + Ay)^2]\, dt + \beta^2 y^2(T) = \min$$

with $y(0) = y_0$ and $y(T)$ unspecified. What (in words) might be the motivation for including the $\beta^2 y^2(T)$ term? Show that the resulting Euler equation and boundary conditions are

$$\ddot{y} - \left(\frac{1 + \alpha^2 A^2}{\alpha^2}\right) y = 0;$$

$$y(0) = y_0, \qquad \alpha^2 \dot{y}(T) + (\alpha^2 A + \beta^2) y(T) = 0.$$

10.33. Make up two examination-type problems on this chapter.

ADDITIONAL REFERENCES FOR PART I

E. K. Blum, *Numerical Analysis and Computation: Theory and Practice*, Addison-Wesley, Reading, Mass., 1972. [See Section 5.7 on Aitken's method and the acceleration of convergence of sequences.]

R. C. Buck, *Advanced Calculus*, 2nd ed., McGraw-Hill, New York, 1965. [A rigorous treatment of functions, limits, continuity, integration, series, curves, surfaces, volumes, and vector field theory.]

B. Carnahan, H. Luther, and J. Wilkes, *Applied Numerical Methods*, Wiley, New York, 1969. [See Chapter 2 on Numerical Integration. This is an excellent general reference for numerical methods and includes many applications from the engineering literature.]

H. Carslaw and J. Jaeger, *Operational Methods in Applied Mathematics*, 2nd ed., Clarendon Press, Oxford, 1948.

G. Doetsch, *Guide to the Applications of the Laplace Transform*, Van Nostrand, Princeton, N.J., 1963.

C. Fox, *An Introduction to the Calculus of Variations*, Oxford University Press, Oxford, 1950.

P. Franklin, *An Introduction to Fourier Methods and the Laplace Transformation*, Dover, New York, 1949.

G. H. Hardy, *Divergent Series*, Clarendon Press, Oxford, 1949. [Contains a detailed discussion of the Euler transformation and other aspects of the general theory of divergent series.]

L. B. W. Jolley, *Summation of Series*, 2nd ed., Dover, New York, 1961. [A compendium of over 1000 series and their (closed form) sum.]

K. S. Kunz, *Numerical Analysis*, McGraw-Hill, New York, 1957. [See Chapter 6 for methods of summing series and accelerating their convergence—for example, the Kummer and Euler transformations, and the summation formulas of Laplace, Euler, and Gauss.]

M. J. Lighthill, *Introduction to Fourier Analysis and Generalized Functions*, Cambridge University Press, New York, 1958.

L. Lyusternik and A. Yanpal'skii, *Mathematical Analysis, Functions, Limits*, Pergamon Press, New York, 1964. [Contains additional techniques for summing and accelerating the convergence of series.]

H. Margenau and G. Murphy, *The Mathematics of Physics and Chemistry*, 2nd ed., Van Nostrand, Princeton, N.J., 1956. [See especially Chapter 1 on the Mathematics of Thermodynamics (including differentials, functions of several variables, line integrals, and Jacobians), and Chapter 5 on Coordinate Systems, Vectors, Curvilinear Coordinates (including 13 important curvilinear systems, such as parabolic, bipolar, and toroidal coordinates).]

W. Rudin, *Principles of Mathematical Analysis*, 2nd ed. McGraw-Hill, New York, 1964.

I. N. Sneddon, *Fourier Transforms*, McGraw-Hill, New York, 1951.

D. V. Widder, *Advanced Calculus*, 2nd ed., Prentice-Hall, Englewood Cliffs, 1961. [A good general reference for most of Part I.]

D. V. Widder, *The Laplace Transform*, Princeton University Press, Princeton, N.J., 1941.

PART II

COMPLEX VARIABLES

Even in Part I we began to feel a need for complex variable theory: in trying to understand fully Taylor series (Section 2.4), in connection with complex Fourier series, in the evaluation of various definite integrals (as in Example 6.1), and in coping with the Laplace inversion formula. Chapters 11 and 12 of Part II introduce basic notions, COMPLEX NUMBERS and FUNCTIONS OF A COMPLEX VARIABLE, including the concept of analyticity. The line of development from Chapters 12 through 15, that is, analyticity, Cauchy's theorem, the Cauchy integral formula, Taylor series, Laurent series, and the Residue Theorem,—reveals not only the power of complex variable theory but its formal elegance as well. Chapter 16 on CONFORMAL MAPPING departs from the logical chain of the previous chapters and develops the important application of complex variable theory to the solution of problems in two-dimensional potential theory (i.e., Laplace's equation in two dimensions).

Chapter 11

Complex Numbers

Considering that *negative* roots of algebraic equations were often held to be "fictitious" as late as the sixteenth century, it is not hard to imagine that "solutions" $x = 2 \pm \sqrt{-2}$ of $x^2 - 4x + 6 = 0$, for instance, were not entirely welcome. In fact, following the introduction of a consistent theory of complex numbers by Bombelli in 1572, it was not until the nineteenth century—particularly the work of Cauchy—that complex numbers gained full acceptance.

11.1. THE ALGEBRA OF COMPLEX NUMBERS

With $\sqrt{-1} \equiv i$, the general complex number is of the form

$$z = a + bi,$$

where a and b are real numbers, called the *real* and *imaginary* parts of z, respectively; that is, $a = \operatorname{Re} z$ and $b = \operatorname{Im} z$. (Note that $\operatorname{Im} z$ is simply b, not bi.)

If $z_1 = a + bi$ and $z_2 = c + di$, we *define* their sum and product as

$$z_1 + z_2 \equiv (a + c) + (b + d)i \tag{11.1}$$

and

$$z_1 z_2 \equiv (ac - bd) + (ad + bc)i. \tag{11.2}$$

These definitions are "rigged" so that the commutative and associative properties of addition and multiplication carry over from real numbers, as well as the distributive property of multiplication.

Having defined addition and multiplication, we also define corresponding inverse operations, subtraction and division. First, let us define *subtraction* according to

$$z_1 - z_2 \equiv z_1 + (-1)z_2. \tag{11.3}$$

Finally, consider the *division* of z_2 into z_1, where $z_2 \neq 0$.[1] We say that

$$\frac{z_1}{z_2} = z_3 \quad \text{if } z_1 = z_2 z_3. \tag{11.4}$$

That this expression uniquely determines z_3 can be seen by writing out $z_1 = z_2 z_3$:

$$a_1 + b_1 i = (a_2 + b_2 i)(a_3 + b_3 i) = (a_2 a_3 - b_2 b_3) + (a_2 b_3 + a_3 b_2)i.$$

Thus

$$a_2 a_3 - b_2 b_3 = a_1, \qquad b_2 a_3 + a_2 b_3 = b_1,$$

which can be solved uniquely for a_3 and b_3, provided that the determinant $a_2^2 + b_2^2 \neq 0$ or, equivalently, provided that $z_2 \neq 0$. Doing so, we have

$$z_3 = a_3 + b_3 i = \left(\frac{a_1 a_2 + b_1 b_2}{a_2^2 + b_2^2}\right) + \left(\frac{a_2 b_1 - a_1 b_2}{a_2^2 + b_2^2}\right)i. \tag{11.5}$$

In practice, it is more direct to proceed simply by multiplying both the numerator and denominator by $a_2 - b_2 i$:

$$\begin{aligned} z_3 = \frac{z_1}{z_2} &= \frac{a_1 + b_1 i}{a_2 + b_2 i} = \frac{a_1 + b_1 i}{a_2 + b_2 i}\frac{a_2 - b_2 i}{a_2 - b_2 i} \\ &= \frac{(a_1 a_2 + b_1 b_2) + (a_2 b_1 - a_1 b_2)i}{a_2^2 + b_2^2} \\ &= \left(\frac{a_1 a_2 + b_1 b_2}{a_2^2 + b_2^2}\right) + \left(\frac{a_2 b_1 - a_1 b_2}{a_2^2 + b_2^2}\right)i. \end{aligned}$$

The quantity $a_2 - b_2 i$ is called the **complex conjugate**, or simply the **conjugate**, of z_2, and it is denoted as $\bar{z}_2$. More generally, if $z = a + bi$, then

$$\bar{z} \equiv a - bi.$$

We leave it for the reader (Exercise 11.2) to show that

$$\overline{z_1 + z_2} = \bar{z}_1 + \bar{z}_2, \qquad \overline{z_1 z_2} = \bar{z}_1 \bar{z}_2, \qquad \overline{\left(\frac{z_1}{z_2}\right)} = \frac{\bar{z}_1}{\bar{z}_2}, \tag{11.6}$$

and that

$$z + \bar{z} = 2 \operatorname{Re} z, \qquad z - \bar{z} = 2i \operatorname{Im} z. \tag{11.7}$$

11.2. THE COMPLEX PLANE

Besides the notation $z = a + bi$, it is common to denote $z = (a, b)$—that is, as an ordered pair of real numbers—with the definitions

$$(a, b) + (c, d) \equiv (a + c, b + d) \qquad \text{[see (11.1)]}$$
$$(a, b)(c, d) \equiv (ac - bd, ad + bc) \qquad \text{[see (11.2)]}$$

and so on. One advantage of this notation is that the mysterious letter i need never

[1]By $z = 0$ we mean $0 + 0i$.

appear; another is that it lends itself more readily to linear vector space concepts, as discussed in Part III.

We will stick to the $a + bi$ notation, but note how the ordered-pair representation (a, b) suggests that we consider z as a point in a cartesian x, y plane with a as abscissa and b as ordinate. This is the so-called **Argand diagram** or, simply, the **z plane** (Fig. 11.1). The length of the radius vector from the origin to the point (x, y) —the "z vector"—is called the **modulus** of z and is denoted as mod z, or $|z|$; clearly,

$$|z| = \sqrt{x^2 + y^2} \quad \text{or} \quad \sqrt{z\bar{z}}. \tag{11.8}$$

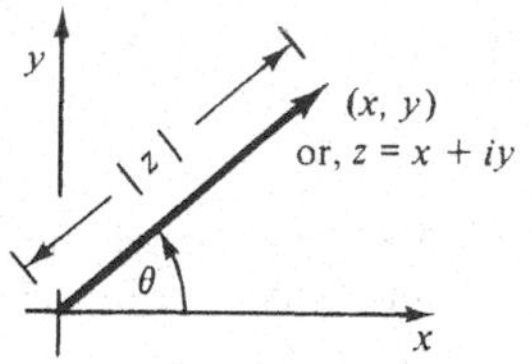

Figure 11.1. The Argand diagram.

The angle θ (Fig. 11.1) is called the **argument** of z and is denoted as arg z. Since $x = |z| \cos\theta$ and $y = |z| \sin\theta$,

$$z = x + iy = |z|(\cos\theta + i\sin\theta). \tag{11.9}$$

Next, recalling **Euler's formula**[2]

$$e^{i\theta} = \cos\theta + i\sin\theta \tag{11.10}$$

and noting that $|z| = r$, where r, θ are the polar coordinates corresponding to x, y, (11.9) becomes

$$z = x + iy = re^{i\theta}; \tag{11.11}$$

these are the alternative *cartesian* and *polar* representations of z, respectively. Furthermore,

$$\bar{z} = x - iy = |z|(\cos\theta - i\sin\theta) = re^{-i\theta}. \tag{11.12}$$

It is important to see that definitions (11.1) and (11.3) of addition and subtraction of complex numbers coincide with the parallelogram law of addition and subtraction of vectors, as illustrated for the case of addition in Fig. 11.2.

Interpretation of multiplication and division in the complex plane is simplest in terms of the polar representation of the complex quantities. Thus

$$z_1 z_2 = (r_1 e^{i\theta_1})(r_2 e^{i\theta_2}) = (r_1 r_2)e^{i(\theta_1+\theta_2)}, \tag{11.13a}$$

that is,

$$|z_1 z_2| = |z_1||z_2| \quad \text{and} \quad \arg(z_1 z_2) = \arg z_1 + \arg z_2. \tag{11.13b}$$

Similarly for division,

$$\frac{z_1}{z_2} = \frac{r_1}{r_2} e^{i(\theta_1-\theta_2)}; \tag{11.14a}$$

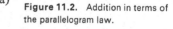

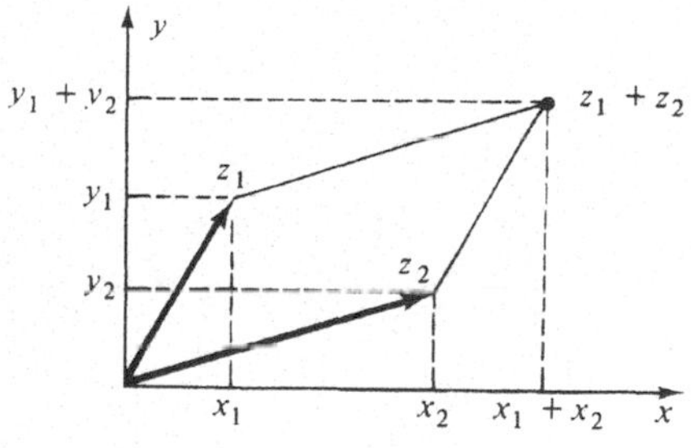

Figure 11.2. Addition in terms of the parallelogram law.

that is,

$$\left|\frac{z_1}{z_2}\right| = \frac{|z_1|}{|z_2|} \quad \text{and} \quad \arg\frac{z_1}{z_2} = \arg z_1 - \arg z_2. \tag{11.14b}$$

[2]We state (11.10) here as the *definition* of $e^{i\theta}$. We won't be in a position to appreciate the *motivation* behind this definition, however, until the discussion of analytic continuation in Chapter 14.

Finally, *note carefully* that the complex numbers are not "ordered" as the real numbers are; for instance, although $-6 < 2$, $10 > 9$, and so on, analogous statements like $z < 0$, $z > 4$, and $3i > 1 + i$ are not meaningful. Of course, $|z| > 4$ and $|3i| > |1 + i|$ *do* make perfect sense because the *modulus* of a complex number is a real number.

EXERCISES

11.1. Verify that the addition and multiplication of complex numbers are both commutative and associative as claimed in Section 11.1.

11.2. Verify equations (11.6) and (11.7).

11.3. Find $|z|$, arg z ($0 < \arg z \leq 2\pi$, say), Re z, and Im z for

(a) -5, (b) $2 - 3i$,

(c) $\dfrac{i}{1 - i}$, (d) $(1 + i)^{50}$,

(e) $\left(\dfrac{1 + i}{1 - i}\right)^4$.

11.4. What point sets in the z plane are defined by the following:

(a) $|z| \leq 1$ (b) $|z| = 1$ (c) $|z| > 1$

(d) $|z - 1| < |z - 2|$ (e) $|z| \leq |z + i - 2|$ (f) $|z + 1| = |z| + 1$

11.5. Derive the equation of a straight line through the points z_1 and z_2 in terms of z_1, z_2, and z.

11.6. (a) Derive the so-called **triangle inequality**, $|z_1 + z_2| \leq |z_1| + |z_2|$ and interpret it geometrically. Under what conditions does the equality hold?

(b) Show further that

$$|z_1 + z_2 + \cdots + z_n| \leq |z_1| + |z_2| + \cdots + |z_n|.$$

Chapter 12

Functions of a Complex Variable

Starting with the notion of a function $f(z)$ as a mapping, we go as far as differentiation and the classical Cauchy–Riemann conditions for analyticity; integration is a longer story, one that will require the next three chapters to complete. In examining various elementary functions toward the end of the chapter, we become acquainted with the idea of singularities, multivaluedness, and the need for branch cuts.

12.1. BASIC NOTIONS

In Part I we considered real-valued functions of one or more real variables, say $f(x_1, \ldots, x_n)$. In contrast, the classical complex variable theory, the subject of Part II, considers complex-valued functions of a single complex variable, say $f(z)$.[1] Denoting the real and imaginary parts of f as u and v, respectively, we write

$$f(z) = u(x, y) + iv(x, y). \tag{12.1}$$

Thus f is a **mapping** from some domain D in the z plane to some range R in the f plane, as indicated in Fig. 12.1. As usual, we absolutely insist that f be *single valued*, although not necessarily *one-to-one*.

[1]For the theory of functions of *several* complex variables, with applications to quantum field theory, function theory, and differential equations with constant coefficients, see V. Vladimirov, *Methods of the Theory of Functions of Many Complex Variables*, MIT Press, Cambridge, Mass., 1966.

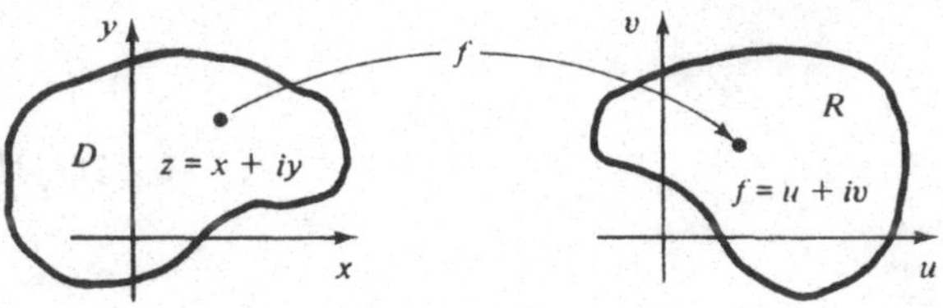

Figure 12.1 f as a mapping.

Example 12.1. Consider the simple function $f(z) = z^2$, where the domain D is the upper half-plane $y > 0$. Then

$$f(z) = z^2 = (x + iy)^2 = (x^2 - y^2) + (2xy)i,$$

so that we identify

$$u(x, y) = x^2 - y^2, \qquad v(x, y) = 2xy. \tag{12.2}$$

What does $f = z^2$ look like? Is it a "parabola"? The fact is that we cannot plot f versus z because, in order to do so, we would need a four-dimensional space—two dimensions for z and two for f. [Of course, we can always plot the two surfaces $u(x, y)$ and $v(x, y)$, but this is not the same as plotting f versus z.]

As a poor substitute for a graph, we could label any number of points in D, say $z_1, \ldots, z_n$, and also their image points $f_1, \ldots, f_n$ in R. Or we could label any number of *curves* in D, as well as their images in R. For instance, consider the curve $y = c$ in D. Then

$$u = x^2 - c^2, \qquad v = 2cx$$

and eliminating x,

$$u = \frac{v^2}{4c^2} - c^2, \tag{12.3}$$

which is a parabola. By sketching a few such curves (Fig. 12.2), it is not hard to see that D (i.e., $y > 0$) maps into the entire f plane except for the positive u axis, which

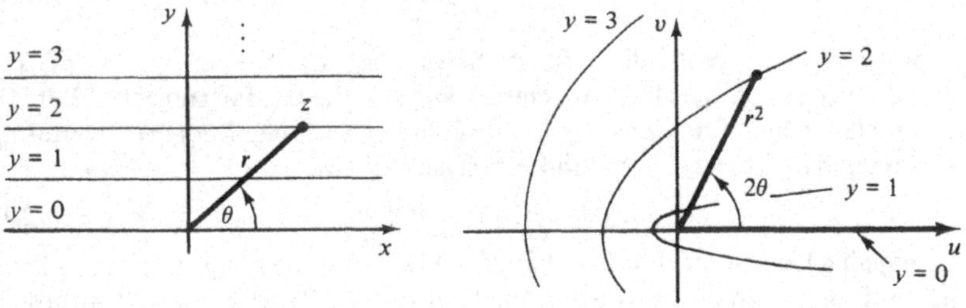

Figure 12.2. The mapping $f(z) = z^2$.

is the image of $y = 0$. Thus R is the entire f plane but with the nonnegative u axis "cut out" or deleted. Furthermore, f is not only single valued but also one to one.

If, on the other hand, D were the region $y \geq 0$, then R would be the entire f plane; f would still be single valued but no longer one to one, since the points $z =$

$\pm x$ (i.e., on $y = 0$) both map into the same point $f = x^2$. Note carefully the way that the upper half of the z plane "fans out" into all of the f plane. This is due to the doubling of angles,

$$z = re^{i\theta}, \qquad f(z) = z^2 = r^2 e^{i2\theta}, \tag{12.4}$$

so that the x axis "folds" into the nonnegative half of the u axis.

Finally, observe from (12.3) that the curves $y = \pm c$ both get mapped into the *same* parabola. So if D is allowed to be *all* of the z plane, then $y > 0$ maps into all of the f plane minus the nonnegative u axis, $y = 0$ maps into the nonnegative u axis, and $y < 0$ maps into all of the f plane minus the nonnegative u axis *again*. There is nothing wrong with this since f is still single valued; it just happens not to be one to one. ▮

Next, we must establish a limit concept if we are to talk about continuity, differentiation, infinite series, and so on. As in Part I,

$$\lim_{z \to z_0} f(z) = L \tag{12.5}$$

will mean that *to each* $\epsilon > 0$ (*no matter how small*) *there corresponds a* $\delta(\epsilon, z_0)$ *such that* $|f(z) - L| < \epsilon$ *whenever* $0 < |z - z_0| < \delta$, as shown in Fig. 12.3. (The radii ϵ and δ are *real* numbers.) Note, of course, that if z_0 happens to be a boundary point of D—for instance, the point z_1 in Fig. 12.3—then "... whenever $0 < |z - z_0| < \delta$" is understood to mean "... whenever $0 < |z - z_0| < \delta$ and z is in D."

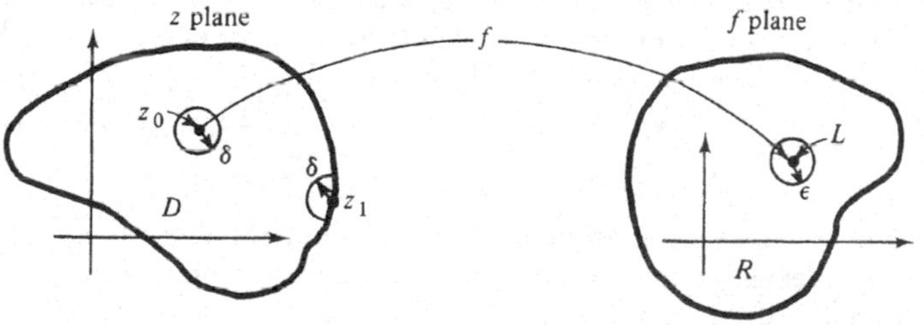

Figure 12.3. I shot an arrow into the air.

If in addition to having $\lim_{z \to z_0} f(z) = L$ we also have $f(z_0) = L$, we say that $f(z)$ is **continuous** at z_0. How would you define **uniform** continuity of f over D (Exercise 12.3)?

12.2. *DIFFERENTIATION AND THE CAUCHY-RIEMANN CONDITIONS FOR ANALYTICITY*

Following the real-variable definition of the derivative, we define

$$f'(z) = \lim_{\Delta z \to 0} \frac{f(z + \Delta z) - f(z)}{\Delta z}. \tag{12.6}$$

Observe that the domain D is a *two*-dimensional region and that $z + \Delta z$ must be allowed to approach z from *any* direction within D.[2]

In general, $\Delta z = \Delta x + i\,\Delta y$. Suppose, however, that we set $\Delta y = 0$ so that $\Delta z = \Delta x$—in other words, $z + \Delta z$ approaches z parallel to the x axis. Then (12.6) can be written

$$\begin{aligned} f'(z) &= \lim_{\Delta x \to 0} \frac{[u(x+\Delta x, y) + iv(x+\Delta x, y)] - [u(x, y) + iv(x, y)]}{\Delta x} \\ &= \lim_{\Delta x \to 0} \frac{u(x+\Delta x, y) - u(x, y)}{\Delta x} + i \lim_{\Delta x \to 0} \frac{v(x+\Delta x, y) - v(x, y)}{\Delta x} \\ &= u_x + iv_x. \end{aligned} \tag{12.7}$$

Alternatively, suppose that $\Delta x = 0$ so that Δz is $i\,\Delta y$—that is, $z + \Delta z$ approaches z parallel to the y axis. Then (12.6) becomes

$$\begin{aligned} f'(z) &= \lim_{\Delta y \to 0} \frac{[u(x, y+\Delta y) + iv(x, y+\Delta y)] - [u(x, y) + iv(x, y)]}{i\,\Delta y} \\ &= \frac{1}{i} u_y + v_y = v_y - iu_y. \end{aligned} \tag{12.8}$$

Since the limit is to be independent of the path of approach, it follows from (12.7) and (12.8) that $u(x, y)$ and $v(x, y)$ must be such that

$$u_x = v_y, \qquad u_y = -v_x \tag{12.9}$$

at a given point z if f' is to exist there; these are the well-known and important **Cauchy–Riemann** conditions. They are clearly *necessary* for differentiability but perhaps not sufficient, since, after all, we've considered only two specific paths of approach; see Exercise 12.4. More generally,

$$\begin{aligned} df &= du + i\,dv = (u_x\,dx + u_y\,dy) + i(v_x\,dx + v_y\,dy) \\ &= (u_x + iv_x)\,dx + (v_y - iu_y)\,i\,dy \end{aligned} \tag{12.10}$$

if u_x, u_y, v_x, v_y are continuous, and thus the chain differentials are justified. If the Cauchy–Riemann conditions are satisfied in addition, then $u_x + iv_x = v_y - iu_y \equiv F$, say, and (12.10) becomes

$$df = (F)(dx + i\,dy) = F\,dz;$$

it follows that

$$\frac{df}{dz} = F = u_x + iv_x = v_y - iu_y$$

for *any* direction of approach. Therefore we have the following *sufficient* conditions.

THEOREM 12.1. (*Sufficient Conditions for Differentiability*) *If the Cauchy–Riemann conditions $u_x = v_y$ and $u_y = -v_x$ are satisfied at (x, y) and u_x, u_y, v_x, v_y are continuous there, then f is differentiable at z and $f'(z) = u_x + iv_x = v_y - iu_y$.*

---

[2] Sometimes it is said that $f'(z)$ is the limit of $[f(z + \Delta z) - f(z)]/\Delta z$ as $z + \Delta z \longrightarrow z$ from all directions. But the idea of approaching a point "from all directions" leaves something to be desired, notwithstanding Stephen Leacock's Lord Ronald, who "... said nothing; he flung himself from the room, flung himself upon his horse and rode madly off in all directions."

Suppose that $f(z)$ is differentiable at z_0. If, in addition, it is differentiable throughout some neighborhood of z_0 (i.e., $|z - z_0| < \epsilon$ for some $\epsilon > 0$), we say that it is **analytic** at z_0; the terms *regular* and *holomorphic* are commonly used, too. You may wonder how f could possibly be differentiable at z_0 and yet *not* be differentiable throughout some neighborhood of z_0. Well, consider the following example.

Example 12.2. Consider

$$f(z) = |z|^2 = z\bar{z}. \tag{12.11}$$

Then $f = (x^2 + y^2) + 0i$, and so $u = x^2 + y^2$ and $v = 0$. Moreover,

$$u_x = 2x, \qquad v_x = 0$$

$$u_y = 2y, \qquad v_y = 0,$$

which means that $u_x = v_y$ on the line $x = 0$ and $u_y = -v_x$ on the line $y = 0$. Although u, v and the four partials are continuous everywhere, the Cauchy–Riemann conditions are satisfied only at $x = y = 0$. Thus $f = |z|^2$ is differentiable *only* at the point $z = 0$ and hence analytic *nowhere*. ∎

Such behavior, however, is hardly typical of the functions that we are apt to encounter, most of which are analytic everywhere or almost everywhere.

12.3 HARMONIC FUNCTIONS

If $f(z) = u + iv$ is analytic throughout some region D, then

$$u_x = v_y \quad \text{and} \quad u_y = -v_x \tag{12.12}$$

in D. Taking $\partial/\partial x$ of the first and $\partial/\partial y$ of the second,

$$u_{xx} = v_{yx} \quad \text{and} \quad u_{yy} = -v_{xy}.$$

It will turn out that if f is analytic, then the partial derivatives of u and v of *all* orders exist and are continuous. Accepting this statement on faith, it follows from page 127 of Chapter 7 that $v_{yx} = v_{xy}$; hence $u_{xx} + u_{yy} = 0$. Similarly, $\partial/\partial y$ of the first of (12.12) and $\partial/\partial x$ of the second reveal that v, too, satisfies Laplace's equation. Consequently,

$$\nabla^2 u = 0 \tag{12.13a}$$

and

$$\nabla^2 v = 0. \tag{12.13b}$$

Since u and v have continuous second-order partials and satisfy Laplace's equation, we call them **harmonic functions**. In fact, we call them *conjugate* harmonic functions because they constitute the real and imaginary parts of the same analytic function f. (The word "conjugate" here has nothing to do with the complex conjugate $\bar{z}$ of z.) Note that $\nabla u \cdot \nabla v = u_x v_x + u_y v_y = 0$ because of the Cauchy–Riemann conditions. Since ∇u is orthogonal to the $u =$ constant curves and ∇v is orthogonal to the $v =$ constant curves, it follows that *the $u =$ constant and $v =$ constant curves are mutually orthogonal.*

Given one of two conjugate harmonic functions, the other can be found by integrating the Cauchy–Riemann relations.

12.4. *SOME ELEMENTARY FUNCTIONS AND THEIR SINGULARITIES*

Among the following examples will be the functions e^z, $\sin z$ and $\cos z$. But what is the sine, for instance, of a complex number? The key is the *definition*

$$e^z = e^{x+iy} \equiv e^x(\cos y + i \sin y), \tag{12.14}$$

where e^x, $\cos y$ and $\sin y$ are the familiar functions from real-variable theory. Contained in (12.14), for the case where $x = 0$, is Euler's formula

$$e^{iy} = \cos y + i \sin y. \tag{12.15}$$

Replacing y by $-y$, we have

$$e^{-iy} = \cos y - i \sin y.$$

Adding and subtracting yield the results

$$\cos y = \frac{e^{iy} + e^{-iy}}{2} \quad \text{and} \quad \sin y = \frac{e^{iy} - e^{-iy}}{2i}, \tag{12.16}$$

which appeared in Part I.

Next, we *define*

$$\begin{aligned} \cos z &\equiv \frac{e^{iz} + e^{-iz}}{2}, \qquad \sin z \equiv \frac{e^{iz} - e^{-iz}}{2i} \\ \cosh z &\equiv \frac{e^{z} + e^{-z}}{2}, \qquad \sinh z \equiv \frac{e^{z} - e^{-z}}{2}, \end{aligned} \tag{12.17}$$

where the complex exponentials are meaningful as a result of (12.14); for example, $e^{iz} = e^{-y+ix} = e^{-y}(\cos x + i \sin x)$.

It should be emphasized that the *motivation* behind definition (12.14) is not particularly obvious. Let us accept it for the moment; later on (Chapter 14) we will see that (12.14) actually amounts to defining e^z as the so-called *analytic continuation* of e^x off the real axis and into the rest of the z plane. Similarly, $\cos z$ will simply be the analytic continuation of $\cos x$ off the real axis and so on.

Let us turn now to a few of the standard elementary functions.

Example 12.3.

$$f(z) = e^z. \tag{12.18}$$

Then

$$f(z) = e^{x+iy} = e^x \cos y + ie^x \sin y,$$

and therefore

$$\begin{aligned} u &= e^x \cos y, & v &= e^x \sin y, \\ u_x &= e^x \cos y, & v_x &= e^x \sin y, \\ u_y &= -e^x \sin y, & v_y &= e^x \cos y. \end{aligned}$$

These partials are continuous everywhere, and the conditions $u_x = v_y$ and $u_y = -v_x$ are satisfied everywhere; consequently, f is differentiable and thus analytic in the entire z plane; as a result, we say that it is an **entire** function. What is $f'(z)$? It is given by any of the expressions $u_x + iv_x$, $u_x - iu_y$, $v_y + iv_x$, or $v_y - iu_y$, which are all equivalent because of the Cauchy–Riemann relations. So

$$f'(z) = e^x \cos y + ie^x \sin y = e^x(\cos y + i \sin y) = e^{x+iy} = e^z. \tag{12.19}$$

Notice that along the x axis $e^z = e^x$ is purely real and coincides with the usual exponential function of real-variable theory, whereas on the y axis $e^z = e^{iy} = \cos y + i \sin y$, which is complex and oscillatory! ▮

Observe that the result (12.19) is exactly the same as the real-variable formula $d(e^x)/dx = e^x$. This situation is not surprising because the definition (12.6) of $f'(z)$ is identical to that for the real-variable case except that the x's are changed to z's. Similarly, it turns out that

$$\frac{d(\sin z)}{dz} = \cos z, \qquad \frac{d(\cos z)}{dz} = -\sin z, \tag{12.20}$$

and so on.

Example 12.4.

$$f(z) = \sin z. \tag{12.21}$$

Taking the "long route," we have

$$f(z) = \sin(x + iy) = \sin x \cos iy + \sin iy \cos x \tag{12.22a}$$

$$f(z) = (\sin x \cosh y) + i(\sinh y \cos x), \tag{12.22b}$$

so that

$$u = \sin x \cosh y, \qquad v = \sinh y \cos x,$$
$$u_x = \cos x \cosh y, \qquad v_x = -\sinh y \sin x,$$
$$u_y = \sin x \sinh y, \qquad v_y = \cosh y \cos x.$$

They are continuous everywhere and the Cauchy–Riemann conditions are satisfied everywhere; consequently, $\sin z$ is an entire function, and

$$f'(z) = \cos x \cosh y - i \sinh y \sin x = \cos x \cos iy - \sin iy \sin x$$
$$= \cos(x + iy) = \cos z$$

as claimed above in (12.20).

COMMENT 1. Note that on the x axis $\sin z$ reduces to $\sin x$, whereas on the y axis it is $\sin iy$ or $i \sinh y$. In particular, $\sin z$ is *not* bounded over the entire z plane in the way that $\sin x$ is bounded over the entire x axis, since on the y axis (for example) $\sin z = i \sinh y \longrightarrow i\infty$ as $y \longrightarrow \infty$. In fact, later we will see that *every* entire function, which is not simply a constant, is necessarily unbounded.

COMMENT 2. You may have accepted the step (12.22a) without question because it looks like the familiar trigonometric identity $\sin(A + B) = \sin A \cos B + \sin B \cos A$. Actually, we need to show that it *follows* from definitions (12.17) and (12.14). Instead let us verify the final result (12.22b):

$$\sin z = \frac{e^{iz} - e^{-iz}}{2i} = \frac{e^{-y+ix} - e^{y-ix}}{2i} = \frac{e^{-y}(\cos x + i \sin x) - e^{y}(\cos x - i \sin x)}{2i}$$
$$= \sin x \left(\frac{e^y + e^{-y}}{2}\right) + i \cos x \left(\frac{e^y - e^{-y}}{2}\right)$$
$$= \sin x \cosh y + i \sinh y \cos x$$

as claimed. ▮

Example 12.5.

$$f(z) = z^{-n} \qquad (n \text{ a positive integer}). \tag{12.23}$$

In this case, the derivative is

$$f'(z) = -\frac{n}{z^{n+1}} \tag{12.24}$$

except at $z = 0$, where the right-hand side "blows up"—that is, becomes infinite in magnitude as $z \longrightarrow 0$. Thus z^{-n} is analytic everywhere except at $z = 0$, where it is said to be **singular** (i.e., not analytic); we say that $z = 0$ is a **singular point** of the function. ∎

Just as the "usual" results (12.20) hold, so do the familiar real-variable rules for differentiating sums, products, and quotients carry over to analytic functions; for instance,

$$(f + g)' = f' + g', \tag{12.25a}$$

$$(fg)' = f'g + fg', \tag{12.25b}$$

$$\left(\frac{f}{g}\right)' = \frac{f'g - fg'}{g^2}, \tag{12.25c}$$

and

$$\frac{d}{dz}f[g(z)] = \frac{df}{dg}\frac{dg}{dz} \tag{12.25d}$$

whenever f and g are analytic functions (Exercise 12.11). It follows from (12.25a) that the sum of analytic functions (e.g., $e^z + \sin z$) is analytic, from (12.25*b*) that the product of analytic functions (e.g., $6e^z \sin z$) is analytic, and from (12.25*d*) that an analytic function of an analytic function (e.g., $e^{\sin z}$) is analytic. Yet it is *not* necessarily true that the *quotient* of analytic functions is analytic; for instance, even though $f = 1$ and $g = z$ are analytic everywhere, $f/g = 1/z$ is singular at $z = 0$.

Example 12.6.

$$f(z) = \frac{\cos z}{z}. \tag{12.26}$$

Here

$$f'(z) = -\frac{\cos z}{z^2} - \frac{\sin z}{z}, \tag{12.27}$$

which exists everywhere except perhaps at $z = 0$, where the denominators vanish. To see how the $(\sin z)/z$ term behaves at $z = 0$, note that

$$\sin z = \sin(x + iy) = \sin x \cosh y + i \sinh y \cos x \sim x + iy = z \tag{12.28}$$

as $x \longrightarrow 0$ and $y \longrightarrow 0$, since $\sin x = x - x^3/3! + \cdots$, $\cosh y = 1 + y^2/2! + \cdots$, and so on. Thus $(\sin z)/z \sim 1$ as $z \longrightarrow 0$, which is fine. Turning to the $(\cos z)/z^2$ term, however, we find [as in (12.28)] that $(\cos z)/z^2 \sim 1/z^2$, which "blows up" as $z \longrightarrow 0$. Thus the right-hand side of (12.27) exists for all z except $z = 0$, which is therefore a singular point of $f(z)$.

The reader may wonder why all the fuss in (12.28); in other words, why not just say

$$\sin z = z - \frac{z^3}{3!} + \cdots \sim z \tag{12.29}$$

directly instead of expressing $\sin z$ in terms of $\sin x$, $\cos x$, $\sinh y$, $\cosh y$ and then expanding them in Taylor series? The reason is that we have not yet established a Taylor series for functions of a *complex* variable. We *will*, and (12.29) will be found to hold, but not until Chapter 14. ▮

12.5. *MULTIVALUEDNESS AND THE NEED FOR BRANCH CUTS*

Here we continue our examination of a few of the elementary functions.

Example 12.7.

$$f(z) = \ln z. \tag{12.30}$$

As in the real-variable case, we *define* (12.30) to mean that $z = e^f$, where e^f has already been defined by (12.14). Differentiating $z = e^f$ with respect to z, with the help of (12.25d) and (12.19), we have $1 = e^f f'$ or[3]

$$f'(z) = e^{-f} = \frac{1}{z}. \tag{12.31}$$

This function blows up at $z = 0$, and so we are tempted to conclude that $\ln z$ is analytic everywhere except at $z = 0$. But $\ln z$ turns out to be a bit more devious than that.

Using the polar form for z, let us express

$$\ln z = \ln (re^{i\theta}) = \ln r + \ln e^{i\theta} = \ln r + i\theta, \tag{12.32}$$

where $\ln r$ is understood to be the "usual" logarithm from real-variable theory—for instance, $\ln 2 \doteq 0.693$. Next, for any given point z in the complex plane, r is uniquely defined but θ is *not*! For example, if $z = 1 + i$, then $\theta = \pi/4 + 2n\pi$ for *any* integer n (positive, negative, or zero). It follows from (12.32) that $\ln z$ is not single valued; in fact, it has infinitely many possible values. Since functions are always to be single valued, we definitely have a problem. We straighten matters out by *limiting the domain* such that θ, and hence $\ln z$, *is* single valued.

To illustrate, suppose that we delete the nonnegative real axis, $0 \leq x < \infty$ and $y = 0$—as might be accomplished picturesquely by cutting the plane along that line with a pair of scissors—and define θ to be zero, say, at any point Q on the upper edge of the cut, as shown in Fig. 12.4. Such a procedure uniquely determines θ, and hence $\ln z$, everywhere in the cut plane. For instance, what is θ at some point R on the bottom edge of the cut? To answer, we start at Q, where θ is *known* (namely,

---

[3]Here we use the little formula $1/e^f = e^{-f}$, which is certainly true for the case where f is real, but here f is complex and so the formula needs to be justified directly from the definition (12.14). Doing so is not hard:

$$e^{-f} = e^{-u-iv} = e^{-u}(\cos v - i \sin v)\left(\frac{\cos v + i \sin v}{\cos v + i \sin v}\right)$$

$$= \frac{e^{-u}(\cos^2 v + \sin^2 v)}{\cos v + i \sin v} = \frac{1}{e^u(\cos v + i \sin v)} = \frac{1}{e^f}.$$

The need to verify a formula like $1/e^f = e^{-f}$ directly from the definition of the complex exponential was also emphasized in Comment 2 following Example 12.4. Practically speaking, however, the usual identities do carry over to the complex case, and from here on we will simply use them without becoming involved in their verification. In any case, some of this verification is covered in Exercise 12.12.

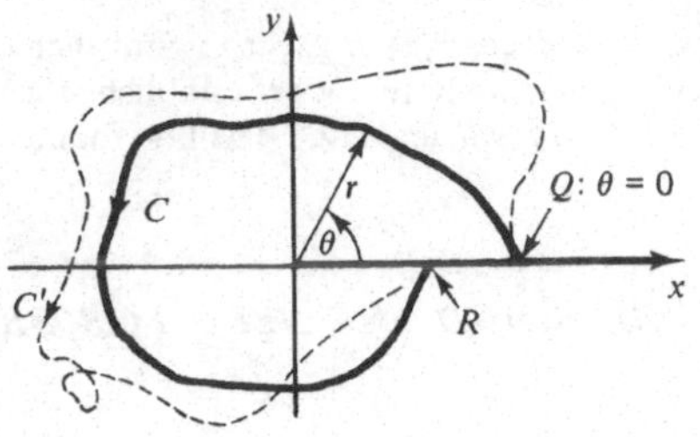

Figure 12.4. The cut domain for ln z.

$\theta = 0$), and take a walk from Q to R along the path C. As the representative point P moves along C from Q to R, we see that θ increases by 2π, and so $\theta = 2\pi$ at R. This result is independent of the shape of the path C, provided that we remain within the domain. In particular, we are *not allowed to cross the cut*, which constitutes a boundary curve; we must go around the origin! If, for definiteness, $x = 3$ at Q and $x = 2$ at R, then $\ln z = \ln 3 + 0i = \ln 3$ at Q and $\ln z = \ln 2 + 2\pi i$ at R.

Note carefully that $\ln z$ is *discontinuous* across the cut, since at any x location $\ln z$ equals $\ln x$ on the top edge and $\ln x + 2\pi i$ on the bottom edge. Indeed, if it were not, there would be no need for the cut!

COMMENT 1. Note also that we need not define $\theta \equiv 0$ at Q (i.e., on the top of the cut); we could have chosen θ to be any multiple of 2π there. With $\theta \equiv 2\pi$ on the upper edge, for instance, it would follow that $\theta = 4\pi$ on the bottom, and therefore $2\pi \le \theta \le 4\pi$ in the cut plane. Or with $\theta \equiv -6\pi$ on the upper edge it would follow that $\theta = -4\pi$ on the bottom.

COMMENT 2. Neither is it necessary to cut along the positive real axis. Several other suitable arrangements are shown in Fig. 12.5. The important point is that in each case the cut extends from infinity into the origin. If instead of going off to infinity,

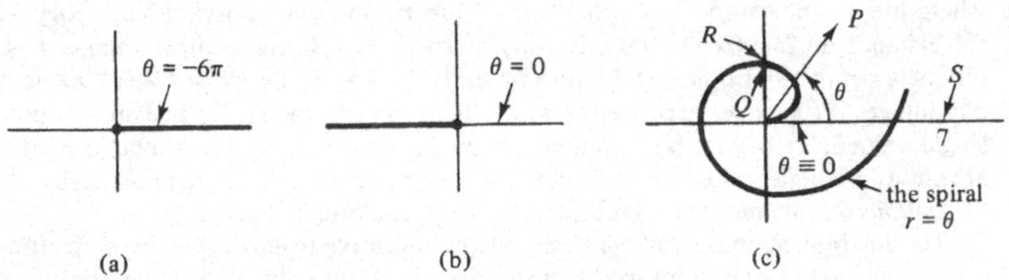

Figure 12.5. Several other definitions of the domain.

the cut were only of finite length, we could sneak around the end of it and encircle the origin any number of times; θ would then retain its multivaluedness and the cut would be useless. As a further illustration, refer again to Fig. 12.5 and note that at the points Q, R, and S in part (c) we have $\ln z = \ln(\pi/2) - 3\pi i/2$, $\ln z = \ln(\pi/2) + \pi i/2$, and $\ln z = \ln 7 + 2\pi i$, respectively. We emphasize that it does not matter at all whether the position vector to P crosses the cut (as it does in the figure). The important factor is that the representative point P itself does not cross the cut, which may therefore be thought of as a "barrier"; that is why θ must be 2π at S, *not* 0.

In addition, note that cuts do not necessarily limit θ to a maximum variation of 2π; for instance, in parts (a) and (b) we have $-6\pi \le \theta \le -4\pi$ and $-\pi \le \theta \le \pi$, but in part (c) we have $-2\pi < \theta < \infty$.

Finally, notice that the function values at a given location in the z plane are, in general, different for the three cases (a), (b), and (c). For example, at $z = i$ we have $\ln z = -11\pi i/2$, $\pi i/2$, and $-3\pi i/2$, respectively; and at $z = 2i$ we have $\ln z = \ln 2 - 11\pi i/2$, $\ln 2 + \pi i/2$, and $\ln 2 + \pi i/2$, respectively. Thus the different cuts define different functions! We call these functions **branches** of $\ln z$ and the cuts **branch cuts.** The origin, the encircling of which is the source of the multivaluedness, is known as a **branch point** of $\ln z$.

Clearly, there are an infinite number of possible branches of $\ln z$. We will call the branch defined by Fig. 12.4 the **principal value** of $\ln z$.

In summary, we see that it is necessary when talking about "$\ln z$" to say what we mean, that is, to specify a branch cut. (The cut selected is up to us. Subsequent examples in Part II will help to clarify this point.) Once that is done, we also see that the resulting function $\ln z$ is analytic throughout the cut plane. (Note that the singular point $z = 0$ is removed by the cut.) ▮

Example 12.8.

$$f(z) = \ln (z - a). \tag{12.33}$$

The story here is essentially identical to the one for $\ln z$. The only difference is that the branch point is now at $z = a$. It is therefore more convenient to use the polar variables ρ, ϕ (Fig. 12.6) rather than r, θ. One suitable branch cut is shown in Fig. 12.6. To illustrate, suppose that $a = 3 + 2i$. Then at $z = 3$, for example, $\ln (z - a) = \ln \rho + i\phi = \ln 2 + 3\pi i/2$. ▮

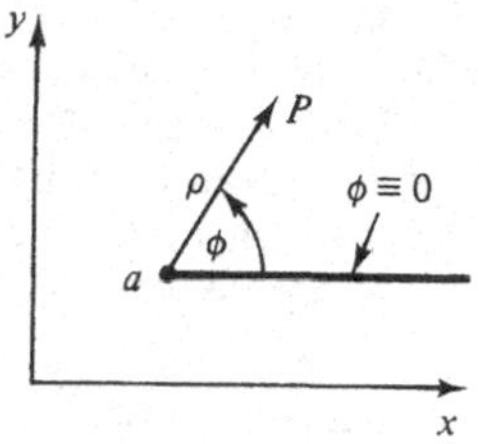

Figure 12.6. A branch cut for $\ln (z - a)$.

Before continuing with several other examples of functions with branch points, we might well wonder whether our earlier examples [(12.3) to (12.6)] involved branch points, too, which we somehow overlooked. The answer is: they did not. As representative, consider $f(z) = z^2 = r^2 e^{i2\theta}$. In the absence of a cut it is true that θ is multivalued, but the crucial point is that this does not lead to multivaluedness of the *function*, since

$$e^{i2(\theta + 2n\pi)} = e^{i2\theta} e^{i4n\pi} = e^{i2\theta}$$

for all n's. The same is true for $\sin z$, e^z, $(\cos z)/z$, and so on. The only cases that we encounter where the *function* is multivalued, and hence where a branch cut is needed, are $\ln (z - a)$, noninteger powers of $(z - a)$, and inverse trigonometric functions (Exercises 12.22 and 12.23). Having already looked at $\ln (z - a)$, we end our discussion by considering some noninteger powers of $(z - a)$.

Example 12.9.

$$f(z) = \sqrt{z}. \tag{12.34}$$

To illustrate the multivaluedness of $\sqrt{z}$, let us compute its values at $z = 1 + i$, say. We have

$$\sqrt{z} = \sqrt{r}\, e^{i\theta/2}, \tag{12.35}$$

where $\sqrt{r}$ is the "usual" square root—for instance, $\sqrt{4} = 2$. At $z = 1 + i$ we have $r = \sqrt{2}$ and $\theta = \pi/4, 9\pi/4, 17\pi/4, \ldots$ (as well as $-7\pi/4, -15\pi/4, \ldots$), and so (12.35) yields

$$\begin{aligned}\sqrt{1+i} &= 2^{1/4}e^{\pi i/8},\quad 2^{1/4}e^{9\pi i/8},\quad 2^{1/4}e^{17\pi i/8},\ldots\\ &= 2^{1/4}\left(\cos\frac{\pi}{8} + i\sin\frac{\pi}{8}\right),\\ &\quad -2^{1/4}\left(\cos\frac{\pi}{8} + i\sin\frac{\pi}{8}\right),\quad 2^{1/4}\left(\cos\frac{\pi}{8} + i\sin\frac{\pi}{8}\right),\ldots\end{aligned} \tag{12.36}$$

Thus if we start at $z = \sqrt{1+i}$ with $\theta = \pi/4$, we have $\sqrt{1+i} = 2^{1/4}(\cos \pi/8 + i \sin \pi/8)$. If we then encircle the origin once, say counterclockwise, we have $\theta = 9\pi/4$ and the function value $\sqrt{1+i} = -2^{1/4}(\cos \pi/8 + i \sin \pi/8)$. Encircling the origin repeatedly, we simply keep cycling between these two function values.

The same situation holds at *any* point z (except $z = 0$, where $\sqrt{z} = 0$, of course); therefore we say that $f(z) = \sqrt{z}$ is *double valued.* This situation is in contrast to $\ln z$, which never comes back to the same value and is infinite valued.

In any case, a branch cut is needed—for example, any of the cuts shown in Figs. 12.4 and 12.5. The one in Fig. 12.4 defines the so-called principal value of $\sqrt{z}$. Notice the discontinuity in $\sqrt{z}$ across the cut; for instance, at any point on the positive x axis, $\sqrt{z} = +\sqrt{x}$ on the upper edge of the cut ($\theta = 0$) and $-\sqrt{x}$ on the bottom edge ($\theta = 2\pi$). ∎

Example 12.10.

$$f(z) = z^{-4/3}. \tag{12.37}$$

Then

$$z^{-4/3} = (re^{i\theta})^{-4/3} = r^{-4/3}\left(\cos\frac{4\theta}{3} - i\sin\frac{4\theta}{3}\right),$$

and we find that this quantity cycles through *three* distinct values (for $z \neq 0$) as θ is repeatedly increased or decreased by 2π. Consider, for example, $z = i$ and 5:

$$\begin{aligned}f(i) = i^{-4/3} &= (1e^{\pi i/2})^{-4/3},\quad (1e^{5\pi i/2})^{-4/3},\quad (1e^{9\pi i/2})^{-4/3}\\ &= -\frac{1}{2} - \frac{\sqrt{3}}{2}i,\quad -\frac{1}{2} + \frac{\sqrt{3}}{2}i,\quad 1,\end{aligned}$$

and

$$\begin{aligned}f(5) = 5^{-4/3} &= (5e^{0i})^{-4/3},\quad (5e^{2\pi i})^{-4/3},\quad (5e^{4\pi i})^{-4/3}\\ &= 5^{-4/3},\quad 5^{-4/3}\left(-\frac{1}{2} - \frac{\sqrt{3}}{2}i\right),\quad 5^{-4/3}\left(-\frac{1}{2} + \frac{\sqrt{3}}{2}i\right),\end{aligned}$$

where $5^{-4/3}$ on the right-hand side is understood to be the "usual" value, $5^{-4/3} \doteq 0.1170$.

Hence a branch cut is needed, and $z^{-4/3}$ will be analytic everywhere in the cut plane. ∎

Example 12.11.

$$f(z) = \sqrt{(z-1)(z-2)}. \tag{12.38}$$

In this case, it is convenient to introduce *two* sets of polar variables, ρ_1, ϕ_1 and ρ_2, ϕ_2, as shown in Fig. 12.7. Thus

$$f(z) = \sqrt{\rho_1 \rho_2}\, e^{i\phi_1/2} e^{i\phi_2/2}. \tag{12.39}$$

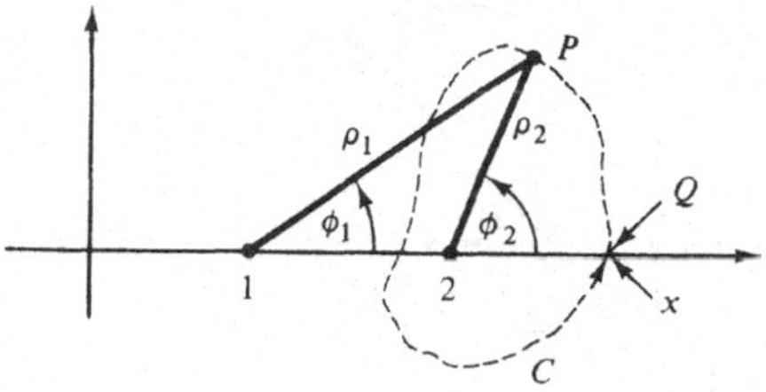

Figure 12.7. Polar variables for $\sqrt{(z-1)(z-2)}$.

Suppose that we start at some point Q, at a location x on the x axis for example, with $\phi_1 = \phi_2 = 0$. Then

$$f = \sqrt{(x-1)(x-2)}\,e^{i0}e^{i0} = \sqrt{(x-1)(x-2)} \tag{12.40}$$

at Q. Suppose next that we encircle the point $z = 2$ by following the counterclockwise path C (Fig. 12.7). Then ϕ_1 starts out as 0, increases to roughly 40°, decreases to roughly $-30°$, and increases back to 0 as we return to Q. So there is no net change and the final value is $\phi_1 = 0$. But ϕ_2 increases to 2π; consequently, on completion of the circuit C we have

$$f = \sqrt{(x-1)(x-2)}\,e^{i0}e^{i2\pi/2} = -\sqrt{(x-1)(x-2)}, \tag{12.41}$$

which is the negative of (12.39). Repeated encirclements of $z = 2$ will cause f at Q to cycle between the two values $\pm\sqrt{(x-1)(x-2)}$. Encirclements of $z = 1$ have the same effect.

Thus f has *two* branch points, one at $z = 1$ and one at $z = 2$. To render f single valued, we introduce cuts to *both* points and define ϕ_1 and ϕ_2 at some point in the plane, for example as shown in Fig. 12.8. To illustrate, what is $f(z)$, then, at $z = 0$

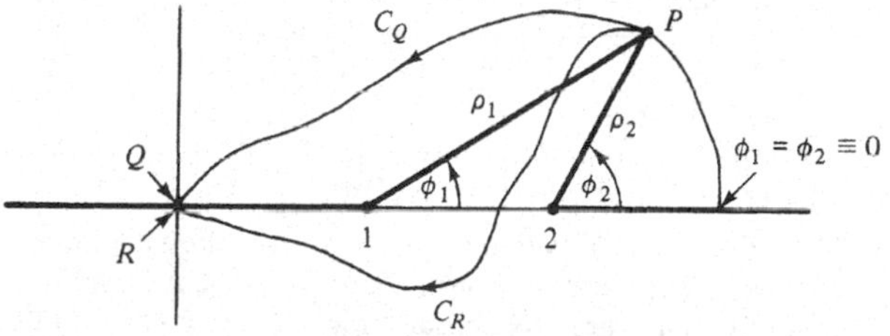

Figure 12.8. A branch cut for $\sqrt{(z-1)(z-2)}$.

on the top and bottom of the left-hand cut—that is, at Q and R? Following paths C_Q to Q and C_R to R, we have $\phi_1 = \phi_2 = \pi$ at Q and $\phi_1 = -\pi$, $\phi_2 = \pi$ at R. Also, $\rho_1 = 1$ and $\rho_2 = 2$ at Q and R; therefore

$$f(Q) = \sqrt{(1)(2)}\,e^{i\pi/2}e^{i\pi/2} = -\sqrt{2}$$

and

$$f(R) = \sqrt{(1)(2)}\,e^{-i\pi/2}e^{i\pi/2} = +\sqrt{2}.$$

Actually, in this example a *finite* cut, from $z = 1$ to $z = 2$, would also be suitable. Defining ϕ_1 and ϕ_2 as shown in Fig. 12.9, suppose that we start at Q ($x = 4$, for instance) and proceed along C until we return to Q.

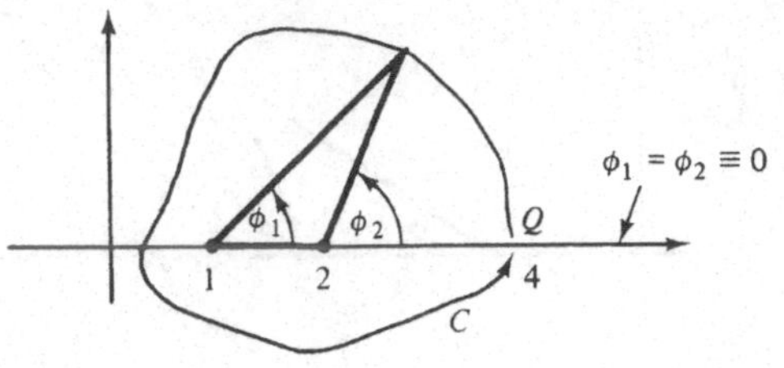

Figure 12.9. A finite branch cut.

$$\text{Initially:}\quad \rho_1 = 3, \qquad \rho_2 = 2, \qquad \phi_1 = \phi_2 = 0$$
$$f = \sqrt{(3)(2)}\, e^{i0} e^{i0} = \sqrt{6}.$$
$$\text{Finally:}\quad \rho_1 = 3, \qquad \rho_2 = 2, \qquad \phi_1 = \phi_2 = 2\pi$$
$$f = \sqrt{(3)(2)}\, e^{i2\pi/2} e^{i2\pi/2} = \sqrt{6}\,(-1)(-1) = \sqrt{6}. \tag{12.42}$$

The reason that f returns to the same value is that we are forced by the cut to encircle *both* branch points. This step causes *two* sign changes [note the $(-1)(-1)$ in (12.42)], the net effect of which is nil.

Note carefully that in this case the cut does *not* render ϕ_1 and ϕ_2 single valued (e.g., at Q we have $\phi_1 = \phi_2 = 2n\pi$ for $n = 0, \pm 1, \pm 2, \ldots$), but it is still a suitable cut because it renders the *function* single valued, which is all that we ask!

For further discussion of this example, see Exercise 12.17. ▮

EXERCISES

12.1. Find (and sketch) the ranges of the following functions $f(z)$ defined over the indicated domains.
(a) z^3 over $0 < \theta < \pi/2$, $0 < r < \infty$.
(b) $2iz$ over $0 \le x \le 2$, $0 \le y \le 1$.
(c) $1/z$ over $y > x$.
(d) $(i/z) + 2$ over $y > x$.
(e) $z^{1/3}$ over the entire z plane.

12.2. Find the images in the f plane of the following lines under the indicated mappings.
(a) The lines $y =$ constant for $0 \le y \le \pi/2$ and $-\infty < x < \infty$, under $f(z) = e^z$.
(b) The lines $x =$ constant for $-\infty < x < \infty$ and $0 \le y \le \pi/2$, under $f(z) = e^z$.
(c) The lines $y =$ constant for $0 \le x < 2\pi$ and $0 \le y < \infty$, under $f(z) = \sin z$.
(d) The lines $x =$ constant for $0 \le x < 2\pi$ and $0 \le y < \infty$, under $f(z) = \sin z$.

12.3. Complete the definitions below in terms of "ϵ, δ language."
(a) $f(z)$ is *continuous* at z_0 if . . . .
(b) $f(z)$ is *uniformly continuous* over D if . . . .

12.4. We've seen that the Cauchy–Riemann conditions are necessary for the existence of $f'(z)$ at a given point z. The example

$$f(z) = \frac{(x^3 - y^3) + i(x^3 + y^3)}{(x^2 + y^2)} \qquad z \neq 0$$
$$f(0) = 0,$$

due to S. Pollard (1928), shows that they are *not* sufficient. That is, show that whereas the Cauchy–Riemann conditions are satisfied at $z = 0$, $f'(0)$ does not exist.

12.5. Prove or disprove the following statements.

(a) If $f(z) = u + iv$ is analytic and not simply a constant, then $\overline{f(z)} = u - iv$ cannot be analytic.

(b) $f(z)$ and $\overline{f(z)}$ can both be analytic, in a given region, if and only if $f(z)$ is a constant.

(c) If $f(z)$ is analytic and not simply a constant, then $|f(z)|$ cannot be analytic.

12.6. Find the derivative $f'(z)$, where it exists, and state where f is analytic.

(a) $f = (1 - 4z^2)^8$ (b) $f = \dfrac{x + iy}{x^2 + y^2}$

(c) $f = \dfrac{1}{z^2 + 3iz - 2}$ (d) $f = \sin \dfrac{1}{z}$

(e) $f = |z| \sin z$ (f) $f = x + y + i(\sin x + \cos y)$

12.7. (*Cauchy–Riemann equations in polar coordinates*) (a) Expressing $f = u(r, \theta) + iv(r, \theta)$, where $z = re^{i\theta}$, show that

$$\lim \frac{f(z + \Delta z) - f(z)}{\Delta z} = e^{-i\theta}\left(\frac{\partial u}{\partial r} + i\frac{\partial v}{\partial r}\right)$$

as $\Delta z \longrightarrow 0$ along the constant-θ line through z and that it equals $e^{-i\theta}(-i\, \partial u/\partial\theta + \partial v/\partial\theta)/r$ as $\Delta z \longrightarrow 0$ along the constant-r line through z. Thus deduce the polar form of the Cauchy–Riemann conditions,

$$u_r = \frac{v_\theta}{r}, \qquad v_r = -\frac{u_\theta}{r}. \tag{12.43}$$

(b) Derive (12.43), alternatively, directly from (12.9) by chain differentiation.

12.8. Test the functions below for differentiability by using the polar Cauchy–Riemann equations (12.43).

(a) $f = z^{27}$ (b) $f = |z| z^2$

(c) $f = i - \sin \dfrac{1}{z}$ (d) $f = r^2 + \theta + 2ir^2\theta$

12.9. Show that if $f(z)$ is differentiable at z, then it must certainly be continuous there.

12.10. Determine whether the following functions $u(x, y)$ are harmonic and (where possible) find the conjugate function $v(x, y)$.

(a) $x^4 - 6x^2y^2$ (b) $e^x \cos y$

(c) $x^3 - 3xy^2$ (d) $3 + r^5 \cos 5\theta$

12.11. (a) Verify (12.25a) and (12.25b).

(b) Verify (12.25c) and (12.25d).

12.12. Using the *definitions* of e^z, $\sin z$, $\cos z$, $\sinh z$, $\cosh z$, and $\ln z$, verify that these formulas hold.

(a) $e^{z_1}e^{z_2} = e^{z_1+z_2}$

(b) $\cos(x + iy) = \cos x \cosh y - i \sin x \sinh y$

(c) $\sin(z_1 + z_2) = \sin z_1 \cos z_2 + \sin z_2 \cos z_1$

(d) $\sin^2 z + \cos^2 z = 1$

(e) $\cosh^2 z - \sinh^2 z = 1$

(f) $\sin iz = i \sinh z$ and $\cos iz = \cosh z$

(g) $\ln(re^{i\theta}) = \ln r + i\theta$

(h) **De Moivre's theorem,** $z^n = r^n(\cos n\theta + i \sin n\theta)$

12.13. (a) Show that if m/n is in simplest form and m and n are nonzero integers, then $z^{m/n}$ is n-valued.
(b) Show that if a is real and irrational, then z^a is infinite valued.

12.14. (a) Compute $\sqrt{(z-1)(z-2)}$, as defined by the branch cut of Fig. 12.9, at $z = 3/2$, both on the upper side of the cut and on the lower side.
(b) Would the branch cut of Fig. 12.9 be suitable for the function $[(z-1)(z-2)]^{1/3}$? Explain.

12.15. Identify the branch points, if any, of the following functions and sketch two suitable branch cuts.

(a) $\sin\sqrt{z}$ (b) $(\sqrt{z})^2$ (c) $\dfrac{1}{\sqrt{z^2-2z}}$
(d) $e^{2\ln z}$ (e) z^z (f) $z^{\pi i}$
(g) $\sqrt{z+1}+\sqrt{z-1}$ (h) $\sqrt{\dfrac{z-1}{z-2}}$ (i) $\cos\sqrt{z}$

12.16. Determine all possible values (in the form $a + ib$) of the given quantities.
(a) 1^i *Hint*: $1^i = e^{i\ln 1}$ (b) $\sin(2+3i)$
(c) $(1-i)^i$ (d) i^i

12.17. (a) Show that the n values of $z^{m/n}$ are equally spaced around the circle of radius $|z|$ with center at the origin; m and n are nonzero integers and m/n is in simplest form.
(b) Find the six values of $1^{1/6}$. *Hint*: One of them is clearly 1.
(c) Find the six values of $1^{7/6}$.

12.18. Find all *zeros* of the following functions $f(z)$—that is, solutions of $f(z) = 0$.
(a) e^z (b) $\sin z$ (c) $\cos z^3$
(d) $\sinh z$ (e) $\cosh z^2$ (f) $\ln z$
(g) $z^2 + 2z$ (h) $\ln(z^2 - i)$ (i) $e^{\sin z}$

12.19. Find all solutions of the equations shown.
(a) $e^z + 1 = 0$ (b) $\sin z = 3$ (c) $\ln z = 3\pi i$
(d) $z^3 + 1 = 0$ (e) $z^3 + i = 0$ (f) $z^4 = 1 + i$

12.20. Define a suitable branch cut for $z^{3/2}$. Now $d(z^{3/2})/dz = 3z^{1/2}/2$, but how are we to interpret $z^{1/2}$; for instance, can its branch cut be selected independently of the cut used for the original $z^{3/2}$ function? Explain.

12.21. To determine which sort of branch cuts are needed for the function $f = \sqrt{1+\sqrt{z}}$, consider f as two successive mappings—that is,

$$g(z) = \sqrt{z}, \qquad f(g) = \sqrt{1+g}.$$

Now $g(z)$ has a branch point at $z = 0$ and $f(g)$ has one at $g = -1$. Introducing cuts as shown in Fig. 12.10, let us compute f at the points P and Q for purposes of illustration. If the curve C, taking us from P to Q, is chosen as the circle $z = 4e^{i\theta}$, then the image C' will be the semicircle $g = \sqrt{z} = 2e^{i\theta/2}$ as shown. Thus P' and Q' are as shown, and the initial and final values of f are

$$\text{Initial:}\quad f = \sqrt{\rho e^{i\phi}} = \sqrt{3e^{i0}} = \sqrt{3}$$
$$\text{Final:}\quad f = \sqrt{\rho e^{i\phi}} = \sqrt{1e^{i\pi}} = i.$$

Observe that since the cut z plane maps into the upper half of the g plane, it would be inappropriate to have the cut in the g plane intrude into the upper half-plane. So

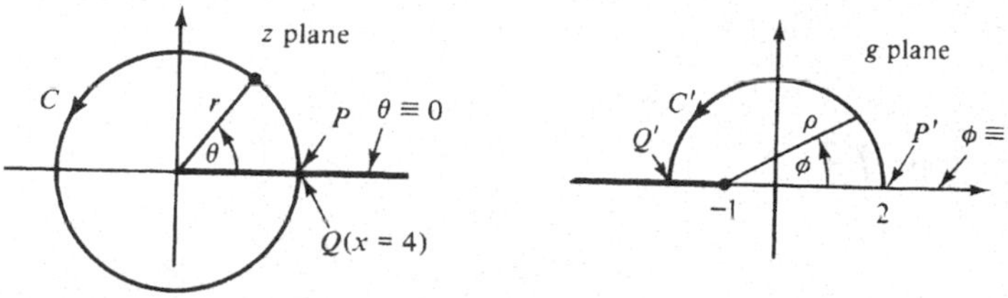

Figure 12.10

our choices for the two cuts are not entirely independent. *Problem.* Discuss (in about the same detail as above) the branch cuts needed for $f = \ln(\sqrt{z} - 1)$.

12.22. If $f = \sin^{-1} z$, then

$$z = \sin f = \frac{e^{if} - e^{-if}}{2i},$$

and so

$$e^{2if} - 2ize^{if} - 1 = 0, \qquad e^{if} = iz \pm \sqrt{1 - z^2}.$$

Moreover,

$$f = \sin^{-1} z = -i \ln (iz + \sqrt{1 - z^2}),$$

where we've replaced the $\pm$ by $+$ for simplicity, since we will need to select a branch of $\sqrt{1 - z^2}$ in any event. In this manner, show also that

(a) $\cos^{-1} z = -i \ln (z + \sqrt{z^2 - 1})$ (b) $\tan^{-1} z = \frac{i}{2} \ln \left(\frac{z + i}{z - i}\right)$

(c) $\sinh^{-1} z = \ln (z + \sqrt{z^2 + 1})$ (d) $\cosh^{-1} z = \ln (z + \sqrt{z^2 - 1})$

(e) $\tanh^{-1} z = \frac{1}{2} \ln \left(\frac{1 + z}{1 - z}\right)$

12.23. Show that $\cos^{-1} z = -i \ln (z + \sqrt{z^2 - 1})$ is infinite valued and discuss the introduction of branch cuts to render it single valued.

12.24. Return to and solve Exercise 9.47 of Chapter 9. (The solution may be more obvious now than before.)

12.25. Make up two examination-type questions on Chapter 12.

Chapter 13

Integration, Cauchy's Theorem, and the Cauchy Integral Formula

In this chapter we meet contour integrals for the first time and obtain two crucial integral theorems, Cauchy's theorem and the Cauchy integral formula.

13.1. INTEGRATION IN THE COMPLEX PLANE

Before continuing, the reader may wish to review the discussion of line integrals, simple closed curves, and so on in Sections 9.1 and 9.2.

In the spirit of definitions (1.16) and (9.2) we *define* the integral of $f(z) = u + iv$ over a curve C in the complex z plane as

$$\int_C f(z)\,dz \equiv \lim_{|P|\to 0} \sum_{k=1}^{n} f(z_{k-1})(z_k - z_{k-1}), \tag{13.1}$$

where $|P| \equiv \max_{1\le k\le n} |z_k - z_{k-1}|$ is called the norm of the partition; C is oriented and is assumed throughout Part II to be piecewise smooth, although not necessarily closed.

Why are such integrals of practical interest? One extremely important application already encountered is the Laplace inversion formula (6.29), which is a contour integral in the complex s plane; C is the infinite straight line shown in Fig. 6.4. Furthermore, *real* definite integrals are often evaluated most conveniently by considering instead a related contour integral in a complex plane. In fact, it is fortunate that apparently difficult contour integrals can often be evaluated with remarkable ease, primarily because of the powerful Residue Theorem.

Definition (13.1) is of no direct help in the practical evaluation of contour integrals, and we are not ready for the Residue Theorem, but two elementary lines of approach can be mentioned here.

First, we can express the integral in terms of *real* line integrals, according to

$$\int_C f(z)\,dz = \int_C (u + iv)(dx + i\,dy) = \int_C (u\,dx - v\,dy) + i\int_C (v\,dx + u\,dy) \quad (13.2)$$

or using polar coordinates,

$$\begin{aligned}\int_C f(z)\,dz &= \int_C (u + iv)[(\cos\theta\,dr - r\sin\theta\,d\theta) + i(\sin\theta\,dr + r\cos\theta\,d\theta)]\\ &= \int_C [(u\cos\theta - v\sin\theta)\,dr - (v\cos\theta + u\sin\theta)\,r\,d\theta]\\ &\quad + i\int_C [(v\cos\theta + u\sin\theta)\,dr + (u\cos\theta - v\sin\theta)r\,d\theta], \qquad (13.3)\end{aligned}$$

since $dz = d(re^{i\theta}) = e^{i\theta}(dr + ir\,d\theta)$. For simple cases, these integrals can be evaluated as in Chapter 9.

Secondly, we have the following analog of the Fundamental Theorem of the Integral Calculus, (1.19).

THEOREM 13.1. (Fundamental Theorem of the Complex Integral Calculus) If $f'(z)$ exists and is continuous in a simply connected region D and if C is a path from z_0 to z lying within D (Fig. 13.1), then

$$\int_C f(z)\,dz \equiv F(z)$$

is analytic and $F'(z) = f(z)$ for all z in D. Conversely, if $F'(z) = f(z)$ is continuous in D, then $\int_C f(z)\,dz = F(z)$.

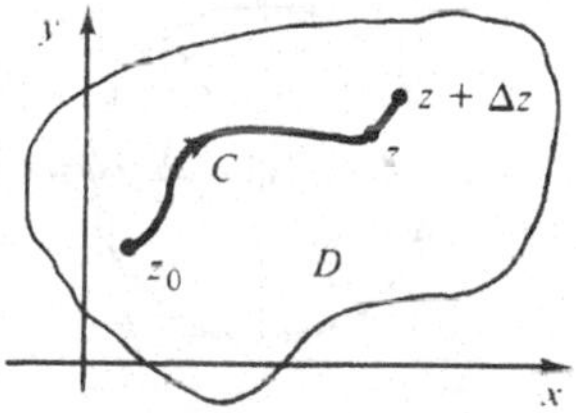

Figure 13.1. Fundamental theorem.

Proof. First, verification is needed that the integral is a single-valued function of z—in other words, that it is independent of the choice of the path C and thus depends only on the endpoints z_0 and z. Considering the form (13.2), note that with $u\,dx - v\,dy \equiv \mathbf{V}\cdot d\mathbf{r}$ and $v\,dx + u\,dy \equiv \mathbf{W}\cdot d\mathbf{r}$, the vector fields $\mathbf{V}$ and $\mathbf{W}$ are irrotational and have continuous first partials, since we've assumed that $f'(z)$ is continuous, and so [recalling (9.64)] the line integrals in (13.2) *are* independent of path. To emphasize this point, we write

$$\int_C f(z)\,dz = \int_{z_0}^{z} f(z)\,dz = F(z).$$

Introducing a dummy integration variable ζ to avoid confusion, we show that $F'(z) = f(z)$.

$$\frac{F(z+\Delta z)-F(z)}{\Delta z}=\frac{1}{\Delta z}\left[\int_{z_0}^{z+\Delta z} f(\zeta)\,d\zeta-\int_{z_0}^{z} f(\zeta)\,d\zeta\right]$$
$$=\frac{1}{\Delta z}\left[\int_{z_0}^{z} f(\zeta)\,d\zeta+\int_{z}^{z+\Delta z} f(\zeta)\,d\zeta-\int_{z_0}^{z} f(\zeta)\,d\zeta\right]$$
$$=\frac{1}{\Delta z}\int_{z}^{z+\Delta z} f(\zeta)\,d\zeta=\frac{1}{\Delta z}\int_{z}^{z+\Delta z}[f(\zeta)-f(z)+f(z)]\,d\zeta$$
$$=\frac{f(z)}{\Delta z}\int_{z}^{z+\Delta z} d\zeta+\frac{1}{\Delta z}\int_{z}^{z+\Delta z}[f(\zeta)-f(z)]\,d\zeta$$
$$=f(z)+I,\text{ say.} \tag{13.4}$$

Then if we take the path from z to $z+\Delta z$ to be a straight line and denote max $|f(\zeta)-f(z)|$ over this path by M, surely

$$\left|\int_{z}^{z+\Delta z}[f(\zeta)-f(z)]\,d\zeta\right|\leq M|\Delta z|, \tag{13.5}$$

so that $|I|\leq M$. But since $f(z)$ is continuous, M must tend to zero as $\Delta z\longrightarrow 0$, and therefore (13.4) becomes $F'(z)=f(z)$. Proof of the converse part of the theorem is easy and will be omitted.

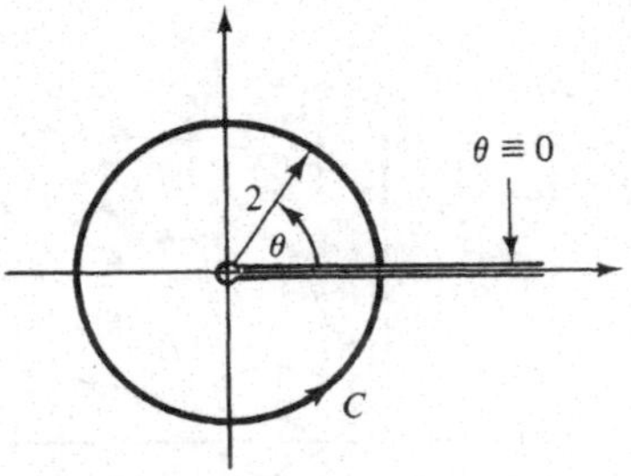

Figure 13.2. Branch cut and contour for Example 13.2.

Example 13.1.

$$\int_C \sqrt{z}\,dz,$$

where C is the circle $z=2e^{i\theta}$ from $\theta=0$ to 2π and $\sqrt{z}$ is defined by the branch cut shown in Fig. 13.2.

Converting to real line integrals, we have

$$\int_C \sqrt{z}\,dz=\int_C \sqrt{2}\,e^{i\theta/2}\,2ie^{i\theta}\,d\theta$$

(since $dr=0$ on C)

$$=-2\sqrt{2}\int_0^{2\pi}\sin\frac{3\theta}{2}\,d\theta$$
$$+\,i2\sqrt{2}\int_0^{2\pi}\cos\frac{3\theta}{2}\,d\theta=-\frac{8\sqrt{2}}{3}. \tag{13.6}$$

Alternatively (and more simply), it follows from Theorem 13.1 that

$$\int_C \sqrt{z}\,dz=\frac{2}{3}z^{3/2}\Big|_{2e^{i0}}^{2e^{i2\pi}}=\frac{2}{3}2^{3/2}(e^{3\pi i}-e^{0i})=-\frac{8\sqrt{2}}{3}. \quad \blacksquare$$

Unfortunately, our integrals are seldom so simple. Consider, for instance, the Laplace inversion

$$\mathbf{L}^{-1}\left(\frac{s}{s^2+a^2}\right)=\frac{1}{2\pi i}\int_{\gamma-i\infty}^{\gamma+i\infty}\frac{se^{st}}{s^2+a^2}\,ds=\cos at \tag{13.7}$$

from Appendix A. An attempt to derive this result by means of the two foregoing methods will convince us that more powerful techniques are needed.

13.2. BOUNDS ON CONTOUR INTEGRALS

It is often important to be able to obtain an upper bound on the magnitude of a contour integral, as in (13.5) for example.

Using the triangle inequality of Exercise 11.6(b), we have from definition (13.1)

$$\left|\int_C f(z)\,dz\right| \equiv \lim_{|P|\to 0}\left|\sum_{k=1}^{n} f(z_{k-1})(z_k - z_{k-1})\right|$$

$$\leq \lim_{|P|\to 0}\sum_{k=1}^{n} |f(z_{k-1})(z_k - z_{k-1})|$$

$$= \lim_{|P|\to 0}\sum_{k=1}^{n} |f(z_{k-1})||z_k - z_{k-1}| = \int_C |f(z)||dz|. \tag{13.8}$$

Furthermore, if $|f(z)| \leq M$ over C, then

$$\left|\int_C f(z)\,dz\right| \leq M\int_C |dz| = ML, \tag{13.9}$$

where L is the length of C. The two bounds (13.8) and (13.9) will be quite useful.

13.3. CAUCHY'S THEOREM

In its simplest and original form Cauchy's theorem states that *if $f'(z)$ exists and is continuous inside and on a closed curve C, then*

$$\int_C f(z)\,dz = 0.$$

For under the stated assumptions (i.e., the Cauchy–Riemann conditions plus the continuity of the partials u_x, u_y, v_x, v_y) we have, with the help of Green's theorem (9.60),

$$\int_C f(z)\,dz = \int_C (u\,dx - v\,dy) + i\int_C (u\,dy + v\,dx)$$

$$= \int_S \left(-\frac{\partial v}{\partial x} - \frac{\partial u}{\partial y}\right) d\sigma + i\int_S \left(\frac{\partial u}{\partial x} - \frac{\partial v}{\partial y}\right) d\sigma = 0,$$

since the two integrands are identically zero over the region S enclosed by C. Actually, Goursat was able to show (in 1900) that it is not necessary to assume the continuity of the partials. Thus[1]

THEOREM 13.2. (Cauchy–Goursat Theorem) If $f(z)$ is analytic inside and on a closed curve C, then

$$\int_C f(z)\,dz = 0. \tag{13.10}$$

Similarly, Theorem 3.1 can also be strengthened; specifically, the words "If $f'(z)$ exists and is continuous" can be reduced to "If $f'(z)$ exists" or, equivalently, "If $f(z)$ is analytic."

---

[1]We state Theorem 13.2 without proof. For a proof and further discussion, see the very readable little book by K. Knopp: *Theory of Functions, Part I*, Dover, New York, 1945, Chapter 4.

Indeed, we will show in Section 13.5 that the continuity of $f'(z)$, as well as its repeated differentiability, is a *consequence* of Theorem 13.2! First, however, we note the following important and immediate consequence of Theorem 13.2.

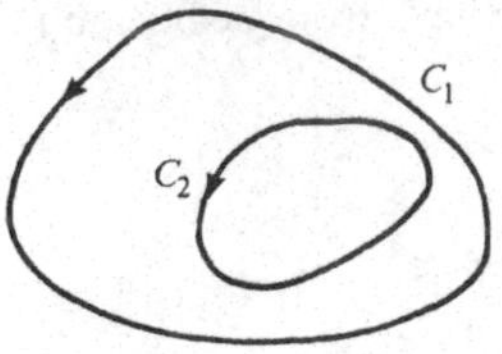

Figure 13.3 Deformation of the contour.

COROLLARY. (*Deformation of the Contour*) *If $f(z)$ is analytic between and on the closed contours C_1 and C_2 (Fig. 13.3), then*

$$\int_{C_1} f(z)\,dz = \int_{C_2} f(z)\,dz. \tag{13.11}$$

In other words, the contour may be deformed continuously without changing the value of the integral, provided that we cross no singularities of $f(z)$. Proof is left for Exercise 13.1.

13.4. *AN IMPORTANT LITTLE INTEGRAL*

Consider the integral

$$I = \int_C (z-a)^n\,dz, \tag{13.12}$$

where n is an integer, C is closed (and counterclockwise, say), and a lies inside (but not on) C. This integral occurs often, and it will be best to evaluate it before proceeding.

First, suppose that n is nonnegative; that is, $n = 0, 1, 2, \ldots$. Then $(z-a)^n$ is analytic everywhere, and so $I = 0$ by Cauchy's theorem. (*By "Cauchy's theorem" we will mean the Cauchy–Goursat Theorem 13.2.*)

Next, suppose that n is negative but not -1; that is, $n = -2, -3, \ldots$. Then

$$I = \frac{(z-a)^{n+1}}{n+1}\bigg|_{z_1}^{z_2},$$

where z_1 and z_2 are the initial and final points. But since C is closed, we have $z_2 = z_1$—that is, they coincide. And since $(z-a)^{n+1}$ is *single valued*, it follows that $I = 0$ again (in spite of the fact that the integrand is not analytic inside C, which means that Cauchy's theorem does not apply).

Finally, suppose that $n = -1$. If we deform the contour to a circle of radius ρ centered at $z = a$ [Fig. 13.4(a)], and denote $z - a = \rho e^{i\phi}$, then $dz = i\rho e^{i\phi}\,d\phi$ on $\rho =$ constant, and so

$$I = \int_C \frac{dz}{z-a} = \int_{C'} \frac{dz}{z-a} = \int_0^{2\pi} \frac{i\rho e^{i\phi}\,d\phi}{\rho e^{i\phi}} = 2\pi i.$$

Alternatively,

$$I = \int_C \frac{dz}{z-a} = \ln(z-a)\bigg|_{z_1}^{z_2} = (\ln\rho + i\phi)\bigg|_{z_1}^{z_2}.$$

Regardless of the choice of the branch cut for $\ln(z-a)$ [Fig. 13.4(b)], we have $\ln\rho\Big|_{z_1}^{z_2} = 0$, since $\rho_1 = \rho_2$, and $i\phi\Big|_{z_1}^{z_2} = i\,\Delta\phi = 2\pi i$, so that $I = 2\pi i$, as found above.

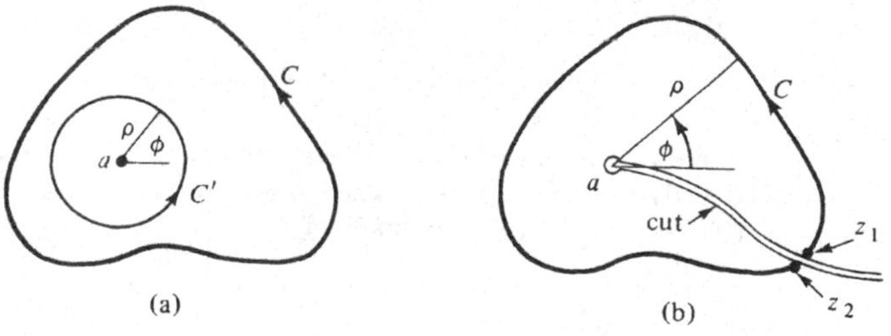

Figure 13.4. The integral (13.12).

Thus for all integer values of n we have

$$\int_C (z-a)^n\,dz = \begin{cases} 0, & n \neq -1 \\ 2\pi i, & n = -1. \end{cases} \tag{13.13}$$

13.5. THE CAUCHY INTEGRAL FORMULA

Suppose that $f(z)$ is analytic inside and on a closed contour C.[2] We will now use Cauchy's theorem to derive the remarkable fact that $f(z)$ is completely determined throughout the interior of C by its values along the boundary curve C.

THEOREM 13.3. (The Cauchy Integral Formula) If $f(z)$ is analytic inside and on a simple closed ccw[3] curve C and if the point a is inside (but not on) C, then

$$\frac{1}{2\pi i}\int_C \frac{f(z)}{z-a}\,dz = f(a). \tag{13.14}$$

Proof. Rewrite

$$\int_C \frac{f(z)}{z-a}\,dz = \int_C \frac{f(a)}{z-a}\,dz + \int_C \frac{f(z)-f(a)}{z-a}\,dz \equiv I + J,$$

say. Since $f(a)$ is a constant, it follows from (13.13) that $I = 2\pi i f(a)$. Thus it remains to show that $J = 0$.

To do so, let us deform C to C', as in Fig. 13.4, where ρ is arbitrarily small. Since f is continuous inside C (recall Exercise 12.9), and hence at a, it follows that for any given $\epsilon > 0$ (no matter how small) we can choose ρ small enough so that $|f(z) - f(a)| < \epsilon$ for all z on C'. According to (13.9), then,

$$|J| < \frac{\epsilon}{\rho}2\pi\rho = 2\pi\epsilon.$$

Since ϵ is arbitrarily small, it must be true that $J = 0$.

[2]Recall that all our contours are assumed to be piecewise smooth unless stated otherwise. In this case, C is also closed so that it is a piecewise smooth closed curve, or according to the terminology introduced in Section 9.2, it is a "simple closed curve."

[3]We will use cw and ccw for clockwise and counterclockwise, respectively. If a direction is not specified, a curve is generally *understood* to be ccw.

Example 13.2. Let us use the Cauchy integral formula (13.14) to evaluate the integral

$$I = \int_C \frac{e^{-z}}{z+2}\, dz, \tag{13.15}$$

where C is the circle $|z| = 3$, taken counterclockwise. Identifying $f(z) = e^{-z}$ (which *is* analytic inside and on C) and $a = -2$ (which *is* inside C), we have, according to (13.14), $I = 2\pi i f(-2) = 2\pi i e^2$. ▮

Let us rewrite (13.14) as

$$f(z) = \frac{1}{2\pi i} \int_C \frac{f(\zeta)}{\zeta - z}\, d\zeta, \tag{13.16}$$

to emphasize that it holds for *every* point z inside C. Forming the difference quotient,

$$\begin{aligned}
\frac{f(z+\Delta z) - f(z)}{\Delta z} &= \frac{1}{2\pi i\, \Delta z} \int_C f(\zeta)\left[\frac{1}{\zeta - z - \Delta z} - \frac{1}{\zeta - z}\right] d\zeta \\
&= \frac{1}{2\pi i} \int_C \frac{f(\zeta)}{(\zeta - z - \Delta z)(\zeta - z)}\, d\zeta \\
&= \frac{1}{2\pi i} \int_C \frac{f(\zeta)}{(\zeta - z)^2}\, d\zeta + \Delta I, \qquad (13.17)
\end{aligned}$$

where

$$\begin{aligned}
\Delta I &= \frac{1}{2\pi i} \int_C \left(\frac{1}{\zeta - z - \Delta z} - \frac{1}{\zeta - z}\right) \frac{f(\zeta)\, d\zeta}{\zeta - z} \\
&= \frac{\Delta z}{2\pi i} \int_C \frac{f(\zeta)\, d\zeta}{(\zeta - z)^2(\zeta - z - \Delta z)}, \qquad (13.18)
\end{aligned}$$

we hope to show that $\Delta I \longrightarrow 0$ as $\Delta z \longrightarrow 0$; then we can conclude from (13.17) that

$$f'(z) = \frac{1}{2\pi i} \int_C \frac{f(\zeta)}{(\zeta - z)^2}\, d\zeta. \tag{13.19}$$

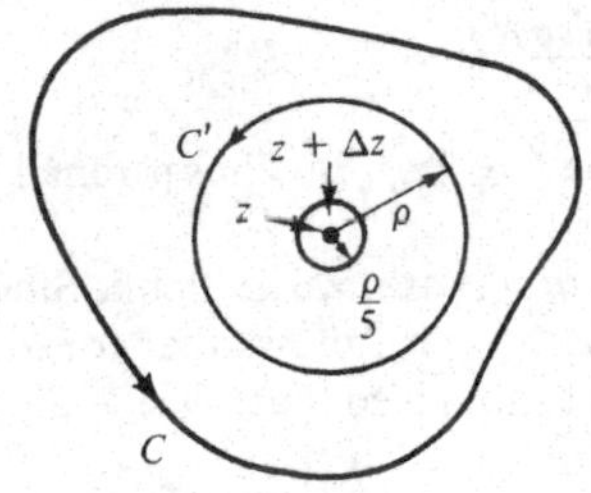

Figure 13.5. Bounding the integral ΔI.

To do so, let us deform C in (13.18), for convenience, to a circle C' with its center at z and its radius ρ small enough so that C' lies entirely inside C (Fig. 13.5). Choosing $|\Delta z| < \rho/5$, say, it follows that $|\zeta - z - \Delta z| > 4\rho/5$ for all ζ on C'. Moreover, there must certainly be some finite constant m such that $|f(\zeta)| < m$ on C', since f is analytic on C'. Therefore applying the bound (13.9),

$$|\Delta I| < \frac{|\Delta z|}{2\pi} \frac{m}{\rho^2(4\rho/5)} 2\pi\rho,$$

and we see that $\Delta I \longrightarrow 0$ as $\Delta z \longrightarrow 0$; hence (13.19) is, in fact, correct.

Similarly, if we form the difference quotient $[f'(z + \Delta z) - f'(z)]/\Delta z$, we can show (with no further assumptions on f) that

$$f''(z) = \frac{2}{2\pi i} \int_C \frac{f(\zeta)}{(\zeta - z)^3}\, d\zeta,$$

and, in fact, that for *all* $n = 0, 1, 2, \ldots,$

$$f^{(n)}(z) = \frac{n!}{2\pi i}\int_C \frac{f(\zeta)}{(\zeta - z)^{n+1}}\,d\zeta. \tag{13.20}$$

Note carefully that *having assumed only that f(z) is once differentiable* (i.e., *analytic inside and on C*), *we have demonstrated that it possesses derivatives of all orders!* This result finally explains our comments following Theorem 13.2, since the existence of $f''(z)$ obviously implies the continuity of $f'(z)$. (Recall Exercise 12.9.[4])

This striking result has no analog in real-variable theory. For instance, observe that $f(x) = x^2H(x)$ (where H is the Heaviside function) is once differentiable but no more; $f(x) = x^3H(x)$ is twice differentiable but no more, and so on.

EXERCISES

13.1. Prove the corollary to the Cauchy–Goursat theorem—that is, equation (13.11).

13.2. Evaluate by two different methods.

(a) $\int_{1-i}^{2+2i} \cos z\,dz$ (b) $\int_{i}^{2i} (z^2 - 2\sin z)\,dz.$

13.3. Consider $I = \int_C dz/z$, where C is the ccw semicircle $|z| = 2$, from $-2i$ to $+2i$. Surely

$$I = \int \frac{d(2e^{i\theta})}{2e^{i\theta}} = \int_{-\pi/2}^{\pi/2} i\,d\theta = \pi i.$$

Alternatively, $I = \ln z\Big|_{-2i}^{2i}$. Is the choice of the branch cut for $\ln z$ completely arbitrary? Discuss.

13.4. Show that if $f(z)$ is analytic inside and on a simple closed ccw curve C, then

$$\int_C \frac{f(\zeta)}{\zeta - z}\,d\zeta = \begin{cases} 2\pi i f(z) & \text{if } z \text{ is inside } C \\ 0 & \text{if } z \text{ is outside } C. \end{cases}$$

13.5. Evaluate $\int_C (z^2 - 3z + 5)\,dz/z$, where C is the cw circle $|z| = 3$.

13.6. Evaluate the following integrals, where C is the ccw circle $|z| = 1$. Repeat for the circle $|z| = 2$.

(a) $\int_C (z^2 - \sin z)\,dz$ (b) $\int_C \frac{\sin z}{z}\,dz$

(c) $\int_C \frac{\sin z}{(z+4)^2}\,dz$ (d) $\int_C \frac{dz}{z^2 + z + 2}$

(e) $\int_C \frac{\sinh z}{z}\,dz$ (f) $\int_C \frac{\cosh z}{z}\,dz$

(g) $\int_C \frac{dz}{z^2(z^2+3)}$ (h) $\int_C \frac{\sin z}{z(z^2+2)}\,dz$

[4] Note further that an additional implication is that *the conditions in Theorem 12.1* (i.e., the Cauchy–Riemann conditions plus continuity of the partials) *are both necessary and sufficient!*

13.7. Evaluate the integrals below with the help of (13.20), where C is the ccw unit circle.

(a) $\displaystyle\int_C \frac{\sin z}{z^2}\,dz$ (b) $\displaystyle\int_C \frac{\sin z}{z^3}\,dz$

(c) $\displaystyle\int_C \frac{z^2 + z}{(2z + 1)^3}\,dz$ (d) $\displaystyle\int_C \frac{e^z}{z^5}\,dz$

13.8. (a) Evaluate the integral

$$\int_C \left(\frac{z + 4}{z - 4}\right)\frac{e^z}{\sin z}\,dz,$$

where C is the ccw rectangle with corners at $-2 - i$, $-2 + i$, $2 + i$, $2 - i$. *Hint:* Write $1/\sin z$ as $(z/\sin z)(1/z)$ and note that the first factor is analytic inside and on C.

(b) Same as part (a), where C's corners are at $3\pi/2 \pm i$ and $\pi/2 \pm i$.

13.9. If $I = \int_C e^z\,dz/(z + 1)^3$, where C is the cw circle $|z| = 3$, use (13.9) to show that $|I| \leq 2\pi(3)e^3/8 = 3\pi e^3/4$.

13.10. (a) If C is a circle of radius ρ with center at z and M is the maximum of $|f(\zeta)|$ on C, then show from (13.20) that

$$|f^{(n)}(z)| \leq \frac{n!\,M}{\rho^n}. \tag{13.21}$$

(b) **(Liouville's theorem)** Use (13.21) with $n = 1$ to prove Liouville's theorem: *If f is entire and bounded for all z, then f is a constant.*

(c) **(Fundamental Theorem of Algebra)** Use Liouville's theorem to prove the Fundamental Theorem of Algebra: *If $P(z)$ is a polynomial of degree one or greater,*

$$P(z) = a_0 + a_1 z + \cdots + a_n z^n \qquad (a_n \neq 0)$$

then $P(z) = 0$ has at least one root. Hint: Suppose that $P(z)$ is nonzero everywhere. Then $f(z) \equiv 1/P(z)$ is analytic everywhere and tends to zero as $z \longrightarrow \infty$ so that it is bounded for all z. Finish the story.

13.11. Given a closed curve C, does

$$\int_C f(z)\,dz = 0$$

imply that f must be analytic inside and on C? Prove or disprove.

***13.12.** **(Morera's theorem)** Prove Morera's theorem: *If f is continuous throughout a simply connected domain D and if*

$$\int_C f(z)\,dz = 0$$

for every *closed contour C within D, then f is analytic in D.*

13.13. Make up two examination-type problems on this chapter.

Chapter 14

Taylor and Laurent Series

We find that the Taylor series of real-variable theory carries over to the complex case $f(z)$. In fact, it turns out to be but a special case of a still more powerful "Laurent series." These results will permit a more precise understanding of the nature of singularities and, at the same time, set the stage for the important Residue Theorem in Chapter 15.

14.1. COMPLEX SERIES

Convergence of series of complex constants $\sum a_n$, or complex functions $\sum a_n(z)$, is defined as the limit of the sequence of partial sums, as for the real case. The theorems and proofs of Chapter 2, carry over virtually intact.

Most of our interest will be in power series.

Example 14.1.

$$\sum_{1}^{\infty} \frac{(z-a)^n}{n}. \tag{14.1}$$

Applying the ratio test,

$$\lim_{n\to\infty} \left| \frac{(z-a)^{n+1}}{n+1} \frac{n}{(z-a)^n} \right| = |z-a|,$$

so that (14.1) converges inside $|z-a| < 1$ and diverges in $|z-a| > 1$; for points on the boundary $|z-a| = 1$, no conclusion is possible from this test. Observe that *the region of convergence is a circular disk centered at* $z = a$; clearly, such will be the

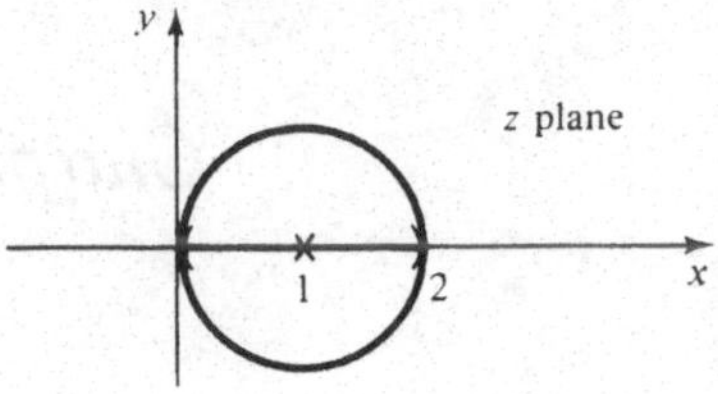

Figure 14.1. Circle of convergence and interval of convergence.

case for the general power series $\sum a_n(z-a)^n$ as well. Suppose, for instance, that $a = 1$ in (14.1). Then the circle of convergence is as shown in Fig. 14.1. For the real case $\sum (x-1)^n/n$—that is, where z is confined to lie on the x axis—we have an *interval* of convergence $|x - 1| < 1$, as discussed in Part I. ∎

Finally, we state (without proof) an important result.[1]

THEOREM 14.1. If $f(z) = \sum a_n(z-a)^n$, then the sum function $f(z)$ is analytic everywhere inside the circle of convergence. Furthermore, the derivatives of f (of all orders) can be found by repeated termwise differentiation of the power series, and each of the resulting series has the same circle of convergence as $\sum a_n(z-a)^n$.

14.2. TAYLOR SERIES

Suppose that $f(z)$ is analytic inside and on the circle C with center at a, as shown in Fig. 14.2. For any point z inside C we have, according to the Cauchy integral formula,

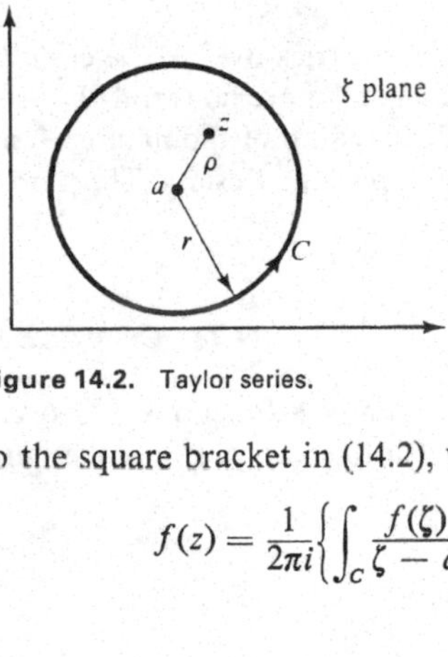

Figure 14.2. Taylor series.

$$f(z) = \frac{1}{2\pi i}\int_C \frac{f(\zeta)}{\zeta - z}\,d\zeta$$

$$= \frac{1}{2\pi i}\int_C \frac{f(\zeta)}{\zeta - a}\left[\frac{1}{1 - \dfrac{z-a}{\zeta - a}}\right] d\zeta. \tag{14.2}$$

Applying the algebraic identity

$$\frac{1}{1-t} = 1 + t + t^2 + \cdots + t^{n-1} + \frac{t^n}{1-t} \qquad \text{(for all } t \neq 1) \tag{14.3}$$

to the square bracket in (14.2), with $(z-a)/(\zeta - a)$ as t, we find

$$f(z) = \frac{1}{2\pi i}\left\{\int_C \frac{f(\zeta)}{\zeta - a}\,d\zeta + (z-a)\int_C \frac{f(\zeta)}{(\zeta - a)^2}\,d\zeta + \cdots + (z-a)^{n-1}\int_C \frac{f(\zeta)}{(\zeta-a)^n}\,d\zeta\right\} + R_n, \tag{14.4}$$

where

$$R_n = \frac{(z-a)^n}{2\pi i}\int_C \frac{f(\zeta)}{(\zeta-a)^n(\zeta - z)}\,d\zeta. \tag{14.5}$$

(Note that there is no question as to the interchange of the order of the summation and the integration; $1 + t + \cdots + t^{n-1}$ is only a *finite* sum, since n has thus far been held

[1] See, for example, E. G. Phillips, *Functions of a Complex Variable*, Oliver and Boyd, Edinburgh, 1957, pp. 18 and 19.

fixed.) With the help of (13.20), (14.4) reduces to

$$f(z) = f(a) + f'(a)(z - a) + \cdots + \frac{f^{(n-1)}(a)}{(n-1)!}(z - a)^{n-1} + R_n, \tag{14.6}$$

which is the complex version of Taylor's formula with remainder.

Then by letting $n \longrightarrow \infty$ in (14.6), we arrive at the **Taylor series** representation

$$f(z) = f(a) + f'(a)(z - a) + \frac{f''(a)}{2!}(z - a)^2 + \cdots, \tag{14.7}$$

provided that we can show that $R_n \longrightarrow 0$ as $n \longrightarrow \infty$. Hopefully, such is the case, for R_n stems from the last term in (14.3), which tends to zero nicely because $|t| = |z - a|/|\zeta - a| = \rho/r < 1$. Denote the maximum of $|f|$ on C as M. Noting that $|\zeta - z| \geq r - \rho$ for all ζ on C, we have from (14.5)

$$|R_n| = \frac{|z - a|^n}{2\pi}\left|\int_C \frac{f(\zeta)\,d\zeta}{(\zeta - a)^n(\zeta - z)}\right| \leq \frac{\rho^n}{2\pi}\frac{M}{r^n}\frac{2\pi r}{(r - \rho)} = \frac{Mr}{r - \rho}\left(\frac{\rho}{r}\right)^n, \tag{14.8}$$

which does tend to zero as $n \longrightarrow \infty$ because $\rho/r < 1$.

Since (14.7) is of exactly the same form as in real-variable theory, we have

$$e^z = 1 + z + \frac{z^2}{2!} + \frac{z^3}{3!} + \cdots$$

$$\sin z = z - \frac{z^3}{3!} + \frac{z^5}{5!} - \cdots$$

$$\cos z = 1 - \frac{z^2}{2!} + \frac{z^4}{4!} - \cdots,$$

and so on, just as in the real-variable case but with the x's changed to z's.

Example 14.2.

$$f(z) = \frac{1}{1 + z^2}. \tag{14.9}$$

Its Taylor series about $z = 0$, say, is obtained easily from (14.7), or recalling the well-known geometric series $1/(1 - x) = 1 + x + x^2 + \cdots$ and changing x to $-z^2$, we have

$$\frac{1}{1 + z^2} = 1 - z^2 + z^4 - \cdots. \tag{14.10}$$

The ratio test tells us that the series converges in the unit disk $|z| < 1$, as shown in Fig. 14.3. The reason for the $|z| < 1$ limitation is that the function $1/(1 + z^2)$ is singular at $z = \pm i$. We can think of the circle of convergence as spreading out (like a ripple on a pond) from the point a—in this case, the origin—until it bumps up against a singularity of the function.

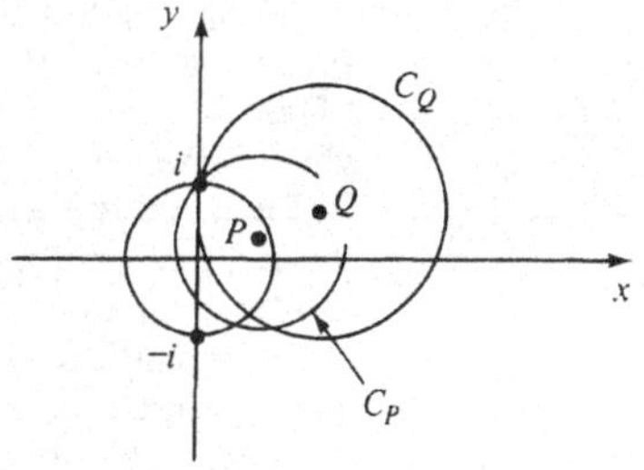

Figure 14.3. Taylor expansions of $1/(1 + z^2)$.

COMMENT 1. If we restrict z to lie on the real axis, (14.10) becomes

$$\frac{1}{1 + x^2} = 1 - x^2 + x^4 - \cdots \quad \text{in } |x| < 1. \tag{14.11}$$

As discussed at the end of Section 2.4, the restriction $|x| < 1$ in (14.11) is strange in the sense that $1/(1 + x^2)$ is beautifully behaved for *all* x; in fact, it is analytic for $-\infty < x < \infty$. But the situation is now transparent; the restriction $|x| < 1$ in (14.11) occurs because of singularities *off the real axis*, at $z = +i$ and $-i$.

COMMENT 2. Note carefully that although $1/(1 + z^2)$ is meaningful everywhere except at the two points $z = \pm i$, the series $1 - z^2 + \cdots$ is meaningful only inside the disk $|z| < 1$ and is therefore only a *partial representation* of the original function. Suppose now that someone hands us the series $1 - z^2 + \cdots$ and we don't recognize it as the Maclaurin expansion of $1/(1 + z^2)$. Computing the circle of convergence $|z| < 1$ by the ratio test, we see that the "original function," whatever it is, has at least one singularity somewhere on $|z| = 1$ (and perhaps others in $|z| > 1$ as well). Nevertheless, we can compute *outside* the unit disk as follows.

Summing the series

$$f(z) = 1 - z^2 + z^4 - \cdots$$

$$f'(z) = -2z + 4z^3 - 6z^5 + \cdots$$

$$f''(z) = -2 + 12z^2 - 30z^4 + \cdots,$$

and so on, at any point P inside the unit disk (Fig. 14.3), we therefore have the values $f(P), f'(P), f''(P), \ldots$ for the Taylor expansion

$$f(z) = f(P) + f'(P)(z - P) + \frac{f''(P)}{2!}(z - P)^2 + \cdots \tag{14.12}$$

about the point P. If we then apply the ratio test to the series (14.12), we will find that it is valid (i.e., convergent) inside the circle C_P (Fig. 14.3). In turn, we use (14.12) to compute f and all its derivatives at some point Q and thus obtain the Taylor expansion

$$f(z) = f(Q) + f'(Q)(z - Q) + \frac{f''(Q)}{2!}(z - Q)^2 + \cdots \tag{14.13}$$

about Q, which will be found to be valid inside C_Q (Fig. 14.3). Repeating the process indefinitely, we can develop a set of expansions that collectively represent $f(z)$ throughout the entire z plane, except for the two singular points $z = \pm i$.

This process is called **analytic continuation**. Although possible *in principle*, it is obviously not too tractable computationally. Pointing toward more reasonable methods of continuation, we note that an Aitken–Shanks transformation (Section 2.5) of the original series $1 - z^2 + z^4 - \cdots$ converges (in one step) to $1/(1 + z^2)$, which is the desired "complete" analytic continuation!

Finally, this discussion brings us back to the remark preceding Example 12.3—namely, that we *define* "e^z" as the analytic continuation of e^x off the real axis and into the complex z plane; similarly for "$\sin z$," "$\cos z$," and so on. You may wonder whether so little data—values on the real axis only—has a *unique* analytic continuation. For instance, surely a function value at a *single point* is not sufficient; the functions $5 \cos z$ and $10/(2 + z)$ both equal 5 at the origin. In fact, the **identity theorem** guarantees that data along a line segment, no matter how short, is sufficient.[2] ∎

[2]See, for example, K. Knopp's *Theory of Functions, Part One*, Dover, New York, 1945, pp. 87–96; also his discussion of *natural boundaries* on p. 101.

In seeking an expansion of $f(z)$ about some point a, let us suppose this time that f is analytic in an *annulus*—that is, between and on concentric circles C_i and C_o with their center at a (Fig. 14.4). To illustrate, suppose that $f = e^z/z(2z+1)^2(z+3i)$, which is singular at 0, $-\frac{1}{2}$, and $-3i$. If we choose $a = 0$, say, then one set of suitable circles C_i and C_o would be $|z| = 1$ and $|z| = 2$, respectively. That f happens to be singular exactly *at* the point a causes no problem; we merely ask that f be analytic in the annulus, and it *is*. On the other hand, if we have $f(z) = \ln z$ and $a = 0$, observe that there exists *no* annulus throughout which f is even single valued let alone analytic.

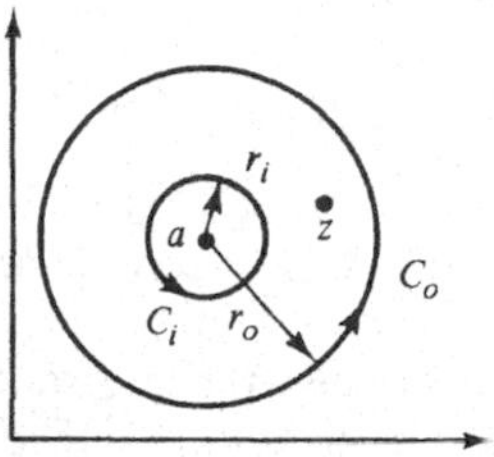

Figure 14.4. Laurent series.

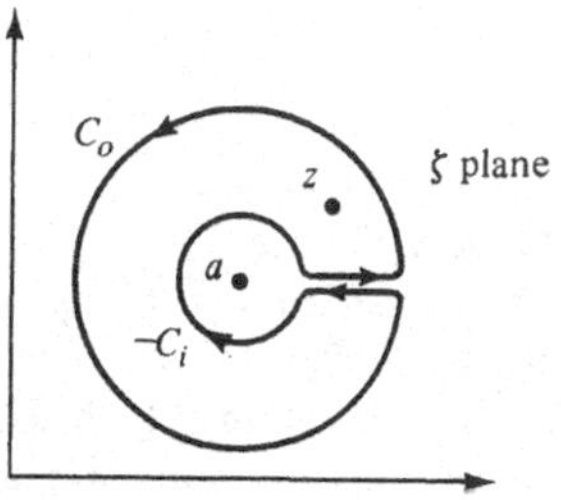

Figure 14.5. The slit contour C.

Next, if C is the "slit" contour shown in Fig. 14.5, then it follows from the Cauchy integral formula that

$$f(z) = \frac{1}{2\pi i}\int_C \frac{f(\zeta)}{\zeta - z}\,d\zeta = \frac{1}{2\pi i}\int_{C_o+(-C_i)+\rightarrow+\leftarrow} \frac{f(\zeta)}{(\zeta - z)}\,d\zeta$$

$$= \frac{1}{2\pi i}\int_{C_o} \frac{f(\zeta)}{\zeta - z}\,d\zeta - \frac{1}{2\pi i}\int_{C_i} \frac{f(\zeta)}{\zeta - z}\,d\zeta, \tag{14.14}$$

since the contributions from the two straight segments exactly cancel. (Actually, the two segments are abutting; we've separated them a little in Fig. 14.5 for clarity.) Following the same lines as in our development of Taylor series, we treat the C_o integral as follows.

$$\frac{1}{2\pi i}\int_{C_o} \frac{f(\zeta)}{\zeta - z}\,d\zeta = \frac{1}{2\pi i}\int_{C_o} \frac{f(\zeta)}{\zeta - a}\left[\frac{1}{1-(z-a)/(\zeta-a)}\right]d\zeta$$

$$= \text{etc.} = \sum_{n=0}^{\infty} a_n (z-a)^n, \tag{14.15}$$

where

$$a_n = \frac{1}{2\pi i}\int_{C_o} \frac{f(\zeta)}{(\zeta - a)^{n+1}}\,d\zeta. \tag{14.16}$$

The only difference is that this time we *cannot* invoke (13.20) to show that the right-hand side of (14.16) reduces to $f^{(n)}(a)/n!$, since f is not necessarily analytic inside of C_o, as assumed in the derivation of (13.20).

Turning next to the C_i integral in (14.14), note that whereas $|(z - a)/(\zeta - a)| < 1$ as ζ runs around on C_o, it will be *greater* than unity as ζ runs around on C_i. Thus we write

$$\frac{1}{\zeta - z} = -\frac{1}{z - a}\frac{1}{1 - (\zeta - a)/(z - a)} \tag{14.17}$$

instead because $|(\zeta - a)/(z - a)|$ *is* less than unity for all ζ's on C_i. Leaving the details for Exercise14.5, we have

$$-\frac{1}{2\pi i}\int_{C_i}\frac{f(\zeta)}{\zeta - z}\,d\zeta = \frac{1}{2\pi i(z - a)}\int_{C_i} f(\zeta)\left[\frac{1}{1 - (\zeta - a)/(z - a)}\right]d\zeta$$

$$= \text{etc.} = \sum_{n=-1}^{-\infty} b_n(z - a)^n, \tag{14.18}$$

where

$$b_n = \frac{1}{2\pi i}\int_{C_i}\frac{f(\zeta)}{(\zeta - a)^{n+1}}\,d\zeta. \tag{14.19}$$

Since $n \leq -1$, $(\zeta - a)^{-n-1}$ is analytic everywhere, but we *cannot* conclude from Cauchy's theorem that the right-hand side of (14.19) is zero, since f is not necessarily analytic inside C_i.

We can change C_i to C_o in (14.19) because the integrand is analytic between and on the two contours. (Recall the corollary to Cauchy's theorem.) Doing so and combining (14.14), (14.15), (14.16), (14.18), and (14.19), we have the so-called **Laurent expansion** of f about a, in the annulus between C_i and C_o,

$$f(z) = \sum_{n=-\infty}^{\infty} c_n(z - a)^n, \tag{14.20}$$

where

$$c_n = \frac{1}{2\pi i}\int_{C_o}\frac{f(\zeta)}{(\zeta - a)^{n+1}}\,d\zeta. \tag{14.21}$$

If, in fact, f has no singularities inside C_i, so that it is analytic everywhere inside (and on) C_o, then Cauchy's theorem tells us that $c_n = 0$ for $n = -1, -2, \ldots$, and (13.20) tells us that $c_n = f^{(n)}(a)/n!$ for $n = 0, 1, 2, \ldots$, and therefore the Laurent expansion (14.20) reduces to the *Taylor* expansion of f about a as it should.

Yet if f *is* singular at points within C_i, then the Laurent series will contain at least one and perhaps an infinite number of negative powers. In this case, the evaluation of the integral (14.21) for both positive and negative n's is not trivial. As subsequent examples will demonstrate, however, it is generally possible to bypass (14.21) completely and compute the c_n's rather easily.

Example 14.3.

$$f(z) = \frac{1}{z - 1}. \tag{14.22}$$

Let us expand about $z = 0$, say. Noting the singularity at $z = 1$, we see that there are two possible expansions, one in $|z| > 1$ (i.e., the annulus $1 < |z| < \infty$) and one in the disk $|z| < 1$. In $|z| < 1$ the function $f(z)$ is, in fact, analytic, and so the Laurent expansion is simply the Taylor series

$$f(z) = \frac{1}{z - 1} = -\frac{1}{1 - z} = -(1 + z + z^2 + \cdots)$$

$$= -1 - z - z^2 - \cdots. \tag{14.23}$$

In $|z| > 1$ the key is to rewrite $f(z)$ as

$$f(z) = \frac{1}{z}\frac{1}{1 - 1/z}. \tag{14.24}$$

Since the $1/z$ factor out in front is already in the desired form—that is, in powers of $(z - a)$, where $a = 0$—we leave it alone. Turning to the second factor, note that $1/z$ is smaller than unity, since we're in the annulus $|z| > 1$. Thus we have the *Taylor* expansion in the variable $1/z \equiv t$,

$$\frac{1}{1 - 1/z} = \frac{1}{1 - t} = 1 + t + t^2 + \cdots$$

since $|t| < 1$. So (14.24) becomes

$$f(z) = \frac{1}{z}\left[1 + \left(\frac{1}{z}\right) + \left(\frac{1}{z}\right)^2 + \cdots\right] = \frac{1}{z} + \frac{1}{z^2} + \frac{1}{z^3} + \cdots, \tag{14.25}$$

which is the desired Laurent expansion in $|z| > 1$. ∎

Example 14.4.

$$f(z) = \frac{1}{z(1 - z)}. \tag{14.26}$$

To illustrate, consider several possible expansions.

(i) About $z = 0$, in $0 < |z| < 1$ [Fig. 14.6(a)].

$$f(z) = \frac{1}{z}\frac{1}{1 - z} = \left(\frac{1}{z}\right)(1 + z + z^2 + \cdots) \tag{14.27}$$

$$f(z) = \frac{1}{z} + 1 + z + z^2 + \cdots. \tag{14.28}$$

(Note the way that the first factor in the right-hand side of (14.27) is valid in $|z| > 0$, the second factor is valid in $|z| < 1$, and their product (14.28) is valid in the "overlap" $0 < |z| < 1$.)

(ii) About $z = 0$, in $|z| > 1$ [Fig. 14.6(b)].

$$f(z) = -\frac{1}{z^2}\frac{1}{1 - 1/z} = -\frac{1}{z^2}\left[1 + \left(\frac{1}{z}\right) + \left(\frac{1}{z}\right)^2 + \cdots\right]$$

$$= -\frac{1}{z^2} - \frac{1}{z^3} - \frac{1}{z^4} - \cdots. \tag{14.29}$$

(iii) About $z = i$, in $|z - i| > \sqrt{2}$ [Fig. 14.6(c)]. Since the expansion that we seek is in powers of $z - i$, it will save some writing if we introduce a temporary

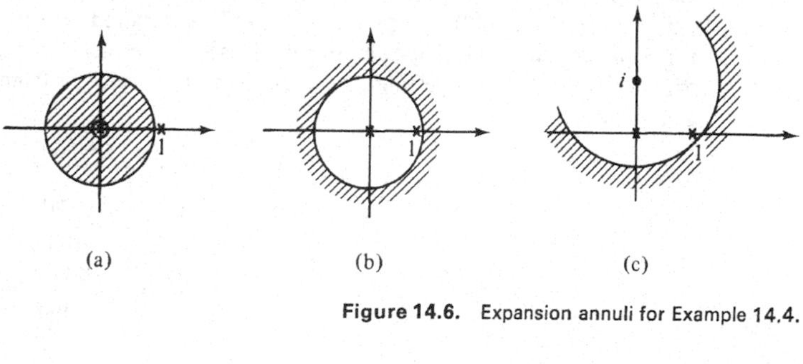

Figure 14.6. Expansion annuli for Example 14.4.

variable $z - i \equiv t$; that is, we "shift the origin" to the point about which we are expanding. Thus we rewrite f as

$$f = \frac{1}{(t+i)(1-i-t)} = -\frac{1}{t(1+i/t)}\frac{1}{t[1-(1-i)/t]},$$

where the last step is motivated by the fact that $|i/t| = 1/|t| < 1$ in $|t| > 1$ (and hence in the expansion annulus) and $|(1-i)/t| = \sqrt{2}/|t| < 1$ in $|t| > \sqrt{2}$. So

$$\begin{aligned} f &= -\frac{1}{t^2}\left[1 - \left(\frac{i}{t}\right) + \left(\frac{i}{t}\right)^2 + \cdots\right]\left[1 + \left(\frac{1-i}{t}\right) + \left(\frac{1-i}{t}\right)^2 + \cdots\right] \\ &= -\frac{1}{t^2}\left[1 + (1-2i)\frac{1}{t} - (2+3i)\frac{1}{t^2} + \cdots\right] \\ &= -\frac{1}{(z-i)^2} - \frac{1-2i}{(z-i)^3} + \frac{2+3i}{(z-i)^4} - \cdots. \quad \blacksquare \end{aligned} \tag{14.30}$$

Example 14.5.

$$f(z) = \frac{\cos z}{z}. \tag{14.31}$$

Expanding about $z = 0$, say, in $|z| > 0$,

$$f(z) = \frac{1}{z}\left(1 - \frac{z^2}{2!} + \frac{z^4}{4!} - \cdots\right) = \frac{1}{z} - \frac{z}{2!} + \frac{z^3}{4!} - \cdots. \quad \blacksquare \tag{14.32}$$

Example 14.6.

$$f(z) = \sec z = \frac{1}{\cos z}. \tag{14.33}$$

This is singular at the zeros of $\cos z$—namely, $z = \pm\pi/2, \pm 3\pi/2, \ldots$. Let us expand about $z = \pi/2$, say, in the annulus $0 < |z - \pi/2| < \pi$. First, expand $\cos z$ in a Taylor series about $\pi/2$,

$$\cos z = -\left(z - \frac{\pi}{2}\right) + \frac{1}{3!}\left(z - \frac{\pi}{2}\right)^3 - \frac{1}{5!}\left(z - \frac{\pi}{2}\right)^5 + \cdots.$$

Since $\cos z \sim -(z - \pi/2)$ as $z \to \pi/2$, it follows that if we rewrite (14.33) as

$$f(z) = \frac{1}{z - \pi/2}\frac{z - \pi/2}{\cos z}, \tag{14.34}$$

then the second factor, $(z - \pi/2)/\cos z \equiv g(z)$ say, should admit a *Taylor* expansion in $|z - \pi/2| < \pi$, where the radius of convergence is limited to π by the neighboring singularities at $-\pi/2$ and $3\pi/2$. The coefficients in the Taylor series $g(z) = g(\pi/2) + g'(\pi/2)(z - \pi/2) + \cdots$ can be computed directly with the help of L'Hospital's rule, but doing so is tedious. It is a little easier to expand $g(z)$ by using ordinary long division. Setting $z - \pi/2 \equiv t$ and canceling t from numerator and denominator, we have

$$\begin{array}{r|l} & -1 - t^2/6 - 7t^4/360 - \cdots \\ \hline -1 + \frac{1}{6}t^2 - \frac{1}{120}t^4 + \cdots & 1 \\ & 1 - t^2/6 + t^4/120 - \cdots \\ \hline & t^2/6 - t^4/120 + \cdots \\ & t^2/6 - t^4/36 + \cdots \\ \hline & 7t^4/360 + \cdots \end{array}$$

so that

$$f(z) = \frac{1}{t}\left[-1 - \frac{1}{6}t^2 - \frac{7}{360}t^4 - \cdots\right]$$

$$= -\frac{1}{z - \pi/2} - \frac{1}{6}\left(z - \frac{\pi}{2}\right) - \frac{7}{360}\left(z - \frac{\pi}{2}\right)^3 - \cdots \qquad (14.35)$$

in $0 < |z - \pi/2| < \pi$. ∎

Example 14.7.

$$f(z) = e^{-1/z^2}. \qquad (14.36)$$

Since $e^\zeta = 1 + \zeta + \zeta^2/2! + \cdots$ for all $|\zeta| < \infty$, it follows that

$$f(z) = 1 - \frac{1}{z^2} + \frac{1}{2!\,z^4} - \cdots \qquad (14.37)$$

for all $|z| > 0$; for z *equal* to zero the series (14.37) diverges. And yet $f(z) = \exp(-1/z^2)$ seems to be well behaved at the origin, for isn't $\lim f(z) = e^{-\infty} = 0$ as $z \longrightarrow 0$? That depends on our path of approach. Approaching along the x axis, for instance,

$$\lim_{z\to 0} e^{-1/z^2} = \lim_{x\to 0} e^{-1/x^2} = 0;$$

but approaching along the y axis,

$$\lim_{z\to 0} e^{-1/z^2} = \lim_{y\to 0} e^{-1/(iy)^2} = \infty!$$

Thus f is indeed badly behaved at $z = 0$. More about this kind of behavior in the next section. ∎

Notice how the Laurent expansions in the preceding examples were all obtained, rather simply, through the use of *Taylor* series and some algebra.

14.4. CLASSIFICATION OF ISOLATED SINGULARITIES

Suppose that z_0 is an isolated singular point of $f(z)$; that is, whereas f is singular at z_0, there is an annulus $0 < |z - z_0| < r$ for some $r > 0$ throughout which f is analytic. It therefore admits a Laurent expansion

$$f(z) = \cdots + \frac{c_{-2}}{(z - z_0)^2} + \frac{c_{-1}}{(z - z_0)} + c_0 + c_1(z - z_0) + \cdots \qquad (14.38)$$

in that annulus. If the series terminates on the left—that is, if it is of the form

$$f(z) = \frac{c_{-m}}{(z - z_0)^m} + \frac{c_{-m+1}}{(z - z_0)^{m-1}} + \cdots + c_0 + c_1(z - z_0) + \cdots, \qquad (14.39)$$

where $c_{-m} \neq 0$—then we classify f's singularity at z_0 as an ***m*th-order pole.** If there are an infinite number of negative powers present, instead, we say that f has an **essential singularity** at z_0.

To illustrate, consider the $f(z)$ of Example 14.4. What kind of singularity does f have at $z = 0$? We have two Laurent expansions about $z = 0$, (14.28) and (14.29). From (14.28) the singularity appears to be a first-order pole, but from (14.29) it looks like an essential singularity. However, since (14.29) holds in $|z| > 1$, it is not relevant in the classification of the singularity at $z = 0$; (14.28) holds in $0 < |z| < 1$—that is, *in an annulus with arbitrarily small inner circle*—and so we conclude that f has a first-order pole at $z = 0$.

More simply,

$$f(z) = \frac{1}{z(1-z)} \sim \frac{1}{z} \quad \text{as } z \longrightarrow 0;$$

consequently, the singularity is clearly a first-order pole.

Example 14.8. Classify the singularity in

$$f(z) = \frac{1}{z^3} + \frac{1}{z^4} + \frac{1}{z^5} + \cdots \tag{14.40}$$

at $z = 0$. Is it obviously an essential singularity? The ratio test reveals that (14.40) holds in $1 < |z| < \infty$. Since this annulus does not have an arbitrarily small inner radius, we can*not* use (14.40) to classify the singularity at $z = 0$! Instead note that

$$\begin{aligned} f(z) &= \frac{1}{z^3}\left(1 + \frac{1}{z} + \frac{1}{z^2} + \cdots\right) = \frac{1}{z^3}\frac{1}{1 - 1/z} = -\frac{1}{z^2}\frac{1}{1-z} \\ &= -\frac{1}{z^2}(1 + z + z^2 + \cdots) = -\frac{1}{z^2} - \frac{1}{z} - 1 - z - z^2 - \cdots \end{aligned} \tag{14.41}$$

in $0 < |z| < 1$; therefore f has a *second-order pole* at $z = 0$. ∎

Example 14.9. Classify all singularities of the function

$$f(z) = \frac{\pi + z + \sin z}{(1 + \cos z)^2}. \tag{14.42}$$

The denominator vanishes at $z = \pm\pi, \pm 3\pi, \ldots$. Consider, for instance, $z = \pi$. Expanding

$$\cos z = -1 + \frac{1}{2}(z - \pi)^2 - \cdots,$$

we see that

$$f(z) \sim \frac{8\pi}{(z - \pi)^4} \quad \text{as } z \longrightarrow \pi,$$

and so f has a fourth-order pole there. The same conclusion holds at $\pm 3\pi, \pm 5\pi, \ldots$, but the point $z = -\pi$ needs to be examined separately because the numerator also vanishes there. Expanding about $-\pi$,

$$\sin z = -(z + \pi) + \frac{1}{6}(z + \pi)^3 - \cdots$$

$$\cos z = -1 + \frac{1}{2}(z + \pi)^2 - \cdots,$$

therefore

$$\begin{aligned} f(z) &\sim \frac{\pi + z - (z + \pi) + (z + \pi)^3/6}{(z + \pi)^4/4} \quad \text{as } z \longrightarrow -\pi \\ &= \frac{2/3}{z + \pi}. \end{aligned}$$

We conclude that the only singularities of f are a first-order pole at $-\pi$ and fourth-order poles at $+\pi, \pm 3\pi, \pm 5\pi, \pm 7\pi, \ldots$. ∎

On the other hand, recalling from Example 14.7 that

$$e^{-1/z^2} = 1 - \frac{1}{z^2} + \frac{1}{2!\,z^4} - \cdots \quad \text{in } 0 < |z| < \infty,$$

we see that e^{-1/z^2} has an *essential* singularity at the origin.

What can we say about the *behavior* of f in the neighborhood of a pole or essential singularity? Suppose first that f has an mth-order pole at z_0. Then $f(z) \sim c_{-m}/(z - z_0)^m$, where $c_{-m} \neq 0$. Or with $z - z_0 = \rho e^{i\phi}$,

$$|f(z)| \sim \frac{|c_{-m}| e^{-im\phi}}{\rho^m |e^{im\phi}|} = \frac{|c_{-m}|}{\rho^m}$$

Now, $c_{-m} \neq 0$, by assumption, so that $|f(z)| \to \infty$ as $z \to z_0$—that is, as $\rho \to 0$; see Fig. 14.7. So we say that f "blows up" at a pole.

The behavior near an essential singularity is somewhat more exotic. For instance, recall that $e^{-1/z^2} \to 0$ as $z \to 0$ along the real axis but tends to ∞ as $z \to 0$ along the imaginary axis. In fact, we have the following striking result, due to Weierstrass.[3]

THEOREM 14.2. Suppose that $f(z)$ has an essential singularity at z_0 and let c be ***any*** *given complex number. Then for each $\epsilon > 0$ and $\delta > 0$ (no matter how small) there will be some point z in $0 < |z - z_0| < \delta$ such that $|f(z) - c| < \epsilon$.*

Thus the behavior near an essential singularity[4] is quite complicated; $|f(z)|$, for instance, hardly resembles the "pole" of Fig. 14.7, and it is not entirely appropriate to say that f "blows up" there because $\lim f(z)$ as $z \to z_0$ *depends on the path of approach.*

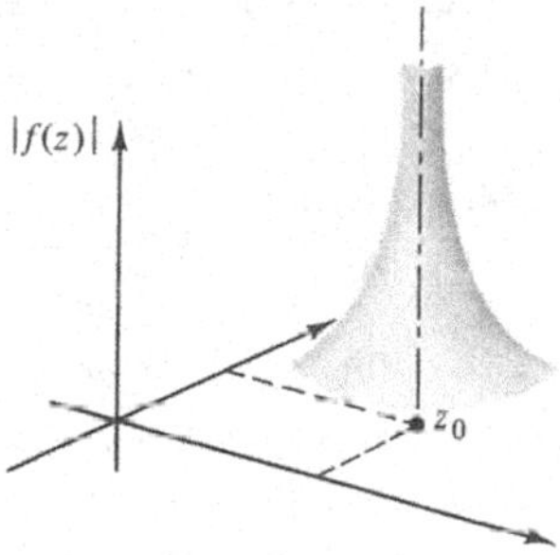

Figure 14.7. "Polelike" behavior.

*

In the first sentence of this section we limited ourselves to *isolated* singular points. According to the definition given there, two examples of *non*isolated singular points are as follows. $\ln z$ is singular at $z = 0$, but it is not an isolated singularity, since there is no annulus $0 < |z| < r$ throughout which $\ln z$ is analytic. $1/\sin(1/z)$ has poles at the zeros of $\sin(1/z)$—namely, at $z = 1/n\pi$ for $n = \pm 1, \pm 2, \ldots$, each of which is an isolated singularity. In addition, $z = 0$ is a singular point but it is not isolated; it is a *limit point* (defined on page 26) of the sequence of poles.

---

[3]See, for example, R. Churchill's *Complex Variables and Applications*, McGraw-Hill, New York, 1960, p. 270.

[4]Strictly speaking, it's not quite accurate to call such a point a singularity. Actually, the function has a *singularity* at a *singular point*, but we will generally use the word singularity to mean either a singularity of a function or the point at which it occurs, as should be clear from the context.

EXERCISES

14.1. Find the regions of convergence of the series below.

(a) $\frac{2}{z} - \frac{2^3 z}{3!} + \frac{2^5 z^3}{5!} - \cdots$ (b) $\sum_{n=1}^{\infty} \frac{1}{(2n)!\,z^{2n}}$

(c) $\sum_{n=1}^{\infty} (-1)^{n+1} n(z-1)^n$ (d) $\sum_{n=1}^{\infty} (-1)^{n+1} n(z-1)^{-n}$

(e) $\sum_{n=0}^{\infty} \frac{z^{n-1}}{4^{n+1}}$

14.2. Expand the principal value of $\ln z$ about $z = -2$. What is the radius of convergence?

14.3. Verify the claim

$$\frac{1}{z^2} = 1 + \sum_{n=1}^{\infty} (n+1)(z+1)^n$$

and determine the region of validity.

14.4. By differentiating the Maclaurin expansion of $1/(1-z)$, obtain the expansion

$$\frac{1}{(1-z)^3} = \frac{1}{2} \sum_{n=1}^{\infty} n(n+1) z^{n-1}.$$

14.5. Fill in the missing steps in the derivation of (14.18) and (14.19).

14.6. Compute the coefficients c_{-1}, c_0, and c_1, say, for the Laurent expansion of $f(z) = 1/(z-1)$ in $|z| > 1$ directly from (14.21) and compare your results with (14.25).

14.7. Expand

(a) $\frac{1}{z^2 - 3z + 2}$ about $z = 0$ in $1 < |z| < 2$.

(b) $\frac{1}{\sin z}$ about $z = 0$ in $\pi < |z| < 2\pi$.

What practical difficulty presents itself in these cases?

14.8. Obtain the first few terms of the Laurent expansions of

(a) $\csc z$ in $0 < |z| < \pi$.

(b) $\sec z$ in $|z| < \frac{\pi}{2}$.

(c) $\csc z$ in $0 < |z - \pi| < \pi$.

(d) $\frac{1}{e^z - 1}$ in $0 < |z| < 2\pi$.

(e) $\frac{1}{e^z + 1}$ in $0 < |z - \pi i| < 2\pi$.

14.9. What are the regions of validity of the following?

(a) $\frac{1}{z^2 \sinh z} = \frac{1}{z^3} - \frac{1}{6}\frac{1}{z} + \frac{7}{360} z - \cdots$

(b) $\frac{1}{z(z^3 + 2)} = \frac{1}{z^4} - \frac{2}{z^7} + \frac{4}{z^{10}} - \cdots$

14.10. Verify.

(a) $\lim_{z \to \pi} \frac{\sin^2 z}{1 + \cos z} = 2$ (b) $\lim_{z \to -\pi i} \frac{z \sinh z}{z + \pi i} = \pi i$

14.11. What kind of singularity (if any) does

$$f(z) = \frac{1}{2z} + \frac{1}{(2z)^2} + \frac{1}{(2z)^3} + \cdots$$

have at $z = 0$?

***14.12.** $f(z) = 1/\sin(1/z)$ has a nonisolated singularity at the limit point of poles $z = 0$. How about the function $f(z) = \sin(1/z)$ for which $z = 0$ is a limit point of *zeros*?

14.13. Find and classify all singular points of the function that is (partially) represented by the series in Exercise 14.1(e).

14.14. Obtain the first four terms of the expansion

$$\frac{e^z}{z(z^2+1)} = \frac{1}{z} + 1 - \frac{1}{2}z - \frac{5}{6}z^2 + \cdots.$$

14.15. Find the complete analytic continuation of the series in Exercise 14.1(c) and (d).

14.16. Using the Taylor series "stepping-stone" method of analytic continuation, compute (approximately) the function that is partially represented by the series $1 + z + z^2 + \cdots$, at the point $z = -\frac{3}{2}$. Compare your result with the exact value—that is, as computed from the complete analytic continuation.

14.17. Find and classify all singularities of the functions below.

(a) $\csc z$ (b) $\sec z$ (c) $\dfrac{1}{e^z - 1}$

(d) $\dfrac{1}{e^z + 1}$ (e) $\dfrac{e^{-z}}{z(z^2+1)}$ (f) $ze^{-1/z}$

14.18. (*The point at infinity*) To study the behavior of a function $f(z)$ as $z \longrightarrow \infty$, we set $z = 1/t$ and consider $f(z) = f(1/t) \equiv g(t)$ as $t \longrightarrow 0$. It is convenient to say that $t = 0$ corresponds to **"the point at infinity"** in the z plane. Furthermore, we say that $f(z)$ is analytic at infinity if $g(t)$ is analytic at $t = 0$ and that if $g(t)$ has a particular kind of singularity at $t = 0$, then $f(z)$ has the same kind of singularity at infinity. Thus determine the behavior at infinity of the functions shown.

(a) $\sin z$ (b) e^z (c) $e^{1/z}$ (d) $e^{-1/z}$

(e) $\dfrac{z}{\sin z}$ (f) $\dfrac{z}{z^3+2}$ (g) $z + \dfrac{3}{z}$

14.19. In Exercise 3.22(c) it is indicated that

$$\int_0^x \cos\frac{1}{\xi}\,d\xi = x + \frac{\pi}{2} - \frac{1}{2x} + \frac{2!}{4!\,3!}\frac{1}{x^3} - \frac{4!}{6!\,5!}\frac{1}{x^5} + \cdots$$

for $x > 0$. Since the left-hand side is simply zero for $x = 0$, the divergence of the series at $x = 0$ may seem strange. Can you reconcile the apparent difference in the behavior of the left- and right-hand sides?

14.20. (**The reflection theorem**) (a) For which of the following functions is it true that $f(\bar{z}) = \overline{f(z)}$: $|z|$, iz, e^{iz}, $\sin z$?

(b) The reflection theorem is as follows. *Let $f(z)$ be analytic in a domain D that includes a segment of the x axis and is symmetric about the x axis. If $f(x)$ is real all along that segment, then $f(\bar{z}) = \overline{f(z)}$ for all z in D; conversely, if $f(\bar{z}) = \overline{f(z)}$ or, equivalently, if $\overline{f(\bar{z})} = f(z)$, then $f(x)$ is real.* Prove the theorem. *Hint:* Denoting $f(z) = u(x, y) + iv(x, y)$ and $F(z) = \overline{f(\bar{z})} = u(x, -y) - iv(x, -y) \equiv U(x, y) + iV(x, y)$, show that analyticity of f implies analyticity of F. Next, show that $F(x) = f(x)$, so that the two functions coincide on the x axis, and then use the principle of analytic continuation. Incidentally, why does the theorem require the shape of D to be symmetric about the x axis?

***14.21.** (*Application to water waves*) First some preliminaries.

(a) If C is a ccw circle of radius R centered at the origin, z is inside C, and $f = u + iv$ is analytic inside and on C, then

$$f(z) = \frac{1}{2\pi i}\int_C \left(\frac{1}{\zeta - z} + \frac{1}{\zeta - R^2/\bar{z}}\right) f(\zeta)\, d\zeta \tag{14.43}$$

by the Cauchy integral formula, where we note that the $\int_C f(\zeta)\, d\zeta/(\zeta - R^2/\bar{z})$ term is zero, since $R^2/\bar{z}$ is outside the circle; that is, with $z = re^{i\theta}$ we have $|R^2/\bar{z}| = R^2/r = R(R/r) > R$. Letting $\zeta = Re^{i\phi}$ in (14.43) and equating real and imaginary parts, derive the relations

$$u(r, \theta) = u(0) + \frac{1}{\pi}\int_0^{2\pi} \frac{Rr \sin(\phi - \theta)}{R^2 + r^2 - 2Rr\cos(\phi - \theta)} v(R, \phi)\, d\phi \tag{14.44a}$$

and

$$v(r, \theta) = v(0) - \frac{1}{\pi}\int_0^{2\pi} \frac{Rr \sin(\phi - \theta)}{R^2 + r^2 - 2Rr\cos(\phi - \theta)} u(R, \phi)\, d\phi \tag{14.44b}$$

between u and v, where $u(0, \theta) \equiv u(0)$ and $v(0, \theta) \equiv v(0)$. Suppose that $f(z)$ is an *entire* function and u is bounded—that is $|u| < M$ for all z. Letting $R \longrightarrow \infty$ in (14.44b), show that this implies that $v =$ constant. Conversely, show that $|v| < M$ implies that $u =$ constant. This result is similar to *Liouville's theorem* [Exercise 13.10(b)] but is a little more informative.

(b) Consider next *small, two-dimensional standing waves on the surface of an infinitely deep ocean.* Let the x axis coincide with the undisturbed water level and let y be measured upward. The governing equations are (see, for example, J. J. Stoker, *Water Waves*, Interscience, New York, 1957)

$$\nabla^2\phi = 0 \quad \text{in } y < 0 \tag{14.45a}$$

$$\phi_{tt} + g\phi_y = 0 \quad \text{on } y = 0, \qquad \phi, \phi_y \longrightarrow 0 \quad \text{as } y \longrightarrow -\infty, \tag{14.45b}$$

where g is gravity, ϕ is the velocity potential (i.e., the velocity $\mathbf{q} = \nabla\phi$), and subscripts denote partials. With ϕ determined, the free surface elevation $y = \eta(x, t) = -\phi_t(x, 0, t)/g$. Seeking standing waves, let us set

$$\phi(x, y, t) = e^{i\omega t}\Phi(x, y), \tag{14.46}$$

so that Φ is governed by

$$\nabla^2\Phi = 0 \quad \text{in } y < 0 \tag{14.47a}$$

$$\Phi_y - \kappa\Phi = 0 \quad \text{on } y = 0, \qquad \Phi, \Phi_y \longrightarrow 0 \quad \text{as } y \longrightarrow -\infty, \tag{14.47b}$$

where $\kappa \equiv \omega^2/g$. Finally, seeking Φ in the product form $\Phi(x, y) = X(x)Y(y)$, it is not hard to show that

$$\Phi = A \sin(\kappa x + B)e^{\kappa y} \tag{14.48a}$$

or

$$\eta(x, t) = \frac{-i\omega A}{g} e^{i\omega t} \sin(\kappa x + B). \tag{14.48b}$$

Since we have assumed $\Phi = X(x)Y(y)$, however, we wonder if there are *other* possible standing waves, not of the assumed form. To answer this question, set $\Psi \equiv \Phi_y - \kappa\Phi$ in $y \leq 0$. Show that $\nabla^2\Psi = 0$ in $y \leq 0$. Letting χ be the conjugate harmonic function, it follows that $F \equiv \chi + i\Psi$ is analytic in $y \leq 0$. Show that F is real valued on the x axis. Continuing F across the x axis into $y > 0$ per $F(z) \equiv \overline{F(\bar{z})}$ and recalling Exercise 14.20, show that F is an entire function. Next, show that Ψ [and hence χ per part (a)] is bounded, and so F is a constant. In particular, Ψ is a constant; and since it is zero on $y = 0$, then it is zero throughout $y \leq 0$. Thus we have *continued the boundary conditions (14.47b) off the x axis*:

$$\Phi_y - \kappa\Phi = 0 \quad \text{in } y \leq 0, \qquad \Phi, \Phi_y \longrightarrow 0 \quad \text{as } y \longrightarrow -\infty.$$

Solving, show that $\Phi = C(x)e^{\kappa y}$; and recalling (14.47a), obtain $\Phi = A \sin(\kappa x + B)e^{\kappa y}$ as before, so that (14.48b) is the *only* possible standing wave of the form (14.46).

14.22. Make up two examination-type problems on this chapter.

Chapter 15

The Residue Theorem and Contour Integration

After completing the derivation of the Residue Theorem, it might be worthwhile to reflect on the logical and elegant sequence of results that culminate in this important theorem: differentiability and analyticity, Cauchy's theorem, the Cauchy integral formula, Taylor and Laurent series, and, finally, the Residue Theorem.

15.1. THE RESIDUE THEOREM

We are interested in the evaluation of a contour integral

$$I = \int_C f(z)\, dz, \tag{15.1}$$

where C is closed and f may have isolated singularities inside C, say at $z_1, z_2, \ldots, z_m$.

First, we deform C to a contour C' as shown in Fig. 15.1(a), where the circles are small enough to be nonintersecting (and where $m = 3$ for definiteness). Since the straight segments cancel in pairs, (15.1) reduces to

$$\begin{aligned} I &= \int_{C_1} f(z)\, dz + \int_{C_2} f(z) dz + \cdots + \int_{C_m} f(z)\, dz \\ &\equiv I_1 + I_2 + \cdots + I_m, \end{aligned} \tag{15.2}$$

where the C_j's encircle the z_j's and are in the same sense as C, as shown (for $m = 3$) in Fig. 15.1(b). To evaluate the I_j's, we need merely return to (14.21). With $n = -1$, we have

$$I_j = \int_{C_j} f(z)\, dz = 2\pi i c_{-1}^{(j)} \tag{15.3}$$

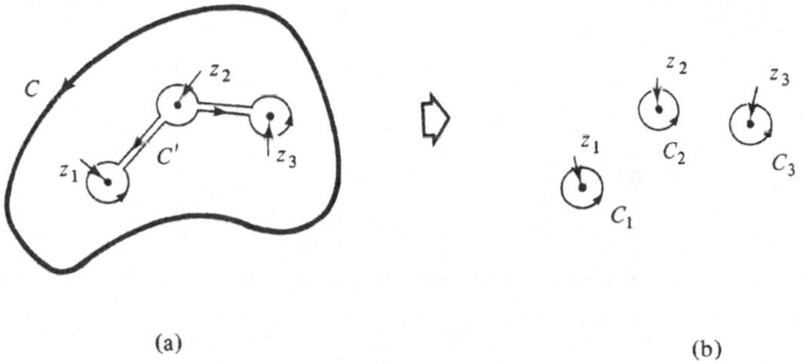

Figure 15.1. Deforming C to circles.

for each j, where $c_{-1}^{(j)}$ is the c_{-1} coefficient in the Laurent expansion of f about z_j in the annulus with arbitrarily small inner radius. Therefore

$$I = \sum_j I_j = 2\pi i \sum_j c_{-1}^{(j)}. \tag{15.4}$$

Since the only Laurent coefficients that survive are the $c_{-1}^{(j)}$'s, we call them the **residues** of $f(z)$ at the points z_j. Thus (15.4) states that *I is $2\pi i$ times the sum of the residues*, and this statement is the so-called **Residue Theorem**. The theorem is actually quite spectacular, for it says that in order to evaluate (15.1), we need only evaluate one coefficient out of the Laurent expansion of f at each of its singularities inside C.

15.2. THE CALCULATION OF RESIDUES

Actually, we already know how to compute residues because we know how to do Laurent expansions. We don't need the whole Laurent series in (15.4), however, only *one* coefficient, and so we wonder if there might be a formula specifically for c_{-1}.

Suppose first that $f(z)$ has a *simple pole* (i.e., a first-order pole) at $z = a$, say. Then its Laurent expansion in an annulus with arbitrarily small inner circle is of the form

$$f(z) = \frac{c_{-1}}{z-a} + c_0 + c_1(z-a) + \cdots. \tag{15.5}$$

It follows that

$$(z-a)f(z) = c_{-1} + c_0(z-a) + c_1(z-a)^2 + \cdots.$$

If we let $z \longrightarrow a$, the right-hand side reduces to the desired coefficient c_{-1}, so that

$$c_{-1} = \lim_{z\to a} [(z-a)f(z)]. \tag{15.6}$$

Example 15.1.

$$f(z) = \frac{2z+3}{z^2-4z}.$$

This function clearly has simple poles at $z = 0$ and 4. At $z = 0$,

$$c_{-1} = \lim_{z\to 0} \frac{z(2z+3)}{z^2-4z} = \lim_{z\to 0} \frac{2z+3}{z-4} = -\frac{3}{4}.$$

At $z = 4$,

$$c_{-1} = \text{etc.} = \frac{11}{4}. \quad ▮$$

Example 15.2.

$$f(z) = \frac{1}{\sin z}.$$

This function has poles at the zeros $z = n\pi$ of $\sin z$. They are simple poles, since

$$\sin z = (-1)^n(z - n\pi) - \frac{(-1)^n}{3!}(z - n\pi)^3 + \cdots \sim (-1)^n(z - n\pi).$$

The residue at $z = n\pi$ is

$$c_{-1} = \lim_{z \to n\pi} \frac{z - n\pi}{\sin z} = \lim_{z \to n\pi} \frac{z - n\pi}{(-1)^n(z - n\pi)[1 + O(z - n\pi)^2]} = (-1)^n.$$

Or we could have simply applied L'Hospital's rule (which holds in the complex case as well):

$$c_{-1} = \lim_{z \to n\pi} \frac{z - n\pi}{\sin z} = \lim_{z \to n\pi} \frac{1}{\cos z} = (-1)^n. \quad ▮$$

Next, suppose that $f(z)$ has a *second-order pole* at $z = a$. Then

$$f(z) = \frac{c_{-2}}{(z-a)^2} + \frac{c_{-1}}{(z-a)} + c_0 + c_1(z-a) + \cdots,$$

$$(z-a)^2 f(z) = c_{-2} + c_{-1}(z-a) + c_0(z-a)^2 + \cdots,$$

$$\frac{d}{dz}[(z-a)^2 f(z)] = c_{-1} + 2c_0(z-a) + \cdots,$$

and so

$$c_{-1} = \lim_{z \to a} \frac{d}{dz}[(z-a)^2 f(z)]. \tag{15.7}$$

Generalizing to a *pole of order k*, it is not hard to show that

$$c_{-1} = \frac{1}{(k-1)!} \lim_{z \to a} \frac{d^{k-1}}{dz^{k-1}}[(z-a)^k f(z)]. \tag{15.8}$$

Example 15.3.

$$f(z) = \frac{z^2}{(z^2 + 4)^2}.$$

This function has second-order poles at $z = \pm 2i$. At $z = 2i$,

$$c_{-1} = \lim_{z \to 2i} \frac{d}{dz} \frac{(z - 2i)^2 z^2}{(z + 2i)^2(z - 2i)^2} = \text{etc.} = -\frac{i}{8}.$$

Alternatively, it's not much longer to expand directly—that is, with $z - 2i \equiv \zeta$,

$$f = \frac{(\zeta + 2i)^2}{\zeta^2(\zeta + 4i)^2} = -\frac{(\zeta + 2i)^2}{16\zeta^2[1 + (\zeta/4i)]^2}$$

$$= -\frac{1}{16\zeta^2}(\zeta^2 + 4i\zeta - 4)\left(1 - \frac{2\zeta}{4i} + \cdots\right),$$

and picking out the coefficient of the $1/\zeta$ term,

$$c_{-1} = -\frac{1}{16}\left(4i + \frac{2}{i}\right) = -\frac{i}{8}$$

as before. We leave it for the reader to show that the other residue, at $z = -2i$, is $c_{-1} = i/8$. ∎

15.3. APPLICATIONS

Consider next some representative applications of the Residue Theorem.

Example 15.4.

$$I = \int_0^\infty \frac{x^2}{(x^2+4)^2}\, dx = ?$$

This is, in fact, a contour integral, where the contour is the positive real axis in the complex plane. Nevertheless, we need a *closed* contour to apply the residue theorem. First, note that

$$I = \frac{1}{2}\int_{-\infty}^{\infty} \frac{x^2}{(x^2+4)^2}\, dx$$

since the integrand is an even function. In order to apply the residue theorem, we consider the integral

$$J = \int_C \frac{z^2}{(z^2+4)^2}\, dz,$$

where C is as shown in Fig. 15.2. The integrand is singular only at $z = \pm 2i$, where it has second-order poles. Of these poles, only $+2i$ lies within C, and (from Example 15.3) the residue there is $-i/8$. According to the residue theorem, then,

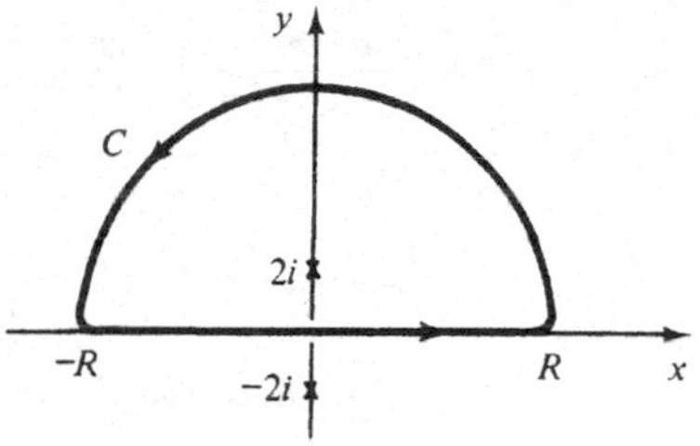

Figure 15.2. The closed contour C.

$$J = 2\pi i \frac{-i}{8} = \int_{-R}^{R} \frac{x^2}{(x^2+4)^2}\, dx + \int_{\frown}. \tag{15.9}$$

In estimating the last integral in (15.9)—the integral over the semicircle—note that for large R we have $z^2 + 4 \sim z^2$ on the semicircle. Thus

$$\begin{aligned} \left|\int_{\frown}\right| &= \text{order of magnitude of integrand} \\ &\quad \text{times } \pi R \text{ (i.e., the length of the contour)} \\ &= O\left(\frac{1}{R^2}\right)\cdot \pi R = O\left(\frac{1}{R}\right) \longrightarrow 0 \quad \text{as } R \longrightarrow \infty. \end{aligned}$$

So by letting $R \longrightarrow \infty$ in (15.9), we have

$$\frac{\pi}{4} = 2I + 0$$

or $I = \pi/8$.

COMMENT. Note that we could also have closed the contour in the *bottom* half-plane, as shown in Fig. 15.3. The contribution from the semicircle is still $O(1/R)$

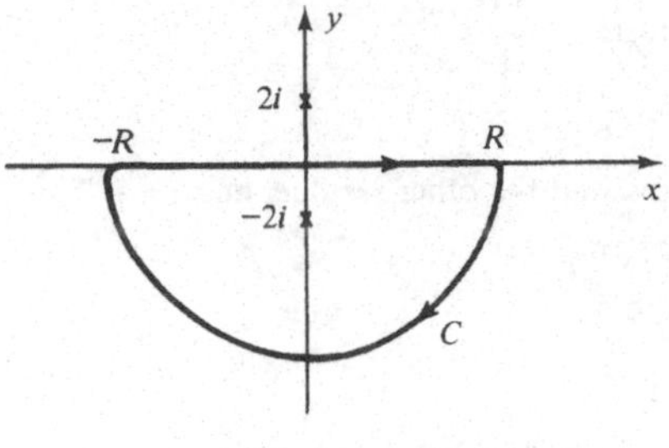

Figure 15.3. An alternative contour.

$\longrightarrow 0$. This time the relevant residue is $+i/8$, but remember that C is now clockwise, whereas it is assumed ccw in the residue theorem. Hence we have

$$J = -2\pi i \frac{+i}{8} = \frac{\pi}{4} = 2I,$$

so that $I = \pi/8$ as before. ∎

Example 15.5.

$$I = \int_0^\infty \frac{\cos ax}{x^2+4}\,dx = ? \qquad (a \geq 0)$$

Proceeding as in the previous example, we could consider

$$J = \int_C \frac{\cos az}{z^2+4}\,dz,$$

where C is the same as in Fig. 15.2. The problem is that $\cos az = (e^{iaz} + e^{-iaz})/2$ is so large on the semicircle that we can no longer claim that the integral over the semicircle tends to zero as $R \longrightarrow \infty$; for instance, at the top of the circle, where $z = iR$, $\cos az = (e^{-aR} + e^{aR})/2 \sim e^{aR}/2$. (If we use the contour of Fig. 15.3 instead, then the e^{-iaz} portion is fine, but the e^{+iaz} part is not.)

Let us try retaining the contour of Fig. 15.2 but considering instead

$$J = \int_C \frac{e^{iaz}}{z^2+4}\,dz.$$

Since

$$|e^{iaz}| = |e^{iax-ay}| = |e^{iax}||e^{-ay}| = |e^{-ay}| \leq 1$$

on the semicircle, we can say that the contribution from the semicircle is

$$O\left(\frac{1}{R^2}\right) \cdot \pi R = O\left(\frac{1}{R}\right) \longrightarrow 0 \quad \text{as } R \longrightarrow \infty.$$

Therefore

$$J = (2\pi i)(\text{Res. @ } 2i) = \int_{-\infty}^\infty \frac{\cos ax}{x^2+4}\,dx + i\int_{-\infty}^\infty \frac{\sin ax}{x^2+4}\,dx \tag{15.10}$$

(where Res. @ $2i$ means the residue at $2i$). But

$$\text{Res. @ } 2i = \lim_{z\to 2i} \frac{(z-2i)e^{iaz}}{z^2+4} = \lim_{z\to 2i} \frac{e^{iaz}}{z+2i} = \frac{e^{-2a}}{4i}.$$

Finally, equating real and imaginary parts in (15.10), we obtain

$$\int_0^\infty \frac{\cos ax}{x^2+4}\,dx = \frac{\pi}{4}e^{-2a}, \tag{15.11}$$

as well as the "bonus"

$$\int_{-\infty}^\infty \frac{\sin ax}{x^2+4}\,dx = 0. \tag{15.12}$$

[Of course, (15.12) is not particularly exciting, since it follows from the antisymmetry of the integrand.] ∎

Example 15.6.

$$I = \int_0^\infty \frac{x^{a-1}}{1+x}\,dx - ? \qquad (0 < a < 1, \text{ for convergence}) \tag{15.13}$$

In evaluating this integral, it looks reasonable to consider

$$J = \int_C \frac{z^{a-1}}{1+z}\,dz, \tag{15.14}$$

where C is a suitably chosen, closed contour, one piece of which is the positive real axis. First, however, it is necessary to say what we mean by z^{a-1}, which is multi-valued. Let us define it according to the (principal value) branch cut shown in Fig. 15.4, say.

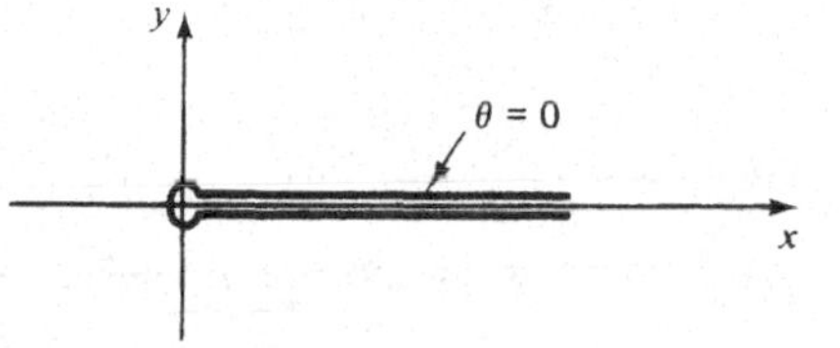

Figure 15.4. Branch cut for z^{a-1}.

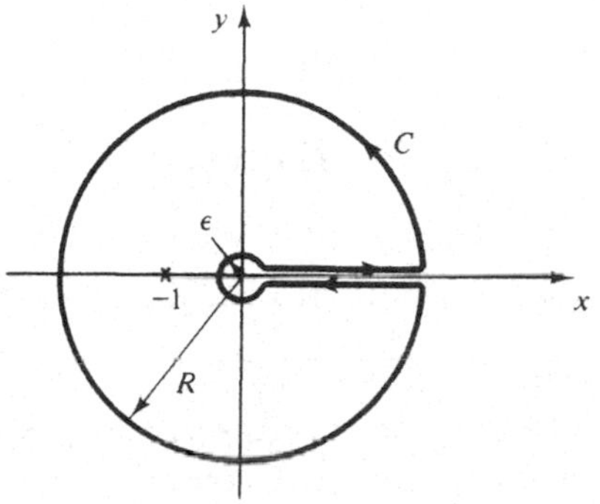

Figure 15.5. Contour C in (15.14).

A reasonable choice for C is therefore as shown in Fig. 15.5. It consists of a large circle of radius R (R will $\to \infty$), a small circle of radius ϵ (ϵ will $\to 0$), and two straight segments along the top and bottom edges of the branch cut. (We do *not* expect the contributions from the two straight segments to cancel, since z^{a-1} will have different values on the two lines as we will see.)

The only singularity inside C is the simple pole at $z = -1$, with

$$\text{Res. @ } -1 = \lim_{z\to -1} \frac{(z+1)z^{a-1}}{1+z} = (-1)^{a-1}.$$

To evaluate $(-1)^{a-1}$, we must be consistent with our choice of the branch cut (Fig. 15.4)! Thus

$$(-1)^{a-1} = (e^{\pi i})^{a-1} = e^{(a-1)\pi i},$$

so that (with obvious shorthand notation)

$$J = 2\pi i e^{(a-1)\pi i} = \int_{\rightarrow} + \int_{R \text{ circle}} + \int_{\leftarrow} + \int_{\epsilon \text{ circle}}. \tag{15.15}$$

We now look at each of the four integrals. Since $1 + z$ is $O(1)$ on the ϵ circle, and $O(R)$ on the R circle,

$$\left| \int_{\epsilon \text{ circle}} \right| = O(\epsilon^{a-1}) \cdot 2\pi\epsilon = O(\epsilon^a) \longrightarrow 0 \quad \text{as } \epsilon \longrightarrow 0,$$

and

$$\left|\int_{R\text{ circle}}\right| = O\left(\frac{R^{a-1}}{R}\right)\cdot 2\pi R = O(R^{a-1}) \longrightarrow 0 \quad \text{as } R \longrightarrow \infty.$$

Finally,

$$\int_{\rightarrow} = \int_{\epsilon}^{R} \frac{(xe^{i0})^{a-1}}{1+x}\,dx \longrightarrow \int_0^{\infty} \frac{x^{a-1}}{1+x}\,dx = I$$

and

$$\int_{\leftarrow} = \int_{R}^{\epsilon} \frac{(xe^{2\pi i})^{a-1}}{1+x}\,dx \longrightarrow \int_{\infty}^{0} \frac{x^{a-1}e^{(a-1)2\pi i}}{1+x}\,dx = -e^{(a-1)2\pi i}I.$$

Consequently, (15.15) becomes

$$2\pi i e^{(a-1)\pi i} = (1 - e^{(a-1)2\pi i})I,$$

or

$$I = \frac{\pi}{\sin \pi(1-a)}. \quad \blacksquare$$

Example 15.7.

$$I = \int_0^{2\pi} \frac{d\theta}{\cos\theta + 2} = ? \tag{15.16}$$

This integral is of the form

$$\int_0^{2\pi} F(\sin\theta, \cos\theta)\,d\theta, \tag{15.17}$$

where F is a rational function (i.e., a polynomial divided by a polynomial) of its two arguments.

Let $e^{i\theta} \equiv z$. Then

$$ie^{i\theta}\,d\theta = dz \quad \text{or} \quad d\theta = \frac{dz}{iz},$$

$$\sin\theta = \frac{e^{i\theta} - e^{-i\theta}}{2i} = \frac{z - z^{-1}}{2i},$$

and

$$\cos\theta = \frac{e^{i\theta} + e^{-i\theta}}{2} = \frac{z + z^{-1}}{2};$$

as a result, (15.17) becomes the contour integral

$$\int_C F\left(\frac{z - z^{-1}}{2i}, \frac{z + z^{-1}}{2}\right)\frac{dz}{iz}, \tag{15.18}$$

where C is the unit circle, traversed ccw. (Recall that $z = e^{i\theta}$, so that $|z| = |e^{i\theta}| = 1$.) The integrand is a rational function of z; and since rational functions can have no singularities other than poles (which are isolated), (15.18) can be evaluated by the residue theorem.

Returning to (15.16), we therefore have

$$I = \int_C \frac{1}{\frac{z + z^{-1}}{2} + 2}\,\frac{dz}{iz} = \frac{2}{i}\int_C \frac{dz}{z^2 + 4z + 1}. \tag{15.19}$$

Factoring the denominator in order to locate the poles,

$$z^2 + 4z + 1 = (z - z_+)(z - z_-),$$

where

$$z_+ = -2 + \sqrt{3} \quad \text{and} \quad z_- = -2 - \sqrt{3}.$$

Of these poles, only z_+ lies within C (Fig. 15.6), and

$$\text{Res. @ } z_+ = \lim_{z \to z_+} (z - z_+) \frac{1}{(z - z_+)(z - z_-)}$$

$$= \frac{1}{z_+ - z_-} = \frac{1}{2\sqrt{3}};$$

hence (canceling an i for an i and a 2 for a 2),

$$I = 2\pi i \left(\frac{2}{i}\right)\left(\frac{1}{2\sqrt{3}}\right) = \frac{2\pi}{\sqrt{3}}. \quad \blacksquare$$

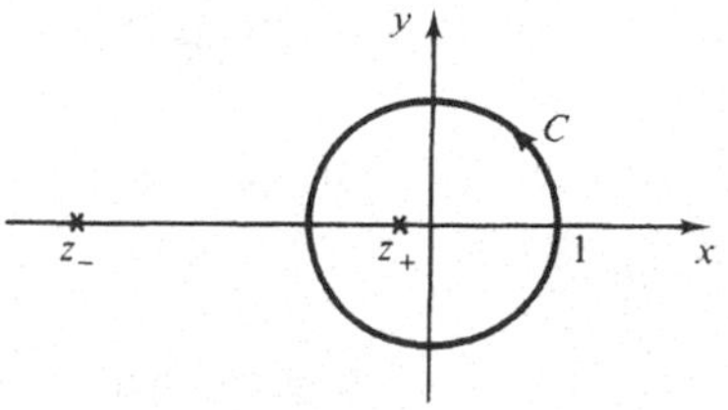

Figure 15.6. The integration (15.19).

Finally, consider the application of the residue theorem to the *Laplace inversion formula*

$$f(t) = \frac{1}{2\pi i} \int_{\gamma - i\infty}^{\gamma + i\infty} \bar{f}(s) e^{st}\, ds. \tag{15.20}$$

(Before continuing, you may want to reread Section 6.2 up to Example 6.2.)

Example 15.8. Find the inverse of the Laplace transform

$$\bar{f}(s) = \frac{1}{s - a},$$

where, for definiteness, a is real and positive. According to (15.20),

$$f(t) = \frac{1}{2\pi i} \int_{\gamma - i\infty}^{\gamma + i\infty} \frac{e^{st}}{s - a}\, ds. \tag{15.21}$$

Recall that γ is an arbitrary real number, except that we ask it to be sufficiently positive so that $e^{-\gamma t}$ "clobbers" $f(t)$ as $t \longrightarrow \infty$—specifically, such that

$$\int_0^\infty e^{-\gamma t} |f(t)|\, dt < \infty. \tag{15.22}$$

Unfortunately, we don't know $f(t)$ yet and so are unsure how to choose γ. Cheating, we know from (6.33) of Chapter 6 that $f(t) = e^{at}$. Condition (15.22) then dictates the choice $\gamma > a$.

Note that this choice places the straight-line contour ($\gamma - i\infty$ to $\gamma + i\infty$) to the *right* of $\bar{f}$'s singularity—namely, the pole at $s = a$. In fact, this is the general rule: *choose γ sufficiently positive so that all singularities of $\bar{f}(s)$ lie to the left of the line $\gamma - i\infty$ to $\gamma + i\infty$*. (Even if skeptical, you can always follow this rule tentatively and check a posteriori.)

Closing the contour (so that we can apply the residue theorem) with a large semicircle on the left, as shown in Fig. 15.7, we therefore consider

$$I = \frac{1}{2\pi i} \int_C \frac{e^{st}}{s - a}\, ds. \tag{15.23}$$

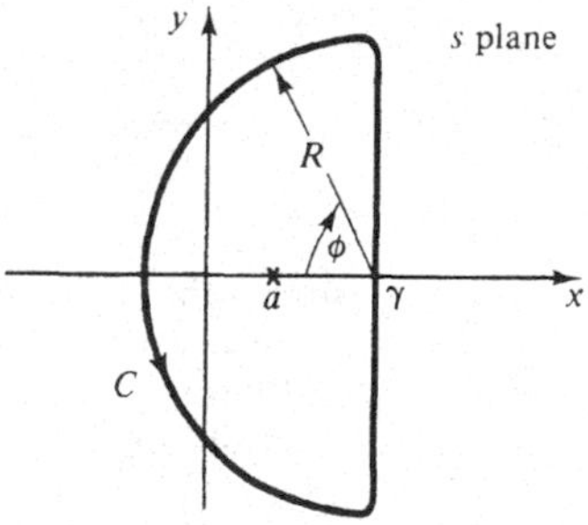

Figure 15.7. Laplace inversion contour for (15.23).

According to the residue theorem,

$$I = (2\pi i)(\text{Res. @ } a) = 2\pi i \frac{e^{at}}{2\pi i} = e^{at}. \tag{15.24}$$

On the other hand,

$$I = \frac{1}{2\pi i} \int_{\gamma - iR}^{\gamma + iR} \frac{e^{st}}{s - a}\, ds + \frac{1}{2\pi i} \int_{C} \frac{e^{st}}{s - a}\, ds. \tag{15.25}$$

In particular, as $R \longrightarrow \infty$, the first integral tends to $f(t)$ according to (15.21), and (as we will soon show) the second integral tends to zero. So

$$I = f(t), \tag{15.26}$$

and comparison of (15.24) and (15.26) tells us that

$$f(t) = e^{at}.$$

As promised, we must now verify that the semicircle integral $\longrightarrow 0$ as $R \longrightarrow \infty$. We have

$$|e^{st}| = |e^{(x+iy)t}| = e^{xt} \leq e^{\gamma t}. \tag{15.27}$$

Since t is regarded as fixed in the integration, we conclude that e^{st} is *bounded* on the semicircle. And since $s - a = O(R)$ on the semicircle, we have

$$\left| \int_C \right| = O\left(\frac{1}{R}\right) \cdot \pi R = O(1). \tag{15.28}$$

Unfortunately, we cannot conclude that the integral tends to zero as $R \longrightarrow \infty$. But behind (15.28) is the inequality (15.27), which is quite conservative. Introducing the polar variable ϕ (Fig. 15.7), (15.27) becomes

$$|e^{st}| = e^{xt} = e^{(\gamma - R\cos\phi)t} = e^{\gamma t} e^{-Rt\cos\phi},$$

but this time we will *not* say that this quantity is $\leq e^{\gamma t}$ on the semicircle, since we will need some help from the $\exp(-Rt\cos\phi)$ factor. Also,

$$\left|\frac{1}{s - a}\right| \leq \frac{1}{R - (\gamma - a)}$$

because the $s - a$ "vector" is shortest when $\phi = 0$ and is $R - (\gamma - a)$ then. Therefore

$$\left| \int_C \right| \leq \frac{e^{\gamma t}}{2\pi(R - \gamma + a)} \int_{-\pi/2}^{\pi/2} e^{-Rt\cos\phi} R\, d\phi. \tag{15.29}$$

As $R \longrightarrow \infty$,

$$\frac{R}{R - \gamma + a} \longrightarrow 1$$

and (Exercise 15.12)

$$\int_{-\pi/2}^{\pi/2} e^{-Rt\cos\phi}\, d\phi \longrightarrow 0, \tag{15.30}$$

so that the semicircle integral does tend to zero as claimed.

So far we have tacitly assumed that t is positive. For $t < 0$, (15.30) does not hold, and it is not difficult to see that now we need to close the contour on the *right*, as shown in Fig. 15.8. Since the singular point $s = a$ is *outside* C, (15.23) becomes

$$0 = \frac{1}{2\pi i} \int_{\gamma - iR}^{\gamma + iR} \frac{e^{st}}{s - a}\, ds + \frac{1}{2\pi i} \int_{C} \frac{e^{st}}{s - a}\, ds,$$

and as $R \longrightarrow \infty$ this becomes

$$0 = f(t)$$

since (remember that t is *negative*)

$$\left|\int_{\gamma}\right| \leq \frac{e^{\gamma t}}{2\pi R}\int_{-\pi/2}^{\pi/2} e^{Rt\cos\varphi} R\, d\varphi \longrightarrow 0.$$

Summarizing,

$$f(t) = \begin{cases} e^{at}, & t > 0 \\ 0, & t < 0. \end{cases}$$

The fact that the inverse is identically zero for $t < 0$ is in keeping with the discussion in Section 6.2.

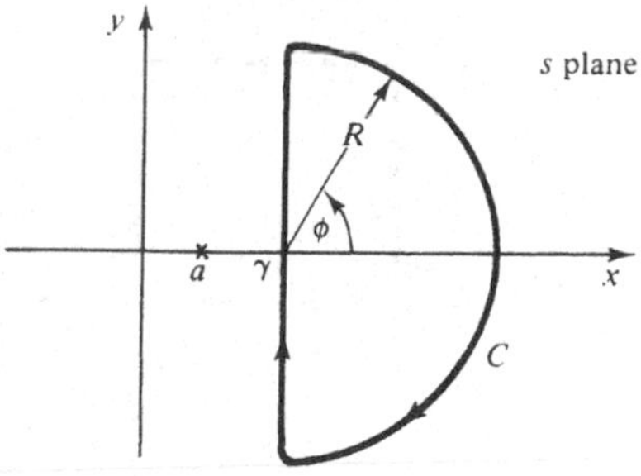

Figure 15.8. Laplace inversion contour for $t < 0$.

COMMENT. The slightly more refined bounding procedure used to show that the semicircle integral tends to zero (a useful generalization of which is contained in Exercise 15.14) is not always needed in Laplace inversions. With $\bar{f}(s) = 1/(s^2 + a^2)$, for instance, the simple bound (15.27) is sufficient, and we have

$$\int_C \frac{e^{st}}{s^2 + a^2}\, ds = O\left(\frac{1}{R^2}\right) \cdot \pi R = O\left(\frac{1}{R}\right) \longrightarrow 0. \blacksquare$$

15.4. CAUCHY PRINCIPAL VALUE

As might be expected, there is a generalization of the **Cauchy principal value,** defined by equation (3.14), to contour integrals. Specifically, suppose that $f(z)$ is singular at some point z_0 on the path of integration, as sketched in Fig. 15.9. Then

$$⨍_A^B f(z)\, dz \equiv \lim_{\epsilon \to 0} \left\{\int_A^a f(z)\, dz + \int_b^B f(z)\, dz\right\}, \tag{15.31}$$

provided, of course, that the limit exists.

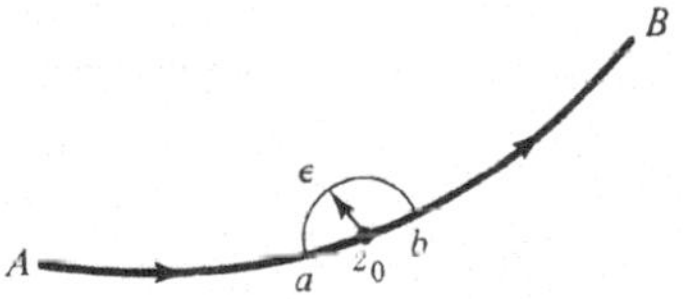

Figure 15.9. Cauchy principal value.

Example 15.9. Evaluate

$$I = ⨍_C \frac{dz}{\sin z}, \tag{15.32}$$

where C is the ccw unit square with vertices at 0, 1, $1 + i$, and i. The only singularity inside or on C is the simple pole at $z = 0$. According to the principal value interpretation called for in (15.32), the neighborhood of this point, from ϵi to ϵ, is deleted from C, as shown in Fig. 15.10(a).

In order to apply the residue theorem, C must be *closed*; so let us close it with a three-quarter circle of radius ϵ, as shown in Fig. 15.10(b). Then the integral from ϵ to 1 around to ϵi, plus the integral around the three-quarter circle, must equal $2\pi i$

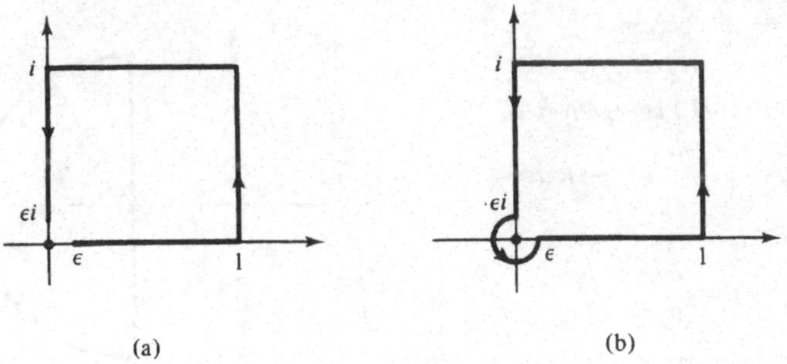

Figure 15.10. Contour of Example 15.9.

times the residue at $z = 0$:

$$\int_{\epsilon}^{\epsilon i} \frac{dz}{\sin z} + \int_{C} \frac{dz}{\sin z} = (2\pi i)(\text{Res. @ } z = 0) \tag{15.33}$$

or

$$I_1 + I_2 = (2\pi i)(1) = 2\pi i. \tag{15.34}$$

Now as $\epsilon \longrightarrow 0$, $I_1 \longrightarrow I$ and

$$I_2 = \int_{C} \left[\frac{1}{z} + O(1)\right] dz = \ln z \Big|_{\epsilon i}^{\epsilon} + O(\epsilon) = \frac{3\pi}{2} i + O(\epsilon) \longrightarrow \frac{3\pi}{2} i,$$

which means that (15.34) becomes $I + 3\pi i/2 = 2\pi i$ or $I = \pi i/2$.

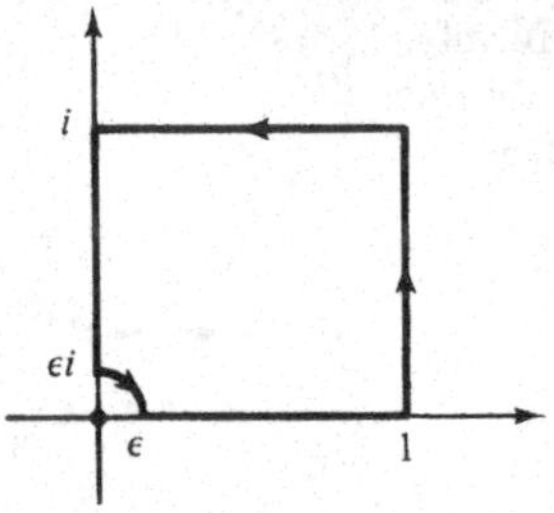

Figure 15.11. Alternative contour.

COMMENT. Suppose that we close the contour with a one-quarter circle (Fig. 15.11) instead. This time the pole at $z = 0$ is outside the contour, and so the integral from ϵ to 1 around to ϵi, plus the integral over the one-quarter circle, must equal zero:

$$\int_{\epsilon}^{\epsilon i} \frac{dz}{\sin z} + \int \frac{dz}{\sin z} = 0$$

or

$$I_1 + I_2 = 0. \tag{15.35}$$

Now as $\epsilon \longrightarrow 0$, $I_1 \longrightarrow I$ and $I_2 \longrightarrow (-\pi/2)i$, and therefore (15.35) becomes $I - \pi i/2 = 0$ or $I = \pi i/2$ as before. ∎

EXERCISES

15.1. Verify the following values by means of contour integration and the residue theorem.

(a) $\int_0^{2\pi} \sin^2 \theta \, d\theta = \pi$

(b) $\int_0^{2\pi} \cos^2 \theta \, d\theta = \pi$

(c) $\int_0^{\infty} \frac{x^2 \, dx}{x^6 + 1} = \frac{\pi}{6}$

(d) $\int_0^{\infty} \frac{dx}{x^4 + a^4} = \frac{\pi}{2\sqrt{2}\, a^3} \quad (a > 0)$

(e) $\displaystyle\int_0^{2\pi} \sin^6 t\, dt = \frac{5\pi}{8}$ (f) $\displaystyle\int_0^{\infty} \frac{\cos x\, dx}{(x^2+1)^2} = \frac{\pi}{2e}$

(g) $\displaystyle\int_0^{2\pi} \frac{\sin^2 x}{a + b\cos x}\, dx = \frac{2\pi}{b^2}(a - \sqrt{a^2 - b^2}) \qquad (a > b > 0)$

(h) $\displaystyle\int_0^{\pi} \frac{dx}{a - \cos x} = \frac{\pi}{\sqrt{a^2 - 1}} \qquad (a > 1)$

(i) $\displaystyle\int_0^{\infty} \frac{x \sin ax}{x^2 + b^2}\, dx = \frac{\pi}{2} e^{-ab} \qquad (a > 0, b > 0)$

(j) $\displaystyle\int_{-\infty}^{\infty} \frac{dx}{4x^2 + 2x + 1} = \frac{\pi}{\sqrt{3}}$

(k) $\displaystyle\int_0^{\pi} \frac{\cos x\, dx}{1 - 2a\cos x + a^2} = \frac{\pi a}{1 - a^2} \qquad (a^2 < 1)$

15.2. Derive equation (6.17) of Chapter 6 by carrying out the indicated integration.

15.3. Evaluate

$$\int_0^{\infty} \frac{dx}{x^2 + x + 1}.$$

Hint: Consider the integral of $(\ln z)/(z^2 + z + 1)$ around the contour shown in Fig. 15.5. Note how the inclusion of the $\ln z$ bails us out of an otherwise difficult situation.

15.4. Evaluate (by contour integration).

(a) $\displaystyle\int_0^{\infty} \frac{dx}{x^2 + 3x + 2}$ (b) $\displaystyle\int_0^{1} \frac{dx}{x^2 + 4}$

Hint: See the previous exercise.

15.5. Show that

$$\int_0^{\infty} e^{-x^2} \cos 2ax\, dx = \frac{\sqrt{\pi}\, e^{-a^2}}{2}$$

by integrating e^{-z^2} around a rectangle with vertices at 0, R, $R + ia$, and ia and recalling that

$$\int_0^{\infty} e^{-x^2}\, dx = \frac{\sqrt{\pi}}{2}. \tag{15.36}$$

15.6. Show that

$$\int_0^{\infty} \cos x^2\, dx = \int_0^{\infty} \sin x^2\, dx = \frac{1}{2}\sqrt{\frac{\pi}{2}} \tag{15.37}$$

by integrating e^{iz^2} around the contour shown in Fig. 15.12 and recalling (15.36) above. More generally,

$$C(x) \equiv \int_0^{x} \cos \frac{\pi t^2}{2}\, dt \quad \text{and} \quad S(x) \equiv \int_0^{x} \sin \frac{\pi t^2}{2}\, dt$$

are the so-called cosine and sine **Fresnel integrals**, and it follows from (15.37) that

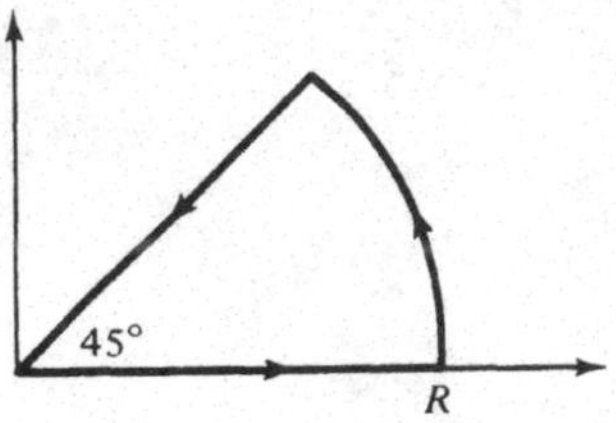

Figure 15.12.

$C(\infty) = S(\infty) = \frac{1}{2}$. *Hint*: You will probably wish to show that $J = R\int_0^{\pi/2} e^{-R^2 \sin\theta}\, d\theta \longrightarrow 0$ as $R \longrightarrow \infty$. To do so, note that $J < R\int_0^{\pi/2} e^{-2R^2\theta/\pi}\, d\theta$ (Why?), which is easily evaluated.

15.7. Show that

(a) $\displaystyle ⨍_0^\infty \frac{x^{a-1}}{1-x}\, dx = \pi \cot (a-1)\pi \qquad (0 < a < 1).$

(b) $\displaystyle ⨍_{-\infty}^\infty \frac{x\, dx}{x^3+1} = \frac{\pi}{\sqrt{3}}.$

15.8. Evaluate

$$⨍_C \frac{dz}{z(1-z)},$$

where C is the ccw unit circle.

15.9. Show that

$$\int_0^\infty \frac{\sin x}{x}\, dx = \frac{\pi}{2}$$

by integrating e^{iz}/z around a suitable contour.

15.10. Show that

$$⨍_0^\pi \frac{\cos\theta\, d\theta}{\cos\theta - \cos\alpha} = \pi.$$

(This is a special case of the so-called **airfoil** or **Glauert integral**

$$⨍_0^\pi \frac{\cos n\theta\, d\theta}{\cos\theta - \cos\alpha} = \pi\frac{\sin n\alpha}{\sin\alpha}.)$$

15.11. Find the inverse of these Laplace transforms directly from the inversion formula.

(a) $\dfrac{1}{s^2+a^2}$ (b) $\dfrac{s}{s^2+a^2}$

*(c) $\dfrac{1}{\sqrt{s}}$ *Hint:* Use the "indented" contour shown in Fig. 15.13, plus Exercise 15.14 below.

(d) $\dfrac{1}{s^3}$ (e) $\dfrac{e^{-as}}{s} \qquad (a \geq 0)$

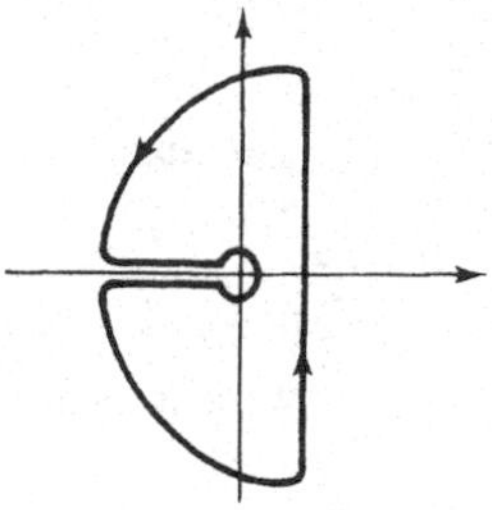

Figure 15.13.

15.12. Show carefully why (15.30) is true.

15.13. Was it necessary to define the branch cut and integration contour for Example 15.6 the way that we did? Try a few other possibilities and see what happens.

15.14. Show that

$$\int_C e^{st}\bar{f}(s)\,ds \longrightarrow 0$$

as $R \longrightarrow \infty$ if $\bar{f}(s) = O(R^{-\alpha})$ on C for $\alpha > 0$, where C is the semicircle on the left for $t > 0$ and on the right for $t < 0$ (Fig. 15.14). *Hint:* Recalling Exercise 4.7 may help.

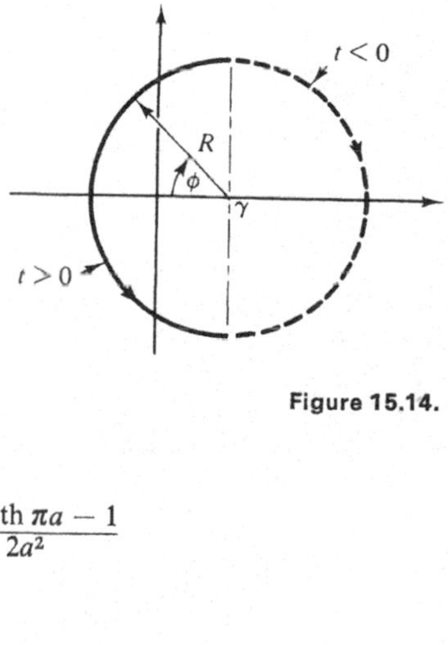

Figure 15.14.

***15.15.** (*The summing of certain series*) Show that

$$\sum_{n=1}^{\infty} \frac{1}{n^2 + a^2} = \frac{\pi a \coth \pi a - 1}{2a^2}$$

by considering the contour integral

$$\int_C \frac{1}{z^2 + a^2}\pi \cot \pi z\,dz,$$

where C is the square with corners at $\pm(N + \frac{1}{2}) \pm (N + \frac{1}{2})i$ and $N \longrightarrow \infty$. More generally, suppose that $f(z)$ is a rational function, none of whose zeros or poles is an integer, such that $zf(z) \longrightarrow 0$ as $|z| \longrightarrow \infty$, and having simple poles at $z_1, z_2, \ldots, z_k$

with residues $b_1, b_2, \ldots, b_k$. By considering

$$\int_C f(z)\, \pi \cot \pi z \, dz,$$

show that

$$\sum_{n=-\infty}^{\infty} f(n) = -\pi(b_1 \cot \pi z_1 + \cdots + b_k \cot \pi z_k).$$

[For alternating series, $\sum (-1)^n f(n)$, we use $\pi \operatorname{cosec} \pi z$ in place of $\pi \cot \pi z$.]

15.16. What is the residue at $z = 0$ of the function

$$f = \frac{1}{z^3} + \frac{1}{z^4} + \frac{1}{z^5} + \cdots ?$$

15.17. Make up two examination-type problems on this chapter.

Chapter 16

Conformal Mapping

The fundamental problem of conformal mapping is to find an analytic function $w = f(z)$ that maps a given region of the z plane into some desired region of the w plane. We will see how such a procedure provides us with a powerful method of solution of Laplace's equation in two dimensions as encountered in connection with steady-state heat transfer, the irrotational flow of an incompressible fluid, membrane theory, the gravitational potential, and numerous other important physical settings.

16.1. THE FUNDAMENTAL PROBLEM

Consider the so-called **Dirichlet problem**—that is,

$$\nabla^2\Phi = \Phi_{xx} + \Phi_{yy} = 0 \tag{16.1}$$

in some region D of the x, y plane with Φ prescribed on the boundary C of D; for instance, Φ might be the steady-state temperature within D due to the maintenance of a known temperature distribution along the boundary C.

We will see in Part V that the degree of difficulty in solving the Dirichlet problem varies considerably with the shape of the region D; some regions, such as circular disks and rectangles, are relatively simple. It is therefore reasonable to consider a change of variables

$$u = u(x, y), \qquad v = v(x, y) \tag{16.2}$$

in the hope that the new region—that is, in the u, v plane—will be simpler than the original one. But what does (16.2) do to the governing equation (16.1)? Using chain differentiation, it is not hard to show that

$$\Phi_{xx} + \Phi_{yy} = (u_x^2 + u_y^2)\Phi_{uu} + 2(u_x v_x + u_y v_y)\Phi_{uv}$$
$$+ (v_x^2 + v_y^2)\Phi_{vv} + (u_{xx} + u_{yy})\Phi_u + (v_{xx} + v_{yy})\Phi_v,$$

which, in general, is not the Laplacian $\Phi_{uu} + \Phi_{vv}$. So in simplifying the region, we inadvertently change the partial differential equation to

$$(u_x^2 + u_y^2)\Phi_{uu} + \cdots + (v_{xx} + v_{yy})\Phi_v = 0, \tag{16.3}$$

which, in general, has nonconstant coefficients and is probably considerably less tractable than the original Laplace equation.

However, note that if we *restrict* the transformation (16.2) so that

$$u_x = v_y \quad \text{and} \quad u_y = -v_x \tag{16.4}$$

everywhere in D, and also ask u and v to be sufficiently well-behaved so that $u_{xy} = u_{yx}$ and $v_{xy} = v_{yx}$ everywhere in D, then (16.3) reduces to

$$(u_x^2 + u_y^2)(\Phi_{uu} + \Phi_{vv}) = 0$$

or

$$\Phi_{uu} + \Phi_{vv} = 0 \tag{16.5}$$

at all points where

$$u_x^2 + u_y^2 \neq 0.^1 \tag{16.6}$$

Thus far there has been no reference to complex variable theory. However, the fact that (16.4) coincides with the Cauchy–Riemann equations suggests that it might be helpful to regard the x, y and u, v planes as complex z and w planes, respectively—that is, $z = x + iy$ and $w = u + iv$. In fact, noting that $u_x^2 + u_y^2 = |w'(z)|^2$, we conclude from the preceding discussion that *if $w(z)$ is analytic and $w'(z) \neq 0$ at all points inside D (although not necessarily at all points on its boundary*[2]*), then the Laplace equation on Φ remains intact; in other words, (16.5) holds everywhere inside the new region.*

16.2. CONFORMALITY

An interesting geometric consequence of these conditions on $w(z)$ is that the mapping turns out to be **conformal**: *angles are preserved in magnitude and sense* as will now be shown.

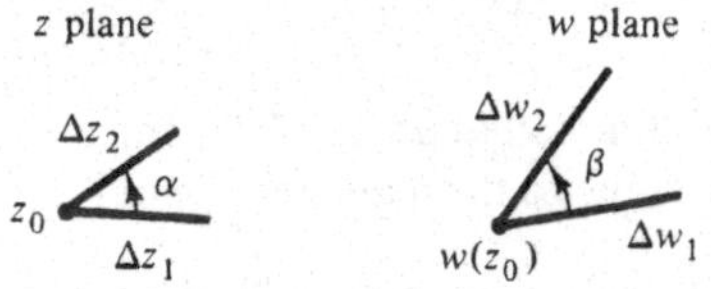

Figure 16.1. Rotation and magnification of line elements.

Consider any two infinitesimal line segments Δz_1 and Δz_2 stemming from the point in question, z_0. Their image in the w plane, Δw_1 and Δw_2, will, in general, be stretched (or contracted) and rotated, perhaps by different amounts (Fig. 16.1). To clarify this statement, note that w' is assumed to exist at z_0, so that

$$\Delta w \sim w'(z)_0 \, \Delta z. \tag{16.7}$$

[1] Recall Exercise 7.12.

[2] We only require (16.1) to be satisfied *inside D*, not necessarily on its boundary.

In particular,

$$\Delta w_1 \sim w'(z_0)\,\Delta z_1, \qquad \Delta w_2 \sim w'(z_0)\,\Delta z_2.$$

Dividing, we can cancel $w'(z_0)$, since $w'(z_0) \neq 0$ by assumption. Thus

$$\frac{\Delta w_2}{\Delta w_1} \sim \frac{\Delta z_2}{\Delta z_1},$$

and it follows (from polar representation of these quantities) that

$$\arg \Delta w_2 - \arg \Delta w_1 \sim \arg \Delta z_2 - \arg \Delta z_1$$

or

$$\beta \sim \alpha. \tag{16.8}$$

Not only does (16.8) say that $|\beta| \sim |\alpha|$, it says that $\beta \sim +\alpha$, and so angles are preserved both in magnitude and sense, as claimed. In addition, observe that $|\Delta w_1|/|\Delta z_1| \sim |w'(z_0)|$ and $|\Delta w_2|/|\Delta z_2| \sim |w'(z_0)|$; hence there is a unique magnification at z_0, independent of the orientation of Δz—namely,

$$\text{magnification} = |w'(z_0)|. \tag{16.9}$$

Example 16.1. To illustrate these ideas, consider the mapping

$$w = z^2$$

of the region consisting of the first quadrant of the z plane. As shown in Fig. 16.2, the image in the w plane is the upper half-plane. First, note that w is analytic everywhere and that $w' = 0$ only at $z = 0$, which is acceptable, since $z = 0$ is a boundary point, not an interior point, of the region under consideration. Not surprisingly, conformality breaks down there, and we see that angles get doubled there (e.g., $\pi/2 \longrightarrow \pi$, as shown in Fig. 16.2). (The fact that angles get *doubled* at $z = 0$ and that $w(z)$ has a *second*-order zero there is not a coincidence; see Exercise 16.1.)

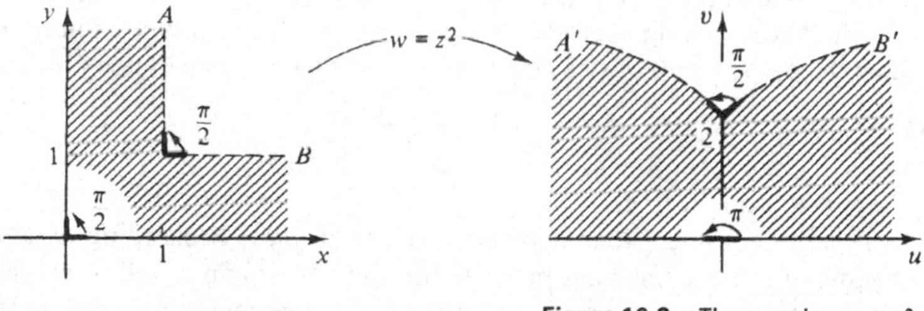

Figure 16.2. The mapping $w = z^2$.

If angles are conserved (except at $z = 0$), however, we might well wonder how the quarter-plane opens out into a half-plane. A look at what happens to the curve AB, for instance, explains how this occurs. The key is that the conformality argument is a *local* one. For example, conformality merely implies that the 90° corner in AB is preserved (as indeed it is; Fig. 16.2). The fanning out of the image $A'B'$ *away* from the corner in no way violates conformality. In other words, recall that the Δz's and Δw's in Fig. 16.1 were *infinitesimal*, not finite curves like AB and $A'B'$. ∎

16.3. THE BILINEAR TRANSFORMATION AND APPLICATIONS

Before we turn to the so-called bilinear transformation, consider the mapping

$$w = \frac{1}{z}, \tag{6.10}$$

which is singular only at the origin.

The most important feature of this mapping will be revealed if we examine how it transforms *circles*. The general equation of a circle in the z plane is

$$(x - a)^2 + (y - b)^2 = c^2, \tag{16.11}$$

where a, b, c are real. If we expand the squared terms and replace $x^2 + y^2$ by $z\bar{z}$, x by $(z + \bar{z})/2$, and y by $(z - \bar{z})/2i$, then (16.11) becomes

$$z\bar{z} + Az + \bar{A}\bar{z} + B = 0, \tag{16.12}$$

where $A = -a + bi$ and $B = a^2 + b^2 - c^2$. Setting $z = 1/w$, (16.12) becomes

$$\frac{1}{w\bar{w}} + \frac{A}{w} + \frac{\bar{A}}{\bar{w}} + B = 0 \tag{16.13}$$

or if $B \neq 0$,

$$w\bar{w} + Cw + \bar{C}\bar{w} + D = 0, \tag{16.14}$$

where $C = \bar{A}/B$ and $D = 1/B$. Then since (16.12) and (16.14) are of exactly the same form and (16.12) is the general equation of a circle in the z plane, it follows that the image in the w plane must also be a circle! This statement is not in the least surprising for the case of circles centered at the origin, since $|w| = 1/|z|$ implies that a circle $|z| = R$ goes into a circle $|w| = 1/R$. What is remarkable is that it holds even if the circle is *not* centered at $z = 0$!

Consider next the case where $B = 0$—that is, $a^2 + b^2 = c^2$. (This condition implies that the circle cuts through the origin $z = 0$. Surely the image cannot be a circle then, since the point $z = 0$ on the circle gets sent off to $w = \infty$.) Then (16.13) becomes

$$1 + A\bar{w} + \bar{A}w = 0,$$

or

$$2au - 2bv = 1,$$

which is a straight line.

Finally, consider what happens to *straight lines*. Proceeding as above, it is easy to show that if the line does not pass through the origin $z = 0$, it is sent into a circle; and if it does pass through $z = 0$, it is sent into another straight line (Exercise 16.3).

In summary, if we regard straight lines as special cases of circles (i.e., of infinite radius), then the conclusion is that *the transformation* $w = 1/z$ *sends circles into circles*!

Consider an application of this mapping to a problem in steady-state heat transfer.

Example 16.2. We seek the temperature $T(x, y)$ in the domain shown in Fig. 16.3—that is, the right half-plane with a circular hole cut out, with the straight boundary maintained at 100° and the circular boundary at 50°. (Whether the temperature exactly *at* $z = 0$ is understood to be 100°, 50°, 75°, or whatever, is unimportant.)

The $1/z$ mapping seems promising because the boundary of the region consists of two "circles." Both pass through the origin and so get sent into straight lines.

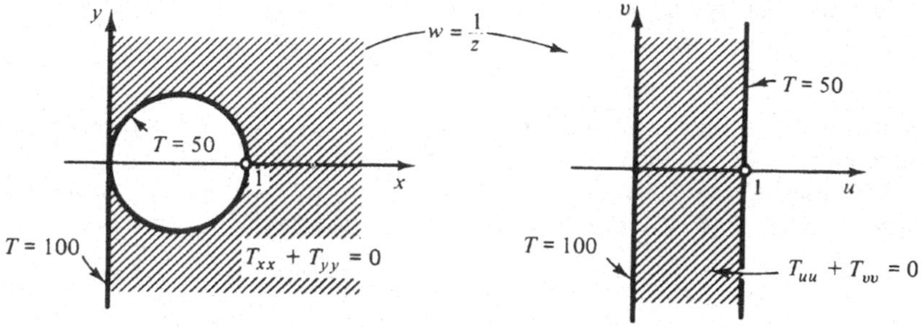

Figure 16.3. Steady state heat conduction, Example 16.2.

Since a straight line is determined by two points, the simplest way to determine the two straight lines in the w plane is to map two points from the $x = 0$ boundary and two from the $|z - 1/2| = 1/2$ boundary. For instance, $i \longrightarrow -i$, $-i \longrightarrow i$, $1 \longrightarrow 1$, and $1/2 + i/2 \longrightarrow 1 - i$; therefore the mapped region is the infinite strip shown in Fig. 16.3.

The boundary value problem in the w plane is simple! Specifically, note that there is no reason why T should vary with v. Thus $T = T(u)$, and the Laplace equation reduces to

$$T_{uu} = 0 \quad \text{or} \quad T = Au + B.$$

But $T(0) = 100$ and $T(1) = 50$, so that $B = 100$ and $A = -50$:

$$T = 100 - 50u. \tag{16.15}$$

Finally, we return to the "physical" plane by replacing u and v in (16.15) (actually, only u appears) by $u(x, y)$ and $v(x, y)$. Since

$$w = u + iv = \frac{1}{x + iy}\frac{x - iy}{x - iy} = \frac{x}{x^2 + y^2} - i\frac{y}{x^2 + y^2},$$

we have $u = x/(x^2 + y^2)$, and (16.15) becomes

$$T(x, y) = 100 - \frac{50x}{x^2 + y^2}. \quad \blacksquare$$

Consider next the problem shown in Fig. 16.4. We would like to map the eccentric annulus into a strip as above or a concentric annulus, either of which would admit a simple solution. The $w = 1/z$ map does neither, for it sends D into a half-plane with a circular hole. Even if we relocate the x, y coordinate system with respect to D, we are still unable to map into a strip, for example. (If the two bounding circles met at some point, we could make that point the origin and the result would be a strip, as in Fig. 16.3. See Exercise 16.4.)

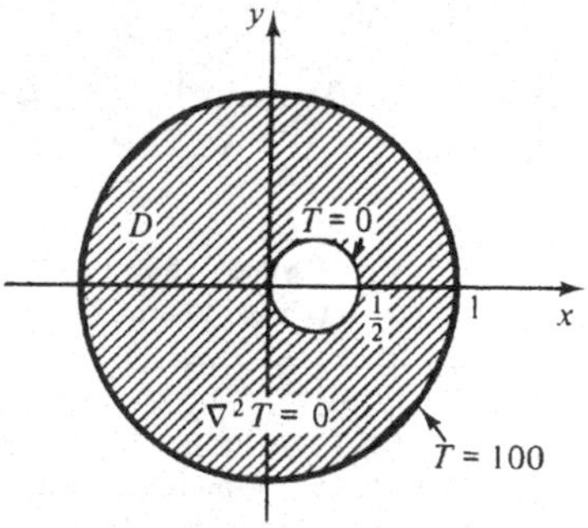

Figure 16.4. Another steady-state heat conduction problem.

This need brings us to a generalization of the $1/z$

mapping—namely, the **bilinear transformation**[3]

$$w = \frac{\alpha z + \beta}{\gamma z + \delta}, \tag{16.16}$$

which contains the four "adjustable" parameters $\alpha, \beta, \gamma, \delta$ and is therefore appreciably more powerful. In general, α, β, γ, and δ will be complex.

It turns out that (16.16) also sends circles into circles. The easiest way to see this is to decompose (16.16) into three successive transformations as follows:

$$\zeta = \gamma z + \delta, \qquad \eta = \frac{1}{\zeta}, \qquad w = \left(\frac{\beta\gamma - \alpha\delta}{\gamma}\right)\eta + \left(\frac{\alpha}{\gamma}\right), \tag{16.17}$$

that is, from the z plane to a ζ plane, from the ζ plane to an η plane, and then from the η plane to the w plane. Certainly the "linear transformation" $\zeta = \gamma z + \delta$ sends circles into circles, since it is merely a scaling by γ and a shift by δ. We've already seen that $\eta = 1/\zeta$ sends circles into circles and the third and final transformation is of the same form as the first. Since *each* of the three transformations sends circles into circles, so must the composite transformation (16.16).

Note carefully that this time it is the point $z = -\delta/\gamma$ rather than $z = 0$ that is sent to infinity and that circles through *this* point are sent into straight lines.

Example 16.3. Returning to the problem of Fig. 16.4, let us try to solve it by means of the bilinear transformation (16.16). Since three points determine a circle, we could require that three selected points on the outer circle $|z| = 1$ get sent into three selected points on some circle with center at $w = 0$ and that three selected points on the inner circle $|z - \frac{1}{4}| = \frac{1}{4}$ get sent into three selected points on some other *concentric* circle in the w plane. Such a requirement would lead to a number of simultaneous, linear, algebraic equations in the desired parameters $\alpha, \beta, \gamma, \delta$.

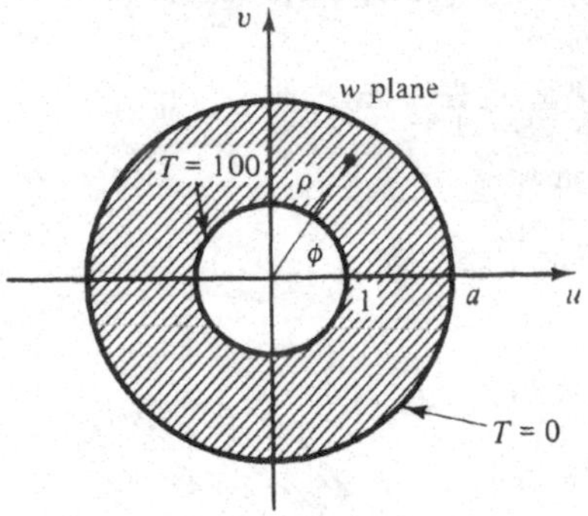

Figure 16.5. Application of (16.18) to D of Fig. 16.4.

However, it is simpler to look in a *table* of transformations in the hope of finding one that suits our needs. One suitable bilinear transformation,

$$w = \frac{z - a}{az - 1}, \qquad a = 2 + \sqrt{3}, \tag{16.18}$$

given in Churchill,[4] maps D into the annulus shown in Fig. 16.5. Observe that $w(z)$ is singular only at $z = 1/a \doteq 0.27$, which is not in D, and that $w'(z) = (a^2 - 1)/(az - 1)^2$ is nonzero in D (everywhere, in fact), so that (16.18) meets our requirements.

Next, we solve the boundary value problem in

---

[3] Studied first by A. Möbius (1790–1868), it is sometimes called **Möbius' transformation.**

[4] See R. V. Churchill's *Complex Variables and Applications,* 2nd ed., McGraw-Hill, New York, 1960, p. 287. A more complete compilation of conformal transformations can be found in the *Dictionary of Conformal Representation,* Dover, New York, 1952, by H. Kober. See also Z. Nehari, *Conformal Mapping,* McGraw-Hill, 1952, and L. Bieberbach, *Conformal Mapping,* Chelsea Publishing Co., New York, 1953.

the w plane. In terms of the polar variables ρ, ϕ we have

$$\nabla^2 T = T_{\rho\rho} + \frac{1}{\rho}T_\rho + \frac{1}{\rho^2}T_{\phi\phi} = 0. \tag{16.19}$$

But since the region is symmetric about $w = 0$ and the boundary conditions are independent of ϕ, clearly T depends only on ρ; therefore (16.19) becomes

$$T_{\rho\rho} + \frac{1}{\rho}T_\rho = 0, \tag{16.20}$$

or[5]

$$T = A + B \ln \rho.$$

Evaluating A and B from the boundary conditions on $\rho = 1$ and $\rho = a$, we have

$$T = 100\left(1 - \frac{\ln \rho}{\ln a}\right) = 100\left[1 - \frac{\ln (u^2 + v^2)}{2 \ln a}\right]. \tag{16.21}$$

Finally, equating real and imaginary parts of (16.18) gives

$$u(x, y) = \frac{(ax - 1)(x - a) + ay^2}{(ax - 1)^2 + a^2y^2}, \qquad v(x, y) = \frac{(a^2 - 1)y}{(ax - 1)^2 + a^2y^2},$$

and $T(x, y)$ follows upon inserting these expressions into (16.21).

COMMENT. Actually, we did not *prove* that T must be independent of ϕ. Logically, the procedure is as follows. Assuming *tentatively* that $T = T(\rho)$, we obtain (16.21). If it satisfies the full equation (16.19) plus the boundary conditions, then it is a bonafide solution. Whether it is the *only* solution involves the question of "uniqueness," which is deferred until Part V. ∎

Before continuing, it might be appropriate to mention a fundamental question that has so far been avoided. Does there necessarily *exist* a conformal transformation $w(z)$ that maps a given region D in the z plane into some specified region D' in the w plane? It suffices to consider whether it is possible to map any given D into the unit disk $|w| < 1$, for if $\zeta = f(z)$ maps D into $|\zeta| < 1$ and $\zeta = F(w)$ [with inverse transformation $w = \mathfrak{F}(\zeta)$, say] maps D' into $|\zeta| < 1$, then $w = \mathfrak{F}[f(z)]$ maps D into D'.

In fact, the famous **Riemann mapping theorem** assures us that any region D with a "decent" boundary can be mapped conformally into the unit disk.[6] However, the theorem provides no assistance in actually finding a suitable transformation. What is generally needed is some familiarity with the mappings provided by the so-called elementary functions, such as e^z and $\sin z$, as well as important special transformations like the *bilinear*, *Joukowsky*, and *Schwarz–Christoffel* transformations. Here, however, we will simply illustrate the *technique* through our foregoing discussion of the bilinear transformation. Additional mappings and aspects of technique are discussed in the exercises, which should be read even by those who do not plan to work them out.[7]

---

[5]We omit details in the solution of simple ordinary differential equations, since an elementary knowledge of the subject is assumed here. However, Part IV does begin with a concise review of the standard methods of solution in the event that equations like (16.20) cause difficulty.

[6]Detailed discussion of this theorem is beyond the scope of this book. If interested, see, for instance, C. Caratheodory's *Conformal Representation*, Cambridge University Press, New York, 1932.

[7]For more complete discussions, including many applications, see Churchill's book or the more advanced one by G. Carrier, M. Crook, and C. Pearson, *Functions of a Complex Variable*, McGraw-Hill, New York, 1966.

EXERCISES

16.1. If $w(z)$ has an nth-order zero at z_0—that is $w(z) \sim \alpha(z - z_0)^n$, where n is an integer greater than 1—show that conformality breaks down and angles increase by n times.

16.2. Determine the image $A'B'$ of AB in Fig. 16.2.

16.3. Show that $w = 1/z$ sends straight lines through $z = 0$ into straight lines through $w = 0$ and straight lines not through $z = 0$ into circles.

16.4. Show that two circles that intersect only at $z = 0$ are sent into *parallel* straight lines by the transformation $w = 1/z$. (Can be done in one sentence.) What if they intersect at another point as well?

16.5. Verify that the simple "scaling transformation" $w = az$, where a is, in general, complex, sends circles into circles, as claimed following equations (16.17).

16.6. Show that the inverse $z = z(w)$ of the bilinear transformation (16.16) is also a bilinear transformation.

16.7. Solve the Dirichlet problem shown in Fig. 16.6 by the mapping $w = z^2$.

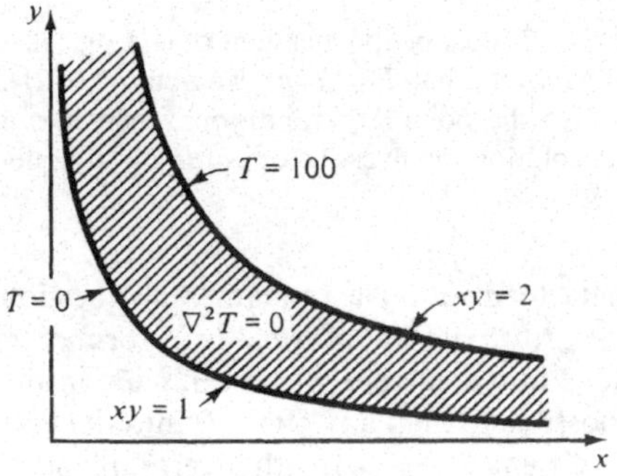

Figure 16.6. Dirichlet problem of Exercise 16.7.

16.8. (*Poisson integral formula*) Just as the Cauchy integral formula

$$f(z) = \frac{1}{2\pi i}\int_C \frac{f(\zeta)}{\zeta - z}\,d\zeta \tag{16.22}$$

expresses an analytic function $f(z) = u + iv$ in terms of its boundary values, we would expect there to be a similar integral formula expressing a harmonic function $u(x, y)$ in terms of its boundary values. Let C be the ccw circle $|\zeta| = R$. If we seek the desired expression for u by equating real parts of (16.22) we find that the right hand side involves both u *and* v, whereas v is not welcome. The reason that v enters is that $\zeta/(\zeta - z)$ is not purely real. However, with $\zeta = Re^{i\phi}$ we can rewrite (16.22) as

$$\begin{aligned} f(z) &= \frac{1}{2\pi i}\int_C \left(\frac{1}{\zeta - z} - \frac{1}{\zeta - R^2/\bar{z}}\right) f(\zeta)\,d\zeta \\ &= \frac{1}{2\pi}\int_0^{2\pi}\left[\frac{\zeta}{\zeta - z} + \frac{\bar{z}}{\bar{\zeta} - \bar{z}}\right] f(\zeta)\,d\phi \end{aligned}$$

(Why?), where the quantity in the square brackets *is* real. In particular, show that

$$\frac{\zeta}{\zeta - z} + \frac{\bar{z}}{\bar{\zeta} - \bar{z}} = \frac{R^2 - r^2}{|\zeta - z|^2}$$

and hence deduce the famous **Poisson integral formula,**

$$u(r, \theta) = \frac{1}{2\pi} \int_0^{2\pi} \frac{(R^2 - r^2)u(R, \phi)\, d\phi}{R^2 - 2Rr \cos(\phi - \theta) + r^2}, \tag{16.23}$$

where $z = re^{i\theta}$ and $\zeta = Re^{i\phi}$.

16.9. Solve the half-plane Dirichlet problem with the following boundary values by mapping onto the unit disk (see Fig. 16.7) and using (16.23).
(a) $T(x, 0) = 100$ for $|x| < 1$ and 0 for $|x| > 1$.
(b) $T(x, 0) = x/(x^2 + 1)$.
You may leave the results in integral form.

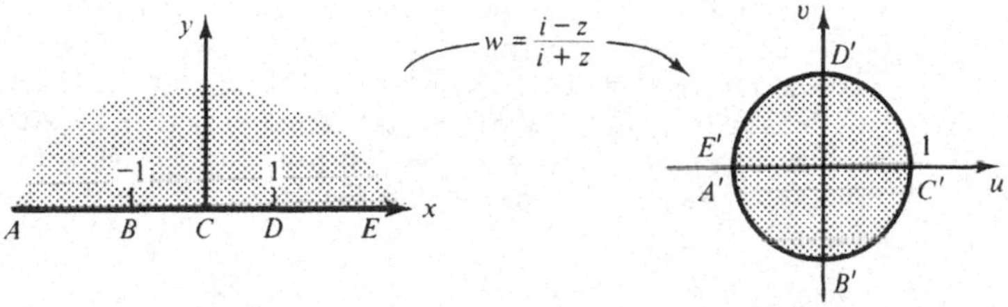

Figure 16.7. Dirichlet problem of Exercise 16.9.

16.10. **(Schwarz–Christoffel transformation)** The Schwarz–Christoffel transformation provides a means of mapping the upper half-plane into a given *polygon* (or, through its inverse, vice versa). To see how it works, consider the mapping of the upper half-plane by a $w(z)$ such that

$$\frac{dw}{dz} = A(z - a)^{-\alpha}, \tag{16.24}$$

where a is a point on the x axis, α is real, and A may be complex. We will have conformality everywhere except perhaps at $z = a$. Thus the x axis to the left of a will go into a straight line, and the x axis to the right of a will go into a straight line, but, in general, there will be a kink where they meet at $w(a)$.

(a) To examine the kink, show from (16.24) that

$$\Delta[\arg(dw) - \arg(dz)] = -\alpha(-\pi)$$

as we move from A to B along the dashed curve in Fig. 16.8. Conclude that $\Delta[\arg(dw)] = \pi\alpha$, and so the kink is as shown in the figure.

(b) To illustrate, suppose that we wish the image to be a semi-infinite strip with $z = -1$ and $z = +1$ mapping into the two 90° corners. Then

$$\frac{dw}{dz} = A(z + 1)^{-1/2}(z - 1)^{-1/2} = \frac{A}{\sqrt{z^2 - 1}};$$

therefore

$$w = A \sin^{-1} z + B. \tag{16.25}$$

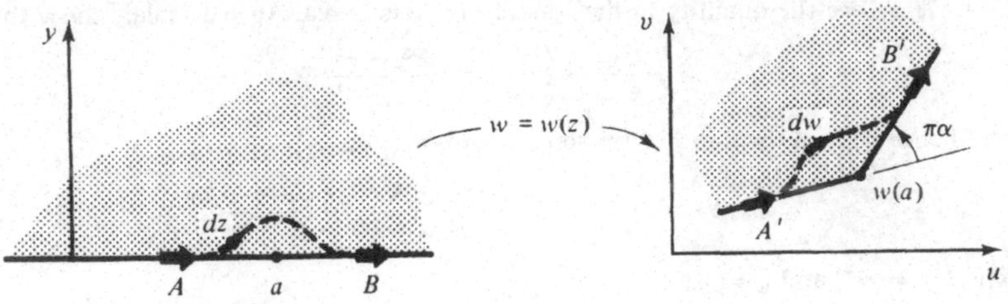

Figure 16.8. Schwarz–Christoffel.

(In this case, the integration yielded an elementary function. Often, however, the result is an *elliptic function.*) Before discussing the choice of A and B, it is important to observe that $\sin^{-1} z$ is *multivalued.* Specifically, $\sin^{-1} z$ maps the upper half-plane into the upper half-strip $-\pi/2 < u < \pi/2$ (see Fig. 16.9). But it also maps the upper half-plane into the upper half-strips $\pi/2 < u < 3\pi/2$, $-3\pi/2 < u < -\pi/2$, $3\pi/2 < u < 5\pi/2$, and so on. Thus we need to specify which branch of $\sin^{-1} z$ is intended in (16.25). Let us choose the "principal" branch, say, whose real part lies between $-\pi/2$ and $+\pi/2$. Returning to (16.25), then, suppose that we wish $z = 0$ to go into $w = 0$. Then $B = 0$. With $A = 1$, say, $w = \sin^{-1} z$, as illustrated in Fig. 16.9.

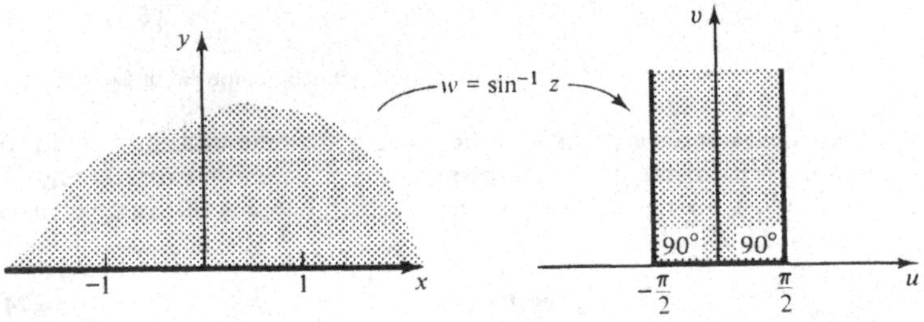

Figure 16.9. The $\sin^{-1} z$ mapping.

(c) Find A and B in (16.25) such that the semi-infinite strip is $u > 0$, $0 < v < 1$ instead. **Important.** *In conformal mapping we go "both ways," from the z plane to the w plane and then back again. So both w(z) and its inverse must be single valued —that is, the mapping needs to be one-to-one. For multivalued functions we therefore need to specify which branch is intended!* This difficulty did not arise in the text, since we treated only the bilinear transformation, which happens to be one-to-one. In fact, the bilinear transformation can be shown to be the *most general* one-to-one mapping!

16.11. Solve the Dirichlet problem for the half-strip $|x| < \pi/2$, $y > 0$, where $T(\pm\pi/2, y) = 0$ and $T(x, 0) = 1$, by means of the successive transformations $\zeta = \sin z$ and $w = \ln [(\zeta - 1)/(\zeta + 1)]$. *Hint:* Be careful about one to oneness, as discussed in the previous exercise.

16.12. Suppose that $\nabla^2 T = 0$ in the z plane with the circular holes $|z| < 1$ and $|z - 5/2| < 1/2$ with $T = 50$ on $|z| = 1$ and $T = 75$ on $|z - 5/2| = 1/2$. Find $T(x, y)$ by means of the bilinear mapping $w = (z - a)/(az - 1)$, where $a = (7 + 2\sqrt{6})/5$.

16.13. (*Two-dimensional fluid flow*) (a) Solve for the velocity potential ϕ of the two-dimensional, irrotational, incompressible corner flow shown in Fig. 16.10 by means of the mapping $\zeta = z^2$. *Hint:* The flow in the ζ plane is simply a uniform stream $\phi_\xi = U$ or $\phi = U\xi$. Thus show that the x, y velocity components are $u(x, y) = 2Ux$, $v(x, y) = -2Uy$. (Note that $x = y = 0$ is a "stagnation point"; that is $u = v = 0$ there.)

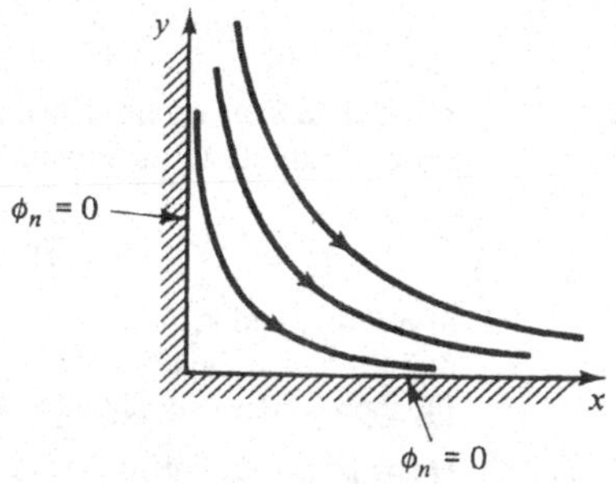

Figure 16.10. Potential flow in a corner.

(b) Generally, however, people work with the *complex potential* $w \equiv \phi + i\psi$, where ψ is the stream function (introduced in Exercise 9.34) rather than just ϕ or ψ. Since the velocity components are $u = \phi_x = \psi_y$ and $v = \phi_y = -\psi_x$, it follows that $w(z)$ is analytic and that $dw/dz = u - iv$ (Note that the letters u, v are usually the coordinates of the complex w plane. Here, however, u and v are the velocity components and $w = \phi + i\psi$ is the complex potential; so we use $\zeta(z) = \xi + i\eta$, say, for the mapping function.) Re-solve the problem of part (a), using the complex potential w. *Hint:* Just as dw/dz is the complex velocity $u - iv$ in the z plane, $dw/d\zeta$ is the complex velocity in the ζ plane. Thus $dw/d\zeta = U$, and therefore $w = U\zeta = Uz^2$.

16.14. (**Joukowsky transformation**) Show that the so-called Joukowsky transformation

$$\zeta = z + \frac{a^2}{z} \qquad (a > 0) \tag{16.26}$$

maps the circles $x^2 + y^2 = c^2$ into the confocal ellipses

$$\frac{\xi^2}{(c + a^2/c)^2} + \frac{\eta^2}{(c - a^2/c)^2} = 1$$

with foci at $\zeta = \pm 2a$. Thus show that the circle $|z| = a$ folds onto the line $|\xi| < 2a$, $\eta = 0$ (where $\zeta = \xi + i\eta$). Show that the region $|z| > a$ maps into the entire ζ plane minus this line segment and that $|z| < a$ does also. Hence the inverse transformation is double valued. In fact, show that

$$z = \frac{\zeta + \sqrt{\zeta^2 - 4a^2}}{2} \tag{16.27}$$

and indicate suitable branch cuts for $\sqrt{\zeta^2 - 4a^2}$ that correspond to the mapping of the slit ζ plane into $|z| > a$ and $|z| < a$, respectively.

16.15. (*Flow over a cylinder*) Solve for the flow [i.e., the velocities $u(x, y)$, $v(x, y)$] of a uniform stream U over a circular cylinder of radius a, as shown in Fig. 9.21 of

Chapter 9, by means of the Joukowsky transformation (16.26). *Important.* Note that the complex velocities in the z plane and ζ plane, say $(u - iv)|_z$ and $(u - iv)|_\zeta$, are related per

$$\frac{dw}{dz} = \frac{dw}{d\zeta}\frac{d\zeta}{dz}.$$

That is, they differ only by the scale factor $d\zeta/dz$:

$$(u - iv)\Big|_z = \frac{d\zeta}{dz}(u - iv)\Big|_\zeta.$$

For the *Joukowsky* transformation, in fact, $d\zeta/dz \sim 1$ at infinity, so that the conditions at infinity in the z plane carry over intact to the ζ plane. Thus

$$\frac{dw}{d\zeta} = (u - iv)\Big|_\zeta \sim (u - iv)\Big|_z \sim U$$

at infinity, and it follows that $w \sim U\zeta$. In fact, $w = U\zeta$ satisfies both this condition at infinity *and* the boundary condition on the "body" (namely, the slit $|\xi| < 2a, \eta = 0$); therefore we simply have $w = U\zeta$.

16.16. (*Flow perpendicular to a plate*) Compute the (two-dimensional) fluid velocity field around a thin flat plate that is perpendicular to a uniform stream V, say (see Fig. 16.11); that is, $dw/dz = u - iv \sim iV$ as $z \to \infty$. In particular, show that $w =$

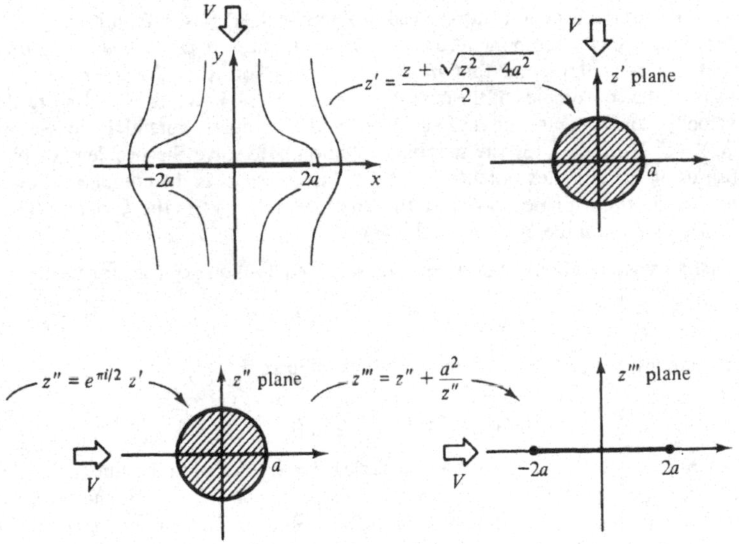

Figure 16.11. Flow perpendicular to a flat plate.

$iV\sqrt{z^2 - 4a^2}$, where the square root is defined by a suitable cut. For instance, on the top of the plate $u = Vx/\sqrt{4a^2 - x^2}$, $v = 0$; on the bottom $u = -Vx/\sqrt{4a^2 - x^2}$, $v = 0$; and on $y = 0, |x| > 2a$ we have $u = 0, v = -Vx/\sqrt{x^2 - 4a^2}$. *Hint:* Use the sequence of mappings indicated. Also, recall Exercises 16.14 and 16.15.

16.17. Solve the Dirichlet problem shown in Fig. 16.12 by the indicated Joukowsky transformation, using the Poisson integral formula

$$T(u, v) = \frac{v}{\pi} \int_{-\infty}^{\infty} \frac{T(u', 0)\, du'}{(u' - u)^2 + v^2} \tag{16.28}$$

for the solution in the w plane. [(16.28) is the *half-plane* version of (16.23) and will be derived in Part V.]

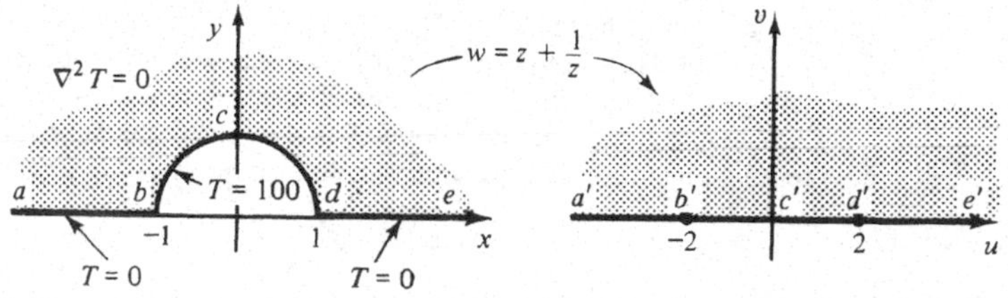

Figure 16.12. Dirichlet problem.

16.18. Solve the heat conduction problem shown in Fig. 16.13 by means of the mapping $w = \ln z$.

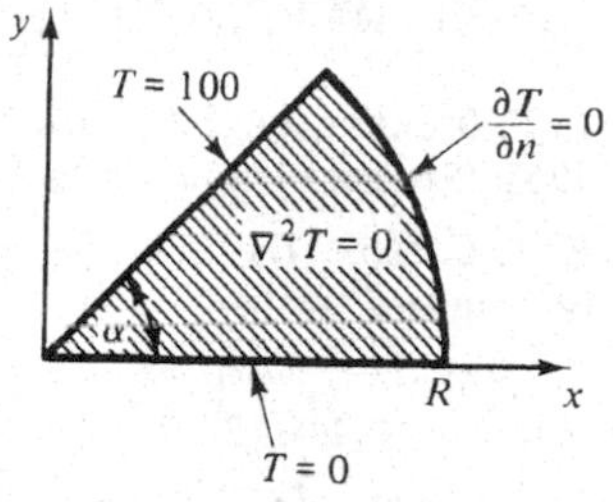

Figure 16.13. Heat conduction problem.

16.19. (*Transformation of boundary conditions*) So far we have considered only Dirichlet boundary conditions (i.e., T prescribed on the boundary) of the simple form $T =$ constant.

(a) *Nonconstant Dirichlet condition.* Show that the transformed boundary condition in Fig. 16.14 is $T = (\sin \phi)/2$.

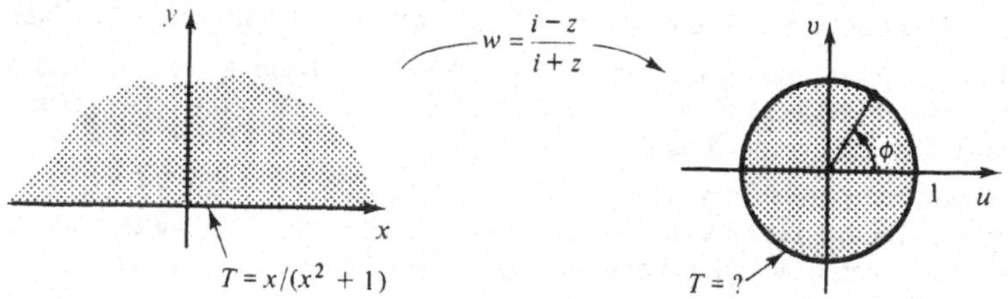

Figure 16.14. Nonconstant Dirichlet conditions.

(b) *Neumann conditions* ($\partial T/\partial n$ prescribed on boundary). Show that the normal derivative $\partial T/\partial n$ transforms from the z plane to the w plane according to

$$\left(\frac{\partial T}{\partial n}\right)_w = \frac{1}{|w'|}\left(\frac{\partial T}{\partial n}\right)_z. \tag{16.29}$$

In particular, observe that $(\partial T/\partial n)_z = 0$ implies that $(\partial T/\partial n)_w = 0$. In the context of heat condition it means that an insulated boundary in the z plane transforms into an insulated boundary in the w plane. To illustrate (16.29), consider again the case shown in Fig. 16.14 but with "$T = x/(x^2 + 1)$" changed to $\partial T/\partial n = x/(x^2 + 1)$ and show that the transformed boundary condition is $(\partial T/\partial n)_w = (\sin \phi)/2(1 + \cos \phi)$.

Dirichlet and Neumann boundary conditions are not the only type encountered. Sometimes we have a **mixed** boundary condition of the form

$$\alpha T + \beta\frac{\partial T}{\partial n} = \gamma, \tag{16.30}$$

where α, β, γ are prescribed along the boundary. Transformation of (16.30) should be apparent from the foregoing discussion.

16.20. Make up two examination-type problems on this chapter.

ADDITIONAL REFERENCES FOR PART II

E. F. Beckenbach (Ed.), *Modern Mathematics for the Engineer*, Second Series, McGraw-Hill, 1961. [See Part 1, Chapter 5, Asymptotic Formulas and Series by J. B. Rosser.]

E. T. Copson, *Theory of Functions of a Complex Variable*, Clarendon Press, Oxford, 1935. [A standard treatise.]

P. Franklin, *Functions of Complex Variables*, Prentice-Hall, Englewood Cliffs, N.J., 1958. [Clear introductory treatment.]

B. Friedman, *Lectures on Applications-Oriented Mathematics*, Holden-Day, San Francisco, 1969. [See Chapter 5 on complex integration.]

A. Kyrala, *Applied Functions of a Complex Variable*, Wiley-Interscience, New York, 1972. [Methods of steepest descent and stationary phase, Laplace and Fourier transforms and control systems, dispersion relations, Hilbert transforms, Plemelj formulas, and special functions.]

G. Moretti, *Functions of a Complex Variable*, Prentice-Hall, Englewood Cliffs, N.J., 1964. [Includes discussion of higher-transcendental functions, Fourier series, and Laplace and Fourier transforms.]

W. F. Osgood, *Functions of a Complex Variable*, Hafner Publishing Co., New York, 1948.

E. G. Phillips, *Some Topics in Complex Analysis*, Pergamon Press, Oxford, 1966. [A continuation of his book on Functions of a Complex Variable into special topics, such as elliptic and other special functions.]

E. Whittaker and G. Watson, *A Course of Modern Analysis*, 4th ed., Cambridge University Press, London, 1958. [A comprehensive treatise on real and complex analysis with emphasis on the transcendental functions and their properties.]

For applications in the areas of solid and fluid mechanics, see, for example,

L. M. Milne–Thomson, *Theoretical Hydrodynamics*, 5th ed., Macmillan, New York, 1968.

N. I. Muskhelishvili, *Some Basic Problems of the Mathematical Theory of Elasticity*, Noordhoff, Groningen, 1953.

S. Timoshenko and J. Goodier, *Theory of Elasticity*, 3rd ed., McGraw-Hill, New York, 1970. [See Chapter 6.]

PART III

LINEAR ANALYSIS

We try, in Part III, to present a unified *theory of linear analysis, although still at an introductory level; careful reading and attention to the exercises will be especially important. In Chapter 17 we introduce the notion of an abstract LINEAR SPACE and get as far as the expansion of a given vector in terms of a set of base vectors, for both finite and infinite-dimensional spaces. In Chapter 18 we introduce LINEAR OPERATORS on those spaces and, in particular, define the adjoint of an operator. We are ready, in Chapter 19, to consider the central problem of linear analysis, THE LINEAR EQUATION $Lx = c$. After a look at the questions of the existence and uniqueness of solutions, we pursue the inverse operator L^{-1} and hence the solution*

$x = L^{-1}c$. The alternative eigenvector expansion approach is developed in Chapter 20, which opens with a discussion of THE EIGENVALUE PROBLEM $Lx = \lambda x$. The case where L is a matrix operator receives the heaviest attention throughout, with corresponding material on integral operators contained in starred sections and important introductory material on differential operators, such as the Sturm–Liouville theory, included, too. Detailed discussion of ordinary and partial differential operators is reserved for Parts IV and V.

Chapter 17

Linear Spaces

A favorite question asked of graduate students in qualifying examinations is "What is a vector?" The response generally takes one of two directions, depending on the student's background and gambling instinct. A reasonable and relatively safe reply would follow the lines of Chapter 8. That is, a vector is an arrowlike quantity having both a magnitude and a direction; vectors can be scaled by multiplicative factors, added according to the parallelogram law; they have a length; dot and cross product operations are defined between two vectors; the angle between two vectors is defined; any vector in a two- or three-dimensional space can be expressed as a linear combination of base vectors, and so on.

A more sophisticated reply would be that a vector is "an element of a linear vector space." The hope that that elegant statement will suffice is always in vain, and the examiner can be counted on to follow with "So what is a vector space?"

17.1. INTRODUCTION; EXTENSION TO N TUPLES

Earlier we distinguished between vectors and scalars by using boldface and lightface type, respectively. In Part III, however, it will be more convenient to dispense with boldface. Instead we will generally use Roman letters, such as u, v, w, x, y, z for vectors and lowercase Greek letters for scalars. Which quantities are vectors and which are scalars should also be clear from the context.

Recall the cartesian representation of vectors in two and three dimensions,

$$x = \xi_1\hat{\imath} + \xi_2\hat{\jmath} \quad \text{and} \quad x = \xi_1\hat{\imath} + \xi_2\hat{\jmath} + \xi_3\hat{k}. \tag{17.1a}$$

They may be expressed in the different but equivalent *n-tuple* notation

$$x = (\xi_1, \xi_2) \quad \text{and} \quad x = (\xi_1, \xi_2, \xi_3), \tag{17.1b}$$

and within this notation it is not hard to imagine an extension to vectors of the form $x = (\xi_1, \xi_2, \ldots, \xi_n)$ for n's greater than 3 as well. For instance, the configuration of the structure shown in Fig. 17.1, consisting of five pinned links restrained by lateral springs, may be defined by the vertical displacements $\xi_1, \ldots, \xi_4$, and it may be convenient to let them be the components of a single vector $x = (\xi_1, \xi_2, \xi_3, \xi_4)$.

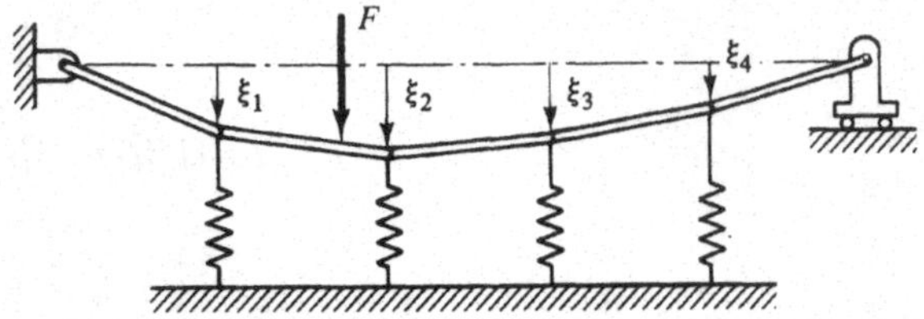

Figure 17.1. Example leading to a four-tuple.

Suppose that x and y are any two n-tuple vectors $x = (\xi_1, \ldots, \xi_n)$, $y = (\eta_1, \ldots, \eta_n)$ and that α is any scalar. Generalizing from the two-and three-dimensional cases discussed in Chapter 8, let us define the *vector sum*

$$x + y \equiv (\xi_1 + \eta_1, \ldots, \xi_n + \eta_n), \tag{17.2a}$$

the *multiplication of a vector by a scalar*

$$\alpha x \equiv (\alpha\xi_1, \ldots, \alpha\xi_n), \tag{17.2b}$$

the *length or norm of a vector*[1]

$$\|x\| \equiv \sqrt{\sum_1^n |\xi_j|^2}, \tag{17.2c}$$

and the *dot* or *inner product*

$$(x, y) \equiv \sum_1^n \xi_j \bar{\eta}_j; \tag{17.2d}$$

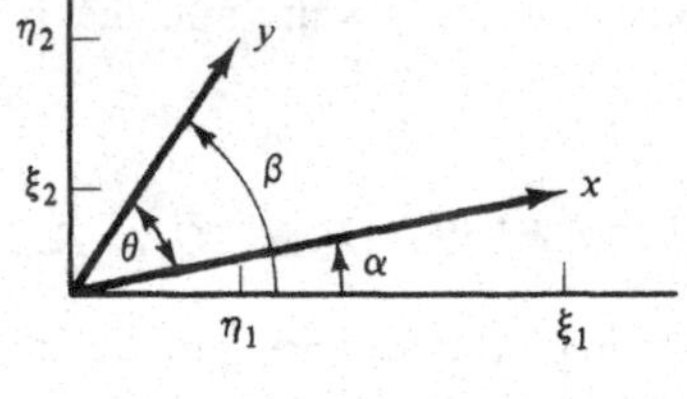

Figure 17.2. The dot product in two dimensions.

the latter is also called the *scalar product* and may be written as $x \cdot y$, (x, y), or $\langle x, y \rangle$.

In the event that n is 2 or 3 (and all quantities are real), surely (17.2a) reduces to the usual parallelogram rule and (17.2c) to the Pythagorean Theorem. To see that (17.2d) reduces to our old dot product $x \cdot y = \|x\| \|y\| \cos\theta$, consider, for simplicity, the two-dimensional case and observe (Fig. 17.2)

[1]Generally the components of our vectors (e.g., the ξ_j's), as well as the scalars that multiply the vectors (e.g., α), will be real. In some applications, however, they may be *complex*, and we shall allow for that possibility here. As Section 17.3 will show, we will therefore need the absolute value braces in (17.2c) and the complex conjugate bar in (17.2d).

that

$$
\begin{aligned}
(x, y) &= \xi_1 \eta_1 + \xi_2 \eta_2 \\
&= \|x\| \cos \alpha \, \|y\| \cos \beta + \|x\| \sin \alpha \, \|y\| \sin \beta \\
&= \|x\| \, \|y\| (\cos \alpha \cos \beta + \sin \alpha \sin \beta) \\
&= \|x\| \, \|y\| \cos (\beta - \alpha) = \|x\| \, \|y\| \cos \theta \qquad (17.3)
\end{aligned}
$$

as claimed.

Note carefully that for $n > 3$ definitions (17.2a) to (17.2d) provide *abstract extensions* of these familiar ideas.

Example 17.1. If $x = (2, 1, -3, 0)$, $y = (1, 3, 3, -2)$, and $z = (1, -2, 0, 5)$, then $x + y = (3, 4, 0, -2)$, $\|x\| = \sqrt{14}$, $3x = (6, 3, -9, 0)$, $(x, y) = -4$, $(x, z) = 0$, and so on. Again, we do not attempt to draw x, y, z as "arrows" as we did in two and three dimensions. When we say that the length of x is $\sqrt{14}$, we mean the length in the sense of (17.2c), which is an abstract extension of the Pythagorean Theorem. Similarly, we would say that x is *orthogonal* to z simply in the sense that $(x, z) = 0$, not in the sense that we could actually measure the angle between them as 90°. ∎

17.2. THE DEFINITION OF AN ABSTRACT VECTOR SPACE

Let us provide a rather broad framework for these ideas. First, we will call a set $\mathcal{S}$ of vectors a linear vector space, or simply a **vector space**, if the following requirements are satisfied.

(i) An operation between any two vectors in $\mathcal{S}$, which will be called *addition* and denoted by $+$, is defined in such a way that if x and y are in $\mathcal{S}$, then $x + y$ is also (i.e., $\mathcal{S}$ is *closed under addition*). Furthermore,

$$x + y = y + x \qquad \text{(addition is } \textit{commutative}\text{)}$$

and

$$(x + y) + z = x + (y + z) \qquad \text{(addition is } \textit{associative}\text{)}.$$

(ii) $\mathcal{S}$ contains a *zero vector* $\mathbb{O}$ such that $x + \mathbb{O} = x$ for each x in $\mathcal{S}$.

(iii) For each x in $\mathcal{S}$ there is also a vector "$-x$" such that $x + (-x) = \mathbb{O}$. We denote $x + (-y)$ as $x - y$ but emphasize that it is actually the $+$ operation between x and $-y$.

(iv) Another operation, called *scalar multiplication*, is defined between any vector x in $\mathcal{S}$ and any scalar α in such a way that if x is in $\mathcal{S}$ and α is any scalar, then αx is in $\mathcal{S}$ also (i.e., $\mathcal{S}$ is *closed under scalar multiplication*). In addition, we require that

$$
\begin{aligned}
\alpha(\beta x) &= (\alpha\beta)x, \\
(\alpha + \beta)x &= \alpha x + \beta x, \\
\alpha(x + y) &= \alpha x + \alpha y,
\end{aligned}
$$

and

$$1x = x, \qquad 0x = \mathbb{O}.$$

Observe that the two plus signs in the equation $(\alpha + \beta)x = \alpha x + \beta x$, for instance, are quite different. The plus on the left is the usual addition of scalars (e.g., $2 + 3 = 5$), whereas the plus on the right is the more exotic addition operation defined between the two vectors αx and βx; the latter must satisfy the conditions under requirement (i), but beyond that its definition is entirely up to whoever is defining the space.

Finally, note carefully that the preceding requirements say nothing about the *nature* of the vectors; they may be "arrows," n tuples, or whatever, as long as these requirements are met.

Example 17.2. Consider, a space where the vectors are *n tuples*. . that is, of the form $x = (\xi_1, \ldots, \xi_n)$. With $y = (\eta_1, \ldots, \eta_n)$, suppose that we define $x + y$ as in (17.2a), αx as in (17.2b), $\mathbb{0} \equiv (0, 0, \ldots, 0)$, and $-x \equiv (-\xi_1, \ldots, -\xi_n)$. Using these definitions, it is not hard to verify that the vector space requirements are satisfied, so that (for any positive integer n) the space of n tuples is a bonafide linear vector space. For instance, note that $x + y = (\xi_1 + \eta_1, \ldots, \xi_n + \eta_n) = (\eta_1 + \xi_1, \ldots, \eta_n + \xi_n) = y + x$ because the addition of *scalars* is commutative, and therefore $\xi_j + \eta_j = \eta_j + \xi_j$ for each j.

Yet we would *not* have a vector space if, for example, we changed the rule of addition (17.2a) to $x + y = (\xi_1 + 2\eta_1, \ldots, \xi_n + 2\eta_n)$, say, since then we would not have $x + y = y + x$ for all x's and y's, nor would $(x + y) + z = x + (y + z)$ in fact.

Again, although the pluses in the right-hand side of (17.2a) denote the "everyday" addition of scalars, the plus on the left denotes a certain operation between vectors, as spelled out by the right-hand side. Similarly, $\alpha\xi_j$ in (17.2b) is the everyday multiplication of scalars, whereas αx denotes a certain operation between the scalar α and the vector x, as spelled out by the right-hand side. ∎

Introducing a Norm and Inner Product. In order to provide our space with a sense of length and distance, let us define for any given vector x a number $\|x\|$, called the **norm** or **length** of x. Staying as close as possible to our usual notions of length, let us ask that $\|x\|$ satisfy the following requirements.

(i)
$$\|\alpha x\| = |\alpha| \|x\|. \tag{17.4a}$$

(ii) *Positiveness:*
$$\|x\| > 0 \quad \text{for all } x \neq \mathbb{0}, \qquad (17.4b)$$
$$\|x\| = 0 \quad \text{for } x = \mathbb{0}. \qquad (17.4b)$$

(iii) *Triangle inequality:*
$$\|x + y\| \leq \|x\| + \|y\|. \tag{17.4c}$$

For instance, (17.4a) simply says that αx is $|\alpha|$ times as long as x. For the space of 2-tuples (or 3-tuples), where x and y can be depicted as "arrows," the triangle inequality (17.4c) amounts to the Euclidean proposition that the length of any one side of a triangle cannot exceed the sum of the lengths of the other two sides (see Fig. 17.3). For more general vector spaces it is an abstract extension of this proposition. Equation (17.4c) is also known as the *Minkowski inequality*.

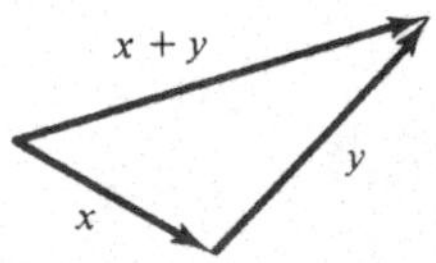

Figure 17.3. Triangle inequality.

Example 17.3. For an n-tuple vector space the choice

$$\|x\| \equiv \sqrt{\sum_{1}^{n} |\xi_j|^2}, \tag{17.5}$$

already given in (17.2c) and known as the *Euclidean* norm, is easily seen to satisfy requirements (17.4a) and (17.4b); that is also satisfies (17.4c) is hereby stated without proof (Exercise 17.5). Although it is the norm that we shall use, note that it is not the *only* suitable norm; for instance, another is (Exercise 17.1)

$$\|x\| \equiv \max_{1 \leq j \leq n} |\xi_j|. \tag{17.6}$$

If $x = (2, -1, 0, -3)$, for example, then (17.5) gives $\|x\| = \sqrt{14}$ and (17.6) gives $\|x\| = 3$. By the way, both (17.5) and (17.6) are special cases of the norm

$$\|x\| \equiv (\sum_{1}^{n} |\xi_j|^p)^{1/p};$$

(17.5) corresponds to $p = 2$ and (17.6) corresponds to the limiting case where $p \longrightarrow \infty$. ∎

Having introduced the "length" of a vector x, it seems reasonable to define the "distance between x and y" as $\|x - y\|$—that is, as the length of their difference. Hence $d(x, y) \equiv \|x - y\|$; that $\|x - y\|$ satisfies the requirements (1.6) to (1.8) of Chapter 1 follows from the requirements (17.4) placed on the norm (Exercise 17.2).

We are thus in a position to talk about the *convergence of a vector sequence*. Specifically, we say that x_n **converges** to x as $n \longrightarrow \infty$ if to each $\epsilon > 0$ (no matter how small) there corresponds an $N(\epsilon)$ such that $\|x - x_n\| < \epsilon$ for all $n > N$. We call x the *limit* of the sequence and write

$$\lim_{n \to \infty} x_n = x \quad \text{or (more simply)} \quad x_n \longrightarrow x. \tag{17.7}$$

(It almost goes without saying that in order for the limit to exist, x must itself be a member of the space $\mathcal{S}$.)

The same is true for the convergence of an infinite series of vectors $\sum_{1}^{\infty} x_n$; with $x_1 + \cdots + x_n = s_n$, we say that $\sum_{1}^{\infty} x_n$ converges to s if the partial sum $s_n \longrightarrow s$.

Finally, we generally introduce a so-called **inner product** (x, y), which is scalar valued and satisfies the conditions below.

(i) *Conjugate symmetry:*

$$(x, y) = \overline{(y, x)}. \tag{17.8a}$$

(ii) *Linearity:*

$$(\alpha x + \beta y, z) = \alpha(x, z) + \beta(y, z). \tag{17.8b}$$

(iii) *Positiveness:*

$$\begin{aligned} (x, x) &> 0 \quad \text{for all } x \neq \mathcal{O} \\ (x, x) &= 0 \quad \text{if } x = \mathcal{O}, \end{aligned} \tag{17.8c}$$

where the bar in (17.8a) denotes the complex conjugate. Note that the linearity condition (17.8b) is equivalent to the two conditions $(\alpha x, y) = \alpha(x, y)$ and $(x + y, z) = (x, z)$

$+ (y, z)$. In contrast to the first of these two, note that

$$(x, \alpha y) = \overline{(\alpha y, x)} = \overline{\alpha(y, x)} = \bar{\alpha}\overline{(y, x)} = \bar{\alpha}(x, y); \tag{17.9}$$

that is, α comes out of the first position *without* a bar and out of the second position *with* a bar.

Next, let us derive a useful result known as the **Schwarz inequality**[2]

$$|(x, y)| \leq \sqrt{(x, x)}\sqrt{(y, y)} \tag{17.10}$$

for all vectors x and y in a given space $\mathcal{S}$. We start with the inequality

$$(x + \alpha y, x + \alpha y) \geq 0,$$

which is guaranteed by (17.8c). Using (17.8a), (17.8b), and (17.9) to expand the left-hand side, we obtain

$$(x, x) + \bar{\alpha}(x, y) + \alpha(y, x) + \alpha\bar{\alpha}(y, y) \geq 0. \tag{17.11}$$

Equation (17.11) is valid for all α's, but we can make it as informative as possible by choosing α so as to minimize the size of the left-hand side, thereby coming as close to an equality as possible. So denoting the left-hand side as $f(a, b)$, where $\alpha = a + bi$, we set $\partial f/\partial a = 0$ and $\partial f/\partial b = 0$. Solving for a and b, we find that $\alpha = a + bi = -(x, y)/(y, y)$; and inserting this expression into (17.11) yields the desired result (17.10) (Exercise 17.4).

Observe further that

$$\begin{aligned}(x + y, x + y) &= (x, x) + (x, y) + (y, x) + (y, y) \\ &= (x, x) + 2 \operatorname{Re}(x, y) + (y, y) \qquad \text{per (17.8a)} \\ &\leq (x, x) + 2|(x, y)| + (y, y) \\ &\leq (x, x) + 2\sqrt{(x, x)(y, y)} + (y, y) \qquad \text{per (17.10)} \\ &= [\sqrt{(x, x)} + \sqrt{(y, y)}]^2;\end{aligned}$$

therefore

$$\sqrt{(x + y, x + y)} \leq \sqrt{(x, x)} + \sqrt{(y, y)}.$$

Comparing this result with the triangle inequality (17.4c) suggests that the quantity $\sqrt{(x, x)}$ can be used as a norm. That (17.4a) and (17.4b) are also satisfied by this choice follows (Exercise 17.7) from (17.8). In fact, *in an inner product space* (i.e., a vector space with an inner product defined) *we will always elect to use the norm*

$$\|x\| \equiv \sqrt{(x, x)}. \tag{17.12}$$

In this case, the **Schwarz inequality** (17.10) can be written in its more usual form

$$|(x, y)| \leq \|x\| \|y\|. \tag{17.13}$$

Example 17.4. For our n-tuple space the inner product

$$(x, y) = \sum_1^n \xi_j \bar{\eta}_j, \tag{17.14}$$

put forward in (17.2d), *is* seen (Exercise 17.8) to be a legitimate inner product. In

---

[2]The names Cauchy and Bunyakovsky are also associated with this well-known inequality.

other words, it satisfies the conditions (17.8). It *follows* from (17.12) that the norm is

$$\|x\| = \sqrt{(x, x)} = \sqrt{\sum_1^n \xi_j \bar{\xi}_j} = \sqrt{\sum_1^n |\xi_j|^2}, \tag{17.15}$$

which is the same as (17.2c). For instance, if $x = (-3, 5i)$ and $y = (1 + i, 2)$, then

$$(x, x) = (-3)(-3) + (5i)(-5i) = 9 + 25 = 34 \qquad (\text{so } \|x\| = \sqrt{34}),$$
$$(y, y) = (1 + i)(1 - i) + (2)(2) = 2 + 4 = 6 \qquad (\text{so } \|y\| = \sqrt{6}),$$
$$(x, y) = (-3)(1 - i) + (5i)(2) = -3 + 13i,$$
$$(y, x) = (1 + i)(-3) + (2)(-5i) = -3 - 13i.$$

Note how $(x, y) = \overline{(y, x)}$, in accordance with (17.8a). This result would *not* be true if the conjugate bar in (17.14) were omitted; indeed, then we would also have $(x, x) = (-3)(-3) + (5i)(5i) = -16$, in violation of (17.8c). Finally, observe that $|(x, y)| = \sqrt{9 + 169} = \sqrt{178}$, which is less than $\|x\|\,\|y\| = \sqrt{34}\sqrt{6} = \sqrt{204}$, in accordance with the Schwarz inequality (17.13). ∎

In this book our vector spaces will always have an inner product defined and a norm derived from the inner product according to $\|x\| = \sqrt{(x, x)}$. However, we should emphasize that it is not necessary to define an inner product, or even a norm, for a given space. Nevertheless, we find that a norm is virtually indispensible in applications, since we are almost always interested in the "length" of vectors, the "closeness" of two vectors, and the convergence of vector sequences. For instance, Struble[3] introduces the so-called *taxicab norm* $\|x\| \equiv \sum_1^n |\xi_j|$ for his n tuples but does *not* introduce an inner product.

What does an inner product do for us? It provides a measure of the "angle" between two vectors. To illustrate, for the case of 2-tuples (where all quantities are real), we recall from (17.3) that $\cos\theta = (x, y)/\|x\|\,\|y\|$. For instance, if $(x, y) = 0$, then $\cos\theta = 0$ and so $\theta = 90°$. If $y = \alpha x$ (so that x and y are aligned), then $\cos\theta = \alpha(x, x)/\|x\|\,\|\alpha x\| = 1$ and so $\theta = 0°$. Generalizing to *any* normed inner product space, it is customary to say that if $(x, y) = 0$, then x and y are *orthogonal*. Moreover, if $(x, y) = \|x\|\,\|y\|$, then x and y are aligned.

17.3. LINEAR DEPENDENCE, DIMENSION, AND BASES

Vectors $x_1, \ldots, x_n$ are said to be **linearly dependent** (LD for short) if there exist scalars α_j, not all zero, such that

$$\alpha_1 x_1 + \alpha_2 x_2 + \cdots + \alpha_n x_n = \mathbb{O}; \tag{17.16}$$

otherwise they are **linearly independent** (LI). If they are LD, then it follows that at least one of the x_j's can be expressed as a linear combination of the others. Suppose, for instance, that $\alpha_2 \neq 0$; it follows from (17.16) that

$$x_2 = -\left(\frac{\alpha_1}{\alpha_2}\right)x_1 - \left(\frac{\alpha_3}{\alpha_2}\right)x_3 - \cdots - \left(\frac{\alpha_n}{\alpha_2}\right)x_n.$$

[3] R. A. Struble, *Nonlinear Differential Equations*, McGraw-Hill, New York, 1962, pp. 35–40.

Example 17.5. Consider the 3-tuples, $x_1 = (1, 0, 0)$, $x_2 = (1, 1, 0)$, $x_3 = (1, 1, 1)$. Then (17.16) becomes

$$\alpha_1(1, 0, 0) + \alpha_2(1, 1, 0) + \alpha_3(1, 1, 1) = (0, 0, 0)$$

or

$$(\alpha_1 + \alpha_2 + \alpha_3, \alpha_2 + \alpha_3, \alpha_3) = (0, 0, 0),$$

so that $\alpha_1 + \alpha_2 + \alpha_3 = 0$, $\alpha_2 + \alpha_3 = 0$, and $\alpha_3 = 0$. It follows that α_1, α_2, and α_3 are necessarily zero, and so x_1, x_2, x_3 are LI. On the other hand, observe that if x_3 were replaced by (0, 1, 0), then the set would be LD; $x_1 - x_2 + x_3 = \Theta$ then. ▮

Furthermore, we say that a vector space $\mathcal{S}$ is **n dimensional** if it contains a set of n LI vectors but no set of $n + 1$ LI vectors. If n LI vectors can be found in $\mathcal{S}$ for *each* n (i.e., no matter how large), we say that $\mathcal{S}$ is *infinite* dimensional.

There is great interest in finding a **basis** for a given space $\mathcal{S}$—that is, a set of LI vectors $e_1, \ldots, e_n$ in $\mathcal{S}$ such that any x in $\mathcal{S}$ can be "expanded in terms of them"—expressed as a linear combination of them,

$$x = \alpha_1 e_1 + \cdots + \alpha_n e_n. \tag{17.17}$$

Such a representation is unique, for if x also admits a representation $x = \beta_1 e_1 + \cdots + \beta_n e_n$ say, then on subtracting

$$(\alpha_1 - \beta_1)e_1 + \cdots + (\alpha_n - \beta_n)e_n = \Theta,$$

and since the e_j's are LI, it follows that $\alpha_j - \beta_j = 0$ for each j—that is, $\alpha_j = \beta_j$. We also say that the e_j's *span* the space $\mathcal{S}$.

Observe that *if a basis for* $\mathcal{S}$ *consists of n vectors, then the dimension of* $\mathcal{S}$ *must be n and vice versa* (Exercise 17.14).

Example 17.6. Clearly, the vectors

$$\begin{aligned} e_1 &= (1, 0, 0, \ldots, 0) \\ e_2 &= (0, 1, 0, \ldots, 0) \\ &\ \ \vdots \\ e_n &= (0, 0, \ldots, 0, 1) \end{aligned} \tag{17.18}$$

in our n-tuple space are LI, since

$$\alpha_1 e_1 + \alpha_2 e_2 + \cdots + \alpha_n e_n = \Theta$$

becomes

$$(\alpha_1, \alpha_2, \ldots, \alpha_n) = (0, 0, \ldots, 0)$$

so that $\alpha_1 = \alpha_2 = \cdots = \alpha_n = 0$. Moreover, it is a *basis* because any n tuple $x = (\xi_1, \xi_2, \ldots, \xi_n)$ can be expanded as $x = \xi_1 e_1 + \cdots + \xi_n e_n$. Since there are n members of the basis, it follows that *n-tuple space is n dimensional.*

If our space consists of vectors $x = (\xi_1, \xi_2, \ldots)$—that is, with an *infinite* number of components—it is infinite dimensional, for we can demonstrate an arbitrarily large number of LI vectors,

$$e_1 = (1, 0, 0, \ldots), \qquad e_2 = (0, 1, 0, \ldots), \ldots$$

Certainly (17.18) (which is the n-dimensional version of the familiar $\hat{\imath}, \hat{\jmath}, \hat{k}$ basis) is not the *only* possible basis for n-tuple space; for $n = 2$, for instance, some other bases are $e_1 = (1, 1), e_2 = (1, -1)$ and $e_1(1, 0), e_2 = (1, 1)$, and so on (Exercise 17.15). In fact, there are an infinite number of such bases. How do we decide which to use? In discussing the *eigenvalue problem* in Chapter 20, we will find that generally one particular basis is most "natural," and its identity will be dictated by the particular problem under investigation; we need not worry about it now. ▮

Suppose that we have a basis $e_1, \ldots, e_n$ (of some n-dimensional space $\mathcal{S}$) in hand and that we wish to expand a given x in $\mathcal{S}$ in terms of this basis, $x = \alpha_1 e_1 + \cdots + \alpha_n e_n$. How do we compute the *expansion coefficients*—the α_j's? Let us "dot" both sides with $e_1, \ldots, e_n$ in turn. This procedure yields the simultaneous linear algebraic equations

$$\begin{aligned} (e_1, e_1)\alpha_1 + (e_2, e_1)\alpha_2 + \cdots + (e_n, e_1)\alpha_n &= (x, e_1) \\ (e_1, e_2)\alpha_1 + (e_2, e_2)\alpha_2 + \cdots + (e_n, e_2)\alpha_n &= (x, e_2) \\ &\vdots \\ (e_1, e_n)\alpha_1 + (e_2, e_n)\alpha_2 + \cdots + (e_n, e_n)\alpha_n &= (x, e_n) \end{aligned} \tag{17.19}$$

in the unknown α_j's. The coupling between these equations is rather unfortunate from a computational point of view, especially if n is large. However, suppose that the basis is *orthogonal*—that is, every e_j is orthogonal to every other one,

$$(e_i, e_j) = 0 \quad \text{for } i \neq j.$$

Then (17.19) reduces to the uncoupled system

$$\begin{aligned} (e_1, e_1)\alpha_1 &= (x, e_1) \\ (e_2, e_2)\alpha_2 &= (x, e_2) \\ &\vdots \end{aligned} \tag{17.20}$$

so that

$$\alpha_j = \frac{(x, e_j)}{(e_j, e_j)}. \tag{17.21}$$

Partly because of this great simplification, we shall use *only* orthogonal bases. Sometimes we go a step further and *normalize* our base vectors so that each is of unit length. Then $(e_j, e_j) = \|e_j\|^2 = 1$, and (17.21) simplifies slightly to the form $\alpha_j = (x, e_j)$. To "normalize" a vector e_j, we simply divide it[4] by its length $\|e_j\|$. For instance, the normalized version of the basis $(1, -1), (2, 2)$ is $(1/\sqrt{2}, -1/\sqrt{2}), (1/\sqrt{2}, 1/\sqrt{2})$. A set $e_1, \ldots, e_n$ that is both orthogonal *and* normalized is said to be **orthonormal** or ON for short. In this case, $(e_i, e_j) = \delta_{ij}$, where

$$\delta_{ij} \equiv \begin{cases} 1, & i = j \\ 0, & i \neq j \end{cases} \tag{17.22}$$

is the well-known **Kronecker delta.**

[4]Perhaps we should say "multiply it by $1/\|e_j\|$" instead, since only scalar *multiplication* of vectors has been defined.

17.4. FUNCTION SPACE

So far the only vector space discussed is the (very important) n-tuple space. Of equal importance are the various *function spaces*—that is, where the vectors are functions.

Example 17.7. Consider the space $\mathcal{C}(a, b)$ of all *continuous functions* $x(t)$ defined over $a \leq t \leq b$, where vector addition and scalar multiplication are defined in the "obvious" way (e.g., if x and y are the functions t^2 and $3 \sin t$, then $x + y$ is the function $t^2 + 3 \sin t$, and αx is the function αt^2), the vector $-x$ is the function with values $-x(t)$, and $\mathcal{O}$ is the function $x(t) = 0$. It is not difficult to verify that it is indeed a bonafide vector space (Exercise 17.16). Sometimes it is called a "function space" because the vectors happen to be functions.

How do we define an inner product for this space? Before answering, let us point out that (17.14) is a special case of the more general inner product

$$(x, y) = \sum_1^n \kappa_j \xi_j \bar{\eta}_j, \tag{17.23}$$

where the κ_j's are positive constants (Exercise 17.17).

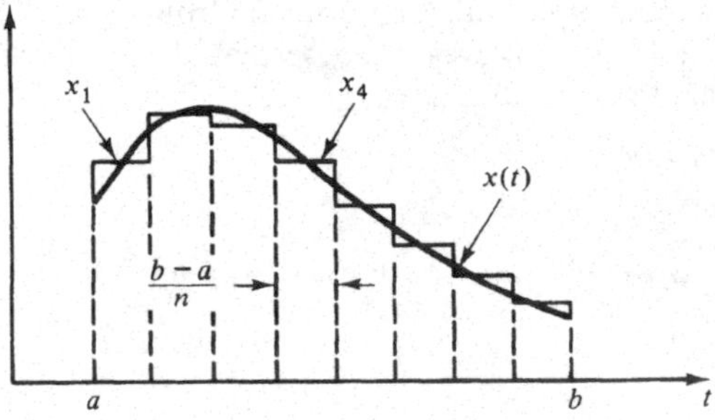

Figure 17.4. "Staircase" approximation of $x(t)$

By way of motivation, let us break the interval $[a, b]$ into n equal parts and introduce a piecewise-constant approximation to $x(t)$, as shown in Fig. 17.4. If we denote by x_j the value of the approximating function over the jth interval, then this function may be represented by an n tuple $(x_1, \ldots, x_n)$. Doing the same for another function $y(t)$ and using (17.23) with κ_j = constant $= (b - a)/n$, we have the Riemann-type sum

$$\sum_1^n \left(\frac{b-a}{n}\right) x_j \bar{y}_j$$

or as $n \longrightarrow \infty$, the integral

$$\int_a^b x(t)\bar{y}(t)\, dt.$$

This formal argument suggests that we define

$$(x, y) \equiv \int_a^b x(t)\bar{y}(t)\, dt \tag{17.24}$$

and we will. [Checking, it is easy to verify that this definition is consistent with the requirements (17.8).] As with n tuples, the bar is needed if we are to admit complex quantities; for instance, $y(t) = e^{it}$ is complex valued.

To illustrate, suppose that $a = 0$ and $b = 1$. If $x = 4t - 3$ and $y = t^2$, then

$$||x|| = \sqrt{(x, x)} = \sqrt{\int_0^1 (4t - 3)^2\, dt} = \sqrt{\frac{7}{3}},$$

$$||y|| = \sqrt{(y, y)} = \sqrt{\int_0^1 t^4\, dt} = \frac{1}{\sqrt{5}},$$

and $$(x, y) = \int_0^1 (4t - 3)t^2\, dt = 0.$$

Thus x and y happen to be orthogonal. Normalizing, $x/\|x\| = \sqrt{3/7}(4t - 3)$ and $y/\|y\| = \sqrt{5}\,t^2$ are orthonormal.

What is the dimension of $\mathcal{C}(a, b)$? Consider the functions $1, (t - a), (t - a)^2, \ldots, (t - a)^{n-1}$ in $\mathcal{C}(a, b)$. Writing

$$\alpha_1 1 + \alpha_2(t - a) + \cdots + \alpha_n(t - a)^{n-1} = 0$$

over $a \leq t \leq b$, we see that setting $t = a$ gives $\alpha_1 = 0$. Differentiating both sides and then setting $t = a$, we find that $\alpha_2 = 0$. Repeating this process, we find that $\alpha_1 = \alpha_2 = \cdots = \alpha_n = 0$, and thus the set $1, (t - a), \ldots, (t - a)^{n-1}$ is LI. Since this is true for *each* n (i.e., no matter how large), it follows that $\mathcal{C}(a, b)$ is *infinite* dimensional. Intuitively, this result should not be surprising because the "staircase" approximation in Fig. 17.4, which is equivalent to an n-tuple representation, becomes exact only in the limit as $n \longrightarrow \infty$. ▮

An unfortunate circumstance surrounds the space $\mathcal{C}(a, b)$. For instance, consider the sequence

$$x_n(t) = \begin{cases} 0, & -1 \leq t \leq 0 \\ nt, & 0 \leq t \leq \dfrac{1}{n} \\ 1, & \dfrac{1}{n} < t \leq 1 \end{cases} \tag{17.25}$$

in $\mathcal{C}(-1, 1)$, as shown in Fig. 17.5. This sequence apparently converges, as $n \longrightarrow \infty$, to the Heaviside function $H(t)$ and yet $H(t)$ is not in $\mathcal{C}(-1, 1)$ because it is not continuous. We must therefore conclude that the sequence (17.25) is divergent.

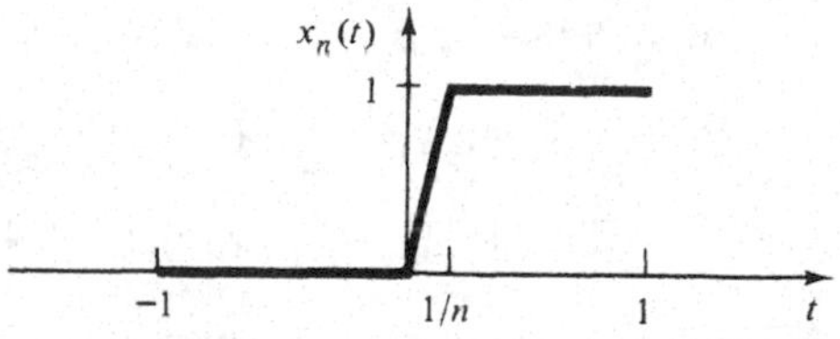

Figure 17.5. The sequence (17.25).

More precisely, the sequence (17.25) is a *Cauchy sequence* [s_n is a Cauchy sequence if to each $\epsilon > 0$ (no matter how small) there corresponds an $N(\epsilon)$ such that $\|s_m - s_n\| < \epsilon$ for all $m, n > N$]. Every convergent sequence is necessarily a Cauchy sequence, but the converse is true only if the space is *closed*—if it contains all its *limit points*. (These ideas were introduced in Section 2.1 in the context of point set theory. Detailed understanding is not essential here, but the reader should at least feel comfortable with the "spirit" of this discussion.) In the present example we see that $\mathcal{C}(-1, 1)$ is not closed, since, for instance, it does not contain the limit point $H(t)$ of the sequence (17.25). This situation is unfortunate, for closed spaces turn out to be more tractable in analysis.

More graphically, we can think of $\mathcal{C}(a, b)$ as having "holes" in it. If we can add vectors to $\mathcal{C}(a, b)$ so as to plug all the holes, then the resulting set will be closed. It turns out that this step *is* possible,[5] and the resulting space is the important and well-known *Lebesgue space* $\mathcal{L}_2(a, b)$—namely, the space of all functions $x(t)$ [with the inner product (17.24)] that are "square-integrable" over the interval—that is, such that[6]

$$\int_a^b |x(t)|^2 \, dt < \infty. \tag{17.26}$$

For example, $x(t) = 6t^2$ is in $\mathcal{L}_2(-1, 1)$ but $x(t) = 1/t$ is not, since $\int_{-1}^{1} 1/t^2 \, dt = \infty$. Observe that (17.26) simply requires that all vectors $x(t)$ be of finite length. (In fact, we will always exclude vectors of infinite length from our spaces.)

$\mathcal{L}_2(a, b)$ is an example of a **Hilbert space**—a *closed inner product space* with the norm $\|x\| \equiv \sqrt{(x, x)}$.

Turning to the expansion of members of $\mathcal{L}_2(a, b)$ in terms of a set of base vectors, consider for definiteness, $\mathcal{L}_2(-1, 1)$. Is the LI set $1, t, t^2, \ldots$ a basis? The answer is not obvious; in fact, Dennery and Krzywicki[7] say that it is, whereas Stakgold[8] says that it is not. The apparent inconsistency is due to a subtle difference in the authors' definitions of the term "basis." In any case, we decided (at the end of Section 17.3) to stick to orthogonal bases, and the set $1, t, t^2, \ldots$ is not orthogonal. Nevertheless, we can generate an orthogonal set from it as follows.

Define $\Phi_1(t) = 1$. Next, seek $\Phi_2(t) = a1 + bt$ such that Φ_2 is orthogonal to Φ_1:

$$(\Phi_2, \Phi_1) = \int_{-1}^{1} (a + bt)(1) \, dt = 2a = 0;$$

so $a = 0$ and b is arbitrary. With $b = 1$, say, $\Phi_2(t) = t$. Next, we seek $\Phi_3(t) = a1 + bt + ct^2$ such that Φ_3 is orthogonal to both Φ_1 and Φ_2:

$$(\Phi_3, \Phi_1) = \int_{-1}^{1} (a + bt + ct^2)(1) \, dt = 2a + \tfrac{2}{3}c = 0,$$

$$(\Phi_3, \Phi_2) = \int_{-1}^{1} (a + bt + ct^2)(t) \, dt = \tfrac{2}{3}b = 0;$$

therefore $b = 0$ and $c = -3a$. With $a = 1$, say, $\Phi_3(t) = 1 - 3t^2$. Continuing in this fashion, we generate an orthogonal set $\Phi_1, \Phi_2, \ldots$. This procedure is called **Gram-Schmidt orthogonalization**.

---

[5]The story is not simple. See, for instance, B. Z. Vulikh, *Functional Analysis for Scientists and Technologists*, Pergamon Press, New York, 1963, Theorem 7.2.1.

[6]The integral in (17.26) is to be understood in the so-called *Lebesgue sense*. We need not elaborate on Lebesgue integration in this introductory treatment, however, for in virtually all practical applications the Lebesgue and Riemann definitions yield the same value.

[7]P. Dennery and A. Krzywicki, *Mathematics for Physicists*, Harper & Row, New York, 1967, p. 202.

[8]I. Stakgold, *Boundary Value Problems of Mathematical Physics*, Vol. I, Macmillan, New York, 1967, p. 119.

Dividing each $\Phi_j(t)$ by its length, $\|\Phi_j(t)\|$, we arrive at the ON set $\phi_j(t) = \Phi_j(t)/\|\Phi_j(t)\|$—that is,

$$\begin{aligned}\phi_1(t) &= \frac{\Phi_1}{\|\Phi_1\|} = \frac{1}{\sqrt{2}},\\ \phi_2(t) &= \frac{\Phi_2}{\|\Phi_2\|} = \sqrt{\frac{3}{2}}\,t,\\ \phi_3(t) &= \frac{\Phi_3}{\|\Phi_3\|} = \sqrt{\frac{5}{8}}(1 - 3t^2),\end{aligned} \tag{17.27}$$

and so on.

Alternatively, if we had scaled the Φ_j's (arbitrarily) so that each one is unity at $t = 1$, say, then we would have the orthogonal set

$$P_0(t) = 1, \qquad P_1(t) = t, \qquad P_2(t) = \frac{3t^2 - 1}{2}, \tag{17.28}$$

and so on; these are the well-known **Legendre polynomials,** which arose earlier in connection with Gauss integration; as mentioned in Exercise 1.24, they satisfy the recursion formula

$$P_{j+1}(t) = \left(\frac{2j+1}{j+1}\right)tP_j(t) - \left(\frac{j}{j+1}\right)P_{j-1}(t) \qquad (j = 1, 2, \dots). \tag{17.29}$$

It turns out (as we shall see later on) that the ON set $\phi_j(t)$ [as well as the orthogonal set $P_j(t)$] *is* a basis for $\mathfrak{L}_2(-1, 1)$. That is, we can expand any $x(t)$ in $\mathfrak{L}_2(-1, 1)$ as

$$x(t) = \sum_1^\infty \alpha_j \phi_j(t). \tag{17.30}$$

To evaluate the α_j's, we "dot" $\phi_k(t)$ into both sides of (17.30):

$$(x, \phi_k) = \left(\sum_1^\infty \alpha_j \phi_j, \phi_k\right) = \sum_1^\infty \alpha_j (\phi_j, \phi_k) = \alpha_k, \tag{17.31}$$

since $(\phi_j, \phi_k) = \delta_{jk}$, where δ_{jk} is the **Kronecker delta**; $\delta_{jk} \equiv 1$ for $j = k$ and 0 for $j \neq k$ (Exercise 17.22). Note that k is fixed in the preceding calculation, whereas j is the dummy summation index. Using (17.31), we have

$$x(t) = \sum_1^\infty (x, \phi_j)\phi_j(t). \tag{17.32}$$

We mentioned, following (17.29), that it "turns out" that the infinite ON set $\phi_j(t)$ is a basis. Is it not obvious, since the dimension of the space is infinite and the number of ϕ_j's is infinite? No! For suppose some infinite orthogonal set $\psi_1, \psi_2, \dots$ *is* a basis of an infinite-dimensional space $\mathcal{S}$. If we remove ψ_6, say, with a tweezers, then the remaining set $\psi_1, \dots, \psi_5, \psi_7, \dots$ is surely no longer a basis, even though it *still* contains an infinite number of ψ_j's, for the vector ψ_6 itself cannot be expanded in terms of the remaining vectors $\psi_1, \dots, \psi_5, \psi_7, \dots$; that is, the expansion

$$\sum_{\substack{j=1\\(j\neq 6)}}^\infty (\psi_6, \psi_j)\psi_j = 0 + 0 + \cdots$$

obviously converges to $\mathcal{O}$, not ψ_6.

Example 17.8. Expand $x(t) = |t|$ over the interval $[-1, 1]$ in terms of the ϕ_j's. We compute the coefficients in (17.32) as follows.

$$(x, \phi_1) = \int_{-1}^{1} |t| \frac{1}{\sqrt{2}}\, dt = \frac{1}{\sqrt{2}},$$

$$(x, \phi_2) = \int_{-1}^{1} |t| \sqrt{\frac{3}{2}}\, t\, dt = 0, \tag{17.33}$$

$$(x, \phi_3) = \int_{-1}^{1} |t| \sqrt{\frac{5}{8}}(1 - 3t^2)\, dt = -\sqrt{\frac{5}{32}},$$

and so on, so that

$$\begin{aligned} |t| &= \frac{1}{\sqrt{2}}\left(\frac{1}{\sqrt{2}}\right) + 0\left(\sqrt{\frac{3}{2}}\,t\right) - \sqrt{\frac{5}{32}}\left[\sqrt{\frac{5}{8}}(1 - 3t^2)\right] + \cdots \\ &= \frac{1}{2} - \frac{5}{16}(1 - 3t^2) + \cdots. \end{aligned} \tag{17.34}$$

The partial sums $s_1 = s_2 = \frac{1}{2}$ and $s_3 = \frac{1}{2} - \frac{5}{16}(1 - 3t^2)$ are sketched in Fig. 17.6 to suggest how the sequence s_n approaches the given function $|t|$ as n increases.

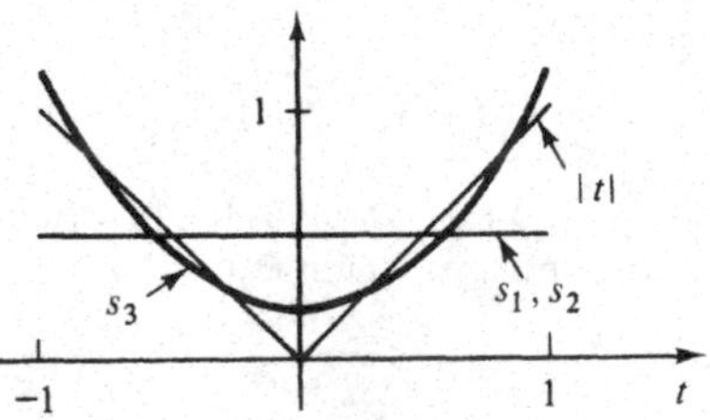

Figure 17.6. Partial sums in expansion of $|t|$.

Expanding in terms of the (orthogonal but not normalized) $P_j(t)$ basis instead, we would write

$$x(t) = \sum_1^\infty \alpha_j P_j(t), \tag{17.35}$$

$$(x, P_k) = \left(\sum_1^\infty \alpha_j P_j, P_k\right) = \alpha_k (P_k, P_k),$$

so that $\alpha_k = (x, P_k)/(P_k, P_k)$ and hence

$$x(t) = \sum_1^\infty \frac{(x, P_j)}{(P_j, P_j)} P_j(t). \tag{17.36}$$

For $x(t) = |t|$, again, we obtain

$$|t| = \frac{1}{2}(1) + 0(t) + \frac{5}{8}\left(\frac{3t^2 - 1}{2}\right) = \frac{1}{2} + \frac{5}{16}(3t^2 - 1) + \cdots,$$

which is identical to (17.34). The point is that the resulting expansion is the same whether the base vectors are normalized or not. ∎

How about the expansion of

$$y(t) = \begin{cases} |t|, & 0 < |t| < 1 \\ 1, & t = 0, \end{cases} \tag{17.37}$$

say, which is identical to $x(t) = |t|$ except for the point discontinuity at $t = 0$? Obviously $y(t)$ is in $\mathfrak{L}_2(-1, 1)$, since $\int_{-1}^{1} |y(t)|^2\, dt < \infty$, and so it must be expandable in terms of the ϕ_j basis. Yet how will the ϕ_j's cope with the point discontinuity? Computing the expansion coefficients, we find that $(y, \phi_1) = 1/\sqrt{2}$, $(y, \phi_2) = 0$, $(y, \phi_3) = -\sqrt{5/32}, \ldots$, just as in (17.33), because the point discontinuity does not affect the value of the integrals. Thus the expansions of $x(t)$ and $y(t)$ are identical. At first this result seems incorrect, for $x(t)$ and $y(t)$ are different functions. However, note carefully that

$$\|x(t) - y(t)\| = \sqrt{\int_{-1}^{1} |x(t) - y(t)|^2 \, dt} = 0,$$

so that x and y are *identical* within the vector space concept! Consequently, they *should* have identical expansions.

Nevertheless, the fact that the expansion of $y(t)$ converges, at $t = 0$, to zero rather than one may still be upsetting. The point to recognize is that we have met two different concepts of convergence. To illustrate, suppose that

$$x(t) = \sum_{1}^{\infty} \alpha_j e_j(t) \tag{17.38}$$

over $a \leq t \leq b$. In Part I this result meant that for a given fixed t

$$x(t) - \sum_{1}^{n} \alpha_j e_j(t) \longrightarrow 0 \quad \text{as } n \longrightarrow \infty, \tag{17.39a}$$

which was the concept of **pointwise convergence**. In Part III, however, (17.38) is understood in the vector sense—that is,

$$\left\| x - \sum_{1}^{n} \alpha_j e_j \right\| \longrightarrow 0 \quad \text{as } n \longrightarrow \infty \tag{17.39b}$$

or, equivalently,

$$\int_a^b \left| x(t) - \sum_{1}^{n} \alpha_j e_j(t) \right|^2 dt \longrightarrow 0 \quad \text{as } n \longrightarrow \infty, \tag{17.39c}$$

which is the so-called **mean square** or **$\mathfrak{L}_2$-convergence**. Thus when we say that $e_1, e_2, \ldots$ is a basis for $\mathfrak{L}_2(a, b)$, what we guarantee is the *mean square* convergence of the expansion. It may or may not be pointwise convergent also; it simply is not relevant. In the present example it turns out that the expansion (17.33) does converge pointwise to $x(t)$ over $-1 \leq t \leq 1$, whereas the expansion of $y(t)$ fails to converge pointwise to $y(t)$ at $t = 0$. (See Exercise 17.24.)

Observe how pointwise convergence is a local or (of course) pointwise concept, whereas vector or mean square convergence is of a *global* nature.

We will see later on that the functions 1, $\cos(\pi t/l)$, $\cos(2\pi t/l), \ldots, \sin(\pi t/l)$, $\sin(2\pi t/l), \ldots$ constitute an orthogonal (although not normalized) basis for $\mathfrak{L}_2(-l, l)$. Expanding a given $x(t)$,

$$x(t) = \sum_{0}^{\infty} \left(a_j \cos \frac{j\pi t}{l} + b_j \sin \frac{j\pi t}{l} \right), \tag{17.40}$$

we compute the b_j's, for example, by dotting $\sin(k\pi t/l)$ into both sides. Noting that $\sin(k\pi t/l)$ is orthogonal to all the other base vectors [i.e., the $\cos(j\pi t/l)$'s, and the $\sin(j\pi t/l)$'s for all j's $\neq k$], we obtain

$$\left(x, \sin \frac{k\pi t}{l} \right) = b_k \left(\sin \frac{k\pi t}{l}, \sin \frac{k\pi t}{l} \right).$$

But

$$\left(x, \sin \frac{k\pi t}{l} \right) = \int_{-l}^{l} x(t) \sin \frac{k\pi t}{l} \, dt$$

and $$\left(\sin\frac{k\pi t}{l}, \sin\frac{k\pi t}{l}\right) = \int_{-l}^{l} \sin^2\frac{k\pi t}{l}\, dt = \text{etc.} = l,$$

so that

$$b_k = \frac{1}{l}\int_{-l}^{l} x(t)\sin\frac{k\pi t}{l}\, dt$$

or changing the k's back to j's,

$$b_j = \frac{1}{l}\int_{-l}^{l} x(t)\sin\frac{j\pi t}{l}\, dt.$$

Evaluating the a_j's in the same way, we find that (17.40) is identical to the Fourier series (5.23). So we are discovering that (5.23) is only one of many possible such series. To distinguish among them, we might call (17.40) a *trigonometric Fourier series*, (17.35) a *Fourier-Legendre series*, and so on. Where do all these bases "come from"? They come from *eigenvalue problems*, in particular, the Sturm–Liouville problem, as we will see in Chapter 20.

*

17.5 CONTINUITY OF THE INNER PRODUCT

We conclude this chapter with a useful theorem.

THEOREM 17.1. (Continuity of the Inner Product) If ω is any vector in the space and $x_n \longrightarrow x$ as $n \longrightarrow \infty$, then $(x_n, \omega) \longrightarrow (x, \omega)$; that is, the inner product is continuous.[9]

Proof. By the Schwarz inequality (17.13), $|(x_n - x, \omega)| \leq \|x_n - x\|\|\omega\|$; and since $\|x_n - x\| \longrightarrow 0$ by assumption, it follows that $|(x_n - x, \omega)| = |(x_n, \omega) - (x, \omega)| \longrightarrow 0$ or $(x_n, \omega) \longrightarrow (x, \omega)$.

EXERCISES

17.1. Show that the norm (17.6) satisfies the triangle inequality (17.4c) as stated in Example 17.3.

17.2. Show that the distance function $d(x, y) = \|x - y\|$ satisfies the requirements (1.6) to (1.8) of Chapter 1, as stated following Example 17.3.

17.3. Using the triangle inequality, show that

$$\|x_1 + \cdots + x_n\| \leq \|x_1\| + \cdots + \|x_n\|.$$

17.4. Fill in the steps between (17.11) and the stated result that

$$\alpha = -\frac{(x, y)}{(y, y)}.$$

[9] With ω fixed, (x, ω) is a function of x, say $f(x)$, and recall that by $f(x)$ being continuous we mean that $\lim_{n\to\infty} f(x_n) = f(\lim_{n\to\infty} x_n)$ or, in the present case, $\lim_{n\to\infty}(x_n, \omega) = (\lim_{n\to\infty} x_n, \omega)$.

17.5. It was stated without proof that the Euclidean norm (17.5) satisfies the triangle inequality (17.4c) as required. Actually, this statement is proved indirectly later in the chapter. Can you show how and where this situation occurs?

17.6. Expand $(\alpha w + \beta x, \gamma y + \delta z)$, where $\alpha, \beta, \gamma, \delta$ are scalars.

17.7. Verify that the choice $\|x\| \equiv \sqrt{(x, x)}$ does satisfy the requirements (17.4a) and (17.4b) as stated just before equation (17.12).

17.8. Show that (17.14) *is* a suitable inner product as claimed.

17.9. Is $(x, y) = \{\max_j |\xi_j|\}\{\max_j |\eta_j|\}$ a suitable inner product for our n-tuple space? Explain.

17.10. Prove that if $(x, z) = (y, z)$ for all z's in the space, then $x = y$ or, equivalently, that if $(u, v) = 0$ for all v's, then $u = \Theta$.

17.11. (*The dual or reciprocal basis*) (a) If $e_1, e_2, \ldots$ is an ON basis and $x = \sum \alpha_j e_j$, we saw, by dotting e_k into both sides, that $\alpha_j = (x, e_j)$, so that $x = \sum (x, e_j)e_j$. If the basis is *not* ON, this result does not hold. However, suppose that we can find a set $e_1^*, e_2^*, \ldots$ such that $(e_j, e_k^*) = \delta_{jk}$. Show that in this case

$$x = \sum (x, e_j^*)e_j. \tag{17.41}$$

The e_j^*'s are called the **dual** or **reciprocal basis** corresponding to the e_j's.

(b) Use (17.41) to expand $x = (3, 1)$ in terms of the basis $e_1 = (1, 0)$, $e_2 = (1, 1)$. Sketch e_1, e_2, e_1^*, e_2^* and x and show that the result is in accord with the usual law of vector addition.

(c) Show that if the e_j's do happen to be ON, then the dual basis coalesces with the original basis—that is, $e_j^* = e_j$.

17.12. Expand $x = (3, 1 - i, 0)$, by any means whatever, in terms of

(a) $e_1 = (1, 0, 0)$, $e_2 = (1, 1, 0)$, $e_3 = (1, 1, 1)$.
(b) $e_1 = (1, 0, 0)$, $e_2 = (0, 1, 0)$, $e_3 = (0, 0, 1)$.
(c) $e_1 = (i, 0, 0)$, $e_2 = (0, i, 1)$, $e_3 = (0, 0, i)$.

17.13. Verify that the following sets are LI.

(a) $e_1 = (i, 0, 0)$, $e_2 = (0, 1, 0)$, $e_3 = (1, 0, 1)$
(b) $e_1 = t$, $e_2 = \sin t$, $e_3 = \cos t$ in $\mathfrak{L}_2(0, \pi)$

Hint: Express $\alpha_1 e_1 + \alpha_2 e_2 + \alpha_3 e_3 = \Theta$. Taking the inner product of both sides with e_1, then e_2, then e_3, recall footnote 3 of Section 7.1 and thus show that the set will be LI if and only if the so-called **Gram determinant**

$$\begin{vmatrix} (e_1, e_1) & (e_2, e_1) & (e_3, e_1) \\ (e_1, e_2) & (e_2, e_2) & (e_3, e_2) \\ (e_1, e_3) & (e_2, e_3) & (e_3, e_3) \end{vmatrix}$$

is nonzero.

17.14. We stated that "if a basis for $\mathfrak{S}$ consists of n vectors, then the dimension of $\mathfrak{S}$ must be n, and conversely." Why is that?

17.15. Show that the following are bases for 2-tuple space.

(a) $e_1 = (1, 1)$, $e_2 = (1, -1)$ (b) $e_1 = (1, 0)$, $e_2 = (1, 1)$

17.16. Verify that the space $\mathfrak{C}(a, b)$, defined in Example 17.7, is a bonafide vector space. Would it still be a vector space if we changed the word "continuous" to "polynomial"? To "discontinuous"?

17.17. (a) Show that the expression (17.23) is a suitable inner product for n-tuple space as claimed.

(b) Show that

$$(x, y) = \int_a^b x(t)\bar{y}(t)w(t)\,dt$$

is a suitable inner product for the spaces $\mathcal{C}(a, b)$ and $\mathcal{L}_2(a, b)$, where the *weighting function* $w(t)$ is positive over $a \leq t \leq b$.

17.18. Show that if $x(t)$ and $y(t)$ are in $\mathcal{L}_2(a, b)$, then so is $x(t) + y(t)$ (as is required of a vector space).

17.19. Is $(x, y) = x(a)y(a)$ a suitable inner product for $\mathcal{L}_2(a, b)$? Explain.

***17.20.** Is the space of all polynomial functions $x(t)$ over $0 \leq t \leq 1$, with the norm $\|x - y\| = \max_{0 \leq t \leq 1} |x(t) - y(t)|$ closed? (Vector addition, scalar multiplication, and so on, are defined in the usual way.) Explain. *Hint:* Recall the Weierstrass approximation theorem (footnote 13, Chapter 1).

17.21. Obtain ON sets from the following LI sets by means of the Gram–Schmidt procedure.

(a) $e_1 = (0, 1, 0)$, $e_2 = (2, 0, 1)$, $e_3 = (1, 1, 1)$

(b) $e_1 = t$, $e_2 = \sin t$, $e_3 = \cos t$ in $\mathcal{L}_2(0, \pi)$

17.22. Regarding the steps in equation (17.31),

*(a) verify the second equality. *Hint:* Write the left-hand side as $\left(\lim_{n\to\infty} \sum_1^n \alpha_j\phi_j, \phi_k\right)$ and then apply Theorem 17.1.

(b) verify the last equality; that is, show that

$$\sum_1^\infty \alpha_j \delta_{jk} = \alpha_k.$$

17.23. Note carefully that mean square convergence of a sequence $s_n(t)$ does *not* imply that it converges pointwise and its pointwise convergence over the interval does *not* imply that it converges in the mean square sense.

(a) Sketch the sequence

$$f_n(t) = \begin{cases} 1 - n|t|, & |t| < \dfrac{1}{n} \\ 0, & \dfrac{1}{n} \leq |t| \leq 1 \end{cases}$$

and show that $f_n(t) \longrightarrow 0$ in the mean square sense, over $-1 \leq t \leq 1$, but not pointwise.

(b) Sketch

$$g_n(t) = \begin{cases} nf_n(t), & t \neq 0 \\ 0, & t = 0 \end{cases} \qquad (f_n \text{ defined above})$$

and show that $g_n(t) \longrightarrow 0$ pointwise but not mean square.

(c) Sketch

$$h_n(t) = \begin{cases} f_n(t), & t \neq 0 \\ 0, & t = 0 \end{cases} \qquad (f_n \text{ defined above})$$

and show that $h_n(t) \longrightarrow 0$ both mean square *and* pointwise.

17.24. Do you think $\sin jt$ ($j = 1, 2, \ldots$) is a basis for $\mathcal{L}_2(-\pi, \pi)$? Explain.

17.25. If $e_1, e_2, \ldots$ is an ON basis for an infinite-dimensional space $\mathcal{S}$, then we can expand any x in $\mathcal{S}$ as

$$x = \sum_1^\infty \alpha_j e_j.$$

Dotting e_k into both sides and recalling that the e_j's are ON, we find that $\alpha_k = (x, e_k)$. The α_k's are called the **Fourier coefficients** of x, and the right-hand side of

$$x = \sum_1^\infty (x, e_j)e_j$$

is called **the Fourier series of x** (with respect to the e_j basis). Some examples are contained in Section 17.4. Suppose instead that we have an ON set $e_1, \ldots, e_n$ that is *not* complete—that is, it is not a basis—and we seek the *best approximation*

$$x \approx \sum_1^n \alpha_j e_j;$$

in other words, we wish to choose the α_j's so that the error $\left\| x - \sum_1^n \alpha_j e_j \right\|$, or, equivalently, its square, is as small as possible.

(a) Denoting $(x, e_j) \equiv a_j$, show that

$$\left\| x - \sum_1^n \alpha_j e_j \right\|^2 \leq \|x\|^2 + \sum_1^n |a_j - \alpha_j|^2 - \sum_1^n |a_j|^2$$

and hence that the optimum choice is $\alpha_j = a_j$—in sum, *the Fourier coefficients do the best job.*

(b) Thus deduce the **Bessel inequality**

$$\sum |a_j|^2 \leq \|x\|^2$$

and in the event that the e_j's *are* complete, the **Parseval equality**

$$\sum |a_j|^2 = \|x\|^2$$

which, if we take the square root of both sides, can be regarded as a generalized "Pythagorean Theorem."

(c) In either case, show that it must be true that $a_j \to 0$ as $j \to \infty$; this modest result is known as the **Riemann–Lebesgue lemma.**

17.26. Make up two examination-type problems on this chapter.

Chapter 18

Linear Operators

Having considered linear spaces in Chapter 17, we now introduce the concept of linear operators on these spaces. With this accomplished, we will, in Chapters 19 and 20, be in a position to consider the solution of the general linear equation $Lx = c$ and the so-called eigenvalue problem $Lx = \lambda x$.

18.1. SOME DEFINITIONS

By an **operator** (or **transformation**) L, we mean a *mapping* from one vector space, called the *domain* $\mathfrak{D}$, to another, called the *range* $\mathfrak{R}$, as indicated schematically in Fig. 18.1. It is actually the same definition as for a function except that whereas the domain and range of a function are point sets, for an operator they are vector spaces. As with a function, L must be single valued, although not necessarily one-to-one. That is, for each x in $\mathfrak{D}$ there is a uniquely determined y in $\mathfrak{R}$ given by $Lx = y$, but there may be other x's in $\mathfrak{D}$, say x', such that $Lx' = y$. If to each y in $\mathfrak{R}$ there corresponds exactly one x in $\mathfrak{D}$, then L is *one-to-one*; or L is one-to-one if and only if $Lx = Lz$ implies $x = z$. Two operators, A and B say, are said to be *equal* if they have the same domain and if $Ax = Bx$ for all x's in their domain.

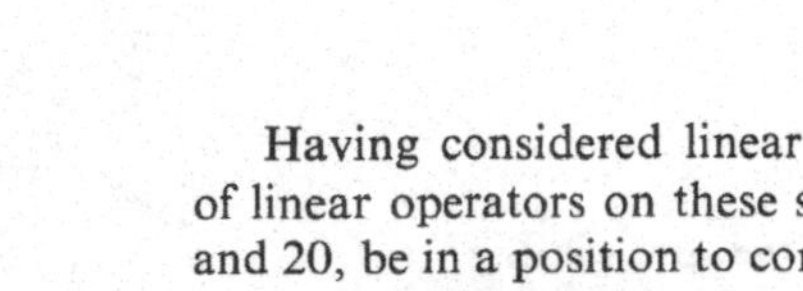

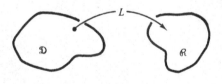

Figure 18.1. L as a mapping.

Finally, we say that I is the **identity operator** if $Ix = x$ for all x and that $\varnothing$ is the **null operator** if $\varnothing x = \mathbf{0}$ for all x.

Example 18.1. *Matrices.* Consider the **matrix** operator

$$A = \begin{pmatrix} a_{11} & a_{12} & \cdots & a_{1n} \\ a_{21} & a_{22} & \cdots & a_{2n} \\ \cdot & & & \cdot \\ \cdot & & & \cdot \\ \cdot & & & \cdot \\ a_{n1} & a_{n2} & \cdots & a_{nn} \end{pmatrix} \tag{18.1}$$

over the domain $\mathfrak{D}$ of n tuples

$$x = \begin{pmatrix} \xi_1 \\ \cdot \\ \cdot \\ \cdot \\ \xi_n \end{pmatrix}, \tag{18.2}$$

where the "elements" a_{ij} of A are real numbers (later we will allow them to be complex) and where the action of A on x is *defined* by

$$Ax = \begin{pmatrix} a_{11} & \cdots & a_{1n} \\ \cdot & & \cdot \\ \cdot & & \cdot \\ \cdot & & \cdot \\ a_{n1} & \cdots & a_{nn} \end{pmatrix} \begin{pmatrix} \xi_1 \\ \cdot \\ \cdot \\ \cdot \\ \xi_n \end{pmatrix} \equiv \begin{pmatrix} a_{11}\xi_1 + \cdots + a_{1n}\xi_n \\ \cdot \\ \cdot \\ \cdot \\ a_{n1}\xi_1 + \cdots + a_{nn}\xi_n \end{pmatrix}; \tag{18.3}$$

for instance,

$$\begin{pmatrix} 2 & 0 & -1 \\ 1 & 2 & 4 \\ 3 & -1 & 0 \end{pmatrix} \begin{pmatrix} 2 \\ 1 \\ 5 \end{pmatrix} = \begin{pmatrix} -1 \\ 24 \\ 5 \end{pmatrix}.$$

The first and second indices on a_{ij} denote the row and column, respectively, of that element; thus we call i the "row index" and j the "column index." Note that the n tuple (18.2) has been arranged in column form for notational compactness in (18.3), as opposed to the row form $(\xi_1, \ldots, \xi_n)$ used in Chapter 17. We will have need for both forms and will refer to them as *column vectors* and *row vectors*, respectively.

The right-hand side of (18.3) is an n-dimensional vector, just like x. Equivalent to (18.3) is the scalar statement

$$(Ax)_j = a_{j1}\xi_1 + \cdots + a_{jn}\xi_n = \sum_{k=1}^{n} a_{jk}\xi_k; \tag{18.4}$$

that is, the jth component of the Ax vector is $\sum_{k=1}^{n} a_{jk}\xi_k$.

What is the range of A? It will clearly be a subset of $\mathfrak{D}$, but whether it is *all* of $\mathfrak{D}$ is not obvious. That's equivalent to the following question. Given any y in $\mathfrak{D}$, is there an x such that $Ax = y$? If the answer is yes, then the range is all of $\mathfrak{D}$. In fact, the general problem of solving

$$Lx = c$$

for x is perhaps *the* central issue in the theory of operators and is the subject of the next chapter.

As a simple illustration, suppose that $n = 2$ and $A = \begin{pmatrix} 1 & 0 \\ 0 & 1 \end{pmatrix}$. Then

$$Ax = \begin{pmatrix} 1 & 0 \\ 0 & 1 \end{pmatrix} \begin{pmatrix} \xi_1 \\ \xi_2 \end{pmatrix} = \begin{pmatrix} \xi_1 \\ \xi_2 \end{pmatrix} = x$$

(so that A happens to be the identity operator I), and so the solution of $Ax = y$, for any y, is simply $x = y$. It follows that the range is $\mathfrak{R} = \mathfrak{D}$. If, on the other hand, $A = \begin{pmatrix} 1 & 2 \\ 1 & 2 \end{pmatrix}$, then $Ax = y$ becomes

$$\begin{pmatrix} \xi_1 + 2\xi_2 \\ \xi_1 + 2\xi_2 \end{pmatrix} = \begin{pmatrix} \eta_1 \\ \eta_2 \end{pmatrix}$$

or

$$\xi_1 + 2\xi_2 = \eta_1$$
$$\xi_1 + 2\xi_2 = \eta_2.$$

Surely a solution for ξ_1, ξ_2 exists only if $\eta_1 = \eta_2$. Thus if we regard the domain $\mathfrak{D}$ as a two-dimensional ξ_1, ξ_2 plane, then we see that the range $\mathfrak{R}$ is only the subset of the η_1, η_2 plane consisting of the line $\eta_1 = \eta_2$. ▮

*

Example 18.2 *Integral Operators.* Consider the integral operator

$$Lx = \int_a^b K(\tau, t)x(\tau)\, d\tau \tag{18.5}$$

over the domain $\mathfrak{L}_2(a, b)$. Suppose, for instance, that $a = 0$, $b = \infty$, and that the *kernel* is $K(\tau, t) = \exp(-t\tau)$. If $x(t) = 1 - H(t - 1)$, say, then

$$Lx = \int_0^\infty e^{-t\tau}[1 - H(\tau - 1)]\, d\tau = \int_0^1 e^{-t\tau}\, d\tau = \frac{1 - e^{-t}}{t}.$$

If we would like Lx to be in $\mathfrak{L}_2(a, b)$, too, we expect that the kernel K must be suitably restricted. In particular,

$$\begin{aligned}\int_a^b |Lx|^2\, dt &= \int_a^b dt \left| \int_a^b K(\tau, t)x(\tau)\, d\tau \right|^2 \\ &\leq \int_a^b dt \int_a^b |K(\tau, t)|^2\, d\tau \int_a^b |\bar{x}(\tau)|^2\, d\tau \\ &= \|x\|^2 \int_a^b \int_a^b |K(\tau, t)|^2\, d\tau\, dt, \end{aligned} \tag{18.6}$$

where the second line follows from the Schwarz inequality and where we have also used the fact that $|\bar{x}| = |x|$ in the last step. Now the $\|x\|^2$ factor in (18.6) is bounded, since x is in $\mathfrak{L}_2$. So we see from (18.6) that a sufficient condition for Lx to be in $\mathfrak{L}_2(a, b)$ is that K be an "$\mathfrak{L}_2$ kernel"—that is,

$$\int_a^b \int_a^b |K(\tau, t)|^2\, d\tau\, dt < \infty, \tag{18.7}$$

for this result will guarantee that $\int_a^b |Lx|^2\, dt < \infty$. In this case, the range will be all or part of $\mathfrak{L}_2(a, b)$. ▮

Given an operator A, we define a new operator $C = \alpha A$, where α is a scalar, according to

$$Cx = (\alpha A)x \equiv \alpha(Ax) \tag{18.8}$$

for all x's in the domain $\mathcal{A}$ of A. Also, given two operators A and B, with domains $\mathcal{A}$ and $\mathcal{B}$, define their "sum" $C = A + B$ according to

$$Cx = (A + B)x \equiv Ax + Bx \tag{18.9}$$

for all x's in both $\mathcal{A}$ and $\mathcal{B}$. Note that the $+$ on the right-hand side denotes the already defined *vector* addition, whereas the $+$ on the left denotes the addition of *operators*, which is being *defined* by (18.9).

Finally, we define the "product" of two operators $C = AB$ according to

$$Cx = (AB)x \equiv A(Bx). \tag{18.10}$$

Similarly, $(ABC)x = A(B(Cx))$ and so on.

Since Ax and Bx are vectors, $Ax + Bx = Bx + Ax$, so that $A + B = B + A$. On the other hand, AB need not equal BA; if it does, we say that A and B **commute**. (Actually, we should say they "commute under multiplication," but the last two words are generally omitted, without confusion, for they *always* commute under addition.)

Example 18.3. For the *matrix* operator defined by (18.3), for instance, if we denote the right-hand side as $\{a_{ij}\}$ for short, it *follows* from (18.8) to (18.10) that if α is a scalar and A and B are $n \times n$ matrices, then

$$\alpha A = \{\alpha a_{ij}\} \tag{18.11}$$

$$A + B = \{a_{ij} + b_{ij}\} \tag{18.12}$$

$$AB = \left\{\sum_{k=1}^{n} a_{ik} b_{kj}\right\}. \tag{18.13}$$

Consider the derivation of (18.12).

$$(A + B)x \equiv Ax + Bx = \begin{pmatrix} a_{11}\xi_1 + \cdots + a_{1n}\xi_n \\ \vdots \\ a_{n1}\xi_1 + \cdots + a_{nn}\xi_n \end{pmatrix} + \begin{pmatrix} b_{11}\xi_1 + \cdots + b_{1n}\xi_n \\ \vdots \\ b_{n1}\xi_1 + \cdots + b_{nn}\xi_n \end{pmatrix}$$

$$= \begin{pmatrix} (a_{11} + b_{11})\xi_1 + \cdots + (a_{1n} + b_{1n})\xi_n \\ \vdots \\ (a_{n1} + b_{n1})\xi_1 + \cdots + (a_{nn} + b_{nn})\xi_n \end{pmatrix}$$

$$= \begin{pmatrix} (a_{11} + b_{11}) \cdots (a_{1n} + b_{1n}) \\ \vdots \\ (a_{n1} + b_{n1}) \cdots (a_{nn} + b_{nn}) \end{pmatrix} \begin{pmatrix} \xi_1 \\ \vdots \\ \xi_n \end{pmatrix},$$

so that $A + B = \{a_{ij} + b_{ij}\}$ as claimed. The crucial step was the next-to-last equality, which follows from our earlier definition of *vector* addition. We leave (18.11) and (18.13) for Exercise 18.3.

Not surprisingly, we also define the "difference" $A - B \equiv A + (-1)B = \{a_{ij} - b_{ij}\}$.

Note that (18.11) to (18.13) are matrix equations. Equivalent to (18.11), for example, is the scalar statement $(\alpha A)_{ij} = \alpha a_{ij}$; that is, the ij element of the αA matrix is given by αa_{ij}. Similarly,

$$(A + B)_{ij} = a_{ij} + b_{ij} \quad \text{and} \quad (AB)_{ij} = \sum_{k=1}^{n} a_{ik}b_{kj}.$$

To illustrate,

$$3\begin{pmatrix} 2 & -1 \\ 0 & 1 \end{pmatrix} = \begin{pmatrix} 3\cdot 2 & 3\cdot -1 \\ 3\cdot 0 & 3\cdot 1 \end{pmatrix} = \begin{pmatrix} 6 & -3 \\ 0 & 3 \end{pmatrix},$$

$$\begin{pmatrix} 2 & 4 \\ 1 & -1 \end{pmatrix} + \begin{pmatrix} 3 & 2 \\ 0 & 1 \end{pmatrix} = \begin{pmatrix} 2+3 & 4+2 \\ 1+0 & -1+1 \end{pmatrix} = \begin{pmatrix} 5 & 6 \\ 1 & 0 \end{pmatrix},$$

and

$$\begin{pmatrix} 2 & 4 \\ 1 & -1 \end{pmatrix}\begin{pmatrix} 3 & 2 \\ 0 & 1 \end{pmatrix} = \begin{pmatrix} 2\cdot 3 + 4\cdot 0 & 2\cdot 2 + 4\cdot 1 \\ 1\cdot 3 - 1\cdot 0 & 1\cdot 2 - 1\cdot 1 \end{pmatrix} = \begin{pmatrix} 6 & 8 \\ 3 & 1 \end{pmatrix}.$$

Since

$$\begin{pmatrix} 3 & 2 \\ 0 & 1 \end{pmatrix}\begin{pmatrix} 2 & 4 \\ 1 & -1 \end{pmatrix} = \begin{pmatrix} 8 & 10 \\ 1 & -1 \end{pmatrix},$$

we see that these two matrices do not commute. ▮

Three more definitions. We say that L is **linear** if

$$L(\alpha x + \beta y) = \alpha Lx + \beta Ly \tag{18.14}$$

for all scalars α, β and all x, y in the domain. As the title of the chapter indicates, we shall consider only linear operators. We leave it for Exercise 18.6 to verify that the matrix and integral operators (18.3) and (18.5) are both linear.

Also, L is said to be **bounded** if for all x in the domain $\mathfrak{D}$ there exists a constant c such that

$$\|Lx\| \leq c\|x\|, \tag{18.15}$$

and we call c "a bound on L."

Example 18.4. Suppose that L is a matrix A with n-tuple domain. Then

$$\begin{aligned} \|Ax\|^2 &= \sum_i |(Ax)_i|^2 = \sum_i |\sum_j a_{ij}\xi_j|^2 \\ &\leq \sum_i (\sum_j |a_{ij}|^2)(\sum_j |\xi_j|^2) = \|x\|^2 \sum_i \sum_j |a_{ij}|^2, \end{aligned} \tag{18.16}$$

where the next to last step follows from the Schwarz inequality (18.13), where we have used the fact that $|\bar{\xi}_j| = |\xi_j|$, and where $\sum_i$ is shorthand for $\sum_{i=1}^{n}$, and similarly for $\sum_j$; that is, if (for i fixed) u is a vector with components a_{ij} ($j = 1, \ldots, n$) and v is a vector with components $\bar{\xi}_j$ ($j = 1, \ldots, n$), then $|\sum_j a_{ij}\xi_j|^2 = |\sum_j a_{ij}\bar{\bar{\xi}}_j|^2 = |\sum_j u_j\bar{v}_j|^2 = |(u, v)|^2 \leq \|u\|^2\|v\|^2 = (\sum_j |a_{ij}|^2)(\sum_j |\bar{\xi}_j|^2)$. From (2.16) we see that

A is bounded and that a bound on A is given by

$$c = \sqrt{\sum \sum |a_{ij}|^2}. \quad \blacksquare \tag{18.17}$$

*

For the integral operator of Example 18.2, (18.6) is the analog of (18.17), and (if K is an $\mathfrak{L}_2$ kernel) we have the bound

$$c = \sqrt{\int_a^b \int_a^b |K(\tau, t)|^2 \, d\tau \, dt}. \tag{18.18}$$

Finally, we call the *smallest* suitable bound the **norm** of L and denote it $\|L\|$. (The norm of an operator $\|L\|$ and the norm of a vector $\|x\|$ are different and must not be confused.) Thus

$$\|L\| \equiv \operatorname*{lub}_{x \neq \theta} \frac{\|Lx\|}{\|x\|}, \tag{18.19}$$

that is, the least upper bound[1] of the "amplification" $\|Lx\|/\|x\|$ for all $x \neq \theta$. In other words, the amplification $\|Lx\|/\|x\|$ depends on x; the *largest* possible amplification (more precisely, the lub) is the norm $\|L\|$.

It can be shown that the following is equivalent to (18.19):

$$\|L\| \equiv \operatorname*{lub}_{\|x\|=1} \|Lx\|. \tag{18.20}$$

If c is a bound on L then surely $\|L\| \leq c$.

Example 18.5. Let us find the norm of the matrix operator $A = \begin{pmatrix} 1 & 1 \\ 0 & 1 \end{pmatrix}$ on real 2-tuple space. Using (18.20), we seek the maximum of $\|Ax\|$, subject to the constraint $\|x\| = 1$. Since

$$Ax = \begin{pmatrix} 1 & 1 \\ 0 & 1 \end{pmatrix} \begin{pmatrix} \xi_1 \\ \xi_2 \end{pmatrix} = \begin{pmatrix} \xi_1 + \xi_2 \\ \xi_2 \end{pmatrix},$$

our problem is as follows.

$$(\xi_1 + \xi_2)^2 + \xi_2^2 = \text{max}, \tag{18.21a}$$

subject to the constraint

$$\xi_1^2 + \xi_2^2 = 1. \tag{18.21b}$$

Solving (18.21) (as discussed in Chapter 10), we find (Exercise 18.15) that $\xi_1 = 2/\sqrt{10 + 2\sqrt{5}}$ and $\xi_2 = (1 + \sqrt{5})/\sqrt{10 + 2\sqrt{5}}$. Consequently,

$$\|A\| = \sqrt{(\xi_1 + \xi_2)^2 + \xi_2^2} = \sqrt{\frac{10 + 4\sqrt{5}}{5 + \sqrt{5}}} \doteq 1.618.$$

By comparison, observe that (18.17) gives $c = \sqrt{3} \doteq 1.732$, which *is* a bound on A, although not the smallest possible bound; the *smallest* possible bound, found to be 1.618, is the *norm* $\|A\|$. $\blacksquare$

---

[1] If a set *has* a maximum, then that will also be its least upper bound. For example, the points $0 \leq t \leq 1$ have lub = max = 1. The points $0 \leq t < 1$, on the other hand, have lub = 1 but no max!

18.3. FURTHER DISCUSSION OF MATRICES

In this section we introduce some additional terminology and information on matrices.

First, matrices need not be "square"—that is, with an equal number of rows and columns. For instance, the n-dimensional column vector (18.2) is considered an $n \times 1$ matrix—it has n rows and one column. More generally, consider $m \times n$ matrices. Suppose that A is $m \times n$ and B is $p \times q$. Then $A + B$ may still be defined per (18.12), provided that A and B have the same "size and shape"—that is, provided that $m = p$ and $n = q$. For example,

$$\begin{pmatrix} 3 & 1 & -2 \\ 1 & 2 & 0 \end{pmatrix} + \begin{pmatrix} 4 & 1 & 1 \\ 0 & 3 & 2 \end{pmatrix} = \begin{pmatrix} 7 & 2 & -1 \\ 1 & 5 & 2 \end{pmatrix}.$$

Similarly, AB may still be defined by (18.13) but only if $n = p$. That requirement follows from the fact that the k index in (18.13) is summed from 1 to n; but k is the column index of A and the row index of B, and so we require A to have as many columns as B has rows. The resulting AB matrix will be $m \times q$. For example,

$$\underset{3 \times 2}{\begin{pmatrix} 1 & 3 \\ 0 & -2 \\ 2 & 1 \end{pmatrix}} \underset{2 \times 4}{\begin{pmatrix} 4 & -1 & 2 & 0 \\ 1 & 3 & 1 & -3 \end{pmatrix}} = \underset{3 \times 4}{\begin{pmatrix} 7 & 8 & 5 & -9 \\ -2 & -6 & -2 & 6 \\ 9 & 1 & 5 & -3 \end{pmatrix}},$$

where the -6 element, for instance, was computed per $(0)(-1) + (-2)(3)$.

A *square* matrix is said to be *diagonal* if $a_{ij} = 0$ for all $i \neq j$—that is, if all elements off the "main diagonal" (running from the upper left corner to the lower right) are zero—*symmetric* if $a_{ij} = a_{ji}$—that is, symmetric about the main diagonal—and *antisymmetric* (or *skew-symmetric*) if $a_{ij} = -a_{ji}$. To illustrate,

$$A = \begin{pmatrix} 2 & 0 & 0 \\ 0 & 0 & 0 \\ 0 & 0 & -1 \end{pmatrix}, \qquad B = \begin{pmatrix} 3 & 2 & -1 \\ 2 & 1 & 0 \\ -1 & 0 & 5 \end{pmatrix}, \qquad C = \begin{pmatrix} 0 & -2 & -3 \\ 2 & 0 & 0 \\ 3 & 0 & 0 \end{pmatrix};$$

A is diagonal (and hence symmetric), B is symmetric, and C is antisymmetric. (Note that the elements on the main diagonal of an antisymmetric matrix *must* be zero, since $a_{jj} = -a_{jj}$.)

The $n \times n$ matrix

$$I = \{\delta_{ij}\} = \begin{pmatrix} 1 & & & \\ & 1 & & \text{\Large 0} \\ & & \cdot & \\ \text{\Large 0} & & & \cdot \\ & & & 1 \end{pmatrix}, \tag{18.22}$$

where the large zeros denote that all off-diagonal terms are zero, is of special importance. It is called the n-dimensional *identity matrix*, since $Ix = x$ for any vector x and $AI = IA = A$ for any $n \times n$ matrix A, as is easily verified.

Corresponding to any $m \times n$ matrix A is an $n \times m$ matrix called the **transpose** of

A and denoted A^T. If $A = \{a_{ij}\}$, then $A^T \equiv \{a_{ji}\}$. For instance, if

$$A = \begin{pmatrix} 2 & 1 & 1-i \\ 0 & 4 & 3 \end{pmatrix}, \text{ then } A^T = \begin{pmatrix} 2 & 0 \\ 1 & 4 \\ 1-i & 3 \end{pmatrix}; \tag{18.23}$$

and if

$$x = \begin{pmatrix} 3 \\ 0 \\ i \end{pmatrix}, \text{ then } x^T = (3 \quad 0 \quad i).$$

Next, suppose that $D = AB$. It follows that

$$d_{ij} = \sum_{k=1}^{n} a_{ik} b_{kj}.$$

Denoting $D^T = \{(d^T)_{ij}\}$, we have

$$(d^T)_{ij} = d_{ji} = \sum_{k=1}^{n} a_{jk} b_{ki} = \sum_{k=1}^{n} (b^T)_{ik} (a^T)_{kj},$$

so that $D^T = (AB)^T = B^T A^T$. Similarly, we find (Exercise 18.16) that $(ABC)^T = C^T B^T A^T$ and so on. Collecting these results, for reference

$$(AB)^T = B^T A^T, \tag{18.24}$$

$$(ABC)^T = C^T B^T A^T, \tag{18.25}$$

and so on.

Furthermore, note that the inner product

$$(x, y) = \sum_{1}^{n} \xi_j \bar{\eta}_j$$

can be expressed compactly as $(x, y) = x^T \bar{y}$ or, equivalently, as $\bar{y}^T x$.

18.4. THE ADJOINT OPERATOR

Finally, we introduce a crucial concept, that of the **adjoint operator**. Specifically, we associate with a linear operator L another operator L^*, called the adjoint of L, such that

$$(Lx, y) = (x, L^* y) \tag{18.26}$$

for all allowable x's and y's. Note that (18.26) serves to *define* L^*.

It is not hard to show (Exercise 18.17) that

$$(L^*)^* = L \tag{18.27a}$$

$$(L_1 + L_2)^* = L_1^* + L_2^* \tag{18.27b}$$

$$(L_1 L_2)^* = L_2^* L_1^* \tag{18.27c}$$

and that L^* is itself linear (Exercise 18.18). In the event that $L^* = L$, we say that L is **self-adjoint** or **Hermitian**.

Example 18.6. What is the adjoint of an $n \times n$ matrix A? Well,

$$(Ax, y) = (Ax)^T \bar{y} = x^T A^T \bar{y} = x^T \overline{\bar{A}^T y} = (x, \bar{A}^T y).$$

According to (18.26), then, the adjoint is

$$A^* = \bar{A}^T. \tag{18.28}$$

For example,

$$\begin{pmatrix} 1 & 3 \\ i & 2-i \end{pmatrix}^* = \begin{pmatrix} 1 & -i \\ 3 & 2+i \end{pmatrix},$$

$$\begin{pmatrix} 3 & 1+2i \\ 1-2i & -1 \end{pmatrix}^* = \begin{pmatrix} 3 & 1+2i \\ 1-2i & -1 \end{pmatrix},$$

and

$$\begin{pmatrix} 2 & 3 \\ 3 & 1 \end{pmatrix}^* = \begin{pmatrix} 2 & 3 \\ 3 & 1 \end{pmatrix}.$$

Here, the last two are seen to be self-adjoint; in fact, every real symmetric matrix is clearly self-adjoint.

In applications, we will find that our operators are often, although not always, self-adjoint and that self-adjoint operators are especially tractable.

COMMENT. Observe that L^* is only defined for an inner product space, of course, since L^* is defined by the formula $(Lx, y) = (x, L^*y)$. Moreover, L^* will depend on the choice of the inner product. Thus when we say that the adjoint of A is $A^* = \bar{A}^T$ in (18.28), it is understood that we mean *with respect to the* (*usual*) *inner product* $(x, y) = x^T\bar{y}$. This point is discussed further in Exercise 18.21. ∎

Example 18.7. Consider the differential operator $L = d/dt + 1$ over the space of real-valued differentiable functions that are defined over $0 \leq t \leq 1$ and that satisfy the boundary condition $x(0) = 0$, with the inner product

$$(x, y) = \int_0^1 x(t)y(t)\, dt. \tag{18.29}$$

(Here the domain is *not* $\mathcal{L}_2$; functions in $\mathcal{L}_2$ are not necessarily differentiable.)

To find the adjoint operator, we start with (Lx, y) and integrate the $x'y$ term by parts in order to throw the result into the form (x, L^*y):

$$\begin{aligned}(Lx, y) &= \int_0^1 [x'(t) + x(t)]y(t)\, dt, \quad \text{where } (\ )' = \frac{d(\ )}{dt} \\ &= x(1)y(1) - x(0)y(0) + \int_0^1 [-y'(t) + y(t)]x(t)\, dt \\ &\equiv (x, L^*y).\end{aligned} \tag{18.30}$$

We see that $L^*y = -dy/dt + y$. According to (18.27), the boundary conditions associated with L^* must be such that the boundary terms (which arose through the integration by parts) all vanish and we are left with an inner product (x, L^*y). Since $x(0) = 0$, the $x(0)y(0)$ term drops out, but $x(1)$ is unspecified and so we must require that $y(1) = 0$.

Consequently, whereas the original operator was $L = d/dt + 1$ with the left end condition $x(0) = 0$, the adjoint operator is $L^* = -d/dt + 1$ with the right end condition $y(1) = 0$. Of course, additional boundary conditions can be imposed on L^*, but it is understood that only those conditions are imposed that are *necessary* for the satisfaction of (18.26).

Recalling the next-to-last sentence before Example 18.1, we conclude that the given operator fails to be self-adjoint for *two* reasons: $L = d/dt + 1$, whereas $L^* =$

$-d/dt + 1$, and *their domains are different* by virtue of the different boundary conditions.

COMMENT. Suppose that the original operator were $L = d/dt + 1$ with *no* boundary condition prescribed. Then from (18.30) we see that the adjoint operator would be $L^* = -d/dt + 1$, together with the *two* boundary conditions $y(0) = y(1) = 0$. Similarly, if the original operator were $L = d/dt + 1$ with the inhomogeneous boundary condition $x(0) = 2$, say, then the adjoint operator would be $L^* = -d/dt + 1$, together with the boundary conditions $y(0) = y(1) = 0$. ▮

Example 18.8. The differential operator $L = d^2/dt^2$ over the space of real-valued, twice-differentiable functions that are defined over $0 \leq t \leq 1$ and that satisfy the boundary conditions $x(0) = x'(0) = 0$, with the inner product (18.29), has the adjoint operator [Exercise 18.19(a)]

$$L^* = \frac{d^2}{dt^2}, \qquad y(1) = y'(1) = 0. \tag{18.31}$$

Since the boundary conditions associated with L^* differ from those associated with L, the operator is *not* self-adjoint. Nevertheless, the "action" of the two operators is the same (i.e., $L = d^2/dt^2 = L^*$), and so we say that the operator is **formally self-adjoint.** ▮

So far we have not worried about the *existence* of the adjoint of a given operator. Indeed, in our examples we were able to *find* the adjoints and this question was of no concern. In fact, it *is* possible that an adjoint operator, as defined by $(Lx, y) = (x, L^*y)$, may not exist, but we will not encounter this exceptional situation in any of our applications.

EXERCISES

18.1. Given the matrices

$$A = \begin{pmatrix} 2 & 0 \\ 1 & 4 \\ 1-i & 3 \\ 0 & -1 \end{pmatrix}, \qquad B = \begin{pmatrix} 3 & 2 \\ 0 & 1 \end{pmatrix}, \qquad x = \begin{pmatrix} 3 \\ i \end{pmatrix},$$

work out whichever of the products AB, BA, Ax, xA, Bx, xB, A^2 (i.e., AA), B^2, x^2 are defined.

18.2. Show that any square matrix A can be decomposed into a sum $B + C$, where B is symmetric and C is antisymmetric. Demonstrate the decomposition for

$$A = \begin{pmatrix} 3 & 4 & -1 \\ 2 & 1 & 0 \\ -1 & 6 & 5 \end{pmatrix} \quad \text{and} \quad A = \begin{pmatrix} i & 1+i \\ 0 & 2 \end{pmatrix}.$$

18.3. Verify (18.11) and (18.13) in the same way that we verified (18.12).

18.4. Show that in order for matrices A and B to commute, they must both be square and of the same size. Show whether $n \times n$ matrices A and B will necessarily commute if both are symmetric.

18.5. Although seemingly trivial, experience shows that this problem is worth discussing. If $(AB)_{ij} = \left\{\sum_{k=1}^{n} a_{ik}b_{kj}\right\}$, what is $\left\{\sum_{k=1}^{n} b_{kj}a_{ik}\right\}$?

18.6. Show that the matrix and integral operators (18.3) and (18.5) are linear.

18.7. If L is linear, what can be said about the linearity or nonlinearity of L^2 (i.e., LL)? Explain.

18.8. Show that if A and B are linear operators, then $(A + B)(C + D)x = (AC + AD + BC + BD)x$.

18.9. If $AB = \varnothing$, does this result imply that $A = \varnothing$ and/or $B = \varnothing$?

18.10. Obtain bounds for the two matrix operators in Exercise 18.2.

18.11. Determine the norm $\|A\|$ of these matrix operators, where all vectors are real.

(a) $A = \begin{pmatrix} -1 & 0 & 0 \\ 0 & 2 & 0 \\ 0 & 0 & 4 \end{pmatrix}$ (b) $A = \begin{pmatrix} 4 & 2 \\ 2 & 1 \end{pmatrix}$ (c) $A = \begin{pmatrix} 0 & 0 \\ 2 & 5 \end{pmatrix}$

18.12. $Lx = (x, a)a$, where a is a fixed unit vector in the domain of L, is called a **projection operator**. Can you explain why that name is appropriate? Show that $\|L\| = 1$.

18.13. Show that (a) $\|L_1 + L_2\| \leq \|L_1\| + \|L_2\|$, (b) $\|L_1 L_2\| \leq \|L_1\| \|L_2\|$, and (c) $\|\alpha L\| = |\alpha| \|L\|$.

***18.14.** If L is the operator $Lx = tx(t)$ over $\mathfrak{L}_2(0, 1)$, show that $\|L\| = 1$.

18.15. Derive the result stated in Example 18.5. Is it an absolute or relative maximum that we've found, or doesn't it matter?

18.16. Recalling the matrix result $(AB)^T = B^T A^T$, show that $(ABC)^T = C^T B^T A^T$. Verify for the case

$$A = \begin{pmatrix} 1 & 2 \\ 3 & 0 \end{pmatrix}, \quad B = \begin{pmatrix} 2 & i \\ 1 & 4 \end{pmatrix}, \quad C = \begin{pmatrix} -1 & 2 \\ 0 & 3 \end{pmatrix}.$$

18.17. (a) Verify equations (18.27).
(b) Make up an illustrative example for each of equations (18.27).

***18.18.** Show that if L is linear, then L^* is, too.

18.19. Find the adjoints of the following operators and indicate whether they are self-adjoint. *Note:* In parts (a) to (f) all quantities are real.

(a) The differential operator of Example 18.8.

(b) $L = d^2/dt^2 + d/dt + 1$, with $x(0) = x(1) = 0$, and the inner product (18.29).

(c) $L = d^3/dt^3$, with $x(0) = 1$, $x'(0) - 2x''(1) = 0$, $x(1) - x'(1) = 0$, and the inner product (18.29).

(d) $L = d/dt$ with $x(0) = 3x(1)$ and the inner product (18.29).

(e) The Laplace operator $L = \nabla^2$ over the rectangle $0 \leq x \leq 1, 0 \leq y \leq 1$, with $u(0, y) = u(x, 0) = u(x, 1) = 0$, $u(1, y) = 100$, and the inner product $(u, v) = \int_S uv\, d\sigma$, where S is the rectangular x, y domain.

(f) The diffusion operator $L = \partial^2/\partial x^2 - \partial/\partial t$ over $0 \leq x \leq l, 0 \leq t \leq T$, with $u(0, t) = u(l, t) = u(x, 0) = 0$ and the inner product $(u, v) = \int_S uv\, d\sigma$.

(g) The projection operator of Exercise 18.12.
(h) The operator of Exercise 18.14.
(i) The matrix operators

$$A = \begin{pmatrix} i & 2 \\ 3 & 0 \end{pmatrix} \quad \text{and} \quad B = \begin{pmatrix} 1 & 4 & 0 \\ 4 & 3 & 0 \\ 0 & 0 & 2 \end{pmatrix}$$

18.20. (a) If

$$L = p(t)\frac{d^2}{dt^2} + q(t)\frac{d}{dt} + r(t)$$

and

$$(x, y) = \int_a^b x(t)y(t)\,dt,$$

where all quantities are real, show that

$$L^* = p\frac{d^2}{dt^2} + (2p' - q)\frac{d}{dt} + (p'' - q' + r)$$

and determine conditions on p, q, r such that L will be formally self-adjoint.

(b) Repeat part (a), using instead the inner product

$$(x, y) = \int_a^b x(t)y(t)w(t)\,dt,$$

where the weighting factor $w(t)$ is positive over $a \leq t \leq b$.

18.21. In Example 18.6 we found that the adjoint of an $n \times n$ matrix A, with respect to the "usual" inner product $(x, y) = x^T\bar{y}$, was $A^* = \bar{A}^T$ or, in terms of the elements, $(A^*)_{ij} = \bar{a}_{ji}$. Show that with respect to the inner product

$$(x, y) = \sum_1^n \kappa_j \xi_j \bar{\eta}_j \qquad (\kappa_j\text{'s} > 0)$$

the adjoint is $(A^*)_{ij} = (\kappa_j/\kappa_i)\bar{a}_{ji}$. With $n = 2$, $\kappa_1 = 1$, and $\kappa_2 = 2$, for example, recompute the adjoints of the three matrices given in Example 18.6.

***18.22.** Show that the adjoint of the integral operator (18.5) is

$$L^*x = \int_a^b \bar{K}(t, \tau)x(\tau)\,d\tau$$

and show how it is analogous to the matrix result $A^* = \bar{A}^T$.

18.23. Show that L^* is unique. *Hint:* Recall Exercise 17.10.

18.24. Make up two examination-type problems on this chapter.

Chapter 19

The Linear Equation $Lx = c$

Here we are concerned with the solution of the linear equation $Lx = c$ for matrix operators and (in starred sections) integral operators. On the whole, we bypass ordinary and partial differential operators, not because they are less important but because they are important enough to warrant separate treatment in Parts IV and V.

19.1. INTRODUCTION

Consider the *matrix* case

$$Ax = c. \tag{19.1}$$

One simple and systematic procedure for the solution of (19.1) is the so-called **Gauss elimination**. Let us look at three examples, not only to illustrate Gauss elimination but also to introduce some ideas regarding the questions of existence and uniqueness.

Example 19.1.

$$\xi_1 + \xi_2 - \xi_3 = 1 \tag{19.2a}$$
$$2\xi_1 - 2\xi_2 + \xi_3 = 6 \tag{19.2b}$$
$$\xi_1 + 3\xi_3 = 0. \tag{19.2c}$$

The first step is to normalize the leading coefficients (i.e., of the ξ_1's). Thus (19.2) becomes

$$\xi_1 + \xi_2 - \xi_3 = 1 \tag{19.3a}$$
$$\xi_1 - \xi_2 + \tfrac{1}{2}\xi_3 = 3 \tag{19.3b}$$
$$\xi_1 + 3\xi_3 = 0. \tag{19.3c}$$

Next, we replace (19.3) by an equivalent system, where the first equation is the same as (19.3a), the second is (19.3a) minus (19.3b), and the third is (19.3a) minus (19.3c):

$$\xi_1 + \xi_2 - \xi_3 = 1 \tag{19.4a}$$

$$2\xi_2 - \tfrac{3}{2}\xi_3 = -2 \tag{19.4b}$$

$$\xi_2 - 4\xi_3 = 1. \tag{19.4c}$$

Again normalizing the leading coefficients,

$$\xi_1 + \xi_2 - \xi_3 = 1 \tag{19.5a}$$

$$\xi_2 - \tfrac{3}{4}\xi_3 = -1 \tag{19.5b}$$

$$\xi_2 - 4\xi_3 = 1. \tag{19.5c}$$

Finally, (19.5c) is replaced by (19.5b) minus (19.5c):

$$\xi_1 + \xi_2 - \xi_3 = 1 \tag{19.6a}$$

$$\xi_2 - \tfrac{3}{4}\xi_3 = -1 \tag{19.6b}$$

$$\tfrac{13}{4}\xi_3 = -2. \tag{19.6c}$$

Since this system is *triangular* (i.e., $a_{ij} = 0$ for $i > j$), its solution is trivial; starting at the bottom, (19.6c) gives $\xi_3 = -8/13$, inserting it into (19.6b) gives $\xi_2 = -19/13$, and inserting both of these into (19.6a) gives $\xi_1 = 24/13$. This process is called "back substitution." The point, then, is the reduction of the original system to one that is triangular. Provided that we're agreed that the systems (19.6) and (19.2) are equivalent (i.e., have the same solution), we see that the solution of (19.2) *exists* and is *unique*; that is, (19.6c) says that $\xi_3 = -8/13$ and *only* $-8/13$, and similarly for ξ_2 and ξ_1. ∎

Example 19.2. Applying the Gauss elimination to the simple system

$$\xi_1 + 2\xi_2 = 1 \tag{19.7a}$$

$$\xi_1 + 2\xi_2 = 0, \tag{19.7b}$$

we obtain, in one step,

$$\xi_1 + 2\xi_2 = 1 \tag{19.8a}$$

$$0 = 1, \tag{19.8b}$$

and it is clear [either from the result (19.8b) or from the original system (19.7)] that *no* solution exists! ∎

Example 19.3. Similarly, we find (Exercise 19.1) that

$$\xi_1 - \xi_2 + 2\xi_3 + \xi_4 = -1 \tag{19.9a}$$

$$2\xi_1 + \xi_2 + \xi_3 - \xi_4 = 4 \tag{19.9b}$$

$$\xi_1 + 2\xi_2 - \xi_3 - 2\xi_4 = 5 \tag{19.9c}$$

$$\xi_1 + \xi_3 = 1 \tag{19.9d}$$

reduces to

$$\xi_1 - \xi_2 + 2\xi_3 + \xi_4 = -1 \tag{19.10a}$$

$$\xi_2 - \xi_3 - \xi_4 = 2 \tag{19.10b}$$

$$0 = 0 \tag{19.10c}$$

$$0 = 0. \tag{19.10d}$$

We say that the system is of "defect two." There is certainly an *infinite* number of solutions this time; for instance, ξ_3 and ξ_4 can be chosen arbitrarily and then ξ_2 and ξ_1 found from (19.10b) and (19.10a) by back substitution. ▮

The Gauss method is actually quite efficient and (with various refinements) is commonly used in digital computer applications. Of course, it is more efficient than we've made it look; we've rewritten (19.2a) four times, whereas this labor is obviously not essential to the calculation.

19.2. *EXISTENCE AND UNIQUENESS*

We start with the question of *existence*—that is, does $Lx = c$ have a solution or is c in the range of L? Notice that if $Lx = c$ *does* have a solution, then c must be orthogonal to all z's such that $L^*z = \Theta$ (i.e., to all z's in the "null space" of L^*), since

$$(z, c) = (z, Lx) = (L^*z, x) = (\Theta, x) = 0. \tag{19.11}$$

THEOREM 19.1. (*Existence*) *A necessary condition for the existence of a solution of* $Lx = c$ *is that* $(c, z) = 0$ *for all* z*'s such that* $L^*z = \Theta$.

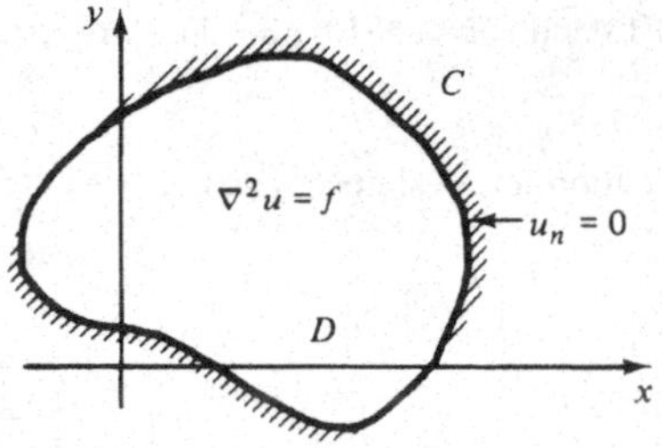

Figure 19.1. The boundary value problem (19.12).

Example 19.4. *Steady-State Heat Conduction.* To illustrate, consider the problem of steady-state, two-dimensional heat conduction in a plane region D with an insulated boundary C (Fig. 19.1). The temperature satisfies the following Poisson equation and boundary condition,

$$\begin{aligned} \nabla^2 u &= f \quad \text{in } D \\ \frac{\partial u}{\partial n} &= 0 \quad \text{on } C, \end{aligned} \tag{19.12}$$

where $f(x, y)$ is a prescribed heat-source distribution. According to Theorem 19.1, we need to examine the adjoint problem $L^*Z = \Theta$. To determine L^*, observe, with the help of the two-dimensional version of Green's second identity, that

$$(Lu, v) = \int_D v\, \nabla^2 u \, d\sigma = \int_C (vu_n - uv_n) ds + \int_D u\, \nabla^2 v \, d\sigma.$$

On C we have $u_n = 0$, and so the vu_n boundary term drops out. But u is *not* prescribed on C; therefore we must require that $v_n = 0$ in order to wipe out the other boundary term, uv_n. Thus the adjoint problem is

$$\begin{aligned} L^*Z = \nabla^2 Z &= 0 \quad \text{in } D \\ \frac{\partial Z}{\partial n} &= 0 \quad \text{on } C, \end{aligned} \tag{19.13}$$

which admits the nontrivial solution $Z = \text{constant} = C$, say. According to Theorem

19.1, then, in order for a solution to (19.12) to exist, it is necessary that

$$(Z, f) = \int_D Cf\, d\sigma = 0 \quad \text{or} \quad \int_D f\, d\sigma = 0. \tag{19.14}$$

With hindsight the condition (19.14) is entirely reasonable. It says that the *net* heat input over D is zero, as must certainly be true if the region is insulated and the temperature distribution is to be *steady state* as assumed. ▮

Obtaining necessary and *sufficient* conditions is more difficult. We state without proof that if, in addition, *the range of* L *is closed*,[1] then $Lx = c$ has a solution if and only if $(c, z) = 0$ for all z's such that $L^*z = \Theta$. The difficulty in applying this result is in determining whether the range of L is closed. However, we are concerned primarily with matrix operators for the present, and it can be shown that every finite-dimensional[2] matrix operator has a closed range. Thus we have the following important theorem:

THEOREM 19.2. (*Existence*) *For finite-dimensional matrix operators, the condition in Theorem* 19.1 *is both necessary and sufficient.*

The question of *uniqueness* is somewhat simpler. Suppose that $Lx = c$ has distinct solutions x_1 and x_2. Then $L(x_1 - x_2) = Lx_1 - Lx_2 = c - c = \Theta$, so that there must be a nontrivial solution of the homogeneous equation $Lx = \Theta$. Conversely, if $Lx = \Theta$ does have a nontrivial solution, say x_0, then the solution of $Lx = c$ (if indeed it exists) must be nonunique, since $L(x + \beta x_0) = Lx + \beta Lx_0 = c + \Theta = c$. Consequently, $x + \beta x_0$ is a solution, too, for *any* β in fact, and so there are an *infinite* number of solutions. Thus

THEOREM 19.3. (*Uniqueness*) *The solution of* $Lx = c$ (*if it exists at all*) *is unique if and only if the "homogenized" equation* $Lx = \Theta$ *has only the trivial solution* ($x = \Theta$). *Furthermore, if* $Lx = \Theta$ *has a nontrivial solution, then there will be infinitely many solutions of* $Lx = c$.

Example 19.5. As a simple illustration, consider the matrix case

$$\begin{pmatrix} 1 & 2 \\ 1 & 2 \end{pmatrix} \begin{pmatrix} \xi_1 \\ \xi_2 \end{pmatrix} = \begin{pmatrix} 1 \\ 0 \end{pmatrix}. \tag{19.15}$$

Now

$$A^* = \bar{A}^T = \begin{pmatrix} 1 & 1 \\ 2 & 2 \end{pmatrix},$$

and $A^*z = \Theta$ is seen to have nontrivial solutions $z = \begin{pmatrix} \alpha \\ -\alpha \end{pmatrix}$ for any α. Computing $(c, z) = (1)(\alpha) + (0)(-\alpha) = \alpha \neq 0$, we conclude from Theorem 19.2 that (19.15) has no solution, which, of course, coincides with our earlier conclusion in Example 19.2.

---

[1]The term "closed" was discussed in Section 17.4 and in Part I.

[2]That is, with a finite number of rows and columns.

On the other hand, if we had $c = \begin{pmatrix} 1 \\ 1 \end{pmatrix}$, then $(c, z) = (1)(\alpha) + (1)(-\alpha) = 0$, and this time Theorem 19.2 tells us that a solution *does* exist. Noting also that the homogenized version of (19.15) has nontrivial solutions—namely, $x = \begin{pmatrix} 2\alpha \\ -\alpha \end{pmatrix}$ for any α—Theorem 19.3 tells us that the solution of

$$\begin{pmatrix} 1 & 2 \\ 1 & 2 \end{pmatrix}\begin{pmatrix} \xi_1 \\ \xi_2 \end{pmatrix} = \begin{pmatrix} 1 \\ 1 \end{pmatrix} \tag{19.16}$$

is nonunique. ∎

19.3. THE INVERSE OPERATOR L^{-1}

We now take a different line of approach in the solution of $Lx = c$. As motivation, suppose that we simply proceed by "algebra"—that is, "dividing both sides by L," we have $x = c/L$ or, more suggestively, $x = L^{-1}c$. The question is, given an L, can we find an operator "L^{-1}" such that $L^{-1}L = LL^{-1} = I$, the identity operator. If so, we call L^{-1} the **inverse** of L, and we have

$$L^{-1}Lx = L^{-1}c$$

and hence the solution

$$x = L^{-1}c. \tag{19.17}$$

To illustrate, recall Example 19.1,

$$Ax = \begin{pmatrix} 1 & 1 & -1 \\ 2 & -2 & 1 \\ 1 & 0 & 3 \end{pmatrix}\begin{pmatrix} \xi_1 \\ \xi_2 \\ \xi_3 \end{pmatrix} = \begin{pmatrix} 1 \\ 6 \\ 0 \end{pmatrix}. \tag{19.18}$$

Without saying how we found it, observe that the operator

$$A^{-1} = \begin{pmatrix} 6/13 & 3/13 & 1/13 \\ 5/13 & -4/13 & 3/13 \\ -2/13 & -1/13 & 4/13 \end{pmatrix} \tag{19.19}$$

is such that $A^{-1}A = AA^{-1} = I$. Therefore

$$x = A^{-1}c = \frac{1}{13}\begin{pmatrix} 6 & 3 & 1 \\ 5 & -4 & 3 \\ -2 & -1 & 4 \end{pmatrix}\begin{pmatrix} 1 \\ 6 \\ 0 \end{pmatrix} = \begin{pmatrix} 24/13 \\ -19/13 \\ -8/13 \end{pmatrix},$$

as we found before.

Although apparent that L^{-1} is nice to have, the question remains, how do we find it? First, note that in the example above, L and L^{-1} are both the same kind of quantity, 3×3 matrices. On the other hand, if L is a *differential* operator, its inverse (if it exists) will be an *integral* operator with the so-called *Green's function* as its kernel. In this case, the calculation of L^{-1} amounts to solution by the method of Green's functions, which is discussed in Parts IV and V.

Here we consider the calculation of L^{-1} for only two (important) cases: first where L is an $n \times n$ matrix A and, secondly, where it is of the form $I + M$, where M is a "small" operator.

The Inverse of a Matrix. To find the inverse A^{-1} (if it exists) of an $n \times n$ matrix

$$A = \begin{pmatrix} a_{11} & \cdots & a_{1n} \\ \cdot & & \cdot \\ \cdot & & \cdot \\ \cdot & & \cdot \\ a_{n1} & \cdots & a_{nn} \end{pmatrix}, \tag{19.20}$$

we must first introduce the **determinant** of A, denoted by various authors in any of the following ways,

$$\det A, \quad |A|, \quad \begin{vmatrix} a_{11} & \cdots & a_{1n} \\ a_{n1} & \cdots & a_{nn} \end{vmatrix}.$$

Note the straight rules used in the last expression, in contrast to the parentheses used in (19.20) to denote A. Now, det A is a *number* associated with any square (i.e., $n \times n$) matrix A and is defined by the well-known **Laplace expansion,**

$$\det A \equiv \sum_{k=1}^{n} a_{jk} A_{jk} \tag{19.21a}$$

for any fixed value of j from 1 to n or, equivalently, by

$$\det A = \sum_{j=1}^{n} a_{jk} A_{jk} \tag{19.21b}$$

for any fixed value of k from 1 to n, where we will define the A_{jk}'s shortly. The fact that (19.21a) yields a unique value, independent of j, and that (19.21b) yields a unique value, independent of k, and that these two values are the same is not difficult to establish, but we choose not to for brevity.

Finally, we define A_{jk}, called the *cofactor* of the a_{jk} element, as

$$A_{jk} = (-1)^{j+k} M_{jk}, \tag{19.22}$$

where M_{jk} is the so-called *minor determinant* of a_{jk}—that is, the determinant of the $(n-1) \times (n-1)$ matrix that survives when the row and column containing a_{jk} (i.e., the jth row and the kth column) are eliminated. As a result of (19.22), the right-hand side of (19.21) amounts to a linear combination of n minor determinants that are one size smaller, $(n-1) \times (n-1)$. Applying (19.21) and (19.22) to each of these $(n-1) \times (n-1)$ determinants in turn, we are then faced with $(n-2) \times (n-2)$ determinants. Repeating the process enough times, we eventually end up with (perhaps a large number of) 1×1 determinants. Finally being forced to commit ourselves, we define the determinant of a 1×1 matrix, say $A = (a)$, simply as det $A \equiv a$. For a 2×2 matrix, it follows from (19.21a), (19.22), and the foregoing definition that

$$\begin{vmatrix} a & b \\ c & d \end{vmatrix} = ad - bc.$$

Example 19.6. Consider the 3×3 matrix in (19.18). Using (19.21a), with $j = 1$ say,

$$\det A = \sum_{k=1}^{3} a_{1k} A_{1k} = (1) \begin{vmatrix} -2 & 1 \\ 0 & 3 \end{vmatrix} - (1) \begin{vmatrix} 2 & 1 \\ 1 & 3 \end{vmatrix} + (-1) \begin{vmatrix} 2 & -2 \\ 1 & 0 \end{vmatrix}$$
$$= (1)(-6) - (1)(5) + (-1)(2) = -13.$$

We call this expression "the Laplace expansion about the first row." Expanding about the second row, we should obtain the same result:

$$\det A = \sum_{k=1}^{3} a_{2k}A_{2k} = -(2)\begin{vmatrix} 1 & -1 \\ 0 & 3 \end{vmatrix} + (-2)\begin{vmatrix} 1 & -1 \\ 1 & 3 \end{vmatrix} - (1)\begin{vmatrix} 1 & 1 \\ 1 & 0 \end{vmatrix}$$

$$= -(2)(3) + (-2)(4) - (1)(-1) = -13$$

as before. Finally, let us expand about the first column, say:

$$\det A = (1)\begin{vmatrix} -2 & 1 \\ 0 & 3 \end{vmatrix} - (2)\begin{vmatrix} 1 & -1 \\ 0 & 3 \end{vmatrix} + (1)\begin{vmatrix} 1 & -1 \\ -2 & 1 \end{vmatrix} = \text{etc.} = -13.$$

Note that if we are doing this calculation by hand, it is easiest to expand about either the third row or second column, for then, because of the 0 element at the 3, 2 location, we have only two 2×2 determinants to evaluate instead of three. ▮

It is possible to prove the following interesting and useful properties of determinants:

D1. The determinants of A and its transpose are the same—that is, $\det A^T = \det A$.
D2. If all the elements of any row or column are zero, then $\det A = 0$.
D3. If any two rows or columns of A are interchanged, to yield a new matrix, say B, then $\det B = -\det A$.
D4. If all the elements of any row or column of A are multiplied by a number α, to yield a new matrix B, then $\det B = \alpha \det A$.
D5. If any two rows or columns are proportional to each other, then $\det A = 0$.
D6. If to the elements of any row (or column) we add α times the corresponding elements of another row (or column) yielding a new matrix B, then $\det B = \det A$.
D7. The determinant of the product of two matrices is equal to the product of their determinants—that is, $\det(AB) = (\det A)(\det B)$.

These properties are not independent of each other; for instance, D5 follows from D3 and D4, D2 follows from D4, and so on.

Finally, we are ready to consider the inverse A^{-1}. Recall that we are trying to solve $Ax = c$ or, in scalar form,

$$\sum_{j=1}^{n} a_{ij}\xi_j = c_i \quad \text{for } i = 1, 2, \ldots, n. \tag{19.23}$$

Multiply both sides by the cofactor A_{ik}. Next, we add the n equations. In other words, summing on i, we have

$$\sum_{i=1}^{n} A_{ik} \sum_{j=1}^{n} a_{ij}\xi_j = \sum_{i=1}^{n} A_{ik}c_i$$

or

$$\sum_{j=1}^{n}\left(\sum_{i=1}^{n} a_{ij}A_{ik}\right)\xi_j = \sum_{i=1}^{n} A_{ik}c_i. \tag{19.24}$$

Then recalling the Laplace expansion (19.21b), we observe that when $j = k$ (j is running from 1 to n, whereas k is fixed), the quantity inside the parentheses in (19.24) is simply the Laplace expansion of A about the kth column, and hence it is equal to $\det A$. When $j \neq k$, it is again the Laplace expansion about the kth column but with the kth column replaced by the jth column. It is therefore the determinant of a matrix with two identical columns that, according to property D5, is zero. That is,

$$\sum_{i=1}^{n} a_{ij}A_{ik} = \begin{cases} \det A & \text{for } j = k \\ 0 & \text{for } j \neq k. \end{cases} \tag{19.25}$$

This result reduces (19.24), to

$$(\det A)\xi_k = \sum_{i=1}^{n} A_{ik}c_i,$$

which yields the desired solution

$$\xi_k = \frac{\sum_{i=1}^{n} c_i A_{ik}}{\det A} \tag{19.26}$$

for each $k = 1, 2, \ldots, n$, *provided that* $\det A \neq 0$. Note that the numerator of (19.26) is also a determinant; it is the Laplace expansion of A about the kth column but with the kth column replaced by the column of c_i's. We'll see what (19.26) has to do with A^{-1} in a minute; first, let us illustrate the calculation (19.26) by applying it to (19.18). Recalling from Example 19.6 that $\det A = -13$, we have

$$\xi_1 = \frac{\begin{vmatrix} 1 & 1 & -1 \\ 6 & -2 & 1 \\ 0 & 0 & 3 \end{vmatrix}}{-13} = \frac{(3)(-8)}{-13} = \frac{24}{13},$$

$$\xi_2 = \frac{\begin{vmatrix} 1 & 1 & 1 \\ 2 & 6 & 1 \\ 1 & 0 & 3 \end{vmatrix}}{-13} = \frac{(1)(7) + (3)(4)}{-13} = -\frac{19}{13},$$

and, similarly, $\xi_3 = -8/13$.

The unique solution of $Ax = c$ for each ξ_k as the ratio of two determinants, as given by (19.26) (if $\det A \neq 0$), is known as **Cramer's rule.**

Returning to our search for A^{-1}, observe that A^{-1} is hiding in (19.26); that is,

$$\xi_k = \sum_{i=1}^{n} \left(\frac{A_{ik}}{\det A}\right) c_i$$

or in matrix form

$$x = \text{“}A^{-1}\text{”}c, \tag{19.27}$$

where

$$A^{-1} = \begin{pmatrix} \frac{A_{11}}{\det A} & \cdots & \frac{A_{n1}}{\det A} \\ \cdot & & \cdot \\ \cdot & & \cdot \\ \cdot & & \cdot \\ \frac{A_{1n}}{\det A} & \cdots & \frac{A_{nn}}{\det A} \end{pmatrix}. \tag{19.28}$$

That is, the i, j element of A^{-1} is $A_{ji}/\det A$; note carefully the switching of the indices. *If* $\det A = 0$, *then* A^{-1} *does not exist and we say that* A *is* **singular**.

Example 19.7. If we compute all the cofactors A_{ij} and det A for the matrix A of (19.18), we do obtain, from (19.28), the A^{-1} matrix (19.19) as claimed (Exercise 19.7) ∎

Some important properties of matrix inverses are as follows (Exercises 19.8 to 19.11):

$$A^{-1}A = AA^{-1} = I \tag{19.29a}$$

$$\det A^{-1} = \frac{1}{\det A} \tag{19.29b}$$

$$(AB)^{-1} = B^{-1}A^{-1} \tag{19.29c}$$

$$(A^T)^{-1} = (A^{-1})^T. \tag{19.29d}$$

So far we have three means of solving $Ax = c$: *Gauss elimination, Cramer's rule* (19.26), and the *inverse matrix* (19.28). Before passing from our discussion of A^{-1} to a consideration of $(I + M)^{-1}$, let us backtrack to the question of existence/uniqueness and give (without proof) a good existence/uniqueness theorem for the problem $Ax = c$, one that is based on the notion of "rank."

First, we define the **augmented matrix** as

$$\begin{pmatrix} a_{11} & \cdots & a_{1n} & c_1 \\ \cdot & & & \cdot \\ \cdot & & & \cdot \\ \cdot & & & \cdot \\ a_{n1} & \cdots & a_{nn} & c_n \end{pmatrix} \tag{19.30}$$

and the **rank** of a matrix (not necessarily square) as *the order of the largest square array within that matrix, formed by deleting certain rows and columns, whose determinant does not vanish.* For instance, the rank of A in (19.18) is three, since the full 3×3 determinant does not vanish (recall that det $A = -13$). As one more example, observe that the rank of

$$\begin{pmatrix} 1 & 2 & 1 \\ 1 & 2 & 0 \end{pmatrix}$$

is two. Obviously it must be less than or equal to two, since the largest possible square array is 2×2; the fact that it *equals* two is seen by deleting either the first or second column and noting that the determinant of the resulting 2×2 array is nonzero.

THEOREM 19.4. (Existence, Uniqueness, and Rank) $Ax = c$ possesses a solution if and only if the rank of the augmented matrix equals the rank of the coefficient matrix A. The solution is unique if and only if the rank of A is n (where A is $n \times n$), i.e. if and only if det $A \neq 0$; if the rank is less than n than there are an infinite number of solutions. The homogeneous case $Ax = 0$ always admits the trivial solution $x = 0$; that solution will be unique if the rank of A is n (i.e. det $A \neq 0$), if the rank of A is less than n then there will be an infinite number of nontrivial solutions as well.

Example 19.8. For the system (19.18) we find that $\text{rank}_{\text{aug}} = \text{rank}_{\text{coeff}} = 3 = n$; so there must exist a unique solution. For the system

$$\begin{pmatrix} 1 & 2 \\ 1 & 2 \end{pmatrix}\begin{pmatrix} \xi_1 \\ \xi_2 \end{pmatrix} = \begin{pmatrix} 1 \\ 0 \end{pmatrix},$$

however, $\text{rank}_{\text{coeff}} = 1$ and $\text{rank}_{\text{aug}} = 2$, and no solution exists. Finally, if

$$\begin{pmatrix} 1 & 2 \\ 1 & 2 \end{pmatrix}\begin{pmatrix} \xi_1 \\ \xi_2 \end{pmatrix} = \begin{pmatrix} 1 \\ 1 \end{pmatrix},$$

then $\text{rank}_{\text{coeff}} = \text{rank}_{\text{aug}} = 1$, whereas $n = 2$, so that a solution exists but is not unique.

*

Inverse of I + M, where M is small; the Neumann series. If L happens to differ only slightly from the identity operator I, it is convenient to express $L = I + M$. By analogy with the Taylor expansion

$$(1 + x)^{-1} = 1 - x + x^2 - \cdots \tag{19.31}$$

for $|x| < 1$, we might anticipate that $(I + M)^{-1} = I - M + M^2 - \cdots$ for $\|M\| < 1$. Indeed, we have the following striking theorem.[3]

THEOREM 19.5. **(Neumann Series)** *If M is linear and* $\|M\| < 1$, *then*

$$(I + M)^{-1} = I - M + M^2 - \cdots. \tag{19.32}$$

Example 19.9. Solve

$$\begin{pmatrix} 1.1 & 0 & -0.2 \\ 0.1 & 0.9 & 0 \\ 0 & 0.1 & 1 \end{pmatrix}\begin{pmatrix} \xi_1 \\ \xi_2 \\ \xi_3 \end{pmatrix} = \begin{pmatrix} 2 \\ 5 \\ 0 \end{pmatrix}. \tag{19.33}$$

Expressing the operator as $I + M$, we see that

$$M = \begin{pmatrix} 0.1 & 0 & -0.2 \\ 0.1 & -0.1 & 0 \\ 0 & 0.1 & 0 \end{pmatrix}.$$

Surely it is a "small" operator. In particular, the bound (18.17) tells us that $\|M\| \leq \sqrt{0.08}$, which is smaller than unity, so that (19.32) can be applied. Thus

$$\begin{aligned} x &= (I - M + M^2 - \cdots)c \\ &= \left\{\begin{pmatrix} 1 & 0 & 0 \\ 0 & 1 & 0 \\ 0 & 0 & 1 \end{pmatrix} - \begin{pmatrix} 0.1 & 0 & -0.2 \\ 0.1 & -0.1 & 0 \\ 0 & 0.1 & 0 \end{pmatrix} + \begin{pmatrix} 0.01 & -0.02 & -0.02 \\ 0 & 0.01 & -0.02 \\ 0.01 & -0.01 & 0 \end{pmatrix} - \cdots\right\}\begin{pmatrix} 2 \\ 5 \\ 0 \end{pmatrix} \\ &= \begin{pmatrix} 0.91 & -0.02 & 0.18 \\ -0.1 & 1.11 & -0.02 \\ 0.01 & -0.11 & 1 \end{pmatrix}\begin{pmatrix} 2 \\ 5 \\ 0 \end{pmatrix} = \begin{pmatrix} 1.72 \\ 5.35 \\ -0.53 \end{pmatrix}. \quad \blacksquare \end{aligned}$$

---

[3]For a proof, see, for example, B. Friedman's *Principles and Techniques of Applied Mathematics*, Wiley, New York, 1957. In fact, this is an excellent general reference for all of Part III.

The Neumann series (19.32) *may* be valid even if $\|M\| > 1$, as we will see in the next example.

Example 19.10. Solve

$$\begin{pmatrix} 1 & 1 & 1 \\ 0 & 1 & 1 \\ 0 & 0 & 1 \end{pmatrix}\begin{pmatrix} \xi_1 \\ \xi_2 \\ \xi_3 \end{pmatrix} = \begin{pmatrix} 4 \\ 3 \\ 2 \end{pmatrix}. \tag{19.34}$$

In this case,

$$M = \begin{pmatrix} 0 & 1 & 1 \\ 0 & 0 & 1 \\ 0 & 0 & 0 \end{pmatrix}$$

does *not* have $\|M\| < 1$. To see this fact, note, for example, that

$$\begin{pmatrix} 0 & 1 & 1 \\ 0 & 0 & 1 \\ 0 & 0 & 0 \end{pmatrix}\begin{pmatrix} 0 \\ 0 \\ 1 \end{pmatrix} = \begin{pmatrix} 1 \\ 1 \\ 0 \end{pmatrix},$$

so that $\|Mx\|/\|x\| = \sqrt{2} > 1$. Nevertheless, if we work out M^2, M^3, and so on, we find that $M^3, M^4, \ldots$ are all identically zero; hence the Neumann series not only converges, it also *terminates* after three terms.

$$\begin{aligned} x &= (I - M + M^2 - \cdots)c \\ &= \left\{\begin{pmatrix} 1 & 0 & 0 \\ 0 & 1 & 0 \\ 0 & 0 & 1 \end{pmatrix} - \begin{pmatrix} 0 & 1 & 1 \\ 0 & 0 & 1 \\ 0 & 0 & 0 \end{pmatrix} + \begin{pmatrix} 0 & 0 & 1 \\ 0 & 0 & 0 \\ 0 & 0 & 0 \end{pmatrix}\right\}\begin{pmatrix} 4 \\ 3 \\ 2 \end{pmatrix} = \begin{pmatrix} 1 \\ 1 \\ 2 \end{pmatrix}, \end{aligned}$$

which is easily verified as the desired solution. The success of the Neumann series in spite of the fact that $\|M\| > 1$ is related to the *triangular* nature of the matrix operator in (19.34). ▮

Next, let us consider the application of the Neumann series to the case of *integral equations*—that is, to the case $Lx = c$, where L is an integral operator, as introduced briefly in Example 18.2.

$$\int_a^b K(\tau, t)x(\tau)d\tau = g(t).$$

More generally, it is standard to consider the so-called **Fredholm equation**

$$f(t)x(t) = g(t) + \Lambda \int_a^b K(\tau, t)x(\tau)\, d\tau, \tag{19.35}$$

where $f(t)$, $g(t)$, $K(\tau, t)$ are given functions and a, b, Λ are given constants. Naturally Λ can be absorbed into the kernel K, but it is often a significant parameter that is best kept in evidence.

If the upper integration limit is the variable t instead of the constant b, so that

$$f(t)x(t) = g(t) + \Lambda \int_a^t K(\tau, t)x(\tau)\, d\tau, \tag{19.36}$$

we call the result a **Volterra equation**. Actually, (19.36) is a special case of (19.35). Note, for instance, that $K(\tau, t)$ for the Fredholm equation (19.35) is defined over the *square ABCD* shown in Fig. 19.2 (i.e., $a \leq \tau \leq b$ and $a \leq t \leq b$), whereas for the Volterra equation (19.36) it is only defined over the *triangle ACD* (i.e., $a \leq \tau \leq t$ and $a \leq t \leq b$). If we replace the t upper limit in (19.36) by b, however, and extend the definition of K from the triangle ACD to the square $ABCD$, with $K \equiv 0$ in the upper portion ABC, we end up with a *Fredholm* equation as claimed.

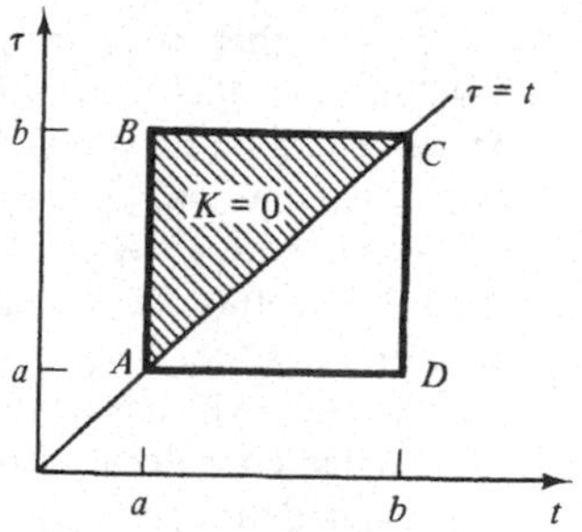

Figure 19.2. Extension of the kernel.

We further classify (19.35) and (19.36) as integral equations of the *first*, *second*, or *third kind*, depending on whether $f(t)$ is zero, unity, or some function of t, respectively.

Not only do integral equations occur naturally in applications, they are also of considerable importance as alternative expressions of *differential* equations. To illustrate, consider two examples.

Example 19.11. Consider the *initial value* problem

$$x'' + 3x = 6t, \qquad x(0) = x'(0) = 0. \tag{19.37}$$

Integrating from 0 to t and using the condition $x'(0) = 0$, we obtain

$$x'(t) + 3\int_0^t x(\tau)\, d\tau = 3t^2.$$

Integrating from 0 to t again and using the condition $x(0) = 0$ give

$$x(t) = t^3 - 3\int_0^t \int_0^{t'} x(\tau)\, d\tau\, dt'.$$

Finally, inverting the order of integration and integrating on t',

$$x(t) = t^3 + 3\int_0^t (\tau - t)x(\tau)\, d\tau. \tag{19.38}$$

Of course, it is not the *solution* of (19.37), since x appears inside the integral. Instead, it is an integral equation—specifically, a *Volterra* equation of the second kind. Observe that there are no initial conditions appended to (19.38), for they have been *built into* the integral equation. ∎

Example 19.12. Now consider a *boundary value* version of (19.37),

$$x'' + 3x = 6t, \qquad x(0) = x(1) = 0. \tag{19.39}$$

Proceeding essentially as above, we obtain the *Fredholm* equation of the second kind (Exercise 19.21)

$$x(t) = (t^3 - t) + 3\int_0^1 K(\tau, t)x(\tau)\, d\tau, \tag{19.40a}$$

where

$$K(\tau, t) = \begin{cases} \tau(1 - t), & \tau < t \\ t(1 - \tau), & \tau > t. \end{cases} \quad ∎ \tag{19.40b}$$

Why bother converting (19.37) and (19.39) to integral equations? Of course, (19.37) and (19.39) are simple examples that can be solved by elementary means. More generally, however, there may be important advantages, from both theoretical and computational points of view, in the conversion of a differential operator to an integral operator.[4] Roughly speaking, the idea is that integral operators are "nicer" than differential operators, since integration is a smoothing process, whereas differentiation has the opposite effect. For instance, the integral of the Heaviside function $H(t)$ is $tH(t)$, which is continuous at $x = 0$, whereas its derivative is the spikelike $\delta(t)$.

Finally, consider the application of the *Neumann series* (19.32) to the solution of integral equations.

Example 19.13. The simple example

$$x(t) = f(t) + \Lambda \int_0^1 x(\tau)\, d\tau \tag{19.41}$$

will permit an important observation. In this case,

$$Lx = x(t) - \Lambda \int_0^1 x(\tau)\, d\tau = (I + M)x$$

so that $Mx = -\Lambda \int_0^1 x(\tau)\, d\tau$. Noting that

$$\begin{aligned}
Mf &= -\Lambda \int_0^1 f(\tau)\, d\tau \\
M^2 f &= M(Mf) = -\Lambda \int_0^1 \left\{-\Lambda \int_0^1 f(\tau)\, d\tau\right\} d\tau' = \Lambda^2 \int_0^1 f(\tau)\, d\tau \\
&\vdots \\
M^n f &= \text{etc.} = (-\Lambda)^n \int_0^1 f(\tau)\, d\tau,
\end{aligned}$$

(19.32) yields

$$\begin{aligned}
x(t) &= (I + M)^{-1} f = f - Mf + M^2 f - \cdots \\
&= f(t) + (\Lambda + \Lambda^2 + \Lambda^3 + \cdots) \int_0^1 f(\tau)\, d\tau,
\end{aligned} \tag{19.42}$$

which is meaningful only if $\Lambda + \Lambda^2 + \cdots$ converges—that is, if $|\Lambda| < 1$. In fact, $\|M\| = |\Lambda|$ (Exercise 19.24), so that Theorem 19.5 guarantees the equality of (19.32) only for $|\Lambda| < 1$ in any case.

Actually, (19.41) can be solved easily by inspection. Observing that the integral is simply a constant, say C, we have

$$x(t) = f(t) + C. \tag{19.43}$$

Next, we substitute (19.43) back into (19.41) and solve for C:

$$C = \frac{\Lambda}{1 - \Lambda} \int_0^1 f(\tau)\, d\tau;$$

[4]As an illustration, see, for example, T. C. Ho, Uniqueness Criteria of the Steady State in Automotive Catalysis, *Chemical Engineering Science*, Vol. 31, 1976, pp. 235–240.

hence the exact solution is

$$x(t) = f(t) + \frac{\Lambda}{1 - \Lambda} \int_0^1 f(\tau)\, d\tau \tag{19.44}$$

for all $\Lambda \neq 1$! Thus note carefully that the Neumann series (19.42) did not give the entire story and that, in order to have it, we need the *analytic continuation* of the Neumann series in terms of Λ! Of course, this deficiency is anticipated by the motivational expression (19.31); $(1 + x)^{-1}$ is analytic for all $x \neq -1$, whereas $1 - x + x^2 - \cdots$ is a valid representation only in $|x| < 1$. ▮

Similarly, the *general* Fredholm equation

$$x(t) = f(t) + \Lambda \int_a^b K(\tau, t) x(\tau)\, d\tau \tag{19.45}$$

has a Neumann series solution of the form

$$x(t) = f(t) + \Lambda \int_a^b \Gamma(\tau, t; \Lambda) f(\tau)\, d\tau. \tag{19.46}$$

According to classical **Fredholm theory**,[5] the desired analytic continuation is provided by the **resolvent kernel** Γ, given by

$$\Gamma(\tau, t; \Lambda) = \frac{D(\tau, t; \Lambda)}{\Delta(\Lambda)}, \tag{19.47a}$$

where both series

$$D = D_0(\tau, t) - \frac{\Lambda}{1!} D_1(\tau, t) + \frac{\Lambda^2}{2!} D_2(\tau, t) - \cdots \tag{19.47b}$$

$$\Delta = A_0 - \frac{\Lambda}{1!} A_1 + \frac{\Lambda^2}{2!} A_2 - \cdots \tag{19.47c}$$

converge for all Λ, and

$$A_0 = 1, \; D_0(\tau, t) = K(\tau, t)$$

$$A_j = \int_a^b D_{j-1}(\tau, \tau)\, d\tau \quad \text{for } j \geq 1$$

$$D_j(\tau, t) = A_j K(\tau, t) - j \int_a^b K(\tau, \tau') D_{j-1}(\tau', t)\, d\tau' \quad \text{for } j \geq 1. \tag{19.47d}$$

Unfortunately, the calculation of the A_js and D_js is often quite difficult.

Example 19.14. For the simple example (19.41), however, we find easily that $A_0 = A_1 = D_0 = 1$ and $A_2 = A_3 = \cdots = D_1 = D_2 = \cdots = 0$; therefore $\Gamma = 1/(1 - \Lambda)$, in agreement with the known solution (19.44). In this example the series (19.47b) and (19.47c) both terminate. This procedure will always happen when the kernel K is **separable**—that is, when K is of the form

$$K(\tau, t) = \sum_1^N h_j(\tau) k_j(t). \; ▮$$

[5] See, for example, W. V. Lovitt, *Linear Integral Equations*, McGraw-Hill, New York, 1924.

Application of the Neumann series to *Volterra* equations is somewhat special, as can be seen from the following example.

Example 19.15.

$$x(t) = t + \Lambda \int_0^t x(\tau)\, d\tau. \tag{19.48}$$

Differentiating with respect to t, it is not hard to see that (19.48) is equivalent to the *differential* system

$$x' - \Lambda x = 1, \qquad x(0) = 0, \tag{19.49}$$

with the solution $x = (e^{\Lambda t} - 1)/\Lambda$.

With M given by $Mx = -\Lambda \int_0^t x(\tau)\, d\tau$, it turns out[6] that $||M|| = 2|\Lambda|/\pi$, and yet the Neumann series solution (Exercise 19.26)

$$x(t) = t + \frac{\Lambda}{2!}t^2 + \frac{\Lambda^2}{3!}t^3 + \cdots = \frac{e^{\Lambda t} - 1}{\Lambda} \tag{19.50}$$

is seen to be valid for *all* Λ, not just for $|\Lambda| < \pi/2$ (i.e., $||M|| < 1$). This fortunate result is due to the triangular nature of the Volterra kernel (recall Fig. 19.2) and is the analog of the matrix Example 19.10. ▮

EXERCISES

19.1. Solve the system (19.9) by Gauss elimination.

19.2. Making minor adjustments in the procedure, if necessary, solve by Gauss elimination:

(a)
$$\begin{aligned} x - 2y + z + 2w &= -11 \\ y - z - w &= 5 \\ x + y + z \qquad &= 4 \\ 3x + 2y \qquad + w &= 7 \end{aligned}$$

(b)
$$\begin{aligned} p + 2q + 3r &= 1 \\ 3p + 2q + r &= 2 \\ p + q \qquad &= 3 \end{aligned}$$

19.3. For what value(s) of α do the following systems have solutions?

(a) $\begin{pmatrix} 1 & 2 & 3 \\ 3 & 2 & 1 \\ 1 & 1 & 1 \end{pmatrix}\begin{pmatrix} x \\ y \\ z \end{pmatrix} = \begin{pmatrix} 2 \\ 0 \\ \alpha \end{pmatrix}$ (b) $\begin{pmatrix} 1 & 2 & 3 \\ 4 & 5 & 6 \\ 9 & 8 & 6 \end{pmatrix}\begin{pmatrix} x \\ y \\ z \end{pmatrix} = \begin{pmatrix} \alpha \\ 4 \\ 1 \end{pmatrix}$

19.4. For what value(s) of λ do the systems shown have nontrivial solutions? (By "trivial" we mean $x = y = 0$.)

(a) $\begin{pmatrix} 1 & 2 \\ 3 & 4 \end{pmatrix}\begin{pmatrix} x \\ y \end{pmatrix} = \lambda\begin{pmatrix} x \\ y \end{pmatrix}$ (b) $\begin{aligned} 2x - y &= \lambda x \\ 4x - 2y &= \lambda y \end{aligned}$

(c) $\begin{aligned} x - y &= 2\lambda y \\ x + y &= \lambda(x - y) \end{aligned}$ (d) $\begin{aligned} 2x + y &= \lambda y \\ x + y &= 0 \end{aligned}$

---

[6]See I. Stakgold, *Boundary Value Problems of Mathematical Physics*, Vol. I, Macmillan, New York, 1967, p. 145.

19.5. What condition(s) need the components of c satisfy if $Ax = c$ is to possess a solution, where

$$A = \begin{pmatrix} 1 & 0 & 2 \\ 2 & -1 & 0 \\ 1 & 1 & 6 \end{pmatrix}?$$

What about uniqueness?

19.6. What condition(s), if any, need f satisfy for the existence of solutions to the problems below?

(a) $\ddot{x} + x = f(t)$, $x(0) = x(\pi) = 0$

(b) $y'''' - y = f(x)$, $y(0) = y(\pi) = y'(\pi) = y''(\pi) = 0$

(c) $\dot{x} = f(t)$, $x(0) = x(1)$

19.7. Compute the inverse of the matrix A of (19.18) and compare your result with (19.19).

19.8. Prove (19.29a).

19.9. Prove (19.29b).

19.10. Prove (19.29c).

19.11. Prove (19.29d).

19.12. Solve by Cramer's rule.

(a)
$$\begin{aligned} x - y &= 6 \\ 2x + y &= 1 \end{aligned}$$

(b)
$$\begin{aligned} x + 2y + 3z &= 1 \\ x + y + z &= 3 \\ 2x + 5y &= -2. \end{aligned}$$

19.13. Show that the inverse of the diagonal matrix $D = \{d_i \delta_{ij}\}$ is simply $D^{-1} = \{\delta_{ij}/d_i\}$.

19.14. (a) Determine the rank of the coefficient matrix and of the augmented matrix for the system (19.9). What do you conclude in regard to existence and uniqueness?

(b) What relationship must exist between the ranks of the coefficient matrix and the augmented matrix before and after Gauss elimination, not only for this example but in general?

19.15. Recall that if $\det A \neq 0$, then $Ax = c$ does have a solution and it is unique. Practically speaking, however, it may happen that $\det A$ is nonzero but *quite small* compared to the size of the products that appear in the Laplace expansion. In this case, we say that the system is *ill-conditioned* because the solution is only weakly implied and may be quite inaccurate due to computer roundoff. For example, the system

$$\begin{aligned} 100x - 50y &= 50 \\ 99x - 49y &= 50 \end{aligned}$$

has as its (unique) solution $x = y = 1$, and yet the values $x = 1.2$, $y = 1.402$ satisfy both equations to within 0.102. Interpret the situation, for this example, graphically.

19.16. A convenient way to compute $\det A$ is provided by the **Crout–Banachiewicz algorithm.** The object is to find matrices G and H of the form

$$G = \begin{pmatrix} g_{11} & & & 0 \\ g_{21} & g_{22} & & \\ \vdots & & \ddots & \\ g_{n1} & g_{n2} & \cdots & g_{nn} \end{pmatrix}, \quad H = \begin{pmatrix} 1 & h_{12} & \cdots & h_{1n} \\ & 1 & & h_{2n} \\ & & \ddots & \vdots \\ 0 & & & 1 \end{pmatrix}$$

such that $GH = A$, where the large zeros denote that $g_{ij} = 0$ for $j > i$ and $h_{ij} = 0$ for $i > j$ because then $\det A = \det(GH) = (\det G)(\det H) = (g_{11}g_{22}\cdots g_{nn})(1) = g_{11}g_{22}\cdots g_{nn}$. Now

$$GH = \begin{pmatrix} g_{11} & g_{11}h_{12} & g_{11}h_{13} & \\ g_{21} & g_{21}h_{12} + g_{22} & g_{21}h_{13} + g_{22}h_{23} & \\ \vdots & \vdots & \vdots & \cdots \\ \vdots & \vdots & \vdots & \\ g_{n1} & g_{n1}h_{12} + g_{n2} & g_{n1}h_{13} + g_{n2}h_{23} + g_{n3} & \end{pmatrix}.$$

Equating it to A gives

$$g_{11} = a_{11}, \quad g_{21} = a_{21}, \ldots, g_{n1} = a_{n1}, \quad g_{11}h_{12} = a_{12},$$
$$g_{21}h_{12} + g_{22} = a_{22}, \ldots, g_{n1}h_{12} + g_{n2} = a_{n2}, \quad g_{11}h_{13} = a_{13}, \text{ etc.,}$$

which can be solved successively for $g_{11}, \ldots, g_{n1}, h_{12}, g_{22}, \ldots, g_{n2}, g_{13}, h_{23}, \ldots,$ and so on. Use this method to determine the determinant of the coefficient matrix for
(a) the system (19.2), (b) the system (19.9),
(c) Exercise 19.2(a), (d) Exercise 19.2(b).

19.17. The Crout–Banachiewicz algorithm for computing $\det A$ was discussed in Exercise 19.16. The idea is easily adapted to the solution of $Ax = c$ as follows. Find G and H, per Exercise 19.16, so that $A = GH$ and then find a C vector such that $c = GC$. Then $GHx = GC$, so that (Why?) $Hx = C$, which is easily solved, since H is triangular. Use this method to solve
(a) the system (19.2), (b) the system (19.9),
(c) Exercise 19.2(a), (d) Exercise 19.2(b).
This method is also called **L-U decomposition**—that is, where A is factored into lower and upper triangular matrices.

***19.18.** An operator of the form

$$Lx = (x, e_1^*)e_1 + \cdots + (x, e_n^*)e_n,$$

where the e_j's are LI and the e_j^*'s are their dual vectors [i.e., $(e_i, e_j^*) = \delta_{ij}$; recall Exercise 17.11], is called a **projection operator**, since Lx is the projection of x onto that portion of the vector space spanned by $e_1, \ldots, e_n$. For instance, consider the three-dimensional space and 3-tuples. If

$$e_1 = \begin{pmatrix} 1/\sqrt{2} \\ 1/\sqrt{2} \\ 0 \end{pmatrix} \quad \text{and} \quad e_2 = \begin{pmatrix} 0 \\ 0 \\ 1 \end{pmatrix},$$

then $Lx = (x, e_1^*)e_1 + (x, e_2^*)e_2$ is the projection of x onto the *plane* spanned by e_1 and e_2. More generally, we will call L an n-term **dyad** if

$$Lx = (x, a_1)e_1 + \cdots + (x, a_n)e_n,$$

where the e_j's, as well as the a_j's, are LI but the a_j's need not be the dual vectors of the e_j's.
(a) Let L be a two-term dyad, $Lx = (x, a_1)e_1 + (x, a_2)e_2$. Solve the linear equation

$$Lx = (x, a_1)e_1 + (x, a_2)e_2 = c$$

and discuss the questions of existence and uniqueness. *Hint:* Seek $x = \alpha_1 a_1 + \alpha_2 a_2 + \kappa$, where κ is orthogonal to both a_1 and a_2 but is otherwise arbitrary.

(b) Next, suppose that L is the identity operator I plus an n-term dyad. For instance, solve

$$Lx = x + (x, a)e = c$$

and discuss existence and uniqueness.

(c) Solve the integral equation

$$x(t) = e^t - 1 + \Lambda \int_0^1 t\tau x(\tau)\, d\tau$$

and discuss existence and uniqueness.

(d) Solve

$$x(t) = e^t - \Lambda \int_0^1 (1 + t\tau^2) x(\tau)\, d\tau$$

and discuss existence and uniqueness.

(e) Although the integral operator in

$$x(t) = 1 + \sin t - \int_0^1 t \cos(t\tau) x(\tau)\, d\tau$$

is not a dyad (since the kernel is not separable), it can be *approximated* by one by expanding $\cos(t\tau)$ in a Maclaurin series and truncating. As the simplest approximation, take $\cos(t\tau) \approx 1$ and solve. Repeat for $\cos(t\tau) \approx 1 - t^2\tau^2/2$. It is easily verified that the exact solution is simply $x(t) = 1$. Compare the two approximate solutions with the exact solution. (An operator L that can be *uniformly approximated by a sequence of n-term dyads*, say L_n, is said to be **completely continuous** or **compact**. Such operators are of great importance in the general theory of linear analysis.)

***19.19.** We motivated the Neumann series (19.32) by the Taylor series (19.31). Obtain it instead by means of a formal iterative approach.[7]

***19.20.** Find the inverse of the matrix $A = \begin{pmatrix} 10 & 2 \\ 1 & 8 \end{pmatrix}$, correct to two decimal places say, with the help of the Neumann series.

***19.21.** Derive the Fredholm equation (19.40).

***19.22.** Convert to equivalent integral equations.

(a) $x'(t) + p(t)x(t) = q(t), \quad x(0) = a$

(b) $x'' + p(t)x = 0, \quad x(0) = 1, \quad x'(0) = 0$

***19.23.** Convert the system

$$x'' + tx = 1, \qquad x(0) = x'(0) = 0$$

to an equivalent integral equation (What type is it? Could you have known in advance?) and develop an iterative solution, through the second iterate, say, taking $x^{(0)} = 0$ as the zeroth iterate.

***19.24.** For Example 19.13, verify that $\|M\| = |\Lambda|$ as claimed.

---

[7]The iterative approach was first used by Liouville in 1837. His work preceded Neumann's contribution, and the name "Liouville–Neumann series" is sometimes used.

***19.25.** Solve

$$x(t) - \Lambda \int_0^1 e^{k(\tau - t)} x(\tau)\, d\tau = f(t)$$

by means of the Fredholm theory. What is the resolvent kernel?

***19.26.** Derive the Neumann series (19.50).

19.27. Make up two examination-type problems on this chapter.

Chapter 20

The Eigenvalue Problem $Lx = \lambda x$

The expansion of a given vector in terms of a set of base vectors is of great importance. The procedure was discussed in Chapter 17, for both finite- and infinite-dimensional spaces, but we did not show where the base vectors came from. In fact, they arise through the so-called *eigenvalue problem*. Besides generating bases, the eigenvalue problem is often of direct interest itself—for instance, in connection with buckling loads, principal stresses, normal coordinates, and convergence of iterative schemes.

20.1. STATEMENT OF THE EIGENVALUE PROBLEM

Certainly $x = \Theta$ is always a solution of the homogeneous equation[1]

$$Lx = \lambda x. \tag{20.1}$$

But we ask, are there *nontrivial* solutions? That is, are there x's such that Lx is *aligned* with x? If so, we call them **eigenvectors** of L. Schematically, the situation is as shown in Fig. 20.1. Given any vector $x \neq \Theta$, Lx might be as shown in part (a); since this Lx is not aligned with x, the chosen x does not satisfy (20.1) and is not an eigenvector. Selecting some other x, suppose that Lx *is* aligned, as shown in part (b). This very special x is an eigenvector, and the ratio of Lx to x is the **eigenvalue** λ corresponding

[1]We say that the equations $Lx = c$ and $Lx = \Theta$ are *inhomogeneous* and *homogeneous*, respectively. Rewriting (20.1) as $(L - \lambda I)x = \Theta$, we see that it is homogeneous. By the way, note that we write $L - \lambda I$ rather than $L - \lambda$, since L is an operator and λ is a scalar, and thus $L - \lambda$ makes no sense.

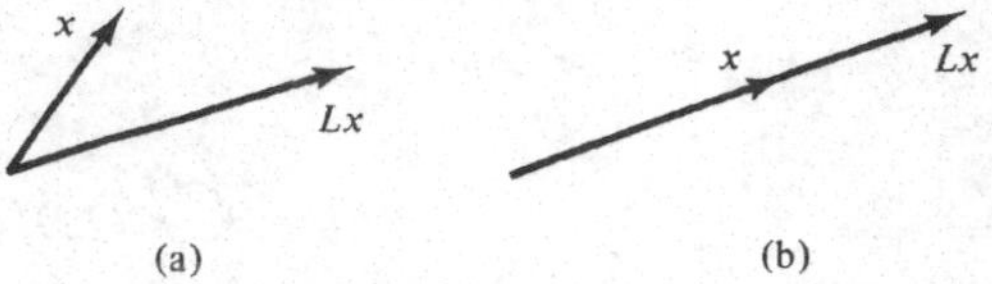

Figure 20.1. Eigenvalue problem.

to that eigenvector. Thus the parameter λ in the eigenvalue problem (20.1) is not specified a priori; it is to be determined together with x in what is uncommonly called (by Professor Ralph Palmer Agnew) an *eigenhunt*. Of course, we sometimes meet the homogeneous problem $Lx = \Lambda x$, where Λ is given. That's fine, but it happens not to be an eigenvalue problem.

20.2. *SOME EIGENHUNTS*

Example 20.1. Consider the matrix case $Ax = \lambda x$, where

$$A = \begin{pmatrix} 2 & 1 \\ 1 & 2 \end{pmatrix}.$$

Rewriting $Ax = \lambda x$ as $(A - \lambda I)x = \mathbb{0}$, we recall from Chapter 19 that it has nontrivial solutions if and only if

$$\det (A - \lambda I) = 0,$$

or

$$\begin{vmatrix} 2-\lambda & 1 \\ 1 & 2-\lambda \end{vmatrix} = (2-\lambda)^2 - 1 = \lambda^2 - 4\lambda + 3 = 0.$$

This is called the **characteristic equation** corresponding to A, and the roots, $\lambda = 1, 3 \equiv \lambda_1, \lambda_2$, respectively, are the characteristic values or "eigenvalues." What are the corresponding eigenvectors? First, consider λ_1. Writing out $Ax = \lambda x$ with λ set equal to λ_1, we have

$$2\xi_1 + \xi_2 = \xi_1, \qquad \xi_1 + 2\xi_2 = \xi_2$$

or $\xi_2 = -\xi_1$. Denoting the eigenvectors as e_j,

$$e_1 = \begin{pmatrix} 1 \\ -1 \end{pmatrix}$$

to within an arbitrary scale factor (as is *always* the case with eigenvectors, for if e_j is a solution of $Lx = \lambda x$, then so is αe_j for any scalar α).

Next, consider λ_2. $Ax = \lambda x$ becomes

$$2\xi_1 + \xi_2 = 3\xi_1, \qquad \xi_1 + 2\xi_2 = 3\xi_2$$

or $\xi_2 = \xi_1$, so that

$$e_2 = \begin{pmatrix} 1 \\ 1 \end{pmatrix},$$

say. Of course, the e_j's can be normalized, if we like, by dividing their components by $\sqrt{2}$. ∎

Observe that the eigenvalues in Example 20.1 happen to be real and that the eigenvectors happen to be mutually orthogonal—that is, $(e_1, e_2) = e_1^T \bar{e}_2 = (1)(1) + (-1)(1) = 0$. This situation is not coincidental. In fact we have the following important theorem.

THEOREM 20.1. If L is self-adjoint, then
(i) *the λ's are real and*
(ii) *eigenvectors corresponding to distinct eigenvalues are mutually orthogonal.*

Proof. (i) Suppose that we have a pair of "eigens"[2] λ_j and e_j; that is, $Le_j = \lambda_j e_j$. "Predotting" and "postdotting" e_j into this equation,

$$(e_j, Le_j) = (e_j, \lambda_j e_j) = \bar{\lambda}_j (e_j, e_j)$$
$$(Le_j, e_j) = (\lambda_j e_j, e_j) = \lambda_j (e_j, e_j).$$

But $(Le_j, e_j) = (e_j, Le_j)$, since L is self-adjoint, and so

$$0 = (\bar{\lambda}_j - \lambda_j)(e_j, e_j).$$

Then $(e_j, e_j) = \|e_j\|^2 \neq 0$, since $e_j \neq \Theta$. Consequently, $\bar{\lambda}_j = \lambda_j$ and λ_j must be real.

(ii) Suppose that we have eigenvectors e_i, e_j corresponding to distinct eigenvalues λ_i, λ_j. Then

$$Le_i = \lambda_i e_i \qquad \text{and} \qquad Le_j = \lambda_j e_j$$
$$(Le_i, e_j) = (\lambda_i e_i, e_j) \qquad (e_i, Le_j) = (e_i, \lambda_j e_j)$$
$$= \lambda_i (e_i, e_j) \qquad = \bar{\lambda}_j (e_i, e_j)$$
$$= \lambda_j (e_i, e_j).$$

Subtracting,

$$0 = (\lambda_i - \lambda_j)(e_i, e_j);$$

and if $\lambda_i \neq \lambda_j$, it follows that $(e_i, e_j) = 0$.

Since $L = A$ was self-adjoint in Example 20.1, the λ's *had* to be real and the e_j's *had* to be mutually orthogonal.

Example 20.2. Consider the matrix example $Ax = \lambda x$, where

$$A = \begin{pmatrix} 2 & 0 & 0 \\ 0 & 1 & 1 \\ 0 & 1 & 1 \end{pmatrix}.$$

The characteristic equation is

$$\begin{vmatrix} 2-\lambda & 0 & 0 \\ 0 & 1-\lambda & 1 \\ 0 & 1 & 1-\lambda \end{vmatrix} = \text{etc.} = -\lambda(\lambda - 2)^2 = 0;$$

therefore $\lambda = 0, 2 \equiv \lambda_1, \lambda_2$, respectively. Note carefully that $x = \Theta$ is never acceptable as an eigenvector because, by definition, an eigenvector is to be a *non*trivial

[2] We will use the term *eigens* as shorthand for eigenvalues and eigenvectors.

solution. There is no objection to $\lambda = 0$, however, as here. The eigenvalue $\lambda_2 = 2$ is said to be of *multiplicity two*, since it is a double root of the characteristic equation.

Proceeding as in Example 20.1, we find that the eigens are as follows.

$$\lambda_1 = 0, \quad e_1 = \begin{pmatrix} 0 \\ 1 \\ -1 \end{pmatrix} \quad \text{and} \quad \lambda_2 = 2, \quad e_2 = \begin{pmatrix} \alpha \\ \beta \\ \beta \end{pmatrix}$$

where α and β are arbitrary scalars (provided, of course, that they're not both zero).

Note that A is self-adjoint and, as predicted by Theorem 20.1, the λ_j's are real, and $(e_1, e_2) = 0$ (for all choices of α and β).

Furthermore, observe that e_2 actually contains *two* orthogonal vectors, for instance,

$$\begin{pmatrix} 1 \\ 0 \\ 0 \end{pmatrix} \quad \text{and} \quad \begin{pmatrix} 0 \\ 1 \\ 1 \end{pmatrix},$$

obtained by setting $\alpha = 1, \beta = 0$ and then $\alpha = 0, \beta = 1$. Since $\lambda = 2$ is an eigenvalue of multiplicity two, let us count it as *two* eigenvalues, λ_2 and λ_3, which just happen to coincide, and let us associate with them the eigenvectors

$$e_2 = \begin{pmatrix} 1 \\ 0 \\ 0 \end{pmatrix} \quad \text{and} \quad e_3 = \begin{pmatrix} 0 \\ 1 \\ 1 \end{pmatrix}.$$

The point to note is that altogether we have three mutually orthogonal eigenvectors and the dimension of the space is three, so that the e_j's constitute an orthogonal basis for the space! [Of course, it is not the *only* such basis; for instance, $e_1 = (1, 0, 0)^T$, $e_2 = (0, 1, 0)^T$, $e_3 = (0, 0, 1)^T$ is one, too. However, we will see that *the basis generated by the eigenvalue problem $Lx = \lambda x$ is particularly helpful in the solution of the inhomogeneous problem $Lx = c$ or, more generally, $Lx = \Lambda x + c$.*] ▮

In fact, it can be shown that

THEOREM 20.2. For any self-adjoint operator L on a finite-dimensional domain $\mathcal{S}$, k mutually orthogonal eigenvectors can be found for each eigenvalue of multiplicity k.

Together with the fact that eigenvectors corresponding to *distinct* eigenvalues are orthogonal (Thorem 20.1), this theorem implies that

THEOREM 20.3. The eigenvectors of any self-adjoint operator L on a finite-dimensional domain $\mathcal{S}$ constitute a basis for $\mathcal{S}$.

Example 20.3. Consider the differential operator $L = -d^2/dx^2$ with the boundary conditions $y(0) = y(\ell) = 0$. The eigenvalue problem is then

$$y'' + \lambda y = 0, \qquad y(0) = y(\ell) = 0. \tag{20.2}$$

The general solution of the differential equation is $A \sin \sqrt{\lambda}\, x + B \cos \sqrt{\lambda}\, x$, where A and B are arbitrary constants. $y(0) = 0$ implies that $B = 0$. Next, we set $y(\ell) = A \sin \sqrt{\lambda}\, \ell = 0$, and thus either $A = 0$ and/or $\sin \sqrt{\lambda}\, \ell = 0$. We cannot accept $A = 0$ because doing so would leave us with $y(x) = 0$, whereas eigenfunctions are to be nontrivial. Hence

$$\sin\sqrt{\lambda}\,\ell = 0 \tag{20.3}$$

[which is the *characteristic equation* for λ, analogous to the polynomial equation $\det(A - \lambda I) = 0$ for the matrix case], so that $\sqrt{\lambda}\,\ell$ must coincide with a zero of the sine function—namely, $n\pi$, where n is an integer. That is, $\lambda = n^2\pi^2/\ell^2$. Setting the arbitrary scale factor A equal to unity, the eigens are therefore as follows:

$$\lambda_n = \frac{n^2\pi^2}{\ell^2}, \qquad y_n = \sin\frac{n\pi x}{\ell} \tag{20.4}$$

for $n = 1, 2, \ldots$; we omit $n = 0$ because $y_0 = \sin 0 = 0$.

Next, L [i.e., $-d^2/dx^2$, together with the boundary conditions $y(0) = y(\ell) = 0$] is found to be self-adjoint, and in accordance with Theorem 20.1, the eigenvalues are all real and the eigenfunctions are mutually orthogonal, since

$$(y_m, y_n) = \int_0^\ell \sin\frac{m\pi x}{\ell}\sin\frac{n\pi x}{\ell}\,dx$$

is found to be zero for $m \neq n$.

We would like to close with a statement that the orthogonal set $\{\sin(n\pi x/\ell)\}$ is **complete**—that is, that it constitutes a basis for the space, but Theorem 20.3 is of no help here because the space in this example is of *infinite* dimension. Let us defer this question until our discussion of the Sturm–Liouville theory in the next section.

COMMENT. We could, if desired, make our derivation of the characteristic equation parallel the matrix case more closely as follows: $y(0) = 0 = 0 + B$ and $y(\ell) = 0 = A\sin\sqrt{\lambda}\,\ell + B\cos\sqrt{\lambda}\,\ell$ or

$$\begin{pmatrix} 0 & 1 \\ \sin\sqrt{\lambda}\,\ell & \cos\sqrt{\lambda}\,\ell \end{pmatrix}\begin{pmatrix} A \\ B \end{pmatrix} = \mathbf{0},$$

and A and B will not both be zero [and hence $y(x) = A\sin\sqrt{\lambda}\,x + B\cos\sqrt{\lambda}\,x$ will be nontrivial] if and only if the determinant of the coefficient matrix is zero. This gives (20.3). ▮

20.3. THE STURM–LIOUVILLE THEORY

The Sturm–Liouville problem is a generalization of Example 20.3. First, let us choose an inner product of the form

$$(f, g) = \int_a^b f(x)g(x)w(x)\,dx, \tag{20.5}$$

where $w(x) > 0$ over $a \leq x \leq b$ (recall Exercise 17.17) and all quantities will be assumed to be real; (20.5) is more general than our usual inner product through the inclusion of the weighting function $w(x)$, which will not necessarily be unity in applications.

It is not hard to show (Exercise 20.7) that the operator

$$L = \frac{1}{w(x)}\left\{\frac{d}{dx}\left[p(x)\frac{d}{dx}\right] + r(x)\right\} \tag{20.6a}$$

$$\alpha y(a) + \beta y'(a) = 0, \qquad \gamma y(b) + \delta y'(b) = 0 \tag{20.6b}$$

is self-adjoint with respect to the inner product (20.5). The eigenvalue problem[3]

$$Ly + \lambda y = \frac{1}{w}\{(py')' + ry\} + \lambda y = 0$$

or

$$(py')' + ry + \lambda wy = 0 \quad \text{over } a \leq x \leq b \tag{20.7a}$$

with homogeneous boundary conditions of the form

$$\alpha y(a) + \beta y'(a) = 0, \qquad \gamma y(b) + \delta y'(b) = 0 \tag{20.7b}$$

is called a **Sturm–Liouville system.**

As mentioned, the operator is self-adjoint, and so Theorem 20.1 applies: (i) the λ's are real, and (ii) eigenfunctions corresponding to distinct eigenvalues are mutually orthogonal with respect to the inner product (20.5).

The question of completeness of the eigenfunctions is complicated. We state, without proof,[4] the following crucial result:

THEOREM 20.4. If both $p(x)$ and $w(x)$ are analytic and positive ($p, w > 0$) over $a \leq x \leq b$, where a and b are finite, then the eigenfunctions of the Sturm–Liouville system (20.7) are complete over $\mathfrak{L}_2(a, b)$.

Completeness is also known to hold for certain other special cases. For instance, if $p(x)$ vanishes at an endpoint, perhaps both, then the boundary condition (20.7b) at that endpoint can be replaced by a condition that y and y' are simply finite there. Also, if it happens that $p(a) = p(b)$, then the "separated" conditions (20.7b) (i.e., one condition at a and the other at b) may be replaced by "periodic" conditions $y(a) = y(b)$ and $y'(a) = y'(b)$ (Exercise 20.8). In this case, the eigenvalues are generally of multiplicity two, whereas for the case of separated conditions there will be no multiplicity.

Note carefully that the domain of the Sturm–Liouville operator (20.6) is the set of twice-differentiable functions $y(x)$ that satisfy the boundary conditions (20.6b), and yet Theorem 20.4 asserts completeness over the much broader space $\mathfrak{L}_2(a, b)$, which includes functions that are not necessarily even continuous and that need not satisfy the boundary conditions (20.6b). The extension to $\mathfrak{L}_2$ space occurs in the conversion of (20.7) to an *integral equation*, in the first step of the proof (which is here omitted), with a self-adjoint $\mathfrak{L}_2$ kernel (as explained in Example 18.2).

The various special forms of the Sturm–Liouville system (20.7) lead to various complete sets for $\mathfrak{L}_2(a, b)$. Expansions in terms of these eigenfunctions are usually called generalized Fourier series—for instance, trigonometric Fourier series, Fourier–Bessel series, and Fourier–Legendre series. Numerous applications of these series are contained in Part V; a very readable account of the theory and its applications can also be found in the book by Churchill.[5]

---

[3] Whether we write $Ly = \lambda y$ or $Ly + \lambda y = 0$ is immaterial. We would prefer the former, for consistency with the rest of our presentation, but the latter is customary in Sturm–Liouville theory.

[4] The classical reference for this material is E. Titchmarsh, *Eigenfunction Expansions Associated With Second-Order Differential Equations*, Oxford University Press, 1946.

[5] R. V. Churchill, *Fourier Series and Boundary Value Problems*, McGraw-Hill, New York, 1956.

Example 20.4. Note that the problem (20.2) of Example 20.3 is a Sturm–Liouville system with $p(x) = w(x) = 1$, $r(x) = 0$, $a = 0$, and $b = \ell$. Thus the eigenfunctions $y_n = \sin(n\pi x/\ell)$ are complete, and any function $f(x)$ in $\mathfrak{L}_2(0, \ell)$ can be expanded as

$$f(x) = \sum_1^\infty f_n \sin \frac{n\pi x}{\ell}. \tag{20.8}$$

To evaluate f_6, for instance, dot $\sin(6\pi x/\ell)$ into both sides. By orthogonality, all terms on the right are zero except for $n = 6$, so that

$$\left(f, \sin \frac{6\pi x}{\ell}\right) = f_6\left(\sin \frac{6\pi x}{\ell}, \sin \frac{6\pi x}{\ell}\right)$$

or

$$f_6 = \frac{(f, \sin(6\pi x/\ell))}{(\sin(6\pi x/\ell), \sin(6\pi x/\ell))}.$$

More generally, for any n we compute

$$f_n = \frac{(f, \sin(n\pi x/\ell))}{(\sin(n\pi x/\ell), \sin(n\pi x/\ell))} \tag{20.9}$$

$$f_n = \frac{\int_0^\ell f(x) \sin(n\pi x/\ell)\, dx}{\int_0^\ell \sin^2(n\pi x/\ell)\, dx}$$

$$= \frac{2}{\ell}\int_0^\ell f(x) \sin \frac{n\pi x}{\ell}\, dx. \tag{20.10}$$

[Analogously, if we wish to expand a given vector $\mathbf{V}$ in three-dimensional space in the form $\mathbf{V} = V_1\hat{\mathbf{i}} + V_2\hat{\mathbf{j}} + V_3\hat{\mathbf{k}}$, we compute the components V_1, V_2, V_3 by dotting $\hat{\mathbf{i}}$, $\hat{\mathbf{j}}$, and $\hat{\mathbf{k}}$ into both sides. For example, $\mathbf{V} \cdot \hat{\mathbf{i}} = V_1\hat{\mathbf{i}} \cdot \hat{\mathbf{i}} + 0 + 0$, and thus $V_1 = (\mathbf{V} \cdot \hat{\mathbf{i}})/(\hat{\mathbf{i}} \cdot \hat{\mathbf{i}})$.]

For example, if $f(x) = 100$, then

$$f_n = \frac{200}{\ell}\int_0^\ell \sin \frac{n\pi x}{\ell}\, dx = \text{etc.} = \begin{cases} \dfrac{400}{n\pi}, & n \text{ odd} \\ 0, & n \text{ even} \end{cases} \tag{20.11}$$

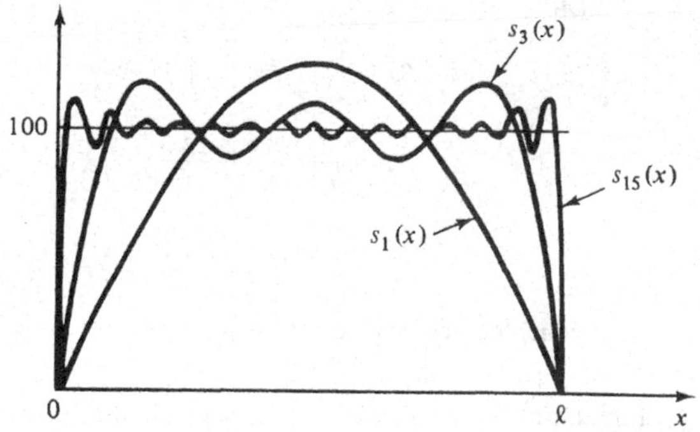

Figure 20.2. Mean square convergence to $f(x) = 100$.

so that

$$f(x) = \sum_{n=1,3,\ldots}^{\infty} \frac{400}{n\pi} \sin \frac{n\pi x}{\ell} = \sum_{n=1}^{\infty} \frac{400}{(2n-1)\pi} \sin \frac{(2n-1)\pi x}{\ell} \tag{20.12}$$

Again, we emphasize that although $f(x) = 100$ does not satisfy the boundary conditions in (20.2), it can still be expanded in terms of the eigenfunctions $\sin(n\pi x/\ell)$; Theorem 20.4 guarantees that *any* function in $\mathfrak{L}_2(0, \ell)$ can! To see how the $\sin(n\pi x/\ell)$'s (which are zero at $x = 0$ and $x = \ell$) cope with $f(x) = 100$ (which is nonzero at $x = 0$ and $x = \ell$), look at Fig. 20.2 and observe how the partial sums $s_n(x) \longrightarrow f(x)$ in the mean square sense, despite the discrepancies at the endpoints. ∎

Example 20.5. Consider the system

$$y'' + \lambda y = 0 \tag{20.13a}$$

$$y(-\ell) = y(\ell), \qquad y'(-\ell) = y'(\ell). \tag{20.13b}$$

Subjecting the solution $y = A \sin \sqrt{\lambda}\, x + B \cos \sqrt{\lambda}\, x$ of (20.13a) to the periodic boundary conditions (20.13b), we obtain

$$A \sin \sqrt{\lambda}\, \ell = 0 \tag{20.14a}$$

$$B\sqrt{\lambda} \sin \sqrt{\lambda}\, \ell = 0; \tag{20.14b}$$

and if A and B are not to be both zero, it follows that $\sin \sqrt{\lambda}\, \ell = 0$ or $\lambda_n = n^2\pi^2/\ell^2$ for $n = 0, 1, 2, \ldots$. With $\lambda = \lambda_n$ in (20.14) we see that A and B are arbitrary, which means that the corresponding eigenfunctions are

$$y_n = A \sin \frac{n\pi x}{\ell} + B \cos \frac{n\pi x}{\ell}. \tag{20.15}$$

In other words, each λ_n (except λ_0) is of multiplicity two, with the orthogonal eigenfunctions $\sin(n\pi x/\ell)$ and $\cos(n\pi x/\ell)$ obtained from (20.15) by setting $A = 1$, $B = 0$ and $A = 0$, $B = 1$, respectively. (An analogous situation arose in the matrix Example 20.2, which you may want to reread at this point.) For $n = 0$, (20.15) yields the single eigenfunction $y_0 = B = 1$, say.

Thus the eigenfunction expansion of a given $f(x)$ in $\mathfrak{L}_2(-\ell, \ell)$ is of the form

$$f(x) = a_0 + \sum_1^{\infty} \left(a_n \cos \frac{n\pi x}{\ell} + b_n \sin \frac{n\pi x}{\ell}\right), \tag{20.16}$$

where

$$a_0 = \frac{(f, 1)}{(1, 1)} = \frac{1}{2\ell} \int_{-\ell}^{\ell} f(x)\, dx \tag{20.17}$$

$$a_n = \frac{(f, \cos n\pi x/\ell)}{(\cos n\pi x/\ell, \cos n\pi x/\ell)} = \frac{1}{\ell} \int_{-\ell}^{\ell} f(x) \cos \frac{n\pi x}{\ell}\, dx \tag{20.18}$$

$$b_n = \frac{(f, \sin n\pi x/\ell)}{(\sin n\pi x/\ell, \sin n\pi x/\ell)} = \frac{1}{\ell} \int_{-\ell}^{\ell} f(x) \sin \frac{n\pi x}{\ell}\, dx, \tag{20.19}$$

which coincides with the "usual" trigonometric Fourier series of Chapter 5. ∎

Incidentally, recall that these eigenfunction expansions, say $f(x) = \sum_1^{\infty} f_n y_n(x)$, are understood in the $\mathfrak{L}_2$ sense—that is, $\left\| f - \sum_1^{N} f_n y_n \right\| \longrightarrow 0$ as $N \longrightarrow \infty$ rather than in the pointwise sense, as discussed in Section 17.4. Nevertheless, positive statements regarding pointwise convergence can also be made. For example, we noted in Section 5.2 that if $f(x)$ is piecewise smooth over $-\ell \leq x \leq \ell$, then the right-hand side of (20.16) will converge to $f(x)$ at each point where f is continuous and to the average value $[f(x+0)+f(x-0)]/2$ at each jump discontinuity.

20.4. ADDITIONAL DISCUSSION FOR THE MATRIX CASE

In the previous section, this one, and the next we consider differential, matrix, and integral operators in turn. The present discussion deals primarily with "quadratic forms" and "diagonalization," but additional material is included in the exercises.

The homogeneous expression of second degree

$$F(x_1, x_2, x_3) = a_{11}x_1^2 + a_{22}x_2^2 + a_{33}x_3^2 + 2a_{12}x_1x_2 + 2a_{13}x_1x_3 + 2a_{23}x_2x_3 \tag{20.20}$$

is called a **quadratic form** in the variables x_1, x_2, x_3. We consider three variables for definiteness; extension to the general case $F(x_1, \ldots, x_n)$ will be obvious. It is easy to verify that F can be expressed more concisely in matrix notation as

$$F = x^T A x, \tag{20.21}$$

where

$$x = \begin{pmatrix} x_1 \\ x_2 \\ x_3 \end{pmatrix} \quad \text{and} \quad A = \begin{pmatrix} a_{11} & a_{12} & a_{13} \\ a_{12} & a_{22} & a_{23} \\ a_{13} & a_{23} & a_{33} \end{pmatrix}. \tag{20.21}$$

(If we did not insert the 2s in (20.20), then the A matrix would end up with $\frac{1}{2}$s in it.)

It is often desirable to find new coordinates x_1', x_2', x_3' as linear combinations of x_1, x_2, x_3 such that F reduces to the *canonical form* $\alpha x_1'^2 + \beta x_2'^2 + \gamma x_3'^2$—that is, to the sum of squares only, with no "mixed" terms. Expressing the linear relationship between the vectors x and x' per

$$x = Qx', \tag{20.22}$$

(20.21) becomes

$$F = (Qx')^T A(Qx') = x'^T(Q^T A Q)x'. \tag{20.23}$$

Clearly, (20.23) will be in canonical form if and only if Q is such that the matrix $Q^T A Q$ is diagonal.

Suppose that the a_{ij} coefficients are all real, both for simplicity and because such is often the case. Then the A matrix is real-symmetric and hence self-adjoint; as a result, ON eigenvectors e_1, e_2, e_3 can be found corresponding to the eigenvalues $\lambda_1, \lambda_2, \lambda_3$. Surely we can arrange it so that all the components of the e_j's are real. We will now show that the so-called **modal matrix**

$$Q = [(e_1)(e_2)(e_3)], \tag{20.24}$$

the columns of which are the normalized eigenvectors, is suitable for the purpose of canonicalization, since

$$Q^T A Q = \begin{bmatrix} \text{———} e_1 \text{———} \\ \text{———} e_2 \text{———} \\ \text{———} e_3 \text{———} \end{bmatrix} \begin{bmatrix} \Big| & \Big| & \Big| \\ Ae_1 & Ae_2 & Ae_3 \\ \Big| & \Big| & \Big| \end{bmatrix}$$

$$= \begin{bmatrix} \text{———} e_1 \text{———} \\ \text{———} e_2 \text{———} \\ \text{———} e_3 \text{———} \end{bmatrix} \begin{bmatrix} \Big| & \Big| & \Big| \\ \lambda_1 e_1 & \lambda_2 e_2 & \lambda_3 e_3 \\ \Big| & \Big| & \Big| \end{bmatrix} = \begin{pmatrix} \lambda_1 & 0 & 0 \\ 0 & \lambda_2 & 0 \\ 0 & 0 & \lambda_3 \end{pmatrix}$$

is diagonal as desired. The last equality follows from the fact that $e_i^T e_j = e_i^T \bar{e}_j$ (since the e_j's have real components) $= (e_i, e_j) = \delta_{ij}$.

Summarizing, if we take Q to be the modal matrix (20.24), we see from (20.23) that the linear transformation $x = Qx'$ reduces F to the canonical form

$$F = \lambda_1 x_1'^2 + \lambda_2 x_2'^2 + \lambda_3 x_3'^2. \tag{20.25}$$

COMMENT 1. As is easily verified, observe that $Q^T Q = I$, so that $Q^T = Q^{-1}$. A matrix with this property is said to be **orthogonal**. Orthogonality of Q permits us to solve (20.22) for x',

$$x' = Q^{-1} x = Q^T x,$$

without having to go through a tedious matrix inversion.

COMMENT 2. Consider the transformation of a given vector x by Q into a new vector y—that is, $Qx = y$. How does the length of y compare with the length of x? Well, $\|y\|^2 = (y, y) = (Qx)^T \overline{Qx} = x^T(Q^T \bar{Q})\bar{x}$. If $Q^T \bar{Q} = I$ (i.e., if $\bar{Q}^T = Q^{-1}$), we say that Q is **unitary** and we see that $\|y\| = \|x\|$. In other words, lengths are preserved under a unitary transformation. If Q is real, the distinction between unitary and orthogonal disappears.

COMMENT 3. A linear operator L is said to be **positive definite** if $(Lx, x) > 0$ for all $x \neq \mathbb{O}$, **positive semidefinite** if $(Lx, x) \geq 0$ for all $x \neq \mathbb{O}$, and similarly for **negative definite** and **negative semidefinite**. In the present (matrix) case, $(Ax, x) = F$. Now $x = \mathbb{O}$ if and only if $x' = \mathbb{O}$ (Exercise 20.15), and so it follows from (20.25) that A is positive definite if and only if all its eigenvalues are positive; the same is true for negative definite and so on. For instance, suppose that $\lambda_1 = 0$, $\lambda_2 = 3$, $\lambda_3 = 7$. Then A is *not* positive definite because of the $\lambda_1 = 0$ eigenvalue. That is, we see from (20.25) that $F = 0$ for $x_2' = x_3' = 0$ even if $x_1' \neq 0$—that is, even if $x' \neq \mathbb{O}$. In this case, then, A is positive *semi*definite; similarly, we say that its quadratic form F is semidefinite, too.

Example 20.6. For

$$F = 4x_1^2 + 2x_1 x_2 + 4x_1 x_3 \tag{20.26}$$

we have

$$A = \begin{pmatrix} 4 & 1 & 2 \\ 1 & 0 & 0 \\ 2 & 0 & 0 \end{pmatrix} \tag{20.27}$$

with the eigens

$$\lambda_1 = 0, \qquad \lambda_2 = 5, \qquad \lambda_3 = -1,$$

$$e_1 = \begin{pmatrix} 0 \\ 2/\sqrt{5} \\ -1/\sqrt{5} \end{pmatrix}, \qquad e_2 = \begin{pmatrix} 5/\sqrt{30} \\ 1/\sqrt{30} \\ 2/\sqrt{30} \end{pmatrix}, \qquad e_3 = \begin{pmatrix} 1/\sqrt{6} \\ -1/\sqrt{6} \\ -2/\sqrt{6} \end{pmatrix}.$$

The foregoing discussion tells us that the linear transformation $x = Qx'$, where

$$Q = \begin{pmatrix} 0 & 5/\sqrt{30} & 1/\sqrt{6} \\ 2/\sqrt{5} & 1/\sqrt{30} & -1/\sqrt{6} \\ -1/\sqrt{5} & 2/\sqrt{30} & -2/\sqrt{6} \end{pmatrix}, \tag{20.28}$$

will reduce (20.26) to the canonical form

$$F = 5x_2'^2 - x_3'^2, \tag{20.29}$$

as is easily verified by substitution. F is neither positive definite nor negative definite.

Of course, Q is not unique. For instance, reordering the e_j's such that

$$Q = \begin{pmatrix} 5/\sqrt{30} & 0 & 1/\sqrt{6} \\ 1/\sqrt{30} & 2/\sqrt{5} & -1/\sqrt{6} \\ 2/\sqrt{30} & -1/\sqrt{5} & -2/\sqrt{6} \end{pmatrix} \tag{20.30}$$

would yield $F = 5x_1'^2 - x_3'^2$ instead. In fact, simply "completing squares," we have

$$\begin{aligned} F &= 4x_1^2 + 2x_1x_2 + 4x_1x_3 \\ &= 4x_1^2 + 2x_1(x_2 + 2x_3) \\ &= [4x_1^2 + 2x_1(x_2 + 2x_3) + \tfrac{1}{4}(x_2 + 2x_3)^2] - \tfrac{1}{4}(x_2 + 2x_3)^2 \\ &= [2x_1 + \tfrac{1}{2}(x_2 + 2x_3)]^2 - \tfrac{1}{4}(x_2 + 2x_3)^2 \\ &= x_1'^2 - \tfrac{1}{4}x_2'^2 \end{aligned} \tag{20.31}$$

under the change of variables

$$\begin{aligned} x_1' &= 2x_1 + \tfrac{1}{2}x_2 + x_3 \\ x_2' &= x_2 + 2x_3 \\ x_3' &= x_3. \end{aligned} \tag{20.32}$$

In this case the Q matrix, which is of the triangular form

$$Q = \begin{pmatrix} 2 & 1/2 & 1 \\ 0 & 1 & 2 \\ 0 & 0 & 1 \end{pmatrix}^{-1} = \begin{pmatrix} 1/2 & -1/4 & 0 \\ 0 & 1 & -2 \\ 0 & 0 & 1 \end{pmatrix}, \tag{20.33}$$

is *not* an orthogonal matrix, and if we regard x_1, x_2, x_3 as a cartesian coordinate system, we find (Exercise 20.16) that the x_1', x_2', x_3' coordinates will *not* be cartesian; in applications, this may be unacceptable or inconvenient. ▮

Example 20.7. *Normal Modes of Vibration.* The equations of motion for the spring-mass system shown in Fig. 20.3 are

$$m\ddot{x}_1 = -kx_1 - k(x_1 - x_2), \qquad m\ddot{x}_2 = -kx_2 - k(x_2 - x_1)$$

or with $m = k = 1$ say,

$$\ddot{x}_1 + 2x_1 - x_2 = 0, \qquad \ddot{x}_2 - x_1 + 2x_2 = 0. \tag{20.34}$$

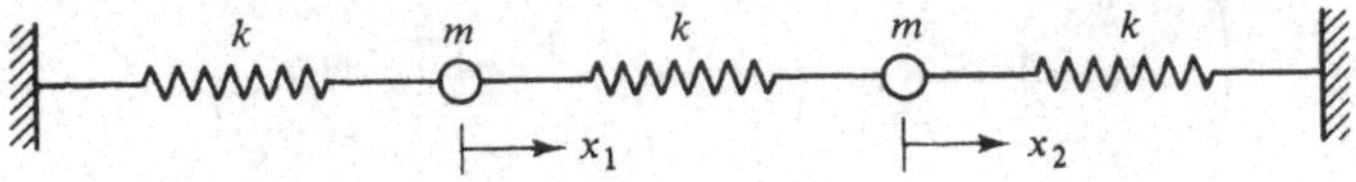

Figure 20.3. Normal modes of vibration.

More compactly,

$$\ddot{x} + Ax = \Theta, \tag{20.35}$$

where
$$x = \begin{pmatrix} x_1 \\ x_2 \end{pmatrix} \quad \text{and} \quad A = \begin{pmatrix} 2 & -1 \\ -1 & 2 \end{pmatrix}.$$

Observe that the "stiffness matrix" A is real-symmetric, with eigens

$$\lambda_1 = 1, \quad e_1 = \begin{pmatrix} 1/\sqrt{2} \\ 1/\sqrt{2} \end{pmatrix}, \qquad \lambda_2 = 3, \quad e_2 = \begin{pmatrix} 1/\sqrt{2} \\ -1/\sqrt{2} \end{pmatrix}.$$

Our object is to *uncouple* the governing equations (20.34). To do so, set $x = Qx'$, where

$$Q = \begin{pmatrix} 1/\sqrt{2} & 1/\sqrt{2} \\ 1/\sqrt{2} & -1/\sqrt{2} \end{pmatrix}$$

is the modal matrix of A. Then (20.35) becomes

$$Q\ddot{x}' + AQx' = \Theta$$

or multiplying through by Q^{-1} $(= Q^T)$,

$$\ddot{x}' + Q^TAQx' = \Theta,$$

and these equations are uncoupled, since

$$Q^TAQ = \begin{pmatrix} \lambda_1 & 0 \\ 0 & \lambda_2 \end{pmatrix} = \begin{pmatrix} 1 & 0 \\ 0 & 3 \end{pmatrix}$$

is diagonal. Thus

$$\ddot{x}_1' + x_1' = 0, \qquad \ddot{x}_2' + 3x_2' = 0, \tag{20.36}$$

where x_1', x_2' are called the **normal coordinates**. Solving (20.36),

$$x_1' = A_1 \sin(t + \phi_1), \qquad x_2' = A_2 \sin(\sqrt{3}\,t + \phi_2),$$

and so

$$\begin{pmatrix} x_1 \\ x_2 \end{pmatrix} = Qx' = \begin{pmatrix} \frac{1}{\sqrt{2}} A_1 \sin(t + \phi_1) + \frac{1}{\sqrt{2}} A_2 \sin(\sqrt{3}\,t + \phi_2) \\ \frac{1}{\sqrt{2}} A_1 \sin(t + \phi_1) - \frac{1}{\sqrt{2}} A_2 \sin(\sqrt{3}\,t + \phi_2) \end{pmatrix}$$

$$= C_1 \begin{pmatrix} \sin(t + \phi_1) \\ \sin(t + \phi_1) \end{pmatrix} + C_2 \begin{pmatrix} \sin(\sqrt{3}\,t + \phi_2) \\ -\sin(\sqrt{3}\,t + \phi_2) \end{pmatrix}.$$

In vibration theory terminology, the solution is thus expressed as a linear combination of the **natural** or **normal modes**. The initial conditions $x_1(0)$, $\dot{x}_1(0)$, $x_2(0)$, $\dot{x}_2(0)$ determine C_1, C_2, ϕ_1, ϕ_2. To illustrate, if $x_1(0) = x_2(0) = 1$ and $\dot{x}_1(0) = \dot{x}_2(0) = 0$, then we find that $C_1 = 1$, $\phi_1 = \pi/2$, and $C_2 = 0$. Since $C_2 = 0$, we say that the system is in its "low mode"—that is, the low frequency of 1 radian per unit time, in which the two masses swing together or are in-phase. On the other hand, if $x_1(0) = -1$, $x_2(0) = 1$, $\dot{x}_1(0) = \dot{x}_2(0) = 0$, say, then we find that $C_1 = 0$, and we say that the system is in its "high mode," where the masses swing 180° out of phase with a frequency of $\sqrt{3}$ radians per unit time. More generally, however, both C_1 and C_2 will be nonzero, and the motion will be a combination of the two natural modes. ▮

*

Suppose that A in (20.35) is *not* self-adjoint. We can still form a modal matrix $X = [(e_1) \cdots (e_n)]$. Then

$$Ax = [(Ae_1) \cdots (Ae_n)] = [(\lambda_1 e_1) \cdots (\lambda_n e_n)]$$

$$= \left[\begin{pmatrix} \\ e_1 \\ \\ \end{pmatrix} \cdots \begin{pmatrix} \\ e_n \\ \\ \end{pmatrix} \right] \begin{bmatrix} \lambda_1 & & 0 \\ & \ddots & \\ 0 & & \lambda_n \end{bmatrix} \equiv X\Lambda,$$

and

$$X^{-1}AX = X^{-1}X\Lambda = \Lambda$$

is diagonal as desired. The catch is that we need X^{-1} to exist—that is, we need $\det X \neq 0$. Such will be the case (we state without proof) if the columns of X are LI—in other words, if the e_j's are LI. Of course if A were self-adjoint then its e_j's would be orthogonal and hence LI. But even if A is *not* self-adjoint we state, without proof that *if the λ_j's of A are distinct, then its e_j's will be LI.* Assuming that such is the case, let $x = Xx'$ in (20.35). Then

$$X\ddot{x}' + AXx' = 0$$

or

$$\ddot{x}' + X^{-1}AXx' = \ddot{x}' + \Lambda x' = 0,$$

where Λ is diagonal as desired.

If the λ_j's are *not* distinct then A can at least be *almost* diagonalized; more specifically, it can be reduced to a so-called **Jordan canonical form** as discussed in Exercises 20.22 and 20.23.

As another application of these ideas, let us consider the so-called inertia tensor, principal axes, and principal inertias encountered in rigid-body mechanics. First, a few words about tensors. Suppose that we simply write down **xy**, where **x** and **y** are vectors. This object is called a *dyad.* (Normally we don't bother with boldface type to denote vectors, but here it will be helpful.) More generally, we call linear combinations of the form $\alpha_1\mathbf{x}_1\mathbf{y}_1 + \cdots + \alpha_n\mathbf{x}_n\mathbf{y}_n$ *dyadics* or **second-order tensors.**

If these second-order tensors are to be of any use, we must decide how to manipulate them, just as we did the vectors in a vector space. As might seem reasonable, we impose the associative rules

$$\alpha(\mathbf{xy}) = (\alpha\mathbf{x})\mathbf{y}, \qquad (\alpha\mathbf{x})(\beta\mathbf{y}) = (\alpha\beta)\mathbf{xy},$$

and the distributive rules

$$\mathbf{x}(\mathbf{y} + \mathbf{z}) = \mathbf{xy} + \mathbf{xz}$$
$$(\mathbf{x} + \mathbf{y})\mathbf{z} = \mathbf{xz} + \mathbf{yz}$$
$$\alpha(\mathbf{wx} + \mathbf{yz}) = \alpha\mathbf{wx} + \alpha\mathbf{yz}.$$

We emphasize, however, that **xy** is *not* the same as **yx**; they are distinct tensors. Finally, we extend the dot or inner product (here the dot notation will be a little more convenient) so that it is defined between a second-order tensor and a vector or between two second-order tensors:

$$(\mathbf{xy}) \cdot \mathbf{z} = \mathbf{x}(\mathbf{y} \cdot \mathbf{z}) = (\mathbf{y} \cdot \mathbf{z})\mathbf{x}$$
$$\mathbf{x} \cdot (\mathbf{yz}) = (\mathbf{x} \cdot \mathbf{y})\mathbf{z}$$
$$(\mathbf{wx}) \cdot (\mathbf{yz}) = (\mathbf{x} \cdot \mathbf{y})\mathbf{wz}.$$

Beyond this, we don't feel obliged to explain the "nature" of **xy** any further. We simply regard it as a fundamental object. The reader should feel more comfortable about this after our discussion of the inertia tensor. First, let us give one additional definition. We call a second-order tensor **I** the *identity element* if it is such that

$$\mathbf{I} \cdot \mathbf{x} = \mathbf{x} \quad \text{and} \quad \mathbf{I} \cdot \mathbf{xy} = \mathbf{xy} \tag{20.37}$$

for all **x** and **y**. Actually, the second follows from the first, since $\mathbf{I} \cdot (\mathbf{xy}) = (\mathbf{I} \cdot \mathbf{x})\mathbf{y} = \mathbf{xy}$. For the three-dimensional vector space with base vectors **i**, **j**, **k**, for example, $\mathbf{I} = \mathbf{ii} + \mathbf{jj} + \mathbf{kk}$, since

$$\mathbf{I} \cdot \mathbf{x} = (\mathbf{ii} + \mathbf{jj} + \mathbf{kk}) \cdot (x_1\mathbf{i} + x_2\mathbf{j} + x_3\mathbf{k}) = \mathbf{ii} \cdot x_1\mathbf{i} + \cdots + \mathbf{kk} \cdot x_3\mathbf{k}$$
$$= x_1\mathbf{i}(\mathbf{i} \cdot \mathbf{i}) + \cdots + x_3\mathbf{k}(\mathbf{k} \cdot \mathbf{k}) = x_1\mathbf{i} + x_2\mathbf{j} + x_3\mathbf{k} = \mathbf{x}.$$

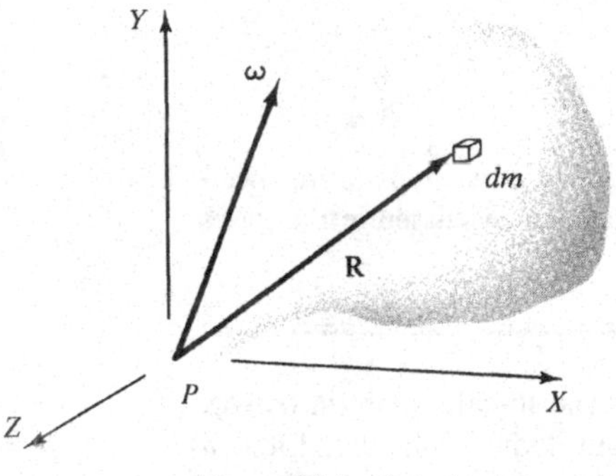

Figure 20.4. Spinning top.

The Inertia Tensor. Consider the motion of a rigid body that is fixed at point P and that has a time-dependent angular velocity $\boldsymbol{\omega}(t)$, as shown in Fig. 20.4—that is, a spinning "top." Of fundamental interest in connection with the dynamics of the body is its so-called *moment of momentum* vector about P, which we will denote as $\mathbf{H}_P$. For the incremental element of mass dm, with position vector $\mathbf{R}$ and velocity $\dot{\mathbf{R}}$, $d\mathbf{H}_P$ is simply $\mathbf{R} \times \dot{\mathbf{R}}\, dm$, since $\dot{\mathbf{R}}\, dm$ is the momentum of dm.[6] Because $\mathbf{R}$ is of fixed length, $\dot{\mathbf{R}} = \boldsymbol{\omega} \times \mathbf{R}$ (Section 8.3), and therefore

$$d\mathbf{H}_P = \mathbf{R} \times \dot{\mathbf{R}}\, dm = \mathbf{R} \times (\boldsymbol{\omega} \times \mathbf{R})\, dm$$
$$= [(\mathbf{R} \cdot \mathbf{R})\boldsymbol{\omega} - (\mathbf{R} \cdot \boldsymbol{\omega})\mathbf{R}]\, dm$$
$$= [R^2\boldsymbol{\omega} - \mathbf{R}(\mathbf{R} \cdot \boldsymbol{\omega})]\, dm = (R^2\mathbf{I} - \mathbf{RR}) \cdot \boldsymbol{\omega}\, dm,$$

[6]Recall from Chapter 8 that the moment of a force $\mathbf{F}$ about the origin is $\mathbf{R} \times \mathbf{F}$; similarly, the "moment of momentum" $d\mathbf{H}_P$ is $\mathbf{R} \times (\dot{\mathbf{R}}\, dm)$.

where we've used the vector identity $\mathbf{A} \times (\mathbf{B} \times \mathbf{C}) = (\mathbf{A} \cdot \mathbf{C})\mathbf{B} - (\mathbf{A} \cdot \mathbf{B})\mathbf{C}$ from Part I and the fact that $\mathbf{I} \cdot \boldsymbol{\omega} = \boldsymbol{\omega}$ per (20.37). Integrating over the whole body,

$$\mathbf{H}_P = \int (R^2\mathbf{I} - \mathbf{RR})\, dm \cdot \boldsymbol{\omega} \equiv \mathscr{I} \cdot \boldsymbol{\omega} \tag{20.38}$$

where $\mathscr{I}$ is a second order tensor!

To proceed—that is, to evaluate the integral in (20.38)—we need to introduce a reference coordinate system. Rather than the X, Y, Z system (Fig. 20.4) which is fixed in space, let us use an x, y, z system (with its origin at P and with base vectors $\mathbf{i}, \mathbf{j}, \mathbf{k}$) that is *fixed in the body*, so that the components of $\mathscr{I}$ will be constant in time. Thus $\mathbf{R} = x\mathbf{i} + y\mathbf{j} + z\mathbf{k}$ and $\mathbf{I} = \mathbf{ii} + \mathbf{jj} + \mathbf{kk}$, and so

$$\begin{aligned}\mathscr{I} &= \int [(x^2 + y^2 + z^2)(\mathbf{ii} + \mathbf{jj} + \mathbf{kk}) - (x\mathbf{i} + y\mathbf{j} + z\mathbf{k})(x\mathbf{i} + y\mathbf{j} + z\mathbf{k})]\, dm \\ &= \mathscr{I}_{xx}\mathbf{ii} - \mathscr{I}_{xy}\mathbf{ij} - \mathscr{I}_{xz}\mathbf{ik} \\ &\quad - \mathscr{I}_{yx}\mathbf{ji} + \mathscr{I}_{yy}\mathbf{jj} - \mathscr{I}_{yz}\mathbf{jk} \\ &\quad - \mathscr{I}_{zx}\mathbf{ki} - \mathscr{I}_{zy}\mathbf{kj} + \mathscr{I}_{zz}\mathbf{kk},\end{aligned} \tag{20.39}$$

where

$$\mathscr{I}_{xx} = \int (y^2 + z^2)\, dm, \quad \mathscr{I}_{yy} = \int (x^2 + z^2)\, dm, \quad \mathscr{I}_{zz} = \int (x^2 + y^2)\, dm, \tag{20.40a}$$

and

$$\mathscr{I}_{xy} = \int xy\, dm = \mathscr{I}_{yx}, \quad \mathscr{I}_{xz} = \int xz\, dm = \mathscr{I}_{zx}, \quad \mathscr{I}_{yz} = \int yz\, dm = \mathscr{I}_{zy}. \tag{20.40b}$$

These integrals occur frequently in mechanics and are called the *moments* and *products of inertia* of the body, respectively. Accordingly, it is reasonable to call $\mathscr{I}$ the **inertia tensor** of the body with respect to the x, y, z coordinate system. We can write it more compactly as

$$\mathscr{I} = \sum_{ij} \mathscr{I}_{ij}\mathbf{t}_i\mathbf{t}_j, \tag{20.41}$$

where $\mathbf{t}_1 = \mathbf{i}, \mathbf{t}_2 = \mathbf{j}, \mathbf{t}_3 = \mathbf{k}, \mathscr{I}_{11} = \mathscr{I}_{xx}, \mathscr{I}_{12} = -\mathscr{I}_{xy}$, and so on. The $\mathscr{I}_{ij}$'s are the nine *components* of the inertia tensor.[7] They are also the nine elements of an inertia matrix $\mathscr{I}$, say, but we emphasize that the tensor $\mathscr{I}$ and the matrix $\mathscr{I}$ are not the same thing.

It turns out that the "Euler differential equations of motion," obtained by setting $d\mathbf{H}_P/dt$ equal to the sum of the applied moments, will be simplest if the $\mathscr{I}$ matrix is diagonal. In general, $\mathscr{I}$ will *not* be diagonal with respect to the x, y, z coordinate sys-

[7]Often, in texts on tensor analysis, the base vectors simply do not appear and the set of numbers $\mathscr{I}_{ij}$ is called "the tensor." However, we prefer to denote $\mathscr{I}$ as the tensor and $\mathscr{I}_{ij}$ as its components. For a given tensor $\mathscr{I}$ the $\mathscr{I}_{ij}$ components will depend on the choice of the reference frame—that is, on the $\mathbf{t}_i$ base vectors.

tem. Let us therefore seek a new basis $\mathbf{t}'_1, \mathbf{t}'_2, \mathbf{t}'_3$, which is related to $\mathbf{t}_1, \mathbf{t}_2, \mathbf{t}_3$ (i.e., **i, j, k**) per

$$\mathbf{t}_i = \sum_k b_{ik}\mathbf{t}'_k \qquad (i = 1, 2, 3). \tag{20.42}$$

Then

$$\mathcal{I} = \sum_{i,j,k,l} \mathcal{I}_{ij} b_{ik} b_{jl} \mathbf{t}'_k \mathbf{t}'_l \equiv \sum_{k,l} \mathcal{I}'_{kl} \mathbf{t}'_k \mathbf{t}'_l,$$

where

$$\mathcal{I}'_{kl} = \sum_{i,j} b_{ik} b_{jl} \mathcal{I}_{ij} \tag{20.43}$$

or in matrix language[8]

$$\mathcal{I}' = B^T \mathcal{I} B. \tag{20.44}$$

From our discussion of quadratic forms and diagonalization of matrices we know that $\mathcal{I}'$ will be diagonal if we choose B as a modal matrix of $\mathcal{I}$. Then the quantities $\mathcal{I}'_{kk}$ are called the **principal inertias**, and the axes defined by $\mathbf{t}'_1, \mathbf{t}'_2, \mathbf{t}'_3$ are called **principal axes.**

Example 20.8. To illustrate the calculation of principal inertias and principal axes, consider a very simple body consisting of unit point masses at (1, 0, 0), (1, 1, 0) and (0, 1, 0), as shown in Fig. 20.5. (If worried about the body not being "rigid," imagine the three masses as joined rigidly by rods that are weightless and hence that do not affect the calculation.)

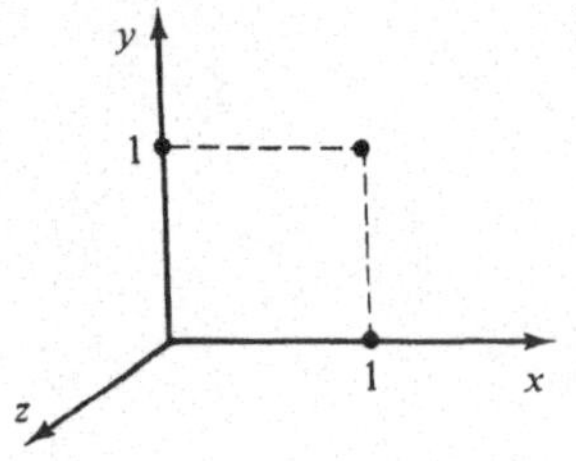

Figure 20.5. The body under consideration.

Since the three masses are point masses, calculation of the inertias is trivial, and we see that[9]

$$\mathcal{I} = 2\mathbf{ii} - \mathbf{ij} - \mathbf{ji} + 2\mathbf{jj} + 4\mathbf{kk}$$

or

$$\mathcal{I} = \begin{pmatrix} 2 & -1 & 0 \\ -1 & 2 & 0 \\ 0 & 0 & 4 \end{pmatrix}. \tag{20.45}$$

Its eigens are found to be

$$\lambda_1 = 1, \quad \lambda_2 = 3, \quad \lambda_3 = 4, \qquad e_1 = \begin{pmatrix} 1/\sqrt{2} \\ 1/\sqrt{2} \\ 0 \end{pmatrix}, \quad e_2 = \begin{pmatrix} 1/\sqrt{2} \\ -1/\sqrt{2} \\ 0 \end{pmatrix}, \quad e_3 = \begin{pmatrix} 0 \\ 0 \\ 1 \end{pmatrix},$$

---

[8] Recalling the definition of matrix multiplication,

$$AB = \{\sum_k a_{ik} b_{kj}\} \quad \text{or} \quad (AB)_{ij} = \sum_k a_{ik} b_{kj},$$

we have

$$\begin{aligned} \mathcal{I}'_{kl} &= \sum_{i,j} b_{ik} b_{jl} \mathcal{I}_{ij} = \sum_j (\sum_i b^T_{ki} \mathcal{I}_{ij}) b_{jl} \\ &= \sum_j (B^T \mathcal{I})_{kj} b_{jl} = (B^T \mathcal{I} B)_{kl} \end{aligned}$$

or

$$\mathcal{I}' = B^T \mathcal{I} B.$$

[9] Since the body consists only of point masses, the integrals in (20.40) reduce to discrete sums over the three masses; for example, $\mathcal{I}_{xx} = (1)^2(1) + (1)^2(1) + (0)(1) = 2$, $\mathcal{I}_{xy} = (0)(1)(1) + (1)(1)(1) + (1)(0)(1) = 1$, etc.

respectively. Choosing

$$B = \begin{pmatrix} 1/\sqrt{2} & 1/\sqrt{2} & 0 \\ 1/\sqrt{2} & -1/\sqrt{2} & 0 \\ 0 & 0 & 1 \end{pmatrix}$$

then, we have

$$\mathscr{I}' = B^T \mathscr{I} B = \begin{pmatrix} 1 & 0 & 0 \\ 0 & 3 & 0 \\ 0 & 0 & 4 \end{pmatrix} \tag{20.46}$$

or

$$\boldsymbol{\mathscr{I}} = 1\mathbf{t}_1'\mathbf{t}_1' + 3\mathbf{t}_2'\mathbf{t}_2' + 4\mathbf{t}_3'\mathbf{t}_3', \tag{20.47}$$

where (since $B^{-1} = B^T$)$\mathbf{t}_k' = \sum_i b_{ik}\mathbf{t}_i$:

$$\mathbf{t}_1' = \frac{1}{\sqrt{2}}\mathbf{i} + \frac{1}{\sqrt{2}}\mathbf{j}, \qquad \mathbf{t}_2' = \frac{1}{\sqrt{2}}\mathbf{i} - \frac{1}{\sqrt{2}}\mathbf{j},$$

$$\mathbf{t}_3' = \mathbf{k},$$

as shown in Fig. 20.6. Thus the principal inertias are $\mathscr{I}_{x'x'}' = 1$, $\mathscr{I}_{y'y'}' = 3$, $\mathscr{I}_{z'z'}' = 4$, and the principal axes x', y', z' are as shown. If the fact that the x', y', z' system is not right-handed, is bothersome, the situation is easily remedied by taking $e_2 = (-1/\sqrt{2}, 1/\sqrt{2}, 0)^T$ instead. Then y' will point in the direction opposite to that shown in Fig. 20.6.

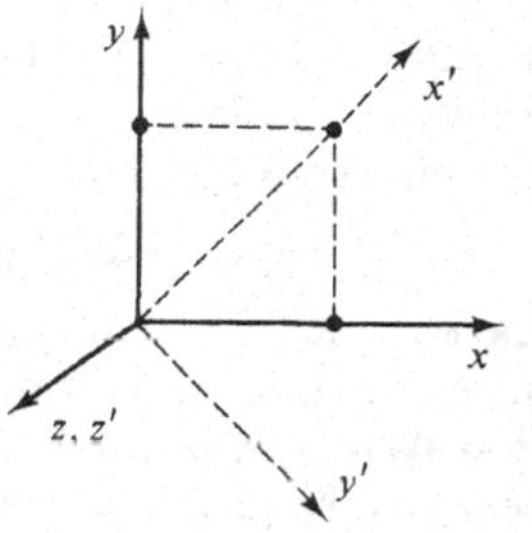

Figure 20.6. The principal axes.

COMMENT. Notice that associated with $\mathscr{I}$ of (20.45) is the quadratic form

$$F = 2x^2 - 2xy + 2y^2 + 4z^2$$

and that $F(x, y, z) =$ constant is a quadric surface. Consider the problem of finding the points on this surface such that the distance from the origin to those points is an extremum. That is,

$$x^2 + y^2 + z^2 - \text{extremum},$$

subject to the constraint $F =$ constant. Taking differentials,

$$x\,dx + y\,dy + z\,dz = 0$$

$$(2x - y)\,dx + (2y - x)\,dy + 4z\,dz = 0;$$

and applying the Lagrange multiplier method (Section 10.2), we arrive at

$$\begin{pmatrix} 2 & -1 & 0 \\ -1 & 2 & 0 \\ 0 & 0 & 4 \end{pmatrix}\begin{pmatrix} x \\ y \\ z \end{pmatrix} = \lambda \begin{pmatrix} x \\ y \\ z \end{pmatrix},$$

where λ is the Lagrange multiplier. It is interesting that this system coincides with the eigenvalue problem that arose in our search for the principal inertias and principal axes and that we can thus interpret the principal axes as the (orthogonal) axes that run from the origin out through the extreme points on the associated quadric surface. ▮

For an introduction to tensor analysis, we recommend the excellent book by Block.[10]

20.5. *THE INHOMOGENEOUS PROBLEM* $Lx = c$

There are two principal approaches to the solution of the general linear equation $Lx = c$. The first is to seek the *inverse operator* L^{-1} and hence to express $x = L^{-1}c$. This procedure was considered in Chapter 19, and we discussed the calculation of L^{-1} for the case where L is a matrix operator. In Parts IV and V we will consider ordinary and partial differential operators, in which case L^{-1} is an integral operator, the kernel of which is the so-called *Green's function.*

The second approach involves *eigenvector expansions*, specifically, the expansion of both x and c in terms of the eigenvectors of L—that is, of the associated eigenvalue problem $Lx = \lambda x$.

Consider, for instance, the inhomogeneous problem

$$Ax = \Lambda x + c, \tag{20.48}$$

where A is an $n \times n$ self-adjoint matrix, Λ is a given constant, and c is a given vector. Of course, the Λx term could be absorbed by the left-hand side, thereby yielding $(A - \Lambda I)x = c$, where $A - \Lambda I$ is still self-adjoint, but Λ is often a significant parameter that is best left in evidence. Since A is self-adjoint, its eigenvectors e_j, considered here as normalized, form an ON basis for the n-dimensional space. Thus we can expand

$$c = \sum_1^n c_j e_j \quad \text{and} \quad x = \sum_1^n \alpha_j e_j,$$

where $c_j = (c, e_j)$ are known and the α_j's are to be found. Inserting them into (20.48) and recalling that $Ae_j = \lambda_j e_j$,

$$\sum_1^n \alpha_j(\lambda_j - \Lambda)e_j = \sum_1^n c_j e_j,$$

so that

$$\alpha_j(\lambda_j - \Lambda) = c_j. \tag{20.49}$$

Next, if Λ does *not* coincide with any of the λ's, then $\alpha_j = c_j/(\lambda_j - \Lambda)$, and the *unique* solution is

$$x = \sum_1^n \left(\frac{c_j}{\lambda_j - \Lambda}\right) e_j. \tag{20.50}$$

On the other hand, suppose that Λ coincides with the kth eigenvalue, $\Lambda = \lambda_k$. Then (20.49) again gives $\alpha_j = c_j/(\lambda_j - \Lambda)$ for all $j \neq k$. For $j = k$ there are two possibilities. (a) If $c_k \neq 0$, then (20.49) is impossible and there is *no solution* of (20.48). (b) If $c_k = 0$ (i.e., if c is orthogonal to the e_k eigenvector), then (20.49) is satisfied for *any* α_k and we have the *nonunique solution*

$$x = \sum_1^{n}{}' \left(\frac{c_j}{\lambda_j - \Lambda}\right) e_j + \alpha e_k, \tag{20.51}$$

where the prime means that the $j = k$ term is omitted and where α is arbitrary.

---

[10] H. D. Block, *Introduction to Tensor Analysis,* Charles E. Merrill Books, Columbus, Ohio, 1962.

These results agree with Theorems 19.1 to 19.3, and the preceding alternative is known as the **Fredholm alternative.**

COMMENT 1. Note that with this approach we find the eigens of A once and for all and then we're "in business." On the other hand, if we chose to reexpress (20.48) as $(A - \Lambda I)x = c$ and to compute $(A - \Lambda I)^{-1}$, then each time we change Λ we would need to recompute $(A - \Lambda I)^{-1}$. Moreover, if $\Lambda = \lambda_k$, say, then $(A - \Lambda I)^{-1}$ *does not exist*, whereas the eigenvector expansion works (provided that $c_k = 0$) and yields (20.51).

COMMENT 2. Why use the eigenvectors of A as our basis? If we did not, we would end up with *coupled* equations for the α_j's in contrast to the simple result (20.49) [Exercise 20.31(b)]. That is, the eigenvectors of A constitute the *natural* basis for the problem $Ax = \Lambda x + c$.

20.6. EIGEN BOUNDS AND ESTIMATES

An extensive literature on eigen bounds and estimates exists. Here we intend to consider only *Rayleigh's quotient* and the *power method.*

Except where noted otherwise, we will assume throughout this section that L is a self-adjoint operator with a complete set of ON eigenvectors $e_1, e_2, \ldots$ with corresponding eigenvalues $\lambda_1, \lambda_2, \ldots,$ which are ordered such that $|\lambda_1| \geq |\lambda_2| \geq |\lambda_3| \geq \cdots$.[11]

For any x we have

$$\begin{aligned} x &= \sum a_j e_j, \\ Lx &= \sum a_j L e_j = \sum \lambda_j a_j e_j, \\ (Lx, x) &= \sum \lambda_j |a_j|^2, \\ (x, x) &= \sum |a_j|^2. \end{aligned} \tag{20.52}$$

Forming $(Lx, x)/(x, x)$, which is called **Rayleigh's quotient**, we easily see that

$$\left|\frac{(Lx, x)}{(x, x)}\right| = \left|\frac{\sum \lambda_j |a_j|^2}{\sum |a_j|^2}\right| \leq \left|\frac{\lambda_1 \sum |a_j|^2}{\sum |a_j|^2}\right| = |\lambda_1|, \tag{20.53}$$

and so $|(Lx, x)/(x, x)|$ provides a lower bound on $|\lambda_1|$ for any x.

Next, suppose that we extremize $(Lx, x)/(x, x) \equiv R$ or, equivalently, suppose that we extremize (Lx, x), subject to the constraint $(x, x) = 1$. Thus, by setting differentials of both (Lx, x) and (x, x) equal to zero, we have

$$\begin{aligned} \lambda_1 \alpha_1\, d\alpha_1 + \lambda_2 \alpha_2\, d\alpha_2 + \cdots &= 0 \\ \alpha_1\, d\alpha_1 + \alpha_2\, d\alpha_2 + \cdots &= 0, \end{aligned} \tag{20.54}$$

where $\alpha_j \equiv |a_j|$. With the help of a Lagrange multiplier we see (Exercise 20.34) that extrema of Rayleigh's quotient occur when $\alpha_1 \neq 0, \alpha_2 = \alpha_3 = \cdots = 0$, when $\alpha_2 \neq 0$, $\alpha_1 = \alpha_3 = \alpha_4 = \cdots = 0$, and so on and that the values of these extrema are λ_1, $\lambda_2, \ldots$ in turn. That is, *the extremals of R occur at $x = e_j$, at which $R = \lambda_j$.*

---

[11] The fact that the eigens are *denumerable* follows from the assumption of completeness.

In particular, (20.53) can be replaced by the more informative statement

$$\max_{x \neq \theta} \left| \frac{(Lx, x)}{(x, x)} \right| = |\lambda_1| \tag{20.55}$$

or recalling the definition of the norm of L,

$$|\lambda_1| = \|L\|, \tag{20.56}$$

and therefore λ_1 is either $+\|L\|$ or $-\|L\|$.

Example 20.9. Consider

$$A = \begin{pmatrix} 2 & 0 & 0 \\ 0 & 1 & 1 \\ 0 & 1 & 1 \end{pmatrix}$$

of Example 20.2. Suppose that we don't know the eigens and would like an estimate of λ_1. If the problem comes from some physical context, say, we may have some intuitive feeling for the eigenvectors. Suppose that our intuition tells us that $e_1 \approx (1, 2, 3)^T$ (not normalized) or that we simply take it as a guess. Inserting it into R gives $R = 27/14 = 1.97$ as our estimate of λ_1. It is striking that this result is so close to the true value $\lambda_1 = 2$ even though our estimate of e_1 is not all that close to $e_1 = (\alpha, \beta, \beta)^T$. ∎

In fact, it turns out (Exercise 20.33) that *if x is within $O(\epsilon)$ of e_k, then $R(x)$ is within $O(\epsilon^2)$ of λ_k!*[12]

Finally, we consider the so-called **power method** for the *iterative* calculation of eigens. Starting with an initial guess $x^{(0)}$, we simply compute $x^{(1)} = Lx^{(0)}$, $x^{(2)} = Lx^{(1)}$, and so on. Before analyzing the situation, however, let us carry out the calculation and see what happens.

Example 20.10. To illustrate, let L be the matrix

$$A = \begin{pmatrix} 0 & 0 & 1 \\ 0 & 0 & 1 \\ 1 & 1 & 1 \end{pmatrix}. \tag{20.57}$$

Starting with $x^{(0)} = (1, 0, 0)^T$, say, we have

$$\begin{matrix} & x^{(0)} & x^{(1)} & x^{(2)} & x^{(3)} & x^{(4)} & x^{(5)} & x^{(6)} & x^{(7)} & x^{(8)} & x^{(9)} \\ \begin{pmatrix} 0 & 0 & 1 \\ 0 & 0 & 1 \\ 1 & 1 & 1 \end{pmatrix} & \begin{pmatrix} 1 \\ 0 \\ 0 \end{pmatrix} & \begin{pmatrix} 0 \\ 0 \\ 1 \end{pmatrix} & \begin{pmatrix} 1 \\ 1 \\ 1 \end{pmatrix} & \begin{pmatrix} 1 \\ 1 \\ 3 \end{pmatrix} & \begin{pmatrix} 3 \\ 3 \\ 5 \end{pmatrix} & \begin{pmatrix} 5 \\ 5 \\ 11 \end{pmatrix} & \begin{pmatrix} 11 \\ 11 \\ 21 \end{pmatrix} & \begin{pmatrix} 21 \\ 21 \\ 43 \end{pmatrix} & \begin{pmatrix} 43 \\ 43 \\ 85 \end{pmatrix} & \begin{pmatrix} 85 \\ 85 \\ 171 \end{pmatrix} \end{matrix} \tag{20.58}$$

and so on. We are apparently converging to an eigenvector because successive $x^{(j)}$'s are tending to multiples of each other. Thus

$$\lambda \approx \frac{85}{43} = 1.9767, \qquad e \approx \begin{pmatrix} 85 \\ 85 \\ 171 \end{pmatrix}.$$

---

[12]By "x within $O(\epsilon)$ of e_k," we mean $a_j = O(\epsilon)$ for all $j \neq k$, and $a_k = O(1)$ in $x = \sum a_j e_j$.

We can probably do much better by applying an Aitken extrapolation (Section 2.5) to the last few iterates:

$$\lambda \approx \frac{\lambda^{(9)}\lambda^{(7)} - \lambda^{(8)2}}{\lambda^{(9)} + \lambda^{(7)} - 2\lambda^{(8)}} = \frac{(\frac{85}{43})(\frac{21}{11}) - (\frac{43}{21})^2}{\frac{85}{43} + \frac{21}{11} - 2(\frac{43}{21})} = 2.0007. \tag{20.59}$$

Scaling $x^{(7)}, x^{(8)}, x^{(9)}$ so that their first components are unity, for convenience, let us extrapolate the resulting vectors,

$$\begin{pmatrix} 1 \\ 1 \\ 43/21 \end{pmatrix}, \begin{pmatrix} 1 \\ 1 \\ 85/43 \end{pmatrix}, \begin{pmatrix} 1 \\ 1 \\ 171/85 \end{pmatrix}.$$

We have

$$1,\ 1,\ 1 \longrightarrow 1$$

$$1,\ 1,\ 1 \longrightarrow 1$$

$$\frac{43}{21},\ \frac{85}{43},\ \frac{171}{85} \longrightarrow \frac{(\frac{171}{85})(\frac{43}{21}) - (\frac{85}{43})^2}{\frac{171}{85} + \frac{43}{21} - 2(\frac{85}{43})} = 2.0002,$$

and hence

$$e \approx \begin{pmatrix} 1 \\ 1 \\ 2.0002 \end{pmatrix}. \tag{20.60}$$

In fact, the exact eigens are

$$\lambda_1 = 2, \quad e_1 = \begin{pmatrix} 1 \\ 1 \\ 2 \end{pmatrix}, \qquad \lambda_2 = -1, \quad e_2 = \begin{pmatrix} 1 \\ 1 \\ -1 \end{pmatrix}, \qquad \lambda_3 = 0, \quad e_3 = \begin{pmatrix} 1 \\ -1 \\ 0 \end{pmatrix}$$

so that the iteration apparently converges to λ_1, e_1. ∎

To understand what has happened, let us expand $x^{(0)}$ in terms of the orthogonal basis $e_1, e_2, \ldots,$

$$x^{(0)} = \sum \alpha_j e_j. \tag{20.61}$$

(Of course, we are not able to evaluate the α_j's, since we don't know what the e_j's are yet, but this fact will not hinder our analysis.) Then

$$x^{(1)} = Lx^{(0)} = \sum \alpha_j L e_j = \sum \lambda_j \alpha_j e_j$$

$$x^{(2)} = Lx^{(1)} = \sum \lambda_j^2 \alpha_j e_j$$

$$x^{(n)} = \sum \lambda_j^n \alpha_j e_j = \lambda_1^n \left[\alpha_1 e_1 + \left(\frac{\lambda_2}{\lambda_1}\right)^n \alpha_2 e_2 + \left(\frac{\lambda_3}{\lambda_1}\right)^n \alpha_3 e_3 + \cdots \right] \sim \lambda_1^n \alpha_1 e_1 \tag{20.62}$$

as $n \to \infty$, since $(\lambda_2/\lambda_1)^n \to 0$, $(\lambda_3/\lambda_1)^n \to 0, \ldots$; consequently, $x^{(n)} \to e_1$ to within some scale factor. The smaller the ratio λ_2/λ_1, the faster the convergence. (See Exercise 20.38.)

Conceivably, $x^{(0)}$ might just happen to be exactly orthogonal to e_1. Then $\alpha_1 = 0$ and we will converge to e_2 instead of e_1. Of course, this situation is unlikely, and it's easy to arrange a procedure that will ensure that we converge to e_1 (Exercise 20.39).

How might λ_2, e_2 and successive eigens be found? Having found e_1 (or at least a close approximation to it), we can start over with an $x^{(0)}$ that is orthogonal to the calculated e_1. This time we will converge to e_2. Actually, even though α_1 is *extremely* small, it is probably not *exactly* zero; therefore after approaching e_2, the iterates will eventually start to move away and head toward $\lambda_1^n \alpha_1 e_1$ again. Such behavior is similar to that of an asymptotic series (Section 2.6), and the idea is to cut off before we start to turn away from e_2.

There are more efficient ways of pursuing the successive eigens, ways that reduce the order of the system by one at each stage. In fact, many important aspects of the eigenvalue problem are beyond our present scope. See, for example, the Additional References for Part III.

EXERCISES

20.1. Find the eigens for each of the following problems, and discuss them in connection with Theorems 20.1 and 20.2.

(a) $Ax = \lambda x$, where $A = \begin{pmatrix} 1 & 3 \\ 3 & 1 \end{pmatrix}$

(b) $Ax = \lambda x$, where $A = \begin{pmatrix} 1 & 1 \\ 0 & 1 \end{pmatrix}$

(c) $Ax = \lambda x$, where $A = \begin{pmatrix} 2 & 1 & 0 \\ 1 & 2 & 0 \\ 0 & 0 & 3 \end{pmatrix}$

(d) $Ax = \lambda x$, where $A = \begin{pmatrix} 1 & -2i \\ i & 0 \end{pmatrix}$. First, use the usual inner product $(x, y) = \xi_1\bar{\eta}_1 + \xi_2\bar{\eta}_2$ and then repeat the calculation using $(x, y) = \xi_1\bar{\eta}_1 + 2\xi_2\bar{\eta}_2$. *Hint:* Recall Exercise 18.21.

*(e) $\int_0^1 t\tau x(\tau)\, d\tau = \lambda x(t)$

*(f) $\int_0^1 x(\tau)\, d\tau = \lambda x(t)$

*(g) $\int_0^\pi \cos(\tau - t) x(\tau)\, d\tau = \lambda x(t)$

*(h) $\int_0^t x(\tau)\, d\tau = \lambda x(t)$

Note: Recall Exercise 18.22.

20.2. Do the eigens of a given operator depend on the choice of the inner product? Explain.

20.3. Do the eigens of the matrix eigenvalue problem $Ax = \lambda x$ stay the same if we change the *order* of the scalar equations? Explain.

20.4. Expand $x = (2, 0, 1)^T$ in terms of the normalized eigenvectors of

$$A = \begin{pmatrix} 0 & 0 & 1 \\ 0 & 0 & 1 \\ 1 & 1 & 1 \end{pmatrix}.$$

20.5. (*Buckling*) Consider the buckling of a column with "stiffness" EI, where E and I are physical constants. Suppose that the column is of length l and is cantilevered at its base; that is, $y'(0) = 0$, where $y(x)$ is the deflected shape (Fig. 20.7). Although we

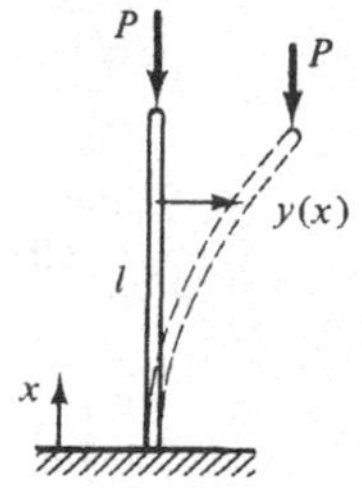

Figure 20.7. Buckling of a column.

could pursue an energy approach, it is easier to use Euler beam theory, which says that

$$EIy'' = P[y(l) - y(x)], \qquad y(0) = y'(0) = 0$$

or differentiating to get rid of $y(l)$,

$$y''' + \lambda y' = 0, \qquad y(0) = y'(0) = y''(l) = 0,$$

where $\lambda \equiv P/EI$. (Actually, it is not of the form $Ly + \lambda y = 0$ but rather $L_1 y + \lambda L_2 y = 0$, and so we call it a *generalized* eigenvalue problem.) Find the eigenvalues and corresponding eigenfunctions or "buckling modes." In particular, show that the critical (i.e., smallest possible) buckling load is $P_{cr} = \pi^2 EI/4l^2$ (which says that it is much easier to buckle a long column than a short one, as we know from experience). Sketch the corresponding buckling mode.

20.6. Consider the same problem as in Exercise 20.5, but let us also constrain the top of the column laterally by a spring of stiffness k that is undeflected when $y(l) = 0$ (Fig. 20.8). According to the Euler beam theory, then,

$$EIy'' = P[y(l) - y(x)] - ky(l)(l - x), \qquad y(0) = y'(0) = 0.$$

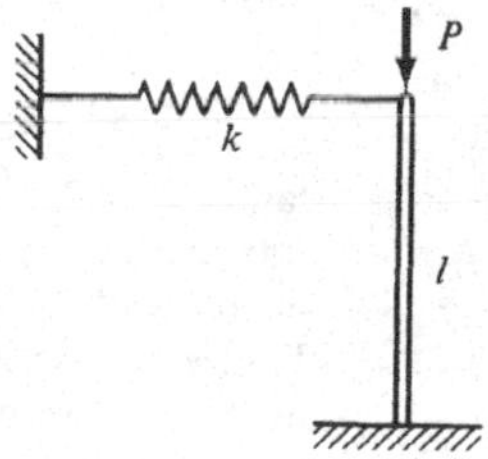

Figure 20.8. Buckling of a constrained column.

Eliminating $y(l)$ from the differential equation, show that

$$y'''' + \lambda y'' = 0$$

$$y(0) = y'(0) = 0, \qquad y''(l) = 0, \qquad y'''(l) = -\lambda y'(l) + \kappa y(l),$$

where $\lambda \equiv P/EI$ and $\kappa \equiv k/EI$. Note that in this generalized eigenvalue problem λ appears in the boundary conditions as well. Show that the eigenvalues are roots of the transcendental equation

$$(\Lambda^2 - \kappa l)\Lambda \cos \Lambda l + \kappa \sin \Lambda l = 0, \qquad (20.63)$$

where $\Lambda \equiv \sqrt{\lambda}$, and that the corresponding buckling modes are (to within an arbitrary scale factor)

$$y = \sin \Lambda x - \tan \Lambda l \cos \Lambda x - \Lambda x + \tan \Lambda l.$$

Clearly, the solution of (20.63) is difficult. But if κ is small (i.e., if the spring stiffness is small compared to the lateral stiffness of the column), we might seek Λ in the form

$$\Lambda = \Lambda_0 + \kappa\Lambda_1 + \kappa^2\Lambda_2 + \cdots, \tag{20.64}$$

where the Λ_j's are found by expanding the left side of (20.63) in powers of κ and equating coefficients of each power of κ to zero. If κ is truly small, then the first two or three terms of (20.64) should suffice. Carry this process out as far as Λ_1 and thus determine the critical buckling load [through $O(\kappa)$]. The expansion procedure (20.64) is known as a *perturbation* method. Perturbation techniques will be developed in Parts IV and V.

20.7. Show that the operator (20.6) is self-adjoint with respect to the inner product (20.5) as claimed.

20.8. Verify that if $p(a) = p(b)$, then the "separated" conditions (20.7b) may be replaced by "periodic" conditions $y(a) = y(b)$ and $y'(a) = y'(b)$ in the Sturm–Liouville theory.

20.9. Show that even if the differential operator $L = a(x)\, d^2/dx^2 + b(x)\, d/dx + c(x)$ is *not* formally self-adjoint, σL *will* be if we choose

$$\sigma = \exp\left\{\int \frac{b - a'}{a}\, dx\right\}. \tag{20.65}$$

Thus convert the system

$$xy'' - y' + y + \lambda y = 0 \qquad (1 \le x \le 2)$$

$$y(1) + 2y'(1) = 0, \qquad y(2) = 0$$

to the Sturm–Liouville form (20.7).

20.10. Show that the eigenfunctions of the Sturm–Liouville system

$$y'' + \lambda y = 0, \qquad y(0) = 0, \quad y(1) - 2y'(1) = 0$$

are $\sin \sqrt{\lambda_n}\, x$, where the eigenvalues are the solutions of the transcendental equation $\tan \sqrt{\lambda} = 2\sqrt{\lambda}$ except for $\lambda = 0$. By a freehand graphical examination of the transcendental equation, show that $\lambda_n \sim (2n - 1)^2\pi^2/4$ as $n \longrightarrow \infty$. Expand $f(x) = x$ in terms of the eigenfunctions. Note that the expansion differs from the "usual" Fourier sine series in that it is not harmonic; that is, the frequencies are not integer multiples of some fundamental frequency.

20.11. Find the eigenfunctions of the Sturm–Liouville system

$$y'' + \lambda y = 0, \qquad y(0) = y'(0), \quad y(1) = 0$$

as well as the transcendental equation for the eigenvalues. Show that $\lambda_n \sim (2n - 1)^2\pi^2/4$ as $n \longrightarrow \infty$. Expand $f(x) = x$ in terms of the eigenfunctions. (The Fourier coefficients may be left in integral form.)

20.12. (a) Determine the eigens for the Sturm-Liouville problem

$$y'' + \lambda y = 0, \qquad y'(0) = y\left(\frac{\pi}{2}\right) = 0.$$

(b) Expand $f(x)$ shown in Fig. 20.9 in terms of the eigenfunctions found in part (a).

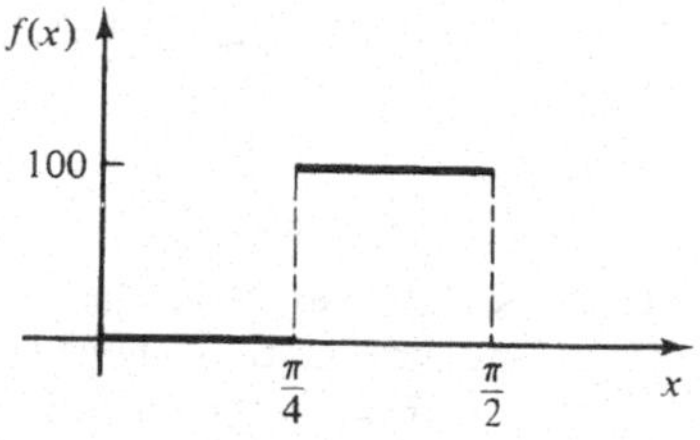

Figure 20.9. $f(x)$ of Exercise 20.12.

(c) Obtain computer plots of the partial sums

$$s_n(x) = \sum_{j=1}^{n} \alpha_j e_j(x)$$

for $n = 4, 16, 64$.

20.13. Recall that for an n-dimensional matrix operator A the characteristic equation is of the polynomial form

$$\det(A - \lambda I) = (-1)^n \lambda^n + a_1 \lambda^{n-1} + \cdots + a_n = 0.$$

According to the *Fundamental Theorem of Algebra*, this equation must have at least one root, and so there *must* exist at least one eigenvalue. For the infinite-dimensional case, this situation is not necessarily true.

(a) Show that

$$y'' + \lambda y = 0, \qquad 2y(0) - y(1) + 4y'(1) = 0, \quad y(0) + 2y'(1) = 0$$

(which is *not* a Sturm–Liouville system) has *no* eigenvalues.

(b) By contrast, show that for

$$y'' + \lambda y = 0, \qquad y(0) - y(1) = 0, \quad y'(0) + y'(1) = 0$$

every λ (real or complex) is an eigenvalue!

20.14. Find the eigens for the problem $y'' + \lambda y = 0$ over $-\infty < x < \infty$, subject to the conditions that $y(-\infty)$ and $y(\infty)$ are both finite. Do you think expansions in terms of the eigenfunctions are possible? Explain.

20.15. In Comment 3 following (20.25) we stated that $x \neq \Theta$ is equivalent to $x' \neq \Theta$. Why is that true?

20.16. The columns of the Q matrix (20.33) are not mutually orthogonal. Show that this fact implies that the x'_1, x'_2, x'_3 coordinates will *not* be cartesian.

20.17. Show that the original cartesian variables x_1, x_2 are related to the rotated variables x'_1, x'_2, in Fig. 20.10, according to

$$\begin{pmatrix} x_1 \\ x_2 \end{pmatrix} = \begin{pmatrix} \cos\alpha & -\sin\alpha \\ \sin\alpha & \cos\alpha \end{pmatrix} \begin{pmatrix} x'_1 \\ x'_2 \end{pmatrix}$$

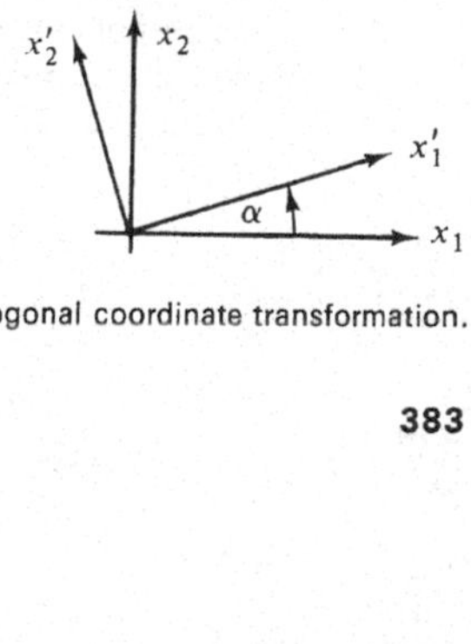

Figure 20.10. Orthogonal coordinate transformation.

or $x = Ax'$, and show that the transformation matrix

$$A = \begin{pmatrix} \cos\alpha & -\sin\alpha \\ \sin\alpha & \cos\alpha \end{pmatrix}$$

is orthogonal.

20.18. Reduce the following quadratic forms to canonical form by means of a modal matrix; are they positive definite?

(a) $F(x_1, x_2) = 4x_1^2 - x_1x_2 + x_2^2$

(b) $F(x_1, x_2, x_3) = x_1^2 - 8x_1x_2 + 4x_1x_3$

(c) $F(x_1, x_2, x_3) = 2x_1x_3 + 2x_2x_3 + x_3^2$

20.19. Show that a unitary transformation Q leaves inner products intact—that is, $(x', y') = (x, y)$ for all x and y, where $x' = Qx$ and $y' = Qy$.

20.20. (*Normal modes*) Show that the equations of motion of the three-mass system shown in Fig. 20.11 are, for $m = k = 1$,

$$\ddot{x}_1 + 2x_1 - x_2 = 0, \qquad \ddot{x}_2 - x_1 + 2x_2 - x_3 = 0, \qquad \ddot{x}_3 - x_2 + 2x_3 = 0.$$

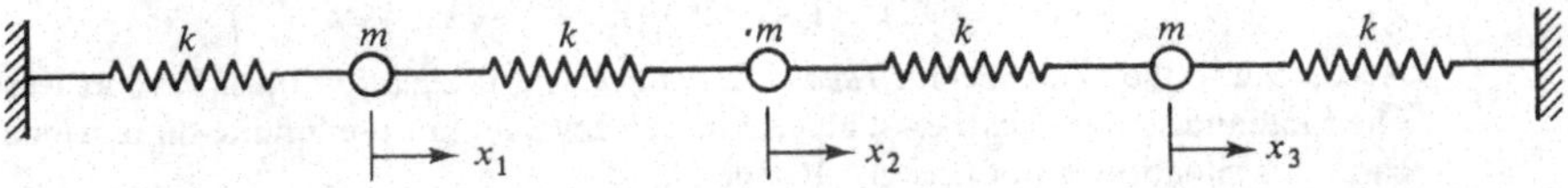

Figure 20.11. Three-mass system.

Uncouple these equations through the introduction of normal coordinates and obtain the solution as a linear combination of normal modes, as for the two-mass system of Example 20.7.

20.21. (*Generalized eigenvectors*) So far we've emphasized matrices that are self-adjoint. Consider the *non*-self-adjoint matrix

$$A = \begin{pmatrix} 1 & 1 & 2 & 0 \\ 0 & 1 & 3 & 0 \\ 0 & 0 & 2 & 2 \\ 0 & 0 & 0 & 1 \end{pmatrix}.$$

(a) Show that $\lambda_1 = \lambda_2 = \lambda_3 = 1$, $\lambda_4 = 2$, and

$$e_1 = \begin{pmatrix} 1 \\ 0 \\ 0 \\ 0 \end{pmatrix}, \qquad e_4 = \begin{pmatrix} 5 \\ 3 \\ 1 \\ 0 \end{pmatrix}.$$

If A were self-adjoint, we would be able to find three orthogonal eigenvectors corresponding to the triple eigenvalue, but it's not and we can't. But we *can* find so-called **generalized eigenvectors** e_2, e_3 such that e_1, e_2, e_3, e_4 are at least *LI*. We introduce e_2, e_3 as follows:

$$(A - \lambda_1 I)e_1 = 0$$
$$(A - \lambda_1 I)e_2 = e_1 \quad \text{or} \quad (A - \lambda_1 I)^2 e_2 = 0$$
$$(A - \lambda_1 I)e_3 = e_2 \quad \text{or} \quad (A - \lambda_1 I)^3 e_3 = 0.$$

(b) Show that $e_1, \ldots, e_4$ are necessarily *LI* (and hence span the space).
(c) Determine e_2, e_3 for this example.

20.22. (*Jordan canonical form*) A self-adjoint matrix A can always be diagonalized by the transformation $\bar{Q}^T A Q$, where Q is a modal matrix of A (and $\bar{Q}^T = Q^{-1}$). But *any* square matrix can be *almost* diagonalized, more specifically, *triangularized*, as follows: Let P be a matrix whose columns are the eigenvectors and generalized eigenvectors (see Exercise 20.21) e_j of A, not necessarily normalized. Since the e_j's are LI, it follows that det $P \neq 0$ so that P^{-1} exists. If we denote the rows of P^{-1} as r_j, then, of course, $(r_i, e_j) = \delta_{ij}$. The matrix $P^{-1}AP$ will be in **Jordan canonical form**; that is, the main diagonal will consist of the λ_j's, and the off-diagonal elements will all be zero except for the elements immediately above repeated λ's, which may be unity. If, for example, $\lambda_5 = \lambda_6 = \lambda_7$, then there will be ones immediately above the λ_6, λ_7 elements. Verify these claims for the A matrix of Exercise 20.21.

***20.23.** Uncouple the following systems of differential equations insofar as possible by means of a linear transformation, using the ideas of Example 20.7, the starred section following Example 20.7, and Exercises 20.21 and 20.22.

(a)
$$\begin{aligned} \ddot{x}_1 + 5x_1 - 2x_2 &= 0 \\ \ddot{x}_2 - 2x_1 + 2x_2 &= 0 \end{aligned}$$

(b)
$$\begin{aligned} \ddot{x}_1 + 2x_1 - x_2 &= 0 \\ \ddot{x}_2 + 4x_1 + 6x_2 &= 0. \end{aligned}$$

20.24. Find the principal inertias and principal axes for the body shown in Fig. 20.12. It consists of a thin bent wire ABC of unit mass per unit length with a concentrated mass $m = 2$ attached at C, all of which lies in the x, y plane.

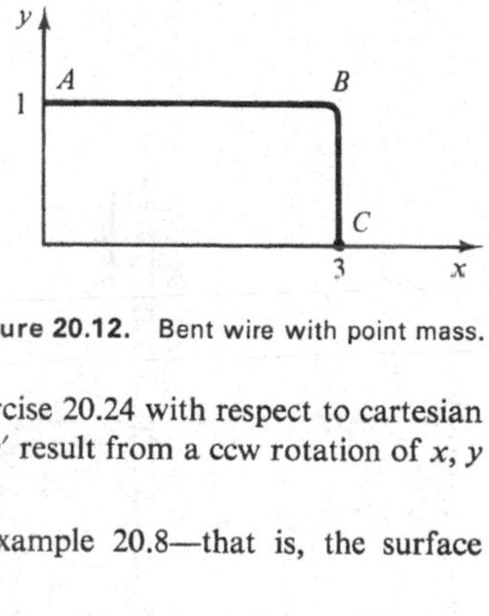

Figure 20.12. Bent wire with point mass.

20.25. Find the inertia components of the body in Exercise 20.24 with respect to cartesian x', y', z' axes, where z' coincides with z and x', y' result from a ccw rotation of x, y through an angle α.

20.26. Sketch the quadric surface associated with Example 20.8—that is, the surface $F(x, y, z) = \text{constant} = 1$, say.

20.27. (*Cayley-Hamilton theorem*) (a) Show that if λ is an eigenvalue of A, then λ^n is an eigenvalue of A^n.

(b) Suppose that A is an $n \times n$ self-adjoint matrix with characteristic equation

$$a_n\lambda^n + \cdots + a_1\lambda + a_0 = 0.$$

Show that

$$(a_nA^n + \cdots + a_1A + a_0I)x = \mathcal{O}$$

for an arbitrary x, so that

$$a_nA^n + \cdots + a_1A + a_0I = \mathcal{O};$$

that is *A satisfies its own characteristic equation*. Known as the **Cayley–Hamilton theorem,** it can be shown to be valid for *any* square matrix.

(c) Use the Cayley–Hamilton theorem to show that if

$$A = \begin{pmatrix} 1 & 2 \\ 2 & 1 \end{pmatrix},$$

then

$$A^2 = 2A + 3I, \qquad A^3 = 7A + 6I,$$

$$A^4 = 20A + 21I, \text{ etc.},$$

and

$$A^{-1} = \tfrac{1}{3}A - \tfrac{2}{3}I.$$

*(d) Show that the Cayley–Hamilton theorem is valid for *any* square matrix. *Hint:* Recall Exercise 20.21.

20.28. For an $n \times n$ matrix A, show that $\lambda_1 + \lambda_2 + \cdots + \lambda_n = \sum_j a_{jj}$ (which is called the *trace* of A) and that $\lambda_1 \lambda_2 \ldots \lambda_n = \det A$.

20.29. Matrices A and B are said to be **similar** if $B = P^{-1}AP$, where P is simply nonsingular. Show that similar matrices have the same characteristic equation (and hence the same eigenvalues). *Hint:* Recall Exercise 20.27.

20.30. (*Generalized eigenvalue problem*) If $B \neq I$, $Ax = \lambda Bx$ is called a **generalized eigenvalue problem.**

(a) Show that if A and B are self-adjoint, then eigenvectors e_i and e_j are orthogonal relative to B—that is, $(e_i, Be_j) = 0$ for $i \neq j$, provided that $\lambda_i \neq \bar{\lambda}_j$.

(b) Three rigid rods, each of length l, are pinned at their ends, with the end A constrained to move in a frictionless vertical slot, as shown in Fig. 20.13. The three lateral springs, each of stiffness k, are unstretched and uncompressed when the configuration is undeflected (i.e., $x = y = z = 0$), and the middle spring is

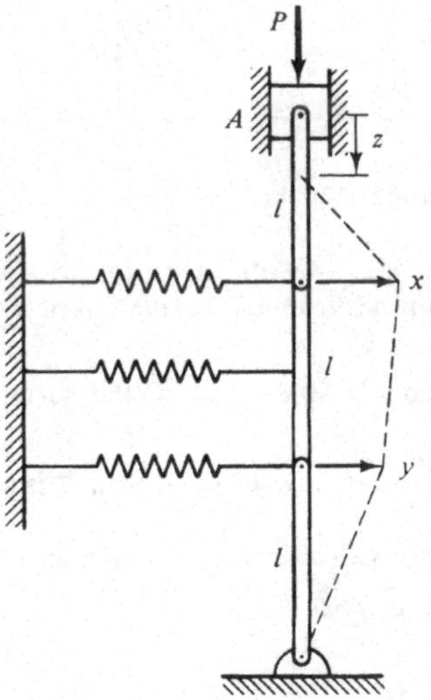

Figure 20.13. Buckling of a system of rods.

attached at the middle of the middle rod. Find the critical buckling load P in terms of k and l and the corresponding buckling mode—that is, the ratio of x to y. Neglect the weight of the rods. *Hint:* Regarding the force P as a *weight* applied to the point A, write down an expression for the total potential energy $V[x, y, z(x, y)]$. Setting $\partial V/\partial x = \partial V/\partial y = 0$ and linearizing in x and y, obtain the generalized eigenvalue problem

$$\begin{pmatrix} 5 & 1 \\ 1 & 5 \end{pmatrix}\begin{pmatrix} x \\ y \end{pmatrix} = \lambda\begin{pmatrix} 8 & -4 \\ -4 & 8 \end{pmatrix}\begin{pmatrix} x \\ y \end{pmatrix}, \tag{20.66}$$

where $\lambda = P/kl$.

(c) Demonstrate the truth of the orthogonality principle $(e_i, Be_j) = 0$ for $i \neq j$ stated in part (a) for the system (20.66).

20.31. (a) Solve $Ax = \Lambda x + c$ (if possible), where

$$A = \begin{pmatrix} 2 & 1 \\ 1 & 2 \end{pmatrix}, \qquad c = \begin{pmatrix} 2 \\ -2 \end{pmatrix}$$

for $\Lambda = 1, 2$, and 3, by means of eigenvector expansions as outlined in Section 20.5.

(b) Re-solve the problem of part (a), using the basis

$$e_1 = \begin{pmatrix} 1 \\ 0 \end{pmatrix}, \qquad e_2 = \begin{pmatrix} 0 \\ 1 \end{pmatrix}$$

say, rather than the basis consisting of the eigenvectors of A. (Recall Comment 2 of Section 20.5.)

20.32. (a) Apply the ideas of Section 20.5 to the problem

$$y'' + y = F(x), \qquad y(0) = y'(\pi) = 0,$$

where $F(x)$ is in $\mathfrak{L}_2(0, \pi)$.

(b) Repeat with the boundary conditions changed to $y(0) = y(\pi) = 0$.

20.33. Show that if x is within $O(\epsilon)$ of e_k, then $R(x)$ is within $O(\epsilon^2)$ of λ_k as claimed following Example 20.9.

20.34. Applying the Lagrange multiplier method to (20.54), show that the extrema of R are $\lambda_1, \lambda_2, \ldots$, as claimed.

20.35. If A is positive definite, show that A^{-1} must be positive definite, too.

20.36. If M is a self-adjoint $n \times n$ matrix, show that the Neumann series $(I + M)^{-1} = I - M + M^2 - M^3 + \cdots$ converges if the largest eigenvalue of M is smaller than unity in magnitude. *Hint:* Recall that if Q is a modal matrix of M, then $Q^{-1}MQ = D$ is diagonal.

20.37. Seek λ_2, e_2 for the matrix (20.57) by means of iteration, by starting with an $x^{(0)}$ that is orthogonal to (20.60), say $x^{(0)} = (0, -2.0002, 1)^T$. If possible, run this calculation on a digital computer for a large number of iterations and observe how $x^{(n)}$ eventually moves away from e_2 and heads back toward e_1.

20.38. The more λ_1 dominates λ_2, the faster the convergence of the iterative method. Show how we can therefore speed convergence by raising A to some power before starting. [Recall Exercise 20.27(a)]. In Example 20.10, for instance, apply the iterative procedure to A^4, starting with $x^{(0)} = (1, 0, 0)^T$ again.

20.39. If $x^{(0)}$ in Example 20.10 happened to be orthogonal to e_1, then (20.60) would be e_2, not e_1. Suggest a simple procedure that will eliminate all doubt.

20.40. What happens to the iterative procedure if $\lambda_2 = \lambda_1$? If $\lambda_2 = -\lambda_1$? Show how these situations can be detected and how $\lambda_1, e_1, \lambda_2, e_2$ can still be found. Illustrate your ideas with suitable 2×2 matrix examples.

20.41. Show that matrices A and A^T have the same eigenvalues. Are their eigenvectors the same, too? How about A and A^{-1}?

20.42. The eigens of the differential operator

$$L = -\frac{d^2}{dx^2}, \qquad y(0) = y(\pi) = 0,$$

that is, the nontrivial solutions of $y'' + \lambda y = 0$, $y(0) = y(\pi) = 0$—are $\lambda = n^2 = 1, 4, 9, \ldots$, $y = \sin nx$. Since there is no largest λ, the iterative scheme

$$y^{(j+1)} = Ly^{(j)}$$

is bound to fail. However, the *inverse* operator has eigenvalues $\lambda = 1/n^2 = 1, 1/4, 1/9, \ldots$ (Exercise 20.41), and so we expect the scheme $y^{(j+1)} = L^{-1}y^{(j)}$—that is, $Ly^{(j+1)} = y^{(j)}$, or

$$-y^{(j+1)''} = y^{(j)}, \qquad y^{(j+1)}(0) = y^{(j+1)}(\pi) = 0$$

—to converge to $\lambda = 1$, $y = \sin x$. Starting with $y^{(0)} = x(\pi - x)$, carry out two iterations and compare your results (numerically) with $\lambda = 1$, $y = \sin x$.

20.43. (*Rayleigh–Ritz*) Recalling Exercise 10.28, note that the eigenvalue problem

$$y'' + \lambda xy = 0, \qquad y(0) = y(1) = 0$$

is equivalent to the variational principle

$$\int_0^1 (y'^2 - \lambda xy^2)\, dx = \text{extremum}$$
$$y(0) = y(1) = 0.$$

Applying the Ritz method (in connection with *eigenvalue* problems, it is generally called the Rayleigh–Ritz method), seek

$$y(x) \approx x(1 - x)(c_1 + c_2x + c_3x^2 + \cdots).$$

(a) Using the one-term approximation $y \approx x(1 - x)c_1$, show that

$$\left(-\frac{1}{3} + \frac{\lambda}{60}\right)c_1 = 0,$$

so that $\lambda = 20$. (The exact eigenvalues are known to be $\lambda_1 \doteq 18.9$, $\lambda_2 \doteq 81.8, \ldots$.)

(b) Using the two-term approximation $y \approx x(1 - x)(c_1 + c_2x)$, show that

$$\left(-\frac{1}{3} + \frac{\lambda}{60}\right)c_1 + \left(-\frac{1}{6} + \frac{\lambda}{105}\right)c_2 = 0$$
$$\left(-\frac{1}{6} + \frac{\lambda}{105}\right)c_1 + \left(-\frac{2}{15} + \frac{\lambda}{168}\right)c_2 = 0$$

and thus obtain an improved estimate of λ_1, as well as a first estimate of λ_2.

20.44. Make up two examination-type problems on this chapter.

ADDITIONAL REFERENCES FOR PART III

A. C. AITKEN, *Determinants and Matrices*, Wiley-Interscience, New York, 1956.

R. ARIS, *Vectors, Tensors and Basic Equations of Fluid Mechanics*, Prentice-Hall, Englewood Cliffs, N.J., 1962. [Application of tensor analysis to fluid mechanics.]

H. D. BLOCK, Periodic Solutions of Forced Systems Having Hysteresis, *IRE Transactions of the Professional Group on Circuit Theory*, Vol. CT–7, No. 4, December 1960, pp. 423–431. [This paper considers the driven mechanical oscillator $\ddot{x} + \mathfrak{F} = P(t)$, where the restoring force $\mathfrak{F}$ entails hysteresis effects and hence is not a (single-valued) function but rather a mapping. Author first introduces the reader to Banach spaces and then recasts problem in terms of mappings on Banach spaces. Good, readable example of the practical application of the methods of functional analysis to an engineering problem.]

L. BRILLOUIN, *Tensors in Mechanics and Elasticity*, Academic, New York, 1964. [Application of tensor analysis to the theory of elasticity.]

B. CARNAHAN, H. LUTHER, and J. WILKES, *Applied Numerical Methods*, Wiley, New York, 1969. [Of interest here because of the considerable overlap of numerical methods and functional analysis.]

L. COLLATZ, *Functional Analysis and Numerical Mathematics*, Academic, New York, 1966. [An extremely readable introduction to functional analysis, together with important applications to iterative methods and monotone operators. Includes numerous illustrative examples.]

J. INDRITZ, *Methods in Analysis*, Macmillan, New York, 1963. [See Chapter 4 on integral operators.]

A LICHNEROWICZ, *Elements of Tensor Calculus*, 4th ed., Methuen, London, 1962.

L. LIUSTERNIK and V. SOBOLEV, *Elements of Functional Analysis*, Frederick Ungar Publishing Co., New York, 1961.

S. G. MIKHLIN, *Integral Equations*, Pergamon Press, New York, 1964. [An excellent book, oriented heavily toward applications in fluid and solid mechanics. Includes singular integral equations as well.]

A. F. MONNA, *Functional Analysis in Historical Perspective*, Wiley, New York, 1973. [An interesting historical account.]

A. NAYLOR and G. SELL, *Linear Operator Theory in Engineering and Science*, Holt, Rinehart and Winston, New York, 1971. [A readable and motivated introduction to functional analysis.]

I. STAKGOLD, *Boundary Value Problems of Mathematical Physics*, Macmillan, New York, 1967. [A clear introduction to linear spaces, linear integral equations, and spectral theory of second-order differential operators.]

PART IV

ORDINARY DIFFERENTIAL EQUATIONS

Chapters 21 and 22 explore some theoretical questions—for instance, existence and uniqueness—but deal mostly with methods of solution of FIRST-ORDER EQUATIONS and HIGHER-ORDER SYSTEMS. Special techniques include the methods of Green's functions, successive approximations, and quasilinearization. Chapter 23 on QUALITATIVE METHODS; THE PHASE PLANE deals with second-order autonomous equations and is complemented by the discussion of QUANTITATIVE METHODS in Chapter 24, which includes the methods of invariant imbedding, weighted residuals, and finite elements. Finally, we look at PERTURBATION TECHNIQUES in Chapter 25, partly for the "regular" case but mostly for "singular" perturbations, including straining and boundary layer concepts.

Chapter 21

First-Order Equations

The first part of this chapter amounts to a brief review of the standard methods of solution of first-order, ordinary differential equations, as discussed (in more detail) in Agnew,[1] Boyce and DiPrima,[2] and Spiegel,[3] for example. The second part is devoted to the fundamental questions of existence and uniqueness.

21.1. STANDARD METHODS OF SOLUTION

The general first-order, ordinary differential equation, with x as independent variable and y as dependent variable, is of the form

$$g(x, y, y') = 0. \tag{21.1}$$

It is an *ordinary* differential equation because there is only one independent variable, and it is of *first order* because the order of the highest derivative that appears is one. A function $y = \phi(x)$ is called a *solution* or *integral* of (21.1) if

$$g[x, \phi(x), \phi'(x)] = 0$$

for all x in the interval under consideration. More generally, if the one-parameter family $\phi(x, C)$ is a solution for every value of the parameter C, then we say that

$$y = \phi(x, C) \tag{21.2}$$

---

[1] R. P. Agnew, *Differential Equations*, McGraw-Hill, New York, 1942.

[2] W. Boyce and R. DiPrima, *Elementary Differential Equations and Boundary Value Problems*, 2nd ed., Wiley, New York, 1969.

[3] M. R. Spiegel, *Applied Differential Equations*, Prentice-Hall, Englewood Cliffs, N.J., 1958.

is the *general solution* of (21.1), and *every* solution (with the possible exception of so-called singular solutions, which will be discussed in the next section) is given by (21.2) for some appropriate choice of C.

Assuming that (21.1) can be solved for y', let us consider instead the equation

$$y' = f(x, y). \tag{21.3}$$

Variables Separable. If $f(x, y)$ is of the separable form $f(x, y) = g(x)h(y)$, we can rewrite (21.3) as

$$\frac{dy}{h(y)} = g(x)\,dx$$

and (hopefully) integrate both sides.

Example 21.1.

$$y' = -3x^2 e^y.$$

Then

$$-\int e^{-y}\,dy = \int 3x^2\,dx,$$

or

$$e^{-y} = x^3 + C,$$

where C is the (arbitrary) constant of integration. Solving for y, we have the general solution

$$y = -\ln(x^3 + C). \quad \blacksquare$$

Homogeneous of Degree Zero. We say that a function $F(x_1, \ldots, x_n)$ is *homogeneous of degree k* if for any λ

$$F(\lambda x_1, \ldots, \lambda x_n) = \lambda^k F(x_1, \ldots, x_n).$$

For instance,

$$F(x, y, z) = \frac{x^5}{y^2 + z^2} \sin\left(\frac{y}{z}\right)$$

is homogeneous of degree three because

$$F(\lambda x, \lambda y, \lambda z) = \frac{(\lambda x)^5}{(\lambda y)^2 + (\lambda z)^2} \sin\left(\frac{\lambda y}{\lambda z}\right) = \lambda^3 F(x, y, z),$$

$G(x, y, z) = (x - 2y)/z$ is homogeneous of degree zero, and $H(x, y, z) = xy + z$ is not homogeneous of *any* degree.

In particular, if $f(x, y)$ in (21.3) is homogeneous of degree *zero*, then $f(\lambda x, \lambda y) = \lambda^0 f(x, y) = f(x, y)$ or, choosing $\lambda = 1/x$,

$$y' = f(x, y) = f(\lambda x, \lambda y) = f\left(1, \frac{y}{x}\right) = F\left(\frac{y}{x}\right), \tag{21.4}$$

which suggests the change of variables $y/x \equiv v$ or $y = xv$. Then $y' = v + xv'$ and (21.4) becomes

$$v + xv' = F(v),$$

or

$$v' = \frac{F(v) - v}{x}, \tag{21.5}$$

which is variable separable and can be dealt with accordingly.

Example 21.2.

$$y' = \frac{y}{x} + 3\sqrt{\frac{x}{y}}.$$

Then

$$v + xv' = v + \frac{3}{\sqrt{v}},$$

$$\sqrt{v}\, dv = \frac{3\, dx}{x},$$

$$v = \left(\frac{9}{2} \ln x + C\right)^{2/3},$$

$$y = x\left(\frac{9}{2}\ln x + C\right)^{2/3}. \quad \blacksquare$$

Exact Differentials and Integrating Factors. Instead of (21.3), consider the equivalent form

$$P(x, y)\, dx + Q(x, y)\, dy = 0. \tag{21.6}$$

Implied is a change in point of view; when we write $y' = f(x, y)$, it is clear (from the y' term) that x is the independent variable and y is the dependent variable, whereas in (21.6) x and y have the same "status" and are both considered independent variables.

We say that $P(x, y)\, dx + Q(x, y)\, dy$ is an **exact differential** if (and only if) there exists a function $\phi(x, y)$ such that

$$P(x, y)\, dx + Q(x, y)\, dy = d\phi. \tag{21.7}$$

If so, (21.6) is simply $d\phi = 0$, which yields

$$\phi(x, y) = \text{constant} = C. \tag{21.8}$$

Returning to our original viewpoint, we see that (21.8) determines y *implicitly* as a function of x and is therefore called an *implicit solution.* The prospects of solving for $y(x)$ explicitly will depend on the form of ϕ.

In pursuing this line of approach, we face two questions: Is $P\, dx + Q\, dy$ exact (i.e., an exact differential)? If so, how do we find ϕ? The first is answered by the following theorem.

THEOREM 21.1. (Exactness) If $P(x, y)$ and $Q(x, y)$ have continuous first partials, then a necessary and sufficient condition that $P\, dx + Q\, dy$ is exact is that

$$\frac{\partial P}{\partial y} = \frac{\partial Q}{\partial x}. \tag{21.9}$$

Proof. This result is essentially proved in Section 9.6, and we leave the details for Exercise 21.7.

To answer the second question, note from (21.7) that $P\, dx + Q\, dy = d\phi = \phi_x\, dx + \phi_y\, dy$. Since x and y are independent variables, the increments dx and dy are independent, and it follows that

$$\frac{\partial \phi}{\partial x} = P \tag{21.10a}$$

$$\frac{\partial \phi}{\partial y} = Q, \tag{21.10b}$$

which can be integrated to yield ϕ.

Example 21.3.

$$\sin y \, dx + (2y + x \cos y) \, dy = 0. \tag{21.11}$$

Then both

$$P = \sin y \quad \text{and} \quad Q = 2y + x \cos y$$

have continuous partials and

$$P_y = \cos y, \qquad Q_x = \cos y$$

which means that (21.9) is satisfied. Consequently, the left side of (21.11) *is* exact. To find ϕ, we use (21.10): (21.10a) says that

$$\frac{\partial \phi}{\partial x} = P = \sin y$$

so that[4]

$$\phi = \int \sin y \, \partial x = x \sin y + A(y).$$

Inserting this result into (21.10b) gives

$$x \cos y + A'(y) = 2y + x \cos y,$$

and then

$$A = y^2 + \text{constant}.$$

Thus the solution of (21.11) is given by

$$\phi(x, y) = x \sin y + y^2 + \text{constant} = C$$

or absorbing the "constant" into C

$$x \sin y + y^2 = C. \quad \blacksquare$$

Even if (21.6) is *not* exact (i.e., the left-hand side is not an exact differential), we can seek to *make* it exact by multiplying through by a suitable "*integrating factor*" $\mu(x, y)$—that is,

$$\mu P \, dx + \mu Q \, dy = 0.$$

According to (21.9), we want to choose μ so that

$$(\mu P)_y = (\mu Q)_x$$

or,

$$P\mu_y - Q\mu_x = (Q_x - P_y)\mu. \tag{21.12}$$

Equation (21.12) is a first-order, *partial* differential equation for μ and is beginning to look more difficult than the original problem (21.6).

---

[4]Recall footnote 15 in Section 9.6.

Backing off a bit, let us hope that a suitable μ can be found that is a function of x alone or y alone. First, try $\mu = \mu(x)$. Then $\mu_y = 0$ and (21.12) becomes

$$\frac{d\mu}{\mu} = \left(\frac{P_y - Q_x}{Q}\right) dx, \tag{21.13}$$

where, for consistency with our assumption that $\mu = \mu(x)$, we need

$$\frac{P_y - Q_x}{Q} = \text{a function of } x \text{ alone;} \tag{21.14}$$

otherwise a suitable $\mu = \mu(x)$ will not exist. Supposing that the consistency condition (21.14) is satisfied, integration of (21.13) yields

$$\mu(x) = \exp\left(\int \frac{P_y - Q_x}{Q}\, dx\right) \tag{21.15}$$

to within an arbitrary multiplicative constant, which is obviously immaterial.

If (21.14) does not hold, we can then try $\mu = \mu(y)$. This time the consistency condition turns out to be

$$\frac{Q_x - P_y}{P} = \text{a function of } y \text{ alone,} \tag{21.16}$$

and if this is fulfilled, then

$$\mu(y) = \exp\left(\int \frac{Q_x - P_y}{P}\, dy\right). \tag{21.17}$$

Example 21.4. Consider the first-order *linear* equation

$$y' + M(x)y = N(x). \tag{21.18}$$

In general, it is neither variable separable nor of homogeneous-of-degree-zero type. Rewriting it as

$$(My - N)\, dx + dy = 0,$$

we see from (21.9) that the left side is exact only in the trivial case where $M = 0$. Seeking an integrating factor, $(Q_x - P_y)/P = -M/(My - N)$ will be a function of y alone only in the unlikely event that $M(x)$ is some constant times $N(x)$. On the other hand, $(P_y - Q_x)/Q = M$ *is* a function of x alone, so that

$$\mu(x) = e^{\int M\, dx}. \tag{21.19}$$

Thus

$$e^{\int M\, dx}(My - N)\, dx + e^{\int M\, dx}\, dy = 0$$

is exact and can be solved as in Example 21.3. The result (Exercise 21.8) is the important formula

$$y = e^{-\int M\, dx}\left(\int N e^{\int M\, dx}\, dx + C\right). \quad ▮ \tag{21.20}$$

Some additional techniques and special equations, such as the **Bernoulli, d'Alembert–Lagrange, Clairaut**, and **Riccati equations**, are contained in the exercises.

21.2. THE QUESTIONS OF EXISTENCE AND UNIQUENESS

In this section we consider the system

$$y' = f(x, y), \qquad y(a) = b, \tag{21.21}$$

that is, where an "initial condition" $y(a) = b$ is adjoined to the differential equation. The fact that *solutions do not always exist* is evident from the following example.

Example 21.5.

$$y' = \frac{y \ln y}{x}, \qquad y(0) = -1.$$

Integrating, we obtain the general solution $y = e^{Cx}$, which cannot satisfy the condition $y(0) = -1$ for any choice of C. ▮

The fact that *more than one solution may exist* is illustrated by the interesting example below.

Example 21.6. *Falling Body.* Suppose that a body of mass m is dropped, from rest, at time $t = 0$. With its displacement $x(t)$ measured downward (Fig. 21.1), the equation of motion is $m\ddot{x} = mg$, where $(\dot{\ }) = d(\ )/dt$ and g is the acceleration of gravity. Thus

$$\ddot{x} = g, \qquad x(0) = \dot{x}(0) = 0, \tag{21.22}$$

which is easily integrated twice to give

$$x = \frac{gt^2}{2}. \tag{21.23}$$

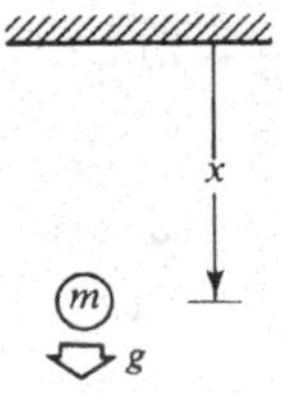

Figure 21.1. Falling body.

Instead of integrating (21.22) by multiplying through by dt, however, let us multiply through by dx. Then

$$\frac{d\dot{x}}{dt}\,dx = g\,dx; \tag{21.24}$$

and formally moving the dt under the dx, we have

$$\dot{x}\,d\dot{x} = g\,dx$$

or

$$\frac{\dot{x}^2}{2} = gx + A, \tag{21.25}$$

where $A = 0$, since $x = \dot{x} = 0$ at $t = 0$. [If sliding the dt differential is bothersome, note that the time derivative of (21.25) does yield the original equation $\ddot{x} = g$.] So we have reduced (21.22) to the *first*-order system

$$\dot{x} = \sqrt{2g}\,x^{1/2}, \qquad x(0) = 0, \tag{21.26}$$

which is the focus of this example. Separating variables,

$$x^{-1/2}\,dx = \sqrt{2g}\,dt,$$

so that

$$2x^{1/2} = \sqrt{2g}\,t + B, \tag{21.27}$$

where $x(0) = 0$ implies that $B = 0$, and therefore $x(t) = gt^2/2$, in agreement with both (21.23) and experiment.

Note, however, that (21.26) also admits the solution $x(t) \equiv 0$! This result sometimes evokes the response "Oh, but $x(t) = 0$ is a trivial solution and so need not be considered." Yet the fact remains that $x(t) = 0$ satisfies both the differential equation and the initial condition (21.26). What more can we ask? We can ask that our solution satisfy the *original* system (21.22) as well. In fact, $x(t) = gt^2/2$ satisfies (21.22), but $x(t) = 0$ does *not*, since it fails to satisfy $\ddot{x} = g$. Consequently, we discard the solution $x(t) = 0$ of (21.26), not because it is "trivial" but because it fails to satisfy the equation of motion $\ddot{x} = g$.

Nevertheless, we may still wonder how this spurious solution of (21.26) crept in. Examining the steps between (21.22) and (21.26), we find the trouble in (21.24); that is, in multiplying both sides by dx, we inadvertantly introduce the additional solution $dx = 0$, or $x =$ constant (in this case, $x = 0$ due to the initial condition).

In more physical terms, observe that (21.25), or

$$\tfrac{1}{2}m\dot{x}^2 + (-mgx) = \text{constant},$$

is a statement of conservation of energy,[5] and certainly energy *is* conserved if the body, on being released, elects not to fall, so that $x(t) = 0$. As we have seen, however, the problem is that this situation violates Newton's second law, which insists that the body accelerate downward at the rate g. ∎

Suppose that the one-parameter family of curves representing the general solution $\phi(x, y, c) = 0$ of the differential equation $F(x, y, y') = 0$ possesses an "**envelope**" Γ, as in Fig. 21.2 for instance. At each point of Γ the values x, y, and y' are such that $F(x, y, y') = 0$ is satisfied, and so Γ must itself be a solution curve, a so-called **singular solution.**

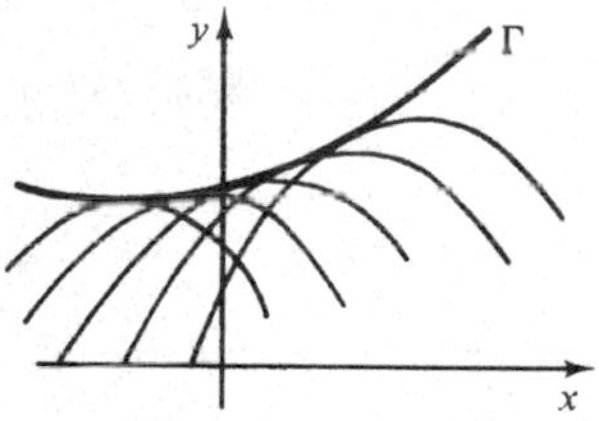

Figure 21.2. Family of curves with an envelope.

Suppose next that we have the general solution $\phi(x, y, c) = 0$ in hand and wish to determine if there are singular solutions as well. We can plot the family of solution curves and *see* if it possesses an envelope, but doing so is tedious and generally leads only to a sketch of Γ rather than its precise equation. Consider, then, the following analytic approach.

The coordinates x, y of point P (Fig. 21.3) must satisfy both $\phi(x, y, c) = 0$ and $\phi(x, y, c + \Delta c) = 0$ or, equivalently,

$$\phi(x, y, c) = 0 \tag{21.28}$$

and
$$\frac{\phi(x, y, c + \Delta c) - \phi(x, y, c)}{\Delta c} = 0. \tag{21.29}$$

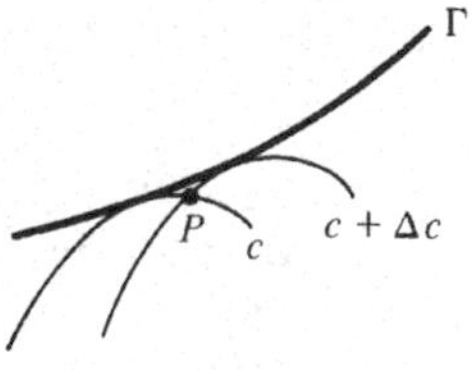

Figure 21.3. Determination of Γ.

Equation (21.29) is valid for Δc arbitrarily small, and so it must hold in the limit as $\Delta c \to 0$, in which case P approaches Γ. Thus (21.28) and (21.29) become

$$\phi(x, y, c) = 0 \tag{21.30a}$$
$$\phi_c(x, y, c) = 0 \tag{21.30b}$$

[5]Not surprising, since multiplication of "$F = ma$" by a *displacement* dx must produce a work-energy relation or, since the gravitational force is conservative, a statement of conservation of energy.

for points x, y on Γ. Eliminating c between (21.30a) and (21.30b) gives the desired equation of Γ.

Envelopes are of importance not only in connection with singular solutions of differential equations but also, for instance, in optics (in connection with "caustics") and in acoustics and the theory of characteristics for hyperbolic partial differential equations.

Example 21.7. Let us return to equation (21.26) in Example 21.6,

$$\dot{x} = \sqrt{2g}\, x^{1/2}. \tag{21.31}$$

Integrating, $2x^{1/2} = \sqrt{2g}\, t + c$, or

$$x = \frac{(\sqrt{2g}\, t + c)^2}{4}. \tag{21.32}$$

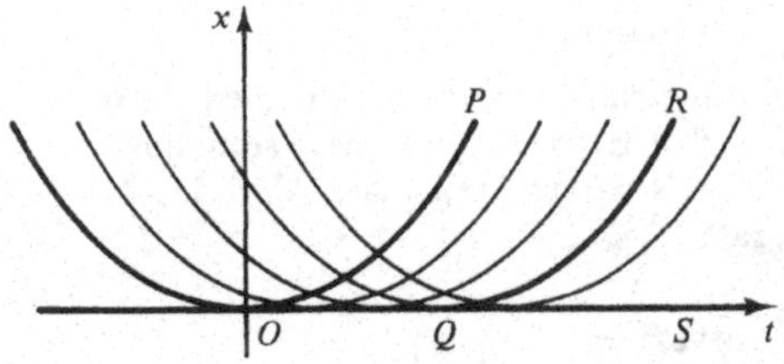

Figure 21.4. Falling body envelope.

Thus

$$\phi(t, x, c) = x - \frac{(\sqrt{2g}\, t + c)^2}{4} = 0$$

$$\phi_c(t, x, c) = -\frac{\sqrt{2g}\, t + c}{2} = 0,$$

and eliminating c yields the envelope $x = 0$. Thus the additional solution $x(t) = 0$, found in Example 21.6, was a singular solution—in particular, the envelope of the one-parameter family of parabolas shown in Fig. 21.4. Besides the curve OP (i.e., $x = gt^2/2$), the singular solution OS (i.e., $x = 0$) also satisfies the initial condition $x(0) = 0$. In fact, there are *infinitely* many solutions, such as OQR, all of which satisfy both the differential equation (21.31) and the initial condition $x(0) = 0$. ∎

Conditions that guarantee the existence and uniqueness of a solution to the system

$$y' = f(x, y), \qquad y(a) = b \tag{21.33}$$

are of fundamental interest in the study of differential equations. Before proceeding, however, a definition is needed.

We say that a function $F(y)$ satisfies a **Lipschitz condition** in a given closed interval I if there is a constant C such that

$$|F(y_2) - F(y_1)| \leq C|y_2 - y_1| \tag{21.34}$$

for every pair of values y_1, y_2 in I; in other words, F has a *bounded difference quotient* over the interval. Certainly the Lipschitz condition will be satisfied if $|F'(y)| \leq C$ over I (Exercise 21.20), but it can be satisfied even if $F'(y)$ does not exist over I. To illustrate, $F(y) = |y|$ satisfies the Lipschitz condition over $-1 \leq y \leq 1$ even though $F'(0)$ fails to exist.

THEOREM 21.2. (Existence and Uniqueness) Suppose that $f(x, y)$ in (21.33) is continuous in some closed region D of the x, y plane and hence is bounded.[6] *In particular,*

[6]The term "closed" was explained in the starred part of Section 2.1. For the reader who has omitted that part, it will suffice to say that D is closed if it includes its boundary curve; for example, the rectangle $x_1 \leq x \leq x_2$, $y_1 \leq y \leq y_2$ is closed, whereas $x_1 < x < x_2$, $y_1 < y < y_2$ is not. That

suppose that

$$|f(x, y)| \leq M \text{ over } D \tag{21.35}$$

and also that f satisfies a Lipschitz condition in the y argument—that is,

$$|f(x, y_2) - f(x, y_1)| \leq C|y_2 - y_1|, \tag{21.36}$$

where the constant C is independent of x. Finally, define a rectangle (Fig. 21.5)

$$|x - a| \leq h, \qquad |y - b| \leq k \tag{21.37}$$

such that $Mh < k$. Then (21.33) has a unique solution $y(x)$ in the shaded part of the rectangle.

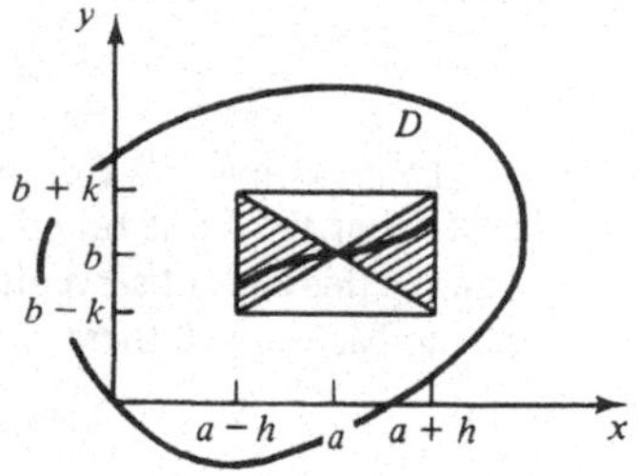

Figure 21.5. Rectangle of existence.

*

Proof: We begin by proving existence. Converting the system (1.32) to an *integral* equation by integrating from a to x,

$$y(x) = b + \int_a^x f[\xi, y(\xi)]\, d\xi, \tag{21.38}$$

we then define the sequence of successive approximations

$$\begin{aligned} y_0(x) &= b, \\ y_n(x) &= b + \int_a^x f[\xi, y_{n-1}(\xi)]\, d\xi \qquad n = 1, 2, \cdots. \end{aligned} \tag{21.39}$$

The next step is to show that $y_n(x) \longrightarrow y(x)$ as $n \longrightarrow \infty$. (Such a proof is called "constructive," since it actually shows *how* to find a solution.)

Observe that the "telescopic" series

$$s(x) = y_0(x) + [y_1(x) - y_0(x)] + [y_2(x) - y_1(x)] + \cdots \tag{21.40}$$

has as its partial sums $s_n(x) = y_n(x)$ for $n = 0, 1, 2, \ldots$, so that convergence of the sequence $y_n(x)$ is equivalent to the convergence of the series (21.40). Consider the bracketed terms in the series. With the help of (21.35), (21.36), and (21.39), we have

$$|y_1(x) - y_0(x)| = \left| \int_a^x f[\xi, y_0(\xi)]\, d\xi \right| \leq M|x - a| \leq Mh, \tag{21.41}$$

$$\begin{aligned} |y_2(x) - y_1(x)| &= \left| \int_a^x \{f[\xi, y_1(\xi)] - f[\xi, y_0(\xi)]\}\, d\xi \right| \\ &\leq \int_a^x C|y_1(\xi) - y_0(\xi)|\, d\xi \leq \int_a^x CM|\xi - a|\, d\xi \\ &= \frac{CM|x - a|^2}{2} \leq \frac{MCh^2}{2} \\ &\ \ \vdots \end{aligned}$$

$$|y_n(x) - y_{n-1}(x)| \leq \frac{MC^{n-1}h^n}{n!}. \tag{21.42}$$

continuity over a closed region implies boundedness is proved in most books on advanced calculus; see, for example, T. M. Apostol, *Mathematical Analysis*, Addison-Wesley, Reading, Mass., 1957, Theorem 4-20, p. 73.

So the series (21.40) is dominated by the convergent series of positive constants

$$b + Mh + M\frac{Ch^2}{2!} + \cdots + M\frac{C^{n-1}h^n}{n!} + \cdots,$$

and hence converges (uniformly) according to the Weierstrass M-test. We must still verify that the limit function $y(x)$ does satisfy the differential equation $y' = f(x, y)$. Toward this end, observe that since the convergence $y_n(x) \longrightarrow y(x)$ is uniform, it follows that to each $\epsilon > 0$ there corresponds an $N(\epsilon)$ such that

$$|y_n(x) - y(x)| \leq \frac{\epsilon}{C} \quad \text{for all } n > N.$$

Recalling (21.36), it follows that

$$|f[x, y_n(x)] - f[x, y(x)]| \leq C|y_n(x) - y(x)| < \frac{C\epsilon}{C} = \epsilon$$

for all $n > N$ and all x in the interval $|x - a| \leq h$, and therefore $f[x, y_n(x)] \longrightarrow f[x, y(x)]$ *uniformly*. Consequently, letting $n \longrightarrow \infty$ in (21.39),

$$\begin{aligned} y(x) &= b + \lim_{n\to\infty} \int_a^x f[\xi, y_{n-1}(\xi)]\, d\xi \\ &= b + \int_a^x \lim_{n\to\infty} f[\xi, y_{n-1}(\xi)]\, d\xi \\ &= b + \int_a^x f[\xi, y(\xi)]\, d\xi, \end{aligned} \tag{21.43}$$

where the second equality follows from the uniformity of convergence [recall (4.5)]. Now $f(x, y)$ is a continuous function of x and y, and y is a continuous function of x [since each term in (21.40) is (Why?), and (21.40) converges *uniformly* to $y(x)$ (Theorem 2.12)]; so $f[\xi, y(\xi)]$ in (21.43) is a continuous function of ξ. Thus the Fundamental Theorem of the Integral Calculus, (1.19) may be applied to (21.43), yielding $y'(x) = f[x, y(x)]$ as desired.

We have therefore shown that (21.33) does have a solution $y(x)$. That it must lie within the shaded part of the rectangle (Fig. 21.5) follows from the fact that

$$|y(x) - b| = \left| \int_a^x f[\xi, y(\xi)]\, d\xi \right| \leq M|x - a| < \frac{k}{h}|x - a|. \tag{21.44}$$

Finally, we wish to establish the *uniqueness* of the solution $y(x)$. Suppose that there *is* another solution of (21.33), say $y = Y(x)$. Surely $|Y(x) - y(x)| \leq 2k$ over $|x - a| \leq h$. Furthermore,

$$\begin{aligned} |Y(x) - y(x)| &= \left| \int_a^x \{f[\xi, Y(\xi)] - f[\xi, y(\xi)]\}\, d\xi \right| \\ &\leq \int_a^x C|Y(\xi) - y(\xi)|\, d\xi \leq 2kC|x - a|. \end{aligned} \tag{21.45}$$

Putting this back into the second integral in (21.45),

$$|Y(x) - y(x)| \leq \int_a^x 2kC^2|\xi - a|\, d\xi = \frac{2kC^2|x - a|^2}{2!}.$$

Repeating again and again, we see that

$$|Y(x) - y(x)| \leq \frac{2kC^n|x - a|^n}{n!} \leq \frac{2k(Ch)^n}{n!},$$

which tends to zero[7] as $n \longrightarrow \infty$, so that $Y(x) = y(x)$ and hence the solution must be unique. This completes the proof.

It has become more fashionable to discuss existence/uniqueness in terms of contraction mappings and the fixed point theorem.

If $\mathfrak{D}$ is a vector space with distance function $d(x, y)$ and L is a mapping of $\mathfrak{D}$ into itself (i.e., if x is in $\mathfrak{D}$, then Lx is, too), then L is a **contraction mapping** if

$$d(Lx, Ly) \leq kd(x, y) \qquad (0 \leq k < 1) \tag{21.46}$$

for all x, y in $\mathfrak{D}$.

THEOREM 21.3. (Fixed Point Theorem) If L is a contraction mapping on a closed vector space $\mathfrak{D}$, then there is one and only one "fixed point" x in $\mathfrak{D}$—that is, such that

$$Lx = x. \tag{21.47}$$

Furthermore, if x_0 is any vector in $\mathfrak{D}$ and $x_n = Lx_{n-1}$ for $n = 1, 2, \ldots$, then $x_n \longrightarrow x$ as $n \longrightarrow \infty$.

In order to apply the theorem to our system (21.33), we reexpress (21.33) as the equivalent integral equation

$$y(x) = b + \int_a^x f[\xi, y(\xi)]\, d\xi$$

or

$$y = Ly$$

and show that L is a contraction mapping. See, for instance, Naylor and Sell.[8]

Example 21.8. Returning again to (21.26), observe that $f(t, x) = \sqrt{2g}\, x^{1/2}$ does *not* satisfy the Lipschitz condition (21.36) in x throughout *any* rectangle "centered" at the initial point $t = 0$, $x = 0$; in fact, the slope $\partial f/\partial x$ is infinite all along the t axis, which passes right through the initial point $t = x = 0$. Thus the conditions of Theorem 21.2 are not met, and as discussed in Examples 21.6 and 21.7, it turns out that we do *not* have uniqueness, although we do have existence. ▮

EXERCISES

21.1. Find the general solution by separation of variables.

(a) $y' = e^{x+2y}$ (b) $y' = \dfrac{ky}{x}$

(c) $y' = \dfrac{ky \ln x}{x}$ (d) $dy = (y^2 - y)e^x\, dx$

---

[7]Recall that the Taylor series $e^x = 1 + x + x^2/2! + \cdots + x^n/n! + \cdots$ converges for all x. *A necessary* condition for $\sum a_n$ to converge is that $a_n \longrightarrow 0$ as $n \longrightarrow \infty$. Thus $x^n/n! \longrightarrow 0$ as $n \longrightarrow \infty$ for all x.

[8]A. Naylor and G. Sell, *Linear Operator Theory*, Holt, Rinehart and Winston, New York, 1971, pp. 125–134.

21.2. (a) Show that the equation $y' = f(ax + by + c)$ is reduced to separable form by the change of variables $ax + by + c = u$ and that among the solutions will be the straight lines $y = (u_0 - c - ax)/b$, where u_0 satisfies $f(u_0) = -a/b$.
(b) Thus solve $y' = x + y + 2$.
(c) Thus solve $y' + (x + y)^2 = 0$.

21.3. Solve $y' = 1 - y^2$ and sketch its solutions throughout the x, y plane.

21.4. Find the general solution of the following homogeneous-of-degree-zero-type equations.
(a) $y' = \dfrac{2y - x}{y - 2x}$ (b) $(2x + y)\,dx + x\,dy = 0$
(c) $y' = \dfrac{xy + 2y^2}{x^2}$ (d) $y^2\,dx + (x^2 + xy + y^2)\,dy = 0$

21.5. Find the general solution of the exact equations below.
(a) $(\sin y - x)\,dy - (e^x + y)\,dx = 0$
(b) $(y - 2x)\,dy - (2y - x)\,dx = 0$
(c) $e^y\,dx + (xe^y - 1)\,dy = 0$
(d) $(x + 2y)\,dx + (y + 2x)\,dy = 0$.

21.6. Find the general solution of these first-order linear equations.
(a) $y' \tan x - 2y = 4$ (b) $y' - 2xy = 1$
(c) $xy' + y = x \sin 2x$ (d) $y' + 4y = x$

21.7. Prove Theorem 21.1.

21.8. Derive the result (21.20).

21.9. Find suitable integrating factors (if necessary) and solve:
(a) $dx + (x - e^{-y})\,dy = 0$ (b) $3y\,dx + dy = 0$
(c) $x\,dy + (y + e^x)\,dx = 0$ (d) $(x^2 - y^2)\,dy = 2xy\,dx$

21.10. The equation

$$y' + M(x)y = N(x)y^n, \tag{21.48}$$

where n is a constant, not necessarily an integer, is known as **Bernoulli's equation** after Jakob Bernoulli (1654–1705).
(a) Solve for the simple cases where $n = 0$ and 1.
(b) For $n \neq 0, 1$ show that the change of variables $v = y^{1-n}$ reduces (21.48) to a linear equation.
(c) Solve $y' = y - y^6$.
(d) Solve $x^2y' - xy = 2y^2$.
(e) As an alternative line of approach, solve (21.48) by seeking an integrating factor of the form $\mu(x)y^\nu$, where ν is a constant.

21.11. The equation

$$y = xf(p) + g(p), \tag{21.49}$$

where $y' = p$, is known as the **d'Alembert–Lagrange equation**.
(a) Differentiating (21.49) with respect to x, obtain the first-order linear equation

$$\frac{dx}{dp} - \frac{f'(p)x}{p - f(p)} = \frac{g'(p)}{p - f(p)} \qquad [f(p) \neq p]$$

for $x(p)$ that, together with (21.49), consists of a solution in parametric form.
(b) If $f(p)$ has a fixed point P_0 [i.e., $f(p_0) = p_0$], show that (21.49) has a linear solution $y = p_0x + g(p_0)$.

(c) Thus solve $(y')^4 + xy' = y - x$.
(d) Solve [according to the discussion in parts (a) and (b)]

$$y = x + y', \qquad y(0) = 1.$$

(e) Repeat part (d) with $y(0) = 1$ changed to $y(0) = 2$.

21.12. For the special case $f(p) = p$, the d'Alembert–Lagrange equation (21.49) becomes

$$y = xp + g(p) \tag{21.50}$$

and is known as the **Clairaut equation.**

(a) Differentiate (21.50) with respect to x and thus discuss its solution(s).
(b) Find and sketch all solutions of $y = xp + p^2$. In particular, find all (real) solutions satisfying the initial condition $y(0) = -1$. Repeat for $y(0) = 0$ and 1.
(c) Find all solutions of $y - xp + e^p = 0$.

21.13. The equation

$$y' = a_1(x) + a_2(x)y + a_3(x)y^2 \tag{21.51}$$

is known as **Riccati's equation** and is of special importance in the study of *optimal control.*
(a) Show that if $y = Y(x)$ is any particular solution of (21.51), then $v = 1/[y - Y(x)]$ satisfies a *linear* differential equation of first order.
(b) Thus find the general solution of $y' = 1 + (y - x)^2$. *Hint:* Note the particular solution $y = x$.
(c) Find the general solution of $y' = 1 - y^2$.
(d) Discuss the utility of a change of variables of the form $y = a(x)u'/u$ [where $a(x)$ is suitably chosen] in solving (21.51).

21.14. (a) Show that

$$\frac{dy}{dx} = \frac{ax + by + c}{dx + ey + f} \qquad (a, b, \ldots, f \text{ constants}) \tag{21.52}$$

can be reduced to homogeneous-of-degree-zero form by the simple change of variables $x = u + h$, $y = v + k$, where h and k are suitably chosen constants, provided that $ae - bd \neq 0$.
(b) Thus solve $y' = (2x - y - 6)/(x - y - 3)$.
(c) Solve $y' = (1 - y)/(x + 4y - 3)$.
(d) Can (21.52) be reduced to the desired form by setting $ax + by + c = u$ and $dx + ey + f = v$ instead? Explain.

21.15. Solve $y' = 1/(e^y + 2x)$. *Hint:* Interchange the dependent and independent variables.

21.16. Solve $y' = y^2$, $y(0) = y_0$ and show how the solution is singular at a value of x that depends on the initial value y_0. Could this happen for a *linear* equation? Explain.

21.17. Solve by any means: (*Hint:* Some of the previous exercises may help.)
(a) $y' + 3y = 0, \quad y(0) = 4$
(b) $y' = \dfrac{2x - 3}{y^2 + y + 4}, \quad y(1) = 2$
(c) $xy' + 2y = 4x^2, \quad y(0) = 0$
(d) $y' = -xy + \sqrt{y}, \quad y(0) = 0$
(e) $y' = -xy + \sqrt{y}, \quad y(0) = 1$
(f) $y = \ln y', \quad y(2) = 0$

21.18. The current $I(t)$ induced in the circuit shown in Fig. 21.6 by an impressed voltage $E(t)$ is governed by the differential equation

$$L\frac{dI}{dt} + RI = E(t).$$

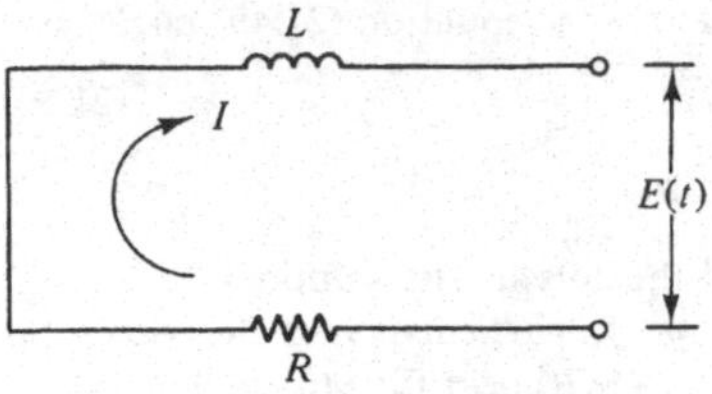

Figure 21.6. Circuit.

Solve for $I(t)$ and sketch it if $I(0) = 0$ and

(a) $E(t) = E_0[H(t - t_1) - H(t - t_2)]$, where E_0, t_1, and t_2 are constants, $t_2 > t_1 > 0$, and H is the Heaviside function.

(b) $E(t) = E_0\,\delta(t - t_1)$, where E_0 and t_1 are constants, $t_1 > 0$, and δ is the delta function. (Be sure that your result is *dimensionally* correct. If it is not, show how to straighten things out.)

21.19. Determine all envelopes (if any) of the following one-parameter families and illustrate with a sketch:

(a) $y = cx + \dfrac{1}{c}$ (b) $(x - c)^2 + y^2 - 9 = 0$

(c) $(x - c)^2 + y^2 = \dfrac{c^2}{2}$ (d) $y^2 = cx - c^{3/2}$

(e) $y^2 - (x - c)^3 = 0$

***21.20.** Prove that the Lipschitz condition (21.34) will be satisfied if $|F'(y)| \leq C$ over I as claimed just before Theorem 21.2.

21.21. To estimate the region of existence and uniqueness of a solution to the equation $y' = 1 + y^2$ through the initial point $y(0) = 0$ (shaded, Fig. 21.7), recall Theorem 21.2. Clearly, we can take M as large as $1 + k^2$, so that $h < k/(1 + k^2)$. For instance, if we choose $k = 4$, say, then h can be as large as $\frac{4}{17}$. Show that the *largest* possible h is $\frac{1}{2}$.

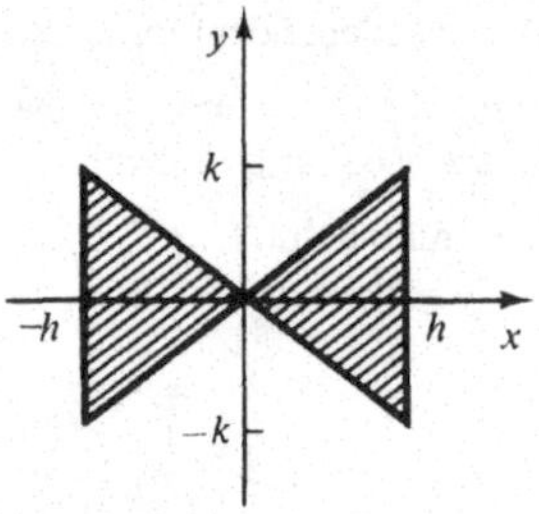

Figure 21.7. Region of existence/uniqueness.

21.22. Actually carry out the first several approximations (21.39) for the problems below and compare with the exact solution.

(a) $y' = 1 + y, \quad y(0) = 0$ (b) $y' = 2y, \quad y(0) = 1$

(c) $y' = y^2, \quad y(0) = 1$ (d) $y' = \sin x, \quad y(0) = 2$

21.23. Give a simple uniqueness proof for the problem

$$y' = f(x), \qquad y(a) = b.$$

21.24. If $f(x) \longrightarrow b$ as $x \longrightarrow \infty$ and if $a > 0$, show that every solution of $y' + ay = f(x)$ tends to b/a as $x \longrightarrow \infty$.

21.25. Determine conditions on $M(t)$ and/or $N(t)$ that are necessary and sufficient to ensure the *asymptotic stability* of the system

$$x' + M(t)x = N(t)$$

with respect to initial conditions—that is, such that $x_1(t) - x_2(t) \longrightarrow 0$ as $t \longrightarrow \infty$ for solutions $x_1(t)$ and $x_2(t)$, independent of the initial conditions $x_1(0)$ and $x_2(0)$.

21.26. Make up two examination-type problems on this chapter.

Chapter 22

Higher-Order Systems

Part of this chapter deals with nonlinear systems—the general existence/uniqueness theory in Section 22.1 and some methods of solution and applications in Section 22.6. Primarily, then, Chapter 22 concerns *linear* systems. This coverage does not imply that most practical problems are essentially linear but rather that linear ones are generally far more tractable, as well as susceptible to a number of powerful lines of approach. And, in fact nonlinear problems can often be *approximated* satisfactorily by linear ones or by a *sequence* of linear ones through the methods of successive approximations or quasilinearization (Section 22.6) or through perturbation techniques (Chapter 25).

*

22.1. NONLINEAR CASE; EXISTENCE AND UNIQUENESS

The second-order equation

$$\ddot{x} + \kappa \sin x = F_0 \sin \omega t, \qquad (\dot{\ }) = \frac{d(\ )}{dt} \tag{22.1}$$

for instance, is nonlinear, since Lx is not of the form $a(t)\ddot{x} + b(t)\dot{x} + c(t)x$—that is, a linear function of x, $\dot{x}$, and $\ddot{x}$. Or return to the defining property of linearity,

$$L(\alpha u + \beta v) = \alpha L u + \beta L v, \tag{22.2}$$

which is to be satisfied for all scalars α, β and all functions u, v in the domain of L, say twice-differentiable functions defined over $0 \leq t < \infty$. If, for example, $\alpha = \beta = 1$

and $u = v = t$, then

$$L(\alpha u + \beta v) = L(2t) = \kappa \sin 2t,$$

whereas

$$\alpha Lu + \beta Lv = \kappa \sin t + \kappa \sin t = 2\kappa \sin t$$

so that L is not linear as stated above.

In considering existence and uniqueness for higher-order systems, whether linear or not, we first convert the system to the *first-order* form (21.33) with the help of vector notation. To illustrate, consider the system

$$\ddot{x} + \kappa \sin x = F_0 \sin \omega t, \qquad x(0) = 3, \quad \dot{x}(0) = 1. \tag{22.3}$$

Define $x_1 = x$ and $x_2 = \dot{x}$. Then (22.3) is equivalent to the two first-order equations

$$\begin{aligned} \dot{x}_1 &= x_2, \qquad x_1(0) = 3 \\ \dot{x}_2 &= F_0 \sin \omega t - \kappa \sin x_1, \qquad x_2(0) = 1 \end{aligned} \tag{22.4}$$

or more compactly

$$\dot{\mathbf{x}} = \mathbf{f}(\mathbf{x}, t), \qquad \mathbf{x}(0) = \mathbf{c}, \tag{22.5}$$

where[1] $\mathbf{x} = \begin{pmatrix} x_1 \\ x_2 \end{pmatrix}, \qquad \mathbf{f}(\mathbf{x}, t) = \begin{pmatrix} f_1(x_1, x_2, t) \\ f_2(x_1, x_2, t) \end{pmatrix} = \begin{pmatrix} x_2 \\ F_0 \sin \omega t - \kappa \sin x_1 \end{pmatrix}$

and $\mathbf{c} = \begin{pmatrix} x_1(0) \\ x_2(0) \end{pmatrix} = \begin{pmatrix} 3 \\ 1 \end{pmatrix}.$

Turning to the questions of existence/uniqueness, consider the system

$$\dot{\mathbf{x}} = \mathbf{f}(\mathbf{x}), \qquad \mathbf{x}(0) = \mathbf{c}. \tag{22.6}$$

That (22.6) is **autonomous**—i.e., $\mathbf{f}$ contains no explicit t-dependence[2]—is not as restrictive as it seems. For instance, with $x_1 = x$, $x_2 = \dot{x}$, and $x_3 = t$, (22.3) can be expressed in the autonomous form (22.6), where

$$\mathbf{x} = \begin{pmatrix} x_1 \\ x_2 \\ x_3 \end{pmatrix}, \qquad \mathbf{f} = \begin{pmatrix} x_2 \\ F_0 \sin \omega x_3 - \kappa \sin x_1 \\ 1 \end{pmatrix}, \qquad \mathbf{c} = \begin{pmatrix} 3 \\ 1 \\ 0 \end{pmatrix}. \tag{22.7}$$

Before stating an existence/uniqueness theorem that is the n-dimensional version of Theorem 21.2, we need to define a vector *norm*. Rather than the Euclidean norm, we choose the algebraically simpler "taxicab norm"[3]

$$\|\mathbf{x}\| = \sum_1^n |x_j|. \tag{22.8}$$

Finally, we say that a vector function $\mathbf{f}(\mathbf{x})$ satisfies a **Lipschitz condition** over a domain D if there is a constant C such that

$$\|\mathbf{f}(\mathbf{x}_2) - \mathbf{f}(\mathbf{x}_1)\| \leq C\|\mathbf{x}_2 - \mathbf{x}_1\| \tag{22.9}$$

for all $\mathbf{x}_1$'s and $\mathbf{x}_2$'s in D. Surely a sufficient, although not necessary, condition that will ensure that $\mathbf{f}$ is a Lipschitz function (i.e., that it satisfies a Lipschitz condition) is

---

[1] If the matrix notation used here is unfamiliar, see Section 18.1.

[2] It is convenient here, although not essential, to regard the independent variable t as "the time."

[3] This norm was mentioned in Section 17.2.

that the partials $\partial f_j/\partial x_k$ exist and are bounded over D for all $j = 1, \ldots, n$ and $k = 1, \ldots, n$ (Exercise 22.1).

THEOREM 22.1. (Existence and Uniqueness) Let D be an n-dimensional neighborhood

$$\|\mathbf{x} - \mathbf{c}\| \leq h$$

of **c**. *For all* **x** *in D, suppose that* **f**(**x**) *is continuous (i.e., each component of* **f** *is), that there is a finite constant M such that*

$$\|\mathbf{f}(\mathbf{x})\| < M,$$

and that **f**(**x**) *satisfies the Lipschitz condition (22.9). If $Mh < k$, then the n-dimensional system*

$$\dot{\mathbf{x}} = \mathbf{f}(\mathbf{x}), \qquad \mathbf{x}(0) = \mathbf{c} \tag{22.10}$$

has a unique solution in D for $0 \leq t \leq h$—namely, $\mathbf{x}(t) = \lim_{j\to\infty} \mathbf{x}_j(t)$*—where*

$$\left.\begin{aligned} \mathbf{x}_0(t) &= \mathbf{c} \\ \mathbf{x}_j(t) &= \mathbf{c} + \int_0^t \mathbf{f}[\mathbf{x}_{j-1}(\tau)]\, d\tau. \end{aligned}\right\} \tag{22.11}$$

Proof follows essentially the same lines as for Theorem 21.2; for details, see the book by Struble.[4] Instead let us illustrate the computational scheme (22.11).

Example 22.1. Consider the system (22.3) or, equivalently, (22.6) and (22.7). According to (22.11), we have the successive approximations

$$\mathbf{x}_0 = \begin{pmatrix} 3 \\ 1 \\ 0 \end{pmatrix}, \qquad \mathbf{x}_1 = \begin{pmatrix} 3 \\ 1 \\ 0 \end{pmatrix} + \begin{pmatrix} \int_0^t 1\, d\tau \\ \int_0^t (0 - \kappa \sin 3)\, d\tau \\ \int_0^t 1\, d\tau \end{pmatrix} = \begin{pmatrix} 3 + t \\ 1 - (\kappa \sin 3)t \\ t \end{pmatrix}, \tag{22.12}$$

and so on. In particular, we have for $x(t)$ (namely, the top element in the column vector **x**) the successive approximations 3 and $3 + t$. After two more iterations we have (Exercise 22.4)

$$x(t) \approx 3 + \left(1 + \frac{F_0}{\omega}\right)t - \frac{F_0}{\omega^2} \sin \omega t + \kappa[\sin(t + 3) - (\cos 3)t - \sin 3], \tag{22.13}$$

which may be a useful approximation for small enough t (and/or small enough κ, since it is seen to be exact for $\kappa = 0$). ▮

Additionally, it can be shown[5] that *under the hypotheses of Theorem 22.1 the solution is a continuous function of the initial values—that is, of* **c**. In particular, we find that if **x**(t) and **x**′(t) (where the prime does *not* denote differentiation) are solutions corresponding to initial conditions **c** and **c**′, then

$$\|\mathbf{x}(t) - \mathbf{x}'(t)\| \leq \|\mathbf{c} - \mathbf{c}'\| e^{Ct}, \tag{22.14}$$

---

[4] R. A. Struble, *Nonlinear Differential Equations*, McGraw-Hill, New York, 1962, pp. 44–47.
[5] *Ibid.*, p. 48.

where C is the Lipschitz constant in (22.9). So to each $\epsilon > 0$ there corresponds a $\delta(\epsilon) = \epsilon e^{-Ch}$ such that $\|\mathbf{x}(t) - \mathbf{x}'(t)\| < \epsilon$ (over $0 \leq t \leq h$) whenever $\|\mathbf{c} - \mathbf{c}'\| < \delta$.

In fact, this result can be used to show that the solution of (22.3) is a continuous function of the parameter κ (or F_0 or ω) in the differential equation. To do so, we append the component $x_4 = \kappa$ to our $\mathbf{x}$ vector. Thus $\dot{x}_4 = 0$ and $\dot{x}_4(0) = \kappa$, and therefore

$$\mathbf{x} = \begin{pmatrix} x_1 \\ x_2 \\ x_3 \\ x_4 \end{pmatrix}, \qquad \mathbf{f} = \begin{pmatrix} x_2 \\ F_0 \sin \omega x_3 - x_4 \sin x_1 \\ 1 \\ 0 \end{pmatrix}, \qquad \mathbf{c} = \begin{pmatrix} 3 \\ 1 \\ 0 \\ \kappa \end{pmatrix},$$

in which κ now appears as an initial condition.

Continuity of a system with respect to input variables and system parameters is of obvious importance in engineering design. For instance, if we use a slightly off-design capacitance, in a given circuit, we would not like the output to be more than slightly off-design.

For analytical purposes, however, we might wish for more than continuity. Specifically, we might like to know if the solution of (22.3), say, is an *analytic* function of κ, for if it is, then we know that (for sufficiently small $|\kappa|$) the solution can be expressed as a power series in κ,

$$x(t) = \sum_0^\infty a_j(t)\kappa^j.$$

This comes under the heading of "perturbation methods," to which we return in Chapter 25.

22.2. THE LINEAR EQUATION; SOME THEORY

Sometimes we meet nth-order systems in the form

$$\begin{aligned} \dot{x}_1 &= f_1(x_1, \ldots, x_n, t) \\ &\vdots \\ \dot{x}_n &= f_n(x_1, \ldots, x_n, t) \end{aligned} \tag{22.15a}$$

or in the more concise vector notation

$$\dot{\mathbf{x}} = \mathbf{f}(\mathbf{x}, t), \tag{22.15b}$$

and sometimes we meet instead a single nth-order equation[6]

$$g(x, y, y', \ldots, y^{(n)}) = 0. \tag{22.16}$$

[6] Why use t as the independent variable in one case and x in the other? Only through habit. In most discussions of (22.16) the variables x, y are used, and in the case of (22.15) most authors use t, $\mathbf{x}$. In the latter case, such usage may result because, in applications, simultaneous first-order equations are often "rate" equations, where t is the time.

[As discussed in the preceding section, it may be convenient to reexpress a given equation (22.16) in the form (22.15)—that is, as a system of n first-order equations.]

Linear versions of (22.15) and (22.16) are as follows: (22.15) becomes

$$\begin{aligned} \dot{x}_1 &= a_{11}(t)x_1 + \cdots + a_{1n}(t)x_n + q_1(t) \\ &\ \vdots \\ \dot{x}_n &= a_{n1}(t)x_1 + \cdots + a_{nn}(t)x_n + q_n(t) \end{aligned}$$

or in matrix form[7]

$$\dot{\mathbf{x}} = A(t)\mathbf{x} + \mathbf{q}(t)$$

where $\mathbf{x} = \begin{pmatrix} x_1(t) \\ \vdots \\ x_n(t) \end{pmatrix}, \quad A = \begin{pmatrix} a_{11}(t) & \cdots & a_{1n}(t) \\ \vdots & & \vdots \\ a_{n1}(t) & \cdots & a_{nn}(t) \end{pmatrix}, \quad \mathbf{q} = \begin{pmatrix} q_1(t) \\ \vdots \\ q_n(t) \end{pmatrix},$

and (22.16) becomes

$$p_0(x)y^{(n)} + p_1(x)y^{(n-1)} + \cdots + p_n(x)y = r(x).$$

In discussing linear equations, we will consider the form (22.16); for a vector version, see Struble[8] for example.

First, we state (it follows from Theorem 22.1) that if the $p_j(x)$'s are all continuous in some interval $a \leq x \leq b$ and $p_0(x)$ does not vanish for any such x—conditions that will be assumed to hold throughout the remainder of this section—then the linear system

$$Ly = r(x), \qquad y, y', \ldots, y^{(n-1)} \quad \text{prescribed at some } x \text{ in the interval,} \tag{22.17}$$

where $\qquad L = p_0(x)D^n + p_1(x)D^{n-1} + \cdots + p_n(x), \qquad D \equiv \dfrac{d}{dx},$

does possess a unique solution $y(x)$ over $a \leq x \leq b$ (Exercise 22.5).[9]

Before proceeding, we should establish some ideas concerning linear dependence and linear independence. We say that *a set of functions* $u_1(x), \ldots, u_n(x)$ *is* **linearly dependent** (*LD*) *over* $a \leq x \leq b$ *if there are constants* $c_1, \ldots, c_n$ *not all zero, such that*

$$c_1u_1(x) + \cdots + c_nu_n(x) \equiv 0 \tag{22.18}$$

over $a \leq x \leq b$*; otherwise it is said to be* **linearly independent** (*LI*).[10] Essentially, this definition is the same as the one for linear dependence and independence of a set of *vectors* in Section 17.3.

---

[7]If the matrix notation used here is new, see Section 18.1.

[8]*Op. Cit.*

[9]Thus singular solutions are out of the question.

[10]In this book the special equal sign $\equiv$ can always be read in one of two ways, as should be clear from the context: *is defined as* (as in $D \equiv d/dx$) or *is identically equal to* [as in (22.18)].

THEOREM 22.2. For a set of functions $u_1(x), \ldots, u_n(x)$, having derivatives through order $n-1$, to be LD over $a \leq x \leq b$, a necessary (although not sufficient) condition is that $W(x) \equiv 0$, where the ***Wronskian***[11]

$$W(x) \equiv \begin{vmatrix} u_1(x) & \cdots & u_n(x) \\ u_1'(x) & & u_n'(x) \\ \cdot & & \cdot \\ \cdot & & \cdot \\ \cdot & & \cdot \\ u_1^{(n-1)}(x) & \cdots & u_n^{(n-1)}(x) \end{vmatrix}. \tag{22.19}$$

Proof. If $u_1, \ldots, u_n$ are to be LD, then (22.18) holds for all x in $a \leq x \leq b$ with the c_j's not all zero. Differentiating $n-1$ times, we have

$$\begin{aligned} c_1u_1(x) \;+ \cdots +\; c_nu_n(x) &= 0 \\ c_1u_1'(x) \;+ \cdots +\; c_nu_n'(x) &= 0 \\ \cdot \qquad\qquad \cdot \qquad & \quad \cdot \\ \cdot \qquad\qquad \cdot \qquad & \quad \cdot \\ \cdot \qquad\qquad \cdot \qquad & \quad \cdot \\ c_1u_1^{(n-1)}(x) + \cdots + c_nu_n^{(n-1)}(x) &= 0. \end{aligned} \tag{22.20}$$

Regarding x as fixed, we must have the determinant $W(x) = 0$ in order for the solution $c_1, \ldots, c_n$ of (22.20) to be nontrivial (Chapter 19). Since this condition must hold for *each* x in $a \leq x \leq b$, it follows that we *must* have $W(x) \equiv 0$ over the interval.

To show that the $W(x) \equiv 0$ condition is not sufficient, however, let $n = 2$ and suppose that

$$u_1 = \begin{cases} x^2, & 0 \leq x \leq 1 \\ 0, & -1 \leq x \leq 0 \end{cases}, \qquad u_2 = \begin{cases} 0, & 0 \leq x \leq 1 \\ x^2, & -1 \leq x \leq 0 \end{cases} \tag{22.21}$$

over $-1 \leq x \leq 1$. Then $c_1u_1 + c_2u_2 = 0$ becomes

$$\begin{aligned} c_1x^2 + 0 = 0 \quad &\text{for} \quad 0 \leq x \leq 1 \\ 0 + c_2x^2 = 0 \quad &\text{for} \; -1 \leq x \leq 0. \end{aligned}$$

The first implies that $c_1 = 0$ and the second that $c_2 = 0$. Consequently, $c_1u_1 + c_2u_2 \equiv 0$ over $-1 \leq x \leq 1$ only if $c_1 = c_2 = 0$, and hence u_1 and u_2 are LI. And yet

$$W(x) = \begin{cases} \begin{vmatrix} x^2 & 0 \\ 2x & 0 \end{vmatrix} = 0, & 0 \leq x \leq 1 \\ \begin{vmatrix} 0 & x^2 \\ 0 & 2x \end{vmatrix} = 0, & -1 \leq x \leq 0, \end{cases}$$

and thus $W(x) \equiv 0$. However,

THEOREM 22.3. If $u_1, \ldots, u_n$ happen to be solutions of an nth-order linear homogeneous differential equation $Ly = 0$, then it can be shown[12] *that $W(x) \equiv 0$ is both necessary and sufficient for the linear dependence of $u_1, \ldots, u_n$. [Recall that we have*

---

[11]The notation $W(u_1, \ldots, u_n)$ is also used.

[12]See J. C. Burkill, *The Theory of Ordinary Differential Equations*, Oliver and Boyd, Edinburgh, 1956, Chapter 2.

assumed, in this section, that the $p_j(x)$ coefficients in L are to be continuous over the interval under consideration, and that $p_0(x)$ does not vanish anywhere in the interval.]

We are now in a position to return to our discussion of the system (22.17). Several steps are involved. First,

THEOREM 22.4. The "homogenized" equation $Ly = 0$ has exactly n LI solutions, say $u_1(x), \ldots, u_n(x)$.

Proof. Suppose that $u_1(x), \ldots, u_{n+1}(x)$ are solutions of $Ly = 0$. Let ξ be some point in $a \le x \le b$. The n equations

$$\begin{aligned} c_1u_1(\xi) + \cdots + c_{n+1}u_{n+1}(\xi) &= 0 \\ &\vdots \\ c_1u_1^{(n)}(\xi) + \cdots + c_{n+1}u_{n+1}^{(n)}(\xi) &= 0 \end{aligned} \tag{22.22}$$

in the $n + 1$ unknowns $c_1, \ldots, c_{n+1}$ clearly have nontrivial solutions. Choosing such a nontrivial set of c_j's, define

$$v(x) \equiv c_1u_1(x) + \cdots + c_{n+1}u_{n+1}(x) \tag{22.23}$$

and observe first that

$$Lv = L(c_1u_1 + \cdots + c_{n+1}u_{n+1}) = c_1Lu_1 + \cdots + c_{n+1}Lu_{n+1} = 0 + \cdots + 0 = 0$$

and also that $v(\xi) = v'(\xi) = \cdots = v^{(n)}(\xi) = 0$. Now one function $v(x)$ that satisfies $Lv = 0$ and $v(\xi) = \cdots = v^{(n)}(\xi) = 0$ is simply $v(x) = 0$. By uniqueness, then, $v(x) \equiv 0$. Recalling that the c_j's in (22.23) are not all zero, it follows that $u_1, \ldots, u_{n+1}$ must be LD. Thus $Ly = 0$ *cannot have more than n* LI solutions.

To show that *there are no less than n* LI solutions of $Ly = 0$, let us simply put forward n such solutions. According to the existence theorem [recall the discussion accompanying equation (22.17)], there must be solutions $Y_1(x), \ldots, Y_n(x)$ of $Ly = 0$, satisfying the initial conditions

$$\begin{array}{llll} Y_1(x_0) = 1, & Y_1'(x_0) = 0, \ldots, & Y_1^{(n-1)}(x_0) = 0 & \\ Y_2(x_0) = 0, & Y_2'(x_0) = 1, & Y_2''(x_0) = 0, \ldots, & Y_2^{(n-1)}(x_0) = 0 \\ \vdots & & & \\ Y_n(x_0) = 0, \ldots, & Y_n^{(n-2)}(x_0) = 0, & Y_n^{(n-1)}(x_0) = 1, & \end{array}$$

where $a \le x_0 \le b$. We call this a **fundamental set** of solutions. That the set is LI follows from the fact that their Wronskian is nonzero at x_0,

$$W(x_0) = \begin{vmatrix} 1 & & & 0 \\ & 1 & & \\ & & \ddots & \\ 0 & & & 1 \end{vmatrix} = 1,$$

and from Theorem 22.3. This completes our proof.

Finally, if $u_1, \ldots, u_n$ are n LI solutions of $Ly = 0$ and y_p is any particular solution of $Ly = r(x)$, then by linearity

$$y = c_1 u_1(x) + \cdots + c_n u_n(x) + y_p(x) \tag{22.24}$$

is also a solution of $Ly = r(x)$, the *general* solution in fact, since $c_1, \ldots, c_n$ can be chosen so as to satisfy any initial conditions $y(x_0), y'(x_0), \ldots, y^{(n-1)}(x_0)$. That is, the equations

$$\begin{aligned}
c_1 u_1(x_0) \quad + \cdots + \quad c_n u_n(x_0) &= \quad y(x_0) - y_p(x_0) \\
c_1 u_1'(x_0) \quad + \cdots + \quad c_n u_n'(x_0) &= \quad y'(x_0) - y_p'(x_0) \\
&\vdots \\
c_1 u_1^{(n-1)}(x_0) + \cdots + c_n u_n^{(n-1)}(x_0) &= y^{(n-1)}(x_0) - y_p^{(n-1)}(x_0)
\end{aligned}$$

can be solved (uniquely) for $c_1, \ldots, c_n$, since the determinant of the coefficient matrix is the Wronskian of $u_1, \ldots, u_n$ at x_0, which cannot vanish because the u_j's are LI.

With these ideas in mind, let us turn to some methods of *solution*—that is, of determining $u_1, \ldots, u_n, y_p$ in (22.24).

22.3. *THE LINEAR EQUATION; SOME METHODS OF SOLUTION*

The examples below provide a *review* of some standard methods of solution of linear equations, which we assume the reader has encountered in a course on differential equations or engineering analysis. If a more systematic and detailed account is needed, the book by Spiegel[13] might be helpful.

Example 22.2. Solve

$$y'' - 9y = 0. \tag{22.25}$$

(When no boundary conditions are prescribed, it will be understood that the general solution is wanted.) Since (22.25) has *constant coefficients*, we will seek $y = e^{\lambda x}$. The idea is that the exponential function has the property that derivatives of all orders have the same form, to within a constant multiple, and so a linear combination of them can be made to cancel by suitable choice(s) of λ. Plugging into (22.25),

$$\lambda^2 e^{\lambda x} - 9e^{\lambda x} = (\lambda^2 - 9)e^{\lambda x} = 0.$$

Now $e^{\lambda x}$ does not vanish for *any* finite x, let alone for *all* x in a given interval; therefore we set the $\lambda^2 - 9$ factor equal to zero. Thus $\lambda = \pm 3$, and we have found *two* solutions, e^{3x} and e^{-3x}. Their Wronskian

$$W(e^{3x}, e^{-3x}) = \begin{vmatrix} e^{3x} & e^{-3x} \\ 3e^{3x} & -3e^{-3x} \end{vmatrix} = -6 \neq 0$$

indicates that they are LI; hence the general solution of (22.25) is

$$y = Ae^{3x} + Be^{-3x}, \tag{22.26}$$

[13]M. R. Spiegel, *Applied Differential Equations*, Prentice-Hall, Englewood Cliffs, N.J., 1958.

where we use the letters A, B in place of "c_1, c_2." Recalling that

$$e^x = \cosh x + \sinh x, \qquad e^{-x} = \cosh x - \sinh x,$$

(22.26) becomes

$$\begin{aligned} y &= (A + B)\cosh 3x + (A - B)\sinh 3x \\ &\equiv C\cosh 3x + D\sinh 3x, \end{aligned} \tag{22.27}$$

which is entirely equivalent to (22.26). Observe that the general solution could also be expressed as $Ee^{3x} + F\sinh 3x$, say, since e^{3x} and $\sinh 3x$ each satisfy (22.25) and are LI, but it is more standard to use (22.26) or (22.27).

Alternatively, note that the operator $L = D^2 - 9$ in (22.25) can be *factored* as $L = (D + 3)(D - 3)$, since

$$\begin{aligned} (D + 3)(D - 3)y &= (D + 3)(y' - 3y) = D(y' - 3y) + 3(y' - 3y) \\ &= y'' - 3y' + 3y' - 9y = y'' - 9y. \end{aligned}$$

Thus we can express (22.25) as

$$(D + 3)(D - 3)y = 0,$$

and if we set $(D - 3)y \equiv u$, this equation reduces to the first-order system

$$u' + 3u = 0 \tag{22.28}$$

$$y' - 3y = u. \tag{22.29}$$

Solving first (22.28) and then (22.29), with the help of (21.20), we obtain $u = Ae^{-3x}$ and $y = C_1e^{3x} + C_2e^{-3x}$ as before. ∎

Example 22.3.

$$y'' + 9y = 0, \qquad y(0) = 1, \quad y'(0) = 2. \tag{22.30}$$

Seeking $y = e^{\lambda x}$, we have $(\lambda^2 + 9)e^{\lambda x} = 0$, so that $\lambda = \pm 3i$, and therefore

$$y = Ae^{i3x} + Be^{-i3x} \tag{22.31}$$

or recalling Euler's formulas,

$$\begin{aligned} y &= A(\cos 3x + i\sin 3x) + B(\cos 3x - i\sin 3x) \\ &= (A + B)\cos 3x + i(A - B)\sin 3x = C\cos 3x + D\sin 3x. \end{aligned} \tag{22.32}$$

The reader may worry that D is inevitably complex, but that need not be so. Specifically, observe that if A and B are complex conjugates, then both $C = A + B$ and $D = i(A - B)$ are purely real.

The form

$$y = E\sin(3x + \phi) \tag{22.33}$$

is also equivalent to (22.21) and (22.32); that is,

$$E\sin(3x + \phi) = E\sin\phi\cos 3x + E\cos\phi\sin 3x,$$

and comparison with (22.32) implies that[14]

$$C = E\sin\phi, \qquad D = E\cos\phi$$

or

$$E = \sqrt{C^2 + D^2}, \qquad \phi = \tan^{-1}\frac{C}{D}.$$

[14] $(C - E\sin\phi)\cos 3x + (D - E\cos\phi)\sin 3x = 0$ implies that $C - E\sin\phi = 0$ and $D - E\cos\phi = 0$, *since* $\cos 3x$ *and* $\sin 3x$ *are* LI.

It is striking that the combination (22.32) should turn out to be a single sine wave (with amplitude E, frequency 3, and phase angle ϕ).

Finally, applying the conditions $y(0) = 1, y'(0) = 2$ to the general solution (22.32), say, we have $1 = C$ and $2 = 3D$, and thus the desired solution of (22.30) is $y = \cos 3x + \frac{2}{3} \sin 3x$. ▮

Example 22.4.

$$y'' + y = 0, \qquad y(0) = y(\pi) = 0. \tag{22.34}$$

The general solution is $y = A \sin x + B \cos x$, and the boundary conditions require that $0 = B$ and $0 = -B$. Thus the solution to (22.34) is $y = A \sin x$ for *any* A. But note carefully that this lack of uniqueness does *not* violate our existence/uniqueness theorem, since (22.34) is a *boundary value* problem, whereas our entire existence/uniqueness discussion was for *initial value problems*—that is, where $y, y', \ldots, y^{(n-1)}$ are all prescribed at one point in the interval. ▮

Example 22.5.

$$y'''' + k^4 y = 0. \tag{22.35}$$

With $y = e^{\lambda x}$ we find that $\lambda^4 = -k^4$. So (Exercise 22.10)

$$\lambda = \frac{k(1+i)}{\sqrt{2}}, \quad \frac{k(-1+i)}{\sqrt{2}}, \quad \frac{k(1-i)}{\sqrt{2}}, \quad \frac{-k(1+i)}{\sqrt{2}},$$

and $y = e^{\alpha x}(A \sin \alpha x + B \cos \alpha x) + e^{-\alpha x}(C \sin \alpha x + D \cos \alpha x), \quad \alpha = \dfrac{k}{\sqrt{2}}$. ▮

Example 22.6.

$$y'' + 2y' + y = 0. \tag{22.36}$$

Seeking $y = e^{\lambda x}$, we have $\lambda^2 + 2\lambda + 1 = 0$ or $(\lambda + 1)^2 = 0$. Instead of obtaining two distinct roots, and hence the general solution of (22.36), we have the repeated root $\lambda = -1, -1$, so that

$$y = Ae^{-x}, \tag{22.37}$$

which is only half the general solution. Apparently the missing solution is not of the assumed form $e^{\lambda x}$. We will seek the missing solution by the method of **variation of parameters**; that is, we let the parameter A in (22.37) vary and thus try

$$y = A(x)e^{-x}. \tag{22.38}$$

Surely the missing solution, whatever it is, can be expressed in this form; for instance, if it is $\sin x$, then $A(x) = e^x \sin x$. The crucial point is not that it *can* be expressed in the form (22.38) but rather that it is *fruitful* to do so, as will now be demonstrated. Plugging (22.38) into (22.36),

$$A''e^{-x} - 2A'e^{-x} + Ae^{-x} + 2A'e^{-x} - 2Ae^{-x} + Ae^{-x} = 0$$

or

$$A'' = 0, \qquad A(x) = C + Dx.$$

Thus the general solution of (22.36) is

$$y = (C + Dx)e^{-x};$$

that is, the "missing solution" is found to be xe^{-x}.

In fact, we state without proof that if λ is repeated, of multiplicity k say, then a solution of the (nth-order linear) differential equation (with constant coefficients) is

$$(C_1 + C_2 x + \cdots + C_k x^{k-1})e^{\lambda x}.$$

To illustrate, if we start with some seventh-order differential equation and find that the λ equation can be factored as

$$\lambda^2(\lambda - 2)(\lambda + 3)^4 = 0,$$

then the general solution is

$$y = (A + Bx) + Ce^{2x} + (D + Ex + Fx^2 + Gx^3)e^{-3x}. \quad \blacksquare$$

Example 22.7.

$$x^2y'' + xy' - 9y = 0. \tag{22.39}$$

Equation (22.39) has *nonconstant coefficients*, which means that the form $y = e^{\lambda x}$ will not work [since the three terms in (22.39) would be of the form $x^2e^{\lambda x}$, $xe^{\lambda x}$, and $e^{\lambda x}$, which are LI and cannot cancel]. Equations with nonconstant coefficients are generally much more unyielding than the constant coefficient case, and we often resort to the powerful but tedious methods of series solution, as will be discussed in the next section.

In the present case, however, we recognize (from a course in differential equations) (22.39) as a **Cauchy–Euler equation**—that is, of the form

$$a_0x^ny^{(n)} + a_1x^{n-1}y^{(n-1)} + \cdots + a_ny = 0, \tag{22.40}$$

where the a_j's are constants. Here we seek

$$y = x^\alpha,$$

for then each term in the equation is proportional to x^α, and we can obtain cancellation by suitable choice(s) of α. Thus (22.39) becomes

$$[\alpha(\alpha - 1) + \alpha - 9]x^\alpha = 0$$

so that $\alpha = \pm 3$ and

$$y = Ax^3 + Bx^{-3}. \quad \blacksquare$$

Example 22.8.

$$x^2y'' - xy' + y = 0. \tag{22.41}$$

Here is another Cauchy–Euler equation, and the form $y = x^\alpha$ yields $\alpha^2 - 2\alpha + 1 = 0$ with the repeated root $\alpha = 1, 1$. Therefore $y = Ax$. To find the missing solution, seek $y = A(x)x$. Plugging into (22.41),

$$A''x^3 + 2A'x^2 - A'x^2 - Ax + Ax = 0 \tag{22.42}$$

or

$$A''x + A' = 0. \tag{22.43}$$

Setting $A' = p$, (22.43) becomes $xp' + p = 0$ or $(dp/p) + (dx/x) = 0$, so that $\ln p + \ln x = \ln C$, $p = C/x$, and $A(x) = C \ln x + D$. Then

$$y = (C \ln x + D)x.$$

Observe that $y = A(x)x$ led to a second-order equation with nonconstant coefficients for A! Fortunately, it was reducible to first order through the substitution $A' = p$ because there were no A terms in (22.43). This situation was not merely coincidental; the A terms in (22.42) *had to* cancel because, if A were constant, the first three terms in (22.42) drop out, and the remaining terms *must* cancel because $y = Ax$ is a solution of (22.41).

Alternatively, Cauchy–Euler equations can always be reduced to constant coefficient equations (Exercise 22.13) through the change of independent variable

$$x = e^t. \tag{22.44}$$

Applying this result to (22.41), we find that

$$\frac{d^2y}{dt^2} - 2\frac{dy}{dt} + y = 0$$

and

$$y = (A + Bt)e^t = (A + B \ln x)x$$

as before. ∎

Example 22.9.

$$y'' - (1 + x^2)y = 0. \tag{22.45}$$

This equation looks difficult, and it is reasonable to expect that we must seek a series solution or settle for an approximate solution by means of the Ritz method (Chapter 10) or the method of weighted residuals (Chapter 24) for example.

Instead let us try to factor the operator $D^2 - (1 + x^2)$ as in Example 22.2. That is, let us seek $a(x)$, $b(x)$ such that $D^2 - (1 + x^2) = (D - a)(D - b)$. Then

$$\begin{aligned}(D - a)(D - b)y &= (D - a)(y' - by) = y'' - (a + b)y' + (ab - b')y \\ &= [D^2 - (a + b)D + (ab - b')]y,\end{aligned}$$

and so we need

$$a + b = 0, \qquad ab - b' = -(1 + x^2),$$

or

$$b' + b^2 = 1 + x^2. \tag{22.46}$$

This, in turn, is a hard equation (specifically, a **Riccati equation**; see Exercise 21.13), and it is simply by good fortune that a solution $b = x$ is seen by inspection. So with $b = x$ and $a = -x$,

$$[D^2 - (1 + x^2)]y = (D + x)(D - x)y = 0, \tag{22.47}$$

and the solution can be found by setting $(D - x)y = u$ and proceeding as indicated in Example 22.2 (Exercise 22.17).

Note carefully that $(D + x)(D - x)$ is *not* equal to $D^2 - x^2$ but rather $D^2 - (1 + x^2)$; the additional term comes from the action of the first D in $(D + x) \cdot (D - x)$ on the second x. In other words, for the case of nonconstant coefficients, factoring may be quite difficult. However, if it does work, it can be very nice, and save a lot of work. (See Exercise 22.18.) ∎

Example 22.10.

$$y'''' - y'' = x^2 - e^{2x}. \tag{22.48}$$

Here is our first *inhomogeneous* equation—that is, $Ly = f(x)$, where f is not zero. Recall from (22.24) that the solution can be found as the sum of the so-called *complementary solution*

$$y_c = c_1u_1(x) + \cdots + c_nu_n(x),$$

that is, the general solution of the "homogenized" equation and a *particular solution* y_p.

It is easily found that

$$y_c = A + Bx + Ce^x + De^{-x}, \tag{22.49}$$

and it remains to determine y_p. In this example we show how to find y_p by the **method of undetermined coefficients**; in the next two another approach is used.

The procedure is as follows: Write down each term on the right-hand side [of (22.48)], together with all its derivatives, ignoring any multiplicative constants.

Thus

$$x^2, x, 1 \quad \text{and} \quad e^{2x}.$$

Next, seek y_p, tentatively, as a linear combination of these terms,

$$y_p = (Ex^2 + Fx + G) + (He^{2x}), \tag{22.50}$$

where it is helpful to enclose each "family" of terms in parentheses. Finally, check each term in y_p for duplication with the terms in y_c. Comparison of (22.50) with (22.49) reveals duplication between the Fx and G terms in y_p and the Bx and A terms in y_c, respectively. We need to multiply the entire family involved in the duplication by the lowest integral power of x that will eliminate all such duplication. So we seek y_p in the revised form[15]

$$y_p = x^2(Ex^2 + Fx + G) + (He^{2x}).$$

Inserting it into (22.48),

$$-12Ex^2 - 6Fx - (24E - 2G) + 12He^{2x} = x^2 - e^{2x},$$

and equating coefficients of LI terms—namely, x^2, x, 1, e^{2x}—we have

$$-12E = 1, \qquad -6F = 0, \qquad 24E - 2G = 0, \qquad 12H = -1,$$

so that $E = -\frac{1}{12}$, $F = 0$, $G = -1$, and $H = -\frac{1}{12}$. Consequently, the general solution of (22.48) is

$$y = y_c + y_p = A + Bx + Ce^x + De^{-x} - \frac{x^2}{12}(x^2 + 12) - \frac{1}{12}e^{2x}.$$

The method is guaranteed, provided that the operator is linear and has constant coefficients and that each family contains only a finite number of terms.[16] It is also understood that the method may not be much fun if a large number of terms are needed.[17] ▮

Example 22.11.

$$\ddot{x} + \omega^2 x = f(t). \tag{22.51}$$

The complementary solution is simply

$$x_c = A \sin \omega t + B \cos \omega t.$$

The method of undetermined coefficients is useless in finding x_p, since the form of $f(t)$ is not specified. We will use instead the method of **variation of parameters,** which is due to Lagrange. According to it, we take x_p to be of the same form as x_c but with variable coefficients

$$x_p = A(t) \sin \omega t + B(t) \cos \omega t. \tag{22.52}$$

This method is similar to the variation of parameter technique of Example 22.6, but here it is being used to obtain the *particular* solution.

Obviously the particular solution can be expressed in the form (22.52);[18] the point is not that it can but rather that it will be fruitful to do so.

---

[15] Note that $y_p = x(Ex^2 + Fx + G) + (He^{2x}) = Ex^3 + Fx^2 + Gx + He^{2x}$ is unsatisfactory because we then have duplication between Gx and Bx—that is, duplication in *form.* The constant factors G and B are not relevant insofar as duplication is concerned.

[16] For instance, $1/x$ generates the family $1/x, 1/x^2, 1/x^3, \ldots$.

[17] For instance, x^{27} generates the family $x^{27}, x^{26}, x^{25}, \ldots, x, 1$.

[18] *Any* function can; that is, any $F(t)$ can be expressed in the form $A(t) \sin \omega t + B(t) \cos \omega t$ by setting $A(t) = F(t)/\sin \omega t$ and $B(t) = 0$ or $A(t) = 0$ and $B(t) = F(t)/\cos \omega t$, and so on.

In order to put (22.52) into the left side of (22.51), we need $\ddot{x}_p$. Taking d/dt of (22.52) first,

$$\dot{x}_p = A\omega \cos \omega t - B\omega \sin \omega t + \dot{A} \sin \omega t + \dot{B} \cos \omega t. \qquad (22.53)$$

Before taking d/dt again, let us look ahead. Substitution of (22.52) into (22.51) will provide us with one equation in the two unknowns $A(t)$ and $B(t)$. Apparently we can afford to add one more relation between A and B,[19] essentially at our pleasure. Consequently, we set

$$\dot{A} \sin \omega t + \dot{B} \cos \omega t = 0. \qquad (22.54)$$

This step deletes the $\dot{A}$ and $\dot{B}$ terms from (22.53), thereby avoiding the entry of $\ddot{A}$ and $\ddot{B}$ terms at the next differentiation. Proceeding,

$$\ddot{x}_p = \dot{A}\omega \cos \omega t - \dot{B}\omega \sin \omega t - A\omega^2 \sin \omega t - B\omega^2 \cos \omega t,$$

so that (22.51) becomes

$$\begin{aligned} &\dot{A}\omega \cos \omega t - \dot{B}\omega \sin \omega t - A\omega^2 \sin \omega t \\ &\quad - B\omega^2 \cos \omega t + A\omega^2 \sin \omega t + B\omega^2 \cos \omega t = f(t). \end{aligned}$$

The A and B terms on the left cancel [as they *must*, since (22.52) satisfies $\ddot{x} + \omega^2 x = 0$ if A and B are constant], and we have the two equations

$$\begin{aligned} \dot{A} \sin \omega t + \dot{B} \cos \omega t &= 0 \\ \dot{A}\omega \cos \omega t - \dot{B}\omega \sin \omega t &= f(t) \end{aligned} \qquad (22.55)$$

or

$$\dot{A} = \frac{f(t) \cos \omega t}{\omega}, \qquad \dot{B} = -\frac{f(t) \sin \omega t}{\omega}.$$

Integrating and putting the results into (22.52), we have[20]

$$\begin{aligned} x_p &= \frac{\sin \omega t}{\omega} \int_0^t f(\tau) \cos \omega \tau \, d\tau - \frac{\cos \omega t}{\omega} \int_0^t f(\tau) \sin \omega \tau \, d\tau \\ &= \frac{1}{\omega} \int_0^t f(\tau) \sin \omega (t - \tau) \, d\tau \end{aligned}$$

and

$$\begin{aligned} x &= x_c + x_p \\ &= A \sin \omega t + B \cos \omega t + \frac{1}{\omega} \int_0^t f(\tau) \sin \omega (t - \tau) \, d\tau. \end{aligned} \qquad (22.56)$$

Note the Laplace-convolution form of the particular solution. In fact, in Example 6.3 we solved essentially the same problem by the Laplace transform.

Observe that the method of variation of parameters is more powerful than undetermined coefficients in that it applies regardless of the form of the "driving term" on the right-hand side [in fact, here $f(t)$ wasn't even specified] and that it works even if the operator L has nonconstant coefficients. ∎

[19] Not surprising in view of the previous footnote; specifically, in view of the nonuniqueness in the determination of $A(t)$ and $B(t)$.

[20] The lower integration limit is immaterial; that is, the difference between $\int_0^t f(\tau) \sin \omega(t - \tau) \, d\tau/\omega$ and $\int_1^t f(\tau) \sin \omega(t - \tau) \, d\tau/\omega$, say, is some constant times $\sin \omega t$ plus some constant times $\cos \omega t$, which can be absorbed into the complementary solution in any case.

Example 22.12.

$$y' + M(x)y = N(x). \tag{22.57}$$

Let us use variation of parameters again. The homogeneous equation is variable separable, and we easily obtain

$$y_c = Ae^{-\int M\,dx}. \tag{22.58}$$

Thus we seek

$$y_p = A(x)e^{-\int M\,dx}. \tag{22.59}$$

Putting it into (22.57),

$$A'e^{-\int M\,dx} - AMe^{-\int M\,dx} + AMe^{-\int M\,dx} = N;$$

therefore

$$A' = Ne^{\int M\,dx}, \qquad A = \int Ne^{\int M\,dx}\,dx,$$

and

$$y = y_c + y_p = e^{-\int M\,dx}\left[\int Ne^{\int M\,dx}\,dx + C\right], \tag{22.60}$$

as obtained in Example 21.4 by means of an integrating factor.

Here we did *not* append an additional relation, analogous to (22.54), since the insertion of (22.59) into (22.57) yielded one equation in the one unknown, $A(x)$.

Like so many of Lagrange's contributions, variation of parameters is simple, powerful, and elegant. ▮

Example 22.13.

$$2\dot{x} - \dot{y} + x + y = t, \qquad \dot{x} + \dot{y} + 4x = 3. \tag{22.61}$$

First, reexpress (22.61) in operator notation,

$$(2D + 1)x - (D - 1)y = t \tag{22.62a}$$

$$(D + 4)x + Dy = 3. \tag{22.62b}$$

To eliminate y, say, operate on (22.62a) with D and on (22.62b) with $D - 1$,

$$D(2D + 1)x - D(D - 1)y = 1$$

$$(D - 1)(D + 4)x + (D - 1)Dy = -3.$$

Adding,

$$(3D^2 + 4D - 4)x = -2, \tag{22.63}$$

so that

$$x = Ae^{2t/3} + Be^{-2t} + \tfrac{1}{2}. \tag{22.64}$$

Similarly, we can eliminate x from (22.62) by operating on (22.62a) with $D + 4$, on (22.62b) with $2D + 1$, and subtracting. This process yields

$$(3D^2 + 4D - 4)y = 2 - 4t, \tag{22.65}$$

and therefore

$$y = Ce^{2t/3} + De^{-2t} + t + \tfrac{1}{2}. \tag{22.66}$$

The fact that the operators in (22.63) and (22.65) are the same is no coincidence; they are the determinant

$$\begin{vmatrix} (2D + 1) & -(D - 1) \\ (D + 4) & D \end{vmatrix}$$

from (22.62).

The integration constants A, B, C, D are *not* independent, since x and y are related through (22.62). Inserting (22.64) and (22.66) into (22.62a) or (22.62b), say the latter, we have

$$\left(\tfrac{14}{3}A + \tfrac{2}{3}C\right)e^{2t/3} + (2B - 2D)e^{-2t} + 3 = 3,$$

so that $C = -7A$ and $D = B$; moreover,

$$y = -7Ae^{2t/3} + Be^{-2t} + t + \tfrac{1}{2}. \tag{22.67}$$

Actually, there was a shortcut for this particular example. Having obtained $x(t)$ according to (22.64), $y(t)$ can be obtained by adding (22.62a) and (22.62b). This step yields

$$y = t + 3 - 3\dot{x} - 5x = -7Ae^{2t/3} + Be^{-2t} + t + \tfrac{1}{2},$$

as before. ▮

22.4. SERIES SOLUTION; BESSEL AND LEGENDRE FUNCTIONS

Equations of the form

$$p(x)y'' + q(x)y' + r(x)y = 0, \tag{22.68}$$

with nonconstant coefficients, are of great importance in applications but are susceptible to the preceding methods only in exceptional cases (e.g., Examples 22.7 and 22.9). In this section we see how *series* solutions can often be obtained and how it is sometimes possible to bypass that approach by finding a change of variables that will throw (22.68) into an equation whose solutions *are* known.

Series Solutions. If both $q(z)/p(z)$ and $r(z)/p(z)$ are analytic at $z = a$, then a is said to be an **ordinary point** of (22.68); otherwise it is a **singular point** of (22.68).[21]

THEOREM 22.5. Suppose that $z = 0$ is an ordinary point of (22.68) and that $q(z)/p(z)$ and $r(z)/p(z)$ are analytic in $|z| < \rho$. Then the general solution of (22.68) can be found as

$$y = \sum_0^\infty c_n x^n = c_0 y_1(x) + c_1 y_2(x) \tag{22.69}$$

where the series representing the LI solutions $y_1(x)$ and $y_2(x)$ have radii of convergence at least as large as ρ. The c_n's are determined through substitution of $\sum c_n x^n$ into (22.68), except for c_0 and c_1, which remain arbitrary.

More generally, we may desire a solution about some point $a \neq 0$—that is, of the form $\sum c_n(x - a)^n$—as will be especially convenient if initial conditions are

---

[21] We found in Section 14.2 that the interval of convergence of the Taylor expansion $f(x) = f(a) + f'(a)(x - a) + \cdots$ is determined not only by the singularities of $f(x)$ on the x axis but by the singularities of $f(z)$ at points off the axis in the complex plane as well. It is not surprising, then, that (in the following theorem) we address the analyticity of q/p and r/p in the *disk* $|z| < \rho$ in the complex plane. If you have not yet read Part II but are familiar with real-variable Taylor series, you may replace the first sentence of Theorem 22.5 by the following: Suppose that the Taylor expansions of $q(x)/p(x)$ and $r(x)/p(x)$ about $x = 0$ converge (to q/p and r/p, respectively) in $|x| < \rho$.

specified at $x = a$. However, there is no loss of generality in Theorem 22.5, since $x = a$ can always be shifted to the origin by the change of variables $x - a = t$.

Example 22.14.

$$y'' - y = 0. \tag{22.70}$$

Here $q(z)/p(z) = 0$ and $r(z)/p(z) = -1$ are analytic for *all* z, and so the form $y = \sum c_n x^n$ will lead to the general solution with the series convergent for all x.

Inserting $y = \sum c_n x^n$ into (22.70),

$$\sum_2^{\infty} n(n-1)c_n x^{n-2} - \sum_0^{\infty} c_n x^n = 0$$

or setting $n - 2 = m$ in the first series,

$$\sum_0^{\infty} (m+2)(m+1)c_{m+2}x^m - \sum_0^{\infty} c_n x^n = 0.$$

Because m is only a dummy index, it can be changed to n. Thus

$$\sum_0^{\infty} [(n+2)(n+1)c_{n+2} - c_n]x^n = 0$$

and so we obtain the *recursion formula*

$$(n+2)(n+1)c_{n+2} - c_n = 0 \qquad (n = 0, 1, 2, \ldots)$$

for the c_n's. Setting $n = 0, 1, 2, \ldots$, we have

$$c_2 = \frac{c_0}{2!}$$

$$c_3 = \frac{c_1}{3!}$$

$$c_4 = \frac{c_2}{4 \cdot 3} = \frac{c_0}{4!}$$

$$c_5 = \frac{c_3}{5 \cdot 4} = \frac{c_1}{5!},$$

and so on, and therefore

$$y = c_0\left(1 + \frac{x^2}{2!} + \frac{x^4}{4!} + \cdots\right) + c_1\left(x + \frac{x^3}{3!} + \frac{x^5}{5!} + \cdots\right)$$

$$= c_0 \cosh x + c_1 \sinh x.$$

Generally, however, identification of the series in terms of known functions is not possible. ∎

If $z = 0$ is a singular point, the situation is more complicated. Nevertheless, if the singularity is "not too strong," results can still be obtained. Specifically, suppose that $z = 0$ is a singular point of (22.68), so that $q(z)/p(z)$ and/or $r(z)/p(z)$ are singular there. If, however, both

$$\frac{zq(z)}{p(z)} \quad \text{and} \quad \frac{z^2 r(z)}{p(z)}$$

are analytic there, then we say that $z = 0$ is a **regular singular point**.

As motivation for the important theorem that follows, suppose that $z = 0$ is a regular singular point and rewrite (22.68) as

$$x^2y'' + x\left[\frac{xq(x)}{p(x)}\right]y' + \left[\frac{x^2r(x)}{p(x)}\right]y = 0.$$

Since the two bracketed quantities are analytic at $x = 0$, they admit Maclaurin expansions; hence

$$x^2y'' + x[a_0 + a_1x + \cdots]y' + [b_0 + b_1x + \cdots]y = 0. \tag{22.71}$$

We expect the solutions of (22.71) and the "reduced" version

$$x^2y'' + a_0xy' + b_0y = 0 \tag{22.72}$$

to coincide in the neighborhood of $x = 0$. But (22.72) is a Cauchy–Euler equation and admits at least one solution of the form $y = x^\alpha$. Consequently, we anticipate that (22.71) will admit at least one solution of the form

$$y = x^\alpha \sum_0^\infty c_nx^n, \tag{22.73}$$

that is, x^α times a regular function; (22.73) is called a **Frobenius type series**, and its determination is called **the method of Frobenius.**

THEOREM 22.6. (Frobenius-Type Solution) Suppose that $z = 0$ *is a regular singular point of* (22.68) *and that* $zq(z)/p(z)$ *and* $z^2r(z)/p(z)$ *are analytic in* $|z| < \rho$. *Then* (22.68) *admits at least one Frobenius-type solution,*

$$y = x^\alpha \sum_0^\infty c_nx^n, \tag{22.74}$$

where the radius of convergence of the series is at least as large as ρ. α *and the* c_n*'s are determined through substitution of* (22.74) *into* (22.68).

How do we interpret x^α if $x < 0$ and $\alpha = \frac{1}{3}$, say? If $x < 0$, we can set $-x = t$ in (22.68). The net result is that the form (22.74) still holds but in the t variable. Since $t^\alpha = (-x)^\alpha = |x|^\alpha$, we see that x^α in (22.74) is to be understood as $|x|^\alpha$. In fact, in a deeper discussion of series solutions it would be best to deal with the *complex* version

$$p(z)\frac{d^2w}{dz^2} + q(z)\frac{dw}{dz} + r(z)w = 0$$

from the start. See, for instance, the treatment in Burkill.[22]

Let us turn to a discussion of some of the important special functions that occur as solutions of equations of the form (22.68).

Bessel Functions. As will be seen in Part V, ***Bessel's equation*** of order ν

$$x^2y'' + xy' + (x^2 - \nu^2)y = 0 \tag{22.75}$$

arises when we separate the heat or wave equation in polar or cylindrical coordinates. This fact alone would be enough to ensure the equation and its solutions a place of great importance. In addition, it appears in a multitude of engineering applications, many of which are described in the book by McLachlan.[23]

---

[22]J. C. Burkill, *The Theory of Ordinary Differential Equations*, Oliver and Boyd, Edinburgh, 1956.

[23]N. W. McLachlan, *Bessel Functions for Engineers*, Oxford University Press, New York, 1934.

The only singular point of (22.75) is $x = 0$, and it is seen to be a regular singular point (so that a Frobenius-type solution must exist). Rather than avoid the singular point $x = 0$, it is precisely that point about which we will expand, since the nature of the solutions will then be most clearly exposed.

Consider first the Bessel equation *of order zero*—that is, for $\nu = 0$,

$$xy'' + y' + xy = 0. \tag{22.76}$$

Inserting (22.74) into this equation,

$$\sum_0^\infty (n + \alpha)(n + \alpha - 1)c_n x^{n+\alpha-1} + \sum_0^\infty (n + \alpha)c_n x^{n+\alpha-1} + \sum_0^\infty c_n x^{n+\alpha+1} = 0.$$

If we set $n = m + 2$ in the first two series and change the dummy index m back to n and the lower summation limit in the third series from 0 to -2, with the understanding that $c_{-1} = c_{-2} = 0$, then the three series can be combined (since the limits are the same in each case, as are the exponents on x) to yield

$$\sum_{-2}^\infty [(n + \alpha + 2)^2 c_{n+2} + c_n]x^{n+\alpha+1} + 0,$$

from which the recursion formula

$$(n + \alpha + 2)^2 c_{n+2} + c_n = 0 \qquad (n = -2, -1, 0, \ldots) \tag{22.77}$$

follows. The *first*—that is, for $n = -2$—will yield a quadratic equation in α, known as the *indicial equation*. Specifically, $\alpha^2 c_0 = 0$; and for $c_0 \neq 0$[24] we have the very simple indicial equation $\alpha^2 = 0$ with the repeated roots $\alpha = 0, 0$.

We state the following results without proof.

(i) *If the indicial roots are equal, only one (Frobenius) solution is obtained.*
(ii) *If they differ by a constant that is not an integer, the general solution is obtained.*
(iii) *If they differ by a nonzero integer, there are two possibilities:*
  (a) *The algebraically smaller root leads to the general solution.*
  (b) *The algebraically smaller root leads to no solution.*
  In either case, the larger root leads to one solution.

In the present case, then, $\alpha = 0$ will lead to one solution, the other solution not being of Frobenius type.

Proceeding, set $\alpha = 0$ in (22.77). For $n = -1, 0, \ldots$ we have

$$c_1 = 0,$$

$$c_2 = -\frac{c_0}{2^2},$$

$$c_3 = -\frac{c_1}{3^2} = 0,$$

$$c_4 = -\frac{c_2}{4^2} = \frac{c_0}{2^2 4^2},$$

---

[24]For definiteness in (22.74), α is understood to be the lowest power of x in the series, and c_0 is its coefficient. Necessarily, then, $c_0 \neq 0$.

and so on; therefore

$$y_1(x) = 1 - \frac{x^2}{2^2} + \frac{x^4}{2^2 4^2} - \cdots \equiv J_0(x), \tag{22.78}$$

where we've normalized $c_0 = 1$ and where

$$J_0(x) = \sum_0^\infty \frac{(-1)^n}{(n!)^2}\left(\frac{x}{2}\right)^{2n} \tag{22.79}$$

is known as the **Bessel function of the first kind and order zero**, the "zero" referring to the fact that $\nu = 0$ in (22.75).

Seeking y_2 by variation of parameters

$$y_2 = A(x)J_0(x), \tag{22.80}$$

and putting this result in (22.76) gives

$$xA''J_0 + 2xA'J_0' + A'J_0 + A(xJ_0'' + J_0' + xJ_0) = 0.$$

Recalling that $xJ_0'' + J_0' + xJ_0 = 0$ and setting $A' = p$, we obtain (Exercise 22.26)

$$A(x) = \int \frac{dx}{xJ_0^2},$$

and hence

$$y_2 = J_0(x) \int \frac{dx}{xJ_0^2}. \tag{22.81}$$

To reduce (22.81), we can observe that $1/J_0^2$ is analytic at $x = 0$, expand it in a Maclaurin series (or else "simply" divide J_0^2 into 1 by long division,) and then integrate termwise.

Alternatively, we can simply use this argument to observe that the *form* of y_2 is

$$J_0(x) \int \left(\frac{1}{x} + a + bx + \cdots\right) dx = J_0(x) \ln x + \text{regular function},$$

and thus seek

$$y_2 = J_0(x) \ln x + \sum_0^\infty c_n x^n, \tag{22.82}$$

where the c_n's are determined by substituting (22.82) into (22.76), except that c_0 is found to remain arbitrary. Setting $c_0 = 0$, the result is

$$y_2 = J_0(x) \ln x + \left(\frac{x}{2}\right)^2 - \frac{(x/2)^4}{(2!)^2}\left(1 + \frac{1}{2}\right) + \frac{(x/2)^6}{(3!)^2}\left(1 + \frac{1}{2} + \frac{1}{3}\right) - \cdots \equiv \mathbf{Y}_0(x), \tag{22.83}$$

which is known as the **Neumann function of order zero**. Next, if J_0 and $\mathbf{Y}_0$ are LI solutions, then so are J_0 and a linear combination of J_0 and $\mathbf{Y}_0$. In particular, at the suggestion of Weber and Schlafli, it is more standard to take

$$\begin{aligned} y_2 &= \frac{2}{\pi}[\mathbf{Y}_0(x) + (\gamma - \ln 2)J_0(x)] \\ &= \frac{2}{\pi}\left[J_0(x)\left(\ln \frac{x}{2} + \gamma\right) + \left(\frac{x}{2}\right)^2 - \frac{(x/2)^4}{(2!)^2}\left(1 + \frac{1}{2}\right)\right. \\ &\quad \left. + \frac{(x/2)^6}{(3!)^2}\left(1 + \frac{1}{2} + \frac{1}{3}\right) - \cdots\right] \\ &\equiv Y_0(x) \end{aligned} \tag{22.84}$$

as the companion solution for $J_0(x)$, where $\gamma \doteq 0.5772157$ is Euler's constant and Y_0 is called the **Bessel function of second kind and order zero.** This form has the advantage of yielding a certain simple asymptotic form as $x \longrightarrow \infty$, as we will see.

Thus we express the general solution of the Bessel equation of order zero as

$$y = AJ_0(x) + BY_0(x). \tag{22.85}$$

Of special interest is the small- and large-argument behavior of J_0 and Y_0. For small x we see immediately from (22.78) and (22.84) that

$$J_0(x) \sim 1 \quad \text{and} \quad Y_0(x) \sim \frac{2}{\pi} \ln x \quad \text{as } x \longrightarrow 0, \tag{22.86}$$

where the logarithmic singularity in $Y_0(x)$ should be carefully noted.

The large x behavior, however, is not directly available from (22.78) and (22.84). Instead let us try to *factor* the Bessel equation in the form

$$\begin{aligned} x^2y'' + xy' + x^2y &= (xD + a)(xD + b)y \\ &= x^2y'' + x(1 + a + b)y' + (xb' + ab)y. \end{aligned}$$

Equating coefficients of y' and y, we see that $a = -b$ and that b satisfies the *Riccatti equation* (recall Exercise 21.13)

$$xb' - b^2 = x^2. \tag{22.87}$$

Then, with $(xD + b)y \equiv u$, the Bessel equation reduces to the system

$$xu' - bu = 0, \qquad xy' + by = u,$$

which is easily solved for u and then y, yielding[25]

$$y = Ce^{-\int (b/x)dx} \tag{22.88}$$

Unfortunately, the Riccatti equation (22.87) is rather intractable insofar as an *exact* solution is concerned. Since we are merely seeking the asymptotic behavior of (22.88), however, we need only the asymptotic behavior of b. Suppose that as $x \longrightarrow \infty$ the xb' term dominates the $-b^2$ term in (22.87) with the result that $xb' \sim x^2$ and hence $b \sim x^2/2$. But then $-b^2 \sim x^4/4$, which is *not* dominated by xb' as assumed. Instead, then, suppose that the $-b^2$ term dominates xb', and thus $-b^2 \sim x^2$ and $b \sim \pm ix$. Then $xb' \sim \pm ix$ *is* dominated by the $-b^2$ term as assumed. This situation suggests the *iterative* solution

$$xb'_n - b_{n+1}^2 = x^2 \quad \text{or} \quad b_{n+1} = \pm\sqrt{-x^2 + xb'_n}. \tag{22.89}$$

With $b_0 = 0$ we have $b_1 = \pm ix$, and

$$b_2 = \pm\sqrt{-x^2 \pm ix} = \pm ix\left(1 \mp \frac{i}{x}\right)^{1/2} \sim \pm ix\left(1 \mp \frac{1}{2}\frac{i}{x}\right),$$

where the upper signs go together and the lower signs go together. Then

$$b(x) \sim \pm ix + \frac{1}{2},$$

[25]Actually, we obtain

$$y = e^{-\int (b/x)dx}\left\{A\int e^{2\int (b/x)dx}\,dx + C\right\},$$

but we can drop the A term if we are able to find two different solutions $b(x)$ of (22.87).

and (22.88) gives

$$y \sim C_1 e^{-\int[-i+(1/2x)]dx} + C_2 e^{-\int[i+(1/2x)]dx}$$

or

$$y \sim C_1 \frac{e^{ix}}{\sqrt{x}} + C_2 \frac{e^{-ix}}{\sqrt{x}}. \tag{22.90}$$

It follows that both $J_0(x)$ and $Y_0(x)$ behave according to (22.90), for different combinations of C_1 and C_2, but since the determination of C_1 and C_2 in each case is beyond our present scope, we merely state the results,

$$J_0(x) \sim \frac{\cos(x - \pi/4)}{\sqrt{\pi x/2}} \tag{22.91a}$$

and[26]

$$Y_0(x) \sim \frac{\sin(x - \pi/4)}{\sqrt{\pi x/2}}. \tag{22.91b}$$

Observe this damped sinusoidal behavior in Fig. 22.1, as well as the small-argument behavior stated in (22.86). Observe also that the *zeros* of these functions are not evenly spaced, although (22.91) indicates that they *tend* to an equal spacing as $x \to \infty$; their precise values are important (as we will see in Part V), and they are tabulated in many places.[27] Finally, observe the "interlacing" of the zeros of J_0 and Y_0. In fact, the interlacing of the zeros of LI solutions of the general second-order linear equation $py'' + qy' + ry = 0$ is well understood and comes under the heading of *oscillation theorems*.[28]

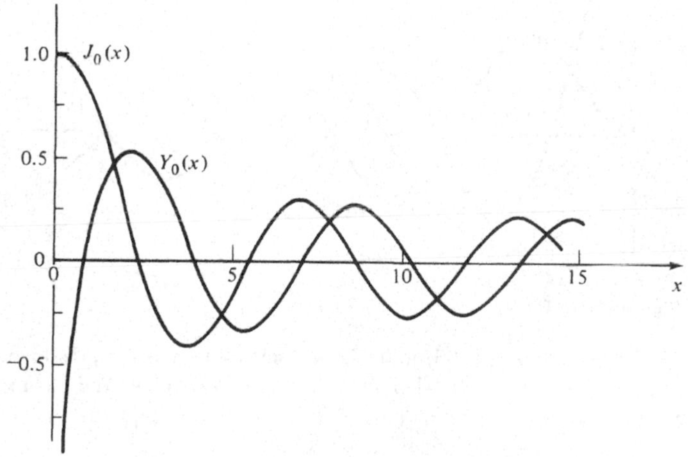

Figure 22.1. The Bessel functions J_0 and Y_0.

[26]The ways that (22.91b) is a natural companion to (22.91a) is due to the Weber–Schlafli definition of Y_0, as mentioned following (22.84).

[27]See, for example, the extensive treatise by G. N. Watson, *Theory of Bessel Functions*, Cambridge University Press, London, 1922.

[28]See, for example, J. C. Burkill, *The Theory of Ordinary Differential Equations*, Oliver and Boyd, Edinburgh, 1956.

We will refer to our method of deriving (22.90) as **approximate factorization**. Its relation to the so-called **WKB method** is discussed in Exercise 22.29.

So far we've considered only the case where $\nu = 0$ in (22.75). If ν is *any integer* instead,[29] then the general solution analogous to (22.85) is found (by the same methods as above) to be

$$y = AJ_\nu(x) + BY_\nu(x), \tag{22.92}$$

where the **Bessel functions of first and second kind and order ν** are given by

$$J_\nu(x) = \frac{(x/2)^\nu}{\nu!}\left[1 - \frac{(x/2)^2}{1(\nu+1)} + \frac{(x/2)^4}{1\cdot 2(\nu+1)(\nu+2)} - \cdots\right] \tag{22.93}$$

and

$$Y_\nu(x) = \frac{2}{\pi}J_\nu(x)\left(\ln\frac{x}{2} + \gamma\right) - \frac{1}{\pi}\sum_{j=0}^{\nu-1}\frac{(\nu-j-1)!}{j!}\left(\frac{x}{2}\right)^{2j-\nu}$$
$$-\frac{1}{\pi}\sum_{j=0}^{\infty}\frac{(-1)^j}{j!(\nu+j)!}\left(\frac{x}{2}\right)^{2j+\nu}\left(1 + \frac{1}{2} + \cdots + \frac{1}{j} + 1 + \frac{1}{2} + \cdots + \frac{1}{j+\nu}\right), \tag{22.94}$$

respectively, where the last quantity in parentheses is $(1 + 1/2 + \cdots + 1/\nu)$ when $j = 0$. The first several J_νs and Y_νs are shown in Figs. 22.2 and 22.3.

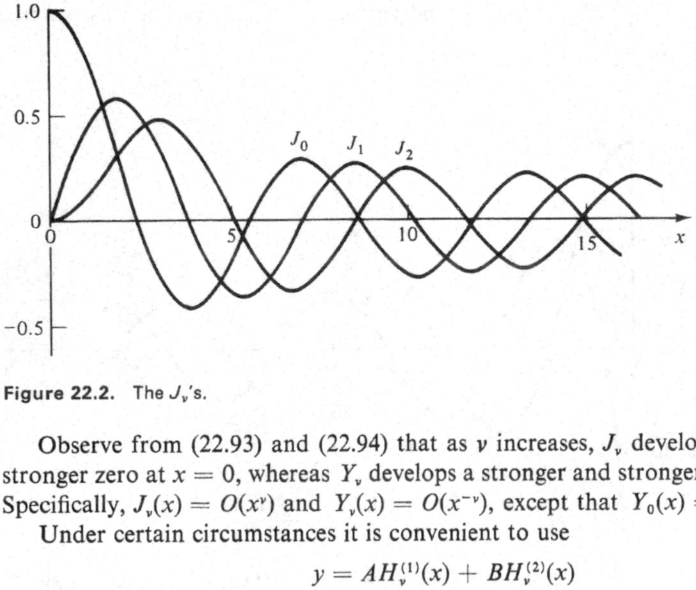

Figure 22.2. The J_ν's.

Observe from (22.93) and (22.94) that as ν increases, J_ν develops a stronger and stronger zero at $x = 0$, whereas Y_ν develops a stronger and stronger singularity there. Specifically, $J_\nu(x) = O(x^\nu)$ and $Y_\nu(x) = O(x^{-\nu})$, except that $Y_0(x) = O(\ln x)$.

Under certain circumstances it is convenient to use

$$y = AH_\nu^{(1)}(x) + BH_\nu^{(2)}(x) \tag{22.95}$$

in place of (22.92), where the **Hankel functions of order ν** are defined as the combinations

$$H_\nu^{(1)}(x) = J_\nu(x) + iY_\nu(x), \qquad H_\nu^{(2)}(x) = J_\nu(x) - iY_\nu(x). \tag{22.96}$$

---

[29] Negative integers need not be considered; (22.75) is insensitive to whether ν is positive or negative.

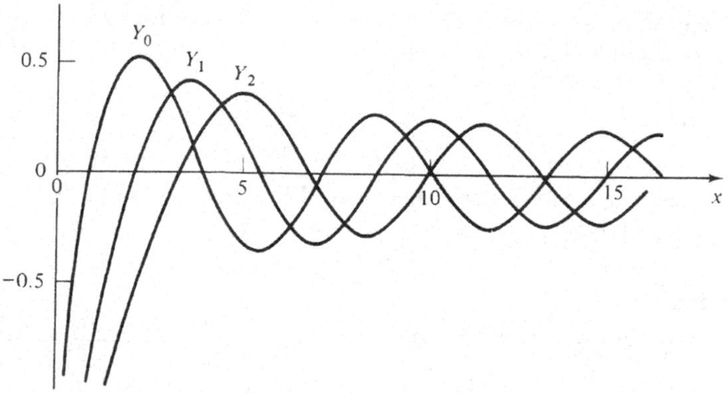

Figure 22.3. The Y_ν's.

That $H_\nu^{(1,2)}$ are LI solutions of the Bessel equation follows easily from their definition (22.96) and the fact that J_ν and Y_ν are LI solutions (Exercise 22.31). $H_\nu^{(1,2)}$ bear the same relation to J_ν and Y_ν as $\exp(\pm i\nu x)$ bear to $\cos \nu x$ and $\sin \nu x$ and are sometimes more convenient than J_ν, Y_ν in the same way that $\exp(\pm i\nu x)$ are sometimes more convenient than $\cos \nu x$, $\sin \nu x$. Observe from (22.91) and (22.96) that

$$H_0^{(1)}(x) \sim \frac{e^{i(x-\pi/4)}}{\sqrt{\pi x/2}} \tag{22.97a}$$

$$H_0^{(2)}(x) \sim \frac{e^{-i(x-\pi/4)}}{\sqrt{\pi x/2}} \quad \text{as } x \longrightarrow \infty. \tag{22.97b}$$

Finally, suppose that ν is *not an integer*. In this case, the indicial equation gives the distinct roots $\alpha = \pm\nu$, and it is found that the method of Frobenius does lead to the general solution—namely,

$$y = AJ_\nu(x) + BJ_{-\nu}(x), \tag{22.98}$$

where J_ν and $J_{-\nu}$ are given by (22.93), with $\nu!$ replaced by the gamma function $\Gamma(\nu + 1)$. Of special note is the case where $\nu = n + 1/2$, where n is any integer; then J_ν and $J_{-\nu}$ turn out to be *elementary functions* (Exercise 22.32).

Just as $y'' + y = 0$ has oscillatory solutions, we've seen that the Bessel equation $x^2y'' + xy' + (x^2 - \nu^2)y = 0$ or

$$y'' + \frac{1}{x}y' + \left(1 - \frac{\nu^2}{x^2}\right)y = 0$$

does, too. And recalling that $y'' - y = 0$ has nonoscillatory solutions, we might expect that the "**modified Bessel equation**"

$$y'' + \frac{1}{x}y' - \left(1 + \frac{\nu^2}{x^2}\right)y = 0 \tag{22.99}$$

will also. The change of variables $ix = \xi$ reduces (22.99) to the Bessel equation

$$y'' + \frac{1}{\xi}y' + \left(1 - \frac{\nu^2}{\xi^2}\right)y = 0, \qquad (\ )' = \frac{d(\ )}{d\xi},$$

and so the solutions of (22.99) are $Z_\nu(ix)$, where Z_ν is any solution of the Bessel equation of order ν. So even though we need no new functions to solve (22.99), it is customary to express its general solution as

$$y = \begin{cases} AI_\nu(x) + BI_{-\nu}(x), & \nu \neq \text{integer} \\ AI_\nu(x) + BK_\nu(x), & \nu = \text{integer} \end{cases} \tag{22.100}$$

where the **modified Bessel functions of first and second kinds of order** ν, I_ν, and K_ν, respectively, are real valued. Specifically,

$$I_\nu(x) = i^{-\nu}J_\nu(ix) \tag{22.101}$$

and

$$\begin{aligned} K_\nu(x) &= (-1)^{\nu+1}I_\nu(x)\left(\ln\frac{x}{2} + \gamma\right) + \frac{1}{2}\sum_{j=0}^{\nu-1}\frac{(-1)^j(\nu-j-1)!}{j!}\left(\frac{x}{2}\right)^{2j-\nu} \\ &\quad + \frac{(-1)^\nu}{2}\sum_{j=0}^{\infty}\frac{1}{j!(\nu+j)!}\left(\frac{x}{2}\right)^{2j+\nu} \\ &\quad \times \left(1 + \frac{1}{2} + \cdots + \frac{1}{j} + 1 + \frac{1}{2} + \cdots + \frac{1}{j+\nu}\right), \end{aligned} \tag{22.102}$$

where the last quantity in parentheses is $(1 + 1/2 + \cdots + 1/\nu)$ when $j = 0$. I_0 and K_0 are shown in Fig. 22.4, where we note the behavior

$$I_0(x) \sim \begin{cases} 1 & \text{as } x \longrightarrow 0 \\ \dfrac{e^x}{\sqrt{2\pi x}} & \text{as } x \longrightarrow \infty \end{cases} \tag{22.103}$$

and

$$K_0(x) \sim \begin{cases} -\ln x & \text{as } x \longrightarrow 0 \\ \dfrac{e^{-x}}{\sqrt{2x/\pi}} & \text{as } x \longrightarrow \infty. \end{cases} \tag{22.104}$$

Not only does the Bessel equation appear frequently, but so do equations that can be *reduced* to the Bessel equation by suitable change of variables. For instance, the

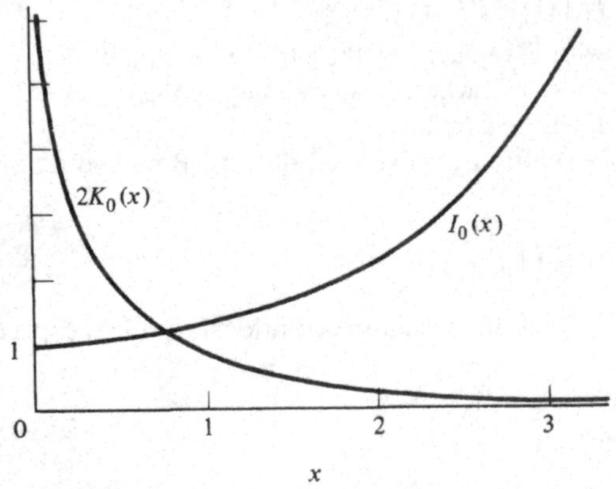

Figure 22.4. The modified Bessel functions I_0 and K_0.

fairly general equation

$$\frac{d}{dx}\left(x^a \frac{dy}{dx}\right) + bx^c y = 0 \tag{22.105}$$

has solutions

$$y = x^{\nu/\alpha} Z_\nu(x^{1/\alpha}\alpha\sqrt{b}), \tag{22.106}$$

where Z_ν is any solution of Bessel's equation of order ν and

$$\nu = \frac{1-a}{c-a+2}, \qquad \alpha = \frac{2}{c-a+2},$$

provided that $c - a + 2 \neq 0$; however, if $c - a + 2 = 0$, then (22.105) is simply a Cauchy–Euler equation, which is easily solved.

Example 22.15. Consider **Airy's equation,**

$$y'' + xy = 0. \tag{22.107}$$

Then $a = 0$ and $b = c = 1$, and so $\nu = \frac{1}{3}$ and $\alpha = \frac{2}{3}$. Thus

$$y = x^{1/2} Z_{1/3}(\tfrac{2}{3}x^{3/2}),$$

which means that the general solution can be expressed as

$$y = Ax^{1/2} J_{1/3}(\tfrac{2}{3}x^{3/2}) + Bx^{1/2} J_{-1/3}(\tfrac{2}{3}x^{3/2}). \quad \blacksquare$$

Finally, it should be mentioned that differences of notation exist among various authors with N_ν sometimes used in place of our Y_ν, $Y^{(\nu)}$ in place of our $\mathbf{Y}_\nu$, and so on. The notation used in this book coincides with that of McLachlan.

Legendre Functions. Another important second-order linear equation is the **Legendre equation**

$$(1 - x^2)y'' - 2xy' + \nu(\nu + 1)y = 0, \tag{22.108}$$

which will be encountered in applications in Part V. We will limit our consideration to the case where ν is a *nonnegative integer*, both for brevity and because it is the one of most interest. For emphasis, let us set $\nu = n$.

Observe that all points x are ordinary points except for $x = \pm 1$, which are regular singular points. Those solutions that are singular at ± 1 will be of less interest to us than certain polynomial solutions; so instead of seeking Frobenius solutions about $x = \pm 1$, it will be convenient to seek power series solutions about the ordinary point $x = 0$. We obtain (Exercise 22.28) the two LI solutions

$$\begin{aligned} y_1 &= 1 - \frac{n(n+1)}{2!}x^2 + \frac{(n-2)n(n+1)(n+3)}{4!}x^4 - \cdots \\ y_2 &= x - \frac{(n-1)(n+2)}{3!}x^3 + \frac{(n-3)(n-1)(n+2)(n+4)}{5!}x^5 - \cdots. \end{aligned} \tag{22.109}$$

Clearly, $y_1(x)$ is even and $y_2(x)$ is odd.

Note that if n is *even*, the y_1 series eventually terminates, due to the factors n, $n-2$, $n-4, \ldots$, and so y_1 is then an nth-degree polynomial and y_2 is an infinite series. Similarly, if n is *odd*, y_1 is an infinite series and y_2 is an nth-degree polynomial.

We denote the polynomial solution, scaled so that it is unity at $x = 0$, as $P_n(x)$ and the series solution (which diverges at $|x| = 1$) as $Q_n(x)$. Thus

$$y = AP_n(x) + BQ_n(x), \tag{22.110}$$

where P_n is called the **Legendre polynomial of degree *n*** or the **Legendre function of the first kind and degree *n***, and Q_n is the **Legendre function of the second kind and degree *n***. Whether n is even or odd, we can express

$$P_n(x) = \sum_{j=0}^{N} \frac{(-1)^j(2n-2j)!}{2^n j!(n-j)!(n-2j)!} x^{n-2j}, \tag{22.111}$$

where $N = n/2$ when n is even and $N = (n-1)/2$ when n is odd.

Other Special Functions. Numerous other equations and their associated functions are of importance in pure and applied mathematics, such as the **hypergeometric equation**

$$x(1-x)y'' + [\gamma - (\alpha + \beta + 1)x]y' - \alpha\beta y = 0, \tag{22.112}$$

the **Hermite equation**

$$y'' - 2xy' + 2\nu y = 0, \tag{22.113}$$

the **Laguerre equation**

$$xy'' + (1-x)y' + \nu y = 0, \tag{22.114}$$

the **Chebyshev equation**

$$(1-x^2)y'' - xy' + \nu^2 y = 0, \tag{22.115}$$

and the **Mathieu equation**

$$y'' + (\alpha + \beta \cos x)y = 0. \tag{22.116}$$

The literature on special functions is quite extensive, and any attempt to do the subject justice here would be a mistake. But the reader who has neither the time nor the inclination to pursue the subject in depth would find even a superficial reading of a book that surveys the subject of special functions and their applications well worth the effort.[30]

22.5. *THE METHOD OF GREEN'S FUNCTIONS*

We now consider the method of Green's functions for the solution of nth-order, linear, ordinary differential equations. The basic idea can be explained through the following example.

Example 22.16. Consider the boundary value problem

$$u'' + k^2 u = f(x), \qquad u(0) = a, \quad u(\pi) = b. \tag{22.117}$$

[30]See, for example, D. Johnson and J. Johnson, *Mathematical Methods in Engineering and Physics; Special Functions and Boundary-Value Problems*, Ronald Press, New York, 1965.

The starting point of the method consists of the integration by parts of the inner product[31]

$$(Lu, G) = \text{boundary terms} + (u, L^*G).$$

Thus

$$\int_0^\pi (u'' + k^2u)G\,d\xi = (Gu' - uG')\Big|_0^\pi + \int_0^\pi (G'' + k^2G)u\,d\xi$$

or

$$\int_0^\pi Gf\,d\xi = (Gu' - uG')\Big|_0^\pi + \int_0^\pi uL^*G\,d\xi. \tag{22.118}$$

So far G is arbitrary; if we choose it cleverly, however, (22.118) can provide us with the solution to our original problem (22.117). Specifically, we require G, the so-called **Green's function** for the problem (22.117), to satisfy the equation $L^*G = \delta(\xi - x)$, where ξ is regarded as the "active" variable and x is regarded as fixed[32]. Since G does depend on the point x, we denote $G = G(\xi, x)$. Hence (22.118) becomes

$$\int_0^\pi G(\xi, x)f(\xi)\,d\xi = G(\pi, x)u'(\pi) - bG_\xi(\pi, x) - G(0, x)u'(0) + aG_\xi(0, x) + u(x),$$

which yields $u(x)$. But note that $u'(\pi)$ and $u'(0)$ are *not prescribed*, and so we eliminate these unwelcome terms by requiring that $G(\pi, x) = G(0, x) = 0$. Then

$$u(x) = bG_\xi(\pi, x) - aG_\xi(0, x) + \int_0^\pi G(\xi, x)f(\xi)\,d\xi, \tag{22.119}$$

where the Green's function G is found as the solution of the system

$$L^*G = G_{\xi\xi} + k^2G = \delta(\xi - x) \tag{22.120a}$$

$$G(0, x) = G(\pi, x) = 0. \tag{22.120b}$$

The point is that we have traded in the original problem (22.117) for the simpler one (22.120), simpler in that $f(x)$ is replaced by a delta function and the boundary conditions become "homogenized."

Let us now determine G. Recalling that $\delta(\xi - x)$ acts only at $\xi = x$, it will be convenient to split the interval into two parts—$0 \le \xi < x$ and $x < \xi \le \pi$—in each of which $L^*G = G_{\xi\xi} + k^2G = 0$. Integrating,

$$G(\xi, x) = \begin{cases} A\sin k\xi + B\cos k\xi, & 0 \le \xi < x \\ C\sin k\xi + D\cos k\xi, & x < \xi \le \pi. \end{cases} \tag{22.121}$$

The boundary conditions (22.120b) yield $B = 0$ and $C\sin k\pi + D\cos k\pi = 0$, so that

$$G(\xi, x) = \begin{cases} A\sin k\xi \\ \dfrac{C}{\cos k\pi}(\sin k\xi\cos k\pi - \sin k\pi\cos k\xi) \end{cases}$$

or

$$G(\xi, x) = \begin{cases} A\sin k\xi, & 0 \le \xi < x \quad (22.122a) \\ E\sin k(\xi - \pi), & x < \xi \le \pi, \quad (22.122b) \end{cases}$$

---

[31]The inner product between two (real-valued) functions $f(x)$ and $g(x)$, over the domain $a \le x \le b$, was defined in Chapter 17 as

$$(f, g) \equiv \int_a^b f(x)g(x)\,dx.$$

[32]L^* is the *adjoint* of L. The ajoint operator is discussed in detail in Sec. 18.4, although all you need to know here is that it is the operator that acts upon G in the integral (u, L^*G), obtained by starting with (Lu, G) and integrating by parts.

where $E \equiv C/\cos k\pi$. To determine A and E, we must somehow "blend" (22.122a) and (22.122b) at $\xi = x$. The presence of the delta function on the right-hand side of (22.120a) implies that the solution G must in some way be singular at $\xi = x$. Since differentiation makes singularities even more singular, we expect the delta function behavior to be borne by the highest derivative on the left-hand side—namely, $G_{\xi\xi}$. Working backward, then, G_ξ should have a finite discontinuity at $\xi = x$, and G should be continuous there but with a finite "kink."

Integrating (22.120a) from "$x - 0$" to "$x + 0$,"

$$\int_{x-0}^{x+0} (G_{\xi\xi} + k^2 G)\, d\xi = \int_{x-0}^{x+0} \delta(\xi - x)\, d\xi$$

$$G_\xi \Big|_{x-0}^{x+0} + k^2 \int_{x-0}^{x+0} G\, d\xi = 1. \tag{22.123}$$

Since G is continuous at $\xi = x$, it is bounded there. Therefore the integral term in (22.123) must be zero, and (22.123) exposes the strength of the kink—that is, the jump in G_ξ. Consequently, our two blending conditions are

$$\text{Continuity:} \quad G\Big|_{x-0}^{x+0} = 0 \tag{22.124a}$$

$$\text{Kink:} \quad G_\xi\Big|_{x-0}^{x+0} = 1, \tag{22.124b}$$

and they yield

$$\begin{aligned} E \sin k(x - \pi) - A \sin kx &= 0 \\ Ek \cos k(x - \pi) - Ak \cos kx &= 1, \end{aligned} \tag{22.125}$$

where G at "$x + 0$," for instance, is the limit of (22.122b) as ξ tends to x from the right—namely, $E \sin k(x - \pi)$. Solving (22.125) for A and E and inserting them into (22.122), we have

$$G(\xi, x) = \begin{cases} \dfrac{\sin k(x - \pi) \sin k\xi}{k \sin k\pi}, & \xi \le x \\[2ex] \dfrac{\sin kx \sin k(\xi - \pi)}{k \sin k\pi}, & \xi \ge x. \end{cases} \tag{22.126}$$

Thus (22.119) gives

$$u(x) = b\frac{\sin kx}{\sin k\pi} - a\frac{\sin k(x - \pi)}{\sin k\pi} + \int_0^\pi G(\xi, x) f(\xi)\, d\xi, \tag{22.127}$$

where G is given by (22.126).

COMMENT 1. Note that the first two terms in the right side of (22.127) are of the form $C_1 \sin kx + C_2 \cos kx$ and amount to the "complementary solution," whereas the integral term is the "particular solution," as could also have been found by variation of parameters.

COMMENT 2. The preceding solution is valid only if k is not an integer. If it is, then (22.126) and (22.127) are meaningless because of the $\sin k\pi$'s in the denominators; or backing up a step, observe that equations (22.125) are then contradictory; or getting to the crux of the matter, the problem (22.120) does not have a solution! We need to modify (22.120) and hence obtain a so-called generalized Green's function. This topic is discussed in Exercise 22.45.

COMMENT 3. Finally, observe from (22.126) that the Green's function is symmetric—that is, $G(\xi, x) = G(x, \xi)$. The source of this symmetry is not obvious, and will now be explored. ▮

*

Let us examine the observed symmetry of G in more generality. Recall that G satisfies a system of the form

$$L^*G(\xi, x) = \delta(\xi - x), \quad \text{plus suitable homogeneous boundary conditions, abbreviated B.C.} \tag{22.128}$$

Denoting the operator of (22.128) (i.e., L^* plus B.C.) as $\mathcal{G}$, we note that associated with $\mathcal{G}$ is its adjoint $\mathcal{G}^*$, consisting of $(L^*)^* = L$ plus certain other homogeneous boundary conditions, say B.C.* So corresponding to $\mathcal{G}^*$ there is a so-called adjoint Green's function, say **G**, that satisfies

$$(L^*)^*\mathbf{G}(\xi, x_0) = L\mathbf{G}(\xi, x_0) = \delta(\xi - x_0), \quad \text{plus suitable homogeneous boundary conditions B.C.*}, \tag{22.129}$$

where we have introduced the zero subscript on x because the x_0's in (22.129) need not be the same as the x's in (22.128).[33]

Next, taking u and v in $(Lu, v) = (u, L^*v)$ to be $\mathbf{G}(\xi, x_0)$ and $G(\xi, x)$, respectively, and using (22.128) and (22.129), we have

$$\int_a^b \delta(\xi - x_0)G(\xi, x)\, d\xi = \int_a^b \mathbf{G}(\xi, x_0)\,\delta(\xi - x)\, d\xi$$

or

$$G(x_0, x) = \mathbf{G}(x, x_0), \tag{22.130}$$

where a, b are the endpoints of the interval under consideration.

In Example 22.16, $\mathcal{G}$ was self-adjoint [although the operator in (22.117) was not], so that (22.130) reduced to the simple symmetry condition $G(x_0, x) = G(x, x_0)$. Symmetry of the Green's function is often related to a known physical principle, such as *Maxwell reciprocity* in structural mechanics (Exercise 22.46).

Example 22.17. Consider the (simple) initial value problem

$$\dot{u} = f(t), \qquad u(0) = a. \tag{22.131}$$

As always, the starting point is $(Lu, G) =$ boundary terms $+ (u, L^*G)$

$$\int_0^\infty G(\tau, t)\dot{u}(\tau)\, d\tau = G(\tau, t)u(\tau)\Big|_0^\infty - \int_0^\infty G_\tau(\tau, t)u(\tau)\, d\tau$$

or

$$\int_0^\infty G(\tau, t)f(\tau)\, d\tau = G(\infty, t)u(\infty) - aG(0, t) + u(t),$$

where $L^*G = -G_\tau(\tau, t) = \delta(\tau - t)$. Since $u(\infty)$ is not prescribed, we require that $G(\infty, t) = 0$. Then

$$u(t) = aG(0, t) + \int_0^\infty G(\tau, t)f(\tau)\, d\tau, \tag{22.132}$$

[33] The name "adjoint Green's function" is ours and is not standard. In fact, often when speaking of "the Green's function" of a problem, people mean **G**, not G.

where G is found per

$$L^*G = -G_\tau(\tau, t) = \delta(\tau - t), \qquad G(\infty, t) = 0. \tag{22.133}$$

Integrating (22.133), $G(\tau, t) = -H(\tau - t) + A$, where H is the Heaviside step function, and $G(\infty, t) = 0$ gives $A = 1$, so that the solution is

$$u(t) = a + \int_0^\infty [1 - H(\tau - t)]f(\tau)\,d\tau = a + \int_0^t f(\tau)\,d\tau. \tag{22.134}$$

Whereas the Green's function for the second-order system in the preceding example was continuous but kinky, the Green's function for this first-order system, $G(\tau, t) = 1 - H(\tau - t)$, has a step discontinuity. ▮

We will also apply Green's functions to *partial* differential equations in Part V, where the Green's function can sometimes be obtained simply by inspection, by means of the *method of images*.[34]

22.6. SOME NONLINEAR PROBLEMS AND TECHNIQUES

Example 22.18. *Nonlinear Pendulum.* We are interested in the motion of a pendulum consisting of a point mass m at the end of a flexible massless string of length l, as shown in Fig. 22.5. Applying Newton's second law, we equate $m\ddot{\mathbf{r}}$ to the applied forces —namely, the weight force and the tension

$$m\ddot{\mathbf{r}} = (mg\cos\theta - T)\hat{\mathbf{e}}_r - mg\sin\theta\,\hat{\mathbf{e}}_\theta, \tag{22.135}$$

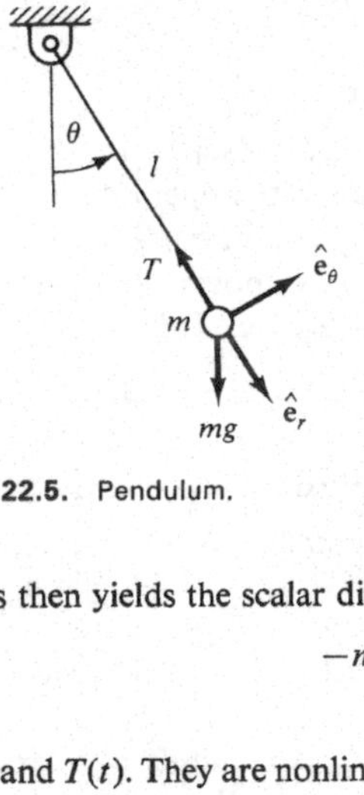

Figure 22.5. Pendulum.

where the force exerted on m by the string is purely radial, since the string is assumed flexible and can support no shear.[35] Furthermore,

$$\mathbf{r} = l\hat{\mathbf{e}}_r$$
$$\dot{\mathbf{r}} = l\dot{\hat{\mathbf{e}}}_r = l\dot{\theta}\hat{\mathbf{e}}_\theta$$
$$\ddot{\mathbf{r}} = l\dot{\theta}\dot{\hat{\mathbf{e}}}_\theta + l\ddot{\theta}\hat{\mathbf{e}}_\theta = -l\dot{\theta}^2\hat{\mathbf{e}}_r + l\ddot{\theta}\hat{\mathbf{e}}_\theta,$$

where the formulas $\dot{\hat{\mathbf{e}}}_r = \dot{\theta}\hat{\mathbf{e}}_\theta$ and $\dot{\hat{\mathbf{e}}}_\theta = -\dot{\theta}\hat{\mathbf{e}}_r$ were derived in Section 8.2. Equating $\hat{\mathbf{e}}_r$ and $\hat{\mathbf{e}}_\theta$ components then yields the scalar differential equations of motion

$$-ml\dot{\theta}^2 = mg\cos\theta - T \tag{22.136a}$$
$$ml\ddot{\theta} = -mg\sin\theta \tag{22.136b}$$

on $\theta(t)$ and $T(t)$. They are nonlinear and coupled, although the coupling presents no problem, since (22.136b) can be solved for θ and then T found (if we wish) from (22.136a).

---

[34]See also M. D. Greenberg, *Application of Green's Functions in Science and Engineering*, Prentice-Hall, Englewood Cliffs, N.J., 1971.

[35]Even if we use a rod (which *can* support shear) in place of the string, the force will still be radial if the mass (and hence inertia) of the rod is negligible, for a shear force at the end would then imply infinite angular accelerations, which we rule out.

Our interest is in the θ equation,

$$\ddot{\theta} + \kappa^2 \sin\theta = 0, \qquad \kappa^2 \equiv \frac{g}{l}, \tag{22.137}$$

and we take $\theta(0) = \theta_0$, $\dot{\theta}(0) = 0$ for definiteness. If θ_0 is small (i.e., if $\theta_0^2/6 \ll 1$), we can linearize (22.137) through the approximation

$$\sin\theta = \theta - \frac{\theta^3}{6} + \frac{\theta^5}{120} - \cdots \approx \theta,$$

in which case (22.137) reduces to the "simple harmonic oscillator"

$$\ddot{\theta} + \kappa^2\theta = 0 \tag{22.138}$$

with the solution $\theta = \theta_0 \cos \kappa t$.

If linearization is unacceptable, the problem is more difficult. To obtain a first integral of (22.137)—that is, a statement of conservation of energy—set $\dot{\theta} = p$. Then $\ddot{\theta} = dp/dt = (dp/d\theta)(d\theta/dt) = p\, dp/d\theta$, so that

$$p\,dp + \kappa^2 \sin\theta d\theta = 0$$

$$\frac{\dot{\theta}^2}{2} - \kappa^2 \cos\theta = \text{constant} = -\kappa^2 \cos\theta_0,$$

where $\theta(0) = \theta_0$ and $\dot{\theta}(0) = 0$. Integrating once more,

$$\int_\theta^{\theta_0} \frac{d\theta}{\sqrt{\cos\theta - \cos\theta_0}} = \kappa\sqrt{2} \int_0^t dt. \tag{22.139}$$

It turns out that the integral on the left cannot be expressed in terms of elementary functions, but it can be expressed in terms of the tabulated **elliptic integral of the first kind**, defined by

$$F(k, \phi) = \int_0^\phi \frac{d\phi}{\sqrt{1 - k^2 \sin^2\phi}} \qquad (0 < k < 1), \tag{22.140}$$

where k is known as the *modulus* and ϕ as the *amplitude*. To convert (22.139) to the desired form, we use the identity $\cos\theta = 1 - 2\sin^2(\theta/2)$; then

$$\frac{1}{\sin\theta_0'} \int_{\theta'}^{\theta_0'} \frac{d\theta'}{\sqrt{1 - k^2 \sin^2\theta'}} = \kappa t, \tag{22.141}$$

where $\theta' = \theta/2$, $\theta_0' = \theta_0/2$, and $k = 1/\sin\theta_0'$. Unfortunately, $k > 1$, and so (22.140) does not apply. To remedy this situation, set[36]

$$k \sin\theta' \equiv \sin\phi; \tag{22.142}$$

then (22.141) becomes

$$\int_\phi^{\phi_0} \frac{d\phi}{\sqrt{1 - (\sin\theta_0')^2 \sin^2\phi}} = \kappa t, \tag{22.143}$$

and the solution is given (in implicit form) by

$$F\left[\sin\left(\frac{\theta_0}{2}\right), \sin^{-1}\left(k \sin\frac{\theta_0}{2}\right)\right] - F\left[\sin\left(\frac{\theta_0}{2}\right), \sin^{-1}\left(k \sin\frac{\theta}{2}\right)\right] = \kappa t. \tag{22.144}$$ ▮

[36]It is implicit in (22.140) that $|k \sin\phi| < 1$. Similarly, $|k \sin\theta'| < 1$ in (22.141), so that (22.142) does indeed define a (real-valued) ϕ.

Before proceeding, let us say a bit more about elliptic integrals. If $R(u, v)$ is a rational function (i.e., the ratio of two polynomials) of its two arguments and

$$X = ax^4 + bx^3 + cx^2 + dx + e \tag{22.145}$$

has distinct roots, then

$$\int R(x, \sqrt{X})\, dx \tag{22.146}$$

is, in general, nonelementary and is called an **elliptic integral**, in connection with a special case that arises in the rectification of an ellipse (Exercise 22.47). Elliptic integrals were first treated systematically by Legendre (1752–1833), who showed that any such integral can be expressed in terms of elementary functions and one or more of three fundamental forms

$$F(k, \phi) \equiv \int_0^\phi \frac{d\phi}{\sqrt{1 - k^2 \sin^2 \phi}} \tag{22.147a}$$

$$E(k, \phi) \equiv \int_0^\phi \sqrt{1 - k^2 \sin^2 \phi}\, d\phi \tag{22.147b}$$

$$\Pi(n, k, \phi) \equiv \int_0^\phi \frac{d\phi}{(1 + n \sin^2 \phi)\sqrt{1 - k^2 \sin^2 \phi}} \tag{22.147c}$$

where $0 < k < 1$ and n is any constant (real or complex) or under the change of variables $\sin \phi \equiv x$

$$F(k, \sin^{-1} x) = \int_0^x \frac{dx}{\sqrt{(1 - x^2)(1 - k^2x^2)}} \tag{22.148a}$$

$$E(k, \sin^{-1} x) = \int_0^x \sqrt{\frac{1 - k^2x^2}{1 - x^2}}\, dx \tag{22.148b}$$

$$\Pi\,(n, k, \sin^{-1} x) = \int_0^x \frac{dx}{(1 + nx^2)\sqrt{(1 - x^2)(1 - k^2x^2)}}, \tag{22.148c}$$

which are known as Legendre's normal forms for the **elliptic integrals of the first, second, and third kind**, respectively. They are called **complete** if $\phi = \pi/2$, and we denote $F(k, \pi/2) \equiv K(k)$ and $E(k, \pi/2) \equiv E(k)$.

Although always *possible*, the reduction of (22.146) in terms of F, E, and Π is not always easy [e.g., even in the foregoing pendulum problem the substitution (22.142) was not entirely obvious], and the handbook by Byrd and Friedman can be of considerable help.[37]

Finally, in exploring methods of solution of nonlinear differential equations, it is reasonable to consider the standard methods of solution of nonlinear *algebraic and transcendental* equations in order to identify techniques that might be carried over to the case of differential equations.

Nonlinear Algebraic and Transcendental Equations. Consider the numerical calculation of the real roots, if any, of the equation

$$g(x) = 0, \tag{22.149}$$

[37] P. Byrd and M. Friedman, *Handbook of Elliptic Integrals for Scientists and Engineers*, 2nd ed., Springer-Verlag, New York, 1971.

where $g(x)$ is sufficiently complicated to rule out an exact analytical solution. One simple line of approach is to reexpress (22.149) in the form

$$x = f(x) \tag{22.150}$$

and to iterate according to

$$x_{n+1} = f(x_n). \tag{22.151}$$

This is known as the **method of successive approximations.**

To illustrate, consider the simple example

$$g(x) = x^2 - 3x + 2 = 0 \tag{22.152}$$

(with roots $x = 1$ and 2). Reexpressing (22.152) as $x = (x^2 + 2)/3 \equiv f(x)$, (22.151) becomes

$$x_{n+1} = \frac{x_n^2 + 2}{3}.$$

Starting with $x_0 = 0$, say, we generate the sequence $x_1 = 2/3$, $x_2 = 22/27, \ldots$, which does converge to the root $x = 1$ (although slowly). Starting with $x_0 = 3$ instead, we have $x_1 = 11/3$, $x_2 = 139/27, \ldots$, which diverges. In fact, it is found that if $|x_0| < 2$, then we converge to the root $x = 1$; if $|x_0| > 2$, we diverge and (unless we happen to choose $x_0 = \pm 2$) the root $x = 2$ is inaccessible.

To understand what has happened, let us define the error $e_n = x_n - x$, where x denotes the *exact* solution. According to the mean value theorem (or Taylor's formula with Lagrange remainder),

$$x_{n+1} = f(x_n) = f(x) + f'(\xi_n)(x_n - x) = x + f'(\xi_n)(x_n - x), \tag{22.153}$$

where ξ_n is somewhere between (or at) x_n and x. But $x_{n+1} - x = e_{n+1}$ and $x_n - x = e_n$, so that

$$e_{n+1} = f'(\xi_n)e_n. \tag{22.154}$$

Now if (22.151) *is* convergent and f' is continuous, then

$$e_{n+1} \sim f'(x)e_n \quad \text{as } n \longrightarrow \infty, \tag{22.155}$$

and we observe that the error sequence does tend to zero, which is consistent with our supposition of convergence only if the "amplification" $|f'(x)| < 1$.

Returning to (22.153), observe that $|f'(x)| = |2x/3| = 2/3$ at the $x = 1$ root and 4/3 at the $x = 2$ root, which explains our convergence to $x = 1$ and our failure to converge to $x = 2$. Note, however, that the factorization $x = (x^2 + 2)/3$ of (22.152) is not unique. Specifically, we can also express $x = 3 - 2/x \equiv f(x)$, in which case $|f'(x)| = 2/x^2 = 2$ at $x = 1$ and 1/2 at $x = 2$. Thus $x_{n+1} = 3 - 2/x_n$ will converge to the $x = 2$ root [if x_0 is sufficiently close to 2 (Exercise 22.52)], with $x = 1$ being inaccessible this time.

Keeping the nonuniqueness of the factorization in mind, we wonder about guidelines for choosing an *optimum* factorization. Looking at (22.155), observe that we would like the amplification $f'(x)$ to be as small as possible. We can do no better than require that $f'(x) = 0$. [Recall that x is the desired root of $g(x) = 0$.] If we express

$$g(x) = h(x)[x - f(x)], \tag{22.156}$$

then

$$\begin{aligned} g'(x) &= h'(x)[x - f(x)] + h(x)[1 - f'(x)] \\ &= h'(x)[0] + h(x)[1 - 0], \end{aligned}$$

and so $h(x) = g'(x)$, which specifies the optimum factorization. Putting this result into (22.156), we obtain $f(x) = x - g(x)/g'(x)$ as our optimum, and (22.151) becomes

$$x_{n+1} = x_n - \frac{g(x_n)}{g'(x_n)}. \tag{22.157}$$

This optimized method of successive approximations is, in fact, none other than **Newton's method**. Recalling (from calculus) the significance of $g'(x_n)$ as the slope of the tangent line to the g curve at x_n, (22.157) admits the simple graphical interpretation shown in Fig. 22.6. Observe that we need to choose x_0 "sufficiently close" to x to ensure convergence; for instance, $x_0 = A$ would probably not lead to convergence to the x root shown.

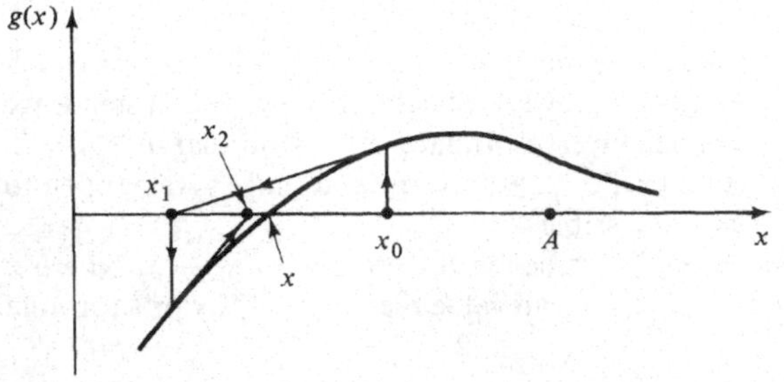

Figure 22.6. Graphical interpretation of Newton's method.

How about the *speed* of convergence? A convergent iteration is said to be *of order* k (not necessarily an integer) if

$$|e_{n+1}| \sim C|e_n|^k \quad \text{as } n \longrightarrow \infty, \tag{22.158}$$

and C is called the *asymptotic error constant.* The larger k is and the smaller C is, the faster the convergence, with the speed being especially sensitive to k (Exercise 22.53). Recalling from (22.155) that $|e_{n+1}| \sim |f'(x)||e_n|$, we see that *the method of successive approximations is a first-order method* (i.e., $k = 1$) with $C = |f'(x)|$. For the optimum Newton factorization, however, $f'(x) = 0$, and so (22.155) does not apply. That is, the leading term on the right is zero, and therefore we need to obtain the first *nonzero* term. In order to find the desired form (22.158), we start with (22.157) or

$$(x_{n+1} - x_n)g'(x_n) = -g(x_n).$$

Expanding $g'(x_n)$ and $g(x_n)$ about x,

$$(x_{n+1} - x_n)[g'(x) + g''(x)(x_n - x) + \cdots]$$
$$= -g(x) - g'(x)(x_n - x) - \frac{g''(x)}{2}(x_n - x)^2 - \cdots.$$

Now,

$$x_n - x = e_n, \qquad x_{n+1} - x_n = (x_{n+1} - x) - (x_n - x) = e_{n+1} - e_n, \qquad g(x) = 0,$$

so that

$$(e_{n+1} - e_n)[g'(x) + g''(x)e_n] \sim -g'(x)e_n - \frac{g''(x)}{2}e_n^2$$

or

$$e_{n+1} \sim \frac{g''(x)}{2g'(x)}e_n^2, \tag{22.159}$$

provided that $g'(x) \neq 0$; the case where $g'(x) = 0$ is left for Exercise 22.57 Thus *Newton's method is a second-order method* and hence is more rapidly convergent than successive approximations.[38]

As suggested by Fig. 22.6, Newton's method amounts to a *step-by-step linearization.* That is, starting at some point x_0, we replace $g = 0$ by the linear tangent line approximation

$$\begin{aligned} 0 = g(x) &= g(x_0) + g'(x_0)(x - x_0) + \frac{g''(x_0)}{2!}(x - x_0)^2 + \cdots \\ &\approx g(x_0) + g'(x_0)(x - x_0), \end{aligned} \tag{22.160}$$

which yields

$$x \approx x_0 - \frac{g(x_0)}{g'(x_0)} \equiv x_1. \tag{22.161}$$

Repeating the process,

$$x_{n+1} = x_n - \frac{g(x_n)}{g'(x_n)}. \tag{22.162}$$

This derivation of Newton's method is readily carried over to nonlinear differential (or integral) equations, as we will now demonstrate.

Application to Nonlinear Differential Equations. Let us illustrate the application of Newton's method, which is also known in the literature as the method of **quasilinearization.**[39]

Example 22.19.

$$yy' + y^2 = 1 \qquad y(0) = 2. \tag{22.163}$$

Paralleling steps (22.160) to (22.162), let us expand the nonlinear terms in (22.163):

$$yy' \approx y_n y_n' + \left.\frac{\partial(yy')}{\partial y}\right|_n (y - y_n) + \left.\frac{\partial(yy')}{\partial y'}\right|_n (y' - y_n') = y_n' y + y_n y' - y_n y_n'$$

$$y^2 \approx y_n^2 + 2y_n(y - y_n) = 2y_n y - y_n^2.$$

[38] By successive approximations we will mean with $f'(x) \neq 0$ in order to distinguish it from Newton's method.

[39] The names generally associated with the development of this extremely powerful method include Newton, Raphson, Kantorovich, Bellman, and Kalaba. For detailed discussion, including the monotone quadratic convergence of the method and its relation to a certain "maximum operation," see R. Bellman and R. Kalaba, *Quasilinearization and Nonlinear Boundary-Value Problems*, American Elsevier Publishing Co., New York, 1965; G. Radbill and G. McCue, *Quasilinearization and Nonlinear Problems in Fluid and Orbital Mechanics*, American Elsevier Publishing Co., New York, 1970; E. S. Lee, *Quasilinearization and Invariant Imbedding*, Academic, New York, 1968.

Putting these terms into (22.163) and denoting the solution y as the next iterate y_{n+1}, our iterative scheme becomes

$$(y_n)y'_{n+1} + (y'_n + 2y_n)y_{n+1} = 1 + y_n^2 + y_n y'_n,$$

which is first order *linear* and yields [recall (21.20)]

$$y_{n+1} = \frac{e^{-2x}}{y_n}\left\{\int (1 + y_n^2 + y_n y'_n)e^{2x}\,dx + \text{constant}\right\} \tag{22.164}$$

for $n = 0, 1, 2, \ldots$, where the constant is chosen so that $y_{n+1}(0) = 2$. Starting with $y_0(x) = 2$, (22.164) yields

$$y_1(x) = \frac{5 + 3e^{-2x}}{4}, \tag{22.165}$$

and so on. (See Exercise 22.60 for further discussion.) ▌

EXERCISES

***22.1.** Show that a sufficient condition for $\mathbf{f}(\mathbf{x})$ to be a Lipschitz function is that the partials $\partial f_j/\partial x_k$ exist and are bounded for all $j = 1, \ldots, n$ and $k = 1, \ldots, n$, as stated just prior to Theorem 22.1.

***22.2.** Use Theorem 22.1 to show that the systems below do have unique solutions:
(a) $\ddot{x} + x = \sin t$, $x(0) = \dot{x}(0) = 0$
(b) $\ddot{x} + x^2 = t$, $x(0) = 1$, $\dot{x}(0) = 3$
(c) $\dddot{x} - x\dot{x} + 2 = 0$, $x(1) = \ddot{x}(1) = 0$, $\ddot{x}(1) = 3$
(d) $\dot{x} + tx^2 = 6$, $x(0) = 2$

***22.3.** (a–d) Apply (22.11) to the systems (a) through (d) of Exercise 22.2 through $\mathbf{x}_2$, say, and thus obtain approximate solutions for $x(t)$ in each case.

***22.4.** Carrying out the necessary iterations, obtain the expression for $x(t)$ given in Example 22.1.

22.5. It was stated that the system (22.17) does possess a unique solution over $a \le x \le b$. Does that follow from Theorem 22.1?

22.6. Show that if $y_1(x)$ is even and $y_2(x)$ is odd and neither is identically zero, then they are LI over any interval $-a < x < a$.

22.7. (*Abel's formula*) If $y_1(x)$ and $y_2(x)$ are two solutions of

$$y'' + p(x)y' + q(x)y = 0$$

over an interval I, where $p(x)$ and $q(x)$ are continuous over I, then for any point ξ in I show that the Wronskian $W(y_1, y_2) = W(x)$ is of the form

$$W(x) = W(\xi)e^{-\int_\xi^x p(t)\,dt}. \tag{22.166}$$

This is known as **Abel's formula**. Since the exponential never vanishes, it follows that W either vanishes identically (over I) or nowhere (over I). *Hint:* Show that $W' + pW = 0$.

22.8. Construct a counterexample like (22.21) for $n = 3$.

22.9. Find the general solution in each case.
(a) $y'' + 2y' = 0$
(b) $y''' - 6y'' + 11y' - 6y = 0$
(c) $y'' - 5y' + 6y = 0$
(d) $y'''' + 4y'' + 4y = 0$
(e) $y''' + y = 0$
(f) $y''' - 3y'' + 3y' - y = 0$

(g) $y'''' - 3y''' = 0$ (h) $y'' + y' + y = 0$

22.10. Fill in the omitted steps in Example 22.5.

22.11. Factor the operator and solve, as indicated at the end of Example 22.2:

(a) $y'' + 2y' = 0$ (b) $y'' - 5y' + 6y = 0$

(c) $y'' + 2y' = e^x$ (d) $y'' - 5y' + 6y = x^2$

(e) $y'' + 9y = f(x)$ (f) $y'' + 2y' + y = 0$

(g) $y'' + (x - 1)y' - xy = (D + x)(D - 1)y = 0$

22.12. Use the fact that $y_1(x) = x$ is a solution to find the general solution (leaving in integral form if necessary):

(a) $y'' + xy' - y = 0$ (b) $xy'' + xy' - y = 0$

(c) $x^2y'' + xy' - y = 0$ (d) $x^3y'' + xy' - y = 0$

22.13. Reduce the Cauchy–Euler equation (22.41) to one with constant coefficients by the change of variables $x = \exp(t)$ and thus obtain the solution $y = (A + B \ln x)x$.

22.14. Find the general solution.

(a) $x^2y'' + y = 0$ (b) $xy'''' + 3y''' = 0$

(c) $(x + 2)^2y'' - y = 0$ (d) $x^2y'' + xy' + 4y = 0$

22.15. Solve $xy'' - y' - 4x^3y = 0$ by setting $x = t^\alpha$ and choosing α suitably.

22.16. If n is a positive integer, then the equation

$$xy'' + 2ny' + kxy = 0$$

is satisfied by

$$y = \left(\frac{1}{x}D\right)^n z,$$

where z is a solution of the equation $z'' + kz = 0$. Thus find the general solution of $xy'' + 4y' + 4xy = 0$.

22.17. Solve (22.47) by setting $(D - x)y = u$ and proceeding as indicated in Example 22.2.

22.18. Solve by factoring the operator:

(a) $y'' - xy' - y = 0$ (b) $y'' - (x + 1)y' + (x - 1)y = 0$.

22.19. Find the general solution for each.

(a) $y'' + 2y' = 4x^2$ (b) $y'' + 2y' + y = \cos 2x - xe^{-x}$

(c) $\ddot{x} + c\dot{x} = f(t)$ (d) $(D^4 - 20D^2 + 4)y = x$

(e) $y'''' - y = \cosh x$ (f) $(D - 1)(2D - 1)y = xe^x + 3$

(g) $xy'''' - y''' = 3$ (h) $x^2y'' - xy' + y = f(x)$

(i) $y'' + y = \tan x$ (j) $x^2y'' + xy' - 4y = x^{30}$

22.20. Solve:

(a) $\ddot{x} + \omega^2x = F_0\delta(t - t_0)$, $x(0) = \dot{x}(0) = 0$ $(t_0 > 0)$

(b) $(D^2 - 1)y = x^2e^x$, $y(0) = 1$, $y'(0) = 0$

(c) $\ddot{x} + kx = F_0H(t - t_0)$, $x(0) = \dot{x}(0) = 0$ $(t_0 > 0)$

22.21. Outline the variation of parameter method for the third-order system $y''' + p(x)y'' + q(x)y' + r(x)y = f(x)$, supposing that we know three LI solutions $y_1(x)$, $y_2(x)$, $y_3(x)$ of the homogenized equation.

22.22. If the complementary solution of $y'' + p(x)y' + q(x)y = f(x)$ is $Ay_1(x) + By_2(x)$, show that its general solution is

$$y(x) = Ay_1(x) + By_2(x) + y_2(x)\int \frac{f(x)y_1(x)}{W(y_1, y_2)}\,dx - y_1(x)\int \frac{f(x)y_2(x)}{W(y_1, y_2)}\,dx,$$

where W is the Wronskian.

22.23. (*Comments on variation of parameters*) (a) Observe that the equations (22.55) do have a solution for $\dot{A}$ and $\dot{B}$—that is, the determinant of the coefficient matrix is nonzero. Generalizing to the case $y'' + p(x)y' + q(x)y = f(x)$ with complementary solution $Ay_1(x) + By_2(x)$, show that the equations analogous to (22.55) will *always* have a solution for $\dot{A}$ and $\dot{B}$.

(b) For fun, replace (22.54) by $\dot{A} \sin \omega t + \dot{B} \cos \omega t = 6$ and show that the end result (22.56) is still obtained.

22.24. Consider a *damped harmonic oscillator* driven by a sinusoidal force $F \sin \omega t$. The governing differential equation is

$$m\ddot{x} + c\dot{x} + kx = F \sin \omega t. \tag{22.167}$$

(a) Show that the particular solution to (22.167) can be found as $\operatorname{Im} v_p$, where v_p is the particular solution to the slightly simpler equation

$$m\ddot{v} + c\dot{v} + kv = Fe^{i\omega t}.$$

Hence find the general solution to (22.167).

(b) Show that the steady-state response (i.e., as $t \longrightarrow \infty$) is

$$x(t) \sim \frac{F}{\sqrt{(k - m\omega^2)^2 + c^2\omega^2}} \sin(\omega t + \phi),$$

where $\phi = \tan^{-1}[c\omega/(m\omega^2 - k)]$. Sketch the response curve—that is, the amplitude $F/\sqrt{(k - m\omega^2)^2 + c^2\omega^2}$ versus the driving frequency ω—and determine the ω at which the peak amplitude occurs. Is it at $\omega = \sqrt{k/m}$?

22.25. Find the general solutions of the following systems of equations:

(a) $\dot{x} - 2\dot{y} - x - y = 1 - t$
$\quad\dot{x} + \dot{y} + 4y = 1 + 4t$

(b) $\dot{x} = y, \dot{y} = z, \dot{z} = x$

(c) $(tD + 2)x - 2y = 0$
$\quad 3x - (tD + 1)y = -t$

(d) $D^2u - 2w = -10$
$\quad (D + 1)v + 3w = 15 - 3x$
$\quad 3Du + 2v = 0$

22.26. Fill in the omitted steps between (22.80) and (22.81).

22.27. Find the general solution of these equations by the *method of Frobenius*:

(a) $y'' - y' = x$

(b) $5x^2y'' + xy' + (x^3 - 1)y = 0$

(c) $4xy'' + 2y' + y = 0$

(d) $xy'' + y = 0$

(e) $x^2y'' - xy' + y = 0$

(f) $x^2y'' + xy' - 9y = 4x^3$

(g) $y'' + xy = 0$ [Compare your result with Example 22.15 and equation (22.93).]

22.28. Derive the power series solutions (22.109).

22.29. (**The WKB method**) If $f(x)$ is such that $|f'/f^{3/2}| \ll 1$, then the approximate solution of

$$y'' + f(x)y = 0 \tag{22.168}$$

is given by the WKB formula[40]

$$y(x) \approx \frac{1}{[f(x)]^{1/4}}\{Ae^{i\int\sqrt{f(x)}dx} + Be^{-i\int\sqrt{f(x)}dx}\}. \tag{22.169}$$

[40]The WKB method was developed by Wentzel, Kramers, and Brillouin in connection with quantum mechanical applications and had been given previously by Jeffreys. Actually, the right-hand side of (22.169) is only a first approximation, that is, the leading term of a systematic expansion. For details and further references see the review article by C. R. Steele, "Application of the WKB Method in Solid Mechanics," *Mechanics Today*, 1976, pp. 243–300.

Obtain (22.169) by means of our method of *approximate factorization* (in Section 22.4).

22.30. In (22.168) (Exercise 22.29) we omit a y' term with no loss of generality. Specifically, show that any equation

$$y'' + 2p(x)y' + q(x)y = 0$$

can be reduced to the "canonical form"

$$v'' + r(x)v = 0$$

by a change of dependent variable $y = a(x)v$; that is, determine $a(x)$ and $r(x)$.

22.31. Show why $H_\nu^{(1,2)}$ must be LI solutions of the Bessel equation of order ν by virtue of their definition (22.96) and the fact that J_ν and Y_ν are LI solutions.

22.32. We stated, following (22.98), that if ν is half an odd integer, then J_ν and $J_{-\nu}$ turn out to be elementary functions. For instance,

$$J_{1/2}(x) = \sqrt{\frac{2}{\pi x}} \sin x, \qquad J_{-1/2}(x) = \sqrt{\frac{2}{\pi x}} \cos x,$$

and

$$J_{3/2}(x) = \sqrt{\frac{2}{\pi x}}\left(\frac{\sin x}{x} - \cos x\right),$$

$$J_{-3/2}(x) = -\sqrt{\frac{2}{\pi x}}\left(\sin x + \frac{\cos x}{x}\right).$$

Derive the preceding formulas for $J_{1/2}(x)$ and $J_{-1/2}(x)$ by the method of Frobenius. Check, by direct substitution, that they do satisfy Bessel's equation of order $\nu = 1/2$.

22.33. Derive the following properties of the Bessel functions of the first kind:
(a) $J_{-n}(x) = (-1)^n J_n(x) \quad (n = \text{integer})$
(b) $xJ_\nu' = \nu J_\nu - xJ_{\nu+1}$
(c) $xJ_\nu' = -\nu J_\nu + xJ_{\nu-1}$
(d) $2\nu J_\nu = xJ_{\nu-1} + xJ_{\nu+1}$
(e) $2J_\nu' = J_{\nu-1} - J_{\nu+1}$
(f) $(x^\nu J_\nu)' = x^\nu J_{\nu-1}$, in particular, $J_0' = J_1$

22.34. Using Exercises 22.33 and 22.32, obtain expressions for $J_{5/2}(x)$ and $J_{-5/2}(x)$.

22.35. If we expand $e^{xt/2}$ and $e^{-x/2t}$ in series of ascending powers of $xt/2$ and $x/2t$, respectively, then it turns out that the coefficients of $t^0, t, t^2, \ldots$ in the product of the two series are equal to $J_0(x), J_1(x), J_2(x), \ldots$; that is,

$$e^{(t-1/t)x/2} = \sum_{-\infty}^{\infty} J_n(x)t^n. \tag{22.170}$$

Thus we call $\exp (t - 1/t)x/2$ the **generating function** for $J_n(x)$.
(a) Verify that the coefficient of t^0 in (22.170) is indeed $J_0(x)$.
(b) Regarding t as complex, (22.170) is a Laurent expansion, valid in $0 < |t| < \infty$. Use (14.21), with $a = 0$ and C_0 the unit circle, to derive the integral representation

$$J_n(x) = \frac{1}{\pi}\int_0^\pi \cos(n\theta - x\sin\theta)\, d\theta \tag{22.171}$$

for integer values of n.
(c) Show, with the help of (22.171) but without using integral tables, that

$$\int_0^\infty e^{-at}J_0(bt)\, dt = \frac{1}{\sqrt{a^2 + b^2}}. \tag{22.172}$$

22.36. Obtain the general solution in terms of Bessel functions or, if possible, elementary functions:

(a) $xy'' + y' + kxy = 0$ (b) $xy'' + y' + kxy = 6x$

(c) $xy'' + y' - kxy = f(x)$ (Recall Exercise 22.22.)

(d) $xy'' + 3y' + 4xy = 0$ (e) $x^2y'' + xy' + (k^2x^2 - \frac{1}{9})y = 0$

(f) $x^2y'' + xy' + k^2y = 0$ (g) $y'' + x^4y = 0$

(h) Solve $y'' + \omega^2 y = 0$ by recalling (22.105), (22.106), and Exercise 22.32.

22.37. Solve $y'' + xy = x$, $y(0) = y(1) = 0$.

22.38. It can be shown that the Legendre polynomials $P_n(x)$ are given by

$$P_n(x) = \frac{D^n(x^2 - 1)^n}{2^n n!}, \tag{22.173}$$

which is known as **Rodrigues' formula**. Use (22.173) to obtain $P_0(x)$, $P_1(x)$, $P_2(x)$, and $P_3(x)$.

22.39. (a) Show that the potential at a point P (Fig. 22.7) induced by charges $-q$ and $+q$ at Q and R, respectively, $\phi = q/PR - q/QR$, can be expressed as

$$\phi = \frac{q}{b}[(1 - 2xy + y^2)^{-1/2} - (1 + 2xy + y^2)^{-1/2}],$$

where $x = \cos\theta$ and $y = a/b$.

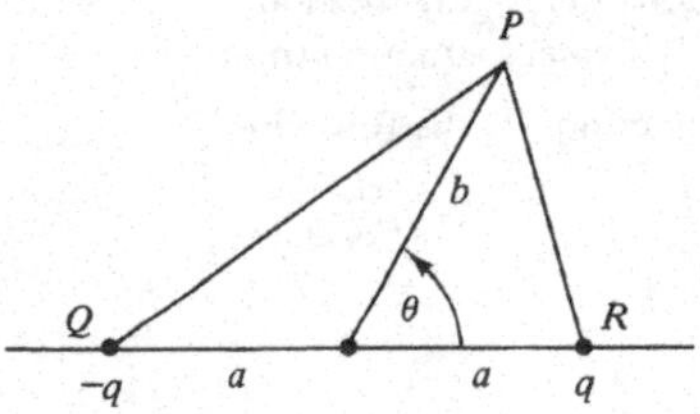

Figure 22.7. Potential due to two charges.

(b) It turns out that if we expand $(1 - 2xy + y^2)^{-1/2}$ in powers of y, then the coefficient of y^n is $P_n(x)$. That is,

$$(1 - 2xy + y^2)^{-1/2} = \sum_0^\infty P_n(x)y^n, \tag{22.174}$$

so that $(1 - 2xy + y^2)^{-1/2}$ is the **generating function** of the $P_n(x)$'s. Verify this result through $n = 3$.

(c) Thus show that

$$\phi = \frac{q}{b}\sum_0^\infty [P_n(\cos\theta) - P_n(-\cos\theta)]\left(\frac{a}{b}\right)^n.$$

22.40. Differentiating (22.174) with respect to y and equating coefficients of powers of y, obtain the *recurrence relation*

$$(n + 1)P_{n+1}(x) - (2n + 1)xP_n(x) + nP_{n-1}(x) = 0.$$

22.41. Setting $x = \cos\theta$, show that Legendre's equation becomes

$$\frac{d^2y}{d\theta^2} + \cot\theta\frac{dy}{d\theta} + n(n + 1)y = 0.$$

22.42. Show that the solution of the initial value problem

$$\ddot{x} + \omega^2 x = f(t) \quad \text{for } t \geq 0 \qquad [x(0) \text{ and } \dot{x}(0) \text{ prescribed}]$$

by the method of Green's functions is

$$x(t) = \frac{\dot{x}(0)}{\omega} \sin \omega t + x(0) \cos \omega t + \int_0^\infty G(\tau, t) f(\tau)\, d\tau$$

where $G(\tau, t) = [\sin \omega(t - \tau)]/\omega$ for $\tau < t$ and 0 for $\tau > t$.

22.43. Find the Green's function, and hence the solution, of the system

$$u'''' = f(x), \qquad u(0) = u''(0) = 0, \quad u'(0) = 2u'(1), \quad u(1) = a.$$

***22.44.** Show that the solution of

$$(xu')' + k^2 xu = f(x), \qquad u(0) \text{ finite}, \quad u(b) = 0$$

is

$$u(x) = \int_0^b G(\xi, x) f(\xi)\, d\xi,$$

where

$$G(\xi, x) = \begin{cases} [J_0(kb) Y_0(kx) - J_0(kx) Y_0(kb)] \dfrac{\pi J_0(k\xi)}{2J_0(kb)}, & 0 \leq \xi \leq x \\ [J_0(kb) Y_0(k\xi) - J_0(k\xi) Y_0(kb)] \dfrac{\pi J_0(kx)}{2J_0(kb)}, & x \leq \xi \leq b \end{cases}$$

(provided that kb does not coincide with a zero of the Bessel function J_0).

22.45. (*Generalized Green's function*) Recall Comment 2 following Example 22.16. Suppose that $k = 1$, in which case G fails to exist. This situation occurs because the right-hand side of (22.120a) is not orthogonal to the nontrivial solution $\sin \xi$ of the homogenized adjoint problem

$$LG = G_{\xi\xi} + G = 0, \qquad G(0, x) = G(\pi, x) = 0.$$

(Recall Theorem 19.1.) That is,

$$(\delta(\xi - x), \sin \xi) = \int_0^\pi \delta(\xi - x) \sin \xi\, d\xi = \sin x \neq 0.$$

To patch things up, let us require that G satisfy instead

$$L^*G = G_{\xi\xi} + G = \delta(\xi - x) + \mathfrak{F}, \qquad G(0, x) = G(\pi, x) = 0,$$

where $\mathfrak{F}$ is chosen so that $\delta(\xi - x) + \mathfrak{F}$ *is* orthogonal to $\sin \xi$. Show that a suitable form for $\mathfrak{F}$ is $\mathfrak{F} = \alpha \sin x \sin \xi$ with α determined by the condition $(\delta(\xi - x) + \mathfrak{F}, \sin \xi) = 0$ and complete the determination of the modified Green's function $G(\xi, x)$ and the solution $u(x)$.

22.46. (*Loaded String*) The problem

$$u'' = \phi(x), \qquad u(0) = u(1) = 0$$

governs the (small) static deflection $u(x)$ of a string stretched under unit tension and subjected to a loading $\phi(x)$ pounds per unit length (see Fig. 22.8).

(a) Determine the Green's function $G(\xi, x)$ and show that it can be interpreted physically as the deflection due to a point unit load at x.

(b) Show that $G(\xi, x) = G(x, \xi)$—that is, that *the deflection at ξ due to a unit load*

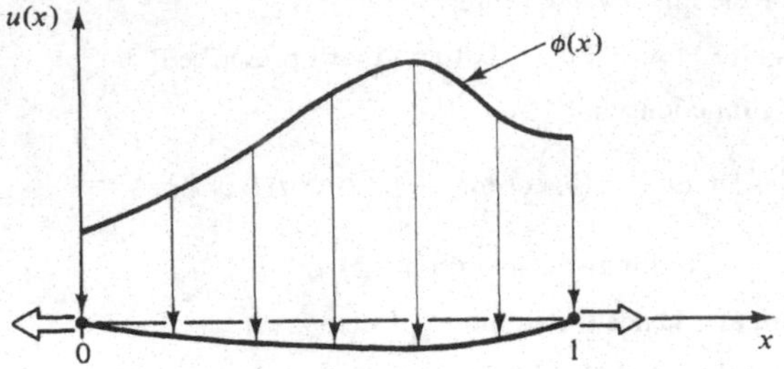

Figure 22.8. Loaded string.

at x is equal to the deflection at x due to a unit load at ξ (Fig. 22.9). This is often referred to as "Maxwell reciprocity."

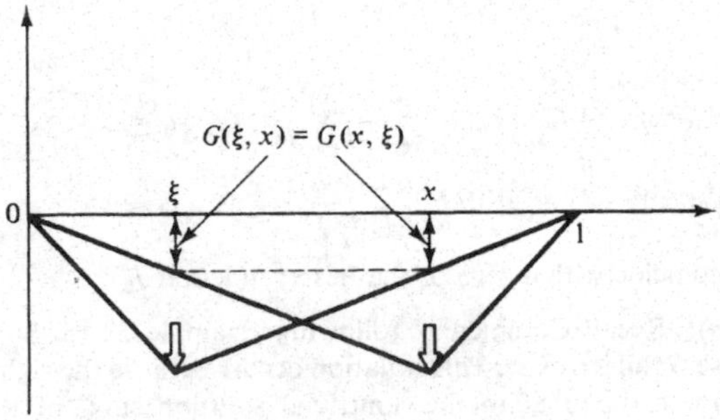

Figure 22.9. Maxwell reciprocity.

(c) Obtain the solution in the form $\int_0^1 G(\xi, x)\phi(\xi)\,d\xi$ or, due to the noted reciprocity,

$$u(x) = \int_0^1 G(x, \xi)\phi(\xi)\,d\xi. \tag{22.175}$$

Since $G(x, \xi)\phi(\xi)\,d\xi$ is the deflection at x due to an incremental load $\phi(\xi)\,d\xi$ at ξ, observe that (22.175) amounts to a *superposition principle*.

22.47. (*Rectification of ellipse*) Show that the length of an ellipse with semimajor axis a and semiminor axis b is given by $L = 4aE(\sqrt{a^2 - b^2}/a)$, where E is the complete elliptic integral of the second kind.

22.48. The integral

$$\int_0^\phi \sqrt{1 - k^2 \sin^2 \phi}\, d\phi$$

was defined as $E(k, \phi)$ when $0 < k < 1$. For $k > 1$ and $k \sin \phi \leq 1$, show that

$$\int_0^\phi \sqrt{1 - k^2 \sin^2 \phi}\, d\phi = kE\left(\frac{1}{k}, x\right) + \frac{1 - k^2}{k} F\left(\frac{1}{k}, x\right),$$

where $x = \sin^{-1}(k \sin \phi)$.

22.49. (a) Compute $E(0.2)$ to four significant figures without using tables.
(b) Compute $K(0.2)$ to four significant figures without using tables.

22.50. Evaluate in terms of elliptic integrals.

(a) $\displaystyle\int_0^{\phi} \frac{\sin^2 \phi \, d\phi}{\sqrt{1 - k^2 \sin^2 \phi}} \qquad (0 < k < 1)$

(b) $\displaystyle\int_0^{\phi} \frac{\cos^2 \phi \, d\phi}{\sqrt{1 - k^2 \sin^2 \phi}} \qquad (0 < k < 1)$

(c) $\displaystyle\int_0^{\pi/2} \frac{dx}{\sqrt{\cos x}}$

(d) $\displaystyle\int_0^{\pi/4} \frac{dx}{\sqrt{5 + \cos x}}$

Hint: $\cos x = 2 \cos^2 \frac{x}{2} - 1.$

22.51. Derive the expansions:

(a) $E(k) = \dfrac{\pi}{2}\left(1 - \dfrac{1}{2^2} k^2 - \dfrac{1^2 \cdot 3}{2^2 \cdot 4^2} k^4 - \dfrac{1^2 \cdot 3^2 \cdot 5}{2^2 \cdot 4^2 \cdot 6^2} k^6 - \cdots\right)$

(b) $K(k) = \dfrac{\pi}{2}\left(1 + \dfrac{1^2}{2^2} k^2 + \dfrac{1^2 \cdot 3^2}{2^2 \cdot 4^2} k^4 + \dfrac{1^2 \cdot 3^2 \cdot 5^2}{2^2 \cdot 4^2 \cdot 6^2} k^6 + \cdots\right).$

22.52. Refer to the paragraph following equation (22.155). Within what interval need x_0 lie in order to obtain convergence to the $x = 2$ root?

22.53. Following (22.158) it is stated that the speed of convergence depends on both C and k but is especially sensitive to k. Can you elaborate on this statement?

22.54. Find the root of $x = 2 \sin x$ that is in the vicinity of $x = 2$ by both successive approximations and Newton's method. Start with $x_0 = 2$ and compute through x_5, say.

22.55. Find the root of $x = 4 \tan x$ that is in the vicinity of $x = 3\pi/2$, by both successive approximations and Newton's method, to seven significant figures. (Suitable factoring will be necessary in order to make the successive approximations method converge to the desired root.)

22.56. Comment on the utility of Aitken extrapolation in speeding the convergence for the method of successive approximations (22.151) and for Newton's method (22.157).

22.57. (a) Show that if $g'(x) = 0$ [but $g''(x) \neq 0$], then we have, in place of (22.159), $e_{n+1} \sim \frac{1}{2} e_n$, so that Newton's method is only a first-order method. Give a simple graphical interpretation of the reduction from second order to first.
(b) What if $g'(x) = g''(x) = \cdots = g^{(k)}(x) = 0$, $g^{(k+1)}(x) \neq 0$?

22.58. (*Extension of Newton's method to higher dimension*) Extend Newton's method to two dimensions and use it to solve

$$x - \sin(x + y) = 0, \qquad y - \cos(x - y) = 0,$$

starting with $x_0 = y_0 = 1$. [This same example was treated by the method of steepest descent in Exercise 10.5(c).]

22.59. (*Duffing's equation*) The equation

$$\ddot{x} + c\dot{x} + \alpha x + \beta x^3 = F \cos \omega t \tag{22.176}$$

governing a driven, damped, nonlinear oscillator is known as **Duffing's equation** and occupies an important place in nonlinear mechanics. Taking $c = 0$ for simplicity and β to be small so that (22.176) is only mildly nonlinear, let us seek a so-called harmonic solution—that is, a periodic solution having the same period $(2\pi/\omega)$ as the driving force. (That such a solution exists seems plausible and is borne out by experiment. In fact, so-called subharmonic and superharmonic solutions are also observed but will not be discussed here.)

(a) Pursuing an iterative approach, show that quasilinearization yields the linear equation

$$\ddot{x}_{n+1} + (\alpha + 3\beta x_n^2)x_{n+1} = F\cos\omega t + 2\beta x_n^3. \tag{22.177}$$

(b) The scheme (22.177) is difficult to carry out, however, due to the nonconstant coefficient multiplying x_{n+1}. Thus let us back off to the simpler, although more slowly convergent, method of successive approximation. As with the algebraic case, our choice of factorization will be crucial; for instance, do we choose

$$x_{n+1} = \frac{-\ddot{x}_n - \beta x_n^3 + F\cos\omega t}{\alpha}, \tag{22.178}$$

or

$$\ddot{x}_{n+1} = -\alpha x_n - \beta x_n^3 + F\cos\omega t, \tag{22.179}$$

or some other form? Recalling our rough notion that "differentiation makes things worse" and "integration makes things better," let us tentatively choose (22.179), which yields x_{n+1} through two *integrations*. Since we are regarding β as small, let us try out (22.179) for the case $\beta = 0$, for then $\ddot{x} + \alpha x = F\cos\omega t$ is linear and easily yields the harmonic solution

$$x = \frac{F}{\alpha - \omega^2}\cos\omega t, \tag{22.180}$$

for comparison with the results of (22.179). Starting with

$$x_0 = A\cos\omega t \tag{22.181}$$

say, show that (22.179) yields

$$x_1 = \frac{\alpha A - F}{\omega^2}\cos\omega t \tag{22.182}$$

$$\vdots$$

$$x_n = \left\{\left(\frac{\alpha}{\omega^2}\right)^n A - \frac{F}{\omega^2}\left[1 + \left(\frac{\alpha}{\omega^2}\right) + \cdots + \left(\frac{\alpha}{\omega^2}\right)^{n-1}\right]\right\}\cos\omega t \tag{22.183}$$

$$\longrightarrow -\frac{F}{\omega^2}\frac{1}{1-(\alpha/\omega^2)}\cos\omega t \quad \text{as } n \longrightarrow \infty,$$

provided that $|\alpha/\omega^2| < 1$, and this result does coincide with (22.180). In fact, observe that if we equate the coefficients of $\cos\omega t$ in (22.181) and (22.182), we obtain $A = F/(\alpha - \omega^2)$ and that putting this result into (22.181) gives the exact answer (22.180) in one step; that is, it provides the "analytic continuation" of (22.183) and bypasses the unwelcome $|\alpha/\omega^2| < 1$ restriction.

In view of this success for $\beta = 0$, we are encouraged to expect good results for the case where β is nonzero but small. Again starting with $x_0 = A\cos\omega t$, use these ideas and (22.179) to obtain

$$x_1 = A\cos\omega t + \frac{\beta A^3}{36[\alpha + (3/4)\beta A^2 - F/A]}\cos 3\omega t, \tag{22.184}$$

together with the amplitude-frequency relation

$$\omega^2 = \alpha + \frac{3}{4}\beta A^2 - \frac{F}{A}. \tag{22.185}$$

22.60. Compute $y_2(x)$ in Example 22.19. Solve (22.163) *exactly* and plot $y_0(x)$, $y_1(x)$, $y_2(x)$ and the exact solution over $0 \leq x \leq 2$, say.

22.61. Consider the nonlinear, coupled differential equations

$$\gamma'' + \left(1 - \frac{1}{T^2}\right)\frac{T}{\gamma} = 1 \tag{22.186a}$$

$$T'' + \gamma T^2 = 0 \tag{22.186b}$$

in $\gamma(x)$, $T(x)$. Use quasilinearization to obtain a linear system that can be solved iteratively. *Hint:* The quasilinearized version of (22.186b) is

$$T''_{n+1} + 2\gamma_n T_n T_{n+1} + T_n^2 \gamma_{n+1} = 2\gamma_n T_n^2.$$

Do the same for (22.186a).

22.62. Make up two examination-type questions on this chapter.

Chapter 23

Qualitative Methods; The Phase Plane

The emphasis in this chapter is on second-order, autonomous, nonlinear equations and the use of the so-called phase plane in portraying and studying their solutions. We will see that the presence of singularities, as well as their type, is of central importance. Often we wish to know whether the solution of a nonlinear system becomes periodic as $t \longrightarrow \infty$, how the solutions change due to the variation of a physical parameter, and so on. Such questions are largely qualitative in nature and frequently involve geometric or topological aspects of the phase plane.

Although much of the early work in nonlinear mechanics dates back to the nineteenth century and Poincaré (1854–1912), in connection with his study of celestial mechanics, it was not until the 1930s that the subject received considerable impetus through the efforts of a number of Russian scientists and mathematicians. In general, this work did not reach the West until the publication of Minorsky's *Introduction to Non-Linear Mechanics*[1] in 1947 and the translations by Lefschetz of the books by Kryloff and Bogoliuboff[2] and Andronow and Chaikin.[3] Since then the subject has attracted the widespread attention that it deserves.

---

[1] N. Minorsky, *Introduction to Non-Linear Mechanics*, J. W. Edwards, Inc., Ann Arbor, Michigan, 1947.

[2] N. Kryloff and N. Bogoliuboff, *Introduction to Non-Linear Mechanics*, Princeton University Press, Princeton, N.J., 1943.

[3] A. Andronow and C. Chaikin, *Theory of Oscillations*, Princeton University Press, Princeton, N.J., 1949.

Consider, for instance, the (linear) simple harmonic mechanical oscillator

$$m\ddot{x} + kx = 0. \tag{23.1}$$

Instead of seeking $x = Ae^{\lambda t}$, let us pursue a different approach. First, we reexpress (23.1), equivalently, as

$$\frac{dx}{dt} = y \tag{23.2a}$$

$$\frac{dy}{dt} = -\frac{kx}{m}, \tag{23.2b}$$

where the letter y for the velocity will be more convenient than the more customary v. Dividing (23.2b) by (23.2a),

$$\frac{dy}{dx} = -\frac{k}{m}\frac{x}{y} \quad \text{or} \quad my\,dy + kx\,dx = 0, \tag{23.3}$$

so that we have as the "first integral" of (23.1),

$$my^2 + kx^2 = C. \tag{23.4}$$

If we solve for y (i.e., for dx/dt), separate variables, and integrate again, we obtain the solution

$$x(t) = A \sin(\omega t + \phi), \tag{23.5}$$

where $\omega = \sqrt{k/m}$ and the amplitude A and phase angle ϕ are integration constants.

Instead let us take the first integral (23.4) as our end result and plot the one-parameter family of ellipses that it defines (Fig. 23.1). It is customary to speak of the x, y plane as the **phase plane** and the integral curves as phase trajectories or simply as **trajectories.** Each trajectory represents a possible motion of the system, and each *point* on a given trajectory represents an instantaneous state of the system. The time t enters only as a parameter—that is, through the parametric representation $x = x(t)$, $y = y(t)$. So we can visualize the representative point $x(t)$, $y(t)$ as moving along a given trajectory, as suggested by the arrows in Fig. 23.1. The *direction* of the arrows (i.e., clockwise) is implied by the fact that $y = \dot{x}$, so that $y > 0$ implies that $x(t)$ is increasing and $y < 0$ implies that $x(t)$ is decreasing.

By way of graphical presentation, note that whereas (23.5) is a *two*-parameter family (amplitude A and phase ϕ), (23.4) is only a *one*-parameter family (C, which in this case is proportional to the total energy) and thus has the advantage of conciseness.[4]

[4]This conciseness is due to the fact that the original system (23.1) is **autonomous**—that is, there is no *explicit* t-dependence in the equation. Considering the more general autonomous case $\ddot{x} = f(x, \dot{x})$, suppose that the general solution is $x = x(t; a, b)$, where a, b are the constants of integration. Then $\dot{x} = y = y(t; a, b)$. If we can eliminate t between these two relations, we obtain $y = \phi(x; a, b)$, say. This is a *two*-parameter family. However, since $\ddot{x} = f(x, \dot{x})$ is *autonomous*, its solution must be of the form $x = x(t - a; b)$, $y = y(t - a; b)$ (Exercise 23.1), so that elimination of $t - a$ yields the *one*-parameter family $y = \phi(x; b)$, say.

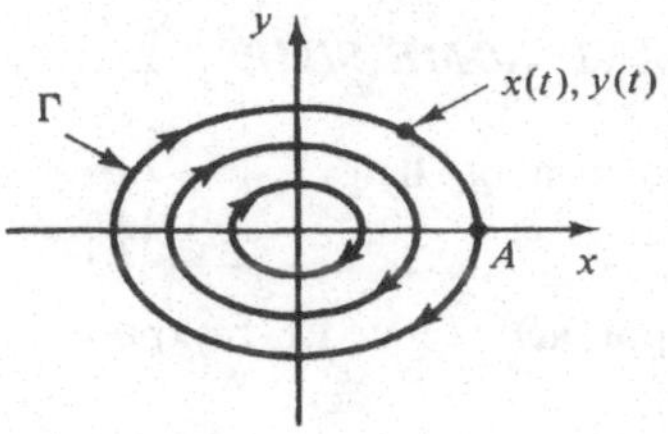

Figure 23.1. Phase portrait of (23.1).

For instance, the single phase trajectory Γ (Fig. 23.1) corresponds to the entire family of oscillations of amplitude A shown in Fig. 23.2, since *any* point on Γ can be designated as the "initial" point (i.e., corresponding to $t = 0$).

More generally, we will consider the system

$$\dot{x} = P(x, y), \qquad \dot{y} = Q(x, y). \tag{23.6}$$

To illustrate, suppose that two species of fish coexist in a lake. The smaller species x feeds on vegetation, which is available in unlimited quantity, and the larger species y feeds exclusively on species x. If the two species were separated from each other, their populations could be assumed to be governed approximately by the rate equations

$$\dot{x} = \alpha x, \qquad \dot{y} = -\beta y,$$

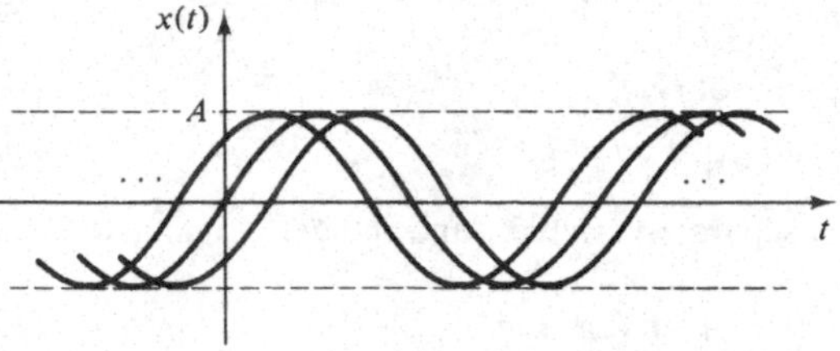

Figure 23.2. Solutions $x(t)$ corresponding to the trajectory Γ.

where x, y are the populations of species x, y, respectively, considered to be large enough to be taken as continuous (rather than integer) variables, and α, β are (presumably positive) constants that reflect net birth/death rates. The species, however, are *not* separated, and so we expect the effective α to decrease as y increases and the effective β to decrease as x increases. An approximate revised model might then be expressed as

$$\dot{x} = (\alpha - \gamma y)x, \qquad \dot{y} = -(\beta - \delta x)y, \tag{23.7}$$

which is precisely of the form (23.6). This ecological problem is known as *Volterra's problem*. We return to it in Exercise 23.4.

Dividing equations (23.6), we have

$$\frac{dy}{dx} = \frac{Q(x, y)}{P(x, y)}, \tag{23.8}$$

the integration of which provides the integral curves $y = \phi(x, c)$ except at points where $P(x, y)$ vanishes. But if $Q \neq 0$ at such a point, we can interchange the roles of x and y and consider $dx/dy = P(x, y)/Q(x, y)$ instead. If *both* P and Q vanish at a point x_0, y_0 of the phase plane, then that point is said to be a **singular point** of (23.8), for then a unique slope is not defined there. In terms of the original system (23.6), however, x_0, y_0 is best described as an **equilibrium point** because both x and y are stationary there.

As mentioned earlier, the notion of a phase plane trajectory holds only in the autonomous case or if there is a t-dependence in Q and P that cancels on dividing Q by P in (23.8); otherwise the slope dy/dx at a given point varies with time. This situation is analogous to two-dimensional fluid flow, where the term "streamline" makes sense only for *steady* flow $\mathbf{q} = u(x, y)\hat{\mathbf{i}} + v(x, y)\hat{\mathbf{j}}$ (or if there is a t-dependence in both u and v that cancels on dividing v by u.)

23.2. SINGULAR POINTS

There are four types of *elementary singularity*, each of which will now be illustrated through a simple linear system.

Consider first the system

$$\dot{x} = x, \qquad \dot{y} = ay \tag{23.9}$$

with a singular point at $x = y = 0$. It is easily found that the integral curves are given by

$$y = C|x|^a. \tag{23.10}$$

If $a > 0$, the curves are as sketched in Fig. 23.3 and the singular point at $x = y = 0$ is classified as a **node**, the definitive feature being that *every integral curve has a limiting direction at the singular point*. Observe that the system

$$\dot{x} = -x, \qquad \dot{y} = -ay \tag{23.11}$$

has the same trajectories as (23.9) except that all the arrows in Fig. 23.3 need to be reversed.

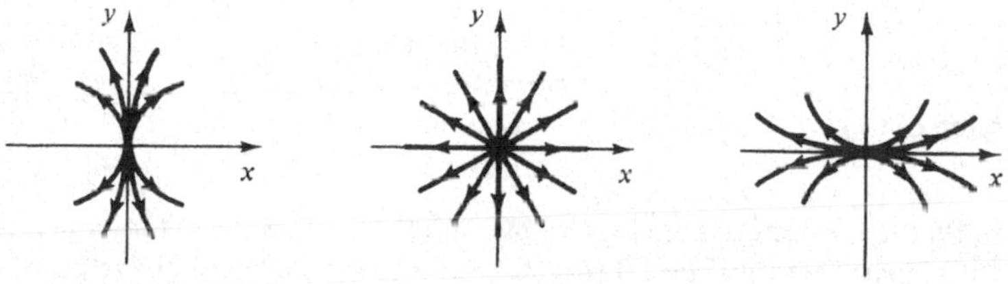

Figure 23.3 Unstable node at $x = y = 0$; system (23.9).

Before continuing, let us introduce a notion of *stability*. For brevity, we denote the representative point $x(t)$, $y(t)$ of any trajectory by the letter $\mathbf{P}(t)$. We will say that *the solution $x = x(t)$, $y = y(t)$ is* ***stable*** *in the sense of* ***Liapunov*** *if to each $\epsilon > 0$ (no matter how small) and t_0 there corresponds a $\delta(\epsilon, t_0)$ such that for any solution $x_1(t)$, $y_1(t)$ whose distance $d[\mathbf{P}_1(t_0), \mathbf{P}(t_0)] < \delta$ we have $d[\mathbf{P}_1(t), \mathbf{P}(t)] < \epsilon$ for all $t \geq t_0$; otherwise it is* ***unstable***. (See Fig. 23.4.) If, in addition, $d[\mathbf{P}_1(t), \mathbf{P}(t)] \to 0$ as $t \to \infty$, then we say that the solution is *asymptotically* Liapunov-stable.

In the present section we are considering singular points, which are actually "point trajectories," since the entire solution curve $x = x(t)$, $y = y(t)$ consists of a single

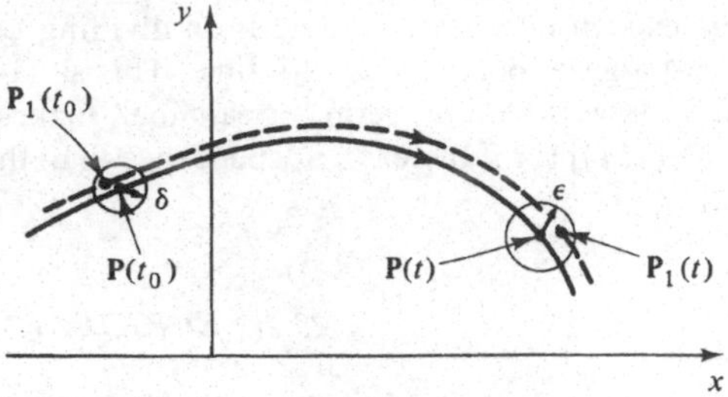

Figure 23.4. Liapunov stability.

point—that is, $\mathbf{P}(t) = \text{constant} = \mathbf{P}(t_0)$. It should be clear from Fig. 23.3 that the nodal point $x = y = 0$ of the system (23.9) is an *unstable* solution in the sense of Liapunov, since points $\mathbf{P}_1(t)$ that are initially close to the origin do not remain close. In fact, it is not hard to show that $d[\mathbf{P}_1(t), \mathbf{P}(t)] \to \infty$ as $t \to \infty$. On the other hand, (23.11)'s node is *stable*.

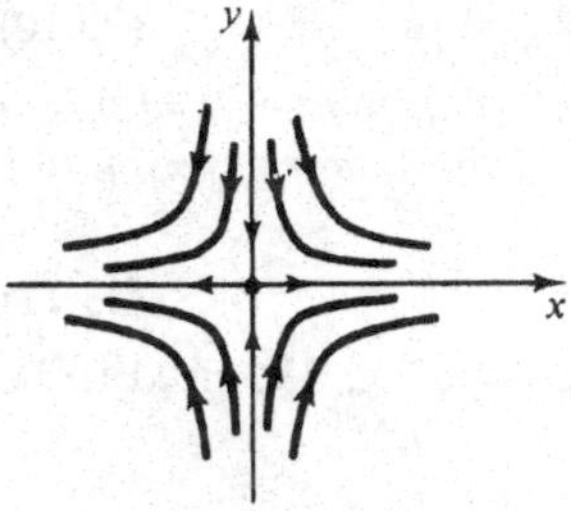

Figure 23.5. Saddlepoint at $x = y = 0$; system (23.9) for $a < 0$.

So far we have considered $a > 0$ in (23.9). If $a < 0$ instead, then the curves (23.10) are hyperbolas, as sketched in Fig. 23.5, and the singular point at $x = y = 0$ is called a **saddlepoint**. Observe that only four trajectories approach the singular point, two as $t \to +\infty$ (y axis) and two as $t \to -\infty$ (x axis). A saddlepoint is always *unstable*.

Next, consider the system

$$\dot{x} = -ax + y, \qquad \dot{y} = -x - ay. \tag{23.12}$$

Here the change $x = r\cos\theta$, $y = r\sin\theta$ to polar variables is convenient, and we find that (23.12) reduces to $\dot{r} = -ar$ and $\dot{\theta} = -1$. Solving,

$$r = Ae^{-at}, \qquad \theta = -t + B,$$

so that the trajectories are logarithmic spirals that approach the singular point if $a > 0$ and depart from it if $a < 0$ (Fig. 23.6). We call such a singular point a **focus**, *stable* if we spiral inward and *unstable* if we spiral outward. Note the absence of a limiting direction as the trajectories approach the singular point.

Finally, return to system (23.2) and consider the singular point at $x = y = 0$. The integral curves are shown in Fig. 23.1, and the singular point is called a **center**. Centers are always stable.

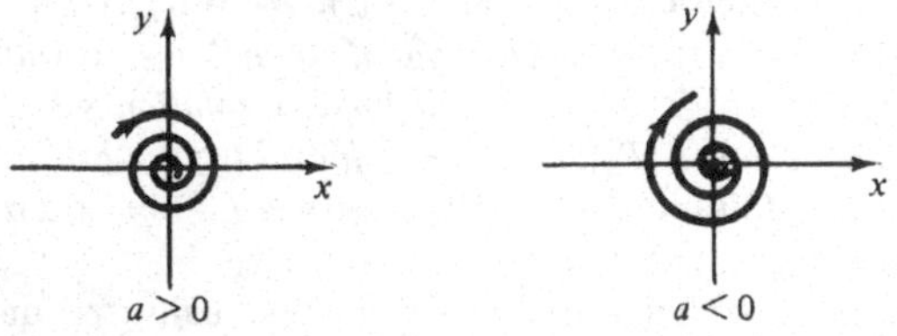

Figure 23.6. Focus at $x = y = 0$; system (23.12).

Thus far we have examined only the three linear systems (23.2), (23.9), and (23.12). Let us now consider the *general linear system*

$$\dot{x} = ax + by, \qquad \dot{y} = cx + dy \tag{23.13}$$

with an isolated singular point at the origin.[5] It can be shown (Exercise 23.3) that the singularity will be one of the four types just considered, depending on the coefficients a, b, c, d, according to the following criteria:

$$\left.\begin{array}{l} D > 0 \begin{cases} \text{Node if } cb - ad < 0 \begin{cases} \text{Stable if } a + d < 0 \\ \text{Unstable if } a + d > 0 \end{cases} \\ \text{Saddle if } cb - ad > 0 \end{cases} \\ D < 0 \begin{cases} \text{Center if } a + d = 0 \\ \text{Focus if } a + d \neq 0 \begin{cases} \text{Stable if } a + d < 0 \\ \text{Unstable if } a + d > 0 \end{cases} \end{cases} \\ D = 0 \text{ Node} \begin{cases} \text{Stable if } a + d < 0 \\ \text{Unstable if } a + d > 0, \end{cases} \end{array}\right\} \tag{23.14}$$

where $D \equiv (a - d)^2 + 4bc$.

Example 23.1. Consider the familiar damped, harmonic oscillator $m\ddot{x} + c\dot{x} + kx = 0$ $(m, c, k > 0)$ or, equivalently,

$$\dot{x} = y,$$
$$\dot{y} = -\frac{k}{m}x - \frac{c}{m}y$$

say. Then $D = (c^2 - 4km)/m^2$, "$cb - ad$" $= -k/m < 0$, and "$a + d$" $= -c/m < 0$. Thus

$$c \geq \sqrt{4km} \text{ implies stable node}$$
$$c < \sqrt{4km} \text{ implies stable focus.}$$

It is customary to call $\sqrt{4km} \equiv c_{cr}$ the *critical damping*. If the system is "overdamped" or "critically damped" (i.e., $c \geq c_{cr}$), then the motion decays without oscillation, hence the node. If it is "underdamped" ($c < c_{cr}$), the motion dies out *with* oscillation, hence the focus.

In order to sketch some typical phase trajectories let us express

$$\frac{dy}{dx} = -\frac{kx + cy}{my},$$

obtained by dividing the above expressions for $\dot{y}$ and $\dot{x}$. Integrating this first order differential equation will produce a relation of the form $f(x, y) = C$, and this will yield the desired integral curves. For purposes of sketching, however, it may be simpler to use the **method of isoclines**. This amounts to finding those curves (iso-

---

[5]Observe that both $\dot{x}$ and $\dot{y}$ vanish at the intersection of the two straight lines $ax + by = 0$ and $cx + dy = 0$ through the origin, so that $x = y = 0$ is surely a singular point. However, if $ad - bc = 0$, then the two lines coincide and *all* points on that line are (nonisolated) singular points. To preclude this situation, we will require that $ad - bc \neq 0$. Note also that the system $\dot{x} = ax + by + c_1$, $\dot{y} = cx + dy + c_2$, where c_1 and c_2 are constants, is actually the most general linear system. However, if the singular point is denoted by x_0, y_0, then the simple translation $x' = x - x_0$, $y' = y - y_0$ reduces this system to $\dot{x}' = ax' + by'$, $\dot{y}' = cx' + dy'$, which has a singular point at $x' = y' = 0$. Thus we may consider the form (23.13) with no loss of generality.

clines) along which the slope dy/dx is a constant, say M. For definiteness let us take $m = k = 1$ and $c = 3$, which is *overdamped* since $c_{cr} = 2$. Then $dy/dx = -(x + 3y)/y = M$ along the curve

$$y = -\frac{1}{M + 3}x.$$

For instance the $M = \infty$ isocline is the curve $y = 0$, the $M = 0$ isocline is the curve $y = -x/3$, and so on. Drawing some of these isoclines in the x, y plane and indicating the slope M along each isocline by a number of short line segments, observe that it is possible to use this direction field to guide us in sketching the solution curves, as indicated in Fig. 23.7(a). Suppose, for example, that the initial conditions are $x(0) = 0$, and $\dot{x}(0) = y(0) = 2$. Then the representative point in the phase plane starts at the point A, moves along the trajectory labeled $\mathcal{C}$, and approaches the origin as $t \longrightarrow \infty$. For comparison, plots of $x(t)$ versus t are included in the figure, corresponding to initial points A and A'. For the *underdamped* case, $c < c_{cr}$, representative curves are as sketched in Fig. 23.7(b). ▮

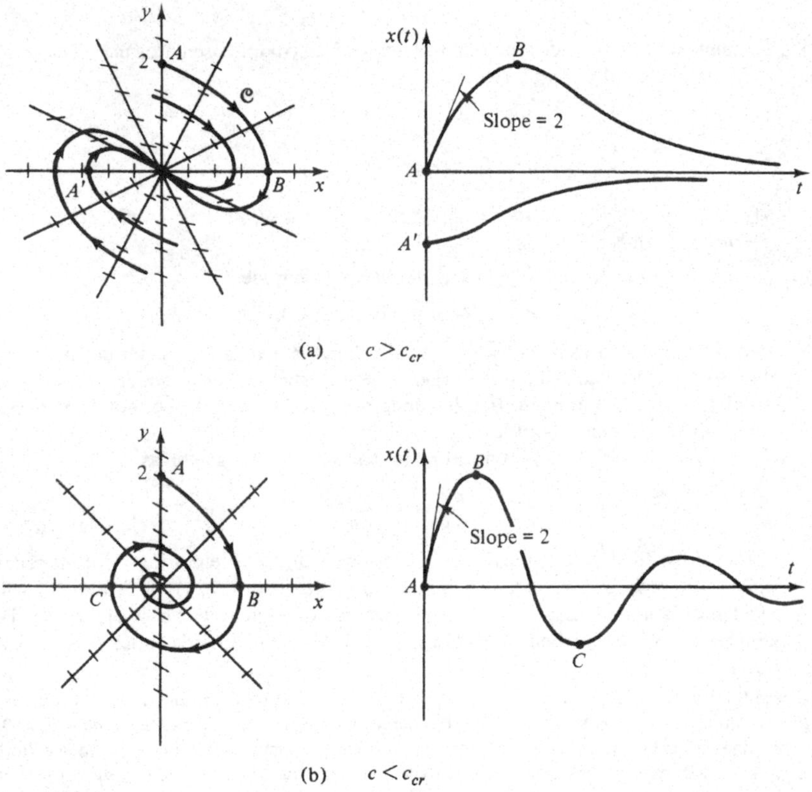

Figure 23.7. Damped harmonic oscillator.

Of course, our principal interest is in the *nonlinear* system (23.6). However, suppose that we can express

$$P(x, y) = ax + by + P_2(x, y)$$
$$Q(x, y) = cx + dy + Q_2(x, y), \tag{23.15}$$

where $ad - bc \neq 0$ (recall footnote 5) and where P_2 and Q_2 are second order or higher (e.g., power series in x and y beginning with terms of degree at least 2). It was shown by Poincaré that *the singularities of the system* $\dot{x} = P(x, y)$, $\dot{y} = Q(x, y)$ *are then identical to those of the abridged (i.e., linearized) version* $\dot{x} = ax + by$, $\dot{y} = cx + dy$ except for the special case where $D < 0$ and $a + d = 0$, which may be a center *or* a focus, depending on the higher-order terms.[6]

Example 23.2. For instance, $\ddot{x} + \epsilon\dot{x}^3 + x = 0$ ($\epsilon > 0$) or

$$\dot{x} = y, \qquad \dot{y} = -x - \epsilon y^3 \tag{23.16}$$

has the abridged system $\dot{x} = y, \dot{y} = -x$, corresponding to the singular point $x = y = 0$. According to (23.14), the abridged system's singularity is a center. However, observe that $D = -4 < 0$ and $a + d = 0$, so that the full system (23.16) may have either a center *or* a focus. In fact, we will find in Chapter 25 (Exercise 25.12) that $\ddot{x} + \epsilon\dot{x}^3 + x = 0$, with $x(0) = 0$ and $\dot{x}(0) = 1$ say, has the approximate solution

$$x(t) \approx \frac{1}{\sqrt{1 + \frac{3}{4}\epsilon t}} \sin t \tag{23.17}$$

for $\epsilon \ll 1$. Thus the abridged system has a center, whereas the full system clearly has an unstable focus. ▮

Although this result covers most cases of practical interest, occasionally we encounter *higher-order* singularities—for instance, if $a = b = c = d = 0$ in (23.15). Such cases can be quite complicated and should be considered individually. You may recall from fluid mechanics (or electrostatics) the way that the coalescing of a source and a sink can lead to a doublet (or dipole), two doublets can coalesce to yield a still higher-order singularity, and so on. The situation here, where higher-order singularities can be thought of as arising from the coalescing of lower-order ones, is not too dissimilar.

23.3. *ADDITIONAL EXAMPLES*

Example 23.3. Consider the undamped, unforced *Duffing equation* (recall Exercise 22.59)

$$\ddot{x} + \alpha x + \beta x^3 = 0, \tag{23.18}$$

[6]See N. Minorsky, *Nonlinear Oscillations*, Van Nostrand, Princeton, N.J., 1962, Sections 6 and 8 of Chapter 1. We will see how to distinguish between the center and focus at least in some cases; see Exercises 23.15 and 23.22(b).

where $\alpha > 0$ and β may be positive (corresponding to a "hard spring") or negative ("soft spring"). With $\dot{x} = y$ we have

$$\dot{x} = y, \qquad \dot{y} = -\alpha x - \beta x^3; \tag{23.19}$$

then

$$\frac{dy}{dx} = -\frac{\alpha x + \beta x^3}{y},$$

and integration yields the curves

$$y^2 + \alpha x^2 + \frac{\beta x^4}{2} = \text{constant} \equiv C, \text{ say}, \tag{23.20}$$

where C is actually twice the total energy.

Before looking at these curves, let us find the singular points. Setting

$$y = 0, \qquad \alpha x + \beta x^3 = 0,$$

we see that, if $\beta > 0$, there is only one singular point, at $x = y = 0$, whereas for $\beta < 0$ two more occur, at $x = \pm\sqrt{\alpha/|\beta|}$ and $y = 0$.

Case 1: $\beta > 0$.

Then the abridged system is

$$\dot{x} = y, \qquad \dot{y} = -\alpha x,$$

and it follows from (23.14) that its singularity at $x = y = 0$ is a center, although [recall the paragraph following (23.15)] the singularity of the full system (23.18) may be either a center *or* a focus. From (23.20), however, it is clearly a center, since, for motions in the vicinity of the origin, the x^4 term is negligible and what's left is the equation of a family of ellipses. (See also Exercise 23.15.) For larger motions the x^4 term becomes important, and the curves are essentially *distorted* ellipses (Fig. 23.8).

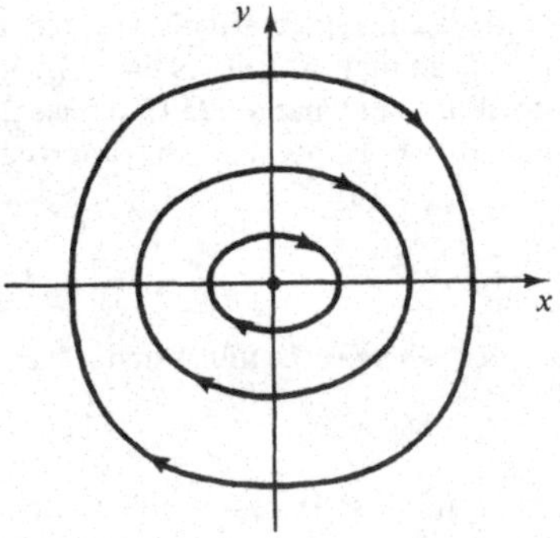

Figure 23.8. Hard spring ($\beta > 0$).

Case 2: $\beta < 0$.

As in Case 1, the singularity at $x = y = 0$ is a center. Consider, then, either of the other two singularities, say at $x = \sqrt{\alpha/|\beta|}$ and $y = 0$. The abridged system is

$$\dot{x} = y \tag{23.21a}$$

$$\dot{y} = 2\alpha\left(x - \sqrt{\frac{\alpha}{|\beta|}}\right). \tag{23.21b}$$

where the right-hand side of (23.21b) is obtained by expanding $-\alpha x - \beta x^3$ in a Taylor series about $x = \sqrt{\alpha/|\beta|}$ and keeping terms only through first degree. Thus $a = 0, b = 1, c = 2\alpha, d = 0$, and it follows from (23.14) that the singularity at $x = \sqrt{\alpha/|\beta|}, y = 0$ is a *saddlepoint*. Similarly, $x = -\sqrt{\alpha/|\beta|}, y = 0$ is found to be a saddlepoint, too.

Plotting the integral curves (23.20) in Fig. 23.9, notice the way that the three singularities determine the essential structure of the entire phase plane. The curves (heavy ink) through the singular points $x = \pm\sqrt{\alpha/|\beta|}, y = 0$ are of special impor-

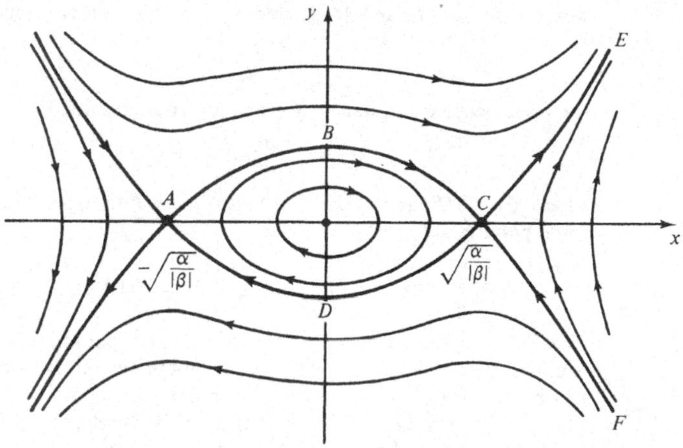

Figure 23.9. Soft spring ($\beta < 0$).

tance because they separate qualitatively different kinds of motions—for instance, periodic and nonperiodic ones. Such a trajectory is called a **separatrix**.

Observe very carefully the highly nonlinear behavior revealed by the phase portrait. For instance, if $x(0) = 0$, $y(0) = y_0$ yields the solution $x = X(t)$, $y = Y(t)$, and the system is *linear*, then $x(0) = 0$, $y(0) = \alpha y_0$ yields the solution $x = \alpha X(t)$, $y = \alpha Y(t)$. Thus there is a simple scaling, and all the phase trajectories of a *linear* system have the same shape and differ only in scale! Returning to Fig. 23.9, note that near $x = y = 0$ the curves are essentially elliptical, but as the amplitude is increased, the ellipses begin to distort and become more pointed at the ends (like a football). The degree of distortion is a measure of the effects of nonlinearity. Once outside the separatrix, the elliptical shape is totally lost and nonlinear effects predominate. Consequently, although the phase plane is not too exciting for linear systems, it *is* for nonlinear ones, since it can capsulize beautifully all the different kinds of solution admitted by the system, as in Fig. 23.9.

Finally, observe that the separatrix $ABCDA$ is itself *not* a periodic motion. To see this fact, first let us introduce the **phase velocity**

$$\dot{s} = \sqrt{\dot{x}^2 + \dot{y}^2},$$

that is, the speed of the representative point $x(t)$, $y(t)$ in the phase plane. From Fig. 23.10 we see, from similar triangles, that

$$\frac{\dot{s}}{\dot{x}} = \frac{a}{y}, \tag{23.22}$$

where a is the length of the perpendicular dropped from P to the x axis. But $\dot{x} = y$ (in this example), and so (23.22) yields the simple result $\dot{s} = a$; that is, $\dot{s}$ may be identified graphically as the length of the perpendicular dropped from P to the x axis. Thus as P approaches C along BC (Fig. 23.9)

$$\dot{s} \sim M(X - x), \tag{23.23}$$

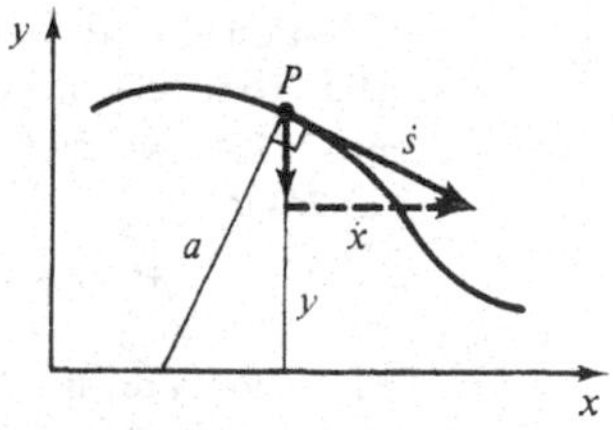

Figure 23.10. The phase velocity for the case where $\dot{x} = y$.

where $X = \sqrt{\alpha/|\beta|}$ and M is some positive constant (Exercise 23.7). But

$$\dot{s} = \frac{ds}{dx}\frac{dx}{dt} \sim N\dot{x} \quad \text{as } P \longrightarrow C$$

for some positive constant N. Combining these results,

$$\frac{dx}{dt} \sim K(X - x),$$

where $K = M/N$, and so the time that it takes P to move along the separatrix from some point P_0 (close to C) to C is

$$\Delta t \sim \frac{1}{K}\int_{x_0}^{X} \frac{dx}{X - x} = -\frac{1}{K}\ln(X - x)\Big|_{x_0}^{X} = \infty. \tag{23.24}$$

That is, P *approaches* C as $t \longrightarrow \infty$ but does not reach it in finite time, and this situation results from the fact that the phase velocity $\dot{s}$ = perpendicular distance to x axis tends to zero as $M(X - x)$. Similarly, P approaches C along FC (Fig. 23.9) as $t \longrightarrow \infty$, along EC as $t \longrightarrow -\infty$, and so on. Surely, then, the separatrix $ABCDA$ does not correspond to a periodic motion.

Although the preceding demonstration was rather specific, it is not hard to show that *if a trajectory of (23.6) starts at a regular (i.e., nonsingular) point, then it cannot reach* any *singular point in finite time (Exercise 23.8).* ▮

Example 23.4. *The van der Pol Equation.* Consider the vacuum tube circuit shown in Fig. 23.11. By Kirchhoff's law,

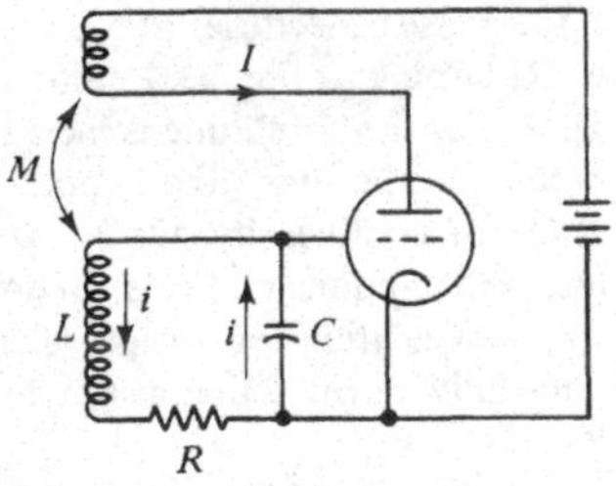

Figure 23.11. Vacuum tube circuit.

$$L\dot{i} + Ri + \frac{1}{C}\int i\,dt = M\dot{I},$$

where L, R, C, M are inductance, resistance, capacitance, and mutual inductance, respectively. It is found empirically that the plate current I depends on the grid voltage x according to

$$I \approx \alpha x - \beta x^3$$

for x not too large, where α and β are positive empirical constants. Noting that

$$x = \frac{1}{C}\int i\,dt \quad \text{or} \quad i = C\dot{x},$$

we obtain $LC\ddot{x} + RC\dot{x} + x = M(\alpha\dot{x} - 3\beta x^2\dot{x})$ or

$$LC\ddot{x} - [(M\alpha - RC) - 3M\beta x^2]\dot{x} + x = 0. \tag{23.25}$$

Equation (23.23) contains three parameters, the combinations LC, $M\alpha - RC$, and $M\beta$. To reduce the number, let us scale x and t per $x = A\bar{x}$, $t = B\bar{t}$. Plugging them into (23.25), we find that the choice $B = \sqrt{LC}$, $A = \sqrt{(M\alpha - RC)/3M\beta}$ reduces (23.25) to

$$\bar{x}'' - \left(\frac{M\alpha - RC}{\sqrt{LC}}\right)(1 - \bar{x}^2)\bar{x}' + \bar{x} = 0, \qquad (\ )' = \frac{d(\ )}{d\bar{t}},$$

which contains the *single* parameter $(M\alpha - RC)/\sqrt{LC} \equiv \epsilon$, say. Finally, dropping the bars and replacing the primes by dots, for simplicity, we arrive at the well-known **van der Pol equation**

$$\ddot{x} - \epsilon(1 - x^2)\dot{x} + x = 0. \tag{23.26}$$

Setting $\dot{x} = y$, $\dot{y} = \epsilon(1 - x^2)y - x$, we see that the only singular point is $x = y = 0$ and that it is an *unstable focus* if $\epsilon < 2$ and an *unstable node* if $\epsilon > 2$ [Exercise 23.5(a)]; we will always take ϵ as positive in (23.26). Next, consider the phase plane (Fig. 23.12) and observe that the "damping coefficient" $c = \epsilon(x^2 - 1)$ is negative in the strip $|x| < 1$ and positive in $|x| > 1$. So as the integral curves spiral outward from the focus, they eventually cut through the regions of positive damping (Fig. 23.12), and we may anticipate that there exists a limiting closed curve over which the positive and negative damping are exactly in balance. In fact, we have the following theorem, which is due to Levinson and Smith.

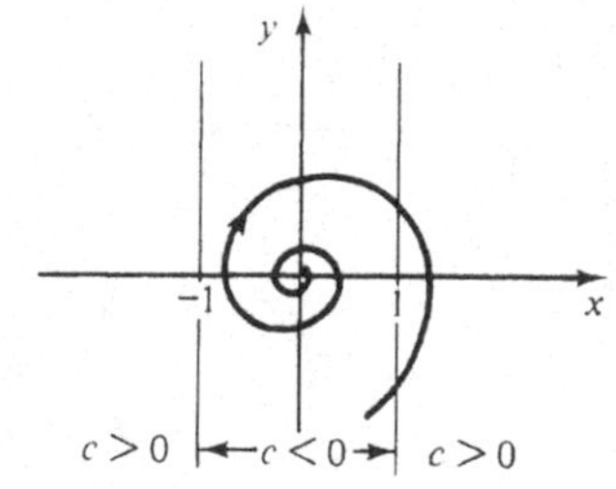

Figure 23.12. Phase plane for (23.26).

THEOREM 23.1. Let $f(x)$ be even and continuous for all x. Let $g(x)$ be odd with $g(x) > 0$ for all $x > 0$ and $g'(x)$ continuous for all x. With

$$\int_0^x f(\xi)\,d\xi \equiv F(x) \quad \text{and} \quad \int_0^x g(\xi)\,d\xi = G(x),$$

suppose that

(i) *$G(x) \longrightarrow \infty$ as $x \longrightarrow \infty$ and*

(ii) *there is an $x_0 > 0$ such that $F(x) < 0$ for $0 < x < x_0$, $F(x) > 0$ for $x > x_0$, and $F(x)$ is monotonically increasing for $x > x_0$ with $F(x) \longrightarrow \infty$ as $x \longrightarrow \infty$. Then the "generalized Liénard equation"*

$$\ddot{x} + f(x)\dot{x} + g(x) = 0 \tag{23.27}$$

has a single periodic solution, the trajectory of which is a closed curve encircling the origin in the x, $\dot{x}$ phase plane; all other trajectories (except the trajectory consisting of the single point at the origin) spiral toward the closed trajectory as $t \longrightarrow \infty$.

Applying this theorem to the van der Pol equation (23.26), $F(x) = -\epsilon(x - x^3/3)$, which is less than zero for $0 < x < \sqrt{3}$, greater than zero for $x > \sqrt{3}$, and increases monotonically to infinity as $x \longrightarrow \infty$. Similarly, the other hypotheses are satisfied, and we conclude that the van der Pol equation does admit a closed trajectory—that is, a periodic solution.

The results are shown in Fig. 23.13 for $\epsilon = 0.1$, 1, and 5, and $x(t)$ for the trajectories labeled C is shown in Fig. 23.14. Following Poincaré, the closed trajectories (heavy ink) are called **limit cycles**. They are clearly of crucial interest because *every* trajectory of (23.26) [except $x(t) \equiv 0$] approaches the limit cycle as $t \longrightarrow \infty$—that is, independent of the initial conditions. In each case (Fig. 23.13), we see that the limit cycle lies partly in $|x| < 1$ and partly in $|x| > 1$, as discussed above. In Chapter 25 we will actually determine the van der Pol limit cycle. We will find that it tends to a circle of radius 2 as $\epsilon \longrightarrow 0$ and becomes more and more kinky as ϵ becomes large. Notice that when ϵ *equals* zero, the phase plane is filled with concentric circles, which are said to have bifurcated from the limit cycle. Thus $\epsilon = 0$ is said to be a **bifurcation point** (also called a branch point). Bifurcation theory involves the notion of the so-called *structural stability* of the differential equation; that is, if the phase portrait of a given differential equation containing a parameter ϵ is qualitatively the same

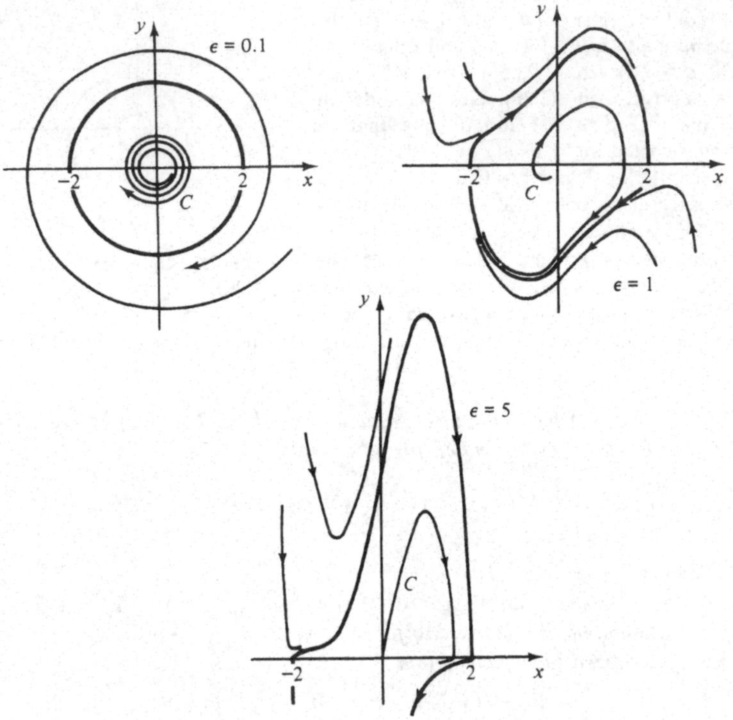

Figure 23.13. The van der Pol limit cycle.

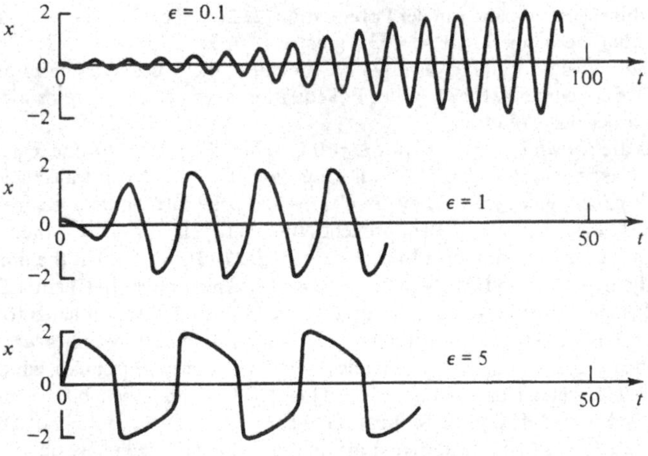

Figure 23.14. The effect of increasing nonlinearity.

for all ϵ within some arbitrarily small neighborhood of ϵ_0, then the system is structurally stable for $\epsilon = \epsilon_0$. For the van der Pol equation we have structural stability for all $\epsilon > 0$ but not for $\epsilon = 0$, which is therefore a bifurcation point.

For additional discussion of the van der Pol limit cycle, see Exercises 25.13 and 25.18. ∎

EXERCISES

23.1. Show that the solution of an *autonomous* differential equation $F(x, \dot{x}, \ddot{x}) = 0$, say, is of the form $x = x(t - a; b)$ or, perhaps more suggestively, $x = x(t - t_0, c)$.

23.2. Sketch the phase portrait of each of these systems (with $\dot{x} = y$ in each case).

(a) $\ddot{x} + \frac{g}{l}\sin x = 0$ (b) $\ddot{x} + c\dot{x} + \frac{g}{l}\sin x = 0$

(c) $\ddot{x} + x^2 = 0$ (d) $\ddot{x} - x^2 = 0$

(e) $\ddot{x} + \dot{x} + x - x^3 = 0$ (f) $m\ddot{x} - kx = 0$

(g) $\ddot{x} - c \operatorname{sgn} \dot{x} + x = 0$—that is, a mechanical oscillator with *Coulomb friction* damping.

23.3. (a) Expressing the system $\dot{x} = ax + by, \dot{y} = cx + dy$ in the matrix form

$$\dot{\mathbf{x}} = A\mathbf{x}, \quad \mathbf{x} = \begin{pmatrix} x \\ y \end{pmatrix}, \quad A = \begin{pmatrix} a & b \\ c & d \end{pmatrix}, \tag{23.28}$$

introduce the linear change of variables $\mathbf{x} = Q\mathbf{x}'$, where Q is a modal matrix (Section 20.4), and thus reduce (23.28) to the "normal" or "canonical" form

$$\dot{x}' = \lambda_1 x'$$
$$\dot{y}' = \lambda_2 y'.$$

Express λ_1 and λ_2 in terms of a, b, c, d. Let $c = b$, for simplicity, so that A is self-adjoint.

(b) Thus obtain the integral curves $y' = Cx'^m$ in the x', y' plane, where $m = \lambda_2/\lambda_1$.

(c) Deduce the $D > 0$ part of the criteria in (23.14).

(d) Deduce the $D = 0$ part of the criteria in (23.14).

(e) Actually, we have thus established the classification in the *transformed* plane x', y'. Why does a node in the x', y' plane correspond to a node in the x, y plane and so on?

23.4. (*Volterra's problem*) Find and classify the singular points of (23.7), show that the axes are trajectories, and hence sketch the phase portrait (for $x \geq 0$ and $y \geq 0$).

23.5. Locate and classify all singularities of the systems below.

(a) $\ddot{x} - \epsilon(1 - x^2)\dot{x} + x = 0$ with $\dot{x} = y$

(b) $\dot{x} = y - x, \dot{y} = y^2 - x$

(c) $\ddot{x} - x = 0$, first with $\dot{x} = y$ and then with $\dot{x} = x + y$

(d) $\ddot{x} + x = 0$, first with $\dot{x} = y$ and then with $\dot{x} = x + y$

(e) $\dot{x} = x + \sin y, \dot{y} = y - x$

(f) $\dot{x} = x + y - 1, \dot{y} = y - x^2 + 1$

23.6. Compare the phase portraits and singular points of the following systems.

(a) $\dot{x} = y, \dot{y} = -x$ (b) $\dot{x} = y^2(x - 1), \dot{y} = -xy(x - 1)$

(c) $\dot{x} = y[y^2 + (x - 1)^2], \dot{y} = -x[y^2 + (x - 1)^2]$

(The singularity at $x = 1, y = 0$ is said to be *removable*.)

23.7. Where did (23.23) come from?

23.8. Show that if a trajectory of (23.6) starts at a regular point, then it cannot reach a singular point x_0, y_0 in finite time. *Hint:* Note that $x = x_0, y = y_0$ is itself a point trajectory and use uniqueness.

23.9. Determine the equation(s) of the separatrix in Fig. 23.8.

23.10. (a) Show that the *period* of oscillation for the Duffing equation (23.18) is (for $\beta > 0$, say)

$$T = 4\int_0^A \frac{dx}{\sqrt{\alpha A^2 + (\beta A^4/2) - \alpha x^2 - (\beta x^4/2)}},$$

where A is the maximum displacement.

(b) Setting $x = A \sin\theta$, show that

$$T = 4\sqrt{2}\int_0^{\pi/2} \frac{d\theta}{\sqrt{2\alpha + \beta A^2 + \beta A^2 \sin^2\theta}}$$
$$= \frac{2\pi}{\sqrt{\alpha}}\left[1 - \frac{3\beta A^2}{8\alpha} + \frac{57\beta^2 A^4}{256\alpha^2} - \cdots\right].$$

23.11. Determine the phase velocity $\dot{s}$ for the oscillator $\ddot{x} + x = 0$ (with $\dot{x} = y$), for the trajectory through the point $x(0) = 0, \dot{x}(0) = A$.

23.12. If a trajectory of the system (23.6) crosses the x axis, at a point P, say, at an angle other than 90°, then is P a singular point? Explain.

23.13. Draw the trajectory of $m\ddot{x} + c\dot{x} + kx = 0$ in the x, y phase plane (where $\dot{x} = y$) for $0 \leq t < \infty$, through the point $x(0) = 1, \dot{x}(0) = 0$, for the limiting case as $m \longrightarrow 0$.

23.14. Show that the system $\ddot{x} + f(\dot{x}) + g(x) = 0$ does not admit periodic solutions if $\dot{x}f(\dot{x})$ is of one sign for all $\dot{x}$. *Hint:* Show that $\dot{x}\,d\dot{x} + f(\dot{x})\,dx + g(x)\,dx = 0$ and, assuming tentatively that a periodic motion *is* possible, integrate over one full period.

23.15. (a) Show that the system

$$\ddot{x} + f(x) = 0 \tag{23.29}$$

is *conservative* (recall Section 9.6).

(b) Supposing that $f(x)$ is analytic and $f(x_0) = 0$, show that the point $x = x_0$, $y = 0$ (where $\dot{x} = y$) is a *center* if $f'(x_0) > 0$ and a *saddle* if $f'(x_0) < 0$. *Hint:* Integrating, obtain the energy equation $y^2 + F(x) = C$ and show that $F(x) \sim f'(x_0)(x - x_0)^2$.

(c) Recall, for the Duffing equation in Example 23.3, that $x = y = 0$ was a center of the abridged system and hence either a center or a focus for the full system. Using part (b) above, show that it is definitely a center. In fact, sketch $f(x)$ (for $\beta < 0$) and discuss the classification of all three singular points.

23.16. (*Dynamic formulation of buckling problem*) Consider the buckling of the system shown in Fig. 23.15, consisting of two rigid massless rods of length l pinned to a mass m and a lateral spring of stiffness k. Derive the equation of motion

$$m\ddot{x} - \frac{2Px}{l}\left[1 - \left(\frac{x}{l}\right)^2\right]^{-1/2} + kx = 0$$

and (with $\dot{x} = y$) examine the singular point $x = y = 0$. Show that it is a center if $P < kl/2$ and a saddlepoint if $P > kl/2$ and thus conclude that the critical buckling load is $P_{cr} = kl/2$.

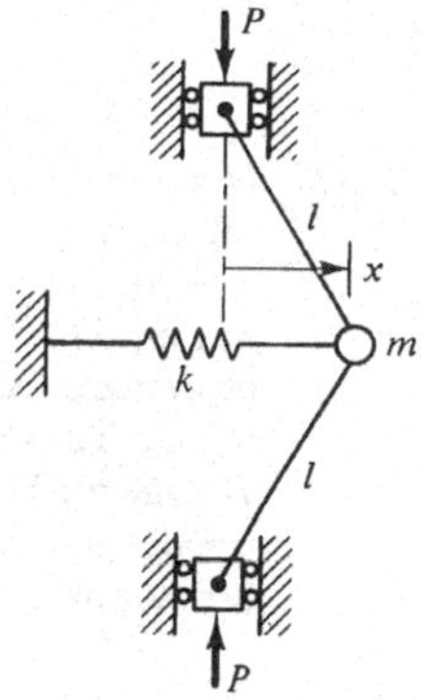

Figure 23.15. Buckling problem.

23.17. A mutual force of attraction is exerted between parallel current-carrying wires. The infinite wire (Fig. 23.16) has current I, and the wire of length l (with leads that are perpendicular to the paper) has current i in the same direction. According to the Biot–Savart law, the mutual force of attraction is $2Iil/(\text{separation}) = 2Iil/(a - x)$, where $x = 0$ is the position at which the spring force is zero. So the equation of motion of the restrained wire is

$$m\ddot{x} + k\left(x - \frac{\lambda}{a - x}\right) = 0 \quad \text{where } \lambda = \frac{2Iil}{k}$$

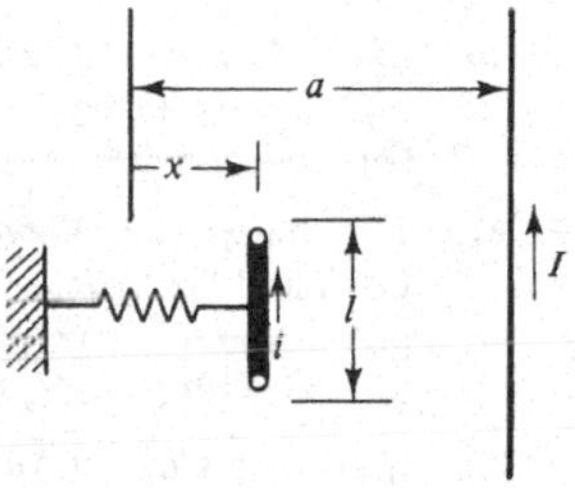

Figure 23.16. Current-carrying wire.

(a) Locate and classify the singularities (with $\dot{x} = y$) and sketch the phase portrait for the case $\lambda < a^2/4$. What is the equation of the separatrix?

(b) Determine the singularities and sketch the phase portrait for the case $\lambda = a^2/4$.

(c) Same as part (b) for $\lambda > a^2/4$.

23.18. (*Negative criterion of Bendixson*) Bendixson has given the following sufficient condition for the *nonexistence* of limit cycles. *If $P(x, y)$ and $Q(x, y)$ have continuous first partials in a bounded simply connected region D of the phase plane and $\partial P/\partial x + \partial Q/\partial y$ does not change sign (or vanish identically) within D, then the system $\dot{x} = P(x, y)$, $\dot{y} = Q(x, y)$ can have no closed trajectory within D.*

(a) Prove the theorem. *Hint:* Use Green's theorem and let the boundary C of D (tentatively) be a closed trajectory.

(b) Use the theorem to show that the van der Pol equation (with $\dot{x} = y$) cannot have a limit cycle in the strip $|x| < 1$.

(c) Make up a nice example illustrating the preceding theorem.

23.19. (*Positive criterion of Poincaré–Bendixson*) Poincaré and Bendixson have established the following sufficient conditions for the *existence* of limit cycles. *If the system* $\dot{x} = P(x, y), \dot{y} = Q(x, y)$ *has no singular points in a finite region D of the phase plane and a half-trajectory C remains within D, then either C is a closed trajectory or it approaches one* (*i.e., a limit cycle*). A difficulty in applying the theorem is in determining a suitable region D or, equivalently, in showing that a given C stays within D. One method of establishing a suitable D is provided by Poincaré in his *curves of contact*, which will not be discussed here.[7]

(a) Applying the Poincaré-Bendixson theorem to the system

$$\dot{x} = y - x(x^2 + y^2 - 4) + \frac{xy}{\sqrt{x^2 + y^2}}$$

$$\dot{y} = -x - y(x^2 + y^2 - 4) + \frac{y^2}{\sqrt{x^2 + y^2}},$$

show that a limit cycle is contained within the annulus $\sqrt{3} \leq r \leq \sqrt{5}$. *Hint:* Show that $dr/dt < 0$ for all $r > \sqrt{5}$ and that $dr/dt > 0$ for all $r < \sqrt{3}$.

(b) How about the system

$$\dot{x} = y + x(x^2 + y^2 - 4) - \frac{xy}{\sqrt{x^2 + y^2}}$$

$$\dot{y} = -x + y(x^2 + y^2 - 4) - \frac{y^2}{\sqrt{x^2 + y^2}}?$$

(c) Construct an example illustrating the foregoing theorem.

23.20. Show that the following equations admit periodic solutions.

(a) $\ddot{x} - (1 - x^2)\dot{x} + x^3 = 0$

(b) $\ddot{x} + x^2(5x^2 - 3)\dot{x} + x = 0$.

23.21. For the systems shown, expressed in polar coordinates, determine all periodic solutions and limit cycles.

(a) $\dot{r} = r^2(1 - r^2), \dot{\theta} = 1$ (b) $\dot{r} = r^2(1 - r^2), \dot{\theta} = -1$

(c) $\dot{r} = r^2(1 - r)^2, \dot{\theta} = 1$ (d) $\dot{r} = r(r - 1)(r - 4), \dot{\theta} = 2$

23.22. (*Liapunov's Direct Method*) Suppose that the (second-order) system

$$\dot{x} = P(x, y), \qquad \dot{y} = Q(x, y) \tag{23.30}$$

has an isolated singular point at $x = y = 0$. We say that $V(x, y)$ is a **Liapunov function** for (23.30) if within some neighborhood of the origin $V(x, y) > 0$ except at the origin where $V(0, 0) = 0$, if it has continuous first partials, and if either

$$V_x(x, y)P(x, y) + V_y(x, y)Q(x, y) \leq 0 \tag{23.31a}$$

or

$$V_x(x, y)P(x, y) + V_y(x, y)Q(x, y) < 0. \tag{23.31b}$$

LIAPUNOV THEOREM. (*Stability*) *If there exists a Liapunov function satisfying (23.31a), then (23.30) is* stable *at the origin; if there exists a Liapunov function satisfying (23.31b) for x, y ≠ 0 then (23.30) is* asymptotically stable *at the origin.*

---

[7] See, for example, Minorsky, *op. cit.*

LIAPUNOV THEOREM. (*Instability*) *If within some neighborhood of the origin a function $W(x, y)$ with continuous first partials satisfies*

$$W_x(x, y)P(x, y) + W_y(x, y)Q(x, y) > 0 \tag{23.32}$$

for $x, y \neq 0$ and vanishes at the origin but is positive at some point of each neighborhood of the origin, then (23.30) is unstable *at the origin.*

Unfortunately, the theorems do not tell us *how* to construct V or W. Nevertheless, a considerable literature exists on the subject. (For a start, see the Additional References for Part IV.)

(a) Give a formal explanation of why the stability theorem should be true.

(b) Show that $V = x^2 + y^2$ is a suitable Liapunov function for the equation $\ddot{x} + \epsilon\dot{x}^3 + x = 0\,(\epsilon > 0)$—that is, for $\dot{x} = y, \dot{y} = -x - \epsilon y^3$—and hence draw conclusions about the stability at $x = y = 0$. Note carefully that the abridged system fails, in this case, to yield conclusive results.

In parts (c) through (e), seek V (or W) of the form $ax^2 + by^2$ and thus draw conclusions as to the stability of the singular point $x = y = 0$.

(c) $\dot{x} = -2x^3 + 3xy^2, \dot{y} = -x^2y - y^3$

(d) $\dot{x} = -x^3 + 3y^3, \dot{y} = -4xy^2$

(e) $\dot{x} = x^3 + y^3, \dot{y} = xy^2 - 2x^2y - y^3$

23.23. Make up two examination-type questions on this chapter.

Chapter 24

Quantitative Methods

Often, in applications, we are faced with one or more ordinary differential equations, together with boundary or initial conditions, that defy our best efforts at analytic solution. In this chapter we explore a number of important techniques for the numerical solution of such problems, from the simple and classical methods of Taylor and Euler to more recent and sophisticated methods like invariant imbedding and finite elements. It turns out that these numerical methods themselves involve considerable analysis—for instance, behind the methods of weighted residuals and finite elements lies a good deal of functional analysis, as can be found in the suggested references.

24.1. *THE METHODS OF TAYLOR AND EULER*

We consider the initial value problem

$$y' = f(x, y), \qquad y(a) = y_0, \tag{24.1}$$

where f is "as nice as necessary," let's say having continuous partials of all orders. According to **Taylor's method**, we seek $y(x)$ (over $a \leq x \leq b$, say) in the form

$$y(x) = y(a) + y'(a)(x - a) + \frac{y''(a)}{2!}(x - a)^2 + \cdots. \tag{24.2}$$

The coefficients are determined systematically as follows. First of all,

$$y(a) = y_0, \qquad y'(a) = f(a, y_0),$$

and subsequent coefficients are evaluated through differentiation of the given equation $y' = f[x, y(x)]$. Thus

$$y'' = \frac{d}{dx} f[x, y(x)] = f_x + f_y y' = f_x + f f_y$$

$$y''' = \frac{d}{dx}(f_x + f f_y) = \text{etc.} = f_{xx} + f_x f_y + f f_y^2 + 2 f f_{xy} + f^2 f_{yy}$$

and so on.

Example 24.1.

$$y' = -y, \qquad y(0) = 1.$$

Then

$$y(0) = 1, \qquad y'(0) = -y(0) = -1, \qquad y''(0) = -y'(0) = 1, \text{ etc.}$$

and so

$$y(x) = 1 - x + \frac{x^2}{2!} - \frac{x^3}{3!} + \cdots. \tag{24.3}$$

There is a serious drawback. If our domain of interest is $0 \leq x \leq 10$ say, we find that a good many terms may be needed to attain a reasonable level of accuracy. For $x = 10$, for instance, about 30 terms are needed for an accuracy of ± 0.001, which is not hard to believe when we compare the (first few) partial sums of (24 3) with the exact solution e^{-x} in Fig. 24.1. ▮

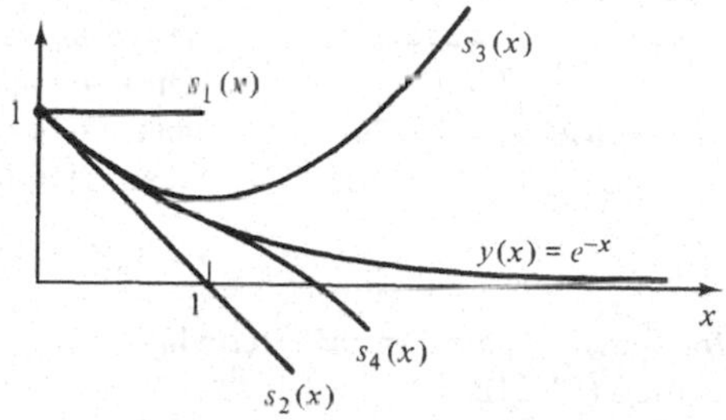

Figure 24.1. Nonuniformity of the approximation.

In fact, it is typical of the Taylor series method that numerous terms are needed to achieve a given level of accuracy *uniformly* over the domain of interest. Moreover, the evaluation of the coefficients $y''(a)$, $y'''(a)$, . . . is, in general, extremely tedious [the present simple example notwithstanding; see Exercise 24.1(e)] and must be done analytically—that is, by *you*.

Seeking to turn the bulk of the labor over to the computer, suppose that we go as far as we can, in (24.2), without incurring the need for hand differentiations—namely,

$$y(x) \approx y(a) + f[a, y(a)](x - a).$$

Generally this result will remain close to the true solution only for x very close to a [e.g., see $s_2(x)$ in Fig. 24.1], and it will be necessary to repeat the calculation many times, in finite steps, in order to generate $y(x)$ over a given domain. Such is the basis of Euler's method, which we consider next.

First, we *discretize* by seeking $y(x)$ only at the points $x_n = a + nh$, where h is the so-called *step size*. Expanding about x_n,

$$y(x_{n+1}) = y(x_n) + y'(x_n)h + \frac{y''(x_n)}{2!}h^2 + \cdots, \tag{24.4}$$

and cutting off after the y' term, as discussed above, we have

$$y(x_{n+1}) \approx y(x_n) + y'(x_n)h = y(x_n) + f[x_n, y(x_n)]h. \tag{24.5}$$

Defining $y(x_n)$ as the *true solution* of (24.1) at x_n and y_n as the *approximate solution*, (24.5) suggests the algorithm

$$y_{n+1} = y_n + f(x_n, y_n)h \qquad (n = 0, 1, 2, \ldots), \tag{24.6}$$

which is known as **Euler's method**. It is illustrated in Fig. 24.2. Since $y' = f(x, y)$, note that $f(x, y)$ defines a field of slopes in the x, y plane. According to Euler's method, we compute the slope $f(x_0, y_0)$ at the initial point $P = (x_0, y_0)$ and then extrapolate out to $x_1 = x_0 + h$ along a straight line with that slope, a process that determines the next point $Q = (x_1, y_1)$. Then we compute the slope $f(x_1, y_1)$ and extrapolate to the next point and so on.

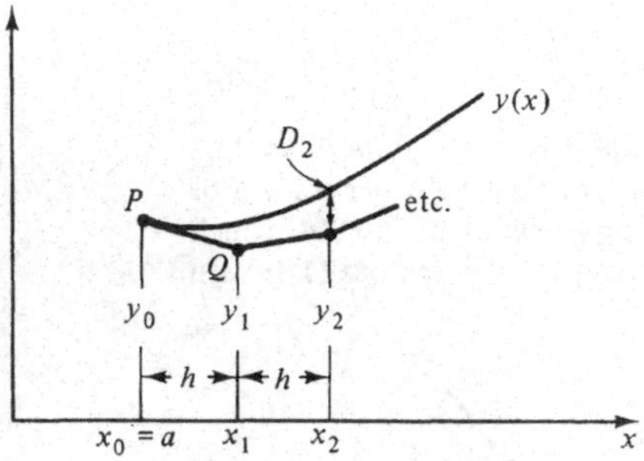

Figure 24.2. Euler's method.

We anticipate that an error will develop between the *computed* points y_n and the *exact* values $y(x_n)$, as indicated in Fig. 24.2. To examine this error, let us replace (24.4) by Taylor's formula with remainder [recall equations (2.25a) and (2.27)],

$$y(x_{n+1}) = y(x_n) + y'(x_n)h + \frac{y''(\xi)}{2}h^2$$

for some suitable ξ in the interval $[x_n, x_{n+1}]$ and observe that the error in the approximation (24.5) is

$$T = \frac{y''(\xi)}{2}h^2. \tag{24.7}$$

This is called the **truncation error** of the Euler method; that is, it is the error incurred in making a *single step* (from x_n to x_{n+1}), due to the truncation of the series (24.4). (There will also be a "roundoff error" incurred by the computer, a subject that will be discussed later. For the present, we will consider our computer to be perfect—that is, entirely free of roundoff.)

Besides the truncation error, we define the **discretization error**[1]

$$D_n \equiv y(x_n) - y_n, \tag{24.8}$$

that is, the total or accumulated effect of all truncation errors incurred up to that point; for instance, D_2 is labeled in Fig. 24.2.

Ultimately we are more interested in the discretization error than in the trunca-

[1] Instead of "truncation error" and "discretization error," some authors use the terms *local truncation error* and *truncation error*, respectively.

tion error. In particular, we would like to know the rate of convergence of a given method—in other words, how fast $D_n \longrightarrow 0$, at a given x, as $h \longrightarrow 0$. To quantify this concept we say that *a method has* **order of convergence** *p if*[2]

$$D_n = O(h^p) \text{ as } h \longrightarrow 0 \tag{24.9}$$

at a fixed x location. Since $x_n = a + nh$ and $h \longrightarrow 0$, then $n \longrightarrow \infty$ such that nh remains constant.

THEOREM 24.1. (*Convergence of Euler's Method*) *Suppose that $y''(x)$ is continuous and $|f_y(x, y)| \leq L$ over the domain of interest. For Euler's method we have the bound*

$$|D_n| \leq \frac{hM}{2L}[e^{(x_n - x_0)L} - 1], \tag{24.10}$$

where $|y''(x)| \leq M$ over $[a, b]$.

Proof. See Exercise 24.2.

According to the big-oh notation, (24.9) means that D_n/h^p is bounded as $h \longrightarrow 0$. The inequality (24.10) indicates that D_n/h is bounded as $h \longrightarrow 0$, and so we can identify $p = 1$. Thus the order of convergence of Euler's method is unity or, more simply, we say that Euler's method is a *first-order method.*

In principle, (24.10) can be used as an a priori error estimate or to determine the size of h that is needed to keep the error $|D_n|$ below some desired value. However, it should be noted that the importance of (24.10) lies instead in its establishment of the order of the method.

That Euler's method is of first order might have been anticipated by a simple order-of-magnitude calculation. Specifically, the accumulated error D_n *at a fixed x location* is, at least in terms of *order of magnitude*, the truncation error per step (T) times the number of steps (n). Since $T = O(h^2)$ per (24.7), this gives

$$D_n = O(h^2)\cdot n = O(h^2)\frac{nh}{h} = O(h^2)\frac{x_n}{h} = O(h)x_n = O(h),$$

where the last equality follows from the fact that x_n is held *fixed* as $h \longrightarrow 0$. Generalizing, we state as a rule of thumb that if $T = O(h^m)$ where $m > 1$, then the order of the method is $p = m - 1$.

Incidentally, it is not necessary for each step h to be of the same size. For instance, the problem $y' = -1000y + 1000$, $y(0) = 0$, has the solution $y = 1 - e^{-1000t}$, which rises *very* sharply initially, and then settles down to a rather slow approach to the asymptote $y(\infty) = 1$. (Sketch it.) Surely, then, we need h to be quite small initially, whereas we can afford to make much larger steps (to reduce the cost) once we get through the region of rapid change. Of course, in practice, we will not know the solution in advance so as to guide our variation in step size. Nevertheless, it is possible to include an algorithm in the program that computes a suitable step size, for the calculation of y_{n+1}, based on the variation in the preceding values, say y_{n-2}, y_{n-1}, and y_n.

---

[2]The "big oh" notation used here was defined in Section 1.5.

24.2. IMPROVEMENTS: MIDPOINT RULE AND RUNGE–KUTTA

Unfortunately, in most practical applications of Euler's method, we find that the step size must be so small, for the desired accuracy, that the computing time is substantial or even prohibitive.[3] It is therefore important to determine methods of higher order that will permit the use of much larger steps for a given level of accuracy.

To illustrate, let us expand about x_n, first with a positive step h and then with a negative step $-h$:

$$y(x_{n+1}) = y(x_n) + y'(x_n)h + \frac{y''(x_n)}{2!}h^2 + \frac{y'''(\xi)}{3!}h^3$$

$$y(x_{n-1}) = y(x_n) - y'(x_n)h + \frac{y''(x_n)}{2!}h^2 - \frac{y'''(\eta)}{3!}h^3$$

where ξ is somewhere in $[x_n, x_{n+1}]$ and η is somewhere in $[x_{n-1}, x_n]$. Subtracting,[4]

$$y(x_{n+1}) - y(x_{n-1}) = 2y'(x_n)h + \frac{y'''(\zeta)}{3}h^3$$

for some ζ in $[x_{n-1}, x_{n+1}]$ or

$$y(x_{n+1}) = y(x_{n-1}) + 2f[x_n, y(x_n)]h + O(h^3). \tag{24.11}$$

Thus the algorithm

$$y_{n+1} = y_{n-1} + 2f(x_n, y_n)h, \tag{24.12}$$

which we shall call the **midpoint rule**, has a truncation error $T = O(h^3)$ and constitutes a *second*-order method, whereas Euler's method was only first order. To understand this improvement in accuracy more clearly, observe that if we solve

$$y(x_{n+1}) \approx y(x_n) + y'(x_n)h \qquad \text{(Euler)}$$

and

$$y(x_{n+1}) \approx y(x_{n-1}) + 2y'(x_n)h \qquad \text{(midpoint)}$$

for $y'(x_n)$ we have the difference quotient approximations

$$y'(x_n) \approx \frac{y(x_{n+1}) - y(x_n)}{h} \qquad \text{(Euler)}$$

and

$$y'(x_n) \approx \frac{y(x_{n+1}) - y(x_{n-1})}{2h} \qquad \text{(midpoint)}$$

and it is apparent from Fig. 24.3 that the latter is appreciably more accurate.

Notice that (24.12) suffers from not being "self-starting"—that is, the first step gives y_1 in terms of x_0, y_0, y_{-1}, where y_{-1} is not defined. We can determine y_1 by using

---

[3] So far we have implicitly assumed that all calculations are performed exactly. In practice, *roundoff* errors enter due to the machine retention of only a finite number of digits. Anticipating that the accumulated effects of roundoff increase as h is diminished, it is possible (even aside from computing time considerations) that when we use an h that is small enough to keep the discretization error within desired limits, the roundoff errors are in excess of these limits. In practice, then, the desired accuracy may be unattainable. For a brief discussion of this point, see C. W. Gear, *Numerical Initial Value Problems in Ordinary Differential Equations*, Prentice-Hall, Englewood Cliffs, N.J., 1971, pp. 18–21.

[4] If $y'''(x)$ is continuous over $[x_{n-1}, x_{n+1}]$, then there must exist a ζ in the interval such that $y'''(\xi) + y'''(\eta) = 2y'''(\zeta)$.

Euler's method, say, and then switch over to (24.12); but in order to maintain consistent second-order accuracy, it will be necessary to subdivide the interval $[x_0, x_1]$ into approximately $1/h$ subintervals so that the Euler calculation of y_1 has an accuracy that is consistent with that of the midpoint rule.

Before looking at a numerical example, consider a different and important line of approach. The low order of Euler's method is undoubtedly related to the fact that (24.6) amounts to an *extrapolation*—that is, based on the slope $f(x_n, y_n)$ at the initial point x_n, y_n—whereas the use of an average slope should be more accurate. Consider, then, a method of the form

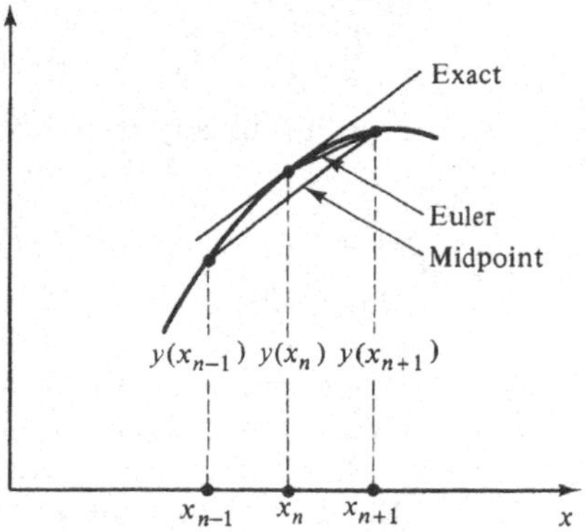

Figure 24.3. Comparison of Euler and midpoint approximations.

$$y_{n+1} = y_n + \{af(x_n, y_n) + bf[x_n + \alpha h, y_n + \beta f(x_n, y_n)h]\}h, \tag{24.13}$$

where the parameters a, b, α, β are to be chosen so that T is of as high an order as possible. First, we expand

$$\begin{aligned} y(x_{n+1}) &= y(x_n) + y'(x_n)h + \frac{y''(x_n)}{2!}h^2 + \frac{y'''(x_n)}{3!}h^3 + \cdots \\ &= y(x_n) + fh + (f_x + ff_y)\frac{h^2}{2} + (f_{xx} + f_xf_y + ff_y^2 \\ &\quad + 2ff_{xy} + f^2f_{yy})\frac{h^3}{6} + O(h^4), \end{aligned} \tag{24.14}$$

where f and its derivatives are understood to be at $x_n, y(x_n)$. To compare (24.14) with (24.13), we need to expand the $f[\ ]$ term in (24.13) in powers of h:

$$\begin{aligned} f[x_n + \alpha h, y_n + \beta f(x_n, y_n)h] &= f + (\alpha f_x + \beta ff_y)h + (\alpha^2 f_{xx} \\ &\quad + 2\alpha\beta ff_{xy} + \beta^2 f^2 f_{yy})\frac{h^2}{2} + \cdots, \end{aligned} \tag{24.15}$$

where f and its derivatives are understood to be evaluated at x_n, y_n. Putting (24.15) into (24.13),

$$\begin{aligned} y_{n+1} &= y_n + (a + b)fh + (\alpha f_x + \beta ff_y)bh^2 \\ &\quad + (\alpha^2 f_{xx} + 2\alpha\beta ff_{xy} + \beta^2 f^2 f_{yy})\frac{bh^3}{2} + O(h^4). \end{aligned} \tag{24.16}$$

Finally, comparing powers of h in (24.14) and (24.16), we have

$$\begin{aligned} h&: \quad a + b = 1 \\ h^2&: \quad \alpha b = \beta b = \tfrac{1}{2}. \end{aligned} \tag{24.17}$$

With four unknowns and only three equations we might hope that agreement through $O(h^3)$ could be obtained as well, but such is not the case.

The outcome is that any method (24.13) with a, b, α, β satisfying (24.17) will have $T = O(h^3)$ and hence be a second-order method. These are the so-called **Runge–Kutta**

methods of second order. For instance, with $a = b = \frac{1}{2}$ and $\alpha = \beta = 1$ we have

$$y_{n+1} = y_n + \{f(x_n, y_n) + f[x_{n+1}, y_n + f(x_n, y_n)h]\}\frac{h}{2} \tag{24.18}$$

or as it is usually expressed,

$$y_{n+1} = y_n + \tfrac{1}{2}(k_1 + k_2), \tag{24.19a}$$

where
$$k_1 = hf(x_n, y_n), \qquad k_2 = hf(x_{n+1}, y_n + k_1). \tag{24.19b}$$

This, (24.19a) and (24.19b), is also known as the *improved Euler method*, and the choice $a = 0$, $b = 1$, $\alpha = \beta = \frac{1}{2}$ is known as the *modified Euler method*. Observe that (24.18) uses a slope that is a simple average of f at the initial point x_n, y_n and at the Euler estimate of the final point x_{n+1}, $y_n + f(x_n, y_n)h$!

In principle, Runge–Kutta methods of any desired order can be derived as above. The derivations however, become extremely tedious. The most commonly used of all methods is probably the following **Runge–Kutta method of order four**,

$$y_{n+1} = y_n + \frac{1}{6}(k_1 + 2k_2 + 2k_3 + k_4), \tag{24.20a}$$

where
$$\begin{aligned} k_1 &= hf(x_n, y_n), \qquad k_2 = hf\left(x_n + \frac{h}{2}, y_n + \frac{1}{2}k_1\right) \\ k_3 &= hf\left(x_n + \frac{h}{2}, y_n + \frac{1}{2}k_2\right), \qquad k_4 = hf(x_{n+1}, y_n + k_3). \end{aligned} \tag{24.20b}$$

Here the effective slope used is a weighted average of the slopes at the four points (x_n, y_n), $(x_n + h/2, y_n + (1/2)k_1)$, $(x_n + h/2, y_n + (1/2)k_2)$ and $(x_{n+1}, y_n + k_3)$ in the x, y plane.

Finally, it is worth noting that, for the special case where f is a *function of x only*, the methods (24.19) and (24.20) amount to the integration of $y' = f(x)$ by the *trapezoidal* and *Simpson's* rules, respectively. For instance, according to Simpson's rule (1.29) (with "h" equal to half of the present h),

$$\int_{y_n}^{y_{n+1}} dy = \int_{x_n}^{x_n+h} f(x)\,dx \approx \frac{(h/2)}{3}\left[f(x_n) + 4f\left(x_n + \frac{h}{2}\right) + f(x_n + h)\right]$$

or
$$y_{n+1} \approx y_n + \frac{h}{6}\left[f(x_n) + 4f\left(x_n + \frac{h}{2}\right) + f(x_n + h)\right],$$

which is the same as (24.20), since $k_2 = k_3$ when f depends only on x.

Example 24.2. As a summary illustration we have solved the problem

$$y' = -y, \qquad y(0) = 1 \tag{24.21}$$

by each of the methods considered, on a Burroughs B6700, using single precision arithmetic and a step size $h = 0.05$. The results are tabulated in Table 24.1, together with the exact solution $y = e^{-x}$ for comparison. The value of y_1 for the midpoint rule method was obtained by Euler's method with a step size of $(0.05)^2 = 0.0025$. We see that the midpoint rule and the second-order Runge–Kutta method yield comparable results initially, but the midpoint rule eventually develops an error that oscillates in sign, from step to step, and grows in magnitude. The reason for this strange behavior will be revealed in Section 24.3. ▮

Note that in the Euler method (24.6) and the Runge–Kutta methods (24.19) and (24.20), only y_n is needed to compute y_{n+1}. Thus they are examples of *single-step* methods. The midpoint rule (24.12), on the other hand, requires both y_n and y_{n-1} and so is a *two-step method.*

24.3. STABILITY

Having developed a number of methods for the calculation of the y_n's, we must realize that the printout obtained from the computer will differ somewhat from these values due to the presence of machine roundoff. Denoting the actual printout as y_n^*, it is instructive to decompose the total error as follows:

$$\begin{aligned}\text{Total error} &= y(x_n) - y_n^* = [y(x_n) - y_n] + [y_n - y_n^*] \\ &= \text{discretization error} + \text{accumulated roundoff.} \qquad (24.22)\end{aligned}$$

TABLE 24.1
Comparison of Results for $y' = -y$, $y(0) = 1$

X	Euler	Midpoint	2nd order R-K	4th order R-K	Exact = e^{-x}
0.00	.1000000000E+01	.1000000000E+01	.1000000000E+01	.1000000000E+01	.1000000000E+01
0.05	.9500000000E+00	.9511698753E+00	.9512500000E+00	.9512294271E+00	.9512294245E+00
0.10	.9025000000E+00	.9048830125E+00	.9048765625E+00	.9048374230E+00	.9048374180E+00
0.15	.8573750000E+00	.8606815741E+00	.8607638301E+00	.8607079834E+00	.8607079764E+00
0.20	.8145062500E+00	.8188148661E+00	.8188015934E+00	.8187307620E+00	.8187307531E+00
0.25	.7737809375E+00	.7788000885E+00	.7788850157E+00	.7788007936E+00	.7788007831E+00
0.30	.7350918906E+00	.7409348462E+00	.7409143712E+00	.7408182327E+00	.7408182207E+00
0.35	.6983372961E+00	.7047066039E+00	.7047947956E+00	.7046881031E+00	.7046880897E+00
⋮					
2.00	.1285121566E+00	.1357351469E+00	.1354524270E+00	.1353352979E+00	.1353352832E+00
2.05	.1220865487E+00	.1285321600E+00	.1288491212E+00	.1287349179E+00	.1287349036E+00
2.10	.1159822213E+00	.1228819309E+00	.1225677266E+00	.1224564422E+00	.1224564283E+00
2.15	.1101831102E+00	.1162439669E+00	.1165925499E+00	.1164841714E+00	.1164841578E+00
2.20	.1046739547E+00	.1112575342E+00	.1109086631E+00	.1108031716E+00	.1108031584E+00
2.25	.9944025699E-01	.1051182135E+00	.1055010650E+00	.1053992374E+00	.1053992246E+00
2.30	.9446824414E-01	.1007457129E+00	.1003586498E+00	.1002588562E+00	.1002588437E+00
2.35	.8974483183E-01	.9504364217E-01	.9546616563E-01	.9536917438E-01	.9536916222E-01
⋮					
5.00	.5920529221E-02	.1262002356E-01	.6752537083E-02	.6737948828E-02	.6737946999E-02
5.05	.5624502760E-02	.2535102060E-03	.6423350900E-02	.6409335204E-02	.6409333446E-02
5.10	.5343277622E-02	.1259467254E-01	.6110212543E-02	.6096748254E-02	.6096746566E-02
5.15	.5076113740E-02	-.1005957048E-02	.5812339682E-02	.5799406349E-02	.5799404727E-02
5.20	.4822308053E-02	.1269526825E-01	.5528988122E-02	.5516565978E-02	.5516564421E-02
5.25	.4581192651E-02	-.2275483873E-02	.5259449952E-02	.5247519895E-02	.5247518399E-02
5.30	.4352133018E-02	.1292281663E-01	.5003051766E-02	.4991595343E-02	.4991593907E-02
5.35	.4134526367E-02	-.3567765536E-02	.4759152993E-02	.4748152379E-02	.4748150999E-02
⋮					
9.65	.5019446397E-04	-.6124855813E+00	.6469508194E-04	.6442560079E-04	.6442556703E-04
9.70	.4768474078E-04	.6440045156E+00	.6154119670E-04	.6128352733E-04	.6128349505E-04
9.75	.4530050374E-04	-.6768860328E+00	.5854106336E-04	.5829469459E-04	.5829466373E-04
9.80	.4303547855E-04	.7116931189E+00	.5568718652E-04	.5545162894E-04	.5545159943E-04
9.85	.4088370462E-04	-.7480553447E+00	.5297243618E-04	.5274722123E-04	.5274719302E-04
9.90	.3883951939E-04	.7864986533E+00	.5039002991E-04	.5017470903E-04	.5017468206E-04
9.95	.3689754342E-04	-.8267052101E+00	.4793351596E-04	.4772765972E-04	.4772763394E-04
10.00	.3505266625E-04	.8691691743E+00	.4559675705E-04	.4539995441E-04	.4539992976E-04

We ask two things of a method: (a) that the discretization error $\longrightarrow 0$ (at any fixed x point) as $h \longrightarrow 0$, which is the question of **convergence** already discussed, (b) that the roundoff error remain "sufficiently small." That is, acknowledging that we must inevitably live with some roundoff error, we ask at least that it does not grow intolerably, and this is the notion of **stability**.

To illustrate, consider the midpoint rule (24.12) applied to the standard "test equation"

$$y' = Ay, \qquad y(0) = 1 \qquad (A \text{ constant}). \tag{24.23}$$

Then

$$y_{n+1} = y_{n-1} + 2Ahy_n, \qquad y_0 = 1. \tag{24.24}$$

In investigating the growth of roundoff errors, it is customary to assume that roundoff enters at one step only, say initially, and that subsequent calculations are all exact. In place of (24.24), then, consider the "perturbed" problem

$$y^*_{n+1} = y^*_{n-1} + 2Ahy^*_n, \qquad y^*_0 = 1 - \epsilon, \text{ say,} \tag{24.25}$$

where ϵ is the roundoff error in the initial condition. Defining the error $e_n \equiv y_n - y^*_n$ and subtracting (24.25) from the system $y_{n+1} = y_{n-1} + 2hAy_n$, $y_0 = 1$, we obtain

$$e_{n+1} = e_{n-1} + 2Ahe_n, \qquad e_0 = \epsilon \tag{24.26}$$

governing the growth of e_n.

Next, $e_{n+1} - 2Ahe_n - e_{n-1} = 0$ is a linear second-order **difference equation** with constant coefficients—second order because $(n + 1) - (n - 1) = 2$ and with constant coefficients because the coefficients do not depend on n. Just as we solve linear (homogeneous) *differential* equations with constant coefficients by seeking "$y = e^{\lambda x}$," the appropriate form for linear (homogeneous) *difference* equations is

$$e_n = \rho^n.$$

With this expression for e_n, $e_{n+1} - 2Ahe_n - e_{n-1} = 0$ becomes

$$\left(\rho - 2Ah - \frac{1}{\rho}\right)\rho^n = 0;$$

and if $\rho^n \neq 0$, then we must have

$$\rho^2 - 2Ah\rho - 1 = 0 \quad \text{or} \quad \rho = Ah \pm \sqrt{1 + A^2h^2}.$$

Consequently, the general solution is[5]

$$e_n = C_1(Ah + \sqrt{1 + A^2h^2})^n + C_2(Ah - \sqrt{1 + A^2h^2})^n. \tag{24.27}$$

Letting $h \longrightarrow 0$,

$$(Ah + \sqrt{1 + A^2h^2})^n \sim (1 + Ah)^n = e^{n \ln (1+Ah)} \sim e^{nAh} = e^{Ax_n}$$

and

$$(Ah - \sqrt{1 + A^2h^2})^n \sim \text{etc.} \sim (-1)^n e^{-Ax_n},$$

[5]If uncomfortable with this abbreviated discussion of difference equations, you might try F. B. Hildebrand, *Methods of Applied Mathematics*, Prentice-Hall, Englewood Cliffs, N.J., 1952, Chapter 3.

so that (24.27) simplifies to

$$e_n = C_1 e^{Ax_n} + C_2(-1)^n e^{-Ax_n}. \tag{24.28}$$

Actually (24.26) is a *second*-order difference equation, and two initial conditions are appropriate whereas we have specified only one, namely $e_0 = \epsilon$. With no great loss, let us specify as a second condition, $e_1 = 0$. Imposing these conditions on (24.28), we have

$$e_0 = \epsilon = C_1 + C_2$$

and

$$e_1 = 0 = C_1 e^{Ax_1} - C_2 e^{-Ax_1}.$$

Solving for C_1, C_2 and inserting these values into (24.28), we obtain

$$e_n = \frac{\epsilon}{2\cosh Ax_1}[e^{A(x_n-x_1)} + (-1)^n e^{-A(x_n-x_1)}]. \tag{24.28a}$$

If $A > 0$, the second term in (24.28a) decays to zero, and *even though the first term grows exponentially*, it remains a tiny fraction of the exact solution $y(x_n) = e^{Ax_n}$ because ϵ is quite small (for example, on the order of 10^{-10}) and all is well. On the other hand, suppose $A < 0$. Then the second term starts out quite small, due to the ϵ factor, but grows exponentially with x_n and oscillates due to the $(-1)^n$ factor, whereas the exact solution is an exponential decay! This is the sort of behavior that is observed in Example 24.2 (where $A = -1$), and we conclude that the midpoint rule is only *weakly stable*. That is, its stability depends on the sign of A in the "test equation" $y' = Ay$; it is stable if $A > 0$ and unstable if $A < 0$. More generally, if $y' = f(x, y)$, we can expect stability if $\partial f/\partial y > 0$ and instability if $\partial f/\partial y < 0$. (See Lieberstein[6] for a brief discussion of this point and Exercise 24.10 for a numerical illustration.)

Notice from (24.24) and (24.26) that y_n and e_n are governed by the same difference equation so that (24.28) holds for y_n as well. That is, letting $h \longrightarrow 0$, the solution of $y_{n+1} - 2Ahy_n - y_{n-1} = 0$ is

$$y_n = B_1 e^{Ax_n} + B_2(-1)^n e^{-Ax_n}.$$

The first of these two terms coincides with the exact solution of the original equation $y' = Ay$, and *the second term (which is associated with the instability for $A < 0$) is an additional or "spurious" solution that enters because we have replaced the original first-order differential equation by a second-order difference equation*! Clearly, then, *single-step methods* (e.g., Euler and Runge–Kutta) are *strongly stable* (i.e., stable independent of the sign of A), since the difference equation is only first order, and so there are no spurious solutions.

We should be aware that these stability claims are based on analyses in which we let $h \longrightarrow 0$, whereas, in practice, h is finite. To illustrate what can happen, let us solve

$$y' = -1000(y - x^3) + 3x^2; \qquad y(0) = 0 \tag{24.29}$$

by Euler's method. The exact solution is simply $y = x^3$, so that at $x = 1$, for instance,

---

[6]H. M. Lieberstein, *A Course in Numerical Analysis*, Harper & Row, New York, 1968, p. 127.

we have $y(1) = 1$. By comparison, computed values (using a Burroughs B6700, which carries 11-digit accuracy) are shown in Table 24.2.

TABLE 24.2
h-Induced Instability

h	Computed $y(1)$
0.2500	2.373756563 10^5
0.1000	8.772506128 10^{14}
0.0100	Exponential overflow
0.0010	0.9999970025
0.0001	0.9999996817

To explain this curious behavior, consider the relevant test equation $y' = -1000y$ (with exponentially decaying solution), that is, $y' = Ay$ where $A = \partial[-1000(y - x^3) + 3x^2]/\partial y = -1000$. Then Euler's method is $y_{n+1} = y_n - 1000hy_n$. Similarly, $y^*_{n+1} = y^*_n - 1000hy^*_n$. Subtracting these two equations, we find that the roundoff error $e_n = y_n - y^*_n$ grows as

$$e_{n+1} = (1 - 1000h)e_n,$$

so that

$$e_n = (1 - 1000h)^n e_0 \tag{24.30}$$

where e_0 is the initial roundoff error. If we let $h \to 0$ as in the preceding discussion, we obtain

$$e_n = (1 - 1000h)^n e_0 = e_0 e^{n \ln(1-1000h)} \sim e_0 e^{-1000nh} = e_0 e^{-1000x_n}$$

which *is* tolerably small and implies stability. This is in agreement with the results shown in Table 24.2; **as h → 0** the scheme is apparently stable. However, for any given calculation, h is *finite*, and it appears, from Table 24.2, there is some critical value, say h_{cr}, such that the guaranteed stability is realized only if $h < h_{cr}$. To see this, let us *not* let $h \to 0$ in (24.30). It is easily seen, from (24.30), that e_n will remain bounded (as $n \to \infty$) only if $|1 - 1000h| \leq 1$, that is, if $h \leq 0.002$. Thus $h_{cr} = 0.002$ for this example, and this agrees with the results shown in Table 24.2. *Finite-h* stability is generally known as **absolute stability**.

A critical h value exists for the Runge-Kutta and other stable (in the $h \to 0$ sense) methods as well. For additional discussion see Exercises 24.21 and 24.22, and the book by Gear.[7]

In closing, we might point out that in the case of *partial* differential equations (Chapter 29) it is standard to consider finite-h stability only, and instead of referring to it as absolute stability (to be consistent with ordinary differential equation nomenclature) it is simply called "stability".

---

[7]C. W. Gear, *Numerical Initial Value Problems in Ordinary Differential Equations*, Prentice-Hall, Englewood Cliffs, N.J., 1971, Sections 1.3.3 and 2.6.1.

So far we've considered only first-order systems of the form

$$y' = f(x, y), \qquad y(a) = y_0. \tag{24.31}$$

Application of the foregoing methods to the higher-order (vector) system

$$\mathbf{y}' = \mathbf{f}(x, \mathbf{y}), \qquad \mathbf{y}(a) = \mathbf{y}_0$$

is straightforward. For instance, consider the second-order system

$$\begin{aligned} y' &= f(x, y, z), & y(a) &= y_0 \\ z' &= g(x, y, z), & z(a) &= z_0. \end{aligned}$$

Then Euler's method, say, takes the form

$$\begin{aligned} y_{n+1} &= y_n + f(x_n, y_n, z_n)h \\ z_{n+1} &= z_n + g(x_n, y_n, z_n)h, \end{aligned} \tag{24.32}$$

and the fourth-order Runge–Kutta method becomes

$$\begin{aligned} y_{n+1} &= y_n + \frac{1}{6}(k_1 + 2k_2 + 2k_3 + k_4) \\ z_{n+1} &= z_n + \frac{1}{6}(l_1 + 2l_2 + 2l_3 + l_4), \end{aligned} \tag{24.33a}$$

where

$$\begin{aligned} k_1 &= hf(x_n, y_n, z_n) \\ l_1 &= hg(x_n, y_n, z_n) \\ k_2 &= hf\left(x_n + \frac{h}{2}, y_n + \frac{1}{2}k_1, z_n + \frac{1}{2}l_1\right) \\ l_2 &= hg\left(x_n + \frac{h}{2}, y_n + \frac{1}{2}k_1, z_n + \frac{1}{2}l_1\right) \\ k_3 &= hf\left(x_n + \frac{h}{2}, y_n + \frac{1}{2}k_2, z_n + \frac{1}{2}l_2\right) \\ l_3 &= hg\left(x_n + \frac{h}{2}, y_n + \frac{1}{2}k_2, z_n + \frac{1}{2}l_2\right) \\ k_4 &= hf(x_{n+1}, y_n + k_3, z_n + l_3) \\ l_4 &= hg(x_{n+1}, y_n + k_3, z_n + l_3). \end{aligned} \tag{24.33b}$$

Example 24.3. Consider the nonlinear second-order system

$$\begin{aligned} \dot{x} &= -4x + y^3, & x(0) &= 1 \\ \dot{y} &= -e^{2t}x, & y(0) &= 1, \end{aligned}$$

which is "rigged" so as to have the simple solution $x = e^{-3t}$, $y = e^{-t}$, for comparison with our computed results. Since there are a number of computer problems in the exercises to this chapter and your programming skills may be "rusty," let us include here our fourth-order Runge–Kutta Fortran IV program for your reference.

```
      READ(5,/) H,JJ,TFIN
      T=0.
      X=1.
      Y=1.
10    WRITE(6,20) T,X,Y
20    FORMAT(1H ,F6.2,2E18.10)
      J=0
30    A1=H*(-4.*X+Y**3)
      B1=H*(-EXP(2.*T)*X)
      A2=H*(-4.*(X+A1/2.)+(Y+B1/2.)**3)
      B2=H*(-EXP(2.*(T+H/2.))*(X+A1/2.))
      A3=H*(-4.*(X+A2/2.)+(Y+B2/2.)**3)
      B3=H*(-EXP(2.*(T+H/2.))*(X+A2/2.))
      A4=H*(-4.*(X+A3)+(Y+B3)**3)
      B4=H*(-EXP(2.*(T+H))*(X+A3))
      X=X+(A1+2.*A2+2.*A3+A4)/6.
      Y=Y+(B1+2.*B2+2.*B3+B4)/6.
      T=T+H
      IF (T.GT.TFIN) STOP
      J=J+1
      IF (J.GE.JJ) GO TO 10
      GO TO 30
      END
```

The "counter" J is inserted to control the amount of printout; for instance, if JJ $= 5$, then t, x, y will be printed only at $t = 0, 5h, 10h, \ldots$. Running this on a Burroughs B6700 (11-digit accuracy), we obtain, for $y(1)$ say:

TABLE 24.3
Fourth-Order Runge–Kutta Results

h	Computed $y(1)$
0.05	0.3678800062
0.02	0.3678794532

compared with the exact value $y(1) \doteq 0.3678794412$.

In some cases, especially in the numerical solution of *partial* differential equations, it is difficult to determine the order of the method analytically. In any case, an *empirical* estimate of the order is possible. To illustrate for the present example, let us express

$$y(x)_{\text{exact}} - y(x)_{\text{computed}} \approx Ch^p,$$

where C and p are constants and p is the order of the method. This equation follows from (24.22) if we assume that the roundoff error is negligible compared to the discretization error—as is reasonable since the method is stable—and if we recall that the discretization error is $O(h^p)$ as $h \longrightarrow 0$. If we fix $x = 1$, then the results in Table 24.3 yield the following two equations in C and p:

$$-0.0000005650 \approx C(0.05)^p$$

$$-0.0000000120 \approx C(0.02)^p.$$

Dividing these results and solving for p, we obtain $p \approx 4.20$, which is approximately correct because the method is known to be of fourth order. ∎

Example 24.4.

$$y'' - xy^2 = 3x, \qquad y(0) = 2, \quad y'(0) = -1. \tag{24.34}$$

In this case we have a second order system consisting of a single second order equation. To integrate numerically we must first convert (24.34) to a system of first-order equations. This may be done by defining an auxiliary dependent variable $u(x)$ according to $y' \equiv u$. Then (24.34) becomes

$$\begin{aligned} y' &= u, & y(0) &= 2 \\ u' &= 3x + xy^2, & u(0) &= -1, \end{aligned}$$

which may be integrated in the same fashion as in the preceding example. ▮

24.5. BOUNDARY VALUE PROBLEMS

Let us begin with an example.

Example 24.5.

$$y''' - x^2 y = -x^4, \qquad y(0) = y'(0) = 0, \quad y(2) = 4. \tag{24.35}$$

Recasting it as a system of *first-order* equations,

$$\begin{aligned} y' &= u, & y(0) &= 0, \quad y(2) = 4 \\ u' &= v, & u(0) &= 0 \\ v' &= x^2 y - x^4. \end{aligned} \tag{24.36}$$

Observe carefully that the initial value $v(0)$ is missing, that it is not prescribed, and so the foregoing methods can not be applied since the first step can not be carried out.[8] Instead of an *initial value* problem, as in the foregoing examples, we have a *boundary value* problem because part of the data is prescribed at one end of the interval and part at the other.

Nevertheless, the linearity of (24.36) saves the day and permits us to work with an initial value version instead. Specifically, suppose that we solve the four *initial value* problems

$$\begin{aligned} LY_1 &= 0, & Y_1(0) &= 1, \quad Y_1'(0) = 0, \quad Y_1''(0) = 0 \\ LY_2 &= 0, & Y_2(0) &= 0, \quad Y_2'(0) = 1, \quad Y_2''(0) = 0 \\ LY_3 &= 0, & Y_3(0) &= 0, \quad Y_3'(0) = 0, \quad Y_3''(0) = 1 \\ LY_p &= -x^4, & Y_p(0) &= Y_p'(0) = Y_p''(0) = 0, \end{aligned} \tag{24.37}$$

where $L = d^3/dx^3 - x^2$ is the operator in (24.35) Each can be expressed in the form (24.31) and solved as discussed in Section 24.4. For instance, the first of (24.37) becomes

$$\begin{aligned} Y_1' &= U_1, & Y_1(0) &= 1 \\ U_1' &= V_1, & U_1(0) &= 0 \\ V_1' &= x^2 Y_1, & V_1(0) &= 0. \end{aligned} \tag{24.38}$$

---

[8] Of course, if y, u, v were all known at the *right* endpoint $x = 2$, then the problem would be of initial value type, since a solution could be "marched out", starting at $x = 2$ and taking the step size h to be *negative*.

In fact, Y_1, Y_2, Y_3 constitute a *fundamental set of linearly independent solutions* of $Ly = 0$ (recall Section 22.2), and Y_p is a *particular solution* of $Ly = -x^4$; therefore the *general solution* of $Ly = -x^4$ can be expressed as

$$y = C_1 Y_1(x) + C_2 Y_2(x) + C_3 Y_3(x) + Y_p(x). \tag{24.39}$$

Finally, we determine C_1, C_2, C_3 by imposing the boundary conditions

$$y(0) = 0 = C_1 + 0 + 0 + 0$$
$$y'(0) = 0 = 0 + C_2 + 0 + 0$$
$$y(2) = 4 = 0 + 0 + C_3 Y_3(2) + Y_p(2);$$

hence $C_1 = C_2 = 0$, and $C_3 = [4 - Y_p(2)]/Y_3(2)$, and

$$y(x) = \frac{4 - Y_p(2)}{Y_3(2)} Y_3(x) + Y_p(x). \tag{24.40}$$

Since $C_1 = C_2 = 0$, there is no need to compute $Y_1(x)$ and $Y_2(x)$; we need only $Y_3(x)$ and $Y_p(x)$ (Exercise 24.15.) ▮

Note that we solved *two* initial value problems [for $Y_3(x)$ and $Y_p(x)$] in order to solve the single boundary value problem (24.35). Alternatively, we could have used the **finite-difference method**, which is discussed briefly in the exercises, but again it is much more cumbersome than the simple marching methods associated with initial value problems. As a rule, then, the numerical solution of boundary value problems is appreciably more unwieldy than for initial value problems.

The superposition method used above requires the system to be linear. Consider next a *nonlinear* case.

Example 24.6.

$$yy' + y'^3 = 1, \qquad y(0) = 2, \quad y(1) = 5. \tag{24.41}$$

One approach would be to use **quasilinearization** (Section 22.6) to reduce (24.41) to a sequence of linear problems, each of which could be solved by superposition, say. We would obtain (Exercise 24.17) the system

$$\begin{gathered} y^{(n)}y^{(n+1)''} + 3(y^{(n)'})^2 y^{(n+1)'} + y^{(n)''}y^{(n+1)} = 2(y^{(n)'})^3 + y^{(n)}y^{(n)''} + 1 \\ y^{(n+1)}(0) = 2, \qquad y^{(n+1)}(1) = 5. \end{gathered} \tag{24.42}$$

As with Newton's method for the iterative solution of algebraic equations, we expect that the initial guess $y^{(0)}(x)$ must be suitably close to the true solution if the method is to converge. In the present case, we might choose $y^{(0)}(x) = 3x + 2$, which at least satisfies the boundary conditions. Of course, any advance knowledge (or "intuition") as to the form of $y(x)$ can be of considerable help in selecting $y^{(0)}$.

Another iterative procedure is to guess at the missing initial value $y'(0)$, "march out" a solution, and compare the computed value $y^{(0)}(1)$ with the required value $y(1) = 5$. If $y^{(0)}(1) < 5$, we increase our estimate of $y'(0)$ and re-solve; similarly, if $y^{(0)}(1) > 5$. This procedure is called *shooting*.[9] Convergence may be a problem, however. ▮

---

[9]See, for example, S. D. Conte, *Elementary Numerical Analysis*, McGraw-Hill, New York, 1965, Section 7.3, for details of the algorithm.

*

Finally, let us turn to an important and elegant method known as **invariant imbedding**. To illustrate, consider the linear, second-order, boundary value problem

$$\begin{aligned} \dot{x} &= p(t)y, \qquad x(0) = 0 \\ \dot{y} &= q(t)x \qquad y(T) = c, \end{aligned} \tag{24.43}$$

and recall that if we knew $x(0)$ and $y(0)$ or $x(T)$ and $y(T)$, then we'd have an *initial* value problem, which would be much simpler. The object of invariant imbedding is the calculation not of the solution to (24.43) but only of the missing end condition $x(T)$.

The starting point is to consider, in place of (24.43),

$$\begin{aligned} \dot{x} &= p(t)y \qquad x(0) = 0 \\ \dot{y} &= q(t)x, \qquad y(a) = c, \end{aligned} \tag{24.44}$$

where a is some point in the interval from 0 to T. Actually, (24.44) is a family of problems, for different values of a, with the original problem (24.43) imbedded in it. To emphasize the role of a, let us rewrite (24.44) as

$$\begin{aligned} x_t(t, a) &= p(t)y(t, a), \qquad x(0, a) = 0 \\ y_t(t, a) &= q(t)x(t, a), \qquad y(a, a) = c, \end{aligned} \tag{24.45}$$

and denote the missing end condition as $x(a, a) = R(a)$. Knowing $R(0) = x(0, 0) = 0$ and desiring the missing condition $R(a) = x(a, a)$, it is reasonable that we will need to know dR/da. So

$$\frac{dR}{da} = x_t(a, a) + x_a(a, a) = p(a)y(a, a) + x_a(a, a) = cp(a) + x_a(a, a), \tag{24.46}$$

where $x_t(a, a)$ and $x_a(a, a)$ mean $x_t(t, a)$ and $x_a(t, a)$ evaluated at $t = a$. But what is $x_a(a, a)$? Differentiating (24.45) with respect to a

$$x_{at}(t, a) = p(t)y_a(t, a), \qquad x_a(0, a) = 0 \tag{24.47a}$$

$$\begin{aligned} y_{at}(t, a) = q(t)x_a(t, a), \qquad y_t(a, a) + y_a(a, a) &= 0 \\ \text{or } y_a(a, a) &= -y_t(a, a) \\ &= -q(a)x(a, a) \\ &= -q(a)R(a) \end{aligned} \tag{24.47b}$$

and comparing systems (24.47) and (24.45), we see that

$$x_a(t, a) = -\frac{q(a)R(a)}{c}x(t, a) \tag{24.48a}$$

$$y_a(t, a) = -\frac{q(a)R(a)}{c}y(t, a). \tag{24.48b}$$

In particular, (24.48a) gives

$$x_a(a, a) = -\frac{q(a)R(a)}{c}x(a, a) = -\frac{q(a)R^2(a)}{c};$$

and putting this result into (24.46), we have the *Riccati equation*

$$\frac{dR}{da} = cp(a) - \frac{q(a)}{c}R^2; \qquad R(0) = 0. \tag{24.49}$$

The idea, then, is that solution of the *initial value* problem (24.49) from $a = 0$ to $a = T$ provides us with the missing end condition $R(T) = x(T)$. With it in hand we then solve (24.43) in backward time steps from T to 0 [unless $x(T)$ was all that we desired].

More generally than (24.43), the linear system

$$\begin{aligned} \dot{x} &= p_1(t)x + p_2(t)y, \qquad x(0) = 0 \\ \dot{y} &= q_1(t)x + q_2(t)y, \qquad y(T) = c \end{aligned} \tag{24.50}$$

leads to

$$\frac{dR}{da} = cp_2(a) + [p_1(a) - q_2(a)]R - \frac{q_1(a)}{c}R^2, \qquad R(0) = 0, \tag{24.51}$$

and the *nonlinear* system

$$\begin{aligned} \dot{x} &= f(x, y, t), \qquad x(0) = 0 \\ \dot{y} &= g(x, y, t), \qquad y(T) = c \end{aligned} \tag{24.52}$$

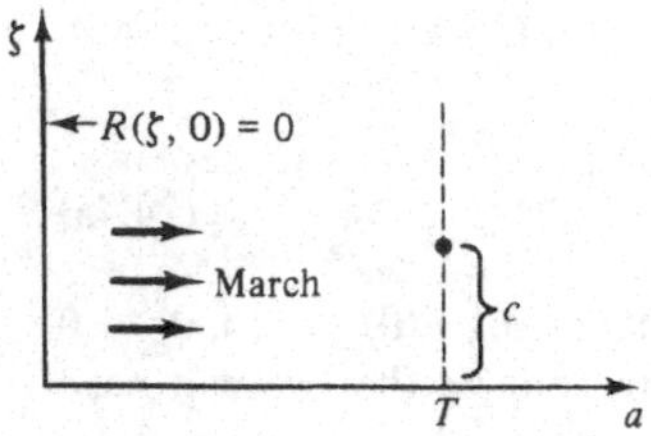

Figure 24.4. Marching out a solution to (24.53).

leads to the *partial* differential equation

$$\frac{\partial R}{\partial a} = f(R, \zeta, a) - g(R, \zeta, a)\frac{\partial R}{\partial \zeta}, \qquad R(\zeta, 0) = 0 \tag{24.53}$$

on $R(\zeta, a)$, where the desired missing end condition is $R(c, T)$; see Fig. 24.4. For details, as well as numerous engineering applications, see the book by Lee.[10]

24.6. THE METHOD OF WEIGHTED RESIDUALS

Consider the system

$$Ly = f(x) \quad \text{over} \quad a \leq x \leq b, \text{ plus homogeneous boundary conditions,} \tag{24.54}$$

where L is an ordinary differential operator. According to the method of weighted residuals (MWR), we seek an approximate solution $y \approx y_N(x)$ of the form

$$y_N(x) = \sum_{j=1}^{N} c_j \phi_j(x), \tag{24.55}$$

where the ϕ_j trial functions are LI (linearly independent) and satisfy the same homogeneous boundary conditions as y. Thus $y_N(x)$ satisfies the given boundary conditions for all choices of the c_j's. Next, we form the error or *residual*

$$R(c_j, x) = Ly_N - f$$

$$\text{or, } R(c_j, x) = \sum_{j=1}^{N} c_j L\phi_j - f \tag{24.56}$$

[10] E. S. Lee, *Quasilinearization and Invariant Imbedding*, Academic, New York, 1968.

if L is linear (for a nonlinear example, see Exercise 24.23) and choose the c_j's so that the weighted integral of the residual

$$(w_i, R) = \int_a^b w_i R \, dx \tag{24.57}$$

is zero for each $i = 1, \dots, N$:

$$(w_i, R) = 0, \qquad (i = 1, \dots, N). \tag{24.58}$$

Finally, imposing (24.58) on (24.56), we obtain the linear algebraic system

$$\sum_{j=1}^{N} (L\phi_j, w_i)c_j = (w_i, f) \tag{24.59}$$

for the c_j's.

So far we have not specified the weight functions w_i. Choice of the w_i's fixes the particular MWR method being used. We will look at three of the most important of these methods. For a detailed account, including numerous applications from the engineering literature, refer to Finlayson.[11]

Collocation. In this case, we employ the delta function

$$w_i(x) = \delta(x - x_i) \tag{24.60}$$

for a prescribed set of points $x_1, \dots, x_N$ in the interval $[a, b]$. Then (24.58) simply amounts to forcing the residual to vanish at the selected points $x_1, \dots, x_N$ since $(w_i, R) = R(x_i)$.

Galerkin. Generally attributed to B. G. Galerkin, who developed the method in 1915, it was also used by I. G. Bubnov in 1913 and is often referred to as the Bubnov–Galerkin method. Here the weight functions $w_i(x)$ are taken to be the same as the trial functions $\phi_i(x)$, and so (24.59) becomes

$$\sum_{j=1}^{N} (L\phi_j, \phi_i)\, c_j = (\phi_i, f) \qquad (i = 1, \dots, N). \tag{24.61}$$

If the set $\phi_1, \phi_2, \dots$ is *complete* and the residual R is orthogonal to $\phi_1, \dots, \phi_N$, then $R \longrightarrow 0$ as $N \longrightarrow \infty$. Of course, we settle for some finite N, but the point is that for an arbitrarily small ϵ we can have the norm $\| R \| < \epsilon$ by taking N sufficiently large.

Least Squares. In the least squares method we choose the c_j's so as to minimize the norm of the residual R or, equivalently, $\| R \|^2$. Since

$$\| R \|^2 = (R, R) = \int_a^b R^2 \, dx, \tag{24.62}$$

then

$$\frac{\partial}{\partial c_j} \| R \|^2 = 2 \int_a^b R \frac{\partial R}{\partial c_j} \, dx = 0; \tag{24.63}$$

and comparing (24.63) with (24.57) and (24.58), we see that the least squares method amounts to the choice

$$w_i = \frac{\partial R}{\partial c_i}. \tag{24.64}$$

---

[11] B. A. Finlayson, *The Method of Weighted Residuals and Variational Principles*, Academic, New York, 1972.

The resulting system for the c_j's is found, from (24.63) and (24.56), to be

$$\sum_{j=1}^{N} (L\phi_j, L\phi_i)c_j = (L\phi_i, f) \qquad (i = 1, \ldots, N). \tag{24.65}$$

Example 24.7.

$$y'' + y = x - 1, \qquad y(0) = -1, \quad y(1) = 1. \tag{24.66}$$

First, let us "homogenize" the boundary conditions by setting $Y = y - (2x - 1)$. Then (24.66) becomes

$$Y'' + Y = -x, \qquad Y(0) = Y(1) = 0. \tag{24.67}$$

Seeking $Y(x) \approx a + bx + cx^2 + dx^3$, say, $Y(0) = 0$ implies $a = 0$, and $Y(1) = 0$ implies $d = -b - c$; so $Y(x) \approx b(x - x^3) + c(x^2 - x^3)$. Consider, then, the two-term approximation

$$Y_2(x) = c_1 x(1 - x^2) + c_2 x^2(1 - x) \equiv c_1\phi_1(x) + c_2\phi_2(x). \tag{24.68}$$

Thus

$$R = LY_2 + x = (-5x - x^3)c_1 + (2 - 6x + x^2 - x^3)c_2 + x. \tag{24.69}$$

Applying collocation, with $x_1 = \frac{1}{3}$ and $x_2 = \frac{2}{3}$ say, yields

$$-1.7037c_1 + 0.0741c_2 = -0.3333$$
$$-3.6296c_1 - 1.8519c_2 = -0.6667.$$

Solving, $c_1 = 0.1947$ and $c_2 = -0.0216$, and so

$$(Y_2)_{\text{col}} = 0.1947x(1 - x^2) - 0.0216x^2(1 - x). \tag{24.70}$$

The Galerkin and least squares systems (24.61) and (24.65) are

$$-0.7238c_1 - 0.2738c_2 = -0.1333$$
$$-0.2738c_1 - 0.1238c_2 = -0.0500$$

and

$$10.4762c_1 + 5.4262c_2 = 1.8667$$
$$5.4262c_1 + 3.7429c_2 = 0.9500,$$

TABLE 24.4
Method of Weighted Residuals, System (24.66)

x	$(y_2)_{\text{Col}}$	$(y_2)_{\text{Gal}}$	$(y_2)_{\text{LS}}$	Exact
0	−1	−1	−1	−1
0.1	−0.7809	−0.7812	−0.7816	−0.7814
0.2	−0.5633	−0.5638	−0.5647	−0.5639
0.3	−0.3482	−0.3489	−0.3499	−0.3488
0.4	−0.1351	−0.1359	−0.1387	−0.1372
0.5	0.0703	0.0681	0.0681	0.0697
0.6	0.2717	0.2707	0.2694	0.2710
0.7	0.4663	0.4655	0.4643	0.4656
0.8	0.6533	0.6526	0.6517	0.6525
0.9	0.8315	0.8311	0.8306	0.8309
1.0	1	1	1	1

respectively, and yield

$$(Y_2)_{\text{Gal}} = 0.1921x(1 - x^2) - 0.0210x^2(1 - x) \tag{24.71}$$

$$(Y_2)_{\text{LS}} = 0.1876x(1 - x^2) - 0.0182x^2(1 - x). \tag{24.72}$$

(The calculations are left for Exercise 24.25.) Finally, $y_2 = Y_2 + 2x - 1$, and this result is compared with the exact solution $y = (\sin x)/(\sin 1) + x - 1$ in Table 24.4.

Calculation of the inner products in (24.59) can be tedious if the functions are not simple and/or N is "large," say greater than three, and it may be appropriate to compute them by numerical integration. Along these lines, the symmetry "a_{ij}" $= (L\phi_j, L\phi_i) = (L\phi_i, L\phi_j) = a_{ji}$ in the least squares system (24.65) is of help. In fact, the same symmetry exists in the Galerkin system (24.61) if L is self-adjoint [as in the present case (24.67)], for then $a_{ij} = (L\phi_j, \phi_i) = (\phi_j, L\phi_i) = (L\phi_i, \phi_j) = a_{ji}$. ∎

24.7. THE FINITE-ELEMENT METHOD

Let us illustrate the main ideas through an example.

Example 24.8.

$$y'' + y = x - 1, \qquad y(0) = -1, \quad y(1) = 1, \tag{24.73}$$

as in Example 24.7. Again we "homogenize" the boundary conditions by setting $Y = y - (2x - 1)$, so that

$$Y'' + Y = -x, \qquad Y(0) = Y(1) = 0, \tag{24.74}$$

and seek

$$Y(x) \approx Y_N(x) = c_1\phi_1(x) + \cdots + c_N\phi_N(x), \tag{24.75}$$

with the c_j's determined by the Galerkin method (24.61) for instance.

This time, however, let us take the ϕ_j's to be the "hat functions" shown in Fig. 24.5—that is,

$$\phi_j(x) = \begin{cases} 1 - j + \dfrac{x}{h}, & x_{j-1} \le x \le x_j \\ 1 + j - \dfrac{x}{h}, & x_j \le x \le x_{j+1} \\ 0, & \text{elsewhere,} \end{cases} \tag{24.76}$$

where $x_j = jh$ and $h = 1/(N + 1)$. Computing the coefficients in (24.61),

$$(L\phi_j, \phi_i) = \int_0^1 (\phi_j'' + \phi_j)\phi_i \, dx$$

$$= (\phi_j'\phi_i)\Big|_0^1 + \int_0^1 (\phi_j\phi_i - \phi_j'\phi_i') \, dx \tag{24.77}$$

or,

$$(L\phi_j, \phi_i) = \int_0^1 (\phi_j\phi_i - \phi_j'\phi_i') \, dx$$

$$= \text{etc.} = \begin{cases} \dfrac{2}{3}h - \dfrac{2}{h}, & i = j \\ \dfrac{h}{6} + \dfrac{1}{h}, & j = i \pm 1 \\ 0, & \text{otherwise,} \end{cases} \tag{24.78}$$

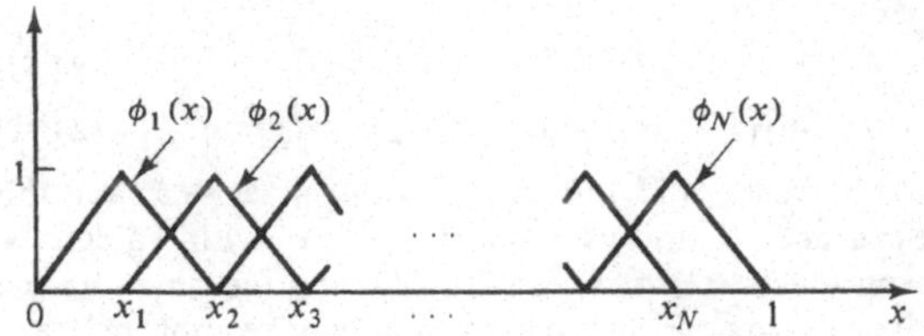

Figure 24.5. The hat functions $\phi_j(x)$.

where we have integrated by parts in (24.77) to avoid the embarrassing ϕ_j'' term. Alternatively, we can forego the integration by parts if we face up to the delta function behavior of ϕ_j'' at x_{j-1}, x_j, x_{j+1} (Exercise 24.27); either way we obtain (24.78). Furthermore,

$$(\phi_i, f) = (\phi_i, -x) = -\int_0^1 \phi_i x \, dx$$
$$= \text{etc.} = -ih^2 \quad \text{for all } i = 1, \ldots, N, \tag{24.79}$$

and so (24.61) becomes

$$\begin{pmatrix} -2+(2/3)h^2 & 1+h^2/6 & & & 0 \\ 1+h^2/6 & -2+(2/3)h^2 & \cdot & & \\ & \cdot & \cdot & \cdot & \\ & & \cdot & \cdot & 1+h^2/6 \\ 0 & & & 1+h^2/6 & -2+(2/3)h^2 \end{pmatrix} \begin{pmatrix} c_1 \\ c_2 \\ \cdot \\ \cdot \\ \cdot \\ c_N \end{pmatrix} = -h^3 \begin{pmatrix} 1 \\ 2 \\ \cdot \\ \cdot \\ \cdot \\ N \end{pmatrix}. \tag{24.80}$$

Before discussing the solution of this system, let us stop a moment and comment on those features that distinguish this *finite-element method* from the "straight Galerkin" procedure already described. The essential finite-element feature is the choice of the ϕ_j's as piecewise polynomials that vanish almost everywhere in the interval. In the present example they are piecewise *linear*; therefore $Y_N(x)$ is as shown in Fig. 24.6. Note that the coefficients c_j have the significance of being the *values*

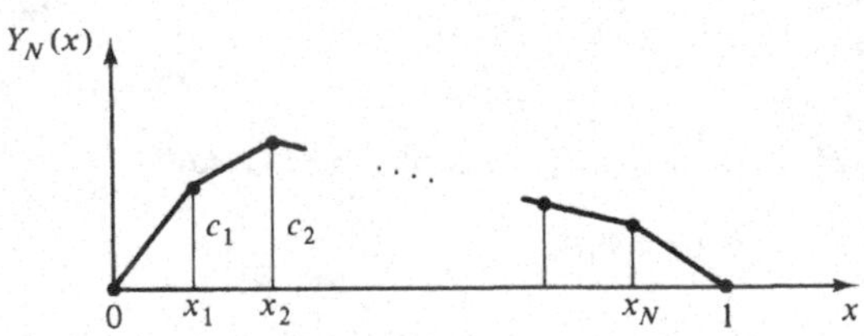

Figure 24.6. Piecewise linear $Y_N(x)$.

of $Y_N(x)$ at the points x_j. [In contrast, the c_j's in the general Galerkin method (24.61) have no such direct and convenient significance.] Furthermore, observe that only ϕ_{j-1} and ϕ_{j+1} overlap ϕ_j, and this condition results in the coefficient matrix in (24.80) being only tridiagonal (with the main diagonal being dominant; i.e., its elements are larger in magnitude than the elements in the adjacent diagonals). The result is to reduce the number of coefficient calculations and permit an efficient iterative solution of (24.81) (which is important because N may be quite large), as we shall see.

We call $Y_N(x)$ of Fig. 24.6 a *piecewise linear Lagrange polynomial of degree one*, since it interpolates Y_N over each (x_j, x_{j+1}) by a first-degree polynomial connecting the values at x_j and x_{j+1}. More generally, higher-order Lagrange or Hermite polynomials can be used, as discussed fully in the book by Prenter.[12] Whereas the error can be shown to be

$$\max_{0 \leq x \leq 1} | Y(x) - Y_N(x) | = O(h) \tag{24.81}$$

for the piecewise linear case, higher-order interpolates permit greater accuracy, as we would expect.[13]

Let us return to the solution of (24.80). The ith equation is of the form

$$ac_{i-1} + bc_i + ac_{i+1} = -h^3 i \qquad (i = 1, \ldots, N), \tag{24.82}$$

where $a = 1 + h^2/6$, $b = -2 + 2h^2/3$, and $c_0 = c_{N+1} \equiv 0$. As mentioned, $|b| > |a|$, which suggests the so-called **Jacobi iteration,**

$$c_i^{(n+1)} = -\frac{a}{b}(c_{i-1}^{(n)} + c_{i+1}^{(n)}) - \frac{h^3 i}{b} \tag{24.83}$$

with $c_1^{(0)} = c_2^{(0)} = \cdots = c_N^{(0)} \equiv 0$ say. That is, with $n = 0$ we set $i = 1, \ldots, N$ in (24.83) so as to obtain $c_1^{(1)}, \ldots, c_N^{(1)}$ in turn. Then with $n = 1$ we set $i = 1, \ldots, N$ and obtain $c_1^{(2)}, \ldots, c_N^{(2)}$ and so on. However, observe that $c_{i-1}^{(n+1)}$ is already known when we are computing $c_i^{(n+1)}$, and this fact suggests changing (24.83) to

$$c_i^{(n+1)} = -\frac{a}{b}(c_{i-1}^{(n+1)} + c_{i+1}^{(n)}) - \frac{h^3 i}{b}, \tag{24.84}$$

which is known as **Gauss–Seidel iteration.** As might be expected, (24.84) converges more rapidly than (24.83) because it uses the most "up-to-date" iterates in the calculation. In fact, additional improvement can be achieved as follows. Rewrite (24.84) as

$$c_i^{(n+1)} = c_i^{(n)} + \left[-\frac{a}{b}(c_{i-1}^{(n+1)} + c_{i+1}^{(n)}) - \frac{h^3 i}{b} - c_i^{(n)}\right] \equiv c_i^{(n)} + \Delta c_i^{(n)}, \tag{24.85}$$

where $\Delta c_i^{(n)}$ is the "correction" at the nth stage. We can seek to speed the convergence by the insertion of a so-called *convergence factor* ω in (24.85), so that

$$c_i^{(n+1)} = c_i^{(n)} + \omega \, \Delta c_i^{(n)}. \tag{24.86}$$

Typically, the optimum ω lies between 1 and 2. For $\omega > 1$, (24.86) is called **successive overrelaxation** or simply **SOR.**

As an illustration, we have solved (24.80) (on a Burroughs B6700), with $h = 0.1$, by each of these three iterative methods. Iterating until $|c_i^{(n+1)} - c_i^{(n)}| < 10^{-5}$ for all i's (from 1 to N), we find that the optimum scheme is SOR, with $\omega \approx 1.6$; see Table 24.5.

Finally, we compare the computed values $Y_N(jh)$—that is, the c_j's—with the exact values $Y(jh)$ in Table 24.6.

Of course, the solution to the original problem (24.73) is then given by $y(x) = Y(x) + 2x - 1 \approx Y_N(x) + 2x - 1$. ▮

---

[12]P. M. Prenter, *Splines and Variational Methods*, Wiley, New York, 1975.
[13]*Ibid.*, p. 221.

TABLE 24.5
Comparative Speed of Convergence

Method		Number of Iterations
Jacobi		127
Gauss–Seidel	$\omega = 1$	73
SOR	$\omega = 1.2$	52
	$\omega = 1.4$	34
	$\omega = 1.6$	19
	$\omega = 1.8$	37

TABLE 24.6
Finite-Element Solution versus Exact; $h = 0.1$

j	x	$Y_N(jh) = c_j$	$Y(jh)$
1	0.1	0.01862	0.01864
2	0.2	0.03606	0.03610
3	0.3	0.05115	0.05119
4	0.4	0.06272	0.06278
5	0.5	0.06967	0.06975
6	0.6	0.07094	0.07102
7	0.7	0.06552	0.06559
8	0.8	0.05245	0.05250
9	0.9	0.03088	0.03090

For detailed discussion of the finite-element method, see the books by Prenter[14] and Strang and Fix.[15]

EXERCISES

24.1. Using Taylor's method, solve through terms of third degree.

(a) $y' = e^{xy}$, $y(0) = 1$ (b) $y' = \sin(x + y)$, $y(0) = 0$

(c) $y' - y = xy^2$, $y(2) = 1$ (d) $xy' = x - y$, $y(2) = 2$

(e) $y' = x^2 \sin[(x - 5y^2)/(1 + xy)]$, $y(1) = 3$

24.2. Let us prove Theorem 24.1. Since

$$y_{n+1} = y_n + (x_n, y_n)\, h$$

and
$$y(x_{n+1}) = y(x_n) + y'(x_n)h + \frac{y''(\xi_n)}{2}h^2 \qquad (x_n \leq \xi_n \leq x_{n+1}),$$

show that

$$D_{n+1} = D_n + f_y(x_n, \eta_n)hD_n - \frac{y''(\xi_n)}{2}h^2,$$

[14] *Op. cit.*

[15] G. Strang and G. Fix, *An Analysis of the Finite Element Method*, Prentice-Hall, Englewood Cliffs, N.J., 1973.

where η_n is between y_n and $y(x_n)$. Recalling the conditions in the theorem, show that

$$|D_{n+1}| \leq (1 + hL)|D_n| + \frac{M}{2}h^2, \qquad D_0 = 0.$$

Next, show by induction that if

$$u_{n+1} = (1 + hL)u_n + \frac{M}{2}h^2, \qquad u_0 = 0,$$

then the u_n's dominate the D_n's—that is, $u_n \geq |D_n|$ for $n = 0, 1, \ldots$. Verify (or derive) the solution

$$u_n = \frac{Mh}{2L}[(1 + hL)^n - 1].$$

Finally, show that $(1 + hL)^n \leq e^{nhL}$, so that

$$|D_n| \leq \frac{Mh}{2L}[e^{(x_n - x_0)L} - 1]$$

as desired.

24.3. Apply the Euler error estimate (24.10) to Example 24.2 for $x = 0.3$, 5, and 10 and compare with the actual error (per Table 24.1).

24.4. Solve the problem $y' = -xy^2$, $y(1) = 2$ on a computer by each of the following methods, with $h = 0.05$, through $x = 3$ and compare with the exact solution.
(a) Euler (b) Improved Euler
(c) Modified Euler (d) Runge–Kutta of order four

24.5. Same as Exercise 24.4 but for the problem

$$(\cos y)y' = e^{-x}, \qquad y(0) = 0.$$

24.6. (*Multistep and predictor-corrector methods*) One means of obtaining a whole class of multistep methods, called *Adams–Bashforth methods*, is to integrate $y' = f(x, y)$ from x_n to x_{n+1},

$$\int_{x_n}^{x_{n+1}} y'\,dx = \int_{x_n}^{x_{n+1}} f[x, y(x)]\,dx$$

or

$$y(x_{n+1}) = y(x_n) + \int_{x_n}^{x_{n+1}} f[x, y(x)]\,dx.$$

To evaluate the integral, we fit $f[x, y(x)]$ approximately with a polynomial of degree m, which interpolates $f[x, y(x)]$ at the $m + 1$ points $x_{n-m}, \ldots, x_{n-1}, x_n$.

*(a) With $m = 3$, show that

$$y(x_{n+1}) = y(x_n) + \{55f[x_n, y(x_n)] - 59f[x_{n-1}, y(x_{n-1})] + 37f[x_{n-2}, y(x_{n-2})] - 9f[x_{n-3}, y(x_{n-3})]\}\frac{h}{24} + \frac{251}{720}y^{v}(\xi)h^5 \qquad (24.87)$$

for some ξ in $[x_{n-3}, x_n]$. This yields the **fourth-order Adams–Bashforth method**

$$y_{n+1} = y_n + (55f_n - 59f_{n-1} + 37f_{n-2} - 9f_{n-3})\frac{h}{24} \qquad (24.88)$$

with

$$T_{AB} = \frac{251}{720}y^{v}(\xi)h^5, \qquad (24.89)$$

where $f_n \equiv f(x_n, y_n)$.

*(b) Suppose that we interpolate instead at $x_{n-m+1}, \ldots, x_n, x_{n+1}$. With $m = 3$ again, derive the **fourth-order Adams–Moulton method**

$$y_{n+1} = y_n + (9f_{n+1} + 19f_n - 5f_{n-1} + f_{n-2})\frac{h}{24} \tag{24.90}$$

with

$$T_{AM} = -\frac{19}{720}y^{v}(\xi)h^5, \tag{24.91}$$

where the ξ's in (24.89) and (24.91) will be identical only by coincidence.

Although both methods are fourth order, note that Adams–Moulton is more accurate because the constant in T_{AM} is roughly an order of magnitude smaller than the constant in T_{AB}. This situation results from the fact that the Adams–Moulton interpolation points $x_{n-2}, x_{n-1}, x_n, x_{n+1}$ are more centered on the interval of integration than the Adams–Bashforth points $x_{n-3}, x_{n-2}, x_{n-1}, x_n$. On the other hand, the term $f_{n+1} = f(x_{n+1}, y_{n+1})$ in (24.90) is embarrassing, since y_{n+1} is not yet known. [If, however, f is linear in y, then (24.90) can be solved for y_{n+1} and there is no problem.] Thus (24.90) is said to be of *closed* type, whereas (24.88) and all our preceding methods are of *open* type.

It is standard to solve closed formulas by iteration. For instance, (24.90) becomes

$$y_{n+1}^{(k+1)} = y_n + [9f(x_{n+1}, y_{n+1}^{(k)}) + 19f_n - 5f_{n-1} + f_{n-2}]\frac{h}{24}. \tag{24.92}$$

To start the iteration, we compute $y_{n+1}^{(0)}$ from a *predictor formula* with subsequent corrections made by the *corrector formula* (24.92). It is recommended that the predictor and corrector formulas be of the same order (obviously the corrector should never be of lower order than the predictor) with the corrector applied only once. Thus the Adams–Bashforth and Adams–Moulton methods constitute a natural **predictor-corrector** pair with "AB" as the predictor and "AM" as the corrector. Why might we choose the fourth-order AB–AM predictor-corrector over the Runge–Kutta method of the same order or vice versa? On the negative side, AB–AM is not self-starting and requires the storage of f_{n-3}, f_{n-2}, and f_{n-1}. On the other hand, it involves only two function evaluations per step (namely, f_n and f_{n+1}) if the corrector is applied only once, whereas Runge–Kutta involves four. So if $f(x, y)$ is reasonably messy, we can expect AB–AM to be almost twice as fast.

24.7. For the problem $y' = -xy^2$, $y(1) = 2$, compute y_1, y_2, y_3 from the exact solution, with $h = 0.05$ say, and use them to determine y_4 by means of the fourth-order AB–AM predictor-corrector scheme given in the preceding exercise, on a hand calculator. Apply the corrector several times and compare $y_4^{(0)}, y_4^{(1)}, \ldots$ with the exact value.

24.8. Find the general solution of each of the following difference equations, by seeking y_n in the form ρ^n.

(a) $y_{n+1} - 3y_n = 0$ (b) $y_{n+1} + y_n = n$

(c) $y_{n+1} + y_{n-1} = 0$ (d) $y_{n+3} + 3y_{n+2} + 2y_{n+1} = 12$

(e) $y_{n+1} - (2\cos\alpha)y_n + y_{n-1} = 0$, α constant

(f) $y_{n+1} - (2\cosh\alpha)y_n + y_{n-1} = 0$, α constant

(g) $y_{n+1} - 2y_n + 2y_{n-1} = 0$

24.9. Derive the last step in (24.27).

24.10. Solve the initial value problem $y' = (4 - x)y$, $y(0) = 1$, using the midpoint rule with $h = 0.05$, through $x = 10$ and compare with the exact solution. Discuss the stability (from your computed results) in terms of the "$\partial f/\partial y$" rule of thumb given near the end of Section 24.3.

***24.11.** (a) Show that the fourth-order Adams–Bashforth method (see Exercise 24.6) is strongly stable.

(b) Show that the fourth-order Adams–Moulton method (Exercise 24.6) is strongly stable.

24.12. Solve the system

$$\ddot{x} + x + x^3 = 0, \qquad x(0) = 1, \quad \dot{x}(0) = 0$$

to five significant figures, over one full period, using whichever numerical method you like.

24.13. The system

$$\ddot{x} = -\frac{Kx}{r^3}, \qquad \ddot{y} = -\frac{Ky}{r^3} \qquad (r = \sqrt{x^2 + y^2})$$

where K is a gravitational constant, describes the orbital motion of one body about another. Integrate the system, using second-order Runge–Kutta through $t = 4$ for $K = 4$, with $h = 0.01$ and the initial conditions $x(0) = 1, \dot{x}(0) = y(0) = 0, \dot{y}(0) = 2$. What is the (exact value of the) period of this particular orbit? *Hint:* Compute r at several different times from your computed x, y values.

24.14. Do an empirical evaluation of the order of the midpoint rule by solving the test equation $y' = y, y(0) = 1$ and comparing with the exact solution at $x = 1$, say, as we did in Example 24.3. *Note:* $e \doteq 2.7182818285$.

24.15. (*Completion of Example* 24.5) (a) Compute $Y_3(x)$ and $Y_p(x)$ in Example 24.5, using Euler integration with $h = 0.05$. Thus compute $y(x)$ at $x = 0.25$, 0.5, 0.75 and compare with the exact solution, which is simply $y(x) = x^2$.

(b) Same as part (a) but use second-order Runge–Kutta instead.

(c) Same as part (a) but use fourth-order Runge–Kutta instead.

24.16. Show how each of these linear boundary value problems can be replaced by an equivalent set of initial value problems through the superposition method outlined in Example 24.5.

(a) $y'' + (\sin x)y = 0, y(0) = 2, y(1) = 3$

(b) $y''' - xy = x^2, y(0) = y(1), y'(1) = 0, y'(0) = 1$

(c) $y' + xz = x, y(0) = 1$
$z' - z + y = 0, z(1) = 3$

24.17. Derive (24.42) by applying quasilinearization to the nonlinear system (24.41).

***24.18.** Solve the boundary value problem

$$\ddot{x} + x = 0, \qquad x(0) = 0, \quad \dot{x}(1) = 2$$

by *invariant imbedding*. Specifically,

(a) obtain the relevant initial value problem (24.51),

(b) solve for the missing end condition by integrating (24.51) by fourth-order Runge–Kutta with $h = 0.01$,

(c) then determine the desired solution $x(t)$ by fourth-order Runge–Kutta integration from $t = 1$ back to $t = 0$ in steps of 0.01 and compare with the exact solution, say at $t = 0.2, 0.4, 0.6, 0.8$.

***24.19.** Same as Exercise 24.18 but for the problem

$$\ddot{x} + tx = 0, \qquad x(0) = 0, \quad \dot{x}(1) = 5.$$

***24.20.** How does (24.53) reduce to (24.49) in the event that $f(x, y, t) = p(t)y$ and $g(x, y, t) = q(t)x$?

24.21. (a) Show that the second-order Runge–Kutta integration of $y' = -Ay$ $(A > 0)$ will be *absolutely stable* if $Ah \leq 2$.

(b) Integrate (24.29) out to $x = 1$ by second-order Runge–Kutta for $h = 0.25, 0.1, 0.01, 0.001$, and 0.0001. Do the results agree with the stability criterion in part (a)?

24.22. (*Stiff systems*) In discussing the system (24.29), we saw that the stability is determined by the $-1000y$ term; in particular, we needed $|1 - 1000h| \leq 1$ for "absolute stability." Suppose that the initial condition were changed to $y(0) = 1$, so that the exact solution became $y = x^3 + e^{-1000x}$. Using Euler's method, we would need $h \leq 0.002$ for stability, independent of x—that is, even for x's that are so large that the e^{-1000x} term (which is at the heart of the stability problem) is quite negligible compared to the x^3 term. In other words, there are two widely different "time scales" (if we think of x as the time), and we are forced to live with the short-time-scale stability limitation even when within the long-time-scale regime. This is the fundamental difficulty presented by so-called **stiff systems.**

(a) Show that if we use the *implicit* Euler method $y_{n+1} = y_n + hf(x_{n+1}, y_{n+1})$ (which is a first-order method) instead of the usual Euler scheme, then the stability condition becomes $|1 + 1000h| \geq 1$ or $h \geq 0$, and thus the stability difficulty is entirely relieved and we can choose h simply on the basis of accuracy (i.e., discretization error).

(b) Recompute the entries in Table 24.2—that is, for the system (24.29)—using the implicit Euler method—and corroborate the claim made in part (a). For a discussion of stiff systems, see Section 11.1 in the book by Gear.[16]

24.23. The steady-state heat conduction across a certain slab with temperature-dependent conductivity $k(T) = 1 + aT$ is governed by the nonlinear system

$$\frac{d}{dx}\left[(1 + aT)\frac{dT}{dx}\right] = 0, \qquad T(0) = 0, \quad T(1) = 10.$$

(a) Seeking $T(x)$ in the form of a polynomial, use a one-term approximation (i.e., containing only one unknown, say c) to obtain an approximate solution by the method of collocation for $a = 0.1$. Compare your result with the exact solution at $x = 0, 0.2, 0.4, 0.6, 0.8, 1$.

(b) Same as part (a) but use Galerkin's method. *Hint:* Be careful; (24.61) does not hold in this case, since the operator is nonlinear.

(c) How do you think the accuracy varies with a, for $0 \leq a < \infty$? Explain.

24.24. In the method of collocation, do the approximate and exact solutions necessarily agree (exactly) at the collocation points? Explain. (You may wish to see if this is true in Example 24.7.)

24.25. Supply the calculations between equations (24.69) and (24.72).

24.26. (a) Find a one-term polynomial approximation for the beam problem

$$y'''' + (1 + x)y = W \qquad (W = \text{constant})$$
$$y(0) = y'(0) = 0, \qquad y(1) = y''(1) = 0$$

by the method of collocation.

---

[16] *Op. cit.*

(b) Same as part (a) but for a *two*-term approximation.

24.27. Express $\phi_j'(x)$ in Example 24.8 in terms of Heaviside functions and hence $\phi_j''(x)$ in terms of delta functions. Use this result to derive (24.78) from (24.77) directly—that is, without integration by parts.

24.28. (a) Solve the boundary value problem

$$y'' - y = x, \qquad y(0) = y(1) = 0$$

using the Galerkin-based finite-element method of Example 24.8 with $h = 0.1$ again.

(b) Compare with the exact solution at $x = 0.1, 0.2, \dots, 0.9$.

24.29. (*Finite-difference method*) To illustrate the **finite-difference method** for solving boundary value problems, consider the example

$$y'' + f(x)y = g(x), \qquad y(0) = y(1) = 0. \tag{24.93}$$

Introducing equally spaced mesh points $x_n = nh$, we replace all derivatives in the differential equation (and boundary conditions) by difference approximations, generally *centered* differences because they are more accurate. For instance (recall Fig. 24.3),

$$y'(x_n) = \frac{y_{n+1} - y_{n-1}}{2h} + O(h^2) \qquad \text{(centered difference),}$$

whereas

$$y'(x_n) = \frac{y_{n+1} - y_n}{h} + O(h) \qquad \text{(forward difference).}$$

(a) Show that

$$y''(x_n) = \frac{y_{n+1} - 2y_n + y_{n-1}}{h^2} + O(h^2).$$

(b) Thus approximate (24.93) by the finite-difference scheme

$$y_{n+1} + (-2 + h^2)f_n y_n + y_{n-1} = h^2 g_n, \qquad y_0 = y_N = 0 \tag{24.94}$$

for $n = 1, 2, \dots, N - 1$, where $N = 1/h$. Write this system in matrix form. [Compare with the first-order finite-element system (24.80), which is for the case where $f(x) = 1$ and $g(x) = -x$.]

(c) Solve (24.94) on a computer for $f(x) = 1$ and $g(x) = -x$, with $h = 0.1$, and compare the results with the first-order finite-element results (third column of Table 24.6) and with the exact solution (fourth column). Use SOR with $\omega = 1.6$. (The calculations can probably be done on a *hand* calculator if you use the *exact* values, from Table 24.6, as the zeroth iterate.) *Note:* If worried about the use of centered differences in view of the stability problem encountered earlier with the midpoint rule, remember that it occurred in the context of *initial* value problems, where we "march out" a solution in (a large number of) successive steps, and that it is not relevant to *boundary* value problems. Finally, we might point out that the finite-difference method amounts to a finite-difference approximation of the differential operator, whereas the finite-element method leaves the operator intact and instead "approximates" the solution space—that is, the space of functions in which the solution is sought.

24.30. Make up two examination-type problems on this chapter.

Chapter 25

Perturbation Techniques

Differential, integral, and algebraic equations often contain a small parameter, and expansions or "perturbations" in terms of this parameter may reduce the problem to a sequence of simpler problems. In the present chapter we provide a short, largely formal introduction to perturbation methods within the context of ordinary differential equations; some applications to partial differential equations are contained in Part V.

If all goes well, a finite number of terms of a straightforward parameter expansion will provide a uniformly valid approximation to the desired solution. Such a case is denoted as *regular*. Perhaps more often than not, however, troublesome nonuniformities occur, and *singular perturbation* methods are needed to eliminate these difficulties.

The development of singular perturbation theory is certainly one of the most important advances in applied mathematics in this century. Although the beginnings go back to the nineteenth century and such names as Lindstedt and Poincaré, it was not until the 1960s that singular perturbation techniques became part of the standard analytical tools of engineers, scientists, and mathematicians. As indicative of this trend, it is interesting to note that of the 440-odd references listed in Nayfeh's book,[1] the breakdown is as follows: 17 in the period 1800–1899, 31 in 1900–1939, 87 in 1940–1959, and 308 in 1960–1971. (And, of course, there is always Euler, 1754.)

[1] A. H. Nayfeh, *Perturbation Methods*, Wiley, New York, 1973. This is a good general reference for the present chapter and quite readable.

The main ideas can be introduced by means of the following example.

Example 25.1.

$$y' + y + \epsilon y^2 = 0, \qquad y(0) = \cos\epsilon \tag{25.1}$$

over $0 \leq x < \infty$, where $0 \leq \epsilon \ll 1$. Suppose that we are unable to solve (25.1) analytically and are willing to settle for a good approximation to the exact solution. According to the perturbation method, we seek $y(x)$ in the form of an expansion in powers of the small parameter ϵ,

$$y(x) = y_0(x) + \epsilon y_1(x) + \epsilon^2 y_2(x) + \cdots. \tag{25.2}$$

First, let us impose the initial condition $y(0) = \cos\epsilon$ or

$$y_0(0) + \epsilon y_1(0) + \epsilon^2 y_2(0) + \cdots = 1 - \frac{\epsilon^2}{2!} + \frac{\epsilon^4}{4!} - \cdots.$$

Equating coefficients of powers of ϵ, we obtain

$$y_0(0) = 1, \qquad y_1(0) = 0, \qquad y_2(0) = -\tfrac{1}{2}, \cdots. \tag{25.3}$$

Next, we put (25.2) into the differential equation (25.1),

$$y_0' + \epsilon y_1' + \epsilon^2 y_2' + \cdots + y_0 + \epsilon y_1 + \epsilon^2 y_2 + \cdots$$
$$+ \epsilon(y_0 + \epsilon y_1 + \cdots)^2 = 0. \tag{25.4}$$

Expanding the last term in powers of ϵ,

$$(y_0 + \epsilon y_1 + \cdots)^2 = y_0^2 + 2\epsilon y_0 y_1 + \cdots,$$

so that (25.4) becomes

$$(y_0' + y_0) + \epsilon(y_1' + y_1 + y_0^2) + \epsilon^2(y_2' + y_2 + 2y_0 y_1) + \cdots = 0.$$

Equating coefficients of powers of ϵ to zero and recalling (25.3), we obtain the system

$$\epsilon^0: \quad y_0' + y_0 = 0, \qquad y_0(0) = 1 \tag{25.5a}$$
$$\epsilon^1: \quad y_1' + y_1 = -y_0^2, \qquad y_1(0) = 0 \tag{25.5b}$$
$$\epsilon^2: \quad y_2' + y_2 = -2y_0 y_1, \qquad y_2(0) = -\tfrac{1}{2} \tag{25.5c}$$
$$\vdots \qquad\qquad \vdots$$

Although these equations are coupled, the coupling is trivial because we can solve (25.5a) for y_0; with y_0 known, we can solve (25.5b) for y_1, and so on. This process yields

$$y_0(x) = e^{-x},$$
$$y_1(x) = -e^{-x} + e^{-2x},$$
$$y_2(x) = \tfrac{1}{2}e^{-x} - 2e^{-2x} + e^{-3x},$$

and so on, and therefore

$$y(x) = e^{-x} + \epsilon(-e^{-x} + e^{-2x}) + \epsilon^2(\tfrac{1}{2}e^{-x} - 2e^{-2x} + e^{-3x}) + \cdots. \tag{25.6}$$

The idea, then, is that the original problem (25.1) was traded in for a system of simpler problems (25.5), in each of which the operator is that of the *reduced equation*—that is, $y' + y + \epsilon y^2$ with ϵ set equal to zero—namely, $y' + y$.

Of course, our procedure has been rather formal and it remains to be seen whether (25.6) is a useful result. In fact, (25.1) admits the exact solution (Exercise 25.1)

$$y(x) = \frac{\cos \epsilon}{(1 + \epsilon \cos \epsilon)e^x - \epsilon \cos \epsilon}, \tag{25.7}$$

and if we expand (25.7) in a Taylor series in ϵ, we do obtain the expansion (25.6).[2] What about the convergence of (25.6)? The easiest way to answer this question is to look at the analytic behavior of (25.7). Regarding x as fixed, the only singularities of (25.7) (in the complex ϵ plane) are the zeros of the denominator—namely, ϵ's—such that

$$\epsilon \cos \epsilon = -\frac{1}{1 - e^{-x}}, \tag{25.8}$$

and it is not hard to show (Exercise 25.2a) that (25.8) has no roots in the disk $|\epsilon| < 1$, say, for all $0 \le x < \infty$. It follows that the expansion (25.6) converges for $|\epsilon| < 1$ for all $0 \le x < \infty$. In fact, we state without proof that the convergence is *uniform* in x; that is, for any $\nu > 0$ (no matter how small) there is an $N(\nu)$ such that $|y(x) - s_n(x)| < \nu$ over $0 \le x < \infty$ for all $n > N$, where $s_n(x) = \sum_0^n y_j(x)\epsilon^j$ is the nth partial sum of (25.2). Since the result (25.6) is *uniformly* valid over the domain of interest ($0 \le x < \infty$), we say that the perturbation is *regular*. Sample results are plotted in Fig. 25.1; $s_2(x)$ and $y(x)$ are virtually identical to within the accuracy of the plot.

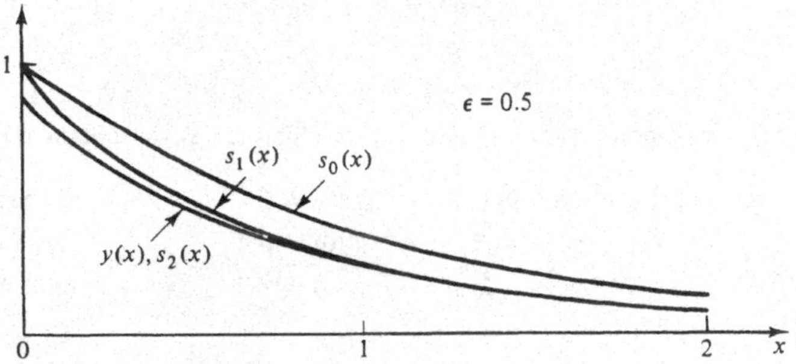

Figure 25.1. The partial sums of (25.6) for $\epsilon = 0.5$.

Unfortunately, perturbation schemes are not always regular. Irregular or *singular* perturbations are the subject of the remaining sections. ∎

25.2. SINGULAR PERTURBATIONS: STRAINING METHODS

Example 25.2. Consider the problem

$$\ddot{x} + x + \epsilon x^3 = 0, \quad x(0) = A, \quad \dot{x}(0) = 0, \quad (0 \le \epsilon \ll 1), \tag{25.9}$$

---

[2] The easiest way to do so is to expand the numerator and denominator of (25.7) in powers of ϵ and then simply divide by "long division."

which we may regard as governing the free vibration of a mass restrained by a (slightly) nonlinear spring. Having already discussed its solution by means of the iterative methods of successive approximation and quasilinearization (Exercise 22.59), we will now explore a perturbation approach. Specifically, we seek $x(t)$ in the form

$$x(t) = x_0(t) + \epsilon x_1(t) + \epsilon^2 x_2(t) + \cdots. \tag{25.10}$$

Imposing the initial conditions

$$x(0) = A = x_0(0) + \epsilon x_1(0) + \cdots$$
$$\dot{x}(0) = 0 = \dot{x}_0(0) + \epsilon \dot{x}_1(0) + \cdots$$

and equating powers of ϵ,

$$\begin{aligned} x_0(0) = A, \qquad x_1(0) = x_2(0) = \cdots = 0 \\ \dot{x}_0(0) = \dot{x}_1(0) = \cdots = 0. \end{aligned} \tag{25.11}$$

Next, we put (25.10) into the differential equation (25.9), expand in powers of ϵ, equate coefficients of powers of ϵ to zero, and thus obtain the system

$$\ddot{x}_0 + x_0 = 0, \qquad x_0(0) = A, \quad \dot{x}_0(0) = 0 \tag{25.12a}$$
$$\ddot{x}_1 + x_1 = -x_0^3, \qquad x_1(0) = \dot{x}_1(0) = 0 \tag{25.12b}$$
$$\ddot{x}_2 + x_2 = -3x_0^2 x_1, \qquad x_2(0) = \dot{x}_2(0) = 0 \tag{25.12c}$$
$$\vdots \qquad\qquad \vdots$$

Although these equations are nonlinear and coupled, observe that the first is linear and yields

$$x_0(t) = A \cos t. \tag{25.13}$$

Putting it into (25.12b),

$$\ddot{x}_1 + x_1 = -A^3 \cos^3 t = -\frac{A^3}{4}(3 \cos t + \cos 3t), \qquad x_1(0) = \dot{x}_1(0) = 0,$$

so that

$$x_1(t) = \frac{A^3}{32}(-\cos t + \cos 3t) - \frac{3A^3}{8} t \sin t. \tag{25.14}$$

Putting (25.14) and (25.13) into (25.12c), we can solve for x_3, and the process can be carried as far as we like, although the calculations become increasingly tedious. Thus

$$x(t) = A \cos t + \epsilon \frac{A^3}{32}(-\cos t + \cos 3t - 12t \sin t) + \cdots.$$

Because of the increasing labor involved in obtaining $x_2, x_3, \ldots$, and the smallness of ϵ, it is reasonable to cut the series off after x_1, say, and express

$$x(t) \approx A \cos t + \epsilon \frac{A^3}{32}(-\cos t + \cos 3t - 12t \sin t). \tag{25.15}$$

Implicit in this description is the assumption that (25.10) is, in fact, either a convergent series or an asymptotic expansion.[3] Besides this question, however, we see that (25.15) falls considerably short of our expectations, since it cannot possibly be

[3] Asymptotic expansions were discussed in Section 2.6, which may have been skipped, since it was a starred section. Although an understanding of asymptotic expansions is needed in perturbation theory, it will not be essential for the present rather formal and introductory treatment of the subject.

uniformly valid over the domain of interest, $0 \leq t < \infty$. That is, we know (from Example 23.3) that the desired solution is a periodic motion, whereas the $t \sin t$ term in (25.15) is nonperiodic and grows in amplitude without bound. The lack of uniform validity is a serious drawback, and we therefore classify this perturbation problem as **singular**.

On the other hand, if the domain were finite, say $0 \leq t \leq T$, then we would expect (25.15) to provide a uniformly valid approximation, and we would classify the perturbation as regular. Yet even here we apparently need $\epsilon T \ll 1$ [from the ϵt combination in (25.15)] for good accuracy, and for T large (say 10 or more) this would require ϵ to be extremely small, namely $\epsilon \ll 0.1$.

Again, it is the appearance of the so-called **secular term** $t \sin t$ that first signals the singular nature of the perturbation, but that is not the end of the story since additional secular terms of the form $t^2 \sin t$, $t^2 \cos t$, $t^3 \sin t, \ldots$ can be expected to arrive in $x_2(t)$, $x_3(t)$, and so on (Exercise 25.8). That is, *successive terms in* (25.10) *become more and more singular.*

In order to expose the source of the difficulty, let us refer back to the approximate solution given by equation (22.184), which was obtained by an iterative procedure. (Note that $\beta \longrightarrow \epsilon$, $\alpha \longrightarrow 1$, and $F \longrightarrow 0$ for the present case.) Consider the $\cos \omega t$ term in (22.184). With $\omega = \sqrt{1 + 3\epsilon A^2/4}$, let us expand in powers of ϵ,

$$\cos \omega t = \cos \sqrt{1 + \frac{3A^2}{4}\epsilon}\, t = \cos t - \frac{3A^2}{8}\epsilon t \sin t + \cdots. \tag{25.16}$$

Specifically, observe the entry of secular terms and the destruction of the uniform validity inherent in (22.184). Here is the source of the difficulty in (25.15)—namely, forcing everything to be expanded in powers of ϵ. On the other hand, if we define a new time variable $\tau = \omega t = \sqrt{1 + 3A^2\epsilon/4}\, t = (1 + 3A^2\epsilon/8 + \cdots)t$, then $\cos \omega t = \cos \tau$, which contains no ϵ's and will no longer lead to secular terms when we expand in powers of ϵ!

Using this motivation, let us seek

$$x = x_0(\tau) + \epsilon x_1(\tau) + \epsilon^2 x_2(\tau) + \cdots \tag{25.17a}$$

in place of (25.10), where τ is a slightly *strained* time defined by[4]

$$t = (1 + a_1\epsilon + a_2\epsilon^2 + \cdots)\tau, \tag{25.17b}$$

and the parameters $a_1, a_2, \ldots$ are to be determined so as to preclude secular terms and hence regularize the perturbation.[5]

Introducing τ in (25.9),

$$\frac{dx}{dt} = \frac{dx}{d\tau}\frac{1}{dt/d\tau} = \frac{dx}{d\tau}(1 + a_1\epsilon + \cdots)^{-1} = \frac{dx}{d\tau}(1 - a_1\epsilon + \cdots) \tag{25.19}$$

and

$$\frac{d^2x}{dt^2} = \text{etc.} = \frac{d^2x}{d\tau^2}(1 - a_1\epsilon + \cdots)^2 = \frac{d^2x}{d\tau^2}(1 - 2a_1\epsilon + \cdots).$$

---

[4]We say "slightly strained" because $t \sim \tau$ as $\epsilon \longrightarrow 0$.

[5]Whether we express $t = (1 + a_1\epsilon + \cdots)\tau$ or $\tau = (1 + a_1\epsilon + \cdots)t$ is immaterial, except that the former turns out to be a little more convenient in the calculations. More generally, we might seek a more powerful procedure by using

$$t = \tau + f_1(\tau)\epsilon + f_2(\tau)\epsilon^2 + \cdots \tag{25.18}$$

in place of (25.17b), where it is expedient to require that $f_1(0) = f_2(0) = \cdots = 0$ so that $t = 0$ corresponds to $\tau = 0$, but the simpler form (25.17b) will suffice in the present example.

Putting these results and (25.17a) into (25.9), we have

$$(x_0'' + \epsilon x_1'' + \cdots)(1 - 2a_1\epsilon - \cdots) + (x_0 + x_1\epsilon + \cdots) + \epsilon(x_0 + x_1\epsilon + \cdots)^3 = 0,$$

where $(\ )' \equiv d(\ )/d\tau$, and equating powers of ϵ,

$$\epsilon^0: \quad x_0'' + x_0 = 0 \tag{25.20a}$$

$$\epsilon: \quad x_1'' + x_1 = 2a_1x_0'' - x_0^3 \tag{25.21a}$$

$$\vdots$$

To obtain the corresponding initial conditions, we observe that $t = 0$ corresponds to $\tau = 0$ and express

$$\begin{aligned} x(0) &= x_0(0) + x_1(0)\epsilon + \cdots \quad \text{per (25.17a)} \\ &= A \\ \dot{x}(0) &= x'(0)(1 - a_1\epsilon - \cdots) \quad \text{per (25.18)} \\ &= [x_0'(0) + x_1'(0)\epsilon + \cdots](1 - a_1\epsilon - \cdots) \\ &= x_0'(0) + [x_1'(0) - a_1x_0'(0)]\epsilon + \cdots = 0, \end{aligned}$$

so that

$$x_0(0) = A, \qquad x_0'(0) = 0 \tag{25.20b}$$

$$x_1(0) = 0, \qquad x_1'(0) = a_1x_0'(0) \tag{25.21b}$$

$$\vdots$$

Solving (25.20), $x_0(\tau) = A\cos\tau$, and so (25.21) becomes

$$x_1'' + x_1 = -2a_1A\cos\tau - A^3\cos^3\tau, \qquad x_1(0) = x_1'(0) = 0$$

or

$$x_1'' + x_1 = -\left(2a_1 + \frac{3A^2}{4}\right)A\cos\tau - \frac{A^3}{4}\cos 3\tau, \qquad x_1(0) = x_1'(0) = 0. \tag{25.22}$$

Given half a chance, the $\cos\tau$ term will lead to a secular (or "resonance") term of the form $\tau\sin\tau$ in the solution. To prevent this occurrence, we choose a_1 so that $2a_1 + 3A^2/4 = 0$. With this choice, the solution of (25.22) is $x_1(\tau) = A^3(-\cos\tau + \cos 3\tau)/32$, and thus through first-order terms we have the *uniformly valid* approximation

$$x \approx A\cos\tau + \frac{A^3}{32}(-\cos\tau + \cos 3\tau)\epsilon$$

$$t \approx \left(1 - \frac{3A^2}{8}\epsilon\right)\tau$$

or

$$x(t) \approx A\cos\left(1 + \frac{3A^2}{8}\epsilon\right)t + \frac{A^3}{32}\left[-\cos\left(1 + \frac{3A^2}{8}\epsilon\right)t + \cos 3\left(1 + \frac{3A^2}{8}\epsilon\right)t\right]\epsilon. \tag{25.23}$$

The straining technique was introduced by Lindstedt in 1882 for the more general equation $\ddot{x} + x = \epsilon f(x, \dot{x})$, which arose in connection with problems in celestial mechanics, and was significantly extended by Lighthill and others, beginning around 1949. The fact that Lindstedt's expansions were asymptotic expansions was established in 1892 by the great mathematician/astronomer Poincaré.[6] Such questions are quite difficult and fall well beyond our present scope. In fact, in applications we often settle for a formal one- or two-term approximation without addressing questions of convergence.

COMMENT. Recall that straightforward expansion led to the expression (25.15), which was not uniformly valid. We then started over with the scheme (25.17) and obtained the uniformly valid result (25.23). It has been pointed out by Pritulo,[7] however, that much of this labor is redundant and can be avoided. Specifically, we need merely insert $t = (1 + a_1\epsilon + a_2\epsilon^2 + \cdots)\tau$ into (25.15), expand in powers of ϵ, and choose the a_j's so as to remove secular terms. Through $O(\epsilon)$,

$$\begin{aligned} x &\sim A\cos(1 + a_1\epsilon + \cdots)\tau + \epsilon\frac{A^3}{32}[-\cos(1 + \cdots)\tau + \cos 3(1 + \cdots)\tau \\ &\qquad - 12(1 + \cdots)\tau \sin(1 + \cdots)\tau] \\ &\sim A\cos\tau - \epsilon a_1 A\tau \sin\tau + \epsilon\frac{A^3}{32}(-\cos\tau + \cos 3\tau - 12\tau\sin\tau), \end{aligned}$$

and so we choose $a_1 = -3A^2/8$. Then

$$\left.\begin{aligned} x &\sim A\cos\tau + \epsilon\frac{A^3}{32}(-\cos\tau + \cos 3\tau) \\ \text{with} \quad \tau &= (1 + a_1\epsilon + \cdots)^{-1}t \sim \left(1 + \frac{3A^2}{8}\epsilon\right)t, \end{aligned}\right\} \tag{25.24}$$

which is identical to (25.23). Pritulo's method is sometimes referred to as *renormalization*. ▮

Example 25.3. Recall from Example 23.4 that the *van der Pol equation*

$$\ddot{x} - \epsilon(1 - x^2)\dot{x} + x = 0 \tag{25.25}$$

admits an isolated periodic solution corresponding to the limit cycle depicted in Fig. 23.12. Let us seek to approximate that solution for the case where $\epsilon \ll 1$.

First, what are our initial conditions? The initial point $x = x(0)$, $y = \dot{x}(0)$ must be *on* the limit cycle Γ, but we do not yet know the precise position of Γ! Let us take $x(0) = 0$, say, and $\dot{x}(0) = c$, where c must be determined later, such that the initial point falls on Γ.

Expanding

$$\dot{x}(0) = c(\epsilon) = c_0 + c_1\epsilon + c_2\epsilon^2 + \cdots, \tag{25.26}$$

we seek x in the form (25.17), that is,

$$x = x_0(\tau) + \epsilon x_1(\tau) + \epsilon^2 x_2(\tau) + \cdots \tag{25.27a}$$

where

$$t = (1 + a_1\epsilon + a_2\epsilon^2 + \cdots)\tau. \tag{25.27b}$$

The idea is to choose $c_0, c_1, \ldots$ and $a_0, a_1, \ldots$ so that x is periodic (since we are *on*

---

[6] A lucid version of this proof can be found in J. J. Stoker, *Nonlinear Vibrations*, Interscience, New York, 1950, Appendix I.

[7] M. F. Pritulo, On the Determination of Uniformly Accurate Solutions of Differential Equations by the Method of Perturbation of Coordinates, *Journal of Applied Mathematics and Mechanics*, Vol. 26, 1962, pp. 661–667.

the limit cycle), and contains no secular terms. Following the same lines as in Example 25.2, we obtain (Exercise 25.9)

$$x \approx 2 \sin \tau + \frac{\epsilon}{4}(\cos \tau - \cos 3\tau) \tag{25.28a}$$

where

$$\tau \approx \left(1 + \frac{\epsilon^2}{16}\right)^{-1} t \approx \left(1 - \frac{\epsilon^2}{16}\right) t. \tag{25.28b}$$

Note carefully how the straining, although only a modest adjustment, is crucial in rendering the solution uniformly valid. ▮

Example 25.4. *Multiple Scales.* Let us again consider the van der Pol equation (25.25) but this time seek solutions other than the periodic (limit cycle) solution. That is, we are interested here in the *approach* to the limit cycle. In this case, we find that even with the help of straining we are unable to obtain a uniformly valid approximation. To illustrate, let us look at the simple linear case

$$\ddot{x} + \epsilon \dot{x} + x = 0, \tag{25.29}$$

because it is similar in form to (25.25) and is easily solved exactly. Specifically,

$$x = e^{-\epsilon t/2}\left[A \sin \sqrt{1 - \frac{\epsilon^2}{4}}\, t + B \cos \sqrt{1 - \frac{\epsilon^2}{4}}\, t\right]. \tag{25.30}$$

Then

$$\sqrt{1 - \frac{\epsilon^2}{4}}\, t = \left(1 - \frac{\epsilon^2}{8} + \cdots\right) t$$

suggests that a strained time variable is needed namely,

$$\tau = \left(1 - \frac{\epsilon^2}{8} + \cdots\right) t \quad \text{or} \quad t = \left(1 + \frac{\epsilon^2}{8} - \cdots\right) \tau.$$

Next, (25.30) becomes

$$x = e^{-(\epsilon + \epsilon^3/8 - \cdots)\tau}[A \sin \tau + B \cos \tau]$$
$$\sim (1 - \epsilon\tau + \cdots)[A \sin \tau + B \cos \tau],$$

so that secular terms now arise through the exponential term.

In physical terms, observe that two time scales are involved in (25.29); a "fast time" that beats at the rate $\omega^{-1} = 1/\sqrt{1 - \epsilon^2/4} \approx 1$ sec/radian and a "slow time" associated with the damping rate. The problem is that *both* times cannot be accommodated simultaneously by a single straining.

So we are led to a so-called **two-variable expansion**

$$x = x_0(\tau, \tilde{\tau}) + \epsilon x_1(\tau, \tilde{\tau}) + \cdots, \tag{25.31a}$$

where $\qquad \tau = (1 + a_1\epsilon + a_2\epsilon^2 + \cdots)t \quad \text{and} \quad \tilde{\tau} = \epsilon t, \qquad (25.31b)$

which is one version of the various **methods of multiple scales.** We leave the details for Exercise 25.10 and state that the result, for the van der Pol equation (25.25), with initial conditions $x(0) = 0$, $\dot{x}(0) = c$ is

$$x = \frac{2c}{\sqrt{c^2 + (4 - c^2)e^{-\tilde{\tau}}}} \sin \tau + O(\epsilon) \tag{25.32a}$$

$$\tau = [1 + O(\epsilon^2)]t \quad \text{and} \quad \tilde{\tau} = \epsilon t, \tag{25.32b}$$

or

$$x = \frac{2c}{\sqrt{c^2 + (4 - c^2)e^{-\epsilon t}}} \sin\,[1 + O(\epsilon^2)]t + O(\epsilon). \tag{25.33}$$

Notice the (slow) approach to the limit cycle as $t \longrightarrow \infty$.

We emphasize that the crucial step in solving a singular perturbation problem is in determining a suitable form for the expansion—(25.17), (25.31), or whatever. Often we can be guided by our experience and the experience of others (i.e., by the literature). In the present example we looked at the problem (25.29), which was much simpler than (25.25) but yet essentially similar, and were then guided by the form of the exact solution of (25.29). ∎

25.3. *SINGULAR PERTURBATIONS: BOUNDARY LAYER METHODS*

In the foregoing examples the nonuniformities that arose were smeared over the entire domain of interest and could be handled by a mild straining of the independent variable. In some problems, however, very thin regions of nonuniformity occur in which the dependent variable undergoes sharp changes. Such cases frequently occur at endpoints of the interval but not always. Independent of the physical context, these regions are often referred to as **boundary layers**, after the classical boundary layer theory of fluid mechanics developed by Prandtl in 1905. The main ideas can be explained through the following example.

Example 25.5. Consider

$$\epsilon \ddot{x} + \dot{x} + x = 0, \qquad x(0) = 0, \quad x(1) = 1, \tag{25.34}$$

where $0 < \epsilon \ll 1$. As a start, let us try a straightforward expansion

$$x = x_0(t) + \epsilon x_1(t) + \cdots.$$

Putting it into (25.34), we easily obtain the system

$$\begin{aligned}
\epsilon^0:&\quad \dot{x}_0 + x_0 = 0, && x_0(0) = 0, \quad x_0(1) = 1\\
\epsilon:&\quad \dot{x}_1 + x_1 = -\ddot{x}_0, && x_1(0) = x_1(1) = 0\\
\epsilon^2:&\quad \dot{x}_2 + x_2 = -\ddot{x}_1, && x_2(0) = x_2(1) = 0\\
&\quad\vdots
\end{aligned}$$

Notice carefully that each of the differential equations is of *first* order and hence not capable (except by coincidence) of coping with *two* boundary conditions. For instance, $x_0 = Ae^{-t}$, and whereas $x_0(0) = 0$ requires that $A = 0$, $x_0(1) = 1$ requires that $A = e$. Or if we simply drop the $\epsilon\ddot{x}$ term in (25.34), then $\dot{x} + x = 0$ and $x = Ae^{-t}$, which can satisfy one but not both boundary conditions. This *reduction of order*, which results from the small parameter multiplying the highest derivative in the differential equation, generally leads to boundary layers (although it is not the only way in which boundary layers occur).

The exact solution

$$x(t) = e^{(1-t)/2\epsilon} \frac{\sinh(\sqrt{1-4\epsilon}\, t/2\epsilon)}{\sinh(\sqrt{1-4\epsilon}/2\epsilon)} \tag{25.35}$$

is plotted in Fig. 25.2 for $\epsilon = 0.05$, for reference. We see that $x(t)$ is closely approximated by e^{1-t} [i.e., the solution of the reduced equation[8] satisfying the right end

---

[8]That is, the differential equation with ϵ set equal to 0, $\dot{x} + x = 0$ in this case.

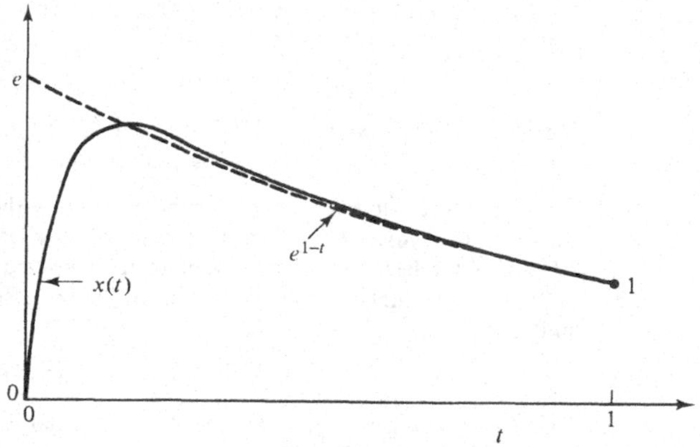

Figure 25.2. The exact solution (25.35), for $\epsilon = 0.05$.

condition $x(1) = 1$] everywhere except in a thin "boundary layer" region near the left endpoint, where $x(t)$ drops sharply to zero.

Following Prandtl's viscous boundary layer concept, it seems natural to think in terms of two adjoining domains, an *inner domain* or boundary layer $0 < t < t_0$ say, where t_0 is small, and an *outer domain* $t_0 < t < 1$. In the outer domain the x, $\dot{x}$, $\ddot{x}$ are all of the same order of magnitude. Thus the $\epsilon\ddot{x}$ term in (25.34) is quite small, and we have a balance between the $\dot{x}$ and x terms. To lowest order, then, the "outer problem" is

$$\dot{x}^o + x^o = 0, \qquad x^o(1) = 1,$$

where the boundary condition $x(0) = 0$ is not relevant and has been dropped. Thus

$$x^o \approx e^{1-t}. \tag{25.36}$$

In the inner domain, however, $\ddot{x}$ is so large that the $\epsilon\ddot{x}$ term is *not* negligible and enters into the balance. In our foregoing calculations, in Examples 25.1 to 25.4, *it was always implicitly understood that all quantities* (such as x, $\dot{x}$, $\ddot{x}$) *were* $O(1)$ *so that the various terms in expansions could be ordered in magnitude by comparing the powers of* ϵ *by which they were multiplied.* To obtain that state of affairs in the boundary layer region, let us *magnify* the t variable according to

$$T = \epsilon^{-\alpha} t \qquad (\alpha > 0) \tag{25.37}$$

where α is sufficiently large so that x' $(= dx/dT)$ and x'' *are* of order unity. Putting (25.37) into (25.34) gives

$$\epsilon^{1-2\alpha} x^{i\prime\prime} + \epsilon^{-\alpha} x^{i\prime} + \epsilon^0 x^i = 0, \qquad x^i(0) = 0, \tag{25.38}$$

where the superscript i denotes the "inner" region and this time the *outer* boundary condition $x(1) = 1$ has been dropped. With x^i, $x^{i\prime}$, and $x^{i\prime\prime}$ all $O(1)$ by assumption, we choose α so that the first term is in balance (i.e., of the same order of magnitude) with one or both of the other terms. In order for all three terms to be in balance, we need $1 - 2\alpha = -\alpha = 0$, which is not possible. In order for the first and third terms to be in balance, we need $1 - 2\alpha = 0$ or $\alpha = 1/2$. This choice fails, however, for then (25.38) becomes $x^{i\prime\prime} + \epsilon^{-1/2} x^{i\prime} + x^i = 0$ or $\epsilon^{1/2} x^{i\prime\prime} + x^{i\prime} + \epsilon^{1/2} x^i = 0$, and so the reduced equation is $x^{i\prime} = 0$ instead of the desired balance $x^{i\prime\prime} + x^i = 0$.

Finally, we try a balance between the first and second terms. Then $1 - 2\alpha = -\alpha$ and so $\alpha = 1$; thus (25.38) becomes

$$x^{i\prime\prime} + x^{i\prime} + \epsilon x^{i} = 0, \qquad x^{i}(0) = 0, \tag{25.39}$$

which is as desired. To lowest order (i.e., for $\epsilon = 0$), it yields

$$x^{i} \approx A(1 - e^{-T}), \tag{25.40}$$

where A is to be determined by somehow blending the inner and outer solutions (25.40) and (25.36) at the edge of the boundary layer. This blending can be accomplished in a number of ways. According to the so-called *primitive matching principle* of Prandtl, we simply set the "inner limit of the outer solution" equal to the "outer limit of the inner solution"—that is,

$$\lim_{t\to 0} x^{o}(t;\epsilon) = \lim_{T\to\infty} x^{i}(T;\epsilon). \tag{25.41}$$

(The $T \longrightarrow \infty$ limit is applied because T is the *magnified* variable; that is, $1/\epsilon \gg 1$ implies that $T = t/\epsilon \gg t$.) Thus

$$\lim_{t\to 0} e^{1-t} = \lim_{T\to\infty} A(1 - e^{-T})$$

$$e = A, \tag{25.42}$$

and so

$$x(t) \approx \begin{cases} e(1 - e^{-t/\epsilon}) & \text{(inner domain)} \quad (25.43a) \\ e^{1-t} & \text{(outer domain),} \quad (25.43b) \end{cases}$$

as shown in Fig. 25.3 for $\epsilon = 0.05$.

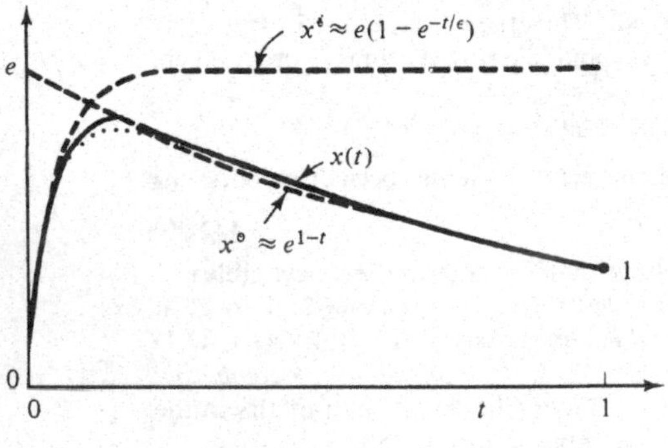

Figure 25.3. x^{i} and x^{o} to lowest order, per (25.43), for $\epsilon = 0.05$.

In applications, we often seek higher-order approximations for both the inner and outer solutions. Thus for the inner problem (25.39) let us set

$$x^{i} = x_0^{i}(T) + \epsilon x_1^{i}(T) + \cdots. \tag{25.44}$$

Putting (25.44) into (25.39) and equating powers of ϵ, we have

$$\begin{aligned} \epsilon^0:&\quad x_0^{i\prime\prime} + x_0^{i\prime} = 0, &\quad x_0^{i}(0) = 0 \\ \epsilon:&\quad x_1^{i\prime\prime} + x_1^{i\prime} = -x_0^{i}, &\quad x_1^{i}(0) = 0 \\ \epsilon^2:&\quad x_2^{i\prime\prime} + x_2^{i\prime} = -x_1^{i}, &\quad x_2^{i}(0) = 0 \\ &\quad\vdots \end{aligned}$$

and it follows that $x_0^{i} = A(1 - e^{-T})$ and $x_1^{i} = B - AT - (B + AT)e^{-T}$, where A, B are constants of integration. Through first order, then,

$$x^{i} \approx A(1 - e^{-T}) + [B - AT - (B + AT)e^{-T}]\epsilon. \tag{25.45}$$

Similarly, for the outer problem

$$\epsilon\ddot{x}^{o} + \dot{x}^{o} + x^{o} = 0, \qquad x^{o}(1) = 1, \tag{24.46}$$

we seek

$$x^{o} = x_0^{o}(t) + x_1^{o}(t)\epsilon + \cdots. \tag{25.47}$$

This produces the system

$$\begin{aligned}
\epsilon^0:&\quad \dot{x}_0^o + x_0^o = 0, &\quad x_0^o(1) = 1\\
\epsilon:&\quad \dot{x}_1^o + x_1^o = -\ddot{x}_0^o, &\quad x_1^o(1) = 0\\
\epsilon^2:&\quad \dot{x}_2^o + x_2^o = -\ddot{x}_1^o, &\quad x_2^o(1) = 0,\\
&\quad\vdots
\end{aligned}$$

which yields $x_0^o = e^{1-t}$, $x_1^o = (1-t)e^{1-t}$, and so on. Through first order,

$$x^o \approx e^{1-t} + [(1-t)e^{1-t}]\epsilon. \tag{25.48}$$

Finally, we *match* (25.45) and (25.48) in order to determine A and B. To do so, we will use the following matching principle proposed by Van Dyke:[9]

The m-term inner expansion (of the n-term outer expansion)

$$= \textit{the n-term outer expansion (of the m-term inner expansion)}, \tag{25.49}$$

where the integers m and n are not necessarily equal. It reduces to the primitive matching principle for the lowest-order case—that is, where $m = n = 1$—and provides an extension for higher-order matching. For instance, let $m = n = 1$. Then

$$\begin{aligned}
&\text{One-term outer expansion:} \quad x^o \sim e^{1-t}\\
&\quad\text{rewritten in inner variable:} \quad = e^{1-\epsilon T}\\
&\quad\text{expanded for small } \epsilon: \quad = e(1 - \epsilon T + \cdots)\\
&\quad\text{one-term inner expansion:} \quad = e
\end{aligned} \tag{25.50}$$

$$\begin{aligned}
&\text{One-term inner expansion:} \quad x^i \sim A(1 - e^{-T})\\
&\quad\text{rewritten in outer variable:} \quad = A(1 - e^{-t/\epsilon}) && (25.51)\\
&\quad\text{expanded for small } \epsilon: \quad = A && (25.52)\\
&\quad\text{one-term outer expansion:} \quad = A, && (25.53)
\end{aligned}$$

and applying (25.49), we set $e = A$, as found above by the primitive matching principle.

We should clarify the expansion of $e^{-t/\epsilon}$ that took us from (25.51) to (25.52). It will suffice to note that, for any fixed $t > 0$, $e^{-t/\epsilon} \longrightarrow 0$ faster than ϵ^N for all positive integers N, no matter how large, and so $e^{-t/\epsilon}$ is simply zero to *any* desired order. It is customary to say that $e^{-t/\epsilon}$ is *transcendentally small* as $\epsilon \longrightarrow 0$.

For $m = n = 2$,

$$\begin{aligned}
&\text{Two-term outer expansion:} \quad x^o \sim e^{1-t} + [(1-t)e^{1-t}]\epsilon\\
&\quad\text{rewritten in inner variable:} \quad = e^{1-\epsilon T} + [(1-\epsilon T)e^{1-\epsilon T}]\epsilon\\
&\quad\text{expanded for small } \epsilon: \quad = e(1 - \epsilon T + \cdots) + e(1 + \cdots)\epsilon\\
&\quad\text{two-term inner expansion:} \quad = e + e(1 - T)\epsilon
\end{aligned} \tag{25.54}$$

$$\text{Two-term inner expansion:} \quad x^i \sim A(1 - e^{-T}) + [B - AT - (B + AT)e^{-T}]\epsilon$$

[9]M. Van Dyke, *Perturbation Methods in Fluid Mechanics*, Academic, New York, 1964.

rewritten in outer variable: $= A(1 - e^{-t/\epsilon})$

$$+ \left[B - A\frac{t}{\epsilon} - \left(B + A\frac{t}{\epsilon}\right)e^{-t/\epsilon}\right]\epsilon$$

expanded for small ϵ: $= A + B\epsilon - At$

two-term outer expansion: $= (A - At) + B\epsilon.$ (25.55)

Applying the matching principle (25.49) gives

$$e + e(1 - T)\epsilon = A - At + B\epsilon$$

or since $T = t/\epsilon$,

$$e(1 - t) + e\epsilon = (A - At) + B\epsilon. \tag{25.56}$$

By equating coefficients of like powers of ϵ, we have $A = B = e$, and so

$$x(t) \approx \begin{cases} e(1 - e^{-t/\epsilon}) + e[\epsilon - t - (\epsilon + t)e^{-t/\epsilon}] & \text{(inner)} \quad (25.57a) \\ e^{1-t} + \epsilon(1 - t)e^{1-t} & \text{(outer).} \quad (25.57b) \end{cases}$$

However, the representations (25.43) and (25.57) suffer from the fact that a precise numerical value of the boundary layer thickness (and hence a definition of the inner and outer domains) is absent. In any event, a *single* representation is generally more desirable, and it can be provided by the so-called **composite solution** x^c, defined as

$$x^c = x^i + x^o - (x^i)^o \tag{25.58}$$

or $x^i + x^o - (x^o)^i$, since $(x^i)^o = (x^o)^i$ according to the matching principle. The idea is that in the inner region (25.58) becomes

$$(x^c)^i = (x^i)^i + (x^o)^i - ((x^o)^i)^i = x^i + (x^o)^i - (x^o)^i = x^i,$$

and in the outer region it becomes

$$(x^c)^o = (x^i)^o + (x^o)^o - ((x^i)^o)^o = (x^i)^o + x^o - (x^i)^o = x^o;$$

that is, x^c agrees with x^i in the inner domain and with x^o in the outer domain (through the desired order) and hence provides a single uniformly valid approximation over the entire domain ($0 \leq t \leq 1$ in the present case).

To see how (25.58) works, let us apply it to (25.43). Then $(x^i)^o$ is the one-term outer expansion of the one-term inner expansion, which, according to (25.53), is A (i.e., e, since $A = e$). So we have[10]

$$x^c \approx e(1 - e^{-T}) + e^{1-t} - e = e(e^{-t} - e^{-t/\epsilon}) \tag{25.59}$$

or, more specifically,

$$x^c = e(e^{-t} - e^{-t/\epsilon}) + O(\epsilon),$$

since x^i and x^o were carried out to within $O(\epsilon)$. x^c is indicated in Fig. 25.3 by the tiny dots (for $\epsilon = 0.05$).

Applying (25.58) to (25.57), we obtain (Exercise 25.14) the more accurate result

$$x^c = [e^{1-t} - (1 + t)e^{1-t/\epsilon}] + \epsilon[(1 - t)e^{1-t} - e^{1-t/\epsilon}] + O(\epsilon^2). \tag{25.60}$$

We've plotted x^i, x^o, x^c, and the exact solution in Fig. 25.4 for $\epsilon = 0.05$. In this case, x^c and the exact solution (solid curve) are virtually indistinguishable!

---

[10] Observe that x^i and x^o have a common part—namely, the part that is matched; in this case, the common part is given by (25.50) or (25.53). The purpose of the $(x^i)^o$ term in (25.58), then, is to subtract off the common part, which would otherwise be counted twice.

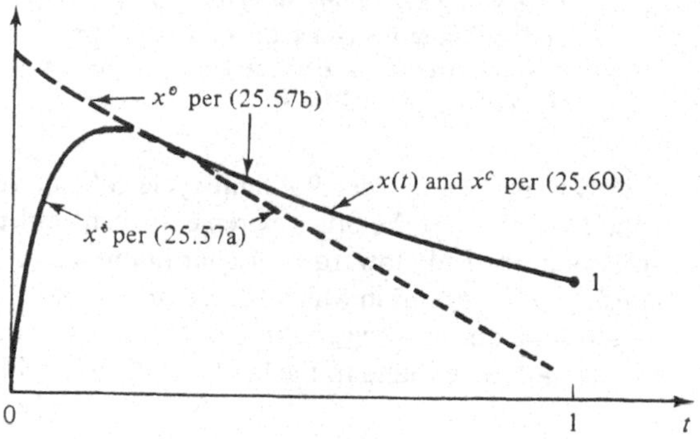

Figure 25.4. First order inner, outer, and composite solutions for $\epsilon = 0.05$.

COMMENT 1. The process of computing inner and outer expansions and matching them is generally known as the **method of matched asymptotic expansions.** In the present example we employed the widely used matching principle of Van Dyke. A number of other matching methods have been proposed—for instance, a more rigorous method due to Kaplun (1967) that involves "intermediate limits" and an "overlap" of the inner and outer domains. In comparing these two procedures, Fraenkel has concluded that Van Dyke's method may not always be correct but that Kaplun's method is more difficult to apply. Rather than cite a long list of articles that elaborate on this and other points in this chapter, we refer the reader to the several books on the subject listed in the footnotes and at the end of Part IV, which, in turn, contain extensive references to the literature.

COMMENT 2. In the present example we allowed for a boundary layer at the left end of the interval, as indicated in Fig. 25.2. More generally, of course, we do not have a picture of the exact solution to guide us, and the location of boundary layer(s) must be determined. This question is covered in Exercise 25.15.

COMMENT 3. Instead of determining an inner expansion, an outer expansion, matching, and then forming a composite expansion, it may be desirable to develop a composite expansion directly. Of course, this procedure hinges on our ability to put forward an appropriate *form* for it. In the present example the form

$$x = \sum_0^\infty \epsilon^n f_n(t) + e^{-t/\epsilon} \sum_0^\infty \epsilon^n g_n(t) \tag{25.61}$$

is appropriate (as can be seen with the benefit of hindsight) and permits a uniformly valid approximation (Exercise 25.16).

COMMENT 4. In reviewing this problem, it may be helpful to identify (25.34) with the standard equation $m\ddot{x} + c\dot{x} + kx = 0$ governing a damped harmonic oscillator. Then $c = k = 1$, and the mass is very small, namely $m = \epsilon$. We will refer to the $m\ddot{x}$, $c\dot{x}$, and kx terms as the inertial, damping, and spring terms, respectively. Now, recalling the equations governing the outer and inner domains,

$$\epsilon\ddot{x}^o + \dot{x}^o + x^o = 0$$

and

$$x^{i\prime\prime} + x^{i\prime} + \epsilon x^i = 0,$$

observe that the outer domain is characterized by a balance between the damping and spring terms (since the $\epsilon\ddot{x}$ inertial term is small), whereas the inner domain is characterized by a balance between the inertial and damping terms (since the ϵx^{*} spring term is small). ▮

Recall that the van der Pol limit cycle calculation for *small* ϵ involved a *straining* of the t variable. At the other extreme, it turns out that, as ϵ becomes *large*, the limit cycle solution tends toward a discontinuous behavior, (Fig. 23.13), containing thin *boundary layer* regions in which x undergoes rapid change. The limiting case as $\epsilon \longrightarrow \infty$ is interesting and is discussed in Exercise 25.18; the boundary layer theory for large but finite ϵ can be found in the books by Cole[11] and Davies and James.[12]

EXERCISES

25.1. Derive the solution (25.7) of (25.1). *Hints Galore:* (25.1) is a Riccati equation (recall Exercise 21.13), a Bernoulli equation (Exercise 21.10), and is even variable separable.

25.2. Show that (25.8) has no roots in the disk $|\epsilon| < 1$. *Hint:* Denoting the right- and left-hand sides as RHS and LHS, show that $|\text{RHS}| > 1$ for all $0 \leq x < \infty$. Next, let $\epsilon = u + iv$ and show that $|\text{LHS}| = (u^2 + v^2)\ |\sin u \sinh v/v| < |\sin 1 \sinh 1/1| \doteq 0.989$ within $|\epsilon| < 1$.

25.3. Modifying Example 25.1 to read $y' + \epsilon y + y^2 = 0$, $y(0) = \cos\epsilon$, solve for $y(x) = y_0(x) + \epsilon y_1(x) + \cdots$ through $y_1(x)$. Sketch $y_0(x)$, $y_0(x) + \epsilon y_1(x)$, and the exact solution $y(x)$. Is the perturbation regular or singular? Explain.

25.4. Find the solution of $\sin x = \epsilon x \cos x$ ($\epsilon \ll 1$) near $x = \pi$, through $O(\epsilon^3)$.

25.5. Return to Exercise 20.6. Expanding $\Lambda = \Lambda_0 + \kappa\Lambda_1 + \cdots$, as suggested, compute Λ_0 and Λ_1 and hence show that the critical buckling load is

$$P_{cr} = \frac{\pi^2 EI}{4l^2} + \frac{8EIl}{\pi^2}\kappa + O(\kappa^2).$$

25.6. Solve for $y(x)$ through $O(\epsilon)$ and classify the perturbation scheme as regular or singular.

(a) $y'' + y = e^{\epsilon \sin x}$, $y(0) = y(1) = 0$

(b) $(1 + \epsilon x)y'' + y = x$, $y(0) = \epsilon$, $y(1) = 1$

(c) $y' - \sqrt{\epsilon}\, y^3 = 0$, $y(0) = \cos\epsilon$, $0 \leq x \leq 1$

25.7. Find the eigens, through $O(\epsilon)$.

$$y'' + \lambda(1 + \epsilon x)y = 0, \qquad y(0) = y(\pi) = 0.$$

25.8. (a) Carry the straightforward expansion (25.10) in Example 25.2 one more step—that is, through $x_2(t)$.

(b) Apply Pritulo's method to the result obtained in part (a).

---

[11] J. D. Cole, *Perturbation Methods in Applied Mathematics*, Blaisdell Publishing Co., Waltham, Mass., 1968. Also an excellent general reference for the present chapter. It is written at a slightly more advanced level than the book by Nayfeh.

[12] T. Davies and E. James, *Nonlinear Differential Equations*, Addison-Wesley, Reading, Mass., 1966, pp. 159–167. In fact, this is a good general reference for Chapters 3 and 5.

25.9. Derive the van der Pol limit cycle result (25.28). You should find, through $O(\epsilon)$, that $c_0 = 2$ and $c_1 = 0$.

25.10. Derive the result (25.32) by means of the two-variable expansion (25.31). *Hint:*

$$\dot{x} = \left(\frac{\partial x_0}{\partial \tau}\frac{d\tau}{dt} + \frac{\partial x_0}{\partial \tilde{\tau}}\frac{d\tilde{\tau}}{dt}\right) + \left(\frac{\partial x_1}{\partial \tau}\frac{d\tau}{dt} + \frac{\partial x_1}{\partial \tilde{\tau}}\frac{d\tilde{\tau}}{dt}\right)\epsilon + \cdots$$

$$= \text{etc.} = \left(\frac{\partial x_0}{\partial \tau}\right) + \left(a_1\frac{\partial x_0}{\partial \tau} + \frac{\partial x_0}{\partial \tilde{\tau}} + \frac{\partial x_1}{\partial \tau}\right)\epsilon + \cdots$$

and similarly for $\ddot{x}$. Thus obtain the system

$$\epsilon^0: \quad \frac{\partial^2 x_0}{\partial \tau^2} + x_0 = 0$$

$$\epsilon^1: \quad \frac{\partial^2 x_1}{\partial \tau^2} + x_1 = -2a_1\frac{\partial^2 x_0}{\partial \tau^2} - 2\frac{\partial^2 x_0}{\partial \tau\, \partial \tilde{\tau}} + (1 - x_0^2)\frac{\partial x_0}{\partial \tau}$$

$$\vdots$$

with initial conditions

$$x_0(0, 0) = 0, \qquad \frac{\partial x_0}{\partial \tau}(0, 0) = c$$

$$x_1(0, 0) = 0, \qquad \frac{\partial x_1}{\partial \tau}(0, 0) = -a_1 c - \frac{\partial x_0}{\partial \tilde{\tau}}(0, 0)$$

$$\vdots$$

You may set $a_1 = 0$ in (25.31) with no loss of generality, since the ϵt combination is already accounted for as $\tilde{\tau}$. (In any event, we can at least set $a_1 = 0$ *tentatively* and see if things work out.) Thus obtain $x_0 = A(\tilde{\tau}) \cos \tau + B(\tilde{\tau}) \sin \tau$, where

$$2A' - A + \frac{A(A^2 + B^2)}{4} = 0, \qquad A(0) = 0$$

$$2B' - B + \frac{B(A^2 + B^2)}{4} = 0, \qquad B(0) = c,$$

and (with the help of "inspection") solve for $A(\tilde{\tau})$, $B(\tilde{\tau})$.

25.11. Apply the two-variable expansion (25.31) to $\ddot{x} + \epsilon\dot{x} + x = 0$, $x(0) = 0$, $\dot{x}(0) = 1$ and obtain the lowest-order result

$$x = e^{-\tilde{\tau}/2} \sin \tau + O(\epsilon)$$
$$= e^{-\epsilon t/2} \sin [1 + O(\epsilon^2)]t + O(\epsilon).$$

Hint: See the hint in Exercise 25.10.

25.12. Apply the two-variable expansion (25.31) to $\ddot{x} + \epsilon\dot{x}^3 + x = 0$, $x(0) = 0$, $\dot{x}(0) = 1$ and obtain the lowest-order result

$$x = \frac{1}{\sqrt{1 + 3\tilde{\tau}/4}} \sin \tau + O(\epsilon)$$

$$= \frac{1}{\sqrt{1 + 3\epsilon t/4}} \sin [1 + O(\epsilon^2)]t + O(\epsilon).$$

Is it surprising that the amplitude decay is only *algebraic* in this case, whereas it was *exponential* in the previous exercise? Explain. *Hint:* See the hint in Exercise 25.10.

25.13. (*Finding limit cycles*) Given the nonlinear equation $\ddot{x} + \beta(x, \dot{x}) = 0$, suppose that we regard the trajectories in the x, $\dot{x}$ phase plane as the streamlines of a plane, steady, compressible fluid flow. Then the x, $\dot{x}$ components of the velocity $\mathbf{V}$ are $\dot{x}$, $\ddot{x}$ or $\dot{x}$, $-\beta$, and so the continuity equation $\nabla \cdot (\rho \mathbf{V}) = 0$ becomes

$$\rho_x \dot{x} - (\rho\beta)_{\dot{x}} = 0, \tag{25.62}$$

where $\rho(x, \dot{x})$ is the fluid mass density. If $\ddot{x} + \beta(x, \dot{x}) = 0$ has a stable limit cycle Γ, show (in words) why ρ must be infinite all along Γ. So in order to determine Γ, we can solve the first-order partial differential equation (25.62) for $\rho(x, \dot{x})$ and seek those curve(s) on which ρ is infinite. To illustrate, consider the van der Pol equation (25.25) for small ϵ. Changing to polar coordinates $x = r\cos\theta$, $\dot{x} = r\sin\theta$, and noting that $\beta = -\epsilon(1 - x^2)\dot{x} + x$, (25.62) becomes

$$\rho_\theta + \epsilon(r^2c^2 - 1)(rs^2\rho_r + sc\rho_\theta + \rho) = 0,$$

where the c, s are short for $\cos\theta$, $\sin\theta$. Introducing the strained variables

$$R = [1 + \epsilon a(\theta) + \cdots]r, \qquad \Theta = \theta,$$

seek

$$\rho = \rho^0(R, \Theta) + \epsilon\rho'(R, \Theta) + \cdots.$$

Thus obtain the system

$$\epsilon^0: \quad \rho^0_\Theta = 0$$
$$\epsilon: \quad \rho^1_\Theta = (1 - R^2c^2)(Rs^2\rho^0_R + \rho^0) - Ra'\rho^0_R$$
$$\vdots$$

and the solution $\rho^0 = f(R)$ and

$$\rho^1 = Rf'\left(\frac{\Theta}{2} - \frac{\sin 2\Theta}{4}\right) + f\Theta - R^3f'\left(\frac{\Theta}{8} - \frac{\sin 4\Theta}{32}\right) - R^2f\left(\frac{\Theta}{2} + \frac{\sin 2\Theta}{4}\right) - Raf'.$$

Requiring that ρ^1 be a periodic function of Θ, of period 2π, show that we need

$$4Rf' + 8f - R^3f' - 4R^2f = 0$$

and thus show that $f(R) = 1/R^2|4 - R^2| = \rho^0$, which is infinite on $R = 0$ (focus) and $R = 2$ (limit cycle). Finally, requiring that ρ^1 be *no more singular* on $R = 2$ and at $R = 0$ than ρ^0, show that we need

$$a = \frac{-2\sin 2\Theta + \sin 4\Theta}{8}$$

and hence conclude that the limit cycle[13] is given by $R = 2$ or

$$r = 2 + \left(\frac{\sin 2\Theta}{2} - \frac{\sin 4\Theta}{4}\right)\epsilon + O(\epsilon^2).$$

25.14. Derive the composite expansion (25.60).

25.15. Suppose, in Example 25.5, that the exact solution was *not* available. In this case, how do we know that a boundary layer does not occur at the *right* endpoint $t = 1$?

---

[13] M. Greenberg, Limit Cycle City, *Quarterly of Applied Mathematics*, April 1976, Vol. 33, pp. 103–105.

25.16. Substituting *Latta's expansion* (25.61) into (25.34) and equating the coefficients of ϵ^n and $\epsilon^n e^{-t/\epsilon}$ to zero for $n = 0, 1, \ldots$, obtain the system

$$\begin{aligned} \dot{f}_0 + f_0 &= 0 & \dot{g}_0 - g_0 &= 0 \\ \dot{f}_1 + f_1 &= -\ddot{f}_0 & \dot{g}_1 - g_1 &= \ddot{g}_0 \\ \dot{f}_2 + f_2 &= -\ddot{f}_1 & \dot{g}_2 - g_2 &= \ddot{g}_1 \\ &\vdots & &\vdots \end{aligned}$$

with the boundary conditions

$$f_0(1) = 1, \qquad f_n(1) = 0 \quad \text{for } n \geq 1,$$
$$f_n(0) + g_n(0) = 0 \quad \text{for } n \geq 0,$$

where the transcendentally small terms $e^{-1/\epsilon} g_n(1)$ are neglected. Thus obtain, through first order,

$$x = \left[1 + \epsilon(1 - t)\right] e^{1-t} - [1 + \epsilon(1 + t)] e^{1+t-t/\epsilon} + O(\epsilon^2).$$

25.17. Obtain three-term inner and outer expansions and then their composite solutions in each case.
(a) $\epsilon\ddot{x} - \dot{x} = 1$, $x(0) = 0$, $x(1) = 2$ $(0 < \epsilon \ll 1)$
(b) $\epsilon\ddot{x} + \dot{x} = -1$, $x(0) = 0$, $x(1) = 1$ $(0 < \epsilon \ll 1)$.

25.18. (*Relaxation oscillation of van der Pol equation*) (a) Show that if we set $x = \dot{z}$ in the van der Pol equation $\ddot{x} - \epsilon(1 - x^2)\dot{x} + x = 0$ and integrate once, we obtain $\ddot{z} - \epsilon(\dot{z} - \dot{z}^3/3) + z =$ constant, say h. Setting $z = \zeta + h$, obtain the so-called **Rayleigh equation** on ζ,

$$\ddot{\zeta} - \epsilon\left(\dot{\zeta} - \frac{\dot{\zeta}^3}{3}\right) + \zeta = 0. \tag{25.63}$$

(This is sometimes called the "van der Pol equation" in the literature.)
(b) To consider (25.63) in the phase plane, set $\dot{\zeta} = v$, scale $\zeta = \epsilon\xi$, and obtain

$$\frac{dv}{d\xi} = \epsilon^2 \frac{(v - v^3/3) - \xi}{v}.$$

As $\epsilon \longrightarrow \infty$, deduce that $dv/d\xi \longrightarrow \infty$ at all points in the v, ξ plane except on the curve $v - (v^3/3) - \xi = 0$. So for any initial point, say P, explain why the solution is as shown in Fig. 25.5; for instance, why is the direction *downward*

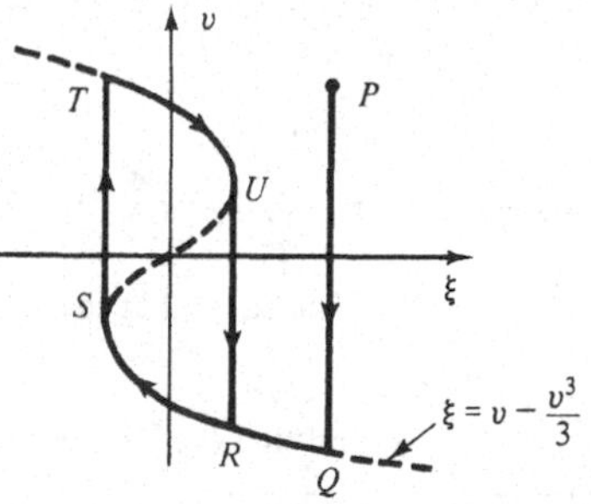

Figure 25.5. Relaxation oscillation of (25.63).

from P to Q and why do we jump from S to T and from U to R? The loop $RSTUR$ is traversed repeatedly and is the *limit cycle.*

(c) Use Fig. 25.5 to sketch $v(t)$ and hence the limit cycle solution of the van der Pol equation (for $\epsilon \longrightarrow \infty$). (The result should be similar to the $\epsilon = 5$ part of Fig. 23.13.) *Hint:* Recall the expression $\dot{s} = a$ for the phase speed from Example 23.3.

(d) Compute the *period* of the limit cycle oscillation.

(e) Why did we bother transforming to the Rayleigh equation (25.63)? *Note:* The motion $x(t)$ is an example of a so-called **relaxation oscillation.** Relaxation oscillations are characterized by a "herkyjerkiness" that results from a repeated slow buildup and sudden discharge type of phenomenon. Quoting van der Pol and van der Mark,[14] examples are

"the Aeolian harp, a pneumatic hammer, the scratching noise of a knife on a plate, the waving of a flag in the wind, the humming noise sometimes made by a water tap, the squeaking of a door, the multivibrator of Abraham and Bloch, the tetrode multivibrator, the periodic sparks obtained from a Wimshurst machine, the Wehnelt interrupter, the intermittent discharge of a capacitor through a neon tube, the periodic reoccurrence of epidemics and economic crises, the periodic density of an even number of species of animals living together with one species serving as food for the other, the sleeping of flowers, the periodic occurrence of showers behind a depression, the shivering from cold, menstruation, and, finally, the beating of the heart."

See also Exercise 25.19.

25.19. (*Prony brake*) Suppose that a disk driven at a constant angular velocity Ω is brought into face-to-face contact with a "Prony brake" consisting of a disk of *very small* inertia I and a restraining torsional spring of stiffness k, as shown in Fig. 25.6(a). Letting the relative angular velocity $\Omega - \dot{\phi} \equiv \zeta$, suppose that the frictional torque supplied to the smaller disk by the larger varies as shown in Fig. 25.6(b).

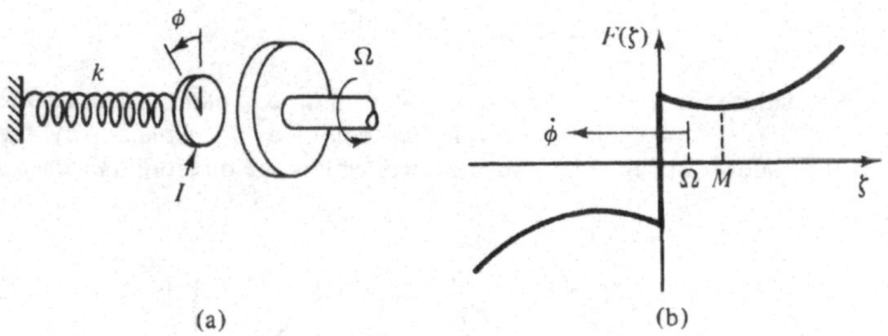

Figure 25.6. Prony brake.

(a) Then show that the equation of motion is

$$I\ddot{\phi} = -k\phi + F(\Omega - \dot{\phi})$$

and that the limit cycle for $I \longrightarrow 0$ is as shown in Fig. 25.7.

[14]The Heartbeat Considered as a Relaxation Oscillation, and an Electrical Model of the Heart, *Philosophical Magazine*, Vol. 6, 1928, pp. 763–775.

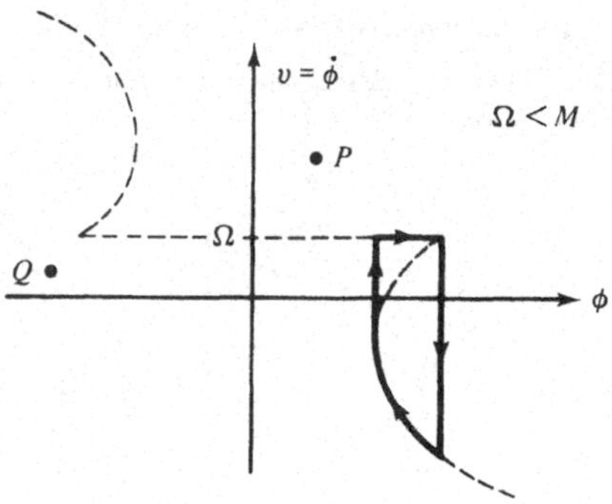

Figure 25.7. Prony brake limit cycle for $I \longrightarrow 0$.

(b) Sketch $\phi(t)$ if the starting point is at P, say.
(c) Sketch $\phi(t)$ if the starting point is at Q.
(d) What if $\Omega > M$ instead? *Hint:* It may help to read through Exercise 25.18.

25.20. Make up two examination-type questions on this chapter.

ADDITIONAL REFERENCES FOR PART IV

W. F. AMES, *Nonlinear Ordinary Differential Equations in Transport Processes*, Academic, New York, 1968.

R. BELLMAN, *Stability Theory of Ordinary Differential Equations*, McGraw-Hill, New York, 1953.

L. BRAND, *Differential and Difference Equations*, Wiley, New York, 1966. [Covers a nice range of topics at an introductory level.]

B. CARNAHAN, H. LUTHER, and J. WILKES, *Applied Numerical Methods*, Wiley, New York, 1969. [Considerable emphasis on application of the methods to problems from the engineering literature. A very nice book.]

E. CODDINGTON and N. LEVINSON, *Theory of Ordinary Differential Equations*, McGraw-Hill, New York, 1955. [Extensive introduction to the theory with some, although not many, applications.]

H. T. DAVIS, *Introduction to Nonlinear Differential and Integral Equations*, Dover Publications, New York, 1962. [Clearly written with many applications to nonlinear mechanics, population dynamics, and problems of pursuit. Includes chapters on the Riccati equation and elliptic integrals.]

M. D. GREENBERG, *Application of Green's Functions in Science and Engineering*, Prentice-Hall, Englewood Cliffs, N.J., 1971.

E. L. INCE, *Ordinary Differential Equations*, Dover Publications, New York, 1956. [A classical reference.]

N. W. MCLACHLAN, *Ordinary Non-Linear Differential Equations in Engineering and Physical Sciences*, Oxford at the Clarendon Press, 1950. [Based largely on applications to nonlinear mechanics.]

M. R. Scott, *Invariant Imbedding and its Applications to Ordinary Differential Equations: An Introduction*, Addison-Wesley, Reading, Mass., 1973.

C. J. Tranter, *Bessel Functions with Some Physical Applications*, Hart Publishing Co., New York, 1968.

PART V

PARTIAL DIFFERENTIAL EQUATIONS

As in Part IV, emphasis here is on applications and methods of solution rather than questions of existence and uniqueness. Of course, the development of suitable methods often depends on some understanding of these and other basic questions, and it is obviously essential to know at the outset whether a problem is of initial value or boundary value type, whether the solution can be expected to be analytic, and so on. This type of information is present in each of the four chapters, although perhaps most heavily in Chapter 27 on CLASSIFICATION, AND THE METHOD OF CHARACTERISTICS. SEPARATION OF VARIABLES AND TRANSFORM METHODS are covered in Chapter 26, GREEN'S FUNCTIONS AND PERTURBATION TECHNIQUES in Chapter 28, and FINITE-DIFFERENCE METHODS in Chapter 29. It is not possible to include the numerous important special techniques from the literature, and it is hoped that the cited references will help to minimize this shortcoming.

Chapter 26

Separation of Variables and Transform Methods

We concentrate in this chapter on *methods of solution* of representative and important second-order, linear, partial differential equations, specifically, the methods of *separation of variables* and *Fourier* and *Laplace transforms*. Considerable attention is paid to the fundamental similarities and differences in the properties of the solutions to the various PDEs (partial differential equations) under consideration, and these ideas are picked up and formalized in Chapter 27.

26.1. INTRODUCTION

The second-order partial differential equation

$$Au_{xx} + 2Bu_{xy} + Cu_{yy} = F \tag{26.1}$$

is categorized according to the "discriminant" $B^2 - AC$.

If $B^2 - AC = 0$, then (26.1) is said to be **parabolic**.

If $B^2 - AC > 0$, it is **hyperbolic**. (26.2)

If $B^2 - AC < 0$, it is **elliptic**.

If A, B, C are functions only of x and y and F is linear in u, u_x, u_y [i.e., $F = a(x, y)u + b(x, y)u_x + c(x, y)u_y + d(x, y)$], then (26.1) is *linear*. If A, B, C depend on u, u_x, u_y as well and/or F depends nonlinearly on u, u_x, u_y, then (26.1) is said to be *quasilinear*—that is, linear at least in the highest (second order) derivatives.

The origin of the criteria (26.2) will be deferred until Chapter 27. For the present, we merely note that it is fundamental insofar as the nature of the solution is concerned and that it may have a bearing on the method of solution as well.

Perhaps the best-known examples of the three types are as follows.

(i) The **diffusion equation** (derived in Section 9.4),

$$\alpha^2 u_{xx} = u_t, \tag{26.3}$$

where t is the time and α is the diffusivity of the material. Then [with $y = t$ in (26.1)]

$$B^2 - AC = 0 - (\alpha^2)(0) = 0,$$

and so (26.3) is *parabolic.*

(ii) The **wave equation**

$$a^2 u_{xx} = u_{yy}, \tag{26.4}$$

where a is a positive constant. Then

$$B^2 - AC = 0 - (a^2)(-1) = a^2 > 0,$$

and so (26.4) is *hyperbolic.*

(iii) The **Laplace equation** (in two dimensions)

$$u_{xx} + u_{yy} = 0. \tag{26.5}$$

Then

$$B^2 - AC = 0 - (1)(1) = -1 < 0,$$

and so (26.5) is *elliptic.*

In (26.3) to (26.5), A, B, C were constants; therefore the equations were of one type throughout the entire plane. More generally, A, B, C may vary such that (26.1) is of different types in different parts of the domain. A well-known example of change of type is the Tricomi equation, which is introduced below.

A basic problem of aerodynamics consists of the calculation of the air disturbance caused by an airfoil moving at a steady speed U through otherwise undisturbed air. It is best to view the flow from a coordinate system that is fixed to the airfoil, for doing so reduces the problem to a steady one (i.e., time independent), as sketched in Fig. 26.1. Neglecting viscosity, it can be shown that the perturbational velocity potential ϕ (i.e., the velocity $\mathbf{q} = U\mathbf{i} + \nabla\phi$) is governed approximately by the PDE

$$(1 - M^2)\phi_{xx} + \phi_{yy} = 0, \tag{26.6}$$

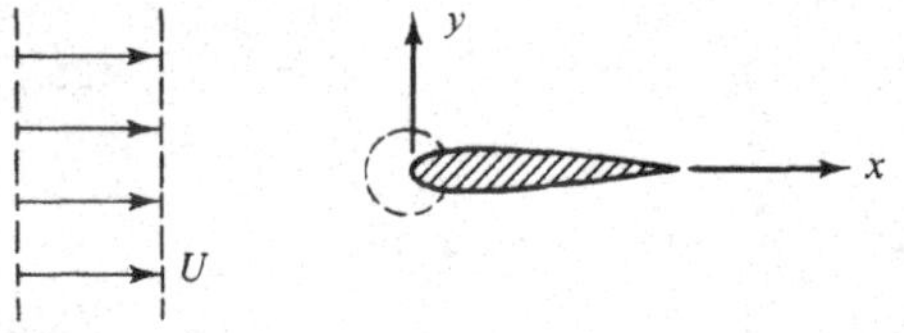

Figure 26.1. Plane flow over airfoil.

where the "Mach number" M is the ratio of U to the speed of sound. For the *subsonic* case $M < 1$, we have $B^2 - AC = M^2 - 1 < 0$ and (26.6) is *elliptic.* In fact, if the fluid is considered incompressible, then the speed of sound is infinite, so that $M = 0$ and (26.6) reduces to the Laplace equation. For the *supersonic* case $M > 1$, we have $B^2 - AC = M^2 - 1 > 0$ and (26.6) is *hyperbolic.*

The case where $M \approx 1$, the so-called *transonic* regime, is especially interesting, as well as difficult. The approximate equation (26.6) is then inaccurate and a nonlinear term, neglected in deriving (26.6), must be retained. Through a complicated change of both dependent and independent variables, Chaplygin[1] was able to suppress the nonlinearity and arrived at what is now called the **Tricomi equation**

$$u_{\xi\xi} + \xi u_{\eta\eta} = 0. \tag{26.7}$$

Now $B^2 - AC = -\xi$, and thus (26.7) is elliptic in the half-plane $\xi > 0$ and hyperbolic in the half-plane $\xi < 0$! This situation reflects the fact that the flow is "mixed" in the original x, y plane, subsonic in some region(s) and supersonic elsewhere. For instance, suppose that M is slightly greater than 1. Then because of the stagnation point at the nose, we expect the flow to be supersonic everywhere except in some neighborhood of the nose, as denoted roughly by the dashed line in Fig. 26.1.[2]

We should emphasize that although the equations that we shall treat are among the most important equations of mathematical physics and engineering, there are many other equations of great importance that are encountered in the literature. To name just a few, at random, we might mention the sixth order (linear) equation

$$E^2\nabla^6 p - 2E\nabla^4 \frac{\partial p}{\partial t} + \nabla^2 \frac{\partial^2 p}{\partial t^2} + 4\frac{\partial^2 p}{\partial z^2} = 0$$

governing the pressure in a rotating viscous fluid (E is the so-called *Ekman number*), the nonlinear *Burgers equation*

$$u_t + uu_x = \lambda u_{xx}$$

used by Burgers in modeling turbulent flow and encountered by others in connection with the theory of shock waves in a real fluid, and the *Korteweg-de Vries equation*

$$u_t + \alpha uu_x + \beta u_{xxx} = 0$$

which is prominent in the theory of nonlinear waves.

Before turning to some concrete examples and their solutions, let us review briefly the subject of eigenfunction expansions (from Part III), which will be needed in most of the problems encountered in this chapter.

26.2. SOME BACKGROUND; THE STURM–LIOUVILLE THEORY

In subsequent sections we will frequently need to expand a given function, say $f(x)$, in the form

$$f(x) = \sum^{\infty} a_n e_n(x) \quad \text{over } a \leq x \leq b \tag{26.8}$$

[1]See, for example, L. D. Landau and E. M. Lifshitz, *Fluid Mechanics*, Addison-Wesley, Reading, Mass., 1959, Chapter 12.

[2]The *one*-dimensional version of (26.7) is the **Airy equation** $y'' + xy = 0$, which has a *turning point* at $x = 0$; that is, y is oscillatory for $x > 0$ and nonoscillatory for $x < 0$ (as we might expect from our knowledge of the simpler equations $y'' + k^2y = 0$, $y'' - k^2y = 0$, where k is a constant).

in much the same way that we expand an "arrow vector" **V** in the form

$$\mathbf{V} = V_1\hat{\mathbf{i}} + V_2\hat{\mathbf{j}} + V_3\hat{\mathbf{k}} \equiv \sum_1^3 V_n\mathbf{e}_n, \tag{26.9}$$

where $\mathbf{e}_1 = \hat{\mathbf{i}}$, $\mathbf{e}_2 = \hat{\mathbf{j}}$, and $\mathbf{e}_3 = \hat{\mathbf{k}}$. To compute the expansion coefficients V_n, we dot $\mathbf{e}_n$ into both sides of (26.9) and use the fact that the $\mathbf{e}_n$'s are mutually orthogonal—that is, $\mathbf{e}_m \cdot \mathbf{e}_n = 0$ for $m \neq n$. Thus

$$\mathbf{V} \cdot \mathbf{e}_m = \sum_1^3 V_n(\mathbf{e}_m \cdot \mathbf{e}_n) = V_m(\mathbf{e}_m \cdot \mathbf{e}_m)$$

so that

$$V_m = \frac{\mathbf{V} \cdot \mathbf{e}_m}{\mathbf{e}_m \cdot \mathbf{e}_m}$$

or reverting to the original notation,

$$V_n = \frac{\mathbf{V} \cdot \mathbf{e}_n}{\mathbf{e}_n \cdot \mathbf{e}_n} \quad \text{for } n = 1, 2, 3. \tag{26.10}$$

Equation (26.10) holds not only for the case where the $\mathbf{e}_n$'s are $\hat{\mathbf{i}}, \hat{\mathbf{j}}, \hat{\mathbf{k}}$ but also for *any* orthogonal set of base vectors. For instance, in a given application, cylindrical base vectors $\mathbf{e}_1 = \hat{\mathbf{e}}_r$, $\mathbf{e}_2 = \hat{\mathbf{e}}_\theta$, $\mathbf{e}_3 = \hat{\mathbf{e}}_z$ or some other set may be more convenient. Of course, $\mathbf{e}_n \cdot \mathbf{e}_n$ in (26.10) is unity if $\mathbf{e}_n$ is of unit length.

Analogously, we can view (26.8) as the expansion of the given $f(x)$ in terms of the "base vectors" $e_n(x)$. Suitable sets of e_n's are generated by the so-called **Sturm–Liouville problem**

$$(py')' + ry + \lambda wy = 0 \quad \text{over } a \leq x \leq b \tag{26.11a}$$

$$\begin{aligned} \alpha y(a) + \beta y'(a) &= 0 \\ \gamma y(b) + \delta y'(b) &= 0, \end{aligned} \tag{26.11b}$$

where $p(x)$, $r(x)$, $w(x)$ are prescribed, as well as the constants α, β, γ, δ. If $p(x)$ vanishes at an endpoint, perhaps both, the boundary condition (26.11b) at that endpoint can be replaced by a condition that y and y' are simply finite there. Also, if it happens that $p(a) = p(b)$, the "separated" conditions (26.11b) (i.e., one condition at a and the other at b) may be replaced by "periodic" conditions $y(a) = y(b)$ and $y'(a) = y'(b)$.

The unknowns in (26.11) are $y(x)$ and λ. Surely $y(x) = 0$ satisfies the homogeneous system (26.11), but the object of the Sturm–Liouville problem is to find *non*trivial solutions. We saw in Section 20.3 that a nontrivial solution exists only if λ is assigned one of a set of special values called **eigenvalues** and that the corresponding nontrivial solutions are called **eigenfunctions.** Analogous to the dot product between two arrow vectors, we defined a dot or *inner* product (f, g) between two functions $f(x)$ and $g(x)$ as

$$(f, g) \equiv \int_a^b f(x)g(x)w(x)\,dx, \tag{26.12}$$

where w is the same w as in (26.11a) and where all functions are assumed to be real valued here. In particular, we found that the Sturm–Liouville eigenfunctions are mutually *orthogonal*—that is, $(y_m, y_n) = 0$ for $m \neq n$.

The principal result (section 20.3) was that virtually any function $f(x)$, defined over $a \leq x \leq b$, can be represented in the form (26.8), where the $e_n(x)$ "base vectors" are the $y_n(x)$ eigenfunctions of a Sturm–Liouville problem. Since the eigenfunctions are orthogonal, we can determine the expansion coefficients a_n by "dotting" $e_m(x)$ into both sides of (26.8)—that is, by multiplying through by $e_m(x)w(x)$ and integrating from a to b. Since $(e_n, e_m) = 0$ for $n \neq m$, this process yields

$$(f, e_m) = \sum^{\infty} a_n(e_n, e_m) = a_m(e_m, e_m),$$

so that $a_m = (f, e_m)/(e_m, e_m)$ or

$$a_n = \frac{(f, e_n)}{(e_n, e_n)}, \tag{26.13}$$

for each n, as in (26.10).

Example 26.1. To illustrate, consider the Sturm–Liouville problem

$$y'' + \lambda y = 0, \qquad y(0) = y(l) = 0, \tag{26.14}$$

where $p(x) = w(x) = 1$ and $r(x) = 0$. Solving, $y = A \sin \sqrt{\lambda}\, x + B \cos \sqrt{\lambda}\, x$. Then $y(0) = 0 = B$ and so B is zero and

$$y(x) = A \sin \sqrt{\lambda}\, x. \tag{26.15}$$

Next, set $y(l) = 0 = A \sin \sqrt{\lambda}\, l$. It follows that either $A = 0$ and/or $\sin \sqrt{\lambda}\, l = 0$. We rule out $A = 0$ because then $y(x) = 0$ per (26.15), whereas we are seeking *non*trivial solutions. Therefore $\sin \sqrt{\lambda}\, l = 0$, and so $\sqrt{\lambda}\, l = n\pi$. Then the eigenvalues are $\lambda_n = n^2\pi^2/l^2$ and, per (26.15), the eigenfunctions are $\sin (n\pi x/l)$ to within an arbitrary scale factor A, which we have set equal to unity.

Thus "virtually any" function $f(x)$ defined over $0 \leq x \leq l$ can be expanded in the form

$$f(x) = \sum_1^{\infty} a_n \sin \frac{n\pi x}{l}, \tag{26.16a}$$

where

$$a_n = \frac{(f, \sin n\pi x/l)}{(\sin n\pi x/l, \sin n\pi x/l)} = \frac{\int_0^l f(x) \sin (n\pi x/l)\, dx}{\int_0^l \sin^2 (n\pi x/l)\, dx}$$

$$= \frac{2}{l}\int_0^l f(x) \sin \frac{n\pi x}{l}\, dx. \tag{26.16b}$$

For instance, if $f(x) = \text{constant} = 100$, then (26.16b) yields $a_n = 400/n\pi$ for n odd and zero for n even, so that

$$f(x) = 100 = \sum_{n=1,3,\ldots}^{\infty} \frac{400}{n\pi} \sin \frac{n\pi x}{l}. \tag{26.17}$$

On the other hand, the (different) Sturm–Liouville problem

$$y'' + \lambda y = 0, \qquad y'(0) = y(l) = 0, \tag{26.18}$$

say, yields the "eigens" $\lambda_n = (2n-1)^2\pi^2/4l^2$ and $y_n(x) = \cos [(2n-1)\pi x/2l]$ and

$$f(x) = \sum_1^{\infty} a_n \cos \frac{(2n-1)\pi x}{2l} \tag{26.19a}$$

where

$$a_n = \frac{(f, \cos (2n-1)\pi x/l)}{(\cos (2n-1)\pi x/2l, \cos (2n-1)\pi x/2l)} = \frac{2}{l}\int_0^l f(x) \cos \frac{(2n-1)\pi x}{2l}\, dx. \tag{26.19b}$$

So observe that for functions defined over a given interval, say $0 \le x \le l$, there are *many* sets of "base vectors" $e_n(x)$ that can be used in (26.8), just as there are many sets of base vectors $\mathbf{e}_n$ in (26.9). How do we know which set to use? In the applications to follow, we will see that the choice of the $e_n(x)$'s is clearly dictated by the context. ▮

26.3. THE DIFFUSION EQUATION

Example 26.2. *Heat Conduction in a Finite Rod.* Suppose that a rod of length l, initially at 100°, has ice (0°) brought in contact with its two ends at time $t = 0$. If the sides of the rod are insulated, we have the one-dimensional, unsteady heat conduction problem

$$\alpha^2 u_{xx} = u_t \tag{26.20a}$$

on the temperature u, together with the initial condition

$$u(x, 0) = 100 \quad \text{over } 0 < x < l \tag{26.20b}$$

and the end conditions

$$u(0, t) = 0, \qquad u(l, t) = 0 \quad \text{for } 0 < t < \infty, \tag{26.20c, d}$$

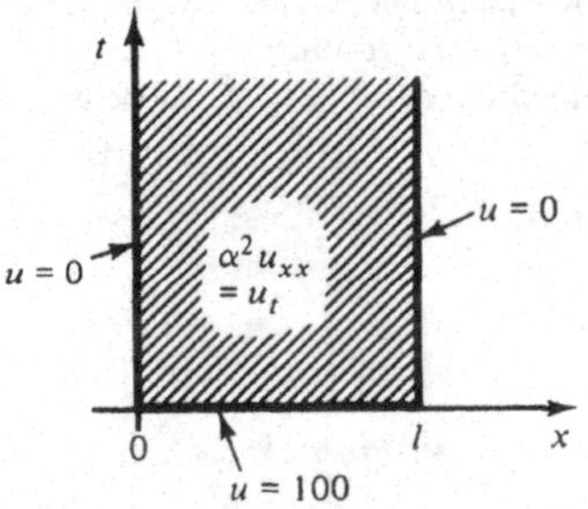

Figure 26.2. The boundary value problem (26.20).

as summarized in Fig. 26.2, and we wish to solve for the temperature history $u(x, t)$.

According to the method of **separation of variables**, we seek a solution in the form

$$u(x, t) = X(x)T(t), \tag{26.21}$$

that is, a function of one independent variable times a function of the other independent variable. Putting (26.21) into (26.20a) yields

$$\alpha^2 X''T = XT'$$

or

$$\frac{X''}{X} = \frac{1}{\alpha^2}\frac{T'}{T}. \tag{26.22}$$

Thus we've "separated the variables." Since the left side is a function only of x, the right side is a function only of t, and x and t are independent variables, it follows[3] that the left- and right-hand sides must each be constant, the *same* constant; that is,

$$\frac{X''}{X} = \frac{1}{\alpha^2}\frac{T'}{T} = \text{constant, say } -k^2. \tag{26.23}$$

The point is that we have reduced the *partial* differential equation (26.20a) to two *ordinary* differential equations

$$X'' + k^2 X = 0, \qquad T' + k^2\alpha^2 T = 0,$$

[3]For instance, $\partial/\partial t$ of (26.22) yields $0 = (\partial/\partial t)(T'/\alpha^2 T) = (d/dt)(T'/\alpha^2 T)$, which implies that $T'/\alpha^2 T$ is a constant. Similarly, $\partial/\partial x$ of (26.22) implies that X''/X is a constant, and according to (26.22), these constants must be identical.

where so far k is arbitrary. Solving, $X = A \sin kx + B \cos kx$ and $T = Ce^{-k^2\alpha^2 t}$; hence

$$u(x, t) = (D \sin kx + E \cos kx)e^{-k^2\alpha^2 t}, \tag{26.24}$$

where $AC \equiv D$ and $BC \equiv E$. Imposing (26.20c),

$$u(0, t) = 0 = Ee^{-k^2\alpha^2 t},$$

so that $E = 0$ and

$$u(x, t) = D \sin kx e^{-k^2\alpha^2 t}. \tag{26.25}$$

Turning to (26.20d), we have

$$u(l, t) = 0 = D \sin kl e^{-k^2\alpha^2 t},$$

and thus $D \sin kl = 0$. If $D = 0$, then (26.25) reduces to $u(x, t) = 0$, which is unacceptable, not because it is "trivial" but because it cannot satisfy the initial condition (26.20b). So $\sin kl = 0$, which implies that the "separation constant" k can have any of the values

$$k = \frac{n\pi}{l} \qquad (n = 1, 2, \ldots).$$

[$n = 0$ is excluded because it contributes nothing; that is, with $k = 0$ the right side of (26.25) is zero. Furthermore, *negative* integers $n = -1, -2, \ldots$ can be omitted with no loss of generality. For instance, if we put $k = 6\pi/l$ and $k = -6\pi/l$ into (26.25), the results differ only in a minus sign out in front, which can be absorbed by the arbitrary constant D.]

Notice that the combination

$$\sin \frac{n\pi x}{l} e^{-n^2\pi^2\alpha^2 t/l^2}$$

satisfies the linear equation (26.20a) and the homogeneous end conditions (26.20c,d) for each $n = 1, 2, \ldots$. Consequently, it follows that any linear combination

$$u(x, t) = \sum_1^\infty D_n \sin \frac{n\pi x}{l} e^{-n^2\pi^2\alpha^2 t/l^2} \tag{26.26}$$

does, too. [You may be worried about interchanging the order of differentiation and summation in verifying (26.20a), but it is best to defer such questions until the (tentative) solution is in hand.]

Finally, we impose the initial condition on (26.26):

$$u(x, 0) = 100 = \sum_1^\infty D_n \sin \frac{n\pi x}{l}. \tag{26.27}$$

Is it *possible* to find D_ns so as to satisfy (26.27); if so *how*? Observe that $u(0, t) = X(0)T(t) = 0$ implies $X(0) = 0$ and that $u(l, t) = X(l)T(t) = 0$ implies $X(l) = 0$. So $X(x)$ satisfies

$$X'' + k^2 X = 0, \qquad X(0) = X(l) = 0. \tag{26.28}$$

Recalling Example 26.1, we see that (26.28) is the Sturm–Liouville problem (26.14), where "λ" is k^2. Therefore $u(x, 0) = 100$ *can* be expanded in terms of the $\sin (n\pi x/l)$ eigenfunctions, and the D_ns in (26.27) are computed as

$$D_n = \left(100, \sin \frac{n\pi x}{l}\right) \Big/ \left(\sin \frac{n\pi x}{l}, \sin \frac{n\pi x}{l}\right)$$

$$= \frac{2}{l} \int_0^l 100 \sin \frac{n\pi x}{l}\, dx = \begin{cases} \dfrac{400}{n\pi}, & n \text{ odd} \\ 0, & n \text{ even}, \end{cases} \tag{26.29}$$

as in Example 26.1. Hence

$$u(x, t) = \sum_{n=1,3,\ldots}^{\infty} \frac{400}{n\pi} \sin \frac{n\pi x}{l} e^{-n^2\pi^2\alpha^2 t/l^2}. \tag{26.30}$$

COMMENT 1. We emphasize that the expression $-k^2$ for the separation constant in (26.23), rather than simply k, is a matter of convenience and foresight; for instance, anticipating taking its square root, we might as well start with k^2. The minus sign is included for two reasons. First, the decay $T = Ce^{-k^2\alpha^2 t}$ looks reasonable, whereas unbounded exponential *growth* $T = Ce^{k^2\alpha^2 t}$ does not. Secondly, the $-k^2$ yields sines and cosines in the x direction, which are capable of handling the end conditions $X(0) = X(l) = 0$, whereas $+k^2$ would yield the less convenient sinh's and cosh's. But even if we used k instead of $-k^2$, the final result would be the same; it would simply turn out that k is negative and a perfect square (Exercise 26.1).

COMMENT 2. It is important to save the initial condition (26.20b) for last because we need to build up the form (26.26) before we're in a position to cope with it.

COMMENT 3. Observe how the relevant Sturm–Liouville system is built right in and is there when we need it.

COMMENT 4. Rigorous verification that (26.30) does indeed satisfy the differential equation (26.20a) is left for Exercise 26.2.

COMMENT 5. Verification of the *uniqueness* of the solution to (26.20) is left for Exercise 26.3. It is sometimes argued from a physical point of view that "of course, the desired $u(x, t)$ exists and is unique, so why bother with these questions?" Generally, however, that query misses the point—whether our *mathematical formulation* of the physical situation [as given in this case by (26.30)] is appropriate—that is, whether our problem is properly posed.

COMMENT 6. Notice that for appreciable values of t the series (26.30) is rapidly convergent thanks to the $e^{-n^2(\cdots)}$. Thus the higher harmonics die out quickly, leaving

$$u(x, t) \sim \frac{400}{\pi} \sin \frac{\pi x}{l} e^{-\pi^2\alpha^2 t/l^2} \quad \text{as } t \longrightarrow \infty. \tag{26.31}$$

On the other hand, for *small* t (26.30) converges with painful slowness; and even though it is "the exact solution," it is not especially useful. For instance, for $t = 0$ the series is only conditionally convergent. A more appropriate form for computation is needed, a topic that will be pursued in Example 26.4.

COMMENT 7. The separation method works frequently but not always. For the **biharmonic equation**

$$\begin{aligned} \nabla^4 u &= \left(\frac{\partial^2}{\partial x^2} + \frac{\partial^2}{\partial y^2}\right)\left(\frac{\partial^2}{\partial x^2} + \frac{\partial^2}{\partial y^2}\right)u \\ &= u_{xxxx} + 2u_{xxyy} + u_{yyyy}, \end{aligned} \tag{26.32}$$

for example, $u = X(x)Y(y)$ yields

$$X''''Y + 2X''Y'' + XY'''' = 0$$

or

$$\frac{X''''}{X} + 2\frac{X''Y''}{XY} + \frac{Y''''}{Y} = 0,$$

and we see that the mixed term $X''Y''/XY$ presents a problem. Besides being able to separate the given PDE, we also need the boundary of the region under consideration to be made up of constant-coordinate lines. In the present example these lines were $x = 0$, $x = l$, and $t = 0$.[4] ▮

Example 26.3. Let us modify the previous problem to read

$$\alpha^2 u_{xx} = u_t \tag{26.33a}$$

$$u(x, 0) = 0 \tag{26.33b}$$

$$u(0, t) = 0 \tag{26.33c}$$

$$u(l, t) = 100. \tag{26.33d}$$

Seeking $u = X(x)T(t)$, we obtain (26.24) as before. Then (26.33c) gives $E = 0$, and (26.33d) becomes

$$u(l, t) = 100 = D \sin kle^{-k^2\alpha^2 t}, \tag{26.34}$$

which is not possible for any choice of D and k. It is NOT correct to solve for D as $D = 100e^{k^2\alpha^2 t}/\sin kl$, since D was assumed to be a *constant* in deriving (26.24)! [Deferring (26.33d) until after (26.33b) is applied does not help.] The difficulty is due to the inhomogeneity of (26.33d).

Consider what happens as $t \longrightarrow \infty$. If a steady state is approached, then $u_t \longrightarrow 0$ and (26.33a) gives $u = Ax + B$ or, since u is 0 at $x = 0$ and 100 at $x = l$,

$$u \sim \frac{100x}{l} \quad \text{as } t \longrightarrow \infty. \tag{26.35}$$

Let us therefore decompose u into a *steady-state* part plus a *transient* part, where the transient part is expressed in product form

$$u(x, t) = \frac{100x}{l} + X(x)T(t). \tag{26.36}$$

Putting it into (26.33a), the $100x/l$ simply drops out; consequently, X and T turn out to be the same as before, and we have

$$u(x, t) = \frac{100x}{l} + (D \sin kx + E \cos kx)e^{-k^2\alpha^2 t}. \tag{26.37}$$

Then

$$u(0, t) = 0 = 0 + Ee^{-k^2\alpha^2 t}$$

and so $E = 0$; moreover,

$$u(l, t) = 100 = 100 + D \sin kle^{-k^2\alpha^2 t} \tag{26.38}$$

and thus the previously disastrous 100 now cancels, and (26.38) implies that $k = n\pi/l$. [Hence the $100x/l$ steady-state term in (26.36) effectively "homogenizes" the end conditions for us—that is, $u(x, t) - 100x/l = 0$ at $x = 0$ and $x = l$.]

With the help of superposition, then,

$$u(x, t) = \frac{100x}{l} + \sum_{1}^{\infty} D_n \sin \frac{n\pi x}{l} e^{-n^2\pi^2\alpha^2 t/l^2},$$

[4]A number of papers on this subject by P. Moon and D. E. Spencer appeared in 1952 and 1953; see, for example, "Separability in a Class of Coordinate Systems," *J. Franklin Institute*, Vol. 254, 1952, pp. 227–242. A second form of separability called "*R*-separability" is discussed in this paper, as well as the "simple separability" discussed here.

and the initial condition becomes

$$u(x, 0) = 0 = \frac{100x}{l} + \sum_1^\infty D_n \sin\frac{n\pi x}{l}$$

or

$$-\frac{100x}{l} = \sum_1^\infty D_n \sin\frac{n\pi x}{l}.$$

Then

$$D_n = \left(-100\frac{x}{l}, \sin\frac{n\pi x}{l}\right)\Big/\left(\sin\frac{n\pi x}{l}, \sin\frac{n\pi x}{l}\right) = \text{etc.} = (-1)^n\frac{200}{n\pi},$$

and

$$u(x, t) = \frac{100x}{l} + \sum_1^\infty (-1)^n \frac{200}{n\pi} \sin\frac{n\pi x}{l} e^{-n^2\pi^2\alpha^2 t/l^2}. \qquad (26.39)$$

Alternatively, suppose that we *Laplace transform* (26.33a) on the t variable—that is,

$$\int_0^\infty \alpha^2 \frac{\partial^2 u}{\partial x^2} e^{-st}\, dt = \int_0^\infty \frac{\partial u}{\partial t} e^{-st}\, dt$$

or

$$\alpha^2 \frac{d^2}{dx^2}\int_0^\infty ue^{-st}\, dt = u(x, t)e^{-st}\Big|_{t=0}^{\infty} + s\bar{u}(x, s)$$

or

$$\alpha^2 \bar{u}_{xx} - s\bar{u} = 0, \qquad (26.40)$$

since $u(x, 0) = 0$. Thus we've reduced the PDE (26.33a) to the ODE (ordinary differential equation) (26.40). Solving,

$$\bar{u} = Ae^{\sqrt{s}\,x/\alpha} + Be^{-\sqrt{s}\,x/\alpha}, \qquad (26.41)$$

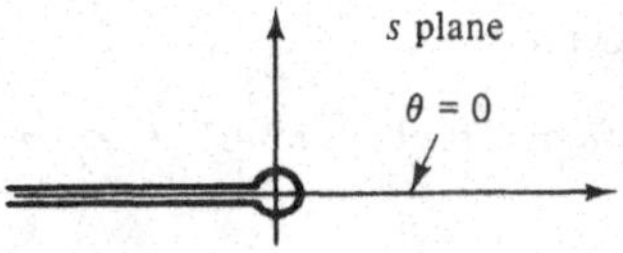

Figure 26.3. Branch cut for $\sqrt{s}$.

where $\sqrt{s}$ is defined by the cut shown in Fig. 26.3, say. Transforming the end conditions (26.33c,d), we have

$$\bar{u}\,|_{x=0} = 0 \qquad (26.42a)$$

and

$$\bar{u}\,|_{x=l} = \frac{100}{s}. \qquad (26.42b)$$

Applying these results to (26.41), we can evaluate A and B (which are allowed to depend on the parameter s) and find that

$$\bar{u} = \frac{100}{s}\frac{\sinh(\sqrt{s}\,x/\alpha)}{\sinh(\sqrt{s}\,l/\alpha)}. \qquad (26.43)$$

Inverting,

$$u(x, t) = \frac{100}{2\pi i}\int_{\gamma-i\infty}^{\gamma+i\infty} \frac{1}{s}\frac{\sinh(\sqrt{s}\,x/\alpha)}{\sinh(\sqrt{s}\,l/\alpha)} e^{st}\, ds. \qquad (26.44)$$

The integration *looks* difficult, partly because of the $\sqrt{s}$'s. However, $\sinh(\sqrt{s}\,x/\alpha)/\sinh(\sqrt{s}\,l/\alpha)$ is an even function of $\sqrt{s}$, which means that only even powers of $\sqrt{s}$ appear in an expansion about $s = 0$. It follows that the square roots disappear, and so $s = 0$ is *not* a branch point of the integrand. In fact, the integrand has only simple poles at $s = 0$ and $s = -n^2\pi^2\alpha^2/l^2$ for $n = 1, 2, \ldots,$ with the following residues (Exercise 26.4):

$$(\text{Res. } @\ s = 0) = \frac{100x}{2\pi i l} \qquad (26.45a)$$

$$\left(\text{Res. } @\ s = -\frac{n^2\pi^2\alpha^2}{l^2}\right) = (-1)^n \frac{100}{n\pi^2 i} \sin\frac{n\pi x}{l} e^{-n^2\pi^2\alpha^2 t/l^2}. \qquad (26.45b)$$

Thus, if we close the straight line contour in (26.44) with a large semicircle to the left (as in Example 15.8) and apply the residue theorem [equation (15.4)], then

$$u(x, t) = 2\pi i \sum \text{Res.} = \frac{100x}{l} + \sum_{1}^{\infty} (-1)^n \frac{200}{n\pi} \sin \frac{n\pi x}{l} e^{-n^2\pi^2\alpha^2 t/l^2} \tag{26.46}$$

as before!

Transform methods provide a powerful line of approach to (linear) PDEs and may work even if separation of variables fails. In addition, the inversion integral [in this case, (26.44)] provides a starting point for the extraction of asymptotic results [in this case, for large time and small time with the small-time case being of great interest because of the already-mentioned slow convergence of the series in (26.46)].[5] ∎

Example 26.4. *Heat Conduction in an Infinite Rod.* This time suppose that the rod is infinite ($-\infty < x < \infty$) and consider the problem

$$\alpha^2 u_{xx} = u_t, \tag{26.47a}$$

$$u(x, 0) = f(x), \tag{26.47b}$$

where the initial temperature $f(x)$ is given.

Recall that for the *finite* rod the x-wise frequencies were $n\pi/l$. As $l \to \infty$, the frequency spectrum tends to a continuum, and the discrete Fourier series solution gives way to a Fourier integral. This situation suggests that separation of variables should give way to a *Fourier transform* on x. (Alternatively, we could Laplace transform on t, as in Example 26.3.) Thus

$$\alpha^2 \int_{-\infty}^{\infty} \frac{\partial^2 u}{\partial x^2} e^{-i\omega x}\, dx = \int_{-\infty}^{\infty} \frac{\partial u}{\partial t} e^{-i\omega x}\, dx$$

or

$$\alpha^2 \left\{ \frac{\partial u}{\partial x} e^{-i\omega x} \Big|_{-\infty}^{\infty} + i\omega \int_{-\infty}^{\infty} \frac{\partial u}{\partial x} e^{-i\omega x}\, dx \right\} = \frac{d}{dt}\hat{u}. \tag{26.48a}$$

Tentatively assuming that $\partial u/\partial x \to 0$ as $|x| \to \infty$, the boundary term drops out. Integrating by parts again, we find that

$$\alpha^2 i\omega \left\{ u e^{-i\omega x} \Big|_{-\infty}^{\infty} + i\omega\hat{u} \right\} = \frac{d}{dt}\hat{u}; \tag{26.48b}$$

and if we assume further that $u \to 0$ as $|x| \to \infty$, then

$$\frac{d\hat{u}}{dt} = -\alpha^2\omega^2\hat{u}. \tag{26.49}$$

At this point we could apply a second transform, a Laplace transform on t, to reduce (26.49) to an *algebraic* equation, but why bother, since (26.49) is already quite simple and easily yields

$$\hat{u} = Ae^{-\alpha^2\omega^2 t}. \tag{26.50}$$

Transforming the initial condition $u(x, 0) = f(x)$,

$$\hat{u}|_{t=0} = \hat{f}(\omega), \tag{26.51}$$

[5] See H. Carslaw and J. Jaeger, *Operational Methods in Applied Mathematics*, 2nd ed., Oxford University Press, London, 1948.

and applying it to (26.50) gives $A = \hat{f}(\omega)$. Inverting, we have

$$u(x, t) = \frac{1}{2\pi}\int_{-\infty}^{\infty} \{\hat{f}(\omega)e^{-\alpha^2\omega^2 t}\}e^{i\omega x}\,d\omega$$

$$= \frac{1}{2\pi}\int_{-\infty}^{\infty} e^{i\omega x-\alpha^2\omega^2 t}\hat{f}(\omega)\,d\omega. \tag{26.52}$$

This result is fine; but since we're given f and not $\hat{f}$, it would be preferable to express the answer in terms of f directly. So

$$u(x, t) = \frac{1}{2\pi}\int_{-\infty}^{\infty} e^{i\omega x-\alpha^2\omega^2 t}\,d\omega \int_{-\infty}^{\infty} f(\xi)e^{-i\omega\xi}\,d\xi \tag{26.53a}$$

$$u(x, t) = \frac{1}{2\pi}\int_{-\infty}^{\infty} f(\xi)\,d\xi \int_{-\infty}^{\infty} e^{i\omega(x-\xi)-\alpha^2\omega^2 t}\,d\omega, \tag{26.53b}$$

where the dummy variable ξ is used to avoid confusion with x. We find (Exercise 26.5) that[6]

$$\frac{1}{2\pi}\int_{-\infty}^{\infty} e^{i\omega(x-\xi)-\alpha^2\omega^2 t}\,d\omega = \frac{e^{-(\xi-x)^2/4\alpha^2 t}}{2\alpha\sqrt{\pi t}} \equiv U(\xi - x, t), \tag{26.54}$$

so that the solution is

$$u(x, t) = \int_{-\infty}^{\infty} f(\xi)U(\xi - x, t)\,d\xi. \tag{26.55}$$

COMMENT 1. Observe that the inversion of $\hat{u}$ could have been carried out more directly with the help of the Fourier convolution theorem [(6.23)] and the Fourier Transform Table A.1 (on the inside front cover of this book); that is,

$$u(x, t) = F^{-1}\{\hat{f}(\omega)e^{-\alpha^2\omega^2 t}\}$$

$$= f(x) * \frac{e^{-x^2/4\alpha^2 t}}{2\alpha\sqrt{\pi t}}$$

$$= \int_{-\infty}^{\infty} f(\xi)\frac{e^{-(\xi-x)^2/4\alpha^2 t}}{2\alpha\sqrt{\pi t}}\,d\xi$$

$$= \int_{-\infty}^{\infty} f(\xi)U(\xi - x, t)\,d\xi$$

as before.

COMMENT 2. As a check on (26.55), let us set $f(\xi) = 1$, since we know that the solution should simply be $u(x, t) \equiv 1$:

$$u(x, t) = \int_{-\infty}^{\infty} \frac{e^{-(\xi-x)^2/4\alpha^2 t}}{2\alpha\sqrt{\pi t}}\,d\xi, \quad \left(\frac{\xi - x}{2\alpha\sqrt{t}} \equiv \zeta\right)$$

$$= \frac{1}{\sqrt{\pi}}\int_{-\infty}^{\infty} e^{-\zeta^2}\,d\zeta = 1. \tag{26.56}$$

In fact, we had no right to expect (26.55) to give the correct result in this case because $\hat{f}$ in (26.51) *does not exist* for $f(x) = 1$. Moreover, the assumption that $u \longrightarrow 0$ as

[6]We write $U(\xi - x, t)$ rather than $U(\xi, x, t)$ because the former is more informative. We call $U(\xi - x, t)$ a *difference kernel.*

$|x| \longrightarrow \infty$ in obtaining (26.49) does not hold. The key lies in the interchange of the order of integration in (26.53); for instance, although (26.53a) is meaningless for $f(\xi) = 1$ (since the ξ integral is divergent), (26.53b) is all right because $U(\xi - x, t)$, which arises from the ω integration, decays very strongly as $|\xi| \longrightarrow \infty$ and helps out in the ξ integration. Of course, the most reasonable approach is to obtain the tentative solution (26.55) working rather formally, as we did, and then to test (26.55) directly to see whether it satisfies the requirements (26.47); this procedure will probably mean placing some restrictions on the allowable f's.

To illustrate, let us verify (26.47b). First, let us be a little more careful about the problem statement. Is the t domain $0 \leq t < \infty$? If so, then in order for $k^2 u_{xx} = u_t$ to hold at $t = 0$, we will need $f(x)$ to be twice differentiable. But this condition is too restrictive, since in applications we generally wish to allow for discontinuous boundary data. Let us therefore take the t domain to be $0 < t < \infty$. In this case, the initial condition $u(x, 0) = f(x)$ needs some repair because $u(x, t)$ is only available in $0 < t < \infty$, not at $t = 0$. One reasonable solution is to replace (26.47b) by the connection $\lim u(x, t) = f(x)$ as $t \longrightarrow 0$. Thus let us restate (26.47) as

$$\alpha^2 u_{xx} = u_t \qquad (-\infty < x < \infty,\ 0 < t < \infty) \tag{26.57a}$$

$$\lim_{t \to 0} u(x, t) = f(x). \tag{26.57b}$$

Next, observe carefully that the kernel $U(\xi - x, t)$ is a *gaussian distribution* centered at $\xi = x$ (which suggests that diffusion may somehow be an essentially statistical phenomenon). Its area is unity for all t [recall (26.56)], and it becomes more and more peaked as $t \longrightarrow 0$ (Fig. 26.4). In fact, it follows from Theorem 4.3 that $U(\xi - x, t)$ is a *delta sequence* as $t \longrightarrow 0$, and so

$$\lim_{t \to 0} \int_{-\infty}^{\infty} f(\xi) U(\xi - x, t)\, d\xi = f(x) \tag{26.58}$$

as desired!

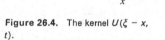

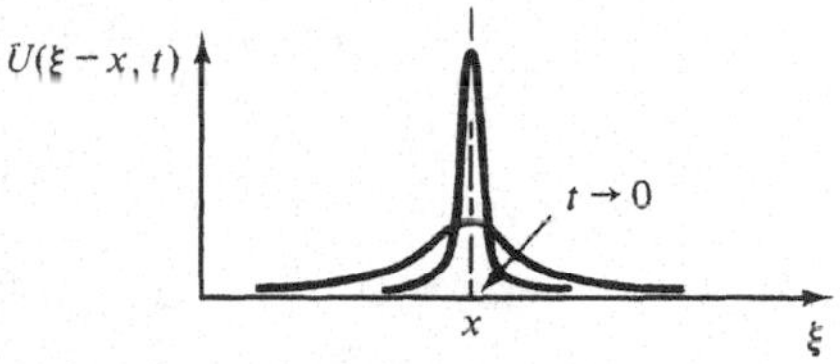

Figure 26.4. The kernel $U(\xi - x, t)$.

COMMENT 3. Looking for a more *physical* interpretation of U, suppose that we let $f(x) = \delta(x - x_0)$. Then (26.55) gives $u(x, t) = U(x_0 - x, t) = U(x - x_0, t)$; that is, the kernel $U(x - x_0, t)$ may be regarded as the temperature response to a δ-function input at $x = x_0$, $t = 0$ (which exposes the *superposition* nature of (26.55)].

COMMENT 4. Next, suppose that $f(x) = H(x)$—that is, a unit step function at $x = 0$. Then (26.55) gives (Exercise 26.6)

$$u(x, t) = \frac{1}{2} + \frac{1}{2} \operatorname{erf}\left(\frac{x}{2\alpha\sqrt{t}}\right), \tag{26.59}$$

as sketched in Fig. 26.5. For each $t > 0$ this is an *analytic* function of x in spite of the fact that $f(x) = H(x)$ is discontinuous![7] Thus diffusion is a *smoothing* process.

Do you recall the difficulty of the slow convergence of the series (26.30) for small t, discussed there in Comment 6? We are now in a position to handle that

[7]To convince yourself, recall the definition $(2/\sqrt{\pi}) \int_0^z e^{-\xi^2}\, d\xi = \operatorname{erf} z$ and hence differentiate (26.59) once or twice with respect to x.

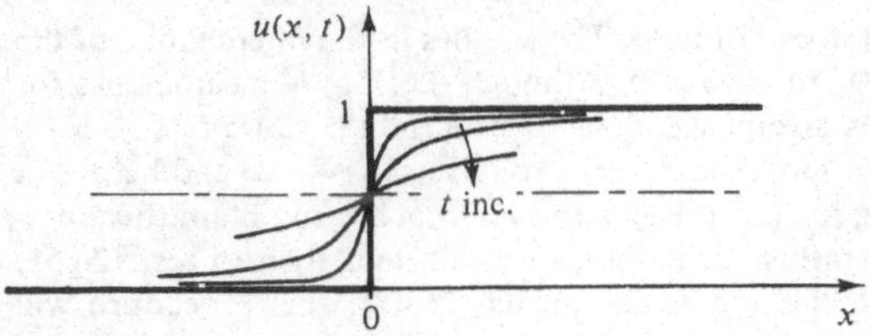

Figure 26.5. Response (26.59) to $f(x) = H(x)$.

difficulty. First, observe that although (26.30) was only intended to apply over $0 \le x \le l$, it can be regarded as the solution for the extended domain $-\infty < x < \infty$, where the initial temperature is the square wave shown in Fig. 26.6(a).[8] Then, for small t, the temperature profile over $0 \le x \le l$ is obviously well approximated by the solution corresponding to the pulselike initial temperature $P(x)$ shown in Fig. 26.6(b), since it takes time for "incorrect information" to arrive at a point in $0 \le x \le l$ from points to the right of $2l$ and to the left of $-l$. Express

$$P(x) = 200[H(x) - H(x - l)] - 100. \tag{26.60}$$

Since the response to a uniform step $H(x)$ is given by (26.59), it follows from linearity of the differential operator that the response to $P(x)$, and hence the small-t behavior of $u(x, t)$, is

$$\begin{aligned} u(x, t) &\sim 200\left[\frac{1}{2} + \frac{1}{2}\operatorname{erf}\left(\frac{x}{2\alpha\sqrt{t}}\right)\right] - 200\left[\frac{1}{2} + \frac{1}{2}\operatorname{erf}\left(\frac{x-l}{2\alpha\sqrt{t}}\right)\right] - 100 \\ &= 100\left[\operatorname{erf}\left(\frac{x}{2\alpha\sqrt{t}}\right) - \operatorname{erf}\left(\frac{x-l}{2\alpha\sqrt{t}}\right) - 1\right] \end{aligned} \tag{26.61}$$

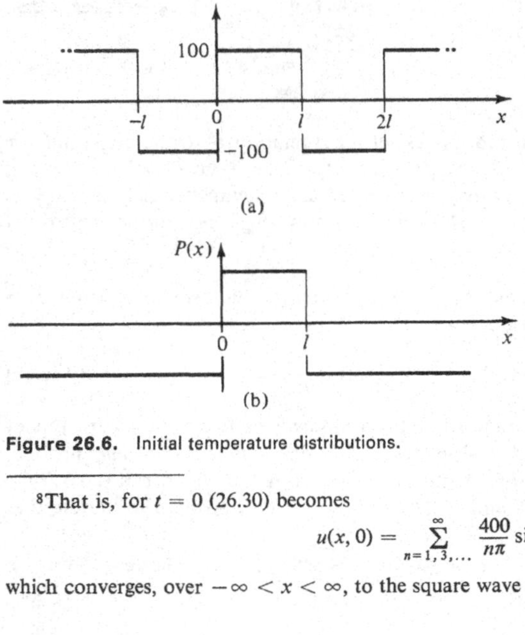

Figure 26.6. Initial temperature distributions.

[8]That is, for $t = 0$ (26.30) becomes

$$u(x, 0) = \sum_{n=1,3,\ldots}^{\infty} \frac{400}{n\pi} \sin\frac{n\pi x}{l},$$

which converges, over $-\infty < x < \infty$, to the square wave shown in the figure.

as $t \longrightarrow 0$. Of course, we need not settle for the approximate form $P(x)$; we can express the entire square wave of Fig. 26.6(a) as an infinite series of Heaviside functions, and (26.61) is the solution corresponding to just the leading term (Exercise 26.7). For small t we can expect the series to converge very rapidly, so that only the leading term or two should suffice.

COMMENT 5. It is interesting to see what (26.55) tells us about *negative* time. Then U is a *positive* exponential; and unless f is of a very special form, the integral diverges and so makes no sense for all $t < 0$. Yet this result is not so surprising in view of the way that diffusion is a *smoothing* process, as mentioned in Comment 4. For instance, if $f(x)$ is kinky (or even more mildly nonanalytic), how can we expect some temperature distribution to exist at $t = -T$, say, that will produce $f(x)$ T units of time later?

COMMENT 6. The result (26.55) for the infinite rod can also be used to provide the solution to the *semi*-infinite problem

$$\alpha^2 u_{xx} = u_t \qquad (0 \leq x < \infty,\ 0 < t < \infty) \tag{26.62a}$$

$$\lim_{t \to 0} u(x, t) = f(x) \tag{26.62b}$$

$$u(0, t) = 0. \tag{26.62c}$$

Specifically, if we extend $f(x)$ over $-\infty < x < \infty$ so as to be antisymmetric [Fig. 26.7(b)], then the initial antisymmetry will ensure that $u(0, t) = 0$ for all $t > 0$

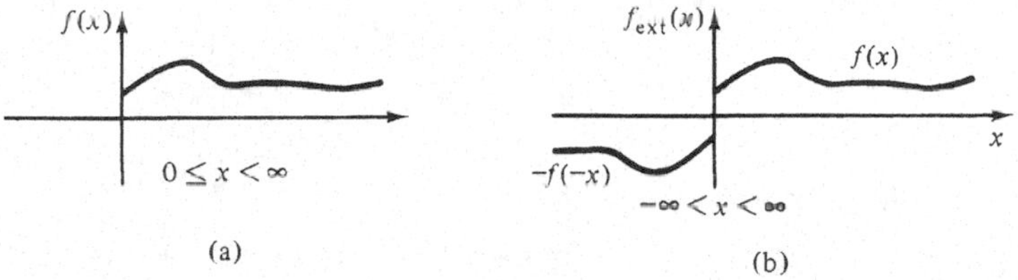

Figure 26.7. Antisymmetric extension of $f(x)$.

[Exercise 26.8(a)], and the solution to (26.62) will be the same as the solution to the extended problem over $0 \leq x < \infty$. Thus

$$\begin{aligned} u(x, t) &= \int_{-\infty}^{\infty} f_{\text{ext}}(\xi) U(x - \xi, t)\, d\xi \\ &= \int_{-\infty}^{0} \{-f(-\xi)\} U(x - \xi, t)\, d\xi + \int_{0}^{\infty} f(\xi) U(x - \xi, t)\, d\xi \\ &= -\int_{0}^{\infty} f(\xi') U(x + \xi', t)\, d\xi' + \int_{0}^{\infty} f(\xi) U(x - \xi, t)\, d\xi \\ &= \int_{0}^{\infty} f(\xi)[U(x - \xi, t) - U(x + \xi, t)]\, d\xi \end{aligned} \tag{26.63}$$

is the solution of (26.62), where the replacement of ξ' by ξ is certainly permissible because ξ' is only a dummy variable of integration. The extension indicated in Fig. 26.7 is an example of the **method of images** with the $x < 0$ portion known as the *image system*.

COMMENT 7. Finally, recall from Comment 3 that $u(x, t) = U(x - x_0, t)$ is the temperature at x, t due to an initial δ-function input $f(x) = \delta(x - x_0)$ at x_0. From Fig. 26.4 we see that for $t > 0$ but *arbitrarily small*, $u(x, t) \neq 0$ for all x. That is, the heat has diffused with an *infinite velocity*! Since propagation of energy at an infinite velocity is impossible, it appears that the diffusion equation can be correct only after a very small time has elapsed. Insofar as practical calculations are concerned, this deficiency is inconsequential and is generally ignored. Interestingly, however, in order to patch things up, the diffusion equation can be modified to read[9]

$$u_{xx} = \frac{1}{\alpha^2}u_t + \frac{1}{c^2}u_{tt}, \tag{26.64}$$

where c is the acoustic velocity in the material. Strictly speaking, (26.64) is *hyperbolic* for all x and t because of the addition of the u_{tt} term and hence is associated with a finite speed of propagation (as we will soon see). However, the $1/c^2$ factor is *extremely small*. So except for the initial instants when u_{tt} is large, the last term becomes negligible, and the character of (26.64) is dominated by the diffusion terms. ∎

Example 26.5. *Unsteady Conduction in a Disk.* Consider next the plane, unsteady heat conduction in an infinite circular cylinder of radius a. By "plane" we mean that the temperature u is independent of the axial variable z. Equivalently, then, we have a thin "coin" whose top and bottom faces are insulated.

Suppose that the initial temperature is $u = f(r)$ over $0 \leq r < a$ and that we quench the edge ($r = a$) to 0° beginning at $t = 0$. Then

$$\alpha^2\nabla^2 u = u_t \tag{26.65a}$$

$$u = f(r) \text{ at } t = 0 \quad \text{for } 0 \leq r < a \tag{26.65b}$$

$$u = 0 \text{ at } r = a \quad \text{for } t \geq 0. \tag{26.65c}$$

In cylindrical coordinates,

$$\nabla^2 u = u_{rr} + \frac{1}{r}u_r + \frac{1}{r^2}u_{\theta\theta} + u_{zz},$$

but $u_{zz} = 0$ by assumption, and apparently $u_{\theta\theta} = 0$ because of the isotropic nature of (26.65a) and the axisymmetry of the region and conditions (26.65b,c); or if you prefer, we can *tentatively* assume $u_{\theta\theta} = 0$ and then see if the resulting solution satisfies the full system (26.65), including the $u_{\theta\theta}$ term. Thus (26.65) becomes

$$u_{rr} + \frac{1}{r}u_r = \frac{1}{\alpha^2}u_t \tag{26.66a}$$

$$u(r, 0) = f(r) \qquad (r < a) \tag{26.66b}$$

$$u(a, t) = 0 \qquad (t \geq 0). \tag{26.66c}$$

Seeking $u(r, t) = R(r)T(t)$, we obtain

$$rR'' + R' + k^2 rR = 0 \tag{26.67a}$$

$$T' + \alpha^2 k^2 T = 0, \tag{26.67b}$$

---

[9]See P. M. Morse and H. Feshbach, *Methods of Theoretical Physics*, Part I, McGraw-Hill, New York, 1953, pp. 865–869. See also the singular perturbation approach (with $\epsilon = 1/c^2$) in A. Nayfeh, *Perturbation Methods*, Wiley, New York, 1973, p. 157, Exercise 4.18.

where the separation constant is $-k^2$. Thus (recall Section 22.4)

$$R = AJ_0(kr) + BY_0(kr) \tag{26.68a}$$

$$T = Ce^{-\alpha^2k^2t}. \tag{26.68b}$$

Then $Y_0 \longrightarrow -\infty$ as $r \longrightarrow 0$, whereas certainly $u(r, t)$ is to be bounded. So we set $B = 0$. With $AC \equiv D$, then,

$$u(r, t) = R(r)T(t) = DJ_0(kr)e^{-\alpha^2k^2t}. \tag{26.69}$$

Applying (26.66c),

$$u(a, t) = 0 = DJ_0(ka)e^{-\alpha^2k^2t}, \tag{26.70}$$

so that ka must coincide with (any of) the zeros of the Bessel function J_0, which we will denote as z_n for $n = 1, 2, \ldots$. Hence acceptable values of k are

$$k_n = \frac{z_n}{a}, \tag{26.71}$$

and by superposition

$$u(r, t) = \sum_1^\infty D_nJ_0(k_nr)e^{-\alpha^2k_n^2t}. \tag{26.72}$$

Finally, the initial condition (26.66b) becomes

$$u(r, 0) = f(r) = \sum_1^\infty D_nJ_0(k_nr). \tag{26.73}$$

Again (as in Example 26.2), we ask: Is it *possible* to find D_ns so as to satisfy (26.73); and if so, *how*? Consider the eigenvalue problem on R,

$$\begin{gathered} \frac{d}{dr}\left(r\frac{dR}{dr}\right) + \lambda rR = 0 \qquad (\lambda \equiv k^2) \\ R(0) = \text{finite}, \qquad R(a) = 0. \end{gathered} \tag{26.74}$$

We see from our discussion in Section 26.2 that this system is of Sturm Liouville type, where "$p(r)$" = "$w(r)$" = r in (26.11a). The eigenfunctions $J_0(k_nr)$ are orthogonal with respect to the inner product

$$(f, g) = \int_0^a f(r)g(r)r\,dr, \tag{26.75}$$

so that the expansion coefficients in (26.73) are given by

$$D_n = \frac{(f(r), J_0(k_nr))}{(J_0(k_nr), J_0(k_nr))} = \frac{\int_0^a f(r)J_0(k_nr)r\,dr}{\int_0^a [J_0(k_nr)]^2r\,dr}$$

or

$$D_n = \frac{2}{a^2[J_1(k_na)]^2}\int_0^a f(r)J_0(k_nr)r\,dr, \tag{26.76}$$

since (Exercise 26.9)

$$\int_0^a [J_0(k_nr)]^2r\,dr = \frac{a^2[J_1(k_na)]^2}{2}. \tag{26.77}$$

The D_n's may be computed from (26.76) once $f(r)$ is specified, by numerical integration if necessary. Therefore the desired solution is given by (26.72), where the k_n's are defined by (26.71) and the D_n's are given by (26.76). ∎

Recall that the equation of (small) lateral motion $y(x, t)$ of a taut string of mass $\rho(x)$ per unit length, under tension τ, was shown in Section 10.5 to be

$$\tau y_{xx} - \rho(x) y_{tt} = 0. \tag{26.78}$$

Example 26.6. Suppose that $\rho(x) =$ constant, that the string extends from $x = 0$ to $x = l$, that it is held fixed at its two ends, and that it is set in motion through some initial displacement $y(x, 0) = F(x)$. Then the boundary value problem becomes

$$y_{xx} = \frac{1}{a^2} y_{tt} \qquad \left(a^2 \equiv \frac{\tau}{\rho}\right) \tag{26.79a}$$

$$y(0, t) = y(l, t) = 0 \tag{26.79b}$$

$$y_t(x, 0) = 0 \tag{26.79c}$$

$$y(x, 0) = F(x), \tag{26.79d}$$

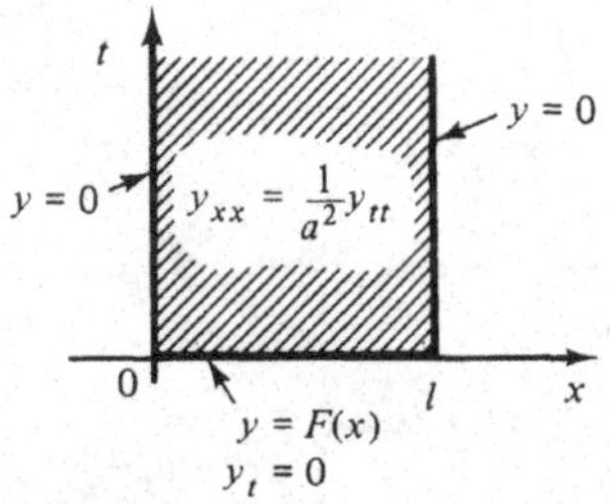

Figure 26.8. Finite string problem (26.79).

as summarized in Fig. 26.8. We expect, on physical grounds, that we will need to know not only the initial displacement but also the initial velocity y_t, and, for definiteness, we have chosen $y_t(x, 0) = 0$. Besides physical arguments, we note that (26.79a) is second order in both x and t, and so we expect that two x conditions [the end conditions (26.79b)] and two t conditions will be appropriate to specify a unique solution.

Pursuing a separation-of-variables approach, we seek $y(x, t) = X(x)T(t)$ and arrive without difficulty at the result

$$y(x, t) = \sum_1^\infty F_n \sin \frac{n\pi x}{l} \cos \frac{n\pi a t}{l}, \tag{26.80a}$$

where

$$F_n = \frac{(F, \sin n\pi x/l)}{(\sin n\pi x/l, \sin n\pi x/l)}$$

$$= \frac{2}{l} \int_0^l F(x) \sin \frac{n\pi x}{l}\, dx. \tag{26.80b}$$

The omitted steps are left for Exercise 26.16. ▮

Suppose that we introduce the simple change of independent variables

$$\xi = x + at, \qquad \eta = x - at \tag{26.81}$$

into the wave equation (26.79a). Using some chain differentiation, it is not hard to show (Exercise 26.17) that (26.79a) reduces to

$$y_{\xi\eta} = 0, \tag{26.82}$$

which is easily solved by integration. Integrating first on η,

$$y_\xi = h(\xi), \tag{26.83}$$

where the "constant" of integration h may depend on ξ, since ξ was held fixed in integrating partially on η; if unhappy about this claim, note that $\partial/\partial\eta$ of (26.83) does give us back (26.82). Integrating again,

$$y = \int h(\xi)\,\partial\xi + g(\eta) \equiv f(\xi) + g(\eta),$$

for arbitrary (twice-differentiable) functions f and g, and

$$y = f(x + at) + g(x - at) \tag{26.84}$$

is the *general solution* of the wave equation (26.79a). Notice how the form is similar to the general solution of a homogeneous, linear, second-order *ordinary* differential equation, but in place of two arbitrary constants we have two arbitrary *functions.*

Example 26.7. *Infinite String.* In order to avoid the complicating effect of end conditions, let us first apply (26.84) to an *infinite* string with prescribed initial displacement and velocity. Specifically,

$$y_{xx} = \frac{1}{a^2}y_{tt} \qquad (-\infty < x < \infty,\ 0 < t < \infty) \tag{26.85a}$$

$$y(x, 0) = F(x) \tag{26.85b}$$

$$y_t(x, 0) = G(x). \tag{26.85c}$$

Since the general solution of (26.85a) is $y = f(x + at) + g(x - at)$, we need worry only about the initial conditions. So

$$y(x, 0) = F(x) = f(x) + g(x) \tag{26.86a}$$

$$y_t(x, 0) = G(x) = af'(x) - ag'(x). \tag{26.86b}$$

Integrating (26.86b) from 0 to x, say,

$$\int_0^x G(\alpha)\,d\alpha = a[f(x) - f(0)] - a[g(x) - g(0)], \tag{26.87}$$

and (26.86a) and (26.87) may be solved for f and g. Thus

$$f(x) = \frac{F(x)}{2} + \frac{1}{2a}\int_0^x G(\alpha)\,d\alpha + \frac{f(0) - g(0)}{2} \tag{26.88a}$$

$$g(x) = \frac{F(x)}{2} - \frac{1}{2a}\int_0^x G(\alpha)\,d\alpha - \frac{f(0) - g(0)}{2}. \tag{26.88b}$$

Replacing x in (26.88a) by $x + at$ and in (26.88b) by $x - at$ and adding, we have

$$y(x, t) = \frac{F(x + at) + F(x - at)}{2} + \frac{1}{2a}\int_{x-at}^{x+at} G(\alpha)\,d\alpha, \tag{26.89}$$

which is the well-known **D'Alembert solution** to (26.85).

As an illustration, suppose that $G(x) = 0$ and that $F(x)$ is as shown in Fig. 26.9. Of course, $F(x)$ should be plotted perpendicular to the paper, but this step is difficult, and so we have laid it on its side for ease of display. Thus we hold the string at A and B, pull up at C, and let go simultaneously at $t = 0$. Consider the two terms in the solution

$$y(x, t) = \frac{F(x + at) + F(x - at)}{2}. \tag{26.90}$$

Observe very carefully that $F(x + at)$ is constant along $x + at =$ constant lines and that $F(x - at)$ is constant along $x - at =$ constant lines. These lines are the

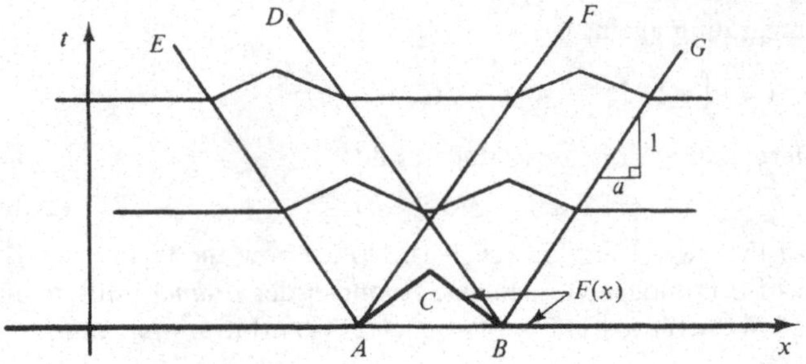

Figure 26.9. The channeling of initial data.

so-called **characteristics.** At $t = 0$ (26.90) becomes $y(x, 0) = [F(x) + F(x)]/2$, so that we can think of it as the sum of two identical "half-waves." For $t > 0$ one half-wave, $F(x + at)/2$, propagates through the channel $EABD$, and the other half-wave, $F(x - at)/2$, propagates through the channel $FABG$.

So if we observe the actual *string*, we see two half-waves moving apart, with no change in form (Fig. 26.10). The fact that the wave speed is a follows from the slope of the channels in Fig. 26.9; for instance, the right edge of the right-running wave moves a distance a per unit time.

Figure 26.10. The left- and right-running waves.

The situation is significantly different from the analogous *diffusion* problem, where even discontinuous initial conditions were smoothed out such that the solution was analytic for all $t > 0$. Furthermore, localized initial disturbances diffused through the entire x, t domain rather than being confined to certain channels. Finally, note the absence of distortion and attenuation of the waves in the case of the wave equation compared to the diffusion and decay associated with the diffusion equation (recall, e.g., Fig. 26.4). ▮

Example 26.8. *Semi-Infinite String.* In incorporating the effect of end conditions, let us consider a *semi*-infinite case, say

$$y_{xx} = \frac{1}{a^2} y_{tt} \qquad (0 \le x < \infty,\ 0 < t < \infty) \tag{26.91a}$$

$$y(x, 0) = F(x) \tag{26.91b}$$

$$y_t(x, 0) = 0 \tag{26.91c}$$

$$y(0, t) = 0. \tag{26.91d}$$

In fact, (26.91) can be solved easily by the method of images introduced in Example 26.4, as indicated in Fig. 26.11, where we have taken $F(x)$ to be a localized "pup tent" for illustratative purposes. That is, we consider instead an *infinite* string and extend the initial condition antisymmetrically (per $ABCDE$). The negative pup tent sends out a left-running half-wave (which is of no interest) and a right-running half-

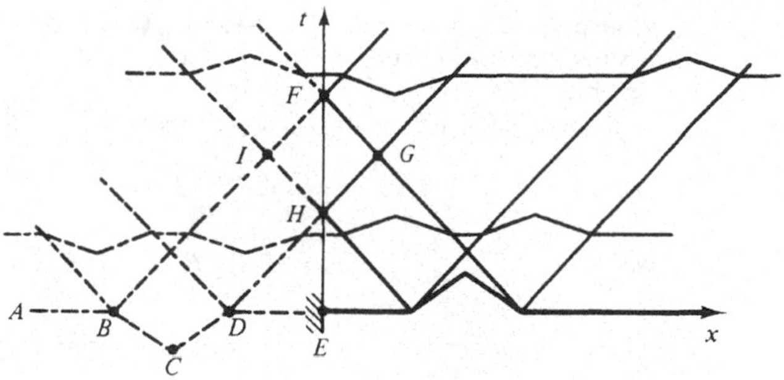

Figure 26.11. Image system for (26.91).

wave. As this negative half-wave and the left-running positive half-wave pass through each other (in *FGHI*), they add to zero along the line $x = 0$; that is, the antisymmetry built into the extended $F(x)$ ensures the satisfaction of the end condition (26.91d). So that portion of the solution of the extended problem that lies in the first quadrant of the x, t plane satisfies *all* conditions (26.91) and is the desired solution. The extension (shown as dashed lines) is the *image system* relevant to this particular problem.

Looking only at the first quadrant, the upshot is that the left-running positive half-wave suffers a *reflection* and an *inversion* off the left end! ∎

Example 26.9. *Finite String Again.* Finally, we return to the finite string problem of Example 26.6, as stated in equations (26.79), but here we will use the D'Alembert solution (26.90) to solve.

Taking $F(x)$ as shown in Fig. 26.12(a), for definiteness, notice that if we extend $F(x)$ over $-\infty < x < \infty$ antisymmetrically [Fig. 26.12(b)], then we guarantee satisfaction of the left end condition $y(0, t) = 0$ but not the right [Exercise 26.8(b)]. Consequently, let us reflect both pulses about $x = l$ [see Fig. 26.12(c)]. Doing so fixes the right end condition but upsets the left. Repeating the reflections, we are

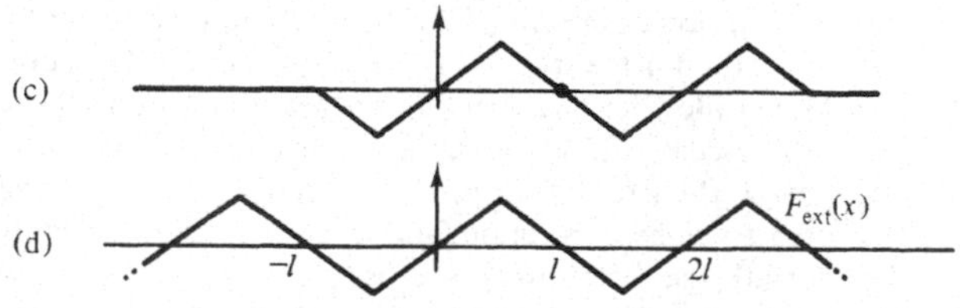

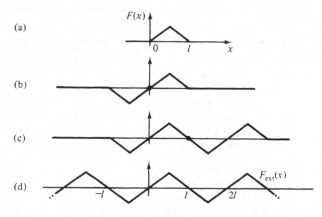

Figure 26.12. Generating the image system.

ultimately led to the *periodic* extension $F_{ext}(x)$ shown in Fig. 26.12(d). Next, $F_{ext}(x)$ is antisymmetric about *both* $x = 0$ and $x = l$ and hence provides the desired image system.

Expanding $F_{ext}(x)$ in a Fourier series gives

$$F_{ext}(x) = \sum_1^\infty F_n \sin \frac{n\pi x}{l}, \tag{26.92}$$

where

$$F_n = \frac{2}{l} \int_0^l F_{ext}(x) \sin \frac{n\pi x}{l}\, dx$$

$$= \frac{2}{l} \int_0^l F(x) \sin \frac{n\pi x}{l}\, dx, \tag{26.93}$$

since $F_{ext}(x) \equiv F(x)$ over $0 \le x \le l$.

Recalling (26.90), we obtain from (26.92)

$$y(x, t) = \sum_1^\infty \frac{F_n\left[\sin \frac{n\pi}{l}(x + at) + \sin \frac{n\pi}{l}(x - at)\right]}{2} \tag{26.94a}$$

$$y(x, t) = \sum_1^\infty F_n \sin \frac{n\pi x}{l} \cos \frac{n\pi at}{l}, \tag{26.94b}$$

which coincides with our separation-of-variables result (26.80)!

COMMENT. Note the different formats of (26.94a,b); (26.94a) is a superposition of leftward and rightward *traveling waves* of various spatial frequencies, whereas (26.94b) is a superposition of *standing waves.* The striking equivalence of the two follows simply from the trigonometric identity $\sin (A \pm B) = \sin A \cos B \pm \sin B \cos A$. ▮

Example 26.10. *A More Complicated Case.* Suppose that we have an infinite string under a tension T and that the density (mass per unit length) is discontinuous at $x = 0$—that is, $\rho(x) = \rho_1$ for $x < 0$ and ρ_2 for $x > 0$. Given some initial rightward moving wave $F(x - a_1 t)$ (Fig. 26.13), where $a_1 = \sqrt{T/\rho_1}$, what happens when

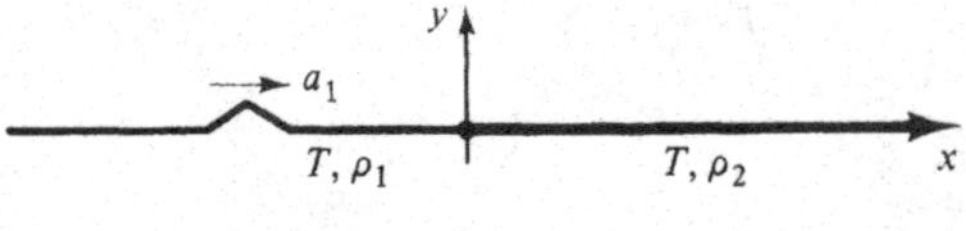

Figure 26.13. Density with a step discontinuity.

this wave reaches the density discontinuity? Turning to the x, t plane, we anticipate a situation as shown in Fig. 26.14, where we have assumed that $\rho_2 > \rho_1$ for definiteness (and hence $a_2 = \sqrt{T/\rho_2} < a_1 = \sqrt{T/\rho_1}$). That is, besides the incoming wave I, we allow for a reflected wave II, and a transmitted wave III. Why? Consider two special cases. If $\rho_2 = \infty$, then the right-hand string is *rigid* and $y(0, t) = 0$ for all t. Next, the left half of the string, with $y(0, t) = 0$, behaves as in Example 26.8, and there will be a reflection (and inversion) of the incoming wave with no transmission at all due to the infinite density of the right-hand string. If, on the other hand, $\rho_2 = \rho_1$, then, of course, "nothing happens"; that is, there is 100% transmission with no reflection. For the general case, then, it is reasonable to allow for both

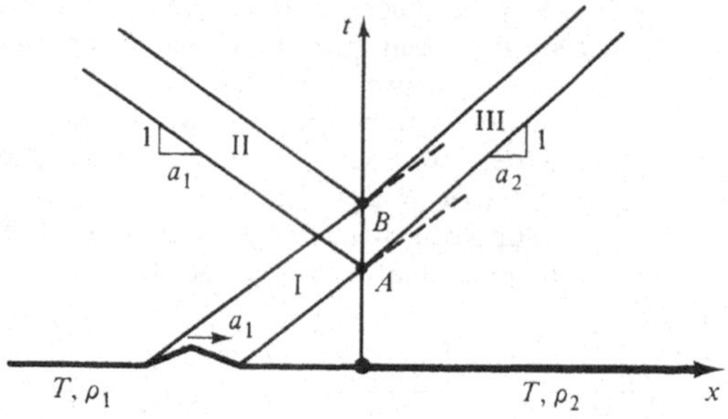

Figure 26.14. Reflection and transmission.

transmitted and reflected waves. So we seek

$$y_1 = F(x - a_1 t) + g(x + a_1 t) \quad \text{in } x < 0 \tag{26.95a}$$

and

$$y_2 = h(x - a_2 t) \quad \text{in } x > 0, \tag{26.95b}$$

where F, g, h are nonzero only in channels I, II, III, respectively. It remains to determine the form of g and h, a step that is accomplished by blending y_1 and y_2 suitably along the t axis or, more specifically, along AB.

Obviously one condition that needs to be enforced is

$$y_1(0, t) = y_2(0, t), \tag{26.96a}$$

so that the string does not break. We might also write $y_{1t}(0, t) = y_{2t}(0, t)$, but this result is already implied by (26.96a). Expecting that a second condition is needed, since we have two unknown functions, g and h, consider a "freebody" diagram of an infinitesimal element of the string located at $x = 0$. Suppose that there is a kink (Fig. 26.15). Then there is a *finite* vertical force on the element, even as $\Delta x \longrightarrow 0$, which implies an infinite vertical acceleration at $x = 0$. If it persists for any finite amount of time, infinite vertical velocities will result. To avoid this situation, we require that the slope be *continuous* at $x = 0$:

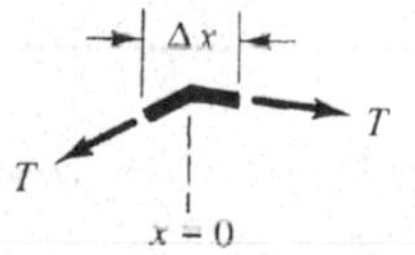

Figure 26.15. Suppose there is a kink at $x = 0$.

$$y_{1x}(0, t) = y_{2x}(0, t), \tag{26.96b}$$

and this is our second condition. Putting (26.95) into (26.96),

$$F(-a_1 t) + g(a_1 t) = h(-a_2 t) \tag{26.97a}$$

and

$$F'(-a_1 t) + g'(a_1 t) = h'(-a_2 t), \tag{26.97b}$$

where (as usual) primes denote differentiation with respect to the argument, whatever that may be. Solving for g and h in terms of F, we obtain (Exercise 26.18)

$$y_1(x, t) = F(x - a_1 t) + \left(\frac{a_2 - a_1}{a_2 + a_1}\right) F(-x - a_1 t) \tag{26.98a}$$

$$y_2(x, t) = \left(\frac{2a_2}{a_2 + a_1}\right) F\left[\frac{a_1}{a_2}(x - a_2 t)\right]. \tag{26.98b}$$

As a partial check, consider the two special cases mentioned above. If $\rho_1 = \rho_2$, then $a_1 = a_2$; thus the "reflection coefficient" $(a_2 - a_1)/(a_2 + a_1) = 0$ and the "transmission coefficient" $2a_2/(a_2 + a_1) = 1$, which, of course, is correct. If $\rho_2 =$

∞ instead, then $a_2 = 0$; as a result; the reflection coefficient is -1 (the minus sign denoting the inversion) and the transmission coefficient is 0, which is correct.

More generally, (26.98) indicates that we have a partial transmission and a partial reflection. The transmitted wave is always right side up, since $2a_2/(a_2 + a_1) \geq 0$, whereas the reflected wave is inverted if $a_2 < a_1$ ($\rho_2 > \rho_1$) and right side up if $a_2 > a_1$ ($\rho_2 < \rho_1$).

Suppose instead that $\rho(x)$ varies *continuously*, as in Fig. 26.16 say. Without even becoming involved in the calculations, we can anticipate the general form of the

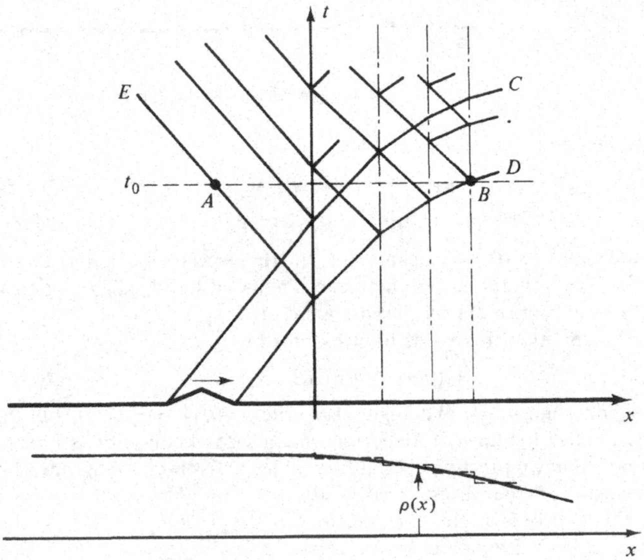

Figure 26.16. Continuous variation in $\rho(x)$.

solution if we approximate $\rho(x)$ by a number of finite steps (dotted lines) and then recall the preceding solution for the single step. For instance, as the pulse moves through the channel between the C and D characteristics, it will be deformed by the sundry reflections. In fact, the disturbance spreads throughout the region between the E and C characteristics as well. At time t_0, for example, the disturbance extends from A to B and is probably of the general form shown in Fig. 26.17, where the dashed line denotes the result if the density had remained constant, for reference.[10]

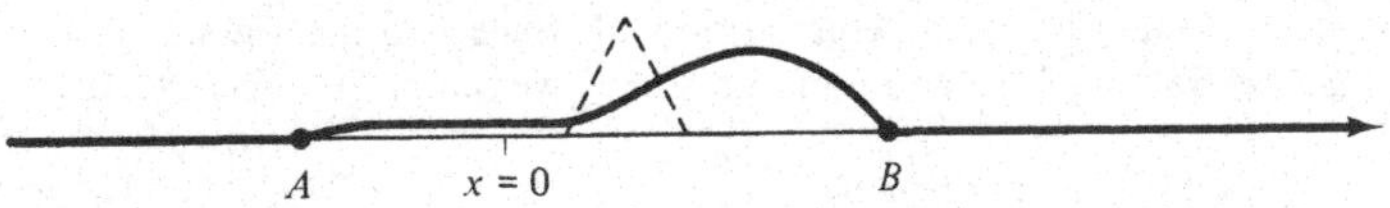

Figure 26.17. The resulting distortion.

[10]If you disagree with the form shown in Fig. 26.17, you may look forward to working Exercise 27.12, which utilizes the more efficient *method of characteristics* to solve this problem. In any event,

As a final comment, note that in the string problem the waves move in the x, t plane, whereas in the two-dimensional supersonic flow governed by (26.6) (with $M > 1$) the waves occur in the physical x, y plane. ▮

26.5. *THE LAPLACE EQUATION*

Example 26.11. *Dirichlet Problem for a Disk.* The **Dirichlet problem** consists of the Laplace equation $\nabla^2 u = 0$ over some region D with boundary B, in either two or three dimensions, together with data $u = f$ prescribed on B. For the two-dimensional case, conformal mapping provides one possible line of approach, as discussed in Chapter 16. In the present section we explore separation of variable and transform methods.

Specifically, consider the Dirichlet problem for a two-dimensional disk of radius a:

$$\nabla^2 u = u_{rr} + \frac{1}{r}u_r + \frac{1}{r^2}u_{\theta\theta} = 0 \qquad (r < a) \tag{26.99a}$$

$$u(a, \theta) = U(\theta), \tag{26.99b}$$

or, more precisely,

$$\lim_{r \to a} u(r, \theta) = U(\theta). \tag{26.99c}$$

Seeking[11]

$$u(r, \theta) = R(r)\Theta(\theta),$$

we obtain

$$\frac{r^2R'' + rR'}{R} = -\frac{\Theta''}{\Theta} = k^2 \tag{26.100}$$

and (noting that the R equation is a Cauchy–Euler equation) hence

$$\Theta = A \sin k\theta + B \cos k\theta \tag{26.101a}$$

$$R = Cr^k + Dr^{-k}. \tag{26.101b}$$

Surely we need 2π-periodicity—that is, $u(r, \theta + 2\pi) = u(r, \theta)$—which implies that $\Theta(\theta + 2\pi) = \Theta(\theta)$. This result, in turn (Exercise 26.25), means that the separation constant k can only take on integer values, say $k = n$, where negative integers can be discarded with no loss of generality. Setting $D = 0$ so that R is bounded as $r \longrightarrow 0$ and employing superposition, we have

$$u(r, \theta) = \sum_0^\infty (E_n r^n \sin n\theta + F_n r^n \cos n\theta). \tag{26.102}$$

Finally, if we impose the boundary condition (26.99b),

$$u(a, \theta) = U(\theta) = \sum_0^\infty (E_n a^n \sin n\theta + F_n a^n \cos n\theta), \tag{26.103}$$

---

the *extent* of the disturbance is determined only by the E and D characteristics, which are easily computed.

[11]There's no point in seeking $u = X(x)Y(y)$, since the boundary does not consist of x = constant and y = constant lines.

and so[12]

$$\left.\begin{aligned} E_n &= \frac{1}{\pi a^n}\int_{-\pi}^{\pi} U(\theta)\sin n\theta\, d\theta \qquad (n = 0, 1, 2, \dots) \\ F_0 &= \frac{1}{2\pi}\int_{-\pi}^{\pi} U(\theta)\, d\theta \\ F_n &= \frac{1}{\pi a^n}\int_{-\pi}^{\pi} U(\theta)\cos n\theta\, d\theta \qquad (n = 1, 2, \dots) \end{aligned}\right\} \tag{26.104}$$

It turns out that if we put (26.104) into (26.102), we can reduce the result to closed form as follows.

$$\begin{aligned} u(r,\theta) &= \frac{1}{2\pi}\int_{-\pi}^{\pi} U(\zeta)\, d\zeta + \sum_1^\infty \left\{\frac{(r/a)^n \sin n\theta}{\pi}\int_{-\pi}^{\pi} U(\zeta)\sin n\zeta\, d\zeta \right. \\ &\qquad \left. + \frac{(r/a)^n \cos n\theta}{\pi}\int_{-\pi}^{\pi} U(\zeta)\cos n\zeta\, d\zeta\right\} \\ &= \frac{1}{\pi}\int_{-\pi}^{\pi}\left\{\frac{1}{2} + \sum_1^\infty \left(\frac{r}{a}\right)^n \cos n(\zeta - \theta)\right\} U(\zeta)\, d\zeta \\ &= \frac{1}{2\pi}\int_{-\pi}^{\pi} \frac{a^2 - r^2}{a^2 - 2ar\cos(\zeta - \theta) + r^2}\, U(\zeta)\, d\zeta, \end{aligned} \tag{26.105}$$

where the key step was the identity (Exercise 26.26)

$$\sum_1^\infty \left(\frac{r}{a}\right)^n \cos n(\zeta - \theta) = \frac{a^2 - ar\cos(\zeta - \theta)}{a^2 - 2ar\cos(\zeta - \theta) + r^2} - 1. \tag{26.106}$$

The nice result (26.105) is known as the **Poisson integral formula**.

Comment 1. Observe that, for $k = n = 0$, (26.101a) reduces to Θ = constant, which is only half the solution; for $k = 0$, the general solution of $\Theta'' + k^2\Theta = 0$ is $\Theta = A_0 + B_0\theta$. Similarly, for $k = 0$, (26.101b) reduces to R = constant, whereas the full solution is $R = C_0 + D_0 \ln r$. But 2π-periodicity implies that $B_0 = 0$ and boundedness implies that $D_0 = 0$; therefore the omission of the θ and $\ln r$ terms in (26.102) caused no harm.

Comment 2. At $r = 0$, (26.105) becomes

$$u|_{r=0} = \frac{1}{2\pi}\int_{-\pi}^{\pi} U(\zeta)\, d\zeta; \tag{26.107}$$

that is, u at the center is the *average* of its boundary values. This striking result must apply "locally" as well; in other words, u at *any* point P within $r < a$ is equal to the average u around the edge of *each* circle (no matter how small) that is centered at P and that lies within $r < a$.

[12]We've omitted the Sturm–Liouville story for brevity. Incidentally, if we impose the more precise condition (26.99c) instead of the $u(a, \theta) = U(\theta)$ version used in (26.103), the result is the same. In this case, we need some help from *Abel's theorem*, which says that $\lim_{r\to a^-} \sum c_n r^n = \sum c_n a^n$ if the latter converges.

COMMENT 3. Consider the *Poisson kernel*

$$P(r, \zeta - \theta) = \frac{1}{2\pi} \frac{a^2 - r^2}{a^2 - 2ar\cos(\zeta - \theta) + r^2}. \tag{26.108}$$

If $U(\zeta) = 1$, then surely $u(r, \theta) \equiv 1$,[13] which implies that

$$\int_{-\pi}^{\pi} P \, d\zeta = 1$$

for each r. Furthermore, (26.99c) says that

$$\lim_{r \to a} \int_{-\pi}^{\pi} P(r, \zeta - \theta) U(\zeta) \, d\zeta = U(\theta),$$

and the conclusion from these two results is that $P(r, \zeta - \theta)$ is a δ sequence—it tends to a delta function at $\zeta = \theta$ as $r \longrightarrow a$. ▮

Example 26.12. *Dirichlet Problem for Half-Plane.* Consider next the Dirichlet problem for the *half-plane*, as sketched in Fig. 26.18. Since y is semi-infinite, we

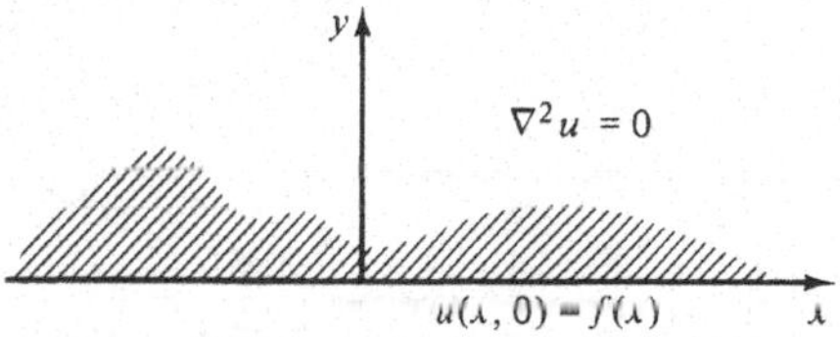

Figure 26.18. Dirichlet problem for half-plane.

might entertain ideas of Laplace transforming on y. However, integration by parts of the $\int_0^\infty e^{-sy} u_{yy} \, dy$ term will introduce both $u(x, 0)$ and $u_y(x, 0)$ boundary terms, of which the latter is not known in advance. Instead let us *Fourier* transform on x:

$$\int_{-\infty}^{\infty} \frac{\partial^2 u}{\partial x^2} e^{-i\omega x} \, dx + \int_{-\infty}^{\infty} \frac{\partial^2 u}{\partial y^2} e^{-i\omega x} \, dx = 0.$$

Integrating the first term by parts twice and assuming tentatively that both $\partial u/\partial x$ and $u \longrightarrow 0$ as $|x| \longrightarrow \infty$, we have

$$-\omega^2 \hat{u} + \frac{d^2 \hat{u}}{dy^2} = 0 \tag{26.109}$$

or

$$\hat{u} = Ae^{|\omega| y} + Be^{-|\omega| y}. \tag{26.110}$$

Just as we assumed that $u \longrightarrow 0$ as $|x| \longrightarrow \infty$, let us assume that

$$u(x, y) \longrightarrow 0 \quad \text{as } y \longrightarrow \infty. \tag{26.111}$$

Fourier transforming (26.111), we obtain the condition $\hat{u} \longrightarrow 0$ as $y \longrightarrow \infty$, which implies that $A = 0$. So far, then,

$$\hat{u} = Be^{-|\omega| y}. \tag{26.112}$$

[13]The question of **uniqueness** is reserved for Exercise 26.27.

Finally, transforming the boundary condition $u(x, 0) = f(x)$, we have

$$\hat{u}|_{y=0} = \hat{f}(\omega), \tag{26.113}$$

and imposing this result on (26.112) yields $B = \hat{f}(\omega)$; therefore

$$\hat{u} = \hat{f}(\omega)e^{-|\omega|y}. \tag{26.114}$$

Inverting, with the help of the Fourier convolution theorem (6.23) and the Fourier Transform Table (inside front cover),

$$u = F^{-1}\{\hat{f}(\omega)e^{-|\omega|y}\} = f(x) * \frac{y}{\pi(x^2 + y^2)}$$

$$= \frac{y}{\pi}\int_{-\infty}^{\infty} \frac{f(\xi)}{(\xi - x)^2 + y^2}\,d\xi \equiv \int_{-\infty}^{\infty} K(\xi - x, y)f(\xi)\,d\xi, \tag{26.115}$$

which is the analog of (26.105) for the circular disk.

COMMENT 1. Why the absolute values in (26.110) rather than

$$\hat{u} = Ce^{\omega y} + De^{-\omega y}? \tag{26.116}$$

First, observe that (26.110) and (26.116) are equivalent, since

$$\hat{u} = Ce^{\omega y} + De^{-\omega y} = \begin{cases} Ce^{|\omega|y} + De^{-|\omega|y}, & \omega > 0 \\ Ce^{-|\omega|y} + De^{|\omega|y}, & \omega < 0 \end{cases}$$

$$\equiv Ae^{|\omega|y} + Be^{-|\omega|y} \quad \text{for } -\infty < \omega < \infty,$$

where $A(\omega) = \begin{cases} C(\omega), & \omega > 0 \\ D(\omega), & \omega < 0 \end{cases}$ and $B(\omega) = \begin{cases} D(\omega), & \omega > 0 \\ C(\omega), & \omega < 0. \end{cases}$

If we use the form (26.116), then $\hat{u} \longrightarrow 0$ as $y \longrightarrow \infty$ implies that $C(\omega) = 0$ for $\omega > 0$ and $D(\omega) = 0$ for $\omega < 0$; whereas if we use (26.110), we simply have $A(\omega) = 0$.

COMMENT 2. As in preceding examples we call attention to the δ-function nature of the kernel; specifically, $K(\xi - x, y)$ is a δ sequence as $y \longrightarrow 0$, so that

$$\lim_{y \to 0} \int_{-\infty}^{\infty} K(\xi - x, y)f(\xi)\,d\xi = f(x)$$

as desired. (In fact, this particular δ sequence was the subject of Exercise 4.8.)

COMMENT 3. And as in preceding examples we note that the final result (26.115) is valid for a larger class of f's than assumed. For instance, $f(\xi) = 1$ in (26.115) yields $u(x, y) = 1$, which is obviously correct, and yet $\hat{f}$ does not exist and $u(x, y)$ does not approach zero at infinity as tentatively assumed. This happy circumstance can be traced to the inversion in the order of integration that lies behind the Fourier convolution theorem. ∎

Example 26.13. *Solution by Superposition.* Any attempt to solve the Dirichlet problem of Fig. 26.19 by separation of variables will show that it simply doesn't work. The difficulty is an insufficient amount of homogeneity in the boundary conditions. Using the principle of superposition, however, we can express

$$u = U(x, y) + V(x, y) + W(x, y), \tag{26.117}$$

where U, V, W are solutions of the Dirichlet problems shown in Fig. 26.20. To verify the equivalence, observe that

$$\nabla^2(U + V + W) = \nabla^2 U + \nabla^2 V + \nabla^2 W$$
$$= 0 + 0 + 0 = 0,$$

so that $U + V + W$ does satisfy Laplace's equation; the crucial step was the first equality, which follows from the *linearity* of the Laplace operator. In addition, it is easy to see that $U + V + W$ satisfies the given boundary conditions on u.

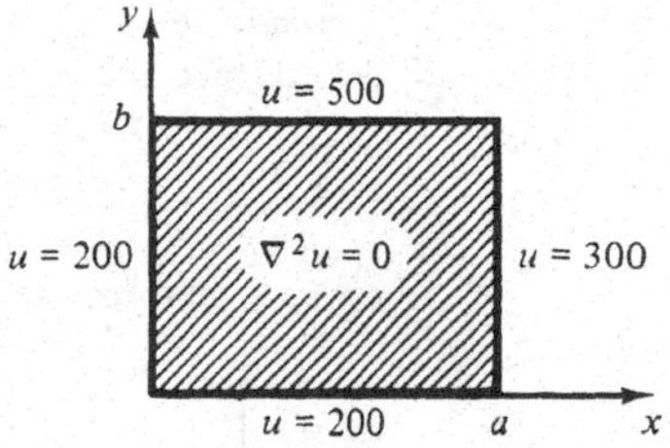

Figure 26.19. Dirichlet problem for rectangle.

The solution of the U problem is trivial. By inspection,

$$U(x, y) = \text{constant} = 200. \tag{26.118a}$$

Turning to the V problem, separation of variables leads to the solution [Exercise 26.34(a)]

$$V(x, y) = \frac{400}{\pi} \sum_{n=1,3,\ldots}^{\infty} \frac{\sinh(n\pi x/b)\sin(n\pi y/b)}{n\sinh(n\pi a/b)}. \tag{26.118b}$$

Actually, it will be more convenient to denote this as $V(x, y; a, b)$ in order to spell out the dependence on the dimensions a and b.

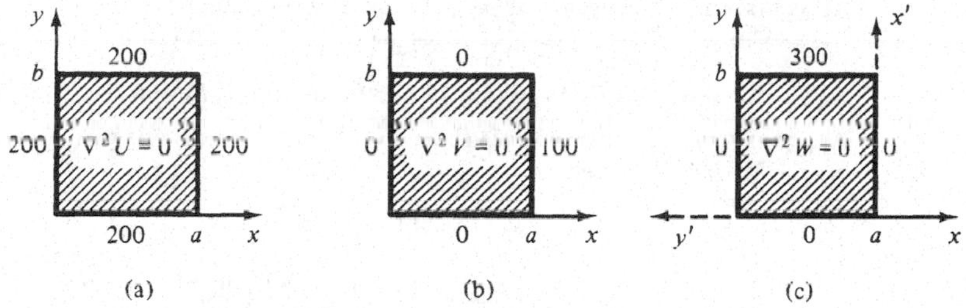

Figure 26.20. Decomposition of u by superposition.

Finally, observe [by comparing Fig. 26.20(b) and (c)] that W is simply

$$W = 3V(x', y'; b, a) = 3V(y, a - x; b, a)$$
$$= \frac{1200}{\pi} \sum_{n=1,3,\ldots}^{\infty} \frac{\sinh(n\pi y/a)\sin[n\pi(a - x)/a]}{n\sinh(n\pi b/a)}, \tag{26.118c}$$

where we have *again* used linearity in scaling V by a factor of 3. That is, $\nabla^2 W = \nabla^2(3V) = 3\,\nabla^2 V = 3 \cdot 0 = 0$ as desired, where the second equality follows from the linearity of ∇^2.

Adding equations (26.118) and noting that

$$\sin[n\pi(a - x)/a] = \sin n\pi \cos(n\pi x/a) - \cos n\pi \sin(n\pi x/a) = \sin(n\pi x/a)$$

for odd n's, we have

$$u(x, y) = 200 + \frac{400}{\pi} \sum_{n=1,3,\ldots}^{\infty} \left\{\frac{\sinh(n\pi x/b)\sin(n\pi y/b)}{n\sinh(n\pi a/b)} + \frac{3\sinh(n\pi y/a)\sin(n\pi x/a)}{n\sinh(n\pi b/a)}\right\} \tag{26.119}$$

as the desired solution. ∎

Example 26.14. *Loaded Membrane.* Recall from Section 10.5 that the normal deflection w of a membrane, under a uniform tension τ and a loading p, is governed by the Poisson equation

$$\nabla^2 w = -\frac{p}{\tau}; \tag{26.120}$$

p was a uniform pressure in Part I, but here we will allow it to be any prescribed function of x and y.

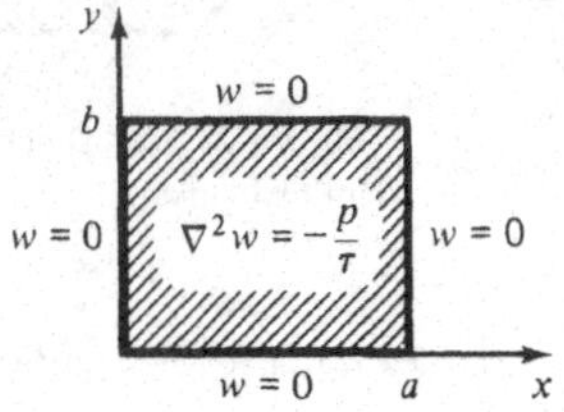

Figure 26.21. Loaded membrane.

Suppose that the membrane is rectangular and that $w = 0$ around its edge, so that the boundary value problem is as indicated in Fig. 26.21. Here is the first example in which the differential equation is *inhomogeneous.* Unless $p(x, y)$ is of a very special form (Exercise 26.29), separation of variables will not work. What to do? Expressing (26.120) in the form

$$Lw = f, \tag{26.121}$$

observe that L (i.e., $L = \nabla^2$ with $w = 0$ on the boundary) is *self-adjoint*, and so its eigenfunctions are mutually orthogonal. If they are also *complete*, we can solve (26.121) by means of eigenfunction expansions, as discussed in Section 20.5. Thus let us first solve the *eigenvalue problem* $Lw = \lambda w$—that is,

$$\nabla^2 w = \lambda w, \tag{26.122a}$$

$$w(0, y) = w(a, y) = w(x, 0) = w(x, b) = 0. \tag{26.122b}$$

Applying separation of variables (i.e., $w = X(x)Y(y)$), it is not hard to show (Exercise 26.30) that the eigens are

$$\phi_{mn}(x, y) = \sin\frac{m\pi x}{a}\sin\frac{n\pi y}{b} \tag{26.123a}$$

$$\lambda_{mn} = -\pi^2\left[\left(\frac{m}{a}\right)^2 + \left(\frac{n}{b}\right)^2\right] \tag{26.123b}$$

for m and $n = 1, 2, \ldots$. So expanding

$$w(x, y) = \sum_1^\infty \sum_1^\infty w_{mn}\phi_{mn}(x, y) \tag{26.124a}$$

$$f(x, y) = \sum_1^\infty \sum_1^\infty f_{mn}\phi_{mn}(x, y) \tag{26.124b}$$

and noting that $L\phi_{mn} = \lambda_{mn}\phi_{mn}$, (26.122) becomes

$$\sum_1^\infty \sum_1^\infty \lambda_{mn}w_{mn}\phi_{mn}(x, y) = \sum_1^\infty \sum_1^\infty f_{mn}\phi_{mn}(x, y),$$

and it follows that

$$w_{mn} = \frac{f_{mn}}{\lambda_{mn}}, \tag{26.125}$$

where

$$f_{mn} = \frac{(f, \phi_{mn})}{(\phi_{mn}, \phi_{mn})} = -\frac{4}{\tau ab}\int_0^b \int_0^a p(x, y)\sin\frac{m\pi x}{a}\sin\frac{n\pi y}{b}\,dx\,dy. \tag{26.126}$$

The solution, then, is given by (26.124a), together with (26.125), (26.123b), and (26.126).

Alternatively, suppose that we return to (26.120) and seek

$$w(x, y) = \sum_1^\infty f_n(x) \sin \frac{n\pi y}{b}, \qquad f_n(0) = f_n(a) = 0 \tag{26.127a}$$

and expand

$$\frac{p(x, y)}{\tau} = \sum_1^\infty p_n(x) \sin \frac{n\pi y}{b}. \tag{26.127b}$$

Then (26.120) becomes

$$\sum_1^\infty \left(f_n'' - \frac{n^2\pi^2}{b^2} f_n\right) \sin \frac{n\pi y}{b} = -\sum_1^\infty p_n(x) \sin \frac{n\pi y}{b},$$

so that the f_n's are determined from the system

$$f_n'' - \frac{n^2\pi^2}{b} f_n = -p_n(x), \qquad f_n(0) = f_n(a) = 0, \tag{26.127c}$$

which can be solved by the method of variation of parameters.

Note that this procedure has an advantage over the foregoing approach in that the solution is only a *single* series.

As emphasized in Part III, it is generally easy to show whether a given operator is self-adjoint and hence whether its eigenfunctions are orthogonal, whereas their completeness is far more difficult to establish (for infinite-dimensional spaces). In the present case, completeness is known.[14] Even in the absence of such information, however, it is not uncommon in engineering applications to go ahead and assume completeness if L is self-adjoint and the domain is bounded. In any case, this point of view may be adopted in working the exercises. ▮

EXERCISES

26.1. Take the separation constant in (26.23) to be k, say, instead of $-k^2$ and show that the same result (26.30) is obtained.

26.2. Verify rigorously that (26.30) does satisfy the differential equation (26.20a).

26.3. (*Heat conduction: uniqueness and maximum principle*) (a) Prove that the solution to the finite-rod heat conduction problem

$$\alpha^2 u_{xx} = u_t \qquad (0 < x < l,\ t > 0) \tag{26.128a}$$

$$u(0, t) = a(t), \qquad u(l, t) = b(t), \qquad u(x, 0) = f(x) \tag{26.128b}$$

is unique. *Hint:* Suppose that $u_1(x, t)$ and $u_2(x, t)$ are any two solutions and define $u_1(x, t) - u_2(x, t) \equiv w(x, t)$. Show that w satisfies the homogenized version of (26.128),

$$\alpha^2 w_{xx} = w_t, \qquad w(0, t) = w(l, t) = w(x, 0) = 0.$$

Next, show by integration by parts that

$$\int_0^l w w_{xx}\, dx = (w w_x)\Big|_0^l - \int_0^l w_x^2\, dx \tag{26.129}$$

[14]See, for example, R. Courant and D. Hilbert, *Methods of Mathematical Physics*, Vol. 1, Interscience, New York, 1953.

and hence that

$$\frac{1}{2\alpha^2}\frac{d}{dt}\int_0^l w^2\,dx = -\int_0^l w_x^2\,dx,$$

$$\frac{1}{2\alpha^2}\int_0^l w^2\,dx = -\int_0^t dt\int_0^l w_x^2\,dx.$$

Finally, use the nonnegativeness of the left side and the nonpositiveness of the right to conlude that $w(x, t) \equiv 0$.

(b) We state without proof the important **maximum principle**. Let $u(x, t)$ be continuous in $0 \le x \le l, 0 \le t \le T$ and satisfy $\alpha^2 u_{xx} = u_t$ throughout the interior. Then the maximum of u is attained at least at one point on the vertical sides ($x = 0, x = l$) or on the bottom ($t = 0$). (Similarly, for *minimum*.) Prove uniqueness for the problem (26.128), using the maximum principle. *Note:* The uniqueness result of parts (a) and (b) does not hold, however, for the *infinite* rod unless we require $u(x, t)$ to be *bounded* (in $-\infty < x < \infty, t > 0$).[15]

(c) Extend the uniqueness proof outlined in part (a) to the problem $\alpha^2 \nabla^2 u = u_t$ in some finite *two-* or *three*-dimensional region D, for $t > 0$, with the *mixed* boundary condition $au + bu_n = c$ on the boundary S of D and the initial condition $u = f$ in D at $t = 0$; a, b, c may vary over S, u_n is the derivative in the direction of the outward normal $\hat{\mathbf{n}}$, and a, b are both nonnegative. *Hint:* The extension of (26.129) is provided by Green's first identity.

26.4. Derive the residues (26.45).

26.5. Derive the expression (26.54). *Hint:* Completing the square in the exponent, obtain

$$\int_{-\infty}^{\infty} e^{i\omega(x-\xi)-\alpha^2\omega^2 t}\,d\omega = \frac{e^{-(\xi-x)^2/4\alpha^2 t}}{\alpha\sqrt{t}}\int_{-\infty-i(x-\xi)/2\alpha\sqrt{t}}^{\infty-i(x-\xi)/2\alpha\sqrt{t}} e^{-\zeta^2}\,d\zeta.$$

Then consider $\int_C e^{-\zeta^2}\,d\zeta$, where C is the rectangle shown in Fig. 26.22 and, letting A and $B \to \infty$, show that

$$\int_{-\infty-i(x-\xi)/2\alpha\sqrt{t}}^{\infty-i(x-\xi)/2\alpha\sqrt{t}} e^{-\zeta^2}\,d\zeta = \int_{-\infty}^{\infty} e^{-\zeta^2}\,d\zeta.$$

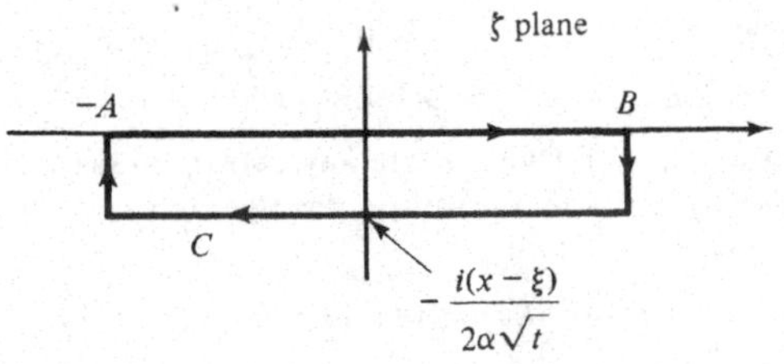

Figure 26.22. Deforming the path.

26.6. Setting $f(x) = H(x)$ in (26.55), derive the result (26.59).

26.7. Express the square wave in Fig. 26.6(a) as a series of Heaviside functions and thus

[15] See B. Epstein, *Partial Differential Equations*, McGraw-Hill, New York, 1962, p. 219.

derive an infinite series of error functions for $u(x, t)$ of which (26.61) is just the leading term.

26.8. (a) Show that if

$$\alpha^2 u_{xx} = u_t \qquad (-\infty < x < \infty,\ t > 0)$$
$$u(x, 0) = f(x),$$

then $f(-x) = f(x)$ for all x implies that $u(-x, t) = u(x, t)$ for all x and all $t > 0$, and similarly for the *anti*symmetric case.

(b) Similarly, show that if $a^2 y_{xx} = y_{tt}$ $(-\infty < x < \infty,\ t > 0)$, $y(x, 0) = f(x)$ and $y_t(x, 0) = 0$, then symmetry (antisymmetry) of $f(x)$ implies symmetry (antisymmetry) of the solution $y(x, t)$.

26.9. Derive (26.77). *Hint:* Recalling that $J_0(\alpha x) \equiv u$ and $J_0(\beta x) \equiv v$ satisfy $xu'' + u' + \alpha^2 xu = 0$ and $xv'' + v' + \beta^2 xv = 0$, show that $(\beta^2 - \alpha^2)xuv = d[x(u'v - uv')]/dx$. Integrating and recalling that $J_0' = -J_1$ and $xJ_1'(x) = -J_1(x) + xJ_0(x)$ from Exercise 22.33, show that

$$\int_0^1 xJ_0(\alpha x)J_0(\beta x)\,dx = \frac{\beta J_0(\alpha)J_1(\beta) - \alpha J_1(\alpha)J_0(\beta)}{\beta^2 - \alpha^2}.$$

If $\beta = \alpha$, each being a root of $J_0(x) = 0$, the right-hand side is indeterminate but can be evaluated by considering α as a root of $J_0(x) = 0$ and letting $\beta \longrightarrow \alpha$. Applying L'Hospital's rule, obtain

$$\int_0^1 x[J_0(\alpha x)]^2\,dx = \tfrac{1}{2}[J_1(\alpha)]^2.$$

26.10. Suppose that we form a ring by taking an insulated heat conductor of length l, bending it into a circle, and joining its ends. Thus we have $\alpha^2 u_{xx} = u_t$ $(0 \le x \le l,\ t > 0)$, together with *periodicity conditions* $u(0, t) = u(l, t)$, $u_x(0, t) = u_x(l, t)$, and some initial condition $u(x, 0) = f(x)$. Solve by separation of variables. What is the steady-state solution $\lim_{t\to\infty} u(x, t)$?

26.11. Solve the heat conduction problem

$$\alpha^2 u_{xx} = u_t \qquad (0 < x < \infty,\ t > 0)$$
$$u(x, 0) = 0, \qquad u(0, t) = 100$$

(a) by a Laplace transform (using Table A.2 in Appendix A)
(b) by means of a **similarity transformation**, as outlined in Exercise. 7.8.

26.12. The following problem emphasizes the importance of a careful formulation of the governing differential equation and boundary conditions. A thin, continuous metal rod of thermal conductivity k, diameter d, specific heat c, and density ρ leaves a furnace that is maintained at a temperature T_0. It travels a distance L at uniform velocity V and enters an extrusion die. During this travel it is exposed to room temperature T_a and the surface heat transfer coefficient is h; that is, the heat loss per unit surface area is $h(T - T_a)$, where the rod temperature $T(x, t)$ may be assumed uniform over each cross section of the rod. Determine the temperature of the metal entering the die. *Hint:* Consider a stationary control volume between x and $x + \Delta x$. Equating the influx of heat to $\partial(\rho A\,\Delta x\,cT)/\partial t$ (where A is the cross-sectional area), derive the governing equation

$$\frac{\partial^2 T}{\partial x^2} = \frac{\rho c}{k}\left(\frac{\partial T}{\partial t} + V\frac{\partial T}{\partial x}\right) + \frac{4h}{kd}(T - T_a) \equiv \frac{1}{\alpha^2}\frac{DT}{Dt} + \sigma(T - T_a).$$

Assume steady-state operation so that $T = T(x)$ and impose suitable boundary conditions at $x = 0$ and $x = L$. Two posibilities at $x = 0$ are $T(0) = T_0$, $T'(0) = 0$. Or how about the following formulation?

$$-\infty < x < 0: \quad T'' - \frac{V}{\alpha^2}T' - \sigma_0 T = -\sigma_0 T_0, \qquad T(-\infty) = T_0$$

$$0 < x < \infty: \quad T'' - \frac{V}{\alpha^2}T' - \sigma T = -\sigma T_a, \qquad T(\infty) = T_a$$

with

$$T(0+) = T(0-) \quad \text{and} \quad T'(0+) = T'(0-).$$

Explain your reasoning.

26.13. (*A sludge problem*) An initially clear stream flows happily with uniform velocity U. Starting at $t = 0$, we put in Q tons/sec of sludge at $x = 0$. The concentration of sludge in the stream $c(x, t)$ is governed by the diffusion-type equation

$$c_t + Uc_x = \epsilon c_{xx} - \kappa c + Q\delta(x),$$

where ϵ is a turbulent diffusivity constant and κ is a chemical decay coefficient. (Recall that the $c_t + Uc_x = Dc/Dt$ combination arose in Exercise 26.12 as well.)

(a) Solve for and sketch the *steady-state* distribution $c_s(x) = c(x, \infty)$. (Use elementary methods, not an integral transform.)

(b) Derive (by Fourier transform) the result

$$c(x, t) = \frac{Q}{2\pi}\int_{-\infty}^{\infty} \frac{1 - e^{-(\epsilon\omega^2+\kappa+i\omega U)t}}{\epsilon\omega^2 + \kappa + i\omega U} e^{i\omega x}\, d\omega.$$

(c) Evaluate the preceding integral by the residue theorem for $t \longrightarrow \infty$ and compare with your answer to part (a). (It's not much harder for $t < \infty$, just more algebra.)

26.14. Show that the diffusion equation $c_t + U(t)c_x = \epsilon c_{xx} - \kappa c$ can be reduced to the form $h_t = \epsilon h_{\xi\xi}$ by the change of variables

$$x = \xi + \int_0^t U(\tau)\, d\tau, \qquad c = he^{-\kappa t}.$$

26.15. Solve the diffusion problem of Example 26.3 with the right-end condition (26.33d) changed to $u(l, t) = F(t)$.

26.16. Fill in the steps between (26.79) and (26.80).

26.17. (a) Show that the change of variables $\xi = x + at$, $\eta = x - at$ reduces the wave equation $a^2 y_{xx} = y_{tt}$ to $y_{\xi\eta} = 0$.

(b) Derive the solution to the *driven* infinite string problem

$$a^2 y_{xx} = y_{tt} + h(x, t), \qquad y(x, 0) = F(x), \quad y_t(x, 0) = G(x).$$

Hint: With $\xi = x + at$ and $\eta = x - at$, obtain

$$4a^2 y_{\xi\eta} = h\left(\frac{\xi + \eta}{2}, \frac{\xi - \eta}{2a}\right) \equiv H(\xi, \eta)$$

and integrate.

(c) If spherical symmetry is present, the wave equation becomes

$$\frac{1}{r^2}\frac{\partial}{\partial r}\left(r^2\frac{\partial u}{\partial r}\right) = \frac{1}{a^2}\frac{\partial^2 u}{\partial t^2},$$

where r is the distance from the origin. Show that it may be reexpressed as

$$\frac{\partial^2}{\partial r^2}(ru) = \frac{1}{a^2}\frac{\partial^2}{\partial t^2}(ru)$$

and hence derive the general solution

$$u = \frac{1}{r}[f(r + at) + g(r - at)].$$

26.18. Solve (26.97) for g and h and thus derive the solution (26.98).

26.19. Solve the *loaded* vibrating string problem

$$y_{tt} = a^2 y_{xx} + x \qquad (0 < x < \pi,\ t > 0)$$
$$y(0, t) = y(\pi, t) = y(x, 0) = y_t(x, 0) = 0.$$

Hint: Seek $y(x, t) = X(x)T(t) + \phi(x)$, where $\phi(x)$ is chosen so as to facilitate the separation of variables. Expansion coefficients may be left in integral form.

26.20. Let the infinite string of Example 26.7 be struck at the origin by a hammer of width $2l$, so that $F(x) = 0$ and $G(x) = V[H(x + l) - H(x - l)]$, where V is the hammer velocity. Then using (26.89), derive the solution and sketch it in the x, t plane.

26.21. Solve the string problem

$$a^2 y_{xx} = y_{tt} \qquad (0 < x < l,\ t > 0)$$
$$y(l, t) = y(x, 0) = y_t(x, 0) = 0,$$

where we "wiggle" the left end according to $y(0, t) = \epsilon \sin \omega t$. *Hint:* Seeking $y = X(x)T(t)$, obtain a form that satisfies all the requirements except that $X(x)T'(0) =$ some function $G(x)$, say, rather than zero. To patch this up, seek $y = X(x)T(t) + g(x, t)$—that is, append a function $g(x, t)$ to the XT that has already been found and require of g that

$$a^2 g_{xx} = g_{tt}, \qquad g(0, t) = g(l, t) = g(x, 0) = 0, \qquad g_t(x, 0) = -G(x).$$

26.22. Suppose in Example 26.10 that the density is uniform ($\rho_1 = \rho_2$), but that there is a bead of mass m at $x = 0$. Modify the formulation (26.95), (26.96) accordingly. (But you need not solve.)

26.23. Recall that the (small-amplitude) vibration of a membrane (neglecting air resistance) is governed by the *two*-dimensional wave equation (10.67) or with the density $\rho(x, y)$ = constant

$$\alpha^2 \nabla^2 w = w_{tt} \qquad \left(\alpha^2 = \frac{T}{\rho}\right).$$

(a) Suppose that the domain is the square $0 < x < a, 0 < y < a$, and $w = 0$ around the edge. Setting $w = \phi(x, y)T(t)$, obtain the equations

$$T'' + k^2\alpha^2 T = 0, \qquad \nabla^2\phi + k^2\phi = 0,$$

where the latter is the well-known **Helmholtz equation**, and thus determine the possible frequencies and their corresponding mode shapes.

(b) Thus, find the lowest frequency of vibration of a *triangular* membrane with vertices at $(0, 0)$, $(a, 0)$, and (a, a). *Hint:* Consider the lowest mode of the *square* membrane that has $y = x$ as a nodal line (i.e., along which $w = 0$ for all t).

(c) Re-solve part (b) approximately by using the Ritz method of Section 10.6. (When applied to *eigenvalue* problems, it is generally called the *Rayleigh*–Ritz method.)

(d) Rework part (a) for a *circular* membrane $r < a$ and solve subject to the initial conditions $w(r, \theta, 0) = 0$, $w_t(r, \theta, 0) = f(r)$.

(e) Comparing the frequencies for the square and circular drums, can you explain why (musical) drums are circular, not rectangular?

26.24. (*Pulsating cylinder; Sommerfeld radiation condition*) The problem

$$\phi_{rr} + \frac{1}{r}\phi_r = \frac{1}{c^2}\phi_{tt} \quad \text{in } a < r < \infty$$

$$\phi_r(a, t) = V \sin \omega t$$

governs (approximately) the velocity potential ϕ for the (compressible) fluid surrounding an infinite circular cylinder of radius a that is pulsating weakly at a rate $V \sin \omega t$.

(a) Considering

$$\Phi_{rr} + \frac{1}{r}\Phi_r = \frac{1}{c^2}\Phi_{tt} \quad \text{in } a < r < \infty \tag{1.130a}$$

$$\Phi_r(a, t) = Ve^{i\omega t} \tag{1.130b}$$

instead, for convenience, show that $\phi(r, t) = \text{Im}\,\{\Phi(r, t)\}$.

(b) Seeking $\Phi = R(r)e^{i\omega t}$, solve for $R(r)$ in terms of Hankel functions. [Recall Section 22.4 and note that $dH_0^{(1,2)}(x)/dx = -H_1^{(1,2)}(x)$.]

(c) Evaluate the integration constants by applying (26.130b), together with the so-called **Sommerfeld radiation condition**—namely, that the velocity Φ_r should consist only of *out*ward-traveling waves as $r \longrightarrow \infty$ and thus show that

$$\phi = -\frac{cV}{\omega}\,\text{Im}\left\{\frac{H_0^{(2)}(\omega r/c)}{H_1^{(2)}(\omega a/c)}e^{i\omega t}\right\}.$$

Note that

$$H_\nu^{(1)}(x) \sim \sqrt{\frac{2}{\pi x}}\,e^{i[x-(\pi/4)-(\nu\pi/2)]},$$

and

$$H_\nu^{(2)}(x) \sim \sqrt{\frac{2}{\pi x}}\,e^{-i[x-(\pi/4)-(\nu\pi/2)]} \quad \text{as } x \longrightarrow \infty.$$

26.25. If $\Theta = A \sin k\theta + B \cos k\theta$ and $\Theta(\theta + 2\pi) = \Theta(\theta)$, show that k must be an integer.

26.26. Derive the identity (26.106). *Hint:* Express

$$\sum_1^\infty \left(\frac{r}{a}\right)^n \cos n(\zeta - \theta) = \text{Re} \sum_1^\infty \left(\frac{r}{a}\right)^n e^{in(\zeta-\theta)}.$$

26.27. (*Poisson equation: uniqueness*) Suppose that D is a finite two- or three-dimensional region with boundary S and that $\nabla^2 u = \phi$ in D, where ϕ is prescribed in D.

(a) If u is subjected to a *Dirichlet* boundary condition $u = f$ on S, show that the solution u is unique.

(b) If u satisfies a *Neumann* boundary condition $u_n = f$ on S, show that u is unique to within an arbitrary additive constant.

(c) Consider the question of uniqueness if u satisfies a *mixed* boundary condition $au + bu_n = c$, where a, b, c may vary over S and $a, b > 0$. *Hint:* In each case, suppose that u_1 and u_2 are solutions, set $u_1 - u_2 \equiv w$, and apply Green's first identity to w.

26.28. (*Bessel functions*) (a) Solve (by separation of variables) for the steady-state temperature $u(r, z)$ in a semi-infinite rod of radius a if $u = 100$ at the end of the rod and $u = 0$ on its lateral surface; that is,

$$\nabla^2 u = u_{rr} + \frac{1}{r}u_r + u_{zz} = 0 \quad \text{in } r < a,\ 0 < z < \infty$$

$$u(r, 0) = 100, \qquad u(a, z) = 0.$$

Hint: Recall Exercise 26.9.

(b) Re-solve the unsteady conduction problem (26.66) with the conditions (26.66b) (26.66c) changed to $u(r, 0) = 0$ and $u(a, t) = 50$.

(c) Solve the two-dimensional unsteady conduction problem

$$\alpha^2 \nabla^2 u = u_t \quad \text{in } 0 \leq r < a,\ 0 < t < \infty$$

$$u(r, \theta, 0) = u_0, \qquad u(a, \theta, t) = u_0 + u_0 \cos\theta$$

for $u(r, \theta, t)$ by separation of variables. Expansion coefficients may be left in integral form.

26.29. Make up a $p(x, y)$ in Example 26.14 such that separation of variables works and thus obtain the solution.

26.30. Derive the eigens (26.123) of the system (26.122).

26.31. Use the Poisson integral formula

$$u(r, \theta) = \int_{-\pi}^{\pi} P(r, \zeta - \theta) U(\zeta)\, d\zeta$$

and the *method of images* to solve the following problems.

(a) $\nabla^2 u = 0$ in $0 < r < a,\ 0 < \theta < \pi$
$u(a, \theta) = f(\theta)$, $u = 0$ on straight edge

(b) $\nabla^2 u = 0$ in $0 < r < a,\ 0 < \theta < \pi$
$u(a, \theta) = f(\theta)$, $u_n = 0$ on straight edge

(c) $\nabla^2 u = 0$ in $0 < r < a,\ 0 < \theta < \pi/2$
$u(a, \theta) = f(\theta)$, $u(r, 0) = 100$, $u_n = 0$ on $\theta = \pi/2$ edge

26.32. In Example 26.11 we solved the *interior* Dirichlet problem $\nabla^2 u = 0$ in $r < a$ with $u(a, \theta) = f(\theta)$. Solve instead the *exterior* problem $\nabla^2 u = 0$ in $r > a$ with $u(a, \theta) = f(\theta)$ (i.e., the domain is the x, y plane with a circular hole of radius a) and obtain the analog of the Poisson integral formula (26.105).

26.33. Solve the steady-state heat conduction problems shown in Fig. 26.23. *Note:* In the circular disk problem (26.99) we enforced the 2π-periodicity condition $u(r, \theta + 2\pi)$

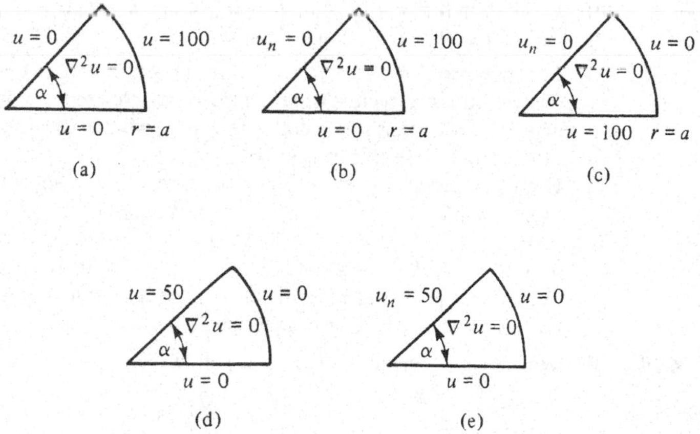

Figure 26.23. Steady-state heat conduction in a wedge.

$= u(r, \theta)$. This condition is neither correct nor relevant for the problems shown in Fig. 26.23, since the θ domain is only $0 < \theta < \alpha$!

26.34. (a) Derive the solution (26.118b) for $V(x, y)$.

(b) Using (26.118b) and the method of *superposition*, write down the solution to the problem shown in Fig. 26.24.

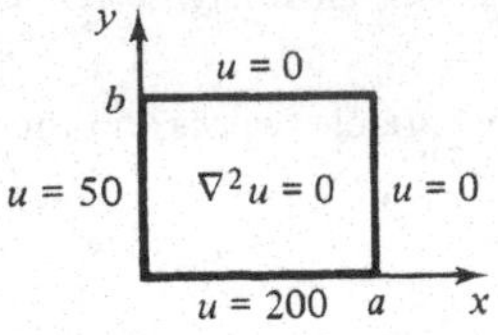

Figure 26.24. Steady-state heat conduction in a rectangle.

26.35. Use (26.118b), *superposition*, and the *method of images* to solve the problem shown in Fig. 26.25.

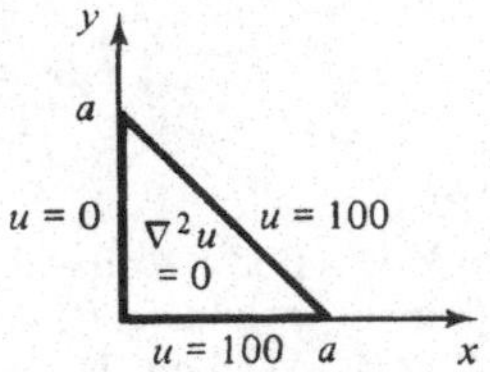

Figure 26.25. Steady-state heat conduction in a triangle.

26.36. Solve the following (two-dimensional) boundary value problems.

(a) $\nabla^2 u = 0$ in $a < r < b$; $u(a, \theta) = f(\theta)$, $u(b, \theta) = 0$

(b) $\nabla^2 u = 0$ in $a < r < b$; $u(a, \theta) = f(\theta)$, $u(b, \theta) = g(\theta)$

(c) $\nabla^2 u = 1$ in $0 \leq r < 1$; $u(1, \theta) = 0$.

26.37. (a) For the problem $\nabla^2 u = 0$ in $y > 0$, $u(x, 0) = H(x)$ (Heaviside function), observe (draw a picture) that there is no "reference length" in the problem, which suggests that u is independent of r, where r, θ are polar coordinates out of the origin. Use this hint to solve.

(b) Use the solution of part (a), plus superposition, to solve the problem $\nabla^2 u = 0$ in $y > 0$, $u(x, 0) = 0$ for $|x| > 1$, and 100 for $|x| < 1$. *Hint:* Express $u(x, 0) = 100H(x + 1) - 100H(x - 1)$. Thus $u = u_1 - u_2$, where $\nabla^2 u_1 = 0$ in $y > 0$, $u_1(x, 0) = 100H(x + 1)$ and $\nabla^2 u_2 = 0$ in $y > 0$, $u_2(x, 0) = 100H(x - 1)$. In expressing the answer in terms of x, y, be careful to choose the branches of the $\tan^{-1}$'s correctly. (See also Exercise 26.44.)

26.38. (a) With $z = x + iy$ and $\bar{z} = x - iy$, show that

$$\frac{\partial^2}{\partial x^2} + \frac{\partial^2}{\partial y^2} = 4\frac{\partial^2}{\partial z\, \partial \bar{z}}.$$

(b) Use part (a) to find the general solution of $\nabla^2 \phi = f(x, y)$. Work out and verify a particular solution for $f = xy$.

(c) Use part (a) to find the general solution of $\nabla^4\phi = f(x, y)$. Work out and verify a particular solution for $f = 1$.

26.39. (*Plane flow over circular cylinder*) (a) Solve the boundary value problem (9.96) for the velocity potential ϕ. From ϕ compute the tangential velocity on the cylinder $r = a$ as a function of θ.

(b) If the stream speed U is a prescribed function of time, then the problem statement (9.96) still holds with U changed to $U(t)$. What is the solution $\phi(r, \theta, t)$ then?

26.40. Solve the steady-state, *three*-dimensional, heat conduction problem

$$\nabla^2 u = 0 \qquad (0 < x < \pi, 0 < y < 2\pi, 0 < z < 2\pi),$$

subject to the boundary conditions $u(\pi, y, z) = 100$ and $u = 0$ on the other five faces. *Hint:* Seek $u = X(x)\phi(y, z)$ and obtain $X'' - k^2 X = 0$, plus the **Helmholtz equation** $\nabla^2\phi + k^2\phi = 0$, with $\phi = 0$ on $y = 0, 2\pi$ and $z = 0, 2\pi$. Solve the ϕ problem, in turn, by separation of variables. Saving the boundary condition $u(\pi, y, z) = 100$ for last, use the completeness of the eigenfunctions of the ϕ problem.

26.41. (a) Find the steady-state temperature u in a sphere of radius a whose upper hemispherical surface is held at 100° while the lower is held at 0°. *Hint:* Use spherical polars, as discussed in Exercise 7.11. Using the axisymmetry, show that $\nabla^2 u = 0$ becomes

$$\rho^2 u_{\rho\rho} + 2\rho u_\rho + u_{\theta\theta} + \cot\theta\, u_\theta = 0.$$

Seeking $u(\rho, \theta) = R(\rho)\Theta(\theta)$, identify the Θ equation as a *Legendre equation* (Section 22.4) through the change of variables $\cos\theta = x$ and obtain

$$u(\rho, \theta) = \sum_0^\infty A_n \rho^n P_n(\cos\theta).$$

Finally, set

$$u(a, \theta) = f(\theta) = \sum_0^\infty A_n a^n P_n(\cos\theta),$$

where $f(\theta) = 100$ for $0 < \theta < \pi/2$ and 0 for $\pi/2 < \theta < \pi$, and show that

$$A_n = \frac{2n+1}{2a^n}\int_0^\pi f(\theta) P_n(\cos\theta)\sin\theta\, d\theta.$$

(b) Repeat part (a) for the case where the domain is $r > a$ instead.

26.42. (*Axisymmetric flow over a sphere*) Reformulate the boundary value problem (9.96) with the circular cylinder replaced by a *sphere* of radius a and derive the result

$$\phi(\rho, \theta) = U\left(\rho + \frac{a^3}{2\rho^2}\right)\cos\theta,$$

where $\theta = 0$ is the axis of symmetry, taken positive downstream. (Recall the preceding exercise.)

26.43. (*Parabolic coordinates*) Suppose that an inviscid fluid flows with a uniform velocity $U = 1$ in the x direction over a flat semi-infinite plate that has no thickness and that coincides with the positive x axis. The fluid is originally (i.e., far upstream) at temperature $T = 0$, say, and becomes heated by the plate, which is held at $T = 1$, say. Then $T(x, y)$ is governed by the PDE

$$\alpha^2\nabla^2 T = \frac{DT}{Dt} = \frac{\partial T}{\partial t} + U\frac{\partial T}{\partial x} = \frac{\partial T}{\partial x},$$

plus the conditions $T = 1$ on $y = 0$ $(x > 0)$ and $T \longrightarrow 0$ far upstream. Changing from x, y to ξ, η according to $x + iy = (\xi + i\eta)^2$, show that the PDE becomes

$$\alpha^2(T_{\xi\xi} + T_{\eta\eta}) = 2(\xi T_\xi - \eta T_\eta).$$

(Exercise 7.12 may help.) In fact, the solution depends only on η and is given by

$$T = \operatorname{erfc}\left(\frac{\eta}{\alpha}\right).$$

Derive this result and give η in terms of x and y. (See Fig. 26.26.)

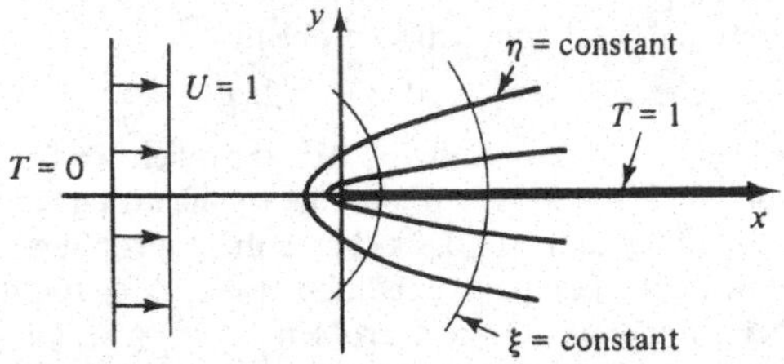

Figure 26.26. Plane flow over a heated plate.

26.44. (*Elliptical coordinates*) Solve the Dirichlet half-plane problem stated in Exercise 26.37(b) by means of elliptical coordinates ξ, η defined by

$$x = a \cosh \xi \cos \eta, \qquad y = a \sinh \xi \sin \eta,$$

where $\xi \geq 0$ and $0 \leq \eta < 2\pi$, as sketched in Fig. 26.27; the constant ξ, η curves are confocal ellipses and hyperbolas. The Laplacian operator is then

$$\nabla^2 = \frac{1}{a^2(\sinh^2 \xi + \sin^2 \eta)}\left(\frac{\partial^2}{\partial \xi^2} + \frac{\partial^2}{\partial \eta^2}\right).$$

What (if any) is the relationship between this approach and conformal mapping?

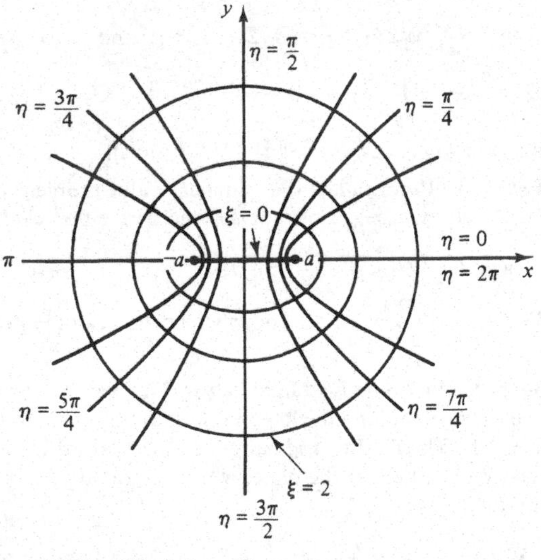

Figure 26.27. Elliptical coordinates.

26.45. (*Vibrating cantilever beam*) Recall [from (10.85) with $\mu = 0$] that the free vibration of a uniform beam is governed by the PDE

$$y_{xxxx} + \frac{\rho}{EI} y_{tt} = 0.$$

Solve this by separation of variables for a cantilever beam of length l (Fig. 26.28) with an initial displacement $y(x, 0) = f(x)$—that is, with the conditions

$$y(0, t) = y_x(0, t) = y_{xx}(l, t) = y_{xxx}(l, t) = 0, \qquad y(x, 0) = f(x), \qquad y_t(x, 0) = 0.$$

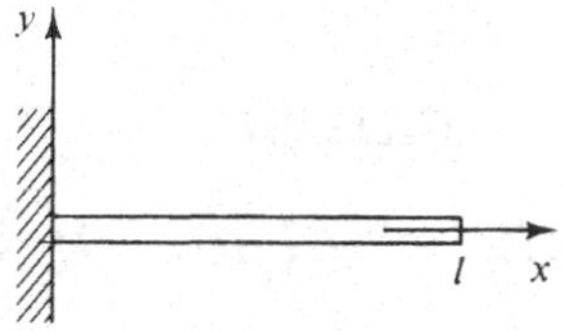

Figure 26.28. Cantilever beam.

Show that the solution is

$$y(x, t) = \sum_1^\infty H_n \Psi_n(x) \cos\left(\sqrt{\frac{EI}{\rho}}\frac{z_n^2}{l^2} t\right),$$

where

$$\Psi_n(x) = \sin\frac{z_n x}{l} - \sinh\frac{z_n x}{l} + \left(\frac{\cos z_n + \cosh z_n}{\sin z_n - \sinh z_n}\right)\left(\cos\frac{z_n x}{l} - \cosh\frac{z_n x}{l}\right)$$

$$H_n = \frac{\int_0^l f(x)\Psi_n(x)\,dx}{\int_0^l \Psi_n^2(x)\,dx},$$

and the z_n's are the roots of the transcendental equation $\cos z \cosh z + 1 = 0$.

26.46. Suppose that $u(x, y)$ satisfies the inhomogeneous **biharmonic equation**

$$\nabla^4 u = \frac{\partial^4 u}{\partial x^4} + 2\frac{\partial^4 u}{\partial x^2 \partial y^2} + \frac{\partial^4 u}{\partial y^4} = \phi(x, y)$$

in a finite two-dimensional region D with boundary C, as well as the boundary conditions $u = f$ and $u_n = g$ on C. Show that the problem admits at most one solution. *Hint:* Suppose there are two solutions, u_1 and u_2, and set $w \equiv u_1 - u_2$. Taking the two functions in Green's second identity to be w and $\nabla^2 w$, show that

$$\int_D [w\nabla^4 w - (\nabla^2 w)^2]\,d\sigma = \int_C \left[w\frac{\partial}{\partial n}(\nabla^2 w) - (\nabla^2 w)\frac{\partial w}{\partial n}\right] ds.$$

Thus show that w must also satisfy *Laplace*'s equation in D and, recalling Exercise 26.27, must vanish everywhere in D.

26.47. Make up two examination-type problems on this chapter.

Chapter 27

Classification and the Method of Characteristics

Most of this chapter deals with hyperbolic systems and their numerical solution by the method of characteristics. At the same time important differences and similarities in the parabolic, hyperbolic, and elliptic versions of the second-order equation $Au_{xx} + 2Bu_{xy} + Cu_{yy} = F$, some of which were already noted in Chapter 26, are discussed.

27.1. CHARACTERISTICS AND CLASSIFICATION

Consider, for instance, the system

$$\begin{aligned} a_1u_x + b_1u_y + c_1v_x + d_1v_y &= f_1 \\ a_2u_x + b_2u_y + c_2v_x + d_2v_y &= f_2 \end{aligned} \tag{27.1}$$

in two dependent variables u, v, where $a_1, a_2, \ldots, f_1, f_2$ are allowed to be functions of x, y, u, v. In general, then, (27.1) is nonlinear; but since it is at least linear in the derivatives u_x, u_y, v_x, v_y, it is customary to call it *quasilinear* (or *planar*, by analogy with the equation of a plane in n dimensions, $a_1x_1 + a_2x_2 + \cdots + a_nx_n =$ constant).

Example 27.1. One-dimensional isentropic flow of a perfect gas:

$$u_t + uu_x + \rho^{\gamma-2}\rho_x = 0 \qquad \text{(equation of motion)}$$

$$\rho_t + u\rho_x + \rho u_x = 0 \qquad \text{(continuity equation)},$$

where $u(x, t)$ is the x velocity and $\rho(x, t)$ is the mass density. ▮

Example 27.2. $\phi_{xx} + \phi_{yy} = 0$. Let $\phi_x \equiv u$ and $\phi_y \equiv v$. Then

$$u_x + v_y = 0, \qquad u_y - v_x = 0. \blacksquare$$

Let us consider the following *fundamental problem*. We wish to determine a solution of (27.1) throughout some region R of the x, y plane that satisfies given data on some curve Γ within R. That is, given u and v on Γ, we wish to extend the solution *off* of Γ. For instance, consider the calculation of u and v at some point Q that is a distance ds from a point P on Γ (Fig. 27.1). Then $u_Q = u_P + (du)_s$ and $v_Q = v_P + (dv)_s$, where u_P and v_P are known and

$$(du)_s = u_x(dx)_s + u_y(dy)_s$$
$$(dv)_s = v_x(dx)_s + v_y(dy)_s \tag{27.2}$$

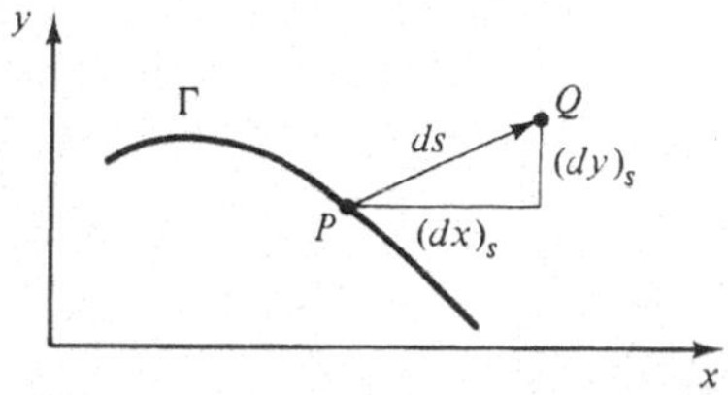

Figure 27.1. Calculation off of Γ.

by chain differentiation. In addition to (27.2), we have the governing equations (27.1), so that

$$\begin{pmatrix} a_1 & b_1 & c_1 & d_1 \\ a_2 & b_2 & c_2 & d_2 \\ (dx)_s & (dy)_s & 0 & 0 \\ 0 & 0 & (dx)_s & (dy)_s \end{pmatrix} \begin{pmatrix} u_x \\ u_y \\ v_x \\ v_y \end{pmatrix} = \begin{pmatrix} f_1 \\ f_2 \\ (du)_s \\ (dv)_s \end{pmatrix}. \tag{27.3}$$

Tentative Scheme: First let ds lie along Γ. Then $(du)_\Gamma$ and $(dv)_\Gamma$ on the right-hand side of (27.3) are known, since u and v are prescribed on Γ. Solve (27.3) for u_x, u_y, v_x, v_y (i.e., at P). With the four partials evaluated, we can compute $(du)_s$ and $(dv)_s$ from (26.2) and, finally,

$$u_Q = u_P + (du)_s \quad \text{and} \quad v_Q = v_P + (dv)_s.$$

Observe that the scheme fails in the event that the Γ direction through P is such that the determinant of the coefficient matrix vanishes—that is, if

$$\alpha(dy)^2 + 2\beta\, dy\, dx + \gamma(dx)^2 = 0, \tag{27.4}$$

where

$$\alpha = c_1 a_2 - a_1 c_2$$
$$2\beta = b_1 c_2 - c_1 b_2 + a_1 d_2 - a_2 d_1$$
$$\gamma = d_1 b_2 - b_1 d_2$$

or (if $\alpha \neq 0$)

$$\frac{dy}{dx} = \frac{-\beta \pm \sqrt{\beta^2 - \alpha\gamma}}{\alpha}. \tag{27.5}$$

These are called **characteristic directions**. Depending on the discriminant $\beta^2 - \alpha\gamma$, there are three possibilities.

$$\text{If } \beta^2 - \alpha\gamma \begin{cases} > 0, & \text{there are } \textit{two} \text{ characteristic directions.} \\ = 0, & \text{there is } \textit{one} \text{ characteristic direction.} \\ < 0, & \text{there are } \textit{no} \text{ (real) characteristic directions.} \end{cases}$$

We classify (27.1) accordingly as **hyperbolic**, **parabolic**, or **elliptic** at P.

Example 27.3. Consider the (important) single, second-order, quasilinear PDE

$$A\phi_{xx} + 2B\phi_{xy} + C\phi_{yy} = F, \tag{27.6}$$

that is, where A, B, C, F are functions of x, y, ϕ, ϕ_x, ϕ_y. Setting

$$\phi_x \equiv u, \qquad \phi_y \equiv v,$$

we have

$$Au_x + 2Bu_y + Cv_y = F, \qquad u_y - v_x = 0, \tag{27.7}$$

which is of the form (27.1) with $a_1 = A$, $b_1 = 2B$, $d_1 = C$, $f_1 = F$, $b_2 = 1$, $c_2 = -1$, $c_1 = a_2 = d_2 = f_2 = 0$, and (27.5) becomes

$$\frac{dy}{dx} = \frac{B \pm \sqrt{B^2 - AC}}{A}. \tag{27.8}$$

Wave equation:

$$a^2\phi_{xx} = \phi_{yy} \qquad (a = \text{constant}). \tag{27.9}$$

Comparing (27.9) with (27.6), we see that $A = a^2$, $B = 0$, $C = -1$, so that $B^2 - AC = 0 - (a^2)(-1) = a^2 > 0$, and thus (27.9) is *hyperbolic* for all x, y.

Diffusion equation:

$$\alpha^2\phi_{xx} = \phi_y \qquad (\alpha = \text{constant}). \tag{27.10}$$

$$B^2 - AC = 0 - (\alpha^2)(0) = 0,$$

and so (27.10) is *parabolic* for all x, y.

Laplace equation:

$$\phi_{xx} + \phi_{yy} = 0. \tag{27.11}$$

$$B^2 - AC = 0 - (1)(1) = -1 < 0;$$

therefore (27.11) is *elliptic* for all x, y. ∎

27.2. *THE HYPERBOLIC CASE*

Suppose that (27.1) is hyperbolic throughout the domain of interest. Then at *each* point x, y there will be two characteristc directions, as defined by (27.4) or (27.5), say $(dy/dx)_+$ and $(dy/dx)_-$. These slopes define two families of curves, so-called +*characteristics* that have the slope $(dy/dx)_+$ at each point and −*characteristics* that have the slope $(dy/dx)_-$ at each point.

Example 27.4. For the wave equation (27.9)

$$\left(\frac{dy}{dx}\right)_+ = \frac{B + \sqrt{B^2 - AC}}{A} = \frac{1}{a}, \qquad \left(\frac{dy}{dx}\right)_- = \frac{B - \sqrt{B^2 - AC}}{A} = -\frac{1}{a}. \tag{27.12}$$

Integrating (27.12), the two families are simply

$$y_+ = \frac{x}{a} + \text{constant}, \qquad y_- = -\frac{x}{a} + \text{constant}, \tag{27.13}$$

as shown in Fig. 27.2. ∎

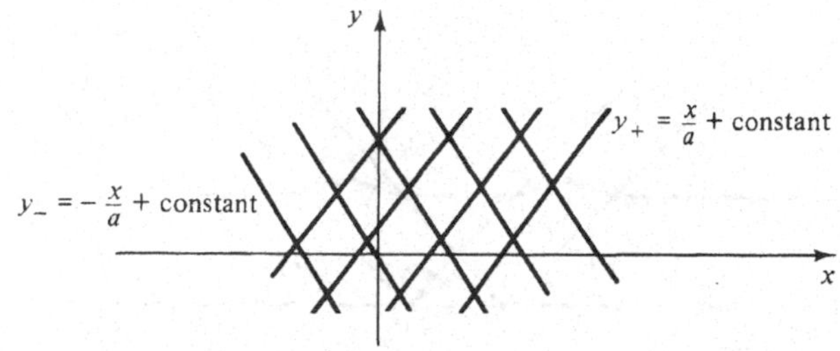

Figure 27.2 ±Characteristics (27.13).

In this example the coefficients A, B, C [or $a_1, a_2, \ldots, d_1, d_2$ in the equivalent system (27.1)] were constants, and so $(dy/dx)_\pm$ were constant; consequently, the characteristics were families of straight parallel lines.

If $a_1, a_2, \ldots, d_1, d_2$ are functions of x, y instead, then $(dy/dx)_\pm$ will be functions of x, y and integration will, in general, yield families of *curved* lines for the characteristics.

Still more generally, suppose that $a_1, \ldots, d_2$ are functions of x, y, u, v. Then we again anticipate curved characteristics, but the catch is that

$$\left(\frac{dy}{dx}\right)_+ = f_+(x, y, u, v), \qquad \left(\frac{dy}{dx}\right)_- = f_-(x, y, u, v) \tag{27.14}$$

cannot be integrated yet because $u(x, y), v(x, y)$ are not yet known throughout the domain!

Let us return to the problem originally posed—computing a solution of (27.1) in R, given u and v on Γ. To fix ideas, consider the "string problem"

$$a^2\phi_{xx} - \phi_{yy}$$

$$\phi(x, 0) = F(x), \qquad \phi_y(x, 0) = 0,$$

where ϕ is the string deflection and y is the time; or with $\phi_x \equiv u$ and $\phi_y \equiv v$,

$$a^2 u_x - v_y = 0, \qquad u_y - v_x = 0, \tag{27.15}$$

with the data $u(x, 0) = F'(x)$ and $v(x, 0) = 0$, so that Γ is the x axis. Taking $F(x)$ to be of "pup tent" shape, we know from Chapter 26 that the solution is as shown in Fig. 27.3. This solution might have been obtained by D'Alembert's method, by an integral transform, or by the method of Green's functions (Chapter 28), but these methods may cease to apply or may become quite unwieldy for more complicated hyperbolic systems, and so we continue in our development of a numerical method of solution. Apparently our *Tentative Scheme* in Section 27.1 should work, since Γ (the x axis) is nowhere tangent to a characteristic. However, observe that if we take our ds's in the y direction and march out a solution from the segment MN of Γ, along which $u = v = 0$, then we will obtain $u \equiv 0, v \equiv 0$ (Right?) even as we cross the characteristic ML (on the other side of which u and v are *not* identically zero). That is, our scheme provides an analytic extension off MN, whereas the solution is *not* analytic

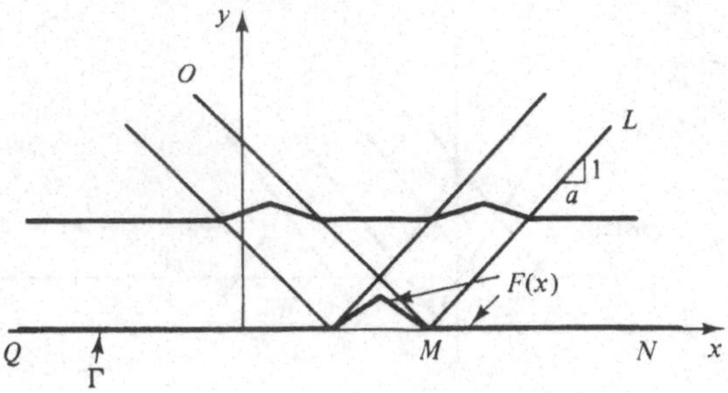

Figure 27.3. Solution to (27.15).

along the ML characteristic. Instead the solution consists of several analytic functions that are "spliced" together along the characteristics ML, OM, and so on.

The message is that *marching across characteristics is to be avoided,* which suggests that we consider marching *along* the characteristics instead. Thus let ds in Fig. 27.1 be along one of the two characteristic directions through P. Then the determinant of the coefficient matrix in (27.3) will be zero and a unique solution for u_x, u_y, v_x, v_y at P will not exist,[1] as we noted earlier. But if equations (27.3) are not to be downright *contradictory* (according to Cramer's rule, Section, Section 19.3), it is necessary and sufficient that the determinant of the coefficient matrix, with any of its columns replaced by the column vector on the right-hand side of (27.3), must vanish too. Thus

$$\begin{vmatrix} a_1 & b_1 & c_1 & f_1 \\ a_2 & b_2 & c_2 & f_2 \\ dx & dy & 0 & du \\ 0 & 0 & dx & dv \end{vmatrix} = 0. \tag{27.16}$$

If, for algebraic simplicity, $f_1 = f_2 = 0$ (this restriction will be relaxed in the water wave example to follow), then (27.16) becomes

$$(a_1b_2 - a_2b_1)\,du + \left[(a_1c_2 - a_2c_1)\frac{dy}{dx} - (b_1c_2 - b_2c_1)\right] dv = 0$$

or

$$\frac{du}{dv} = \frac{(a_1c_2 - a_2c_1)(dy/dx) - (b_1c_2 - b_2c_1)}{a_2b_1 - a_1b_2} \equiv g(x, y, u, v) \tag{27.17}$$

or

$$\left(\frac{du}{dv}\right)_{\pm} = g_{\pm}(x, y, u, v), \tag{27.18}$$

since dy/dx in (27.17) is either $(dy/dx)_+$ or $(dy/dx)_-$. This connection between du and dv along the characteristics is known as the **compatibility relation**.

So we have the four ordinary differential equations

$$\left(\frac{dy}{dx}\right)_{+} = f_{+}(x, y, u, v), \qquad \left(\frac{dy}{dx}\right)_{-} = f_{-}(x, y, u, v) \tag{27.19a}$$

$$\left(\frac{du}{dv}\right)_{+} = g_{+}(x, y, u, v), \qquad \left(\frac{du}{dv}\right)_{-} = g_{-}(x, y, u, v) \tag{27.19b}$$

---

[1] For instance, the values of u_x, u_y, v_x, v_y at M (Fig. 27.3) will depend on whether ML is considered as the edge of the region LMN or of the region LMQ.

together with starting data along Γ, and this system can be integrated by Runge–Kutta or any of the other standard methods discussed in Chapter 24. This is the so-called **method of characteristics**. In general, then, we march out a solution for both the characteristic "net" and the u, v solution simultaneously. Of course, if the $f_\pm$ do not depend on u, v, then the entire net may be computed in advance of u and v.

To illustrate, let us simply use Euler's method to integrate. From data at P and Q (Fig. 27.4) we compute at R as follows:

$$\begin{aligned} y_R &= y_P + f_+(P)(x_R - x_P), \\ y_R &= y_Q + f_-(Q)(x_R - x_Q) \\ u_R &= u_P + g_+(P)(v_R - v_P), \\ u_R &= u_Q + g_-(Q)(v_R - v_Q) \end{aligned} \tag{27.20}$$

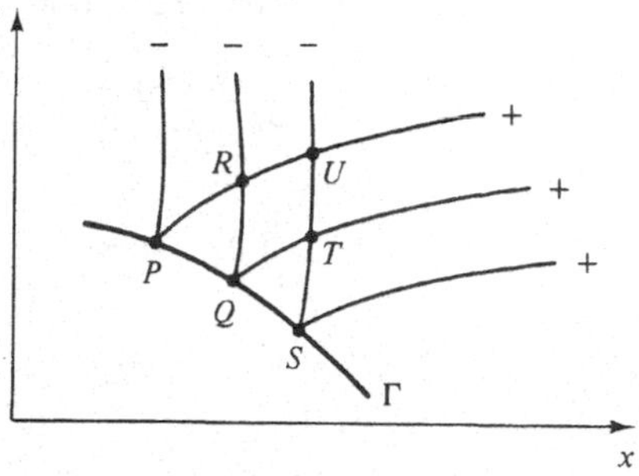

Figure 27.4. Method of characteristics.

or solving for x_R, y_R, u_R, v_R,

$$\begin{aligned} x_R &= \frac{y_Q - y_P + x_P f_+(P) - x_Q f_-(Q)}{f_+(P) - f_-(Q)} \\ y_R &= y_P + f_+(P)(x_R - x_P) \\ v_R &= \frac{u_Q - u_P + v_P g_+(P) - v_Q g_-(Q)}{g_+(P) - g_-(Q)} \\ u_R &= u_P + g_+(P)(v_R - v_P). \end{aligned} \tag{27.21}$$

Similarly, we can use Q and S to compute at T, R and T to compute at U, and so on.

The simple Euler integration amounts to a linear extrapolation scheme. For instance, if the actual characteristics through P and Q are as shown by the solid lines in Fig. 27.5, then the intersection R predicted by the first two of equations (27.21) would be as indicated by the dotted lines.

In applications we would probably use a higher-order integration scheme.

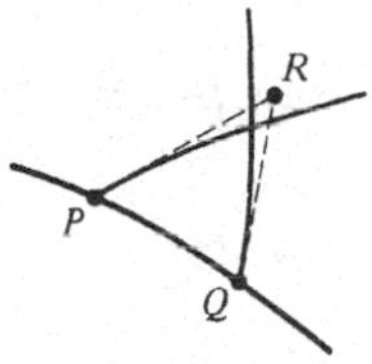

Figure 27.5. Euler integration.

Example 27.5. *Application to Water Waves; Shallow Water Theory.* Consider the two-dimensional, irrotational, inviscid motion of water, as shown in Fig. 27.6. The undisturbed water level coincides with the x axis, the (rigid) bottom is given by $y = -h(x)$, and the free surface is $y = \eta(x, t)$.

Let l be some measure of the length of the waves under consideration and d be a measure of the water depth [e.g., an average value of $h(x)$]. If $\sigma \equiv d/l \ll 1$, we are

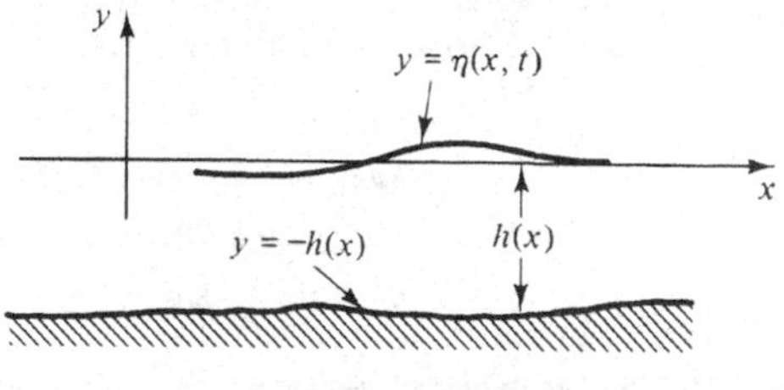

Figure 27.6. Shallow water theory.

within the so-called shallow water theory (also known as long wave theory), which applies to a number of important cases—the tides in the oceans (where the wave length is *many* times greater than the ocean depth), waves on beaches, floods, and open channel flow. If we expand the quantities of interest in powers of σ and plug into the governing equations, we find,[2] to zeroth order in σ, that

$$u_t + uu_x = -g\eta_x, \qquad [u(\eta + h)]_x = -\eta_t, \tag{27.22}$$

where g is the acceleration of gravity and u is the x-velocity component. [Remarkably, the derivation of (27.22) indicates that u does not depend on y—that is, $u = u(x, t)$—so that we have, in effect, a "one-dimensional" flow.]

If we further assume that u, η and their derivatives are all small quantities, say $O(\epsilon)$, then the linearized version of (27.22) is

$$u_t = -g\eta_x, \qquad (uh)_x = -\eta_t. \tag{27.23}$$

If desired, we can eliminate either u or η from (27.23)—for instance,

$$g(h\eta_x)_x = \eta_{tt}. \tag{27.24}$$

Finally, if h is a constant, we have the classical wave equation

$$c^2\eta_{xx} = \eta_{tt}, \tag{27.25}$$

where

$$c = \sqrt{gh} \tag{27.26}$$

is the wave speed.

But let us work with the nonlinear system (27.22). It is convenient and customary to define $c \equiv \sqrt{g(\eta + h)}$, in which case (27.22) becomes

$$u_t + uu_x + 2cc_x = H_x, \qquad 2c_t + 2uc_x + cu_x = 0, \tag{27.27}$$

where $H(x) \equiv gh(x)$. Then (27.3) becomes

$$\begin{pmatrix} u & 1 & 2c & 0 \\ c & 0 & 2u & 2 \\ dx & dt & 0 & 0 \\ 0 & 0 & dx & dt \end{pmatrix} \begin{pmatrix} u_x \\ u_t \\ c_x \\ c_t \end{pmatrix} = \begin{pmatrix} H_x \\ 0 \\ du \\ dc \end{pmatrix}, \tag{27.28}$$

and it follows (Exercise 27.3) that the characteristic and compatibility equations are

$$\left(\frac{dx}{dt}\right)_{\pm} = u \pm c \tag{27.29}$$

and

$$(du)_{\pm} \pm 2(dc)_{\pm} - H_x(dt)_{\pm} = 0. \tag{27.30}$$

Recall that in deriving (27.18), we assumed that $f_1 = f_2 = 0$. In the present case, however, f_1 is H_x, and its inclusion leads to the third term in (27.30). Although this term prevents us from expressing (27.30) in the neat form of (27.18), it does not prevent us from carrying out the integration. In fact, Euler integration of (27.29) and (27.30) gives (recall Fig. 27.4)

$$\left.\begin{aligned} x_R &= x_P + (u_P + c_P)(t_R - t_P) \\ x_R &= x_Q + (u_Q - c_Q)(t_R - t_Q) \\ u_R - u_P &+ 2c_R - 2c_P - H_x(x_P)(t_R - t_P) = 0 \\ u_R - u_Q &- 2c_R + 2c_Q - H_x(x_Q)(t_R - t_Q) = 0. \end{aligned}\right\} \tag{27.31}$$

[2] J. J. Stoker, *Water Waves*, Interscience, New York, 1957; see pp. 22–24 for a short formal derivation and pp. 27–32 for the perturbation approach.

To illustrate, suppose that

$$h = \text{constant} = 1, \qquad u(x, 0) = 0, \qquad \eta(x, 0) = F(x) \tag{27.32}$$

[so that $c(x, 0) = \sqrt{g[F(x) + 1]}$], where $F(x)$ is shown in Fig. 27.7.

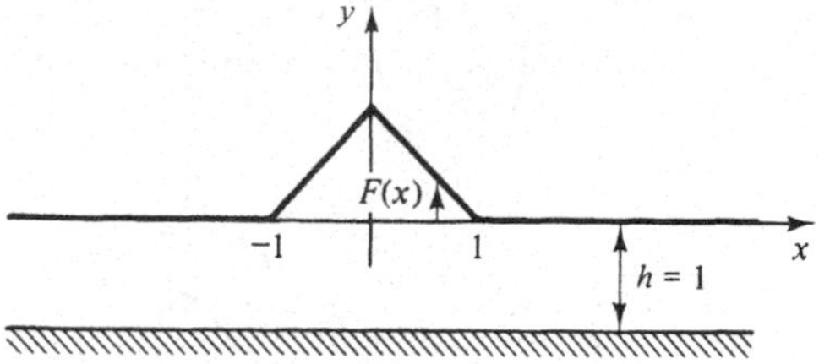

Figure 27.7. Initial free surface shape.

With data given on the x axis (i.e., in the x, t plane), let us march out a solution from the segment $-6 \leq x \leq 6$ say. Define the discrete initial data points

$$x_{j0} = -6 + 0.25(j - 1), \qquad t_{j0} = 0 \quad (j = 1, 2, \dots, 49)$$

say, where x_{jk}, t_{jk} denote the x and t coordinates of the kth mesh point along the jth +characteristic; for instance, the $*$ point in Fig. 27.8 has the coordinates x_{23}, t_{23}.

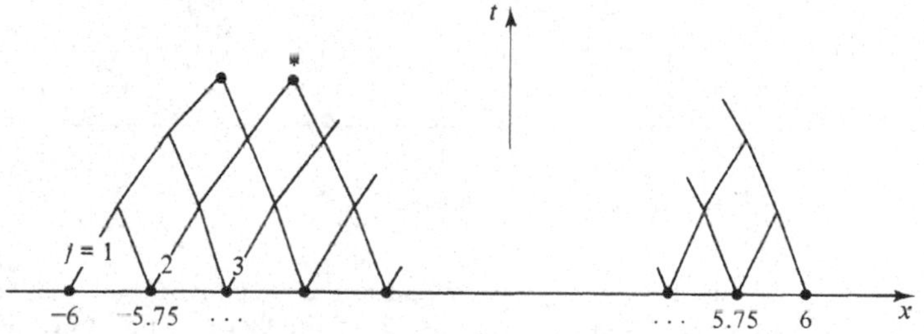

Figure 27.8. Mesh layout.

With $H_x = 0$, since $h(x)$ is constant, solution of (27.31) for x_R, t_R, u_R, c_R gives

$$\left.\begin{aligned} t_R &= \frac{x_Q - x_P - (u_Q - c_Q)t_Q + (u_P + c_P)t_P}{u_P - u_Q + c_P + c_Q} \\ x_R &= x_P + (u_P + c_P)(t_R - t_P) \\ u_R &= c_P - c_Q + \frac{u_P + u_Q}{2} \\ c_R &= c_Q + \frac{u_R - u_Q}{2}, \end{aligned}\right\} \tag{27.33}$$

where $(\ )_P \equiv (\ )_{j,k}$, $(\ )_Q \equiv (\ )_{j+1,k}$, and $(\ )_R \equiv (\ )_{j,k+1}$. That is, starting with

$$t_{j0} = 0, \qquad x_{j0} = -6 + 0.25(j - 1), \qquad u_{j0} = 0,$$

j	22	23	24	25	26	27	28	all other j's
c_{j0}	6.34	6.95	7.51	8.02	7.51	6.95	6.34	5.67

we compute $t_{j1}, x_{j1}, u_{j1}, c_{j1}$ from (27.33) for $j = 1$ through 48, then $t_{j2}, x_{j2}, u_{j2}, c_{j2}$ for $j = 1$ through 47, and so on. Clearly, the farthest we can carry the calculation is through $k = 48$, since our net then terminates at a point.

The results are shown in Fig. 27.9.[3] Only the $x > 0$ portion is shown because the results will obviously be symmetric about $x = 0$. To keep the figure uncluttered, not all the characteristics used in the calculation are drawn.

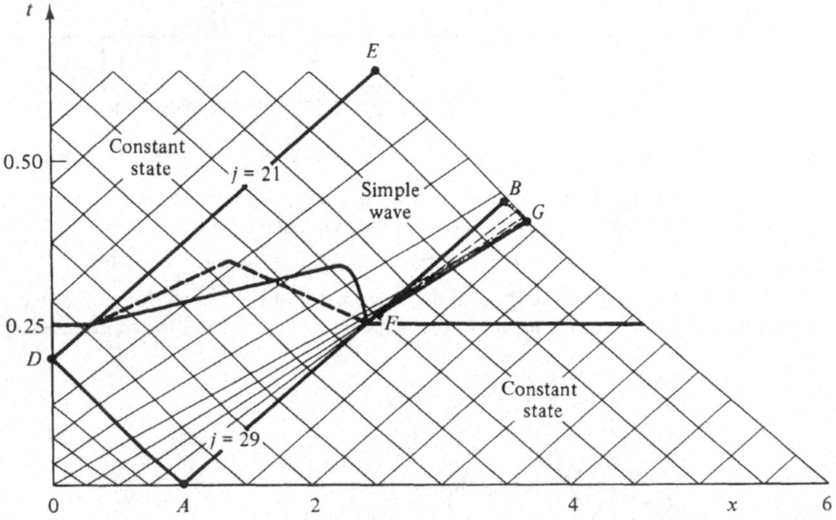

Figure 27.9. Shallow water theory calculation.

COMMENT 1. First, observe that in the region ABC both families of characteristics are parallel straight lines in spite of the fact that the system (27.27) is nonlinear. This result occurs because ABC is a region of *constant state*; that is, u and c are constant there ($u = 0$ and $c = \sqrt{g} \doteq 5.67$), so that, in effect, (27.27) is linear with constant coefficients. Similarly, the region between DE and the t axis is also a region of constant state with $u = c = 0$. In between, in $ABED$, we find that both u and c are constant along the +characteristics, which are straight but nonparallel.

$ABED$ is an example of what is called a *simple wave*, a region in which the characteristics of one family (in this case, the +characteristics) are straight but nonparallel and along which the solution is a constant and where the characteristics of the other family are, in general, curved (here the curvature is very slight) and all have the same slope as they cross a given characteristic of the other family. It can be shown[4] that the region adjoining a constant state *must* be a simple wave. (Note that the corner region, to the left of DA, is not a simple wave.)

---

[3]I am grateful to Ms. Wendy Morrison, who was kind enough to program and run these calculations.

[4]R. Courant and K. O. Friedrichs, *Supersonic Flow and Shock Waves*, Interscience, New York, 1948, p. 59.

COMMENT 2. For comparison with the linearized theory—that is, (27.25)—we've plotted the linearized solution at time $t = 0.25$. Naturally it's a pup tent shape of height 0.5 [i.e., half the original pulse $F(x)$, according to the D'Alembert solution given in Chapter 26], as indicated by the dashed curve. But because of the way that the +characteristics "fan" and the way that η is constant along each one, we see that the nonlinear solution has steepened at the right and flattened at the left. By the time that we reach $t \approx 0.26$ the wave slope has become *infinite* at its right edge, and we might say that the wave is about to "break." Beyond this point the fanlike characteristics cross, and this situation implies a multivalued solution, which is not acceptable; the solution needs to be modified so as to allow for a *bore* along the envelope FG.[5]

Note carefully that the degree to which the nonlinear effects are manifested is directly related to the magnitude of the initial disturbance $F(x)$; for instance, if $F(x)$ were reduced by a factor of 10, we would find that the fanning of the characteristics is very slight. Steepening and developing of a bore would still take place but much more slowly. This situation should not be surprising; as a more familiar example, note that for the pendulum problem

$$\ddot{x} + \frac{g}{\ell}\sin x = 0, \qquad x(0) = x_0, \quad \dot{x}(0) = 0$$

x tends to the "linearized solution" $x_0 \cos\sqrt{g/\ell}\,t$ as the initial condition $x_0 \longrightarrow 0$ and deviates more and more as x_0 is increased.

COMMENT 3. Without spelling out the details of the equivalence,[6] we might at least mention that equations (27.22) also govern a class of flows in one-dimensional gasdynamics. In this case, the development of a bore corresponds to the development of a *shock wave.* Incidentally, equations (27.22) are not the "final word" in modeling long waves or one-dimensional gasdynamics, but they do contain most of the physics and serve as a nice illustration of the method of characteristics.

COMMENT 4. Finally, it is interesting to note, in passing, how our wave is governed by the *elliptic* Laplace equation $\nabla^2\phi = 0$ (since the flow is inviscid and initially irrotational, and hence irrotational for *all* time, as well as incompressible) on the one hand and the *hyperbolic* system (27.22) on the other. ∎

27.3. REDUCTION OF $A\phi_{xx} + 2B\phi_{xy} + C\phi_{yy} = F$ TO NORMAL FORM

Suppose that the characteristics of the quasilinear equation

$$A\phi_{xx} + 2B\phi_{xy} + C\phi_{yy} = F \tag{27.34}$$

are given by

$$\psi^+(x, y) = \text{constant}, \qquad \psi^-(x, y) = \text{constant}. \tag{27.35}$$

If (27.35) is *hyperbolic*, then the $\psi^\pm$ are real and distinct. The method of characteristics revealed that the $\psi^\pm = $ constant curves constitute a more "natural" (cur-

[5]See, for example, Stoker, *op. cit.*, p. 351.
[6]Stoker, *op. cit.*, p. 25.

vilinear) coordinate system than the original x, y system, and this fact suggests actually changing variables from x, y to ξ, η according to

$$\psi^+(x, y) \equiv \xi, \qquad \psi^-(x, y) \equiv \eta. \tag{27.36}$$

In fact, by doing so, we find (after some chain differentiation; Exercise 27.4) that (27.34) reduces to the so-called *normal* (or *canonical* or "simplest") *form*

$$\phi_{\xi\eta} = \mathfrak{F}(\xi, \eta, \phi, \phi_\xi, \phi_\eta). \tag{27.37}$$

Example 27.6. Recall that the characteristics for the wave equation (27.9) are

$$\psi^+(x, y) = x + ay, \qquad \psi^-(x, y) = x - ay.$$

Introducing

$$x + ay \equiv \xi, \qquad x - ay \equiv \eta, \tag{27.38}$$

we find that (27.9) becomes

$$\phi_{\xi\eta} = 0, \tag{27.39}$$

which can be integrated easily to give

$$\phi = F_1(\xi) + F_2(\eta) = F_1(x + ay) + F_2(x - ay). \tag{27.40}$$

Of course, this result coincides with the D'Alembert solution discussed in Chapter 26, but this time the change of variables (27.38) is introduced systematically and not merely "out of the blue." ▮

If (27.35) is *elliptic*, it turns out that the change of variables

$$\psi^+(x, y) \equiv \xi + i\eta, \qquad \psi^-(x, y) \equiv \xi - i\eta \tag{27.41}$$

leads to the normal form

$$\phi_{\xi\xi} + \phi_{\eta\eta} = \mathfrak{F}(\xi, \eta, \phi, \phi_\xi, \phi_\eta) \tag{27.42}$$

if (27.34) is Laplace's equation or Poisson's equation, (27.41) yields $x = \xi$, $y = \eta$—that is, the Laplace and Poisson equations are already in normal form.

Finally, if (27.35) is *parabolic*, it turns out that the change of variables

$$\psi^+(x, y) = \psi^-(x, y) \equiv \xi, \qquad x \equiv \eta \tag{27.43}$$

leads to the normal form

$$\phi_{\eta\eta} = \mathfrak{F}(\xi, \eta, \phi, \phi_\xi, \phi_\eta). \tag{27.44}$$

27.4. *COMPARISON OF HYPERBOLIC, ELLIPTIC, AND PARABOLIC SYSTEMS; SUMMARY*

In summarizing the essential features associated with the three types of systems (from Chapters 26 and 27), let us compare the three representative problems H, E, and P shown in Fig. 27.10. Although by no means the most general problems of their type, they will suffice for our purposes.

For the *hyperbolic* case H the domain is the semi-infinite strip. There are two fami-

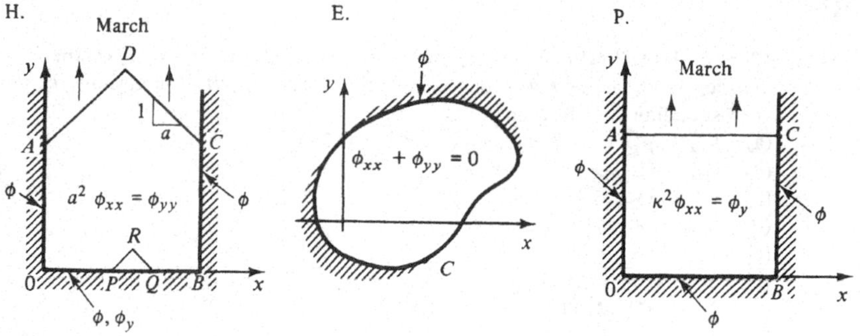

Figure 27.10. Representative hyperbolic, elliptic, and parabolic problems.

lies of characteristics, $x \pm ay =$ constant, and the boundary data indicated on $AOBC$ suffices to permit us to march a solution out as far as ADC. The solution need *not* be analytic, since singularities in boundary data propagate into the region along characteristics.

The domain for the *elliptic* case E is the closed region bounded by C. There are *no* (real) characteristics and no "method of characteristics" as in the hyperbolic case. Although the solution at R is completely determined by data on the segment PQ for case H, the fact is that the solution at any interior point in case E is dependent on *all* the boundary data—that is, over the entire curve C; for instance, recall the Poisson integral formula (26.105). We say that E is of *boundary value* type, whereas H is an *initial value* (i.e., if we think of y as the time) or *marching* problem. Also in sharp contrast with the hyperbolic case, discontinuities of the boundary values on C are *not* continued into the interior; in fact, the solution will be beautifully *analytic* throughout the interior.

The *parabolic* case P turns out to be a sort of cross between the hyperbolic and elliptic systems.

(i) The two families of characteristics coalesce into *one*—namely, the lines $y =$ constant.

(ii) As in the hyperbolic case, the region is open and the solution may be marched out; for instance, ϕ boundary data over $AOBC$ determines the solution up to the line AC.

(iii) The solution will be an analytic function of x for each $y > 0$, but not necessarily of x *and* y. For example, if $\phi = 0$ along the $AOBC$ part of the boundary for the P problem in Fig. 27.10, then the solution within that rectangle will be $\phi(x, y) \equiv 0$. If ϕ were to be an analytic function of x and y throughout the semi-infinite strip domain, it would follow that $\phi \equiv 0$ above the line AC as well, that is, the analytic continuation of the solution $\phi \equiv 0$ within $AOBC$. However, if the boundary data on ϕ is not identically zero above the points A and C, then it will obviously *not* be true that $\phi \equiv 0$ above (the characteristic) AC, and hence the solution will not be analytic in both x and y.

EXERCISES

27.1. Classify the following equations and systems of equations as to whether they are elliptic, parabolic, or hyperbolic. If of mixed type, specify the various regions and corresponding classifications.
(a) $u_{xx} = p(x)u_{tt} + h(x)u_t + f(x)e^{i\omega t}$; $p(x) > 0$
(b) $u_{xx} + (1 - x)^2 u_{yy} = 6$
(c) $u_{xy} + xyu_{yy} - u_y = 0$
(d) $u_{xx} = p(x)u_t + h(x)u + f(x, t)$
(e) $u_x = (1 - u)^2 u_{yy}$
(f) $e_x + Ri = 0$ and $i_x + Ce_t = 0$,
where R and C are positive constants

27.2. We converted the second-order equation (27.6) to the first-order system (27.7) by introducing $\phi_x \equiv u$ and $\phi_y \equiv v$. Is this choice unique? How about $\phi_x \equiv u$ and $\phi_x + \phi_y \equiv v$, for example?

27.3. Derive the characteristic and compatibility equations (27.29) and (27.30).

27.4. Verify that the transformation (27.36) does, in fact, reduce (27.34) to the normal form (27.37) if (27.34) is hyperbolic.

27.5. Reduce to their normal form.
(a) $\epsilon\phi_{xx} + \phi_{yy} = 0$ $(\epsilon > 0)$
(b) $\phi_{xx} - x^2\phi_{yy} - \kappa\phi^3 = 0$
(c) $\phi_{xx} + 2\phi_{xt} + \phi_{tt} = f(x, t)$,
where ϵ and κ are constants

27.6. (a) To obtain the general solution of $A\phi_{xx} + 2B\phi_{xy} + C\phi_{yy} = 0$, seek $\phi = f(x + my)$ and obtain $\phi = f(x + m_1 y) + g(x + m_2 y)$, where m_1, m_2 are the roots of a quadratic equation.
(b) Thus obtain the general solution of $3\phi_{xx} + 8\phi_{xy} - 3\phi_{yy} = 0$.
(c) Suppose that the roots are repeated: $m_1 = m_2$. Proceeding formally, $\lim f(x + my)$ should be a solution as $m \longrightarrow m_1$ and hence so should the combination

$$\lim_{m \to m_1} \frac{f(x + my) - f(x + m_1 y)}{m - m_1} = \frac{\partial}{\partial m} f(x + my)\Big|_{m_1}$$
$$= yf'(x + m_1 y) \equiv yh(x + m_1 y).$$

Verify that it is indeed a solution, and so the general solution is $\phi = f(x + m_1 y) + yh(x + m_1 y)$.
(d) Thus obtain the general solution of $4\phi_{xx} - 4\phi_{xy} + \phi_{yy} = 0$.

27.7. Seeking a solution of $\phi_{xx} - \phi_{yy} - \phi_x - \phi_y = 0$ in the form $\phi = Ae^{\alpha x + \beta y}$, obtain, with the help of superposition,

$$\phi = \sum_\alpha [A(\alpha)e^{\alpha(x-y)} + B(\alpha)e^{-y+\alpha(x+y)}].$$

This expression suggests the form $\phi = f(x - y) + e^{-y}g(x + y)$ for the general solution. Verify this form by substitution into the PDE.

27.8. (*A fourth-order equation*) The equation governing the free vibration of a uniform beam of flexural rigidity EI, mass per unit length ρ, and length L is

$$y_{xxxx} + \frac{\rho}{EI} y_{tt} = 0, \tag{27.45}$$

where $y(x, t)$ is the transverse displacement. In contrast with the apparently analogous vibrating *string* equation, show that (27.45) is *parabolic*; in particular, show that the four families of characteristics coalesce and are the $t =$ constant lines. [In fact,[7] it can be shown that a unique solution is determined up to the characteristic BC (Fig. 27.11) if we prescribe y and y_t on AD, as well as two independent linear relations between y, y_x, y_{xx}, and y_{xxx} on each of the segments AB and DC; for instance,

$$y(0, t) = y_x(0, t) = y_{xx}(L, t) = y_{xxx}(L, t) = 0,$$

that is, clamped at $x = 0$ and free at $x = L$ (i.e., "cantilevered").]

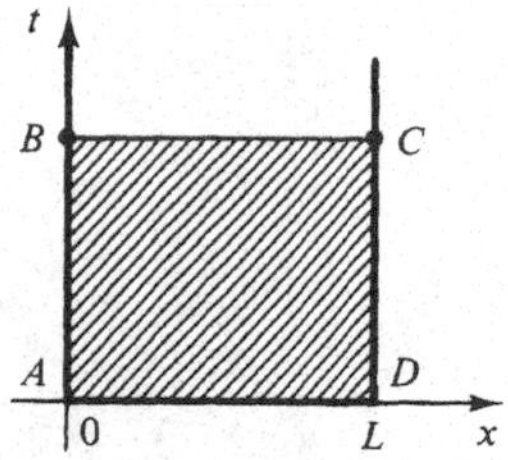

Figure 27.11. Vibrating beam problem.

29.9. (*A first-order equation*) Obtain the solution of

$$u_x + u_y = 0, \qquad u(x, 0) = 3 \sin x, \quad u(0, y) = 3e^{-y}$$

in the quarter plane $x > 0$, $y > 0$ by the method of characteristics.

27.10. Plot $\eta(x, 0.2)$ for the water wave problem of Example 27.5. Note how the steepening of the wave increases with t. (All the data needed is in Fig. 27.9; no calculations should be necessary.)

27.11. Re-solve the water wave example (on a digital computer), using a second-order Runge–Kutta method in place of the Euler integration.

27.12. Set up the variable-density string problem (Fig. 27.12)

$$Ty_{xx} = \rho(x)y_{tt}$$

$$\frac{\rho(x)}{T} = 1 + xH(x) \qquad (H \text{ is Heaviside function})$$

$$y(x, 0) = F(x)$$

$$y_t(x, 0) = 0$$

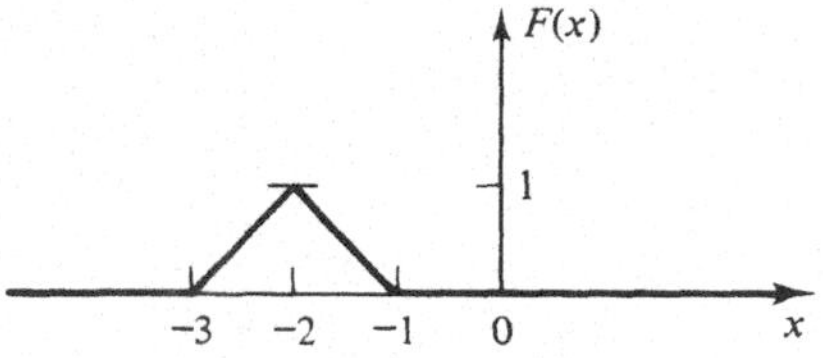

Figure 27.12. Initial deflection.

---

[7]See, for example, R. Courant and D. Hilbert, *Methoden der Mathematischen Physik*, Vol. 2, Springer-Verlag OHG, Berlin, 1937, p. 138.

for the digital computer and run it. Choice of initial x interval, integration scheme, display of results, and so on, is up to you. Of course, our interest is in what happens to the rightward-running wave as it enters the increasingly dense part of the string ($x > 0$).

27.13. If all characteristics of a quasilinear hyperbolic system are straight, does this fact imply that the coefficients are constants? Explain.

27.14. Make up two examination-type questions on this chapter.

Chapter 28

Green's Functions and Perturbation Techniques

Both Green's functions and perturbation techniques were discussed in Part IV, in the solution of ordinary differential equations. In the present chapter we extend these ideas to *partial* differential equations with one or two differences. In the Green's function approach it will be convenient to introduce so-called *principal solutions*, which did not appear in the ODE case because they were not needed. And our perturbation discussion leads up to but does not actually include singular perturbation techniques except for some material on *straining* in the exercises. This decision was made in the interest of "boundedness." Moreover, the straining and boundary layer concepts that arise in the singular case are essentially similar to those already discussed in Part IV; consequently, the reader should be in a good position to pursue this subject more thoroughly in the cited references.

28.1. THE GREEN'S FUNCTION APPROACH

The essential ideas can be explained best through a concrete example, say the two-dimensional Poisson problem shown in Fig. 28.1, with Dirichlet boundary conditions.

As in Part IV, the starting point of the method is the integration by parts of the inner product (Lu, G).[1] Since $L = \nabla^2$,

$$(Lu, G) = \int_D G \nabla^2 u \, d\sigma = \int_C (Gu_n - uG_n) \, ds + \int_D u \nabla^2 G \, d\sigma, \tag{28.1}$$

[1]For a *two*-dimensional region D we define the inner product of any two functions, say $f(x, y)$ and $g(x, y)$, as

$$(f, g) \equiv \int_D fg \, d\sigma.$$

or

$$\int_D G\phi \, d\sigma = \int_C (Gu_n - fG_n)\, ds + \int_D u \nabla^2 G \, d\sigma, \tag{28.2}$$

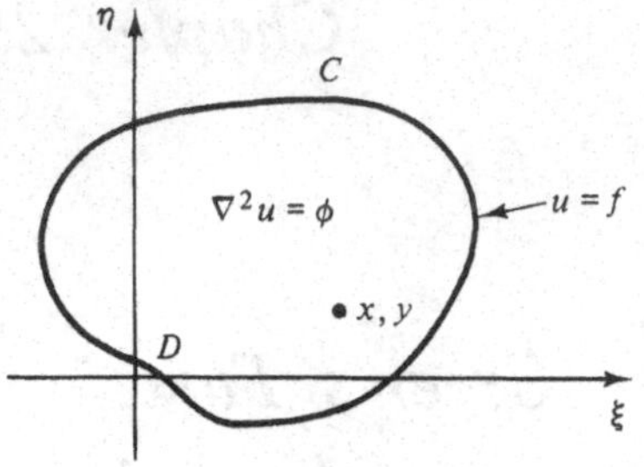

Figure 28.1. Poisson problem.

where the "integration by parts" in (28.1) is provided by the two-dimensional version of Green's second identity (Exercise 28.1). Since u_n is not prescribed on C, we remove the unwelcome u_n term from (28.2) by requiring that $G = 0$ on C. In addition, we set $\nabla^2 G = \delta(\xi - x, \eta - y)$,[2] so that (28.2) becomes

$$u(x, y) = \int_C fG_n \, ds + \int_D G\phi \, d\sigma, \tag{28.3}$$

where the **Green's function** G satisfies the system[3]

$$L^*G = \nabla^2 G = \delta(\xi - x, \eta - y) \text{ in } D \tag{28.4a}$$

$$G = 0 \text{ on } C. \tag{28.4b}$$

The point is that the problem (28.4) on G is simpler than the original one on u; that is, $\nabla^2 G = \delta(\xi - x, \eta - y)$ is simpler than $\nabla^2 u = \phi$, since the δ function is simpler than the perhaps complicated $\phi(\xi, \eta)$, and $G = 0$ (on C) is simpler than $u = f$.

It is generally convenient to seek G in two parts

$$G(\xi, \eta; x, y) = U(\xi, \eta; x, y) + g(\xi, \eta; x, y), \tag{28.5}$$

where U is a *particular* solution of (28.4a) that need not satisfy the prescribed boundary conditions (28.4b) and g is a solution of the homogeneous equation $L^*G = 0$ such that the combination $U + g$ does satisfy the boundary condition. We will refer to U as a **principal solution**, although it is also called a *fundamental solution*, an *elementary solution*, and a *free-space Green's function* in the literature. It contains the basic singularity of the Green's function. The hard part, in constructing the Green's function, is usually in the determination of the "regular" part g so as to satisfy the required boundary conditions.

For the present example U satisfies

$$\nabla^2 U = \delta(\xi - x, \eta - y). \tag{28.6}$$

Since no special boundary conditions are to be imposed on U, we might as well simplify matters by asking U to be symmetric about the field point x, y. That is, introducing polar coordinates r, θ with their origin at x, y, let us ask U to be independent of θ. Then (28.6) simplifies to the *ordinary* differential equation

$$\frac{1}{r}(rU_r)_r = \delta. \tag{28.7}$$

---

[2] Having already discussed the one-dimensional delta function $\delta(\xi - x)$ in some detail (in Section 4.2), we will extend the definition to two dimensions with a minimum of discussion. Specifically, we define $\delta(\xi - x, \eta - y)$ such that

$$\int_D \delta(\xi - x, \eta - y) f(\xi, \eta) \, d\xi \, d\eta = f(x, y)$$

if f is continuous over the region D and x, y is inside D.

[3] If you have read Chapter 18, you will know that the $\nabla^2 G$ that arises (through the integration by parts) in (28.1) is actually L^*G, where L^* is the *adjoint* of L.

Proceeding formally, suppose that $r > 0$, where the δ function is zero. Then (28.7) becomes

$$\frac{1}{r}(rU_r)_r = 0, \qquad (28.8)$$

so that $rU_r = A$ and

$$U = A \ln r + B. \qquad (28.9)$$

Having considered only $r > 0$, we've not yet built into (28.9) the fact that there is a δ function at $r = 0$ in (28.7). Therefore let us integrate (28.7) over an ϵ disk centered at x, y (Fig. 28.2):

$$\int \nabla^2 U \, d\sigma = \int \delta \, d\sigma = 1. \qquad (28.10)$$

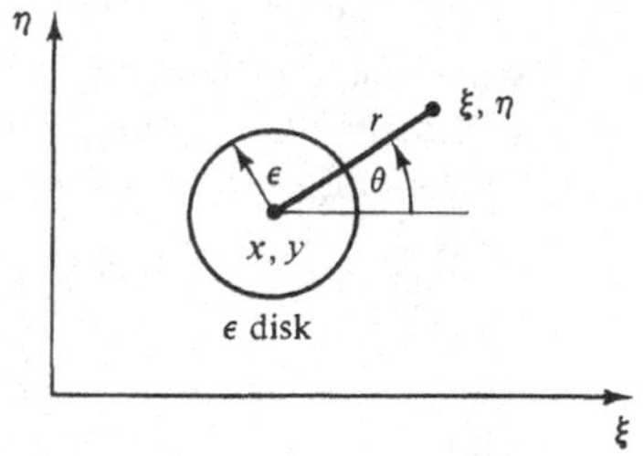

Figure 28.2. ϵ disk at x, y.

Applying the two-dimensional version of the Gauss divergence theorem to the left-hand side (Exercise 28.1), we have

$$\int \nabla \cdot (\nabla U) \, d\sigma = \int \hat{\mathbf{n}} \cdot \nabla U \, ds = \int_0^{2\pi} U_r \Big|_{r=\epsilon} \epsilon \, d\theta$$

$$= \int_0^{2\pi} \frac{A}{\epsilon} \epsilon \, d\theta = 2\pi A, \qquad (28.11)$$

so that (28.10) gives $2\pi A = 1$ or $A = 1/2\pi$. B remains arbitrary and will be set equal to zero because we are merely seeking the *singular* part ($\ln r$) for the principal solution U. Thus

$$U = \frac{1}{2\pi} \ln r \qquad [r = \sqrt{(\xi - x)^2 + (\eta - y)^2}] \qquad (28.12)$$

for the two-dimensional Laplace operator.

*

As mentioned, the preceding derivation is only formal. For instance, we've said that $\delta = 0$ for $r > 0$, whereas we're not supposed to talk about the *values* of generalized functions (Section 4.2), and the singular behavior of U leaves the application of the divergence theorem in (28.11) open to question. However, the argument can be made rigorous as follows.

Instead of dealing directly with $L^* U = \delta$, suppose that we solve

$$L^* U_k = w_k, \qquad (28.13)$$

where w_k is a δ *sequence*. Then we claim that the desired U is

$$U = \lim U_k, \qquad (28.14)$$

where "lim" will mean the limit as $k \longrightarrow \infty$.

Proof. By definition, a δ sequence w_k is such that

$$\lim \int w_k h \, d\sigma = h(x, y), \qquad (28.15)$$

that is,

$$\lim (w_k, h) = h(x, y). \qquad (28.16)$$

But

$$\lim (w_k, h) = \lim (L^*U_k, h) = \lim (U_k, Lh) = (U, Lh) = (L^*U, h), \qquad (28.17)$$

where the third equality in (28.17) follows from the continuity of the inner product (Theorem 17.1). From (28.16) and (28.17) we have $(L^*U, h) = h(x, y)$, and therefore $L^*U = \delta(\xi - x, \eta - y)$ as desired.

Applying this result to the present example, let w_k be the δ sequence

$$w_k = \frac{ke^{-kr^2}}{\pi}. \qquad (28.18)$$

Then

$$\frac{1}{r}\frac{d}{dr}\left(r\frac{dU_k}{dr}\right) = \frac{ke^{-kr^2}}{\pi}. \qquad (28.19)$$

If we multiply through by r, integrate from 0 to r, and require that dU_k/dr be finite at $r = 0$ (which is not to say that dU/dr will be), we obtain

$$\frac{dU_k}{dr} = \frac{1 - e^{-kr^2}}{2\pi r}. \qquad (28.20)$$

Integrating from r to 1 and requiring that $U_k(1) = 0$, say, we have

$$U_k(r) = \frac{1}{2\pi}\ln r + \frac{1}{2\pi}\int_r^1 \frac{e^{-k\rho^2}}{\rho}\,d\rho, \qquad (28.21)$$

and hence [(4.5)]

$$\lim U_k(r) = \frac{1}{2\pi}\ln r, \qquad (28.22)$$

since $e^{-k\rho^2}/\rho$ tends uniformly to 0 over $r \le \rho \le 1$ for each fixed $r > 0$. Thus $U = (1/2\pi)\ln r$ as before.

In order to determine the regular part g, we must first specify the region D in Fig. 28.1.

Example 28.1. Suppose that D is the upper half-plane $\eta > 0$. Then

$$G = U + g = \frac{1}{2\pi}\ln r + g, \qquad (28.23)$$

where g satisfies the conditions

$$L^*g = \nabla^2 g = 0 \quad \text{in } \eta > 0 \qquad (28.24\text{a})$$

$$g = -U = -\frac{1}{2\pi}\ln\sqrt{(\xi - x)^2 + y^2} \quad \text{on } \eta = 0. \qquad (28.24\text{b})$$

In fact, (28.24) may be solved by *inspection*, using *the method of images*. That is, $U = (1/2\pi)\ln r$ is the field induced by a point singularity (i.e., the δ function) at x, y. If we let g be the field induced by a *negative* point singularity at the image point $x, -y$ (Fig. 28.3), then

$$g = -\frac{1}{2\pi}\ln r' = -\frac{1}{2\pi}\ln\sqrt{(\xi - x)^2 + (\eta + y)^2}. \qquad (28.25)$$

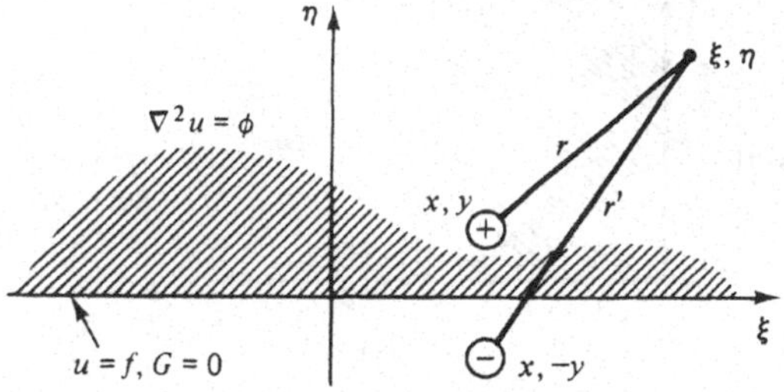

Figure 28.3. The image system.

(28.25) satisfies $\nabla^2 g = 0$ everywhere except at $x, -y$ and hence in $\eta > 0$. In addition, $G = U + g$ vanishes along $\eta = 0$ due to the built-in antisymmetry. So

$$G = \frac{1}{2\pi} \ln \sqrt{(\xi - x)^2 + (\eta - y)^2} - \frac{1}{2\pi} \ln \sqrt{(\xi - x)^2 + (\eta + y)^2}, \tag{28.26}$$

and therefore the solution (28.3) becomes

$$\begin{aligned} u(x, y) &= \int_{-\infty}^{\infty} f(\xi)[-G_\eta(\xi, 0; x, y)]\, d\xi \\ &\quad + \int_0^{\infty} \int_{-\infty}^{\infty} G(\xi, \eta; x, y)\phi(\xi, \eta)\, d\xi\, d\eta \\ &= \frac{y}{\pi} \int_{-\infty}^{\infty} \frac{f(\xi)}{(\xi - x)^2 + y^2}\, d\xi \\ &\quad + \frac{1}{4\pi} \int_0^{\infty} \int_{-\infty}^{\infty} \ln \left[\frac{(\xi - x)^2 + (\eta - y)^2}{(\xi - x)^2 + (\eta + y)^2}\right] \phi(\xi, \eta)\, d\xi\, d\eta. \end{aligned} \tag{28.27}$$

For the case where $\phi = 0$, (28.27) does reduce to the solution (26.115) obtained by means of a Fourier transform.

Obviously we are not always able to construct g by the method of images so easily. Tractibility of the image method depends not only on the *shape* of the region being suitably "simple"[4] but also on the nature of the *boundary conditions.*[5] ∎

Example 28.2. *Quarter-Plane.* Consider next the quarter-plane problem shown in Fig. 28.4. Proceeding as before,

$$\int_D G\, \nabla^2 u\, d\sigma = \int_C (Gu_n - uG_n)\, ds + \int_D u\, \nabla^2 G\, d\sigma \tag{28.28}$$

or with $L^*G = \nabla^2 G = \delta(\xi - x, \eta - y)$,

$$0 = \int_0^{\infty} (Gu_n - uG_n)\Big|_{\xi=0} d\eta + \int_0^{\infty} (Gu_n - uG_n)\Big|_{\eta=0} d\xi + u(x, y). \tag{28.29}$$

[4] See, for instance, M. D. Greenberg, *Application of Green's Functions in Science and Engineering*, Prentice-Hall, Englewood Cliffs, N.J., 1971.

[5] To illustrate, if the Dirichlet boundary condition $u = f$ in the present example is replaced by the *mixed* condition $u_n + \alpha u = f$ (α = constant, say), then the image system is considerably more complicated; see Exercise 28.5.

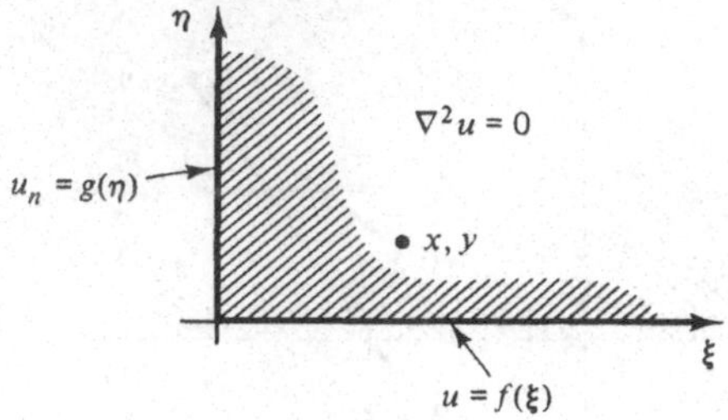

Figure 28.4. Quarter-plane problem.

Next, u_n is prescribed on $\xi = 0$ but u is not. So the u term in the first integral in (28.29) is unwelcome, and we remove it by requiring that $G_n = 0$ on $\xi = 0$. Similarly, we set $G = 0$ on $\eta = 0$ so as to remove the unwelcome u_n term in the second integral.

As in Example 28.1, $G = U + g = (1/2\pi) \ln r + g$. To determine g, recall that U represents the field induced by a point singularity at x, y. Thus if we place a negative singularity at $x, -y$, then the combined field will certainly vanish along the ξ axis (by antisymmetry) as desired. Turning to the condition $G_n = 0$ on the η axis, let us simply reflect the + and − singularities at x, y and $x, -y$ across the η axis, as shown in Fig. 28.5. Then the ξ axis is a line of antisymmetry (of the combined field) and the η axis is a line of symmetry, so that $G = 0$ on $\eta = 0$ and $G_n = 0$ on

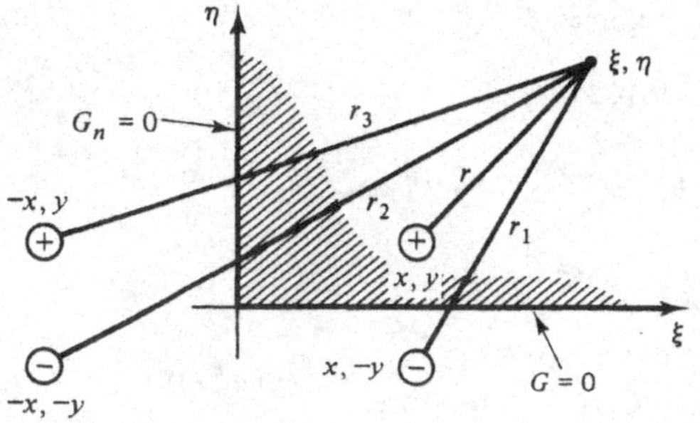

Figure 28.5. The image system.

$\xi = 0$ as desired. Furthermore, $g = -(1/2\pi) \ln r_1 - (1/2\pi) \ln r_2 + (1/2\pi) \ln r_3$ is harmonic everywhere except at the three singular points; hence it is harmonic in the first quadrant as required. Consequently,

$$
\begin{aligned}
G = U + g &= \frac{1}{2\pi}(\ln r - \ln r_1 - \ln r_2 + \ln r_3) \\
&= \frac{1}{4\pi} \ln \frac{[(\xi - x)^2 + (\eta - y)^2][(\xi + x)^2 + (\eta - y)^2]}{[(\xi - x)^2 + (\eta + y)^2][(\xi + x)^2 + (\eta + y)^2]},
\end{aligned}
\tag{28.30}
$$

and therefore the solution is

$$
u(x, y) = -\int_0^\infty G(0, \eta; x, y) g(\eta)\, d\eta - \int_0^\infty G_\eta(\xi, 0; x, y) f(\xi)\, d\xi \tag{28.31}
$$

from (28.29), with G given by (28.30).

COMMENT. You may be disturbed by the direction of integration in the two boundary integrals in (28.29) and puzzled at the "orientation" of the boundary contour C, since ξ runs *counterclockwise* from 0 to ∞, whereas η runs *clockwise* from 0 to ∞. This situation occurs because ds in (28.28) must always be *positive*, since (28.28) actually comes from the divergence theorem

$$\int G\,\nabla^2 u\,d\tau = \int (Gu_n - uG_n)\,d\sigma + \int u\,\nabla^2 G\,d\tau,$$

where $d\tau = 1\,d\sigma$ and $d\sigma = 1\,ds$ as we pass to the two-dimensional case (Exercise 28.1). Obviously the area elements $d\sigma$ are necessarily positive; and since $d\sigma = 1\,ds$, then ds must be positive, too. ∎

28.2. PERTURBATION METHODS

We will limit our discussion to a single example, the classical "thickness problem" of thin airfoil theory, which contains the essential points of the expansion procedure and also ties in well with some other ideas, such as Green's functions and the Cauchy principal value (from Part I).

Example 28.3. *The Thickness Problem.* Consider a thin, symmetric, two-dimensional airfoil moving through otherwise still air or, equivalently, consider the airfoil to be fixed and the air to be streaming by (Fig. 28.6). We emphasize the thinness of

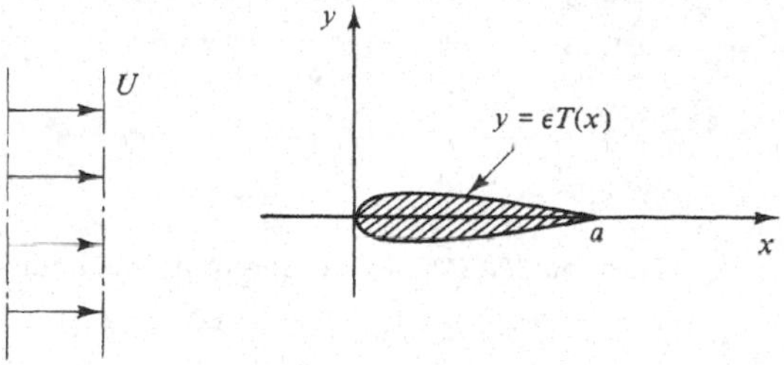

Figure 28.6. Flow over the airfoil.

the airfoil by expressing the semithickness as $\epsilon T(x)$, where $\epsilon \ll 1$ and $T(x)$ is $O(1)$, rather than simply as $T(x)$. We will neglect the effects of viscosity and will also assume that the Mach number $M = U/(\text{speed of sound}) \ll 1$ so that we may consider the fluid to be incompressible.[6] Then there exists a velocity potential $\phi(x, y)$ such that $\mathbf{q} = u\hat{\mathbf{i}} + v\hat{\mathbf{j}} = \nabla\phi$, where u, v are the x, y fluid velocity components and ϕ must satisfy the Laplace equation

$$\nabla^2\phi = 0 \tag{28.32a}$$

plus the boundary condition at infinity $\nabla\phi \sim U\hat{\mathbf{i}}$, or

$$\phi(x, y) \sim Ux \quad \text{as } r \longrightarrow \infty. \tag{28.32b}$$

Finally, there is the tangent-flow boundary condition on the surface of the airfoil. From Fig. 28.7 we see that it can be expressed as

$$\frac{\phi_y[x, \epsilon T(x)]}{\phi_x[x, \epsilon T(x)]} = \epsilon T'(x). \tag{28.32c}$$

[See Exercise 28.10 for a slightly different formulation of (28.32c).]

[6] You may wish to review Sections 9.8 and 26.1 on the relevant fluid mechanics.

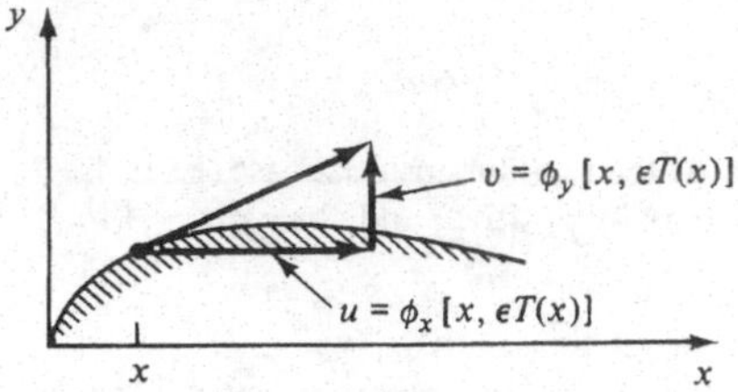

Figure 28.7. Tangent-flow boundary condition.

The resulting boundary value problem (28.32) is *linear* but complicated by the presence of the $y = \pm\epsilon T(x)$ boundary of the region. One possible line of approach is *conformal mapping*, but the success of this method depends on the precise form of $T(x)$, and it may be necessary to settle for only an approximate mapping. Furthermore, the method fails to apply for compressible and/or three-dimensional flows, whereas the perturbation method applies equally well. In any case, conformal mapping was discussed in Part II. Here we intend to take advantage of the smallness of ϵ instead. That is, we will seek ϕ in the form

$$\phi(x, y) = \phi_0(x, y) + \epsilon\phi_1(x, y) + \epsilon^2\phi_2(x, y) + \cdots. \tag{28.33}$$

Now as $\epsilon \longrightarrow 0$, the airfoil creates no disturbance at all,[7] and so the velocity potential should reduce to the free-stream potential Ux. Thus $\phi_0(x, y) = Ux$, and it follows that

$$\phi(x, y) = Ux + \epsilon\phi_1(x, y) + \epsilon^2\phi_2(x, y) + \cdots. \tag{28.34}$$

Putting this result into (28.32a) and (28.32b) and equating powers of ϵ, we see easily that

$$\begin{aligned} \nabla^2\phi_1 &= 0, \qquad \phi_1 \longrightarrow 0 \quad \text{as } r \longrightarrow \infty \\ \nabla^2\phi_2 &= 0, \qquad \phi_2 \longrightarrow 0 \quad \text{as } r \longrightarrow \infty \\ &\ \,\vdots \qquad\qquad \vdots \end{aligned}$$

Next, put (28.34) into the tangent-flow condition (28.32c). First, we need to expand

$$\begin{aligned} \phi_y[x, \epsilon T(x)] &= \epsilon\phi_{1y}(x, \epsilon T) + \epsilon^2\phi_{2y}(x, \epsilon T) + \cdots && \text{per (28.34)} \\ &= \epsilon[\phi_{1y}(x, 0) + \phi_{1yy}(x, 0)\epsilon T + \cdots] \\ &\quad + \epsilon^2[\phi_{2y}(x, 0) + \cdots] + \cdots && \text{per Taylor series} \\ \phi_x[x, \epsilon T(x)] &= U + \epsilon\phi_{1x}(x, \epsilon T) + \cdots \\ &= U + \epsilon[\phi_{1x}(x, 0) + \cdots] + \cdots. && (28.35) \end{aligned}$$

Putting these expressions into (28.32c),

$$\epsilon\phi_{1y}(x, 0) + \epsilon^2[T\phi_{1yy}(x, 0) + \phi_{2y}(x, 0)] + \cdots = \epsilon UT' + \epsilon^2 T'\phi_{1x}(x, 0) + \cdots, \tag{28.36}$$

[7] Actually, a nontrivial circulatory flow can exist even in the event that $\epsilon = 0$. (See, for example, H. Glauert, *Elements of Aerofoil and Airscrew Theory*, 2nd ed., Cambridge University Press, Cambridge, 1948; see also Exercise 28.11.) But then the rear stagnation point would not be at the trailing edge $x = a$, and hence the flow would make a 180° turn around the trailing edge. This is not possible if the fluid has even an infinitesimal amount of viscosity, and so we impose the so-called *Kutta condition* that the flow be smooth at the trailing edge. (See Glauert, Chapter 9.) Thus even though we have ignored viscosity in the equations of motion [more specifically, in applying the Kelvin–Helmholtz theorem (Section 9.8) in order to deduce the existence of the velocity potential ϕ], the effects of viscosity still enter through the application of the Kutta condition. The effect of this condition is to rule out any circulation, so that ϕ_0 is simply Ux. Stated another way, the flow is to be *symmetric* about the x axis.

so that

$$\begin{aligned}
\epsilon:\quad \phi_{1y}(x, 0) &= UT'(x) \\
\epsilon^2:\quad \phi_{2y}(x, 0) &= T'(x)\phi_{1x}(x, 0) - T(x)\phi_{1yy}(x, 0) \\
&= T'(x)\phi_{1x}(x, 0) + T(x)\phi_{1xx}(x, 0) \qquad (28.37) \\
&= [T(x)\phi_{1x}(x, 0)]', \\
&\vdots
\end{aligned}$$

where the (simplifying but noncrucial) substitution $\phi_{1yy} = -\phi_{1xx}$ in (28.37) follows from the fact that $\nabla^2\phi_1 = 0$.

Consequently; the $O(\epsilon)$ problem is

$$\nabla^2\phi_1 = 0 \qquad (28.38a)$$

$$\phi_1 \longrightarrow 0 \quad \text{as } r \longrightarrow \infty \qquad (28.38b)$$

$$\phi_{1y}(x, 0) = UT'(x) \equiv UT_1'(x) \text{ say,} \qquad (28.38c)$$

the $O(\epsilon^2)$ problem is

$$\nabla^2\phi_2 = 0 \qquad (28.39a)$$

$$\phi_2 \longrightarrow 0 \quad \text{as } r \longrightarrow \infty \qquad (28.39b)$$

$$\phi_{2y}(x, 0) = [T(x)\phi_{1x}(x, 0)]' \equiv UT_2'(x) \text{ say,} \qquad (28.39c)$$

and so on.

Notice carefully that whereas the tangent-flow boundary condition (28.32c) is applied on the physical boundary $y = \epsilon T(x)$, the boundary conditions (28.38c), (28.39c), . . . are applied on $y = 0$, which is much simpler! Actually, these conditions apply only for $0 < x < a$, but if we extend the definition of $T(x)$ so that $T(x) = 0$ for $a < x < \infty$ and $-\infty < x < 0$, then they hold for all x. Thus (28.38), (28.39), . . . are a sequence of upper half-plane Neumann problems, (Fig. 28.8) that

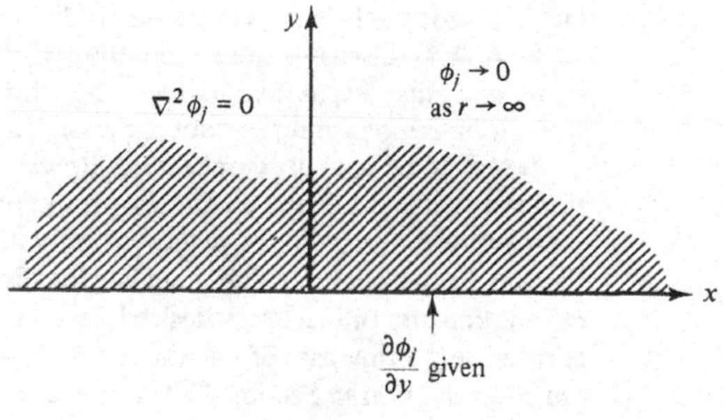

Figure 28.8. The Neumann problems.

are readily solved by the method of Green's functions. We find (Exercise 28.16) that

$$\phi_j(x, y) = \frac{U}{2\pi}\int_0^a T_j'(\xi)\ln[(\xi - x)^2 + y^2]\,d\xi, \qquad (28.40)$$

where $T_1(x) \equiv T(x)$, $T_2(x) = T(x)\phi_{1x}(x, 0)/U$, and so on. Through $O(\epsilon)$, say, the fluid velocity components are therefore given by

$$u(x, y) = \frac{\partial \phi}{\partial x} \approx U + \epsilon \frac{\partial \phi_1}{\partial x} = U + \epsilon \frac{U}{\pi} \int_0^a \frac{x - \xi}{(\xi - x)^2 + y^2} T'(\xi)\, d\xi \tag{28.41a}$$

$$v(x, y) = \frac{\partial \phi}{\partial y} \approx \epsilon \frac{\partial \phi_1}{\partial y} = \epsilon \frac{Uy}{\pi} \int_0^a \frac{T'(\xi)\, d\xi}{(\xi - x)^2 + y^2}. \tag{28.41b}$$

Is the perturbation *regular*? Observe from (28.34) that $\phi = Ux + O(\epsilon)$ or, equivalently, $\mathbf{q} = U\hat{\mathbf{i}} + O(\epsilon)$, so that the deviation of the flow from the uniform stream is assumed to be uniformly small throughout the flow field (because of the thinness of the airfoil). However, note that for a blunt-nose airfoil the velocity $\mathbf{q} \approx 0$ in the immediate neighborhood of the stagnation point $x = y = 0$ (Fig. 28.9), and hence the disturbance velocity is comparable to U—that is, $O(1)$ rather than $O(\epsilon)$. Apparently, then, (28.34) breaks down in this neighborhood, which means that it is *not* uniformly valid and the perturbation is *singular*. In order to expose the details of the nonuniformity, as well as find means of eliminating it, it is helpful to consider a (thin) *elliptic* airfoil, for in this case the exact solution is available (by conformal mapping) and can be compared with the perturbation solution. This situation is discussed in detail by Van Dyke,[8] who shows that the region of non-uniformity is a circular neighborhood of the origin, the radius of which is the same (in order of magnitude) as the nose radius of the airfoil, a result that applies to other blunt-nose bodies as well—for instance, the airfoil shown in Fig. 28.6. In fact, the airfoil also has a stagnation point and hence a second region of nonuniformity at the *trailing* edge. Since the trailing edge is wedge shaped rather than blunt, the region of nonuniformity is extremely small, in fact, of order $e^{-1/\epsilon}$ (i.e., transcendentally small in ϵ).[9]

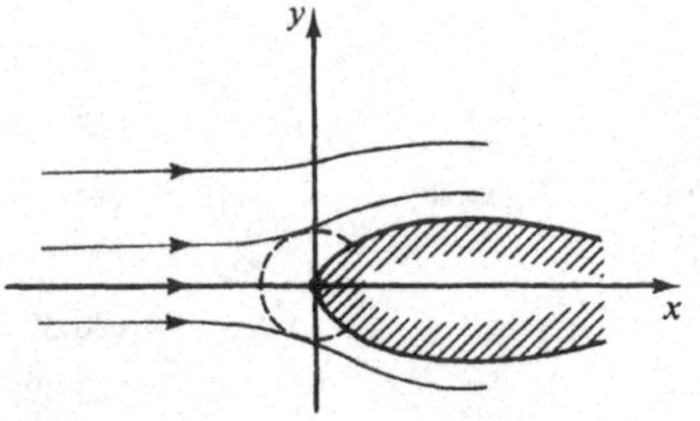

Figure 28.9. Stagnation region for blunt-nose airfoil.

Van Dyke discusses the elimination of these nonuniformities, both by *straining* and *inner-outer expansions* (where the "inner regions" are the above-mentioned stagnation regions and the "outer region" is the rest of the flow field).

It should be strongly emphasized, however, that even a nonuniformly valid solution may suffice! For the airfoil shown in Fig. 28.6, for example, the region of nonuniformity at the trailing edge has virtually no impact on practical calculations. Even the one at the blunt leading edge is often ignored, depending on the object of the calculation and the accuracy desired. In this case, the region(s) of nouniformity constitutes only a tiny part of the domain. In contrast, recall from our discussion of the Duffing equation in Section 25.1 that the nonuniformity for the expansion

$$x(t) = A \cos t + \epsilon \frac{A^3}{32}(-\cos t + \cos 3t - 12t \sin t) + \cdots$$

was quite extensive, extending over $T < t < \infty$ (where T depends on A and ϵ), so that elimination of the nonuniformity in that case was more of a necessity than a refinement.

[8] M. Van Dyke, *Perturbation Methods in Fluid Mechanics*, Academic, New York, 1964, Chapter 4.
[9] *Ibid.*

COMMENT. Recall that we inferred from (28.32b) and (28.34) that $\phi_1, \phi_2, \ldots \to 0$ as $r \to \infty$. Actually, this result does not follow. For instance, if $\phi(x, y) = U(x + \epsilon\sqrt{r})$, say, then $\phi \sim Ux$, whereas $\phi_1 = U\sqrt{r}$ does *not* $\to 0$ as $r \to \infty$. Nevertheless, $\phi_1, \phi_2, \ldots \to 0$ *is* correct and follows instead from the fact that the body is closed. This topic is discussed briefly in Van Dyke.[10] ▮

EXERCISES

28.1. (a) Derive the two-dimensional version of Green's second identity (28.1). *Hint:* Starting with

$$\int_V G\,\nabla^2 u\,d\tau = \int_S (Gu_n - uG_n)\,d\sigma + \int_V u\,\nabla^2 G\,d\tau,$$

let u and G be independent of z and let V be a cylinder of cross section D (Fig. 28.1) and unit thickness in the z direction.

(b) Derive the two-dimensional version of the divergence theorem,

$$\int_S \nabla \cdot \mathbf{Q}\,d\sigma = \int_C \mathbf{Q} \cdot \hat{\mathbf{n}}\,ds.$$

28.2. Apply the splitting ($G = U + g$) to the ODE example

$$u'' = \phi, \qquad u(a) = 3, \quad u(b) = 2.$$

28.3. Using the method of images, determine G, for the problems shown in Fig. 28.10. *Hint:* In part (a) use seven image singularities as indicated; in part (b) you will need an infinite sequence of them.

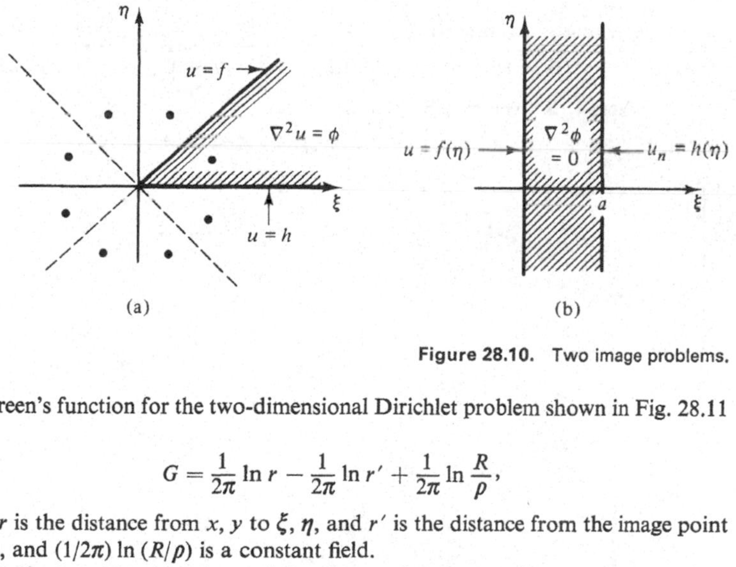

Figure 28.10. Two image problems.

28.4. The Green's function for the two-dimensional Dirichlet problem shown in Fig. 28.11 is

$$G = \frac{1}{2\pi}\ln r - \frac{1}{2\pi}\ln r' + \frac{1}{2\pi}\ln\frac{R}{\rho},$$

where r is the distance from x, y to ξ, η, and r' is the distance from the image point to ξ, η, and $(1/2\pi)\ln(R/\rho)$ is a constant field.

(a) Verify that this expression satisfies the requirements on G.

---

[10] *Ibid.*

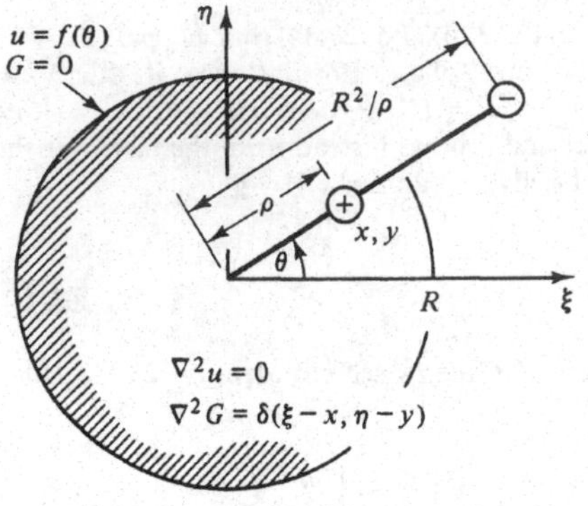

Figure 28.11. Image system for circle.

(b) Use it to derive the *Poisson integral formula* solution,

$$u(\rho, \theta) = \frac{1}{2\pi} \int_0^{2\pi} \frac{R^2 - \rho^2}{R^2 + \rho^2 - 2R\rho \cos(\tilde{\vartheta} - \theta)} f(\tilde{\vartheta})\, d\tilde{\vartheta}.$$

28.5. (*Mixed Boundary Conditions*) Verify that the Green's function for the half-plane problem

$$\nabla^2 u = u_{\xi\xi} + u_{\eta\eta} = 0 \quad \text{in } \eta > 0,$$

with the *mixed* boundary condition

$$\frac{\partial u}{\partial n} + \alpha u = f(\xi) \quad \text{on } \eta = 0$$

(where α is a constant) is

$$G = U(\xi, \eta; x, y) + U(\xi, \eta; x, -y) + \int_{-\infty}^{-y} U(\xi, \eta; x, \zeta) m(\zeta)\, d\zeta,$$

where U is given by (28.12) and

$$m(\zeta) = -2\alpha e^{\alpha(y+\zeta)}.$$

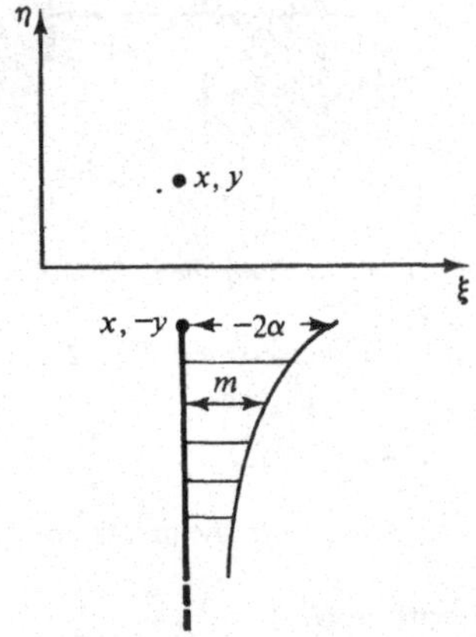

Figure 28.12. Image system.

Interpret G as the field induced by two point singularities, together with a linear distribution of singularities, as suggested in Fig. 28.12. Express the solution $u(x, y)$ in terms of the Green's function given above.

28.6. (a) The principal solution U corresponding to the two-dimensional **Helmholtz operator** $L = \nabla^2 + k^2$ satisfies

$$L^*U = \nabla^2 U + k^2 U = \delta(\xi - x, \eta - y).$$

Proceeding as in equations (28.6) to (28.12) and recalling the small-argument behavior of the Bessel functions J_0, Y_0 from section 22.4, derive the result

$$U = \tfrac{1}{4} Y_0(kr).$$

(b) Thus solve the half-plane problem

$$\nabla^2 u + k^2 u = 0 \quad \text{in } \eta > 0$$
$$u(\xi, 0) = f(\xi).$$

28.7. (a) Integral transform methods are of great use in finding principal solutions. Let us consider, for instance, the *modified Helmholtz operator*, $\nabla^2 - k^2$ ("modified" because of the minus sign). The form of ∇^2 in polar coordinates lends itself to a **Hankel transform.** [Recall from Section 6.3 that the Hankel transform of order n and its inverse are given by

$$\bar{f}(\rho) = \int_0^\infty r J_n(r\rho) f(r)\, dr, \qquad f(r) = \int_0^\infty \rho J_n(r\rho) \bar{f}(\rho)\, d\rho,$$

respectively, where $n > -\frac{1}{2}$.] Starting with

$$\frac{1}{r}(rU_r)_r - k^2 U = \delta(\xi - x, \eta - y), \tag{28.42}$$

compute U by means of a Hankel transform of order zero. *Hint:* Multiply (28.42) by $rJ_0(r\rho)\, dr\, d\theta$ and integrate on r from 0 to ∞ and on θ from 0 to 2π. Note carefully that

$$\iint \delta(\xi - x, \eta - y) r J_0(r\rho)\, dr\, d\theta$$
$$= \iint \delta(\xi - x, \eta - y) J_0(r\rho)\, d\xi\, d\eta = J_0(0) = 1.$$

Thus obtain $\bar{U} = -1/2\pi(\rho^2 + k^2)$ and invert by means of the well-known formula

$$\int_0^\infty \frac{\rho J_0(r\rho)}{\rho^2 + k^2}\, d\rho = K_0(kr).$$

(b) Obtain the solution $U = -K_0(kr)/2\pi$ to part (a) by pursuing a different line of approach. *Hint:* Recall from Section 22.4 that the general solution of $xy'' + y' - k^2 xy = 0$ is $y = AI_0(kx) + BK_0(kx)$. Also, note that you will need to let $\epsilon \longrightarrow 0$ when integrating $\nabla^2 U - k^2 U = \delta$ over an ϵ disk. This situation is in contrast to our calculation (28.11) for the Laplace operator, since (28.11) happens to be exact for all ϵ and there was no need to let $\epsilon \longrightarrow 0$.

28.8. (a) Derive the principal solution $U = -1/4\pi r$ of the *three*-dimensional Laplace operator $L = \nabla^2$.

(b) Use the result of part (a) to solve the problem

$$\nabla^2 u = \phi(x, y, z) \quad \text{in } y > 0,\ -\infty < x < \infty,\ -\infty < z < \infty$$
$$u(x, 0, z) = 0.$$

28.9. Derive the principal solution $U = (1/8\pi)r^2 \ln r$ for the two-dimensional **biharmonic operator** $L = \nabla^4$.

28.10. Derive the tangent-flow condition (28.32c) from $\nabla\phi \cdot \hat{\mathbf{n}} = 0$, where $\hat{\mathbf{n}}$ is proportional to $\nabla[y - \epsilon T(x)]$.

28.11. Consider an inviscid, incompressible flow aligned with the axis of a corrugated infinite cylinder, which is defined in terms of cylindrical coordinates by $r = a(1 + \epsilon \cos \omega z)$, where $\epsilon \ll 1$ (Fig. 28.13). The velocity potential ϕ is governed by $\nabla^2\phi = 0$, plus the

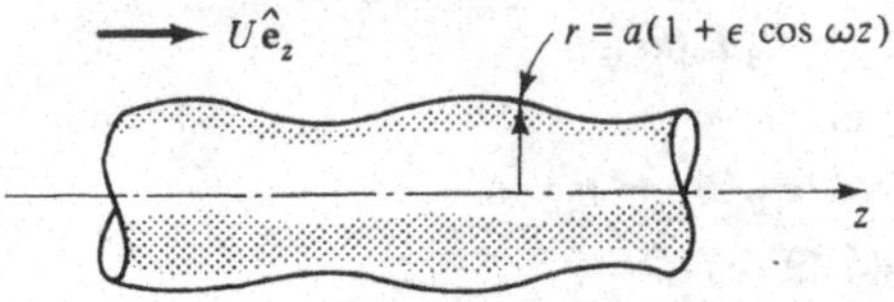

Figure 28.13. Axial flow over corrugated cylinder.

condition $\nabla\phi \sim U\hat{\mathbf{e}}_z$ as $r \longrightarrow \infty$ and a tangent-flow condition on the cylinder. Show that

$$\phi(r, z) = Uz + \epsilon a U \frac{K_0(\omega r)}{K_1(\omega a)} \sin \omega z + O(\epsilon^2),$$

where K_0 and K_1 are Bessel functions of the second kind. *Hint:* Note the identity $K_0' = -K_1$.

28.12. We have already discussed the problem of a plane, uniform, inviscid, incompressible flow over a circular cylinder in terms of a velocity potential ϕ (Example 9.14 and Exercise 26.39). Suppose instead that the incoming flow is slightly *sheared*, (Fig. 28.14)—that is,

$$\mathbf{q} \sim U\left(1 + \epsilon\frac{y}{a}\right)\hat{\mathbf{\imath}}$$

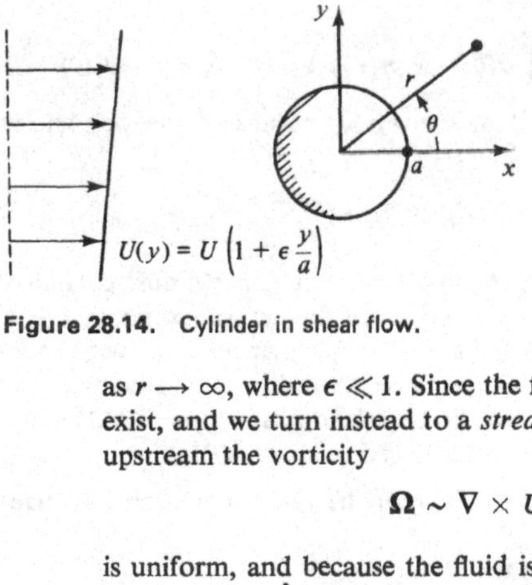

Figure 28.14. Cylinder in shear flow.

as $r \longrightarrow \infty$, where $\epsilon \ll 1$. Since the flow is *rotational*, a velocity potential ϕ does not exist, and we turn instead to a *stream function* ψ, as discussed in Exercise 9.34. Far upstream the vorticity

$$\mathbf{\Omega} \sim \nabla \times U\left(1 + \epsilon\frac{y}{a}\right)\hat{\mathbf{\imath}} = -\epsilon\frac{U}{a}\hat{\mathbf{k}}$$

is uniform, and because the fluid is assumed to be inviscid, we will therefore have $\mathbf{\Omega} = -\epsilon(U/a)\hat{\mathbf{k}}$ everywhere, so that ψ is governed by the PDE

$$\nabla^2\psi = \frac{\epsilon U}{a},$$

together with boundary conditions

$$\psi \sim U\left(r \sin \theta + \frac{\epsilon}{2a} r^2 \sin^2 \theta\right) \quad \text{as } r \longrightarrow \infty$$

$$\psi(a, \theta) = \text{constant, say } C.$$

First, explain how these two boundary conditions were obtained. Then seek $\psi = \psi_0 + \epsilon\psi_1 + \epsilon^2\psi_2 + \cdots$ and obtain the solution

$$\psi_0 = U\left(r - \frac{a^2}{r}\right) \sin \theta + C\frac{\ln r}{\ln a}$$

$$\psi_1 = \frac{U}{4}\left[\frac{r^2}{a}(1 - \cos 2\theta) + \frac{a^3}{r^2} \cos 2\theta - a\right]$$

$$\psi_2 = \psi_3 = \cdots = 0.$$

(Thus the series happens to terminate, and we end up with an exact solution in closed form. Observe that ψ_0, which is the stream function for the case where $\epsilon = 0$, consists of the superposition of three terms: $Ur \sin \theta = Uy$ is simply the stream function of the free stream $\mathbf{q} = U\hat{\mathbf{i}}$; $-Ua^2 (\sin \theta)/r$ can be identified as a *doublet* located at the center of the cylinder (as discussed in H. Glauert, *Elements of Airfoil and Airscrew Theory*, 2nd ed., Cambridge University Press, Cambridge, 1948, Section 3.6); and $C \ln r/\ln a$ can be identified as a counterclockwise point *vortex* of strength $\Gamma = -2\pi C/\ln a$ located at the origin, as discussed in Glauert, Section 4.2. Physically, $\Gamma \neq 0$ implies a nonzero *circulation* around the cylinder that will move the stagnation points to $\theta = \sin^{-1} (\Gamma/4\pi Ua)$ rather than 180° and 0°. We omitted the circulatory term in Exercise 26.39 because there we were solving for the flow in Fig. 9.21, which is pictured as symmetric—that is, with no circulation.)

28.13. For the eigenvalue problem shown in Fig. 28.15, show that the smallest eigenvalue is given by

$$\lambda = 2\left[1 + \frac{8}{3\pi^2}\epsilon + O(\epsilon^2)\right].$$

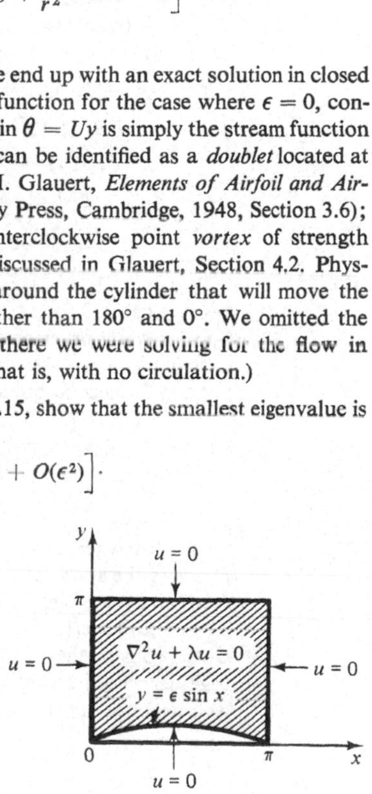

Figure 28.15. Eigenvalue problem.

28.14. The lateral vibration of a certain variable-density string is governed by the PDE

$$a^2(1 + \epsilon x)y_{xx} = y_{tt} \qquad (\epsilon \ll 1)$$

plus the end conditions $y(0, t) = y(\pi, t) = 0$. Show that the frequency of the lowest mode is given by

$$\text{Frequency} = a\left[1 + \frac{\pi}{4}\epsilon + O(\epsilon^2)\right].$$

Hint: It should be clear that you will need to *strain* the time variable.

28.15. For the incompressible case ($M = 0$), the airfoil flow in Fig. 28.6 was governed by the Laplace equation $\nabla^2\phi = 0$, where $\mathbf{q} = \nabla\phi$. If compressibility effects are included, however, then we have

$$\phi_{yy} - B^2\phi_{xx} = M^2\Big[\frac{\gamma-1}{2}(2\phi_x + \phi_x^2 + \phi_y^2)(\phi_{xx} + \phi_{yy}) + (2\phi_x + \phi_x^2)\phi_{xx} + 2(1+\phi_x)\phi_y\phi_{xy} + \phi_y^2\phi_{yy}\Big],$$

where $B^2 = M^2 - 1$, γ is the adiabatic index of the gas (i.e., the ratio of its two specific heats), and this time ϕ is the *disturbance* potential (i.e., $\mathbf{q} = U\hat{\imath} + U\nabla\phi$, following the notation in A. Nayfeh, *Perturbation Methods*, Wiley, New York, 1973, pp. 26–28). In addition, there are the tangent-flow condition on the airfoil and the condition $\phi \longrightarrow 0$ far upstream.

(a) Letting $\phi = \epsilon\phi_1 + \epsilon^2\phi_2 + \cdots$, derive the first- and second-order problems

$$\epsilon:\quad \phi_{1yy} - B^2\phi_{1xx} = 0$$
$$\phi_{1y}(x, 0) = T'(x)$$
$$\phi_1(x, y) \longrightarrow 0 \text{ upstream}$$
$$\epsilon^2:\quad \phi_{2yy} - B^2\phi_{2xx} = M^2[(\gamma+1)\phi_{1x}\phi_{1xx} + (\gamma-1)\phi_{1x}\phi_{1yy} + 2\phi_{1y}\phi_{1xy}]$$
$$\phi_{2y}(x, 0) = \phi_{1x}(x, 0)T'(x) - \phi_{1yy}(x, 0)T(x)$$
$$\phi_2(x, y) \longrightarrow 0 \text{ upstream,}$$

where $T(x) \equiv 0$ for $-\infty < x < 0$ and $a < x < \infty$.

(b) Solve the first-order problem by D'Alembert's method for the supersonic case $B > 0$.

(c) Solving the second-order problem (for $B > 0$), it turns out that

$$\phi_x = -\epsilon\frac{T'}{B} + \epsilon^2\Big[\Big(1 - \frac{(\gamma+1)M^4}{4B^2}\Big)\Big(\frac{T'}{B}\Big)^2 - \frac{\gamma+1}{2}\frac{M^4}{B^3}yT'T'' - TT''\Big] + O(\epsilon^3)$$

(in $y > 0$), where the argument of the T, T', and T'' terms is $\xi \equiv x - By$. Note carefully the appearance of y in the ϵ^2 term. Thus the ϵ^2 term is *not* a small correction to the ϵ term if y is large, and we say that the perturbation is *singular*. The basic idea is that the combination $\xi = x - By$ persists, whereas beyond the $O(\epsilon)$ problem the characteristics would like to curve a bit. This situation, together with our experience with secular terms in the ODE case, suggests that we introduce the *strained* variables X, Y according to

$$\xi = X + \epsilon f(X, Y) + \cdots$$
$$y = Y.$$

Put these into the preceding expression for ϕ_x and (following Pritulo; recall Section 25.1) determine f so as to remove secular terms. Thus obtain the uniformly valid result

$$\phi_x = -\epsilon\frac{T'(X)}{B} + O(\epsilon^2)$$

where
$$\xi = X - \epsilon\frac{\gamma+1}{2}\frac{M^4}{B^2}yT'(X) + O(\epsilon^2).$$

28.16. Derive the solution (28.40) of the Neumann half-plane problem in Fig. 28.8 by the method of Green's functions.

28.17. *(a) Using (28.41a,b), show that the surface velocity components are

$$u(x, \pm\epsilon T) \sim U - \epsilon\frac{U}{\pi} ⨍_0^a \frac{T'(\xi)\,d\xi}{\xi - x} \tag{28.43a}$$

$$v(x, \pm\epsilon T) \sim \pm\epsilon U T'(x) \tag{28.43b}$$

as $\epsilon \longrightarrow 0$. You may assume that T' is reasonably well-behaved, for example, that it satisfies a *Lipschitz condition* $|T'(x_2) - T'(x_1)| \leq C|x_2 - x_1|$ over the interval $0 \leq x \leq a$, for some finite constant C.

(b) Evaluate the right-hand side of (3.43a) by setting

$$\xi = \frac{a}{2} - \frac{a}{2}\cos\theta, \qquad x = \frac{a}{2} - \frac{a}{2}\cos\alpha$$

(Fig. 28.16), expanding $T'(\xi) = \sum_1^\infty a_n \sin n\theta$, and recalling the Glauert airfoil integral in Exercise 15.10.

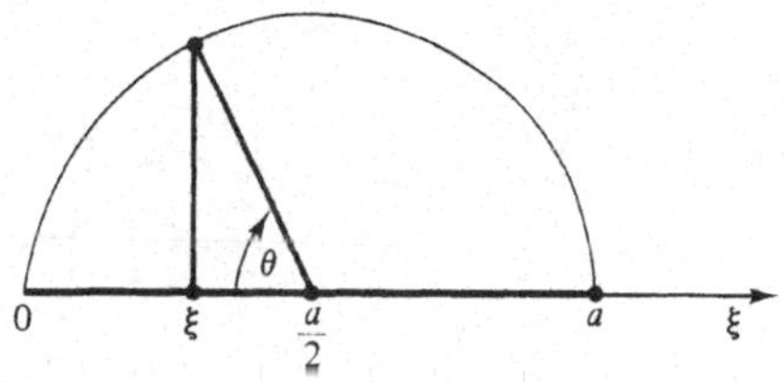

Figure 28.16. Change of variables.

28.18. Make up two examination-type problems on this chapter.

Chapter 29

Finite-Difference Methods

Approximate and numerical methods that have already been discussed for partial differential equations include the Ritz method (Section 10.6), the method of characteristics (Section 27.2), and even perturbation methods (Section 28.2) in the sense that we often settle simply for the first term or two of the desired expansion. The various methods of weighted residuals, finite elements, quasilinearization, and iteration discussed in Part IV for ODEs also bear important application to PDEs, as well as special techniques too numerous to mention (let alone discuss). Many are contained within the cited references.

In this chapter we consider only finite differences, with an emphasis on the parabolic heat equation.

29.1. *THE HEAT EQUATION; AN EXPLICIT METHOD AND ITS CONVERGENCE AND STABILITY*

For definiteness, consider the one-dimensional heat conduction problem shown in Fig. 29.1. The usual diffusivity factor α^2 has been omitted for simplicity; in any case, it can always be absorbed by a simple scaling of the independent variables.

As in the ODE case, we seek $U(x, t)$ only at discrete points, say

$$\begin{aligned} x_i &= i\,\Delta x, \qquad i = 1, 2, \ldots, N-1 \quad \left(\Delta x = \frac{\ell}{N}\right) \\ t_j &= j\,\Delta t, \qquad j = 1, 2, \ldots, \end{aligned} \tag{29.1}$$

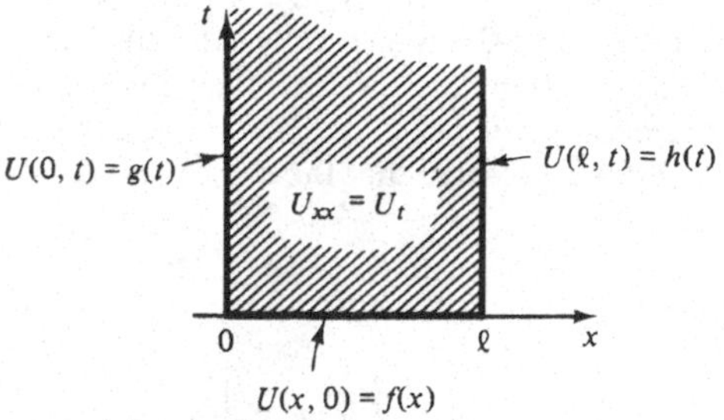

Figure 29.1. A heat conduction problem.

and we replace the derivatives U_{xx} and U_t by finite-difference approximations (as in Exercise 24.29) as follows. From Taylor's formula with Lagrange remainder [recall (2.27)],

$$U(x_i, t_{j+1}) = U(x_i, t_j) + U_t(x_i, t_j)\,\Delta t + \frac{1}{2}U_{tt}(x_i, \tau)(\Delta t)^2,$$

where τ is some point in $[t_j, t_{j+1}]$. Solving for U_t,

$$U_t(x_i, t_j) = \frac{U(x_i, t_{j+1}) - U(x_i, t_j)}{\Delta t} - \frac{1}{2}U_{tt}(x_i, \tau)\,\Delta t. \tag{29.2}$$

Similarly (Exercise 29.1),

$$U_{xx}(x_i, t_j) = \frac{U(x_{i+1}, t_j) - 2U(x_i, t_j) + U(x_{i-1}, t_j)}{(\Delta x)^2} - \frac{1}{12}U_{xxxx}(\xi, t_j)(\Delta x)^2 \tag{29.3}$$

for some ξ in $[x_{i-1}, x_{i+1}]$. Thus our PDE $U_{xx}(x_i, t_j) - U_t(x_i, t_j) = 0$ may be written

$$\frac{U_{i+1,j} - 2U_{ij} + U_{i-1,j}}{(\Delta x)^2} - \frac{U_{i,j+1} - U_{ij}}{\Delta t} = \frac{1}{12}U_{xxxx}(\xi, t_j)(\Delta x)^2 - \frac{1}{2}U_{tt}(x_i, \tau)\,\Delta t; \tag{29.4}$$

where U_{ij} denotes $U(x_i, t_j)$. The right-hand side is said to be $O(\Delta x)^2 + O(\Delta t)$ since we have not specified any a priori relationship between Δx and Δt. If we simply drop these unwieldy terms, then

$$\frac{u_{i+1,j} - 2u_{ij} + u_{i-1,j}}{(\Delta x)^2} - \frac{u_{i,j+1} - u_{ij}}{\Delta t} = 0, \tag{29.5}$$

where we call attention to the so-called **truncation error**

$$\begin{aligned} T(x_i, t_j) &= \frac{1}{12}U_{xxxx}(\xi, t_j)(\Delta x)^2 - \frac{1}{2}U_{tt}(x_i, \tau)\Delta t \\ &= \frac{1}{12}U_{xxxx}(x_i, t_j)(\Delta x)^2 + O(\Delta x)^3 - \frac{1}{2}U_{tt}(x_i, t_j)\Delta t + O(\Delta t)^2 \end{aligned} \tag{29.6}$$

by changing the Us (which refer to the *exact* solution) to u's.

Solving (29.5) for $u_{i,j+1}$,

$$\begin{aligned} u_{i,j+1} &= u_{ij} + \frac{\Delta t}{(\Delta x)^2}(u_{i+1,j} - 2u_{ij} + u_{i-1,j}) \\ &= ru_{i+1,j} + (1 - 2r)u_{ij} + ru_{i-1,j}, \end{aligned} \tag{29.7}$$

where

$$r \equiv \frac{\Delta t}{(\Delta x)^2}. \tag{29.8}$$

Equation (29.7) is generally attributed to E. Schmidt (1924) and L. Binder (1911). With $j = 0$ in (29.7) we can compute $u_{11}, u_{21}, \ldots, u_{N-1,1}$ (Fig. 29.2). Then with $j = 1$ we compute $u_{12}, u_{22}, \ldots, u_{N-1,2}$ and thus continue to march out a solution one line at a time. Observe that the choice $r = \frac{1}{2}$ produces the especially simple "averaging"

$$u_{i,j+1} = \frac{u_{i+1,j} + u_{i-1,j}}{2} \tag{29.9}$$

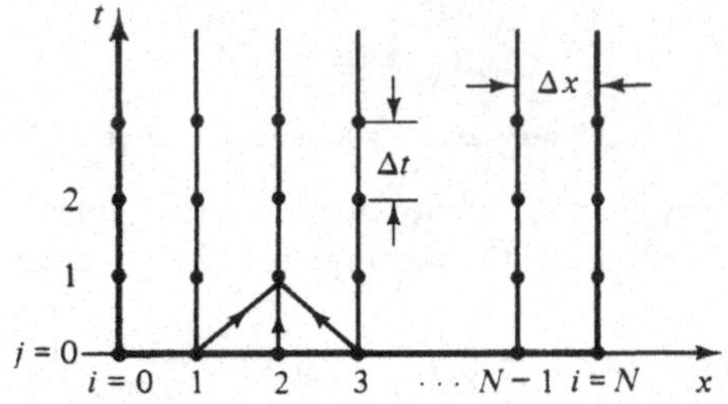

Figure 29.2. The grid.

To illustrate, let us apply (29.7) to the problem shown in Fig. 29.1, with $\ell = 1$, $g(t) = h(t) = 0$, and $f(x) = 1$. Suppose that we choose $\Delta x = 0.02$ and $\Delta t = 0.00018$ (so that $r = 0.45$). The results are plotted as dots in Fig. 29.3 and compared with the exact solution (solid curves)

$$U(x, t) = \sum_{n=1,3,\ldots}^{\infty} \frac{4}{n\pi} \sin n\pi x \, e^{-n^2\pi^2 t}. \tag{29.10}$$

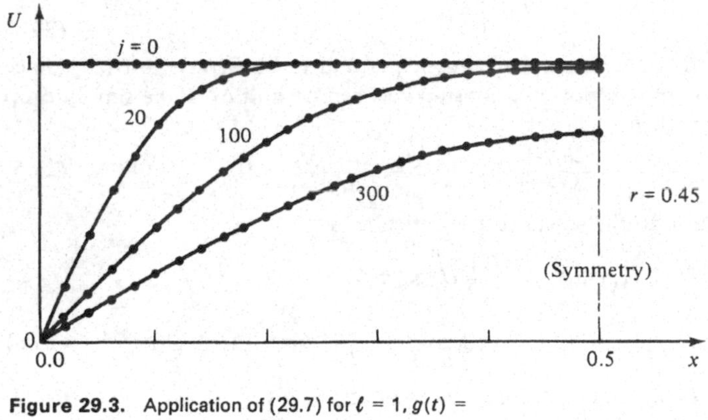

Figure 29.3. Application of (29.7) for $\ell = 1$, $g(t) = h(t) = 0$, $f(x) = 1$, $\Delta x = 0.02$, $\Delta t = 0.00018$.

Before discussing the important questions of convergence and stability, let us introduce the idea of "consistency." We say that a finite-difference version of the PDE is **consistent** with the PDE if the truncation error $T(x, t) \to 0$ for each fixed point x, t in the domain, as the increments Δx and $\Delta t \to 0$. It is clear from (29.6) that (29.7) satisfies this condition.

Denoting the actual machine printout as u_{ij}^*, let us decompose the total error at any given grid point as follows:

$$\begin{aligned}\text{Total error} &= U_{ij} - u_{ij}^* = (U_{ij} - u_{ij}) + (u_{ij} - u_{ij}^*) \\ &\equiv \text{discretization error} + \text{roundoff error,}\end{aligned} \tag{29.11}$$

where the discretization error is the cumulative effect of all truncation errors (including those incurred by finite-difference approximations of derivative boundary conditions). If at each fixed x, t the discretization error $\to 0$ as Δx, $\Delta t \to 0$, we say that the finite-difference scheme is **convergent**. And if the roundoff error stays "small," the scheme is **stable**. (This definition of stability is rather formal but will suffice for our purposes. For more detailed discussion, see, for example, Forsythe and Wasow[1] and Ames.[2])

The following connection between consistency, convergence, and stability is of great importance.

THEOREM 29.1. (*The Lax Equivalence Theorem*) *Given a properly posed linear initialvalue problem and a finite-difference approximation to it that satisfies the consistency condition, stability is necessary and sufficient for convergence.*[3]

By *properly posed* we mean that a solution exists, is unique, and depends continuously on the input data.

As a result of the equivalence between convergence and stability for our scheme (29.7), it will suffice to consider one or the other, say the stability.

Two well-known methods for studying stability are (a) the *Fourier method*,[4] in which we use a finite Fourier representation of the error, and (b) the *matrix method*, in which we express the scheme in matrix form and examine the eigenvalues of an associated matrix.

Although these methods of examining stability rely on linearity of the operators, we emphasize that the finite-difference methods themselves do not (see, for instance, Exercise 29.15)! Unfortunately, however, very little general theory is presently available for nonlinear cases, and we might mention that stability then generally depends not only on the form of the finite-difference scheme but also on the particular solution being obtained.

It is standard, in examining stability, to suppose that roundoff errors are introduced at each point on a given line, say $j = 0$ for definiteness, and to ask: How do these errors propagate (as j increases), assuming that no further errors are introduced?

---

[1]G. Forsythe and W. Wasow, *Finite-Difference Methods for Partial Differential Equations*, Wiley, New York, 1960, Section 5.1. This excellent book is a standard reference for finite-difference methods.

[2]W. F. Ames, *Numerical Methods for Partial Differential Equations*, Barnes & Noble, New York, 1969, Section 1-8.

[3]For proof and a discussion, see R. Richtmeyer and K. Morton, *Difference Methods for Initial-Value Problems*, 2nd ed., Interscience, New York, 1967.

[4]The Fourier stability method, developed by J. von Neumann in the early 1940s, was first discussed in detail in 1951 in an important paper by G. O'Brien, M. Hyman, and S. Kaplan, "A Study of the Numerical Solution of Partial Differential Equations," *Journal of Mathematical Physics*, Vol. 29, 1951, pp. 223–251.

The roundoff error is $e_{ij} \equiv u_{ij} - u_{ij}^*$, with the e_{i0}'s considered as prescribed for $i = 0, 1, \ldots, N$. Since there are no errors committed after $j = 0$, both u_{ij} and u_{ij}^* proceed according to (29.7):

$$u_{i,\,j+1} = ru_{i+1,\,j} + (1 - 2r)u_{ij} + ru_{i-1,\,j}$$
$$u_{i,\,j+1}^* = ru_{i+1,j}^* + (1 - 2r)u_{ij}^* + ru_{i-1,j}^*$$

and, subtracting, we find that e_{ij} proceeds in the same way,

$$e_{i,\,j+1} = re_{i+1,\,j} + (1 - 2r)e_{ij} + re_{i-1,\,j}. \tag{29.12}$$

Motivated by the separation-of-variables form $\sum T_n(t)e^{\mathbf{i}n\pi x/l}$ for the PDE, let us express e_{ij} in the analogous form

$$e_{ij} = \sum_{n=0}^{N} B_n(j)e^{\mathbf{i}n\pi x/l} = \sum_{n=0}^{N} B_n(j)e^{\mathbf{i}\pi in/N}, \tag{29.13}$$

where $\mathbf{i} = \sqrt{-1}$ (Exercise 29.2). Observe that only a finite number of x-wise modes are needed for the discrete (finite-difference) case, whereas an infinite number are normally required in the continuous (PDE) case. The $B_n(0)$s are considered as known, since the initial errors are prescribed, and the question is: How do the $B_n(j)$s grow with time—that is, with j?

Putting (29.13) into (29.12),

$$\begin{aligned}\sum_{n=0}^{N} B_n(j+1)e^{\mathbf{i}\pi in/N} &= \sum_{n=0}^{N} B_n(j)[re^{\mathbf{i}\pi n/N} + 1 - 2r + re^{-\mathbf{i}\pi n/N}]e^{\mathbf{i}\pi in/N} \\ &= \sum_{n=0}^{N} B_n(j)\left(2r\cos\frac{\pi n}{N} + 1 - 2r\right)e^{\mathbf{i}\pi in/N}\end{aligned} \tag{29.14}$$

so that (Exercise 29.3)

$$B_n(j+1) = C_nB_n(j), \tag{29.15}$$

where

$$C_n = 2r\cos\frac{n\pi}{N} + 1 - 2r. \tag{29.16}$$

Thus

$$\begin{aligned}B_n(1) &= C_nB_n(0)\\ B_n(2) &= C_nB_n(1) = C_n^2B_n(0)\\ B_n(3) &= C_nB_n(2) = C_n^3B_n(0)\\ &\;\;\vdots\end{aligned}$$

and it follows that if the B_ns (and hence the e_{ij}'s) are not to grow without bound, we must have

$$|C_n| \le 1 \quad \text{for } n = 0, 1, \ldots, N, \tag{29.17}$$

that is,

$$-1 \le 2r\cos\frac{n\pi}{N} + 1 - 2r \le 1. \tag{29.18}$$

The right-hand inequality is clearly satisfied for *all* (positive) r's. The left-hand inequality reduces to

$$r \le \frac{1}{1 - \cos(n\pi/N)}, \tag{29.19}$$

and since this result needs to hold for all $n = 0, 1, \ldots, N$, we conclude that

$$r \leq \tfrac{1}{2} \tag{29.20}$$

for stability (and, according to the Lax equivalence theorem, also for convergence).[5]

To illustrate, recall that we had $r = 0.45$ for the calculation shown in Fig. 29.3. Increasing Δt to 0.00022, so that $r = 0.55$, the results are as shown in Fig. 29.4. The instability is quite apparent.

Having established convergence only indirectly, we are still lacking an estimate of the actual magnitude of the discretization error $U_{ij} - u_{ij} \equiv E_{ij}$. Leaving the details for Exercise 29.4, we note the result

$$E_{ij} = O(\Delta t) + O(\Delta x)^2, \tag{29.21}$$

which is the same as for the truncation error (29.6).

Before turning to other questions, let us reconsider the stability of the scheme (29.7) by means of the so-called *matrix method.* Expressing (29.7) (plus the boundary conditions at $x = 0$ and $x = \ell$) in matrix form,

$$\begin{pmatrix} u_{1,\,j+1} \\ u_{2,\,j+1} \\ \cdot \\ \cdot \\ \cdot \\ \\ u_{N-1,\,j+1} \end{pmatrix} = \begin{pmatrix} (1-2r) & r & & & 0 \\ r & (1-2r) & r & & \\ & \cdot & & & \\ & & \cdot & & \\ & & & \cdot & \\ & & r & (1-2r) & r \\ 0 & & & r & (1-2r) \end{pmatrix} \begin{pmatrix} u_{1j} \\ u_{2j} \\ \cdot \\ \cdot \\ \cdot \\ \\ u_{N-1,\,j} \end{pmatrix} + \begin{pmatrix} rg_j \\ 0 \\ \cdot \\ \cdot \\ \cdot \\ 0 \\ rh_j \end{pmatrix}$$

or

$$\mathbf{u}_{j+1} = A\mathbf{u}_j + \mathbf{c}_j. \tag{29.22}$$

As above, we suppose that roundoff errors enter along the first line ($j = 0$) and that no further errors are made. Then

$$\mathbf{u}_1 = A\mathbf{u}_0 + \mathbf{c}_0 \qquad \text{(exact)}$$
$$\mathbf{u}_1^* = A\mathbf{u}_0^* + \mathbf{c}_0^* \qquad \text{(actual).}$$

Subtracting,

$$\mathbf{e}_1 = A\mathbf{e}_0 + \mathbf{b}_0,$$

where $\mathbf{e}_j \equiv \mathbf{u}_j - \mathbf{u}_j^*$ and $\mathbf{b}_0 \equiv \mathbf{c}_0 - \mathbf{c}_0^*$. At the next step,

$$\mathbf{u}_2 = A(A\mathbf{u}_0 + \mathbf{c}_0) + \mathbf{c}_1 = A^2\mathbf{u}_0 + A\mathbf{c}_0 + \mathbf{c}_1 \qquad \text{(exact)}$$
$$\mathbf{u}_2^* = A(A\mathbf{u}_0^* + \mathbf{c}_0^*) + \mathbf{c}_1 = A^2\mathbf{u}_0^* + A\mathbf{c}_0^* + \mathbf{c}_1 \qquad \text{(actual).}$$

Subtracting,

$$\mathbf{e}_2 = A^2\mathbf{e}_0 + A\mathbf{b}_0.$$

---

[5]If attempting to correlate the present PDE discussion with the ODE stability discussion in Section 24.3, recall that in the ODE case we distinguished between *stability* and *absolute stability*. The former corresponded to the $h \longrightarrow 0$ limit and the latter to the finite-h case. If, in the present discussion, we were concerned with "stability," then we would have let Δt and hence r tend to zero in (29.16); in this case, $C_n \longrightarrow 1$ and we would have concluded that the method was stable, which is quite reasonable, since we have only a single-step method in terms of the "marching" variable t. But we do *not* let $r \longrightarrow 0$, and so the stability concept adopted in the present chapter actually corresponds to what we called *absolute* stability in the ODE case.

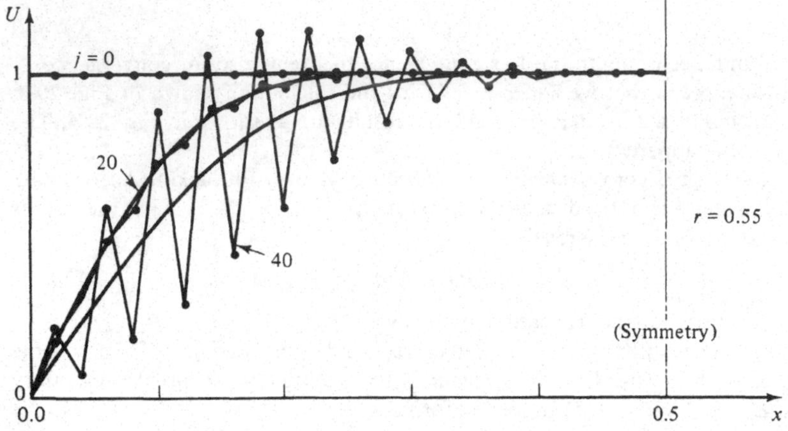

Figure 29.4. Increasing r above $\frac{1}{2}$; $\Delta x = 0.02$, $\Delta t = 0.00022$.

Continuing in this manner, we obtain

$$\mathbf{e}_j = A^j \mathbf{e}_0 + A^{j-1}\mathbf{b}_0. \tag{29.23}$$

Since A is self-adjoint, we know (Chapter 20) that we can expand $\mathbf{e}_0$ and $\mathbf{b}_0$ in terms of the eigenvectors of A,[6] say $\boldsymbol{\phi}_1$ through $\boldsymbol{\phi}_{N-1}$,

$$\begin{aligned} \mathbf{e}_0 &= \alpha_1 \boldsymbol{\phi}_1 + \alpha_2 \boldsymbol{\phi}_2 + \cdots + \alpha_{N-1} \boldsymbol{\phi}_{N-1} \\ \mathbf{b}_0 &= \beta_1 \boldsymbol{\phi}_1 + \beta_2 \boldsymbol{\phi}_2 + \cdots + \beta_{N-1} \boldsymbol{\phi}_{N-1}, \end{aligned} \tag{29.24}$$

so that (29.23) becomes

$$\mathbf{e}_j = (\alpha_1 \lambda_1^j + \beta_1 \lambda_1^{j-1})\boldsymbol{\phi}_1 + \cdots + (\alpha_{N-1}\lambda_{N-1}^j + \beta_{N-1}\lambda_{N-1}^{j-1})\boldsymbol{\phi}_{N-1}. \tag{29.25}$$

For boundedness as $j \to \infty$, it follows that we need $|\lambda_1|, \ldots, |\lambda_{N-1}| \leq 1$ or, equivalently, we need the magnitude of the *largest* eigenvalue of A to be less than or equal to unity.

We state that the "tridiagonal" $L \times L$ matrix

$$\begin{pmatrix} a & b & & & & 0 \\ c & a & b & & & \\ & c & a & b & & \\ & & & \cdot & & \\ & & & & \cdot & \\ & & & & & \cdot \\ & & & c & a & b \\ 0 & & & & c & a \end{pmatrix}$$

[6]Even if A were not self-adjoint, if its $N - 1$ eigenvalues were distinct, then the eigenvectors would be linearly independent (Exercise 29.5), which is all we really need.

has eigenvalues

$$\lambda_n = a + 2\sqrt{bc}\cos\left(\frac{n\pi}{L+1}\right) \quad \text{for } n = 1, \ldots, L. \tag{29.26}$$

In our case [i.e., for the A matrix of (29.22)], $a = 1 - 2r$, $b = c = r$, and $L = N - 1$, and thus

$$\lambda_n = 1 - 2r + 2r\cos\left(\frac{n\pi}{N}\right) \qquad (n = 1, \ldots, N-1)$$

and $|\lambda_n| \leq 1$ implies that

$$r \leq \frac{1}{1 - \cos[(N-1)\pi/N]}. \tag{29.27}$$

This result, for reasonably large N, is essentially the same as the condition (29.20) obtained by the Fourier method. The slight discrepancy can be traced to the absence of the boundary conditions in the Fourier method; specifically, it was assumed that (29.7) applies even at the ends $x = 0$ and $x = \ell$, whereas u_{0j} and u_{Nj} are, in fact, prescribed, not computed.

Generalizing, we note that the Fourier method is easy to apply but does not incorporate boundary conditions and so must be used with caution. The matrix method, on the other hand, automatically includes the boundary conditions but is generally difficult to apply.

To illustrate, suppose that we change the boundary conditions to

$$U_x(0, t) = 0, \tag{29.28a}$$

$$U(\ell, t) = 0. \tag{29.28b}$$

We need to express (29.28a) in finite-difference form—for instance, as

$$U_x(0, t_j) = \frac{U(x_1, t_j) - U(0, t_j)}{\Delta x} - \frac{1}{2}U_{xx}(\xi, t_j)\Delta x = 0 \tag{29.29}$$

for some ξ in $[0, x_1]$. Dropping the last term and changing U to u,

$$\frac{u_{1j} - u_{0j}}{\Delta x} = 0$$

or

$$u_{1j} = u_{0j}. \tag{29.30}$$

The problem with this form is that it contributes to the overall truncation error in the amount $O(\Delta x)$, whereas the U_{xx} term in the PDE was approximated to $O(\Delta x)^2$. Thus we would expect our discretization error to increase from $O(\Delta t) + O(\Delta x)^2$ to $O(\Delta t) + O(\Delta x)$.

To avoid this loss of accuracy, let us use a second-order approximation for (29.28a), say the centered difference

$$\frac{u_{1j} - u_{-1j}}{2\,\Delta x} = 0$$

or

$$u_{-1j} = u_{1j}; \tag{29.31}$$

that is, we extend the interval back to $x_{-1} = -\Delta x$, apply the boundary condition

(29.31) there, and apply (29.7) from $i = 0$ (rather than 1) through $N - 1$. Effectively, then, we have a *false boundary* at x_{-1}. So (Exercise 29.6) in place of (29.22) we have

$$\begin{pmatrix} u_{0,j+1} \\ u_{1,j+1} \\ \cdot \\ \cdot \\ \cdot \\ \\ u_{N-1,j+1} \end{pmatrix} = \begin{pmatrix} (1-2r) & \mathbf{2}r & & & 0 \\ r & (1-2r) & r & & \\ & & \cdot & & \\ & & & \cdot & \\ & & \cdot & & \\ & & r & (1-2r) & r \\ 0 & & & \mathbf{2}r & (1-2r) \end{pmatrix} \begin{pmatrix} u_{0j} \\ u_{1j} \\ \cdot \\ \cdot \\ \cdot \\ \\ u_{N-1,j} \end{pmatrix} \tag{29.32}$$

and we see that the A matrix differs from the one in (29.22) in two ways because it is $N \times N$ rather than $(N-1) \times (N-1)$ and also because of the two boldface **2**s. These changes affect the eigenvalues, and we can expect some change in the resulting stability criterion. Unfortunately, (29.26) does not apply this time. One fruitful approach would be to obtain a *bound* on the largest λ by means of *Gerschgorin's theorem*.[7]

Before closing this section, let us return to (29.6). It has been pointed out by Milne[8] that $U_{xx} = U_t$ implies that $U_{xxxx} = U_{tt}$, so that (29.6) becomes

$$T = \tfrac{1}{12} U_{xxxx}(x_i, t_j)[(\Delta x)^2 - 6\,\Delta t] + O(\Delta x)^3 + O(\Delta t)^2. \tag{29.33}$$

Therefore the special choice $r = \Delta t/(\Delta x)^2 = 1/6$ reduces T to $O(\Delta x)^3 + O(\Delta t)^2$ or, since $\Delta t = (\Delta x)^2/6$, to $O(\Delta x)^3$. As noted, however, consistent accuracy will be needed in derivative boundary conditions if a discretization error of this order is to be realized.

29.2. *THE HEAT EQUATION; IMPLICIT METHODS*

It turns out that the restriction $r = \Delta t/(\Delta x)^2 \leq \frac{1}{2}$ is quite severe in many applications, for if Δx is chosen small for accuracy, then the allowable Δt may be exceedingly tiny, thereby necessitating a great many time steps and considerable computer time.

Obviously, however, (29.7) is not the only possible finite-difference scheme for the heat equation. For instance, observe that for the scheme

$$\frac{u_{i,j+1} - u_{ij}}{\Delta t} = \frac{u_{i+1,j} - 2u_{ij} + u_{i-1,j}}{(\Delta x)^2}, \tag{29.34}$$

which we've been using, the U_{xx} term is approximated entirely at the initial (jth) time, whereas we might expect some weighted average over the time interval to be more suitable—that is,

$$\frac{u_{i,j+1} - u_{ij}}{\Delta t} = (1-\theta)\frac{u_{i+1,j} - 2u_{ij} + u_{i-1,j}}{(\Delta x)^2} + \theta\frac{u_{i+1,j+1} - 2u_{i,j+1} + u_{i-1,j+1}}{(\Delta x)^2}; \tag{29.35}$$

---

[7]See, for instance, G. D. Smith, *Numerical Solution of Partial Differential Equations*, Oxford University Press, New York, 1965.

[8]W. E. Milne, *Numerical Solution of Differential Equations*, Wiley, New York, 1953, p. 122.

where $0 \le \theta \le 1$. The computation points for (29.34) and (29.35) are indicated in Fig. 29.5. Note carefully that (29.34) is an **explicit** formula, since we may solve [in the form (29.7)] explicitly for $u_{i,\,j+1}$. The scheme (29.35), on the other hand, is **implicit** because it involves the three unknowns $u_{i-1,\,j+1}, u_{i,\,j+1}, u_{i+1,\,j+1}$; that is, the equations corresponding to the $(j+1)$ st line are *coupled.*

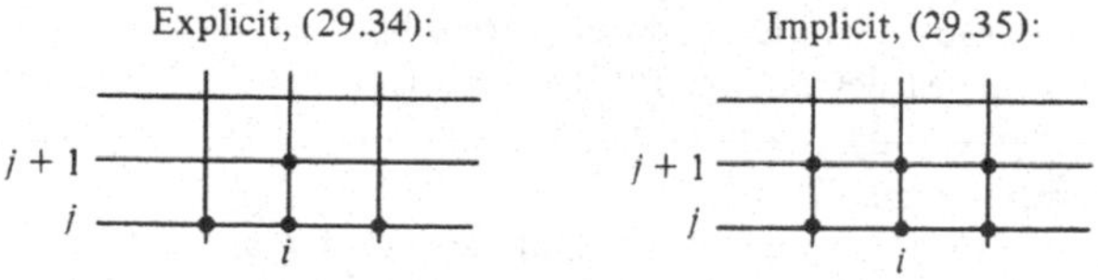

Figure 29.5. The computation points.

It can be shown that the implicit scheme (29.35) satisfies the consistency condition and has a truncation error

$$T = O(\Delta t) + O(\Delta x)^2, \tag{29.36}$$

which is the same as for the explicit scheme (29.34) except for the choice $\theta = \frac{1}{2}$, in which case it becomes $O(\Delta t)^2 + O(\Delta x)^2$ (Exercise 29.7).

A Fourier stability analysis of (29.35) proceeds just as before; again we obtain $B_n(j+1) = C_n B_n(j)$, but this time [Exercise 29.8(a)]

$$C_n = \frac{1 - 2r(1-\theta)(1 - \cos n\pi/N)}{1 + 2r\theta(1 - \cos n\pi/N)}. \tag{29.37}$$

Requiring $|C_n| \le 1$ leads [Exercise 29.8(b)] to the condition

$$r(1 - 2\theta)\left(1 - \cos\frac{n\pi}{N}\right) \le 1 \tag{29.38}$$

on r, and we see that if we choose the constant $\theta \ge \frac{1}{2}$, then (29.38) is satisfied for *all* r; that is, *we have stability (and convergence) for all* r! In particular, the choice $\theta = \frac{1}{2}$ yields the well-known **Crank–Nicolson** scheme.

This impressive result is not entirely surprising in view of the following simpler formal argument. Since this is a "marching" problem, we might expect the x derivatives in $U_{xx} = U_t$ not to be central to the question of stability, which is more fundamentally related to the t variable. As in the ODE case (Section 24.3), then, let us look at the *test equation* $-AU = U_t$ instead. Our scheme becomes

$$\frac{u_{i,\,j+1} - u_{ij}}{\Delta t} = -(1-\theta)Au_{ij} - \theta A u_{i,\,j+1} \tag{29.39}$$

or

$$u_{i,\,j+1} = \left[\frac{1 - A\Delta t(1-\theta)}{1 + A\Delta t\theta}\right]u_{ij} \equiv K u_{ij} \text{ say.}$$

Similarly,

$$u^*_{i,j+1} = K u^*_{ij};$$

and subtracting these results, we see that the roundoff error $e_{ij} \equiv u_{ij} - u^*_{ij}$ propagates according to $e_{i,\,j+1} = K e_{ij}$. Thus $e_{ij} = K^j e_{i0}$, and so for stability we need $|K| \le 1$

or $-1 \leq K \leq 1$. Writing this out, we find that for stability we need

$$A\,\Delta t < \frac{2}{1-2\theta}. \tag{29.40}$$

For $0 \leq \theta < \frac{1}{2}$ the right-hand side is positive, and (29.40) then limits the step size Δt for stability. For $\theta \geq \frac{1}{2}$, however, (29.40) is satisfied for Δt arbitrarily large. Consequently, we have stability for all Δt's and are free to choose Δt simply on the basis of accuracy (i.e., discretization error). Incidentally, this discussion is essentially the same as our earlier one on *stiff systems* (Exercise 24.22) except that in Chapter 24 we considered only the two limiting cases of (29.39), the explicit case $\theta = 0$ and the fully implicit case $\theta = 1$.

Turning to computational aspects, let us write out the Crank–Nicolson scheme for the problem of Fig. 29.1,

$$\begin{pmatrix} 2(1+r) & -r & & & 0 \\ -r & 2(1+r) & -r & & \\ & & \cdot & & \\ & & \cdot & & \\ & & \cdot & & \\ & -r & 2(1+r) & -r & \\ 0 & & & -r & 2(1+r) \end{pmatrix} \begin{pmatrix} u_{1,\,j+1} \\ u_{2,\,j+1} \\ \cdot \\ \cdot \\ \cdot \\ u_{N-1,\,j+1} \end{pmatrix}$$

$$= \begin{pmatrix} rg_j + 2(1-r)u_{1j} + ru_{2j} + rg_{j+1} \\ ru_{1j} + 2(1-r)u_{2j} + ru_{3j} \\ \cdot \\ \cdot \\ \cdot \\ ru_{N-2,\,j} + 2(1-r)u_{N-1,\,j} + rh_j + rh_{j+1} \end{pmatrix}.$$

This is of the form $Ax = c$, where A is *tridiagonal*. The same form arose in our discussion of the finite-element method in Section 24.7, where efficient iterative methods of solution were discussed. For further discussion, refer to Smith[9] and the treatise by Varga.[10]

Besides the schemes (29.34) and (29.35), others are also available. Richtmeyer and Morton,[11] for instance, list *14* finite-difference schemes (mostly implicit) for the heat equation $U_{xx} = U_t$. Of them, the one by *DuFort–Frankel*,

$$\frac{u_{i,\,j+1} - u_{i,\,j-1}}{2\,\Delta t} = \frac{u_{i+1,\,j} - u_{i,\,j-1} - u_{i,\,j+1} + u_{i-1,\,j}}{(\Delta x)^2} \tag{29.41}$$

is especially interesting. Its truncation error is

$$T = \frac{(\Delta x)^2}{12} U_{xxxx}(x_i, t_j) - \frac{(\Delta t)^2}{6} U_{ttt}(x_i, t_j) + \left(\frac{\Delta t}{\Delta x}\right)^2 U_{tt}(x_i, t_j) + \cdots \tag{29.42}$$

---

[9] *Op. cit.*

[10] R. S. Varga, *Matrix Iterative Analysis*, Prentice-Hall, Englewood Cliffs, N.J., 1962.

[11] R. Richtmeyer and K. Morton, *Difference Methods for Initial-Value Problems*, 2nd ed., Interscience, New York, 1967, pp. 189–191.

so that the scheme is consistent only if Δt goes to zero faster than Δx. If they tend to zero "together"—with $\Delta t/\Delta x = \text{constant} = \kappa$—then the scheme (29.41) is consistent not with the heat equation but with the hyperbolic equation

$$u_{xx} = \kappa^2 u_{tt} + u_t!$$

This situation has no counterpart in the case of ODEs.

Some additional considerations for the parabolic case, including nonconstant coefficients, nonlinear cases, singularities, and extension to more than one dimension, are contained in the exercises.

29.3. *HYPERBOLIC AND ELLIPTIC PROBLEMS*

Hyperbolic systems are generally handled most naturally and accurately by the method of characteristics, for it is along the characteristics that discontinuities propagate. If, however, the desired solution is sufficiently smooth, then finite-difference methods offer an attractive alternative.

To illustrate, the wave equation

$$a^2 U_{xx} = U_{tt} \tag{29.43}$$

may be discretized in the form [Exercise 29.9(a)]

$$a^2 \frac{u_{i+1,j} - 2u_{ij} + u_{i-1,j}}{(\Delta x)^2} = \frac{u_{i,j+1} - 2u_{ij} + u_{i,j-1}}{(\Delta t)^2}$$

or

$$u_{i,j+1} = r^2(u_{i+1,j} + u_{i-1,j}) + 2(1 - r^2)u_{ij} - u_{i,j-1} \tag{29.44}$$

where

$$r \equiv \frac{a\,\Delta t}{\Delta x}. \tag{29.45}$$

A typical set of computation points is shown (heavy dots) in Fig. 29.6. Observe that the calculation of u at P from information along QR [Exercise 29.9(b)] involves marching out u_{ij} in the triangle QPR. Suppose that the parameter a is such that the characteristics through P (SP and TP) fall within QR as shown. We recall from Chapter 27 that the solution at P depends on data along ST; therefore the data on QR is more than sufficient and we can be optimistic about the prospects for convergence and stability. If, on the other hand, QR is contained within ST, we may be pessimistic instead.

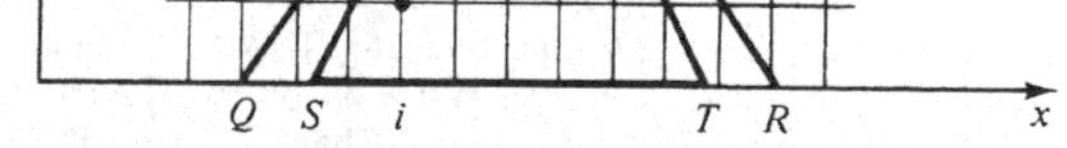

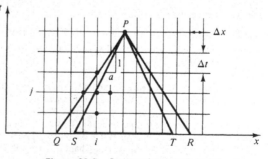

Figure 29.6. Computational mesh and the characteristics.

Summarizing, we anticipate convergence and stability of the scheme (29.44) if and only if the so-called **interval of dependence** *ST* of *P* is contained within the mesh interval *QR*—that is, if and only if

$$\frac{1}{a} \geq \frac{\Delta t}{\Delta x}, \tag{29.46}$$

or if $r \leq 1$. (In the present case, a is a constant; more generally, it may vary, in which case we expect to need $r \leq 1$ *locally*—that is, at each mesh point.)

In fact, this formal claim is correct and was established in an early and pioneering paper by Courant, Friedrichs, and Lewy.[12]

In contrast with the parabolic and hyperbolic marching-type problems, *elliptic* problems are of boundary value type, as discussed in Chapter 27, and often arise in describing equilibrium and steady-state phenomena. Thus the question of stability of a finite-difference scheme does not occur, although, of course, convergence is as relevant as before.

Often in marching problems, although not always, the domain is rectangular and hence fitted perfectly by a simple rectangular grid. The heat equation $U_{xx} = U_t$ discussed in the foregoing sections was of this type [but would cease to be if, for example, the applied temperature $U(0, t) = g(t)$ at the left end were high enough to cause the material to melt away continually[13]]. Boundary shapes for elliptic problems, however, are more varied, a factor that introduces considerable inconvenience in applying boundary conditions. (Of course, curvilinear nets may be used, but they lack the simplicity of rectangular ones and even they may have trouble in fitting irregular boundaries.)

In any case, the finite-difference scheme for an elliptic problem with M interior mesh points ultimately leads to an M-dimensional system of the form $A\mathbf{u} = \mathbf{c}$, where the components of the M-dimensional vector $\mathbf{u}$ are the M unknowns. So we are confronted with systems that are much larger than in marching problems; for instance, even for implicit schemes like (29.35) the $\mathbf{u}$ vector consisted of only *one line* of unknowns. Fortunately, A is generally rather *sparse* and well suited to iterative solution.[14]

EXERCISES

29.1. Derive the expression (29.3).

29.2. Rather than assume the form (29.13) a priori, solve the difference equation (29.12) by seeking e_{ij} in the variable-separable form $X_i T_j$ and proceed to derive the final result $r \leq \frac{1}{2}$.

---

[12] R. Courant, K. Friedrichs, and H. Lewy, Über die Partiellen Differenzengleichungen der Mathematischen Physik, *Mathematische Annalen*, Vol. 100, 1928, pp. 32–74.

[13] As a start, see, for example, W. F. Ames, *Nonlinear Partial Differential Equations in Engineering*, Academic, New York, 1965, pp. 360–361. This book contains numerous special techniques from the engineering literature. Emphasis is not on general theory but rather on methods of solution.

[14] See, for instance, Ames, *op. cit.*, Chapter 7, Part B; Smith, *op. cit.*; and Varga, *op. cit.*

***29.3.** Deducing (29.15) from (29.14) amounts to saying that

$$\sum_{n=0}^{N} a_n e^{inx} = 0 \text{ for } x = 0, \frac{\pi}{N}, \frac{2\pi}{N}, \ldots, \pi$$

implies that $a_n = 0$ for $n = 0, 1, \ldots, N$. Show why this is so.

29.4. (*Convergence*) (a) An outline of the derivation of (29.21) is as follows: Since $E_{ij} \equiv U_{ij} - u_{ij}$, (29.4) and (29.5) yield

$$E_{i,j+1} = r(E_{i+1,j} + E_{i-1,j}) + (1 - 2r)E_{ij} + O(\Delta t)^2 + O[\Delta t(\Delta x)^2].$$

If $r \leq \frac{1}{2}$, for stability, then

$$\max_i |E_{i,j+1}| \leq \max_i |E_{ij}| + K_1(\Delta t)^2 + K_2\,\Delta t(\Delta x)^2,$$

so that for $j\,\Delta t \leq t$, say,

$$\max_i |E_{ij}| \leq t[K_1\,\Delta t + K_2(\Delta x)^2].$$

Fill in whatever reasoning is needed.

(b) A more detailed estimate than (29.21) must take into account the degree of smoothness of the boundary data. For example, if $g(t) = h(t) = 0$ in Fig. 29.1 and the odd periodic extension of $f(x)$ and its first $p - 1$ derivatives are continuous (and the pth derivative has bounded variation over $0 \leq x \leq \ell$), then (29.21) is refined to[15]

$$E_{ij} = \begin{cases} O[(\Delta t)^{p/4}] & \text{for } p \leq 3 \\ O(\Delta t\,|\ln \Delta t\,|) & \text{for } p = 4 \\ O(\Delta t) & \text{for } p > 4, \end{cases} \tag{29.47}$$

where we have used the fact that $r = \Delta t/(\Delta x)^2$ is fixed in omitting the Δx terms. Note carefully that (29.47) does *not* violate (29.21) but is simply a *sharper* estimate in the same way that $\sin x = O(x)$ as $x \longrightarrow 0$ is a sharper statement than $\sin x = O(1)$ as $x \longrightarrow 0$. For instance, in the calculation shown in Fig. 29.3 we have $p = 1$, and so the truncation error can be expected to be $O(\Delta t)^{1/4}$, at least near the ends of the rod and for small time (where the effects of the discontinuity in the boundary conditions at $x = t = 0$ and $x = \ell, t = 0$ are most strongly felt); away from the ends and/or for large t we can expect the truncation error to approach the value $O(\Delta t)$ given by (29.21). [We drop the $(\Delta x)^2$ term in (29.21) because $(\Delta x)^2$ and Δt are of the same order for $r = \Delta t/(\Delta x)^2$ fixed.] Here is the problem. Run the calculation of Fig. 29.3 for $\Delta x = 0.05$, $\Delta t = 0.001125$, for $\Delta x = 0.025$, $\Delta t = 0.00028125$, and then for $\Delta x = 0.01666\ldots$, $\Delta t = 0.000125$ (so that $r = 0.45$ in each case). For each run, obtain u at $x = 0.1$, $t = 0.00225$. Expressing

$$u(0.1, 0.00225) - U(0.1, 0.00225) \approx A(\Delta t)^{\alpha},$$

where U, A, α are unknown, use the three sets of results to evaluate α and see how close the result is to $\frac{1}{4}$. (Of course, the computer output is actually u^*, not u, but it is found that for a stable method the roundoff error is generally neg-

[15] See Richtmeyer and Morton, *op. cit.*, Section 1.7.

ligible compared to the discretization error.) Repeat for $x = 0.5$ and $t = 0.01125$ and see how close α is to 1.[16] Discuss.

***29.5.** Prove that if the eigenvalues of an $M \times M$ matrix A are distinct, then the eigenvectors must be linearly independent (as stated in footnote 5).

29.6. Show how the system (29.32) is obtained.

29.7. Show that (29.36) becomes $T = O(\Delta t)^2 + O(\Delta x)^2$ for the special (Crank–Nicolson) case $\theta = \frac{1}{2}$, as stated in the text.

29.8. (a) Apply the Fourier stability method to the implicit scheme (29.35) and obtain the expression (29.37) for C_n.
(b) From it, deduce the inequality (29.38).

29.9. (a) What is the order of the truncation error T associated with the approximation of the wave equation (29.43) by the finite-difference scheme (29.44)?
(b) Let (29.43) apply in $0 \leq x \leq 1$, $t \geq 0$, with the conditions

$$U_x(0, t) = U(0, t), \qquad U(1, t) = 0$$
$$U(x, 0) = x^2(1 - x)^2, \qquad U_t(x, 0) = 0$$

and with $a = \frac{1}{4}$. Express derivative boundary conditions in finite-difference form, to an accuracy that is consistent with that of (29.44), and introduce "false boundaries" where necessary. With $\Delta x = 0.25$ and $\Delta t = 0.5$, compute $u_{2,\,3}$.

29.10. Apply the Fourier stability method to (29.44) and thus derive the stated stability condition $r \leq 1$.

29.11. (Sample explicit calculation) (a) Suppose that $\ell = 1$, $g(t) = h(t) = 0$, and $f(x) = 2x$ for $0 \leq x \leq \frac{1}{2}$ and $2 - 2x$ for $\frac{1}{2} \leq x \leq 1$ in Fig. 29.1. Taking $\Delta x = 0.1$ and $\Delta t = 0.001$, use (29.7) to compute u at $x = 0.3$ and $t = 0.005$ and compare with the exact solution $u \doteq 0.5966$.
(b) Same as part (a) but with $\Delta t = 0.005$.

29.12. By means of the Fourier method, show that if we use the more accurate centered difference $(u_{i,\,j+1} - u_{i,\,j-1})/2\,\Delta t$ in (29.5) in place of the forward difference $(u_{i,\,j+1} - u_{ij})/\Delta t$, then the resulting scheme is *unstable for all r*.[17]

29.13. Show that

$$a(x)\frac{\partial}{\partial x}\left[b(x)\frac{\partial U}{\partial x}\right] = \frac{\partial U}{\partial t}$$

can be simplified a little to the form

$$\frac{a}{b}\frac{\partial^2 U}{\partial y^2} = \frac{\partial U}{\partial t}$$

by the change of variables $y = \int dx/b(x)$.

---

[16] A number of methods of "desingularizing," to improve the accuracy, are discussed in W. F. Ames, *Nonlinear Partial Differential Equations in Engineering*, Academic, New York, 1965. To the "lowest order," such methods indicate that the average value should be assigned at a jump discontinuity in the boundary data. In the present example it means using $U(0, 0) = U(1, 0) = 0.5$, which is what we did in Figs. 29.3 and 29.4 and what you should do here.

[17] In fact, this scheme was proposed in a pioneering paper by Richardson in 1910 and not shown to be unstable until 1951. See L. F. Richardson, The Approximate Arithmetical Solution by Finite Differences of Physical Problems Involving Differential Equations with an Application to the Stresses in a Masonry Dam, *Trans. Roy. Soc. London,* A210, 1910, pp. 307–357.

29.14. As a *nonlinear* example, with *nonconstant coefficients* as well, consider the PDE

$$a(x, t)U^2U_{xx} = U_t. \tag{29.48}$$

(a) Write out an *explicit* finite-difference scheme for (29.48) and show that it is *linear* in the unknown $u_{i,\,j+1}$!

(b) What if the scheme is *implicit*? Show how to apply the method of *quasilinearization* (Section 22.6). *Note:* In the absence of a definitive stability analysis we might expect the explicit scheme to remain stable, provided that $a(x, t)U^2\,\Delta t/(\Delta x)^2 \leq \frac{1}{2}$. This condition may require refinement of Δt from line to line.[18]

29.15. (*Symmetrical heat conduction in cylinders and spheres*) (a) Cylinders. If $U(r, \theta, z, t)$ is independent of θ and z, then

$$U_{rr} + \frac{1}{r}U_r = U_t.$$

Show that this result reduces to the slightly more convenient form

$$e^{-2R}U_{RR} = U_t$$

under the change of variables $R = \ln r$.

(b) Spheres. If $U(\rho, \theta, \phi, t)$ is independent of θ and ϕ, then

$$U_{\rho\rho} + \frac{2}{\rho}U_\rho = U_t.$$

Show that it reduces to the more familiar form

$$W_{\rho\rho} = W_t$$

under the change of variables $U = W/\rho$.

29.16. For the *two-dimensional* heat equation

$$\alpha^2(U_{xx} + U_{yy}) = U_t$$

with explicit finite-difference representation

$$\frac{u_{ij,\,k+1} - u_{ijk}}{\Delta t} = \frac{\alpha^2}{(\Delta x)^2}(u_{i-1,\,jk} - 2u_{ijk} + u_{i+1,\,jk}) + \frac{\alpha^2}{(\Delta y)^2}(u_{i,\,j-1,\,k} - 2u_{ijk} + u_{i,\,j+1,\,k}),$$

show that the stability condition (independent of boundary conditions)is

$$\alpha^2\,\Delta t\left\{\frac{1}{(\Delta x)^2} + \frac{1}{(\Delta y)^2}\right\} \leq \frac{1}{2}.$$

29.17. Show that the "five-point" difference scheme

$$\frac{u_{i+1,\,j} - 2u_{ij} + u_{i-1,\,j}}{(\Delta x)^2} + \frac{u_{i,\,j+1} - 2u_{ij} + u_{i,\,j-1}}{(\Delta y)^2} = 0 \tag{29.49}$$

for the two-dimensional Laplace equation has a truncation error

$$T = O(\Delta x)^2 + O(\Delta y)^2. \tag{29.50}$$

Observe that if $\Delta x = \Delta y$, then (29.49) takes the well-known form

$$u_{i+1,\,j} + u_{i-1,\,j} + u_{i,\,j+1} + u_{i,\,j-1} - 4u_{ij} = 0,$$

---

[18] You should definitely look at the discussion of the nonlinear example $U_t = (U^5)_{xx}$ in Section 8.6 of Richtmeyer and Morton, *op. cit.*

which indicates that u_{ij} is equal to the *average* $(u_{i+1,j} + u_{i-1,j} + u_{i,j+1} + u_{i,j-1})/4$ of its neighboring values. This result should not be surprising in view of our earlier observation in Comment 2 of Example 26.11.

29.18. (*Torsion of a hollow square cylinder*) The stress function $\phi(x, y)$ for a long cylinder in torsion is governed by a Poisson equation

$$\phi_{xx} + \phi_{yy} = -2 \tag{29.51}$$

over the cross section, where ϕ is constant around the edge. Consider the cross section to be square and hollow (Fig. 29.7) with $\phi = 0$ on the outer edge and $\phi = 1$ on the inner edge.

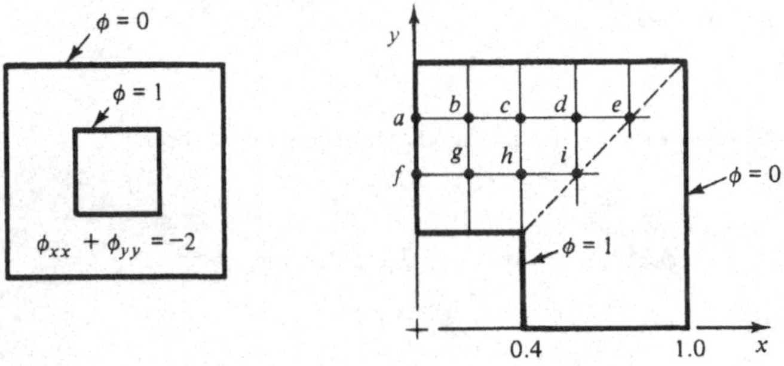

Figure 29.7. A torsion problem.

(a) Using the five-point form in (29.49) for the left-hand side of (29.51), write down the nine equations in the nine unknowns $\phi_a, \phi_b, \ldots, \phi_i$ (using the obvious symmetries). You need not solve.

(b) Show how an *extrapolation* technique can be used, with the help of (29.50), to improve the accuracy substantially.

29.19. Make up two examination-type problems on this chapter.

ADDITIONAL REFERENCES FOR PART V

M. B. ABBOTT, *The Method of Characteristics*, American Elsevier Publishing Co., New York, 1966. [A highly motivated treatment with numerous numerical examples on long water waves, gasdynamics, plasticity, and the mechanics of granular materials.]

G. CARRIER and C. PEARSON, *Partial Differential Equations*, Academic, New York, 1976. [Excellent.]

B. EPSTEIN, *Partial Differential Equations*, McGraw-Hill, New York, 1962. [Intended as an introductory text at the graduate level. Emphasis is on existence theory rather than methods of solution. Banach and Hilbert space concepts are introduced in two chapters and used in the subsequent development.]

F. B. HILDEBRAND, *Advanced Calculus for Applications*, 2nd ed., Prentice-Hall, Englewood Cliffs, N.J., 1976. [Chapters 8 and 9 provide a clear discussion of PDEs that is both detailed and compact. Be sure to look at this book.]

A. G. MACKIE, *Boundary Value Problems*, Hafner Publishing Co., New York, 1965. [Concerned, for the most part, with the application of Green's function, delta function, integral transform, and Riemann–Green function techniques to boundary value problems in ordinary and partial differential equations.]

F. H. MILLER, *Partial Differential Equations*, Wiley, New York, 1941. [A more elementary treatment than the other references in this list. Chapters 4 and 5 discuss *first*-order PDEs, which we have discussed hardly at all.]

A. SOMMERFELD, *Partial Differential Equations in Physics*, Academic, New York, 1949. [Combination of theory and application to problems of mathematical physics.]

A TYCHONOV and A. SAMARSKI, *Partial Differential Equations of Mathematical Physics*, Vols. I and II, Holden-Day, San Francisco, 1964. [Volume I deals with hyperbolic, parabolic, and elliptic equations and covers the physics, careful derivation of the mathematical problems, their solution, and the existence, uniqueness, and properties of solutions. Volume II extends this material to three dimensions and introduces a number of special functions.]

H. F. WEINBERGER, *Partial Differential Equations*, Blaisdell Publishing Co., Waltham, Mass., 1965. [Intended as a text for a two-semester, advanced, undergraduate-level course in PDEs. Nicely written.]

Survey-Type References

G. ARFKEN, *Mathematical Methods for Physicists,* 2nd ed., Academic, New York, 1970.

R. ARIS and N. AMUNDSON, *Mathematical Methods in Chemical Engineering*, Vol. 2, *First-Order Partial Differential Equations*, Prentice-Hall, Englewood Cliffs, N.J., 1973.

E. F. BECKENBACH (Ed.), *Modern Mathematics for the Engineer*, McGraw-Hill, New York, 1961.

R. COURANT and D. HILBERT, *Methods of Mathematical Physics*, Vol. 1, Interscience, New York, 1953.

H. CROWDER and S. MCCUSKEY, *Topics in Higher Analysis*, Macmillan, New York, 1964.

P. DENNERY and A. KRZYWICKI, *Mathematics for Physicists*, Harper & Row, New York, 1967.

J. W. DETTMAN, *Mathematical Methods in Physics and Engineering*, McGraw-Hill, New York, 1962.

P. FRANKLIN, *Methods of Advanced Calculus*, McGraw-Hill, New York, 1944.

B. FRIEDMAN, *Lectures on Applications-Oriented Mathematics*, Holden-Day, San Francisco, 1969.

G. GOERTZEL and N. TRALLI, *Some Mathematical Methods of Physics*, McGraw-Hill, New York, 1960.

F. B. HILDEBRAND, *Methods of Applied Mathematics*, 2nd ed., Prentice-Hall, Englewood Cliffs, N.J., 1965.

J. IRVING and N. MULLINEUX, *Mathematics in Physics and Engineering*, Academic, New York, 1959.

H. Jeffreys and B. Jeffreys, *Methods of Mathematical Physics*, 3rd ed., Cambridge at the University Press, London, 1956.

T. von Kármán and M. Biot, *Mathematical Methods in Engineering*, McGraw-Hill, New York, 1940.

G. Korn and T. Korn, *Mathematical Handbook for Scientists and Engineers*, 2nd ed., McGraw-Hill, New York, 1968.

E. Kreyszig, *Advanced Engineering Mathematics*, 3rd ed., Wiley, New York, 1972.

H. Margenau and G. Murphy, *The Mathematics of Physics and Chemistry*, 2nd ed., Vols. I and II, Van Nostrand, Princeton, N.J., 1956, 1964.

J. Mathews and R. Walker, *Mathematical Methods of Physics*, W. A. Benjamin, New York, 1965.

P. Morse and H. Feshbach, *Methods of Theoretical Physics*, Parts I and II, McGraw-Hill, New York, 1953.

F. D. Murnaghan, *Introduction to Applied Mathematics*, Wiley, New York, 1948.

A. Page, *Mathematical Analysis and Techniques*, Oxford University Press, New York, 1974.

L. Pipes and L. Harvill, *Applied Mathematics for Engineers and Physicists*, 3rd ed., McGraw-Hill, New York, 1970.

I. Sokolnikoff and R. Redheffer, *Mathematics of Physics and Modern Engineering*, 2nd ed., McGraw-Hill, New York, 1966.

I. Stakgold, *Boundary Value Problems of Mathematical Physics*, Vols. I and II, Macmillan, New York, 1967.

C. R. Wylie, Jr., *Advanced Engineering Mathematics*, 4th ed., McGraw-Hill, New York, 1975.

Answers to Selected Exercises

PART I

Chapter 1

1.9. Yes. No.
1.10. $\delta(\epsilon) = \sin^{-1} \epsilon$.
***1.13.** It is.
1.21. $n = 92$ will suffice.
1.25. (b) $I'(x) = 4x^7 \ln x$.
1.29. $-100/289$.

Chapter 2

***2.4.** Closed and bounded with respect to either distance function.
2.9. (a) conv. (c) div. (e) conv. (g) div.
2.10. (a) conv. conditionally (c) conv. absolutely.
2.12. (a) div. for all x (c) conv. for $x > 0$
(e) conv. for all x (g) conv. for all x
(i) conv. for all $x \neq -1, -2, -3, \ldots$
(k) conv. for all $x \neq \pm\pi/2, \pm 3\pi/2, \pm 5\pi/2, \ldots$.
2.13. (a) conv. only for $x = 0$ (c) conv. in $|x| < 1$.
2.16. $N(\epsilon) = -\ln \epsilon/\ln 2$, for example.
2.20. (a) $\sin(3x^2) = 3x^2 - \dfrac{3^3}{3!}x^6 + \dfrac{3^5}{5!}x^{10} - \cdots$ for all x
(c) $\ln x = \ln 3 + \dfrac{1}{3}(x-3) - \dfrac{1}{2 \cdot 3^2}(x-3)^2 + \dfrac{1}{3 \cdot 3^3}(x-3)^3 - \cdots$ for $0 < x \leq 6$.

2.26. $I \doteq 1.3764$.

2.27. $\left| \operatorname{erf} x - \frac{2}{\sqrt{\pi}}\left(x - \frac{x^3}{3} + \frac{x^5}{10} - \frac{x^7}{42} + \frac{x^9}{216}\right)\right| \leq 0.001$.

***2.35.** Conv.

Chapter 3

3.2. (a) $|\alpha| < 3$ (c) $|\alpha| < \infty$ (e) $-4 < \alpha < 2$ (g) none.

3.3. (a) conv. (c) conv. (e) conv. (g) div.

3.4. Yes; no; no; yes.

3.7. $I = \int_0^2 \left\{\frac{1 - H(\xi - 1)}{(2 - \xi)^{3/2}} - \frac{1}{(2 + \xi)^{3/2}}\right\} \frac{d\xi}{\xi}$, which is *non*singular.

3.9. $\sqrt{\pi}/4$.

3.14. -6.

3.17. $\pi/8$.

3.18. $I = \int_0^1 \int_0^{\sqrt{y}} f(x, y)\, dx\, dy - \int_1^4 \int_{\sqrt{y}}^2 f(x, y)\, dx\, dy$.

3.20. Here's half the answer:

$$I = \int_0^1 \int_p^1 \int_?^? \int_?^? f\, ds\, dr\, dq\, dp.$$

3.21. (b) $(2 \ln 2)/3$.

Chapter 4

4.2. No; yes; yes

4.4. $\pi/4$.

Chapter 5

5.9. (b) $G(t) = -100[f(t) - 1] = 500 - \frac{200}{\pi} \sum_{1,3,\ldots}^{\infty} \frac{\sin nt}{n}$.

5.10. (b) $f(x) = \frac{2}{\pi} - \frac{4}{\pi} \sum_1^{\infty} \frac{\cos 2nx}{4n^2 - 1}$

(d) $f(x) = \frac{\pi}{4} - \frac{1}{\pi} \sum_1^{\infty} \left\{\frac{2 \cos (2n - 1)x}{(2n - 1)^2} + \frac{\pi(-1)^n \sin nx}{n}\right\}$

(g) $f(x) = \frac{2}{\pi} - \frac{4}{\pi} \sum_1^{\infty} \frac{(-1)^n \cos 2nx}{4n^2 - 1}$

(i) $f(x) = \frac{50}{3} + \frac{50}{\pi} \sum_1^{\infty} \frac{1}{n}\left\{\sin \frac{2n\pi}{3} \cos \frac{2n\pi x}{3} + \left(1 - \cos \frac{2n\pi}{3}\right) \sin \frac{2n\pi x}{3}\right\}$.

5.17. $u(a, \theta) = 50 + \frac{200}{\pi} \sum_{1,3,\ldots}^{\infty} \frac{\sin n\theta}{n}$.

Chapter 6

6.1. (a) $2\alpha/(\alpha^2 + \omega^2)$.

***6.2.** $\hat{f}(\omega) = (2 \sin \omega)/\omega$.

6.13. (b) $\sin (e^{t^2})$, yes; $(d/dt) \sin (e^{t^2}) = O(te^{t^2})$, no

(e) yes (g) yes.

6.15. $*(\pi/2 - \tan^{-1} s)/s$.

6.18. *(a) $H(t - 1)[\operatorname{erf} \sqrt{3(t - 1)} - \sqrt{3(t - 1)} \operatorname{erf}' \sqrt{3(t - 1)}]/3\sqrt{3}$

(c) $(\sin^2 t)/2$.

6.21. $u(x) = \frac{P}{2EI}\left[\frac{H(x - l/2)(x - l/2)^3}{3} + \frac{l^2 x}{8} - \frac{x^3}{6}\right]$.

6.22. (c) $(1 - e^{-s})/s(1 - e^{-3s})$.
6.23. (a) $x(t) = (\sin \sqrt{k/m}\, t)/\sqrt{km}$.
6.25. (a) $(\sin^2 t)/2$ (d) $-3 - 3t - t^2 + 3e^t$
(f) $(\sinh t + t \cosh t)/2$.
6.26. $I(t) = (2/a) \int_0^t \dot{E}_i(t - \tau)e^{-R\tau/2L} \sinh (a\tau/2L)\, d\tau,\ a = \sqrt{R^2 - 4L/C}$.
6.28. (a) $[2(1 - e^{-2t}) + H(t - 1)(2t - 3 + e^{-2(t-1)})]/4$
(c) $1 + \int_0^t f(t - \tau)(\cosh \tau - 1)\, d\tau$
(e) $x(t) = e^t \cosh 2t,\ y(t) = 2e^t \sinh 2t$
(g) 1.
***6.33.** (b) $p(v) = 2\sqrt{0.0744/\pi}\, e^{-1/(0.2976v^2)}$.

Chapter 7

7.6. (b) $y = AJ_0(2k\sqrt{x}) + BY_0(2k\sqrt{x})$.
7.10. $x_v = y/(y - x),\ x_{uu} = -1/2(x - y)^3$.
7.16. (a) $\sin (x + y) = x + y - \frac{x^3}{6} - \frac{x^2 y}{2} - \frac{xy^2}{2} - \frac{y^3}{6} + \cdots$
(c) $x^2 y = 2 + 4(x - 1) + (y - 2) + 2(x - 1)^2 + 2(x - 1)(y - 2) + (x - 1)^2(y - 2)$.
7.19. (a) $z_y = x/[2z - \cos (x + z)]$ (b) $y_x = e^y/(2y - xe^y)$.
7.20. $y(x) = 1 + \frac{1}{2e}(x - 1) - \frac{e + 2}{8e^2}(x - 1)^2 + \cdots$.
7.21. $z_{\text{approx}} = 2 - (x/2)^2 - (y/2)^2$.
7.22. $t(q, r, s) = (q - 1) + (r - 1) + (s - 1) + (q - 1)^2 + (r - 1)^2 + (s - 1)^2$
$+ (q - 1)(r - 1) + (q - 1)(s - 1) + (r - 1)(s - 1) + \cdots$.

Chapter 8

8.4. No. Yes.
8.6. $\cos^{-1} (4/\sqrt{30})$.
8.9. (a) Not possible.
8.17. $\dot{\mathbf{A}} = (3 \cos 3t + 27t^4)\hat{\mathbf{e}}_r + (3 \sin 3t - 36t^3)\hat{\mathbf{e}}_\theta + 2t\hat{\mathbf{e}}_z$.
8.18. $\mathbf{a} = (2t^2\hat{\mathbf{e}}_r - 4\hat{\mathbf{e}}_\theta)/\sqrt{t^4 + 4}$.
8.23. (a) $s = \sqrt{a^2 + \omega^2}\, \zeta$
(b) $\hat{\mathbf{t}} = (-a \sin \zeta\hat{\mathbf{i}} + a \cos \zeta\hat{\mathbf{j}} + \omega\hat{\mathbf{k}})/\sqrt{a^2 + \omega^2},\ \hat{\mathbf{n}} = -\cos \zeta\hat{\mathbf{i}} - \sin \zeta\hat{\mathbf{j}}$
(c) $\kappa = a/(a^2 + \omega^2),\ \tau = -\omega/(a^2 + \omega^2)$.
8.24. $\rho(0, 1, 0) = 4/\sqrt{6}$.
8.29. (a) $(e - 1)/4$ (c) $\pi/8$.
8.31. $5\pi a^2/4$.
8.32. (a) $F = 2\pi\sigma h(1 - \cos \alpha)$.

Chapter 9

9.1. (a) $-47/4$.
9.2. (a) $-3\pi a^2/2$.
9.4. (a) $2\pi ab$ (b) 0 (f) $-\pi ab$.
9.6. $df/ds = 9/\sqrt{5}$, max. directional derivative = 7 in the direction $(6\hat{\mathbf{i}} + 2\hat{\mathbf{j}} + 3\hat{\mathbf{k}})/7$,
tangent plane: $6x + 2y + 3z = 18$.
9.19. (b) 18.
9.20. (a) 36π.
9.23. (a) Integrals $= -2\pi a^2$.

9.30. (b) $8\sqrt{2}\,(6\pi)^{15/2}$.
9.31. (a) $u = (2x^3/3) - y^2z - 3z + \text{const.}$
(d) $u = (z^4/4) + \text{const.}$
9.32. (b) $\mathbf{w} = xy\hat{\mathbf{j}} + (yz - x^2/2)\hat{\mathbf{k}} + \nabla u$, for arbitrary scalar field u.
9.36. Cylindrical: $\rho_t + \dfrac{1}{r}(r\rho u)_r + \dfrac{1}{r}(\rho v)_\theta + (\rho w)_z = 0$.
9.46. $p < 1$.

Chapter 10

10.2. (b) $(0, 0, \sqrt{2})$ and $(0, 0, -\sqrt{2})$.
10.5. (b) $x = 0$, $y \doteq 0.0201746905841$
(c) $x \doteq 0.9350821$, $y \doteq 0.9980201$.
10.14. (b) The semicircle $(x - \frac{1}{2})^2 + y^2 = \frac{1}{4}$, for $y \geq 0$.
***10.15.** Natural boundary condition is $W + [\rho g y(a) - \lambda][y'(a)/\sqrt{1 + y'^2(a)}] = 0$.
10.21. $m\ddot{r} + k(r - L) - mr\dot{\theta}^2 - mg\cos\theta = 0$, $r\ddot{\theta} + 2\dot{r}\dot{\theta} + g\sin\theta = 0$.
10.25. (b) $w \approx -\frac{9}{16}xy(x - 1)(y - 1)$.
10.28. (b) $y = x - x(x - 1)(2.764 - 2.091x)$.

PART II

Chapter 11

11.3. (d) $|z| = 2^{25}$, $\arg z = \pi/2$, $\text{Re}\, z = 0$, $\text{Im}\, z = 2^{25}$.
11.4. (d) Half-plane $\text{Re}\, z < \frac{3}{2}$.

Chapter 12

12.2. (c) For $y > 0$, image is the ellipse $(u/\cosh y)^2 + (v/\sinh y)^2 = 1$; for $y = 0$, image is the segment $v = 0$, $-1 \leq u \leq 1$.
12.6. (b) Differentiable (and analytic) nowhere.
12.10. (c) $v = 3x^2y - y^3 + \text{const.}$
12.14. (a) At $z = 3/2$ on top of cut $\sqrt{(z - 1)(z - 2)} = i/2$.
12.15. (e) $z = 0$ [as can be seen by expressing $z^z = (re^{i\theta})^{x+iy}$].
12.16. (d) $i^i = e^{-(2n+1/2)\pi}$ for $n = 0, \pm 1, \pm 2, \ldots$.
12.18. (b) $z = 0, \pm\pi, \pm 2\pi, \ldots$ (d) $z = 0, \pm\pi i, \pm 2\pi i, \ldots$.
12.19. (a) $z = (2n + 1)\pi i$ for $n = 0, \pm 1, \pm 2, \ldots$.

Chapter 13

13.2. (a) $(\sin 2\cosh 2 - \sin 1\cosh 1) + i(\cos 1\sinh 1 + \cos 2\sinh 2)$.
13.6. (a) 0 (f) $2\pi i$.
13.7. (d) $\pi i/12$.

Chapter 14

14.1. (d) $|z - 1| > 1$.
14.2. Through first order, $\ln z = \ln 2 + \pi i - \frac{1}{2}(z + 2) + \cdots$.
14.8. (a) $\csc z = (1/z) + (z/6) + (7/360)z^3 + \cdots$.

Chapter 15

15.11. (c) $f(t) = 1/\sqrt{\pi t}$.

Chapter 16

16.7. $T = 100(xy - 1)$.

16.11. $T = (2/\pi)\tan^{-1}(\cos x/\sinh y)$, where $0 < \tan^{-1} < \pi$.

16.12. $T = 50 + \dfrac{25}{2\ln(5 - 2\sqrt{6})}\ln\left[\dfrac{(x-a)^2 + y^2}{(ax-1)^2 + a^2y^2}\right]$.

16.15. $u = U(1 - a^2\cos 2\theta/r^2)$.

16.17. $T = \dfrac{100}{\pi}\left\{\tan^{-1}\left[\dfrac{(2-x)(x^2+y^2) - x}{y(x^2+y^2) - y}\right] + \tan^{-1}\left[\dfrac{(2+x)(x^2+y^2)+x}{y(x^2+y^2)-y}\right]\right\}$,

where $-\dfrac{\pi}{2} < \tan^{-1} < \dfrac{\pi}{2}$.

PART III

Chapter 17

17.12. (c) $x = -3ie_1 - (1+i)e_2 + (1-i)e_3$.

***17.20.** No.

17.24. No.

Chapter 18

18.7. L^2 is linear if L is.

18.11. (b) $\|A\| = 5$ (c) $\|A\| = \sqrt{29}$.

18.19. (c) $L^* = -d^3/dt^3$; $y(0) = y''(0) = y(1) + 2y'(0) = y''(1) - y'(1) = 0$, not self-adjoint

(f) $L^* = \partial^2/\partial x^2 + \partial/\partial t$; $v(0, t) = v(l, t) = v(x, T) = 0$, not self-adjoint.

Chapter 19

19.3. (a) $\alpha = 1/2$.

19.4. (c) $\lambda = (1 \pm \sqrt{5})/2$.

19.5. $3c_1 - c_2 - c_3 = 0$.

19.6. (c) $\int_0^1 f(t)\,dt = 0$.

***19.18.** (c) $x(t) = e^t - 1 + 3\Lambda t/(3 - \Lambda)$ for $\Lambda \neq 3$; no solution if $\Lambda = 3$

(e) $x_1(t) \doteq 1 + \sin t - 0.9731t$

$x_2(t) \doteq 1 + \sin t - 1.0010t + 0.1671t^3$.

***19.22.** (b) $x(t) = 1 - \int_0^t p(\tau)(t - \tau)x(\tau)\,d\tau$.

***19.23.** $x^{(2)}(t) = t^2/2 - t^5/40$

***19.25.** $\Gamma(\tau, t; \Lambda) = e^{k(\tau - t)}/(1 - \Lambda)$

$$x(t) = f(t) + \Lambda \int_0^1 \frac{e^{k(\tau - t)}}{1 - \Lambda} f(\tau)\,d\tau.$$

Chapter 20

20.1. (d) For $(x, y) = \xi_1\bar{\eta}_1 + \xi_2\bar{\eta}_2$, A is not self-adjoint; for $(x, y) = \xi_1\bar{\eta}_1 + 2\xi_2\bar{\eta}_2$, A is self-adjoint

*(g) $\lambda = \pi/2$, $e(t) = \cos t$ and $\sin t$.

20.2. No.

20.5. Buckling mode corresponding to P_{cr} is $y(x) = A(1 - \cos \pi x/2l)$.

20.6. $P_{cr} = \pi^2 EI/4l^2 + (8EIl/\pi^2)\kappa + O(\kappa^2)$.

20.10. $f(x) = x = \sum_1^\infty \alpha_n \sin\sqrt{\lambda_n}\, x;\ \alpha_n = \dfrac{\int_0^1 x \sin\sqrt{\lambda_n}\, x\, dx}{\int_0^1 \sin^2\sqrt{\lambda_n}\, x\, dx}$.

20.18. (a) $F = [(5 + \sqrt{10})/2]x_1'^2 + [(5 - \sqrt{10})/2]x_2'^2$; positive definite.

20.21. (c) $e_2 = (0, 1, 0, 0)^T$, $e_3 = (0, -2/3, 1/3, -1/6)^T$.

20.24. Principal inertias $\doteq$ 39.33, 37.07, 2.27.

20.30. (b) $P_{cr} = kl/3$.

20.32. (a) $y(x) = \sum_1^\infty \dfrac{4f_n}{4 - (2n-1)^2} \sin \dfrac{2n-1}{2} x$.

PART IV

Chapter 21

21.1. (a) $y = -(1/2)\ln(C - 2e^x)$.

21.2. (c) $y = [C(1-x)e^{2x} - 1 - x]/(1 + Ce^{2x})$.

21.4. (c) $y = -x/(C + 2\ln x)$.

21.5. (c) $xe^y - y = C$.

21.6. (a) $y = -2 + C\sin^2 x$.

21.9. (d) $x^2 + y^2 = Cy$.

21.10. (c) $y = (1 + Ce^{-5x})^{-1/5}$

21.11. (c) $x = 24 - 24p + 12p^2 - 4p^3 + Ce^{-p}$
$y = (1 + p)x + p^4$.

21.12. (c) $y = Cx - e^C$ (general solution)
$y = x\ln x - x$ (singular solution).

21.13. (b) $y = x + 1/(C - x)$.

21.14. (b) $y^2 + 2(x-3)^2 - 2(x-3)y = C$.

21.17. (b) $2y^3 + 3y^2 + 24y = 6x^2 - 18x + 88$
(e) $y = e^{x^2/2}[1 - (\sqrt{\pi}/2)\operatorname{erf}(x/2)]^2$.

21.19. (d) $y = \pm 2x^{3/2}/3\sqrt{3}$.

Chapter 22

*22.3. (b) $x(t) \approx 1 + 3t - (t^2/2)$.

22.9. (e) $y = Ae^{-x} + e^{x/2}[B\sin(\sqrt{3}/2)x + C\cos(\sqrt{3}/2)x]$.

22.11. (g) $y = Ae^x + B[\operatorname{erf}[(x+1)/\sqrt{2}] - \operatorname{erf}(1/\sqrt{2})]e^x$.

22.12. (d) $y = Ax\int (e^{1/x}/x^2)\, dx + Bx$.

22.14. (d) $y = A\sin(2\ln x) + B\cos(2\ln x)$.

22.18. (a) $y = Ae^{x^2/2} + Be^{x^2/2}\operatorname{erf}(x/\sqrt{2})$.

22.19. (c) $x = A + Be^{-ct} + (1/c)\int_0^t \{1 - \exp[c(\tau - t)]\}f(\tau)\, d\tau$
(i) $y = A\sin x + B\cos x - \cos x \ln(\sec x + \tan x)$.

22.20. (c) $x = (F_0/k)H(t - t_0)[1 - \cos\sqrt{k}\,(t - t_0)]$.

22.25. (b) $y = Ae^t + e^{-t/2}[B\sinh(\sqrt{3}/2t) + C\cosh(\sqrt{3}/2t)]$
$z = \dot{y} =$ etc., $x = \dot{z} =$ etc.

22.27. (b) $y = Ax\left[1 - \dfrac{x^3}{9\cdot 7} + \dfrac{x^6}{9^2\cdot 2\cdot 7\cdot 12} - \cdots + (-1)^n \dfrac{x^{3n}}{9^n\cdot n!\cdot 7\cdot 12 \cdots (5n+2)} + \cdots\right]$
$+ Bx^{-1/5}\left[1 - \dfrac{x^3}{9\cdot 3} + \dfrac{x^6}{9^2\cdot 2!\cdot 3\cdot 8} - \cdots + (-1)^n \dfrac{x^{3n}}{9^n\cdot n!\cdot 3\cdot 8 \cdots (5n-2)} + \cdots\right]$.

(d) $y = A\left(x - \frac{x^2}{1!2!} + \frac{x^3}{2!3!} - \frac{x^4}{3!4!} + \cdots\right)$

$+ B\left[\left(x - \frac{x^2}{1!2!} + \frac{x^3}{2!3!} - \frac{x^4}{3!4!} + \cdots\right)\ln x - 1 + \frac{x}{2} + \frac{x^2}{2} - \frac{11}{72}x^3 + \cdots\right]$

22.36. (e) $y = AJ_{1/3}(kx) + BJ_{-1/3}(kx)$

(g) $y = A\sqrt{x}J_{1/6}(x^3/3) + B\sqrt{x}J_{-1/6}(x^3/3)$.

22.37. $y = 1 + \left(\frac{\Gamma(2/3)J_{-1/3}(2/3) - 3^{1/3}}{3^{1/3}J_{1/3}(2/3)}\right)x^{1/2}J_{1/3}\left(\frac{2}{3}x^{3/2}\right)$

$- \left(\frac{\Gamma(2/3)}{3^{1/3}}\right)x^{1/2}J_{-1/3}\left(\frac{2}{3}x^{3/2}\right)$.

22.43. $u(x) = a(6x - x^3)/5 + \int_0^1 G(\xi, x)f(\xi)\,d\xi$, where

$$G(\xi, x) = \frac{(\xi - x)^3}{6}H(\xi - x) + \frac{(1-x)(x^2+x-5)}{30}\xi^3 + \frac{x(1-x^2)}{10}\xi^2 + \frac{3x^3 - 5x^2 + 2x}{10}\xi.$$

22.45. $u(x) = K\sin x + \int_0^\pi G(\xi, x)\phi(\xi)\,d\xi$, where

$$G(\xi, x) = \frac{\xi \sin x \cos \xi}{\pi} - \begin{cases} \cos x \sin \xi, & 0 \le \xi \le x \\ \sin x \cos \xi, & x \le \xi \le \pi \end{cases}, \quad \text{and}$$

K is arbitrary.

22.49. (b) $K(0.2) \doteq 1.587$.

22.50. (b) $[E(k, \phi) + (k^2 - 1)F(k, \phi)]/k^2$

(d) $\sqrt{2/3}F(1/\sqrt{3}, \pi/8)$.

22.54. Successive approximations: $x_5 \doteq 1.8837$

Newton's method: $x_5 \doteq 1.8955$.

22.55. $x \doteq 3.9164354$.

22.58. $x \doteq 0.9351$, $y \doteq 0.9980$.

Chapter 23

23.5. (f) Unstable focus at $(1, 0)$; saddle at $(-2, 3)$.

23.9. $y^2 + \alpha x^2 + \beta x^4/2 = \alpha^2/2|\beta|$.

23.17. (a) Separatrix: $my^2 + kx^2 + 2k\lambda \ln(a - x) = kx_+^2 + 2k\lambda \ln(a - x_+)$, where $x_\pm = (a \pm \sqrt{a^2 - 4\lambda})/2$.

23.21. (a) $r = 1$ is stable limit cycle; origin is unstable focus.

23.22. (c) Origin is stable.

Chapter 24

24.1. (c) $y = 1 + 3(x-2) + 8(x-2)^2 + (64/3)(x-2)^3 + \cdots$

24.3. $x = 0.3$: $|E_n| \lesssim 0.0087$, actual $|E_n| = 0.0057$

$x = 10$: $|E_n| \lesssim 550.6$, actual $|E_n| = 0.00001$.

24.7. $AB = 1.388972$, converged AB-$AM = 1.388884$, exact $= 1.388889$.

24.8. (d) $y_n = 2 + A(-1)^n + B(-2)^n$

(f) $y_n = A\sinh n\alpha + B\cosh n\alpha$.

24.12. For example, $x(4) \doteq 0.50263$; period ≈ 4.8.

24.13. For example, $x(4) \doteq -0.14550$, $y(4) \doteq 0.98936$.

24.16. (a) $y(x) = 2Y_1(x) + [3 - 2Y_1(1)]Y_2(x)/Y_2(1)$,

where $Y_1'' + (\sin x)Y_1 = 0$; $Y_1(0) = 1$, $Y_1'(0) = 0$

and $Y_2'' + (\sin x)Y_2 = 0$; $Y_2(0) = 0$, $Y_2'(0) = 1$.

24.23. (a) $T_{\text{col}} = 10x + 3.166(x - x^2)$

(b) $T_{\text{Gal}} = 10x + 3.258(x - x^2)$.

24.26. (a) Collocating at $x = 0.5$, $y(x) \approx (8W/387)x^2(3 - 5x + 2x^2)$.

Chapter 25

25.4. $x = \pi[1 + \epsilon + \epsilon^2 + \epsilon^3(1 - \pi^2/3) + O(\epsilon^4)]$.

25.6. (a) $y = 1 + (\epsilon/2)(\sin x/\tan 1 - x \cos x) + O(\epsilon^2)$; regular.

25.7. $y_n(x) = \sin nx - \epsilon x[\sin nx + n(\pi - x) \cos nx]/4 + O(\epsilon^2)$

$\lambda_n = n^2 - n^2\pi\epsilon/2 + O(\epsilon^2)$.

25.17. (a) $x^o \sim -t + 0\epsilon + 0\epsilon^2$

$x^i \sim -t + 0\epsilon^2 + 3e^{(t-1)/\epsilon}$

$x^c \sim -t + 3e^{(t-1)/\epsilon}$.

PART V

Chapter 26

26.10. $u(x, t) = \sum_0^\infty [D_n \sin (2n\pi x/l) + E_n \cos (2n\pi x/l)]e^{-(2n\pi\alpha/l)^2 t}$,

where $E_0 = (1/l) \int_0^l f(x)\, dx$, $E_n = (2/l) \int_0^l f(x) \cos (2n\pi x/l)\, dx$ $(n \geq 1)$,

and $D_n = 2/l \int_0^l f(x) \sin (2n\pi x/l)\, dx$.

26.11. $u(x, t) = 100 \operatorname{erfc} (x/2\alpha\sqrt{t})$.

26.15. $u(x, t) = (2\pi\alpha^2/l^2) \sum_1^\infty (-1)^{n+1} n\left[\int_0^t F(\tau)e^{(n\pi\alpha/l)^2\tau} d\tau\right] \sin (n\pi x/l)e^{-(n\pi\alpha/l)^2 t}$.

26.19. $y(x, t) = x(\pi^2 - x^2)/6a^2 + \sum_1^\infty H_n \sin nx \cos nat$

$H_n = \int_0^\pi x(x^2 - \pi^2) \sin nx\, dx/3\pi a^2$.

26.21. $y = \epsilon[\cos (\omega x/a) - \cot (\omega l/a) \sin (\omega x/a)] \sin \omega t + g(x, t)$,
where $g(x, t)$ is solution of $a^2 g_{xx} = g_{tt}$; $g(0, t) = g(l, t) = g(x, 0) = 0$,
$g_t(x, 0) = \epsilon[\cot (\omega l/a) \sin (\omega x/a) - \cos (\omega x/a)]\omega$.

26.23. (b) Lowest frequency $= \pi\alpha\sqrt{5}/a$ rad/sec.

(d) $w(r, t) = \sum_1^\infty A_n J_0\left(z_n \frac{r}{a}\right) \sin\left(\frac{z_n \alpha}{a} t\right)$,

where $A_n = \dfrac{2}{\alpha a z_n [J_1(z_n a/\alpha)]^2} \displaystyle\int_0^a f(r) J_0\left(\frac{z_n r}{\alpha}\right) r\, dr$

and the z_n's are the roots of $J_0(z) = 0$.

26.28. (a) $u(r, z) = 200 \sum_1^\infty \dfrac{1}{z_n J_1(z_n)} J_0\left(z_n \dfrac{r}{a}\right) e^{-z_n z/a}$,

where the z_n's are the roots of $J_0(z) = 0$.

26.31. (a) $u(r, \theta) = \int_0^\pi [P(r, \zeta - \theta) - P(r, \zeta + \theta)] f(\zeta)\, d\zeta$.

26.32. $u(r, \theta) = \dfrac{1}{2\pi} \displaystyle\int_{-\pi}^{\pi} \dfrac{(r^2 - a^2)U(\zeta)\, d\zeta}{r^2 - 2ar \cos (\zeta - \theta) + a^2}$.

26.33. (c) $u(r, \theta) = 100 - \dfrac{400}{\pi} \displaystyle\sum_{1,3,\ldots}^\infty \dfrac{1}{n}\left(\dfrac{r}{a}\right)^{n\pi/2\alpha} \sin \dfrac{n\pi\theta}{2\alpha}$.

26.36. (a) $u(r, \theta) = A\left(1 - \dfrac{\ln r}{\ln b}\right) + \displaystyle\sum_1^\infty \left[\left(\dfrac{r}{b}\right)^n - \left(\dfrac{r}{b}\right)^{-n}\right](E_n \sin n\theta + F_n \cos n\theta)$, where

$$A = \int_0^{2\pi} f(\theta)\, d\theta/2\pi\left(1 - \frac{\ln a}{\ln b}\right),\quad E_n = \int_0^{2\pi} f(\theta) \sin n\theta\, d\theta/\pi[(a/b)^n - (a/b)^{-n}]$$

$$F_n = \int_0^{2\pi} f(\theta) \cos n\theta\, d\theta/\pi[(a/b)^n - (a/b)^{-n}].$$

26.37. (b) $u = \frac{100}{\pi}\left[\tan^{-1}\left(\frac{x+1}{y}\right) - \tan^{-1}\left(\frac{x-1}{y}\right)\right]$, where both $\tan^{-1}$'s are between $-\pi/2$ and $+\pi/2$.

26.40. $u = \frac{1600}{\pi^2} \sum_{m=1,3,\ldots}^{\infty} \sum_{n=1,3,\ldots}^{\infty} \frac{\sinh(\sqrt{m^2+n^2}\, x/2) \sin(my/2) \sin(nz/2)}{mn \sinh(\sqrt{m^2+n^2}\, \pi/2)}$

Chapter 27

27.5. (b) $4(\xi + \eta)\phi_{\xi\eta} + \phi_\xi + \phi_\eta - \kappa\phi^3 = 0$; $\xi = \frac{x^2}{2} - y,\ \eta = \frac{x^2}{2} + y.$

27.9. $u = \begin{cases} 3e^{x-y} & \text{in } \pi/4 < \theta < \pi/2 \\ 3\sin(x-y) & \text{in } 0 < \theta < \pi/4. \end{cases}$

Chapter 28

28.6. (b) $u(x, y) = \frac{ky}{2} \int_{-\infty}^{\infty} f(\xi) \frac{Y_0'(k\sqrt{(\xi - x)^2 + y^2})}{\sqrt{(\xi - x)^2 + y^2}}\, d\xi.$

28.17. (b) $u(x, \pm\epsilon T) \sim U - \epsilon U \sum_1^{\infty} a_n \cos n\alpha.$

Chapter 29

29.9. (b) $u_{2,3} \doteq 0.0171.$

29.11. (a) $u \doteq 0.5971$ (b) $u \doteq 0.6000.$

Index

Note: Author citations from the Additional References at the end of Parts I through V, and the Survey-Type References on pages 614 and 615 are *not* included in this index.

Abel:
  formula, 444
  integral equation, 116
  theorem, 548
Abramowitz, M., 11, 15, 59, 60
Absolute convergence:
  integrals, 55
  series, 29
Absolute stability, 482
Acceleration techniques, 38
Adams–Bashforth method, 495
Adams–Moulton method, 496
Adjoint operator, 335
Agnew, R.P., 360, 393
Airfoil integral, 288
Airy's equation, 433, 525
Aitken–Shanks transformation, 38
Ames, W.F., 129, 131, 599, 608, 610
Analytic continuation, 240, 264
Analytic function, 36, 239
Andronow, A., 454
Angular velocity, 139
Apostol, T.M., 28, 54, 67, 197, 401
Arfken, G., 94
Argand diagram, 233
Argument, 233
Asymptotic:
  error constant, 442
  expansion, 43
  series, 44
  stability, 407
Asymptotically equal, 11
Autonomous differential equation, 409, 455

Bachman–Landau big oh, 12
Base vectors:
  cartesian, 136
  cylindrical, 137
Basis, 316, 325
Bellman, R., 220, 443
Bendixson criterion, 469
Bernoulli, 221
  equation, 404
Bessel equation:
  modified, 431
  of order μ, 425
  of order zero, 109, 127

Bessel functions, 425
modified, 432
of first kind, 22, 109
of order zero, 109, 127, 447
Bessel inequality, 327
Beta function, 63
Bieberbach, L., 296
Bifurcation point, 465
Big oh notation, 12
Biharmonic:
equation, 530, 563
operator, 592
Bilinear transformation, 294
Binormal, 147
Biot–Savart law, 188
Block, H.D., 376
Bogoliuboff, N., 454
Boltyanskii, V.G., 220
Bolzano-Weierstrass theorem, 27
Boundary layer method, 508
Boundary value problem, 485
Bounded set, 27
Bounded variation, 10
Bounds on:
contour integrals, 255
eigenvalues, 377, 604
Boyce, W., 393
Brachistochrone problem, 221
Buck, R.C., 67, 175
Burgers equation, 525
Burkill, J.C., 413, 425, 429
Byrd, P., 440

Canonical form, 367, 574
Jordan, 371
Carrier, G., 66, 297
Carslaw, H., 533
Cartesian coordinates, 136
Cauchy:
condensation test, 47
convergence, 25
–Euler equation, 418
–Goursat theorem, 255, 256
integral formula, 257
Cauchy: (*cont.*)
principal value, 54
sequence, 25, 319
theorem, 255, 256
Cauchy–Riemann conditions, 129, 238
in polar coordinates, 249
Cayley–Hamilton theorem, 386
Center, 458
Central force field, 146, 194
Centripetal acceleration, 138
Chaikin, C., 454
Chain differentiation, 121
Characteristic equation, 360
Characteristics, 542, 564, 569
Chartier's test, 55
Chasle's theorem, 140
Chebyshev equation, 434
Churchill, R.V., 112, 271, 296, 297, 364
Ci(x), 43
Clairaut equation, 405
Clegg, J.C., 203
Closed interval $[\alpha, \beta]$, 35
Closed set, 26
Cofactor, 345
Cole, J.D., 514
Collocation, 489
Compact operator, 357
Comparison:
functions, 203
test, 28
Compatibility equation, 568
Complementary error function, 11
Complete, 363
Completely continuous operator, 357
Complex:
conjugate, 232
plane, 232
Composite:
function, 120
solution, 512
Conformal mapping, 291
Conjugate:
complex, 232
function, 239
Connected, 152
Conservative field, 172

Consistency, 598
Constraint, 199, 208
  nonintegral, 223
Conte, S.D., 12, 13, 14, 486
Continuity:
  equation of fluid mechanics, 162
  of a function of a complex variable, 237
  of a function of one real variable, 5
  of a function of several real variables, 121
  uniform, 6
Continuously differentiable, 203
Contraction mapping, 403
Control system:
  discontinuous, 220
  feedback control, 219
  liquid level control system, 218
  proportional control, 219
  terminal process control, 227
Convective derivative, 189
Convergence:
  factor, 493
  of numerical integration, 480, 599
  of vector sequence, 313, 323
Convergence of integrals:
  absolute, 55
  Cauchy, 54
  tests for, 54
Convergence of series:
  absolute, 29
  Cauchy, 25
  Cesàro, 24
  conditional, 30
  ordinary, 24
  tests for, 28, 30, 31, 33, 47
Coriolis acceleration, 138
Cosine integral, 43
Courant, R., 553, 572, 577, 608
Cramer's rule, 347
Crank–Nicolson scheme, 605
Crook, M., 66, 297
Cross product, 134
Crout–Banachiewicz algorithm, 355, 356
Curl, 157, 176, 177
Curvature, 147
Curvilinear coordinates, 141, 175
Cylindrical coordinates, 137, 176
d'Alembert–Lagrange equation, 404
d'Alembert solution, 541
Davies, T., 514
Delta function, 68
  derivatives of, 70
  sequence, 69, 535
  two-dimensional, 580
DeMoivre's theorem, 249
Denn, M., 220
Dennery, P., 320
Derivative:
  complex variable, 237
  real variable, 8
Determinant, 345
  2 X 2, 3 X 3, 124
Difference:
  equation, 480
  kernel, 534
Differential equation (ordinary):
  Airy, 433, 525
  autonomous, 409, 455
  Bernoulli, 404
  Bessel, 109, 127, 425, 431
  Cauchy–Euler, 418
  Chebyshev, 434
  Clairaut, 405
  constant coefficients, 415
  d'Alembert–Lagrange, 404
  Duffing, 451, 461
  existence and uniqueness, 400, 410
  first order, 393
  first order linear, 422
  general solution, 394, 415
  Hermite, 434
  homogeneous of degree zero, 394
  hypergeometric, 434
  Korteweg-de Vries equation, 525
  Laguerre, 434
  Legendre, 433
  Liénard, 465
  linear, 411
  Mathieu, 434
  nonlinear, 438, 443
  Riccati, 405, 419, 428
  series solution, 423
  undetermined coefficients, 419

Differential equation (ordinary): (*cont.*)
  van der Pol, 464, 506, 517
  variables-separable, 394
  variation of parameters, 417, 420
Differentiation of integrals, 17, 163, 185
Diffusion equation, 165, 524, 528
Diffusivity, 165
Dimension of space, 316
DiPrima, R., 393
Dirac, P.A.M., 68
Dirac delta function, 68, 580
Direct method, 217
Directional derivative, 156
Dirichlet:
  conditions, 85
  discontinuous integral, 63
  problem, 215, 291, 547
Dirichlet test:
  integrals, 55
  series, 30
Discretization error, 474, 599
Displacement operator E, 41
Distance function, 7, 121
Diverge, 24, 52
Divergence, 153, 176, 177
  theorem, 161
Domain, 3
Dot product, 134
Double integral, 57
Dual basis, 325
Duffing's equation, 451, 461
DuFort–Frankel scheme, 606
Dyad operator, 325
Dynamic programming, 220

Eigen, 361
Eigenfunction, 364, 526
Eigenvalue:
  bounds, 377, 604
  problem, 359, 526
Eigenvector, 359
  expansion, 376
  generalized, 384
Ei(x), 43
Elliptic:
  coordinates, 562
Elliptic: (*cont.*)
  equation, 523, 565
  integral, 439
Entire function, 240
Envelope, 399
Epstein, B., 554
Equilibrium point, 456
Erdélyi, A., 100
erf(x), 11
erfc(x), 11
Error function, 11
Euclidean distance function, 121
Euler:
  beam theory, 87
  equation, 205
  formulas, 80, 233
  method, 474
  theorem on homogeneous functions, 128
  transformation, 41
Euler's constant, 43
Even function, 78
Exact differential, 395
Existence, for solution of:
  $Lx = c$, 342, 343, 348, 376
  $y' = f(x, y)$, 400, 410
Explicit formula, 605
Exponential:
  integral, 43
  order, 103

Fermat's principle, 222
Feshbach, H., 538
Finite:
  difference method, 486, 499, 596
  element method, 491
Finlayson, B.A., 217, 489
Fix, G., 494
Fixed point theorem, 403
Flanders, H., 164
Fluid mechanics, 177
  continuity equation, 162
  energy equation, 190
  equations of motion, 189
  tangent-flow boundary condition, 585

Focus, 458
Formal, 18
Formally self-adjoint, 337
Forsythe, G., 599
Fourier:
  coefficients, 80, 327
  integral, 88
  integral theorem, 90
  law of heat conduction, 164
  sine and cosine transforms, 112, 114
Fourier series, 80, 327
  complex, 96
  pointwise convergence of, 80
  termwise differentiation of, 86
  termwise integration of, 85
Fourier transform, 98, 533, 549
  convolution, 102, 113
  inversion formula, 98
  properties, 102, 113
  tables, 100
Franklin, P., 185
Fredholm:
  alternative, 377
  equation, 350
  theory, 353
Frenet formulas, 147
Frequency spectrum, 89
Fresnel integrals, 287
Friedman, B., 349
Friedman, M., 440
Frobenius series, 425
Function, 3, 235
  entire, 240
  space, 318
Functional, 4, 202
Fundamental:
  set, 414
  solution, 580
Fundamental theorem:
  of algebra, 260, 383
  of complex integral calculus, 253
  of integral calculus, 10

Galerkin method, 217, 489
Gamma function, 56
Gauss:
  distribution, 75
  divergence theorem, 161
  elimination, 340
  integration, 16
  –Seidel iteration, 493
Gear, C.W., 476, 482, 498
Gelbaum, B., 20
Generalized eigenvalue problem, 381, 386
Generating function, 447, 448
Geodesic, 206, 223
Geometric series, 24, 39
Gerschgorin's theorem, 604
Gibbs' phenomenon, 93
Glauert, H., 586, 593
Glauert integral, 288
Goodier, J.N., 216
Goursat, E., 255
Gradient, 155, 176, 177
Gram determinant, 325
Gram–Schmidt orthogonalization, 320
Gravitation, 182
Greenberg, M.D., 438, 516, 583
Green's:
  function, 101, 435, 449, 579
  identities, 165
  theorem, 168

Hamilton's principle, 209
Hankel function, 430
Hankel transform, 112, 591
  tables, 100
Hardy, G.H., 24
Harmonic:
  function, 199, 239
  series, 27
Hastings, C., Jr., 11
Heat conduction:
  Fourier law of, 164
  maximum principle, 553
  steady state, 165, 199, 294, 342
  uniqueness, 553
  unsteady, 164, 528, 533, 538, 597, 604
Heaviside:
  expansion theorems, 111
  step function, 6, 20

Helmholtz, 177
  equation, 557
  operator, 591
Hermite equation, 434
Hermite operator, 335
Hessian, 198
Hilbert, D., 553, 577
Hilbert space, 320
Hildebrand, F.B., 480
Ho, T.C., 352
Holomorphic, 239
Homogeneous function, 127
Housner, G.W., 224
Hyperbolic equation, 523, 565, 566
Hypergeometric equation, 434
Hyslop, J.M., 28

Identity:
  element, 372
  operator, 328
  theorem, 264
Images, 535, 537, 543, 582, 584
Imaginary part, 231
Implicit:
  function, 124
  methods, 604, 605
Improper integral, 51, 186, 195
Improved Euler method, 478
Incompressible fluid, 163
Indicial equation, 426
Inner:
  domain, 509
  product, 313, 324
Integral:
  airfoil, 288
  double, 57
  Glauert, 288
  in complex plane, 252
  Lebesgue, 320
  line, 150
  Riemann, 9
  singular, 51, 59
Integral equation, 350
  Abel, 116
  Fredholm, 350
Integral equation, (*cont.*)
  singular, 116
  Volterra, 351
Integral form of remainder, 35
Integral transform:
  bilateral Laplace, 112
  Fourier, 98
  Fourier sine and cosine, 112, 114
  Hankel, 112
  Laplace, 103
  Mellin, 112
Integrating factor, 396
Interval:
  of convergence, 32
  of dependence, 608
Invariant imbedding, 487
Inverse:
  matrix, 345
  operator, 314
Irrotational field, 170
Isoclines, 459
Isoperimetric problem, 208, 223
Iterated integral, 57

Jacobi iteration, 493
Jocobian, 125
Jaeger, J., 533
James, E., 514
Jeffery, R.L., 77
Jeffreys, H., 160
Johnson, D., 434
Johnson, J., 434
Jordan canonical form, 371, 385
Joukowsky transformation, 301

Kalaba, R., 443
Kaplan, S., 599
Kaplan, W., 164
Kellogg, O.D., 161
Kernel, 353
  difference, 534
  resolvent, 353
  separable, 353

Knopp, K., 255
Kober, H., 296
Korteweg-deVries equation, 525
Kronecker delta, 317, 321
Krook, M., 66
Kryloff, N., 454
Krylov, V., 104
Krzywicki, A., 320
Kutta condition, 586

Lagrange:
  form of remainder, 35
  identity, 146
  multiplier, 201
Lagrangian, 209
Laguerre equation, 434
Landau, L.D., 525
Laplace:
  equation, 165, 215, 524, 547
  expansion, 345
  operator, 123, 129, 159, 176
Laplace transform, 103, 532
  convolution, 106
  inversion fomula, 104, 283
  partial fraction representation, 110
  properties, 105, 106, 115
  tables, 100
  use in evaluating integrals, 119
Laplacian, 123, 129, 159, 176
Lasalle, J.P., 220
Latta's expansion, 517
Laurent expansion, 266
Lax equivalence theorem, 599
Least squares, 489
Lebesgue space, 320
Lee, E.S., 443, 488
Left-hand derivative, 81
Legendre:
  equation, 433
  function, 434
  polynomial, 15, 21, 321, 434
Leibnitz differentiation, 18, 163, 185
Lewy, H., 608
Liapunov:
  function, 470
Liapunov: (*cont.*)
  stability, 457
  theorem, 470
Lieberstein, H.M., 481
Lifshitz, E.M., 525
Limit, 4
  cycle, 465, 516
  interchange, 66
  point, 26
Line integral, 150
Linear:
  dependence, 315, 412
  independence, 315, 412
  operator, 10, 332
  programming, 202
Linearity:
  derivative, 10
  infinite series, 28
  integral, 10
  lim operator, 19
Linearization, 218
Liouville's theorem, 260, 274
Lipschitz condition, 74, 400, 409, 595
Longman, I.M., 17
Lovitt, W.V., 353
L–U decomposition, 356
Luke, Y.L., 11

McCue, G, 443
Mach number, 163
McLachlan, N.W., 425
Maclaurin expansion, 36
Mapping, 3, 235, 328
  conformal, 292
Matched asymptotic expansion, 513
Mathieu equation, 434
Matrix, 329
  antisymmetric, 334
  augmented, 348
  diagonal, 334
  ill-conditioned, 354
  inverse, 345
  modal, 367
  orthogonal, 368
  rank, 348

Matrix, (*cont.*)
singular, 347
skew-symmetric, 334
symmetric, 334
unitary, 368
Maximum:
absolute, 197
relative or local, 197
Mean square convergence, 323
Mean value theorem:
of differential calculus, 35, 68
of integral calculus, 68, 154
Mellin transform, 112
tables, 100
Membrane:
pumped up, 216, 552
vibrating, 214
Meridian, 142
Metric, 7
Metric coefficients, 175
Midpoint rule, 476
Mikhlin, S.G., 217
Miles, J.W., 112
Milne, W.E., 604
Minimum:
absolute, 197
relative or local, 197
Minkowski inequality, 312
Minor determinant, 345
Minorsky, N., 454, 461
Mixed boundary condition, 554, 583, 590
Möbius:
band, 152
transformation, 296
Modified Euler method, 478
Modulus, 233
Monotone decreasing, 30
Moon, P., 531
Morera's theorem, 260
Morrison, Wendy, 572
Morse, P.M., 538
Morton, K., 599, 606, 609, 611
Multiple scales, 507
Multiply connected, 153

n tuples
Natural boundary condition, 207, 208
Nayfeh, A., 500, 594
Naylor, A., 403
Negative definite, 368
Nehari, Z., 296
Neighborhood, 26
Neumann:
boundary condition, 558
function, 427
series, 349, 352
Neustadt, L.W., 220
Newton–Cotes formulas, 14
Newton's method, 442
Nichols, R.L., 202
Node, 457
Nonlinear:
algebraic equations, 440
transformation of series, 41
Norm of:
operator, 333
partition, 9
vector, 312, 409
Normal:
coordinates, 370
form, 573
modes, 371
Numerical integration, 12

O'Brien, G., 599
Odd function, 78
Olmstead, J., 20
Open interval (α, β), 35
Operator, 328
adjoint, 335
algebra, 331
bounded, 332
compact, 357
completely continuous, 357
dyad, 356
Hermitian, 335
identity, 328
integral, 330
inverse, 344
matrix, 329
norm of, 333
null, 328
projection, 338
Order of convergence, 475

Ordinary point, 423
Orientable, 152
Orthogonal:
funtions, 319
matrix, 368
vectors, 134, 315
Orthonormal, 136, 317
Osculating plane, 147
Outer domain, 509

p integral, 55
p series, 28
Parabolic:
coordinates, 561
equation, 523, 565
Parallelogram law, 133
Parseval equality, 327
Pearson, C., 66, 297
Periodic function, 78
Perturbation method, 382, 500, 585
Phase:
plane, 455
trajectory, 455
velocity, 463
Phillips, E.G., 262
Piecewise:
continuous, 80
smooth, 80, 152
Poincaré, H., 461
–Bondixson criterion, 470
Pointwise convergence, 323
Poisson:
equation, 185, 216
integral formula, 298, 548
Polar coordinates, 122
Pontryagin, L.S., 220
maximum principle, 220
Positive definite, 368
Potential:
flow between cylinders, 180
flow over cylinder, 178
gravitational, 182
logarithmic, 194
scalar, 172
theory, 178
Power method, 377
Power series, 31
differentiation of, 75
interval of convergence, 32
radius of convergence, 32
Predictor-corrector methods, 495
Prenter, P.M., 493, 494
Primitive matching principle, 510
Principal:
axes, 374
inertias, 374
normal, 147
solution, 580
value of integral, 54, 285
Pritulo, M.F., 506
Projection operator, 338, 356

Quadratic form, 367
Quasilinearization, 443, 486

Radbill, G., 443
Radiation condition, 558
Radius of convergence, 32
Range, 3
Rank of matrix, 348
Ratio test, 28
Rayleigh:
equation, 517
quotient, 377
Rayleigh–Ritz method, 388
Real part, 231
Reciprocal basis, 325
Rectangular rule, 12
Reflection theorem, 273
Regular:
function, 239
perturbation, 501
Relaxation oscillation, 517
Residue, 277
theorem, 277
Resolvent kernel, 353
Reynolds number, 163
Riccati equation, 405, 419, 428
Richardson, L.F., 610
Richtmeyer, R., 599, 606, 609, 611

Riemann:
integral, 9
–Lebesgue lemma, 327
mapping theorem, 297
sum, 9, 89
zeta function, 29, 74
Right-hand derivative, 81
Ritz method, 216
Rodrigues' formula, 448
Romberg integration, 14, 21
Roundoff error, 479, 599
Rudin, W., 67
Runge–Kutta method, 477

Saddlepoint, 198, 458
Saint–Venant's principle, 73
Samarski, A.A., 187
Scalar, 133
potential, 172
product, 134
triple product, 135, 139
Schwarz–Christoffel transformation, 299
Schwarz inequality, 314
Sectionally smooth, 152
Secular term, 504
Self-adjoint operator, 335
Sell, G., 403
Separable kernel, 353
Separation of variables, 528
Separatrix, 463
Series:
complex, 261
convergence, 24, 25
geometric, 24, 31
harmonic, 27
Maclaurin, 36
p series, 28
power, 31
solution, 423
summing of, 289
Taylor, 34, 263
uniform convergence, 32, 33
sgn (x), 8
Shames, I., 140
Sharing, *see* W. Morrison
signum (x), 8
Similarity transformation, 128, 555
Simpson's rule, 14
Sine integral, 43
Singular, 242
at infinity, 273
integral, 51, 59
matrix, 347
point, 36, 242, 423, 424, 456, 457
solution, 399
Singularity, 269
branch point, 245
essential, 269
isolated, 271
nonisolated, 271
pole, 269
Si (x), 43
Skoblya, N., 104
Sludge, 556
Smith, G.D., 604, 606, 608
Smooth, 152, 203
Snell's law, 222
Solenoidal field, 173
Sommerfeld radiation condition, 558
Spencer, D.E., 531
Spherical polars, 129, 138, 177
Spiegel, M.R., 393, 415
Stability:
absolute, 482, 601
asymptotic, 407, 457
Liapunov, 457
of numerical integration, 480, 482, 599
Stakgold, I., 320, 354
Stark, R.M., 202
Steele, C.R., 446
Steepest descent, 220
Stegun, I., 11, 15, 60
Stiff systems, 498, 606
Stirling's formula, 66
Stoker, J.J., 274, 506, 570, 573
Stokes, G.G., 32
theorem, 168
Straining, 504
Strang, G., 494
Stream function, 128, 192
Struble, R.A., 315, 410, 412